ADVANCES IN
MOLECULAR AND CELL BIOLOGY
VOLUME 35

Perspectives on Lung Endothelial Barrier Function

ADVANCES IN
MOLECULAR AND CELL BIOLOGY
VOLUME 35

Perspectives on Lung Endothelial Barrier Function

Series Editor:

E. Edward Bittar
University of Wisonsin – Madison
Madison, Wisconsin,
USA

Volume Editor:

Carolyn E. Patterson
Indiana University School of Medicine &
Roudebush VA Medical Center
Indianapolis, Indiana,
USA

2005

ELSEVIER

Amsterdam – Boston – Heidelberg – London – New York – Oxford
Paris – San Diego – San Francisco – Singapore – Sydney – Tokyo

ELSEVIER B.V.
Sara Burgerhartstraat 25
P.O. Box 211, 1000 AE
Amsterdam, The Netherlands

ELSEVIER Inc.
525 B Street
Suite 1900, San Diego
CA 92101-4495, USA

ELSEVIER Ltd.
The Boulevard
Langford Lane, Kidlington,
Oxford OX5 1GB, UK

ELSEVIER Ltd.
84 Theobalds Road
London WC1X 8RR
UK

© 2005 Elsevier B.V. All rights reserved.

This work is protected under copyright by Elsevier B.V., and the following terms and conditions apply to its use:

Photocopying
Single photocopies of single chapters may be made for personal use as allowed by national copyright laws. Permission of the Publisher and payment of a fee is required for all other photocopying, including multiple or systematic copying, copying for advertising or promotional purposes, resale, and all forms of document delivery. Special rates are available for educational institutions that wish to make photocopies for non-profit educational classroom use.

Permissions may be sought directly from Elsevier's Rights Department in Oxford, UK: phone (+44) 1865 843830, fax (+44) 1865 853333, e-mail: permissions@elsevier.com. Requests may also be completed on-line via the Elsevier homepage (http://www.elsevier.com/locate/permissions).

In the USA, users may clear permissions and make payments through the Copyright Clearance Center, Inc., 222 Rosewood Drive, Danvers, MA 01923, USA; phone: (+1) (978) 7508400, fax: (+1) (978) 7504744, and in the UK through the Copyright Licensing Agency Rapid Clearance Service (CLARCS), 90 Tottenham Court Road, London W1P 0LP, UK; phone: (+44) 20 7631 5555; fax: (+44) 20 7631 5500. Other countries may have a local reprographic rights agency for payments.

Derivative Works
Tables of contents may be reproduced for internal circulation, but permission of the Publisher is required for external resale or distribution of such material. Permission of the Publisher is required for all other derivative works, including compilations and translations.

Electronic Storage or Usage
Permission of the Publisher is required to store or use electronically any material contained in this work, including any chapter or part of a chapter.

Except as outlined above, no part of this work may be reproduced, stored in a retrieval system or transmitted in any form or by any means, electronic, mechanical, photocopying, recording or otherwise, without prior written permission of the Publisher.
Address permissions requests to: Elsevier's Rights Department, at the fax and e-mail addresses noted above.

Notice
No responsibility is assumed by the Publisher for any injury and/or damage to persons or property as a matter of products liability, negligence or otherwise, or from any use or operation of any methods, products, instructions or ideas contained in the material herein. Because of rapid advances in the medical sciences, in particular, independent verification of diagnoses and drug dosages should be made.

First edition 2005

Library of Congress Cataloging in Publication Data
A catalog record is available from the Library of Congress.

British Library Cataloguing in Publication Data
A catalogue record is available from the British Library.

ISBN: 0-444-51834-7
ISSN: 1569-2558 (Series)

∞ The paper used in this publication meets the requirements of ANSI/NISO Z39.48-1992 (Permanence of Paper).
Printed in The Netherlands.

Working together to grow
libraries in developing countries

www.elsevier.com | www.bookaid.org | www.sabre.org

ELSEVIER   BOOK AID International   Sabre Foundation

Perspectives on Lung Endothelial Barrier Function

Table of Contents

Foreword viii
Contributors ix-xiv

Chapters:

Foundations

Chapter 1. The Cellular and Molecular Foundations of Pulmonary Edema   1-24
Basic architecture of the lung; pulmonary and bronchial circulation; the alveolar-capillary
membrane and lung fluid balance; injury vs. activation; clinical pulmonary edema.
Carolyn E. Patterson* & Michael A Matthay

Chapter 2. Molecular Architecture of Endothelium   25-64
Cytoskeletal structure and adhesions; cell polarity; relationship of molecular structure
to barrier function; the contractile model and the tensegrity model.
Carolyn E. Patterson* & Dimitrije Stamenovic

Chapter 3. The Activated Endothelial Cell Phenotypes   65-104
Response to mediators, pathogens, and mechanical stimuli; receptors and physiologic
consequences; association of other endothelial functions with barrier changes.
Hazel Lum*

Signaling

Chapter 4. Membrane and Cellular Signaling of Integrity and Acute Activation   105-138
Acute signaling – receptor/membrane events, second messenger events, signaling control
of cytoskeletal/adhesion effectors; lipid and oxidant signaling.
Viswanathan Natarajan*, Peter V. Usatyuk, & Carolyn E. Patterson

Chapter 5. Adenylyl Cyclase and cAMP Regulation of the Endothelial Barrier   139-164
Adenylate cyclase, phosphodiesterases, and calcium regulation.
Sarah L. Sayner & Troy Stevens*

Chapter 6. Signaling of Prolonged Activation   165-204
Pathways of prolonged signaling leading to altered protein synthesis/degradation
and gene induction; upregulation of leukocyte adhesion factors and inflammatory factors;
responses to mediators, hypoxia, oxidants, and shear stress.
Carolyn E. Patterson* & Matthias A. Clauss

Structure-Function Relationship

Chapter 7. Dynamic Microfilaments & Microtubules Regulate Endothelial Function   205-236
Regulation of the cytoskeleton -actin polymerization and organization;
interrelationship with the microtubule cytoskeleton.
Joanna Zurawska, Mabel Sze, Joanne Lee, and Avrum Gotlieb*

Chapter 8. Endothelial-Matrix Interactions in the Lung   237-250
Molecular structure of focal adhesions and their role in cell movement
and barrier function; integrin signaling.
Sunita Bhattacharya, Sadiqa Quadri, & Jahar Bhattacharya*

Chapter 9. Interendothelial Junctions and Barrier Integrity   251-276
Adherens junctions, tight junctions, PECAM, and cell-cell integrins; regulation
and relationship to the actin-cytoskeleton, signaling, and function.
Lopa Leach*, Carolyn E. Patterson, & Donna Carden

Chapter 10. Heterogeneity of Lung Endothelial Cells   277-310
Organ specific differences in endothelium; endothelium in pulmonary and
bronchial circulation; endothelium in arteries, capillaries, and veins;
variations with conditions; endothelial markers and heterogeneity.
Eric Thorin*, Troy Stevens, & Carolyn E. Patterson

Inflammation & Pathology

Chapter 11. Interaction of Pulmonary Endothelial Cells with Blood Elements   311-334
Adhesion of circulating formed blood elements; receptor expression;
leukocyte transmigration; effect of activated leukocytes on endothelium;
influence of endothelium on leukocytes and platelet function;
influence of platelets and albumin on the endothelium.
Qin Wang*, Inkyung Kang, & Claire M. Doerschuk

Chapter12. Endothelial Cell Injury and Defense   335-364
Inflammation, injury, and pathologic consequences of leukocyte products,
oxidants, and bioactive agents; toxic death and apoptosis of endothelium;
recovery/re-endothelialization.
Hedwig Murphy*, James Varani, & Peter A. Ward

Chapter 13. Endothelial Injury Due to Infectious Agents   365-400
Pathogen attack and invasion; inflammation and response; pathophysiologic
and clinical consequences of infection.
Stefan Hippenstiel* & Norbert Suttorp

Chapter 14. Chronic Lung Vascular Hyperpermeability   401-422
Chronic hyperpermeability; mediators, mechanisms, and transcription
in chronic permeability, vascular remodeling, and lung disease.
Geerten P. van Nieuw Amerongen, Victor W.M. van Hinsbergh,
& Bradford C. Berk*

Application in the Clinical Setting

Chapter 15. Advances in Protection of Endothelial Barrier Function   423-470
Experimental advances in establishing the barrier, blocking activation,
reversing dysfunction, and repairing injury.
Carolyn E. Patterson*, Hazel Lum, & A.B. Johan Groeneveld

Chapter 16. Looking to the Future as Keepers of the Dam   471-516
Advances in endothelial biology, activation, and promotion of barrier integrity;
scientific and clinical significance in ARDS and other clinical conditions
based in endothelial dysfunction; projection of new directions
for prevention and treatment of barrier dysfunction and acute lung injury.
Carolyn E. Patterson* & Michael A Matthay

Abbreviations and Acronyms   517-520
Abbreviations and acronyms common amongst the chapters
compiled into one central reference.

Index of Contents   521-525

Foreword

The importance of the primary topic of this volume, endothelial barrier function, comes into sharp focus in clinical practice when dealing with acute lung injury and the development of Acute Respiratory Distress Syndrome. Yet, the fundamental mechanisms of endothelial dysfunction and the potentials described for intercepting the pathologic consequences or in reversing them are highly applicable to many diverse conditions, such as septic shock, diabetic renal failure, and atherosclerosis. While the information provided is state-of-the-art, with the rapid pace of investigation in this exciting area of science and medicine, it is expected that the next decade will see great advancement in both our ability to understand the contribution of endothelial dysfunction to initiation and progression of many diseases and, hopefully, our ability to support endothelial integrity. The information contained in this volume will provide a launching pad from which such discoveries can be made.

These chapters, therefore, are a comprehensive guide to endothelial biology and have been carefully organized and coordinated to give the reader an integrated understanding, rather than assembled as a mere collection of isolated efforts. Thus, the reader is encouraged to actually read this volume as presented to fully appreciate the role of endothelial integrity in pulmonary function, and for that matter, in organ function in general. Now, a word regarding layout mechanics to aid our readers. A brief outline of individual chapter contents can be found at the first page of each chapter and these heading are used within the chapter. We have attempted to minimize the use of abbreviations and acronyms, but in scientific parlance these may often be more recognized and understood than the original word definition. Those abbreviations and acronyms that are common amongst the chapters are contained in a centralized appendix found at the end of the book. Specialized abbreviations found only within a paragraph or two, are defined in the text and are not included in the appendix. An index of key terms has also been prepared for cross-reference.

Finally, it should be said that preparation of this volume has been a labor of love, extending from a long-sustained infatuation with understanding the molecular basis of structure and function in biomedical science. I would like to express appreciation to Dr. Bittar for instigating this project and encouraging me to take on the venture. The chapter contributors were chosen for their expertise and important contributions to the science and it has been a delightful learning experience in working with them and their respective co-authors. Similarly, the people at Elsevier have been very supportive in this endeavor. And, not the least, I must acknowledge the amazing patience of my husband, Jeff.

Carolyn E. Patterson

Ad Marjorem Dei Gloriam

Perspectives on Lung Endothelial Barrier Function
Contributors

Editor:

Carolyn E. Patterson, Ph.D.
Associate Professor & Scientist
Departments of Medicine and Cellular & Integrative Physiology
Indiana University School of Medicine and
Research Investigator, Roudebush VA Medical Center
1481 W. 10th. St. VA 111P
Indianapolis, IN 46202   USA
Email: caepatte@iupui.edu

Chapter Contributors: * denotes corresponding author

Bradford C. Berk, M.D., Ph.D.
Director, Center for Cardiovascular Research
Professor and Chairman, Department of Medicine
University of Rochester School of Medicine
Box MED, 601 Elmwood Ave.
Rochester, NY 14642   USA
Email: Bradford_Berk@URMC.rochester.edu

Jahar Bhattacharya, M.D., Ph.D.*
Professor, Departments of Physiology & Cellular Biophysics and Medicine
Lung Biology Laboratory, College of Physicians and Surgeons
Columbia University
St. Luke's-Roosevelt Hospital Center
New York, New York 10019   USA
Email: Jb39@columbia.edu

Sunita Bhattacharya, M.S.
Associate Professor of Clinical Medicine, Department of Pediatrics
Columbia University
St. Luke's-Roosevelt Hospital
1000 10th Ave.
New York, New York 10019   USA
Email: Sb80@columbia.edu

Donna Carden, M.D.
Professor of Internal and Emergency Medicine,
Department of Emergency Medicine
Louisiana State University Health Sciences Center, Shreveport
1501 Kings Highway
PO Box 33932
Shreveport, LA 71130-3932 USA
Email: dcarde@lsuhsc.edu

Matthias A. Clauss, Ph.D.
Associate Professor, Department of Cellular & Integrative Physiology
Indiana University School of Medicine
IB 433
975 West Walnut Street, IB 441
Indianapolis, IN 46236 USA
Email: mclauss@iupui.edu

Claire M. Doerschuk, M.D.
Professor of Pediatrics and Pathology
Case Western Reserve University
Rainbow Babies and Children's Hospital, Room 787
11100 Euclid Avenue
Cleveland, OH 44106 USA
Email: cmd22@po.cwru.edu

Avrum I. Gotlieb, MD, CM, FRCPC *
Professor and Chair, Department of Laboratory Medicine and Pathobiology,
Faculty of Medicine, University of Toronto,
Vascular Research Laboratory
Toronto General Research Institute,
Department of Pathology, University Health Network
100 College Street, Room 110
Toronto, Ontario, M5G 1L5 Canada
Email: avrum.gotlieb@utoronto.ca

A.B. Johan Groeneveld, M.D., Ph.D., FCCP, FCCM
Professor of Intensive Care Medicine, Department of Intensive Care
Institute for Cardiovascular Research
Vrije Universiteit Medical Centre
De Boeleaan 1117
1081 HV   Amsterdam   The Netherlands
Email: johan.groeneveld@vumc.nl

Stefan Hippenstiel, M.D.*
Department Internal Medicine: Infectious Diseases, Respiratory, and Intensive Care Medicine
Medizinische Klinik m.S. Infektiologie
Charité – University Medicine Berlin,
Augustenburger Platz 1
13353 Berlin   Germany
Email: stefan.hippenstiel@charite.de

Inkyung Kang, M.S.
Division of Integrative Biology, Department of Pediatrics
Case Western Reserve University
Rainbow Babies and Children's Hospital
Cleveland, OH 44106 USA
Email: ikang@cwru.edu.

Lopa Leach, Ph.D.*
Senior Lecturer in Vascular Biology, Faculty of Medicine
Centre for Integrated Systems Biology and Medicine
Institute of Clinical Research, Faculty of Medicine and Health Sciences
University of Queen's Medical Centre
Nottingham, NG 7 2 UH UK
Email: lopa.leach@nottingham.ac.uk

Joanne Lee, MSc
Department of Laboratory Medicine and Pathobiology
University of Toronto
Toronto General Hospital
200 Elizabeth Street, NU 1-126
Toronto, ON M5G 2C4   Canada

Hazel Lum, Ph.D.*
Associate Professor, Department of Pharmacology
Rush University Medical Center
Cohn Research Building, Rm. 416
1735 W. Harrison St.
Chicago, IL 60612 USA
Email hlum@rush.edu

Michael A. Matthay, M.D.
Professor of Medicine and Anesthesia, Cardiovascular Research Institute
Director of the Critical Care Medicine Training and
Associate Director of the Intensive Care Unit
University of California
505 Parnassus Ave., HSW-825
San Francisco, CA 94143-0130 USA
Email: mmatt@itsa.ucsf.edu

Hedwig S. Murphy, M.D., Ph.D.*
Assistant Professor, Department of Pathology
University of Michigan
3510 MSRB I
1301 Catherine Road
Ann Arbor, Michigan 48109-0602   USA
Email: hsmurphy@umich.edu

Viswanathan Natarajan, Ph.D.*
Professor, Department of Medicine
Division of Pulmonary & Critical Care Medicine
Johns Hopkins University School of Medicine
MFL 675 Center Tower
5200 Eastern Avenue
Baltimore, MD 21224   USA
Email: vnataraj@jhmi.edu

Sadiqa K. Quadri, Ph.D.
Associate Research Scientist, Department of Physiology & Cellular Biophysics
Lung Biology Laboratory, College of Physicians and Surgeons
Columbia University
432 W. 58th St., Antenucci 517-520
New York, New York 10019   USA
Email: skq1@columbia.edu

Sarah L. Sayner, B.Sc.
Department of Molecular and Cellular Pharmacology, Center for Lung Biology
MSB 3362
University South Alabama College of Medicine
Mobile, Alabama 36688   USA
Eamil: slsayner@hotmail.com

Dimitrije Stamenovic, Ph.D.
Associate Professor, Department of Biomechanical Engineering
Boston University
44 Cummington St.
Boston, MA 02215   USA
Email: dimitrij@engc.bu.edu

Troy Stevens, Ph.D.*
Director of the Center for Lung Biology
Professor, Department of Molecular and Cellular Pharmacology
MSB 3364
University South Alabama College of Medicine
Mobile, Alabama 36688   USA
Email: tstevens@jaguar1.usouthal.edu

Norbert Suttorp, M.D.
Professor and Director, Department Internal Medicine:
Infectious Diseases, Respiratory and Intensive Care Medicine
Medizinische Klinik m.S. Infektiologie,
Charité – University Medicine Berlin,
Augustenburger Platz 1
13353 Berlin   Germany
Email: norbert.suttorp@charite.de

Mabel Sze, B.Sc.
Department of Laboratory Medicine and Pathobiology
University of Toronto
Toronto General Hospital
200 Elizabeth Street, NU 1-119a
Toronto, ON   M5G 2C4   Canada

Eric Thorin, Ph.D.*
Research Associate Professor, Département de Chirurgie,
Institut de Cardiologie de Montréal
Université de Montréal
5000, rue Bélanger
Montréal (Qc) H1T 1C8   Canada
E-mail: Eric.Thorin@icm-mhi.org

Peter V. Usatyuk, Ph.D.
Research Associate, Department of Medicine
Division of Pulmonary & Critical Care Medicine
Johns Hopkins University School of Medicine
MFL 683 Center Tower
5200, Eastern Avenue
Baltimore, MD 21224   USA
Email: pusatyuk@yahoo.com

Victor W.M. van Hinsbergh
Professor, Faculty of Medicine
VU Medical Center
Laboratory for Physiology, Rm. B150
Boechorststraat 7
1081 BT Amsterdam   The Netherlands
 Email: hinsbergh@physiol.med.vu.nl

Gerard P. van Nieuw Amerongen, Ph.D.*
Assistent Professor, Faculty of Medicine
VU Medical Center
Laboratory for Physiology, Rm. C-162
Boechorststraat 7
1081 BT Amsterdam   The Netherlands
Email: nieuwamerongen@physiol.med.vu.nl

James Varani, Ph.D.
Professor of Pathology, Department of Pathology
University of Michigan
M4224 MSI 0602
1301 Catherine Street
Ann Arbor, Michigan 48109-0602 USA
Email: varani@umich.edu

Qin Wang, Ph.D.*
Assistant Professor, Department of Pediatrics
Case Western Reserve University
Room 787, RB&C
11100 Euclid Avenue
Cleveland, OH 44106 USA
Email: qxw9@po.cwru.edu

Peter A Ward, M.D.
Professor and Chairman, Department of Pathology
University of Michigan
M5240 MSI 0602
1301 Catherine Road
Ann Arbor, Michigan 48109-0602 USA
Email: pward@umich.edu

Joanna Zurawska
Department of Laboratory Medicine and Pathobiology
University of Toronto
Toronto General Hospital
200 Elizabeth Street, NU 1-119a
Toronto, ON M5G 2C4 Canada

Chapter 1

The Cellular and Molecular Foundations of Pulmonary Edema

Carolyn E. Patterson, Ph.D.[1]* & Michael A. Matthay, M.D.[2]

[1]Departments of Medicine & Physiology, Indiana University School of Medicine & Roudebush VA Med. Center, Indianapolis IN., USA 46202, [2] Cardiovascular Research Institute, Univ. of California, San Francisco CA, USA 94143

CONTENTS:

Introduction

Basic Architecture of the Lung and its Circulation
Dual Circulation Systems
Pulmonary Circulation and Gas Exchange
Bronchial Circulation and Endothelial Heterogeneity

The Alveolar-Capillary Membrane and Lung Fluid Balance
Architecture, Area, and Leakage
Forces Influence Leakage
Barrier Permeability Influences Leakage

Endothelial Barrier Dysfunction: Activation & Injury
Weakening and Collapse of the Dam

Clinical Occurrence and Importance of Pulmonary Edema
Victims of the Flood
Pathology Alters the Forces on the Dam
Pathology Alters Barrier Permeability
Saboteurs
Endothelial Dysfunction and Gas Exchange

Multifunctional Endothelium
Interactivity of Barrier and Other Vital Endothelial Functions
Importance of the Endothelium in Multiple Pathologies

Building Better Dams

Figure 1: The Flood and the Fire.
Lithograph of the Johnstown Flood of 1889. Reproduction from The Granger Collection, New York.

Introduction

The rain began in the peaceful little town, nestled among the Pennsylvania hills, shortly after the Memorial Day parade ended. It continued through the night and on into the next day, spilling over the banks of small streams and filling up Lake Conemaugh, a picturesque fishing and hunting playground for prominent wealthy industrialists. By mid-afternoon, the water was pushing over the top of the sagging middle of the earthen South Fork Dam. Suddenly, a section gave way forming a ten-foot gap, followed closely by collapse of the entire center barrier. The forty-foot-high torrential wave thundered down the valley, sweeping away animals, trees, locomotives, tracks, houses, and - - - the town. The deluge somehow precipitated the outbreak of the sinister fire that consumed helpless, trapped survivors of the flood. In the end, over two thousand Johnstown residents perished from the flood and the fire.

Though fragile in construction, that dam had served as an effective barrier separating the waters from the valley below for 37 years. Our endothelium, like that dam, is fragile, but serves as an effective barrier separating the circulating waters from the tissues below and we, like the residents of Johnstown, rely on that dam for our very lives. Amazingly, this vitally important barrier function is largely performed by a single thin layer of the endothelial cells, which line all blood vessels and indeed are the capillaries. The importance of endothelial barrier integrity is particularly apparent in the lung, where the tissue beyond the barrier is also only a single, thin layer of epithelial cells, and a breech in the endothelial dam can cause bursting of the epithelium, rapid flooding of the alveolar air sacs, impairment of gas exchange, ignition of the fires of inflammation, and drowning from the pulmonary edema.

Although once merely thought of as a passive barrier, with advances in cell biology and biochemistry, endothelial cells are known today as dynamic, metabolically active cells that perform a wide variety of critical life functions. Nevertheless, the active responses of endothelium to untoward stimuli resulting in sabotage of the barrier function remain a significant hazard in a number of clinical settings. Therefore, study of this essential role of endothelium as a dam has progressed from the physiologic description of a nearly inert membrane to a fervent endeavor at defining the cellular and biochemical mechanisms that govern maintenance of barrier integrity and recovery from insult. As pathologic stresses resulting in rupture of this dam and concomitant inflammation remains a serious problem in medicine, these efforts to define the causes and consequences of endothelial dysfunction and to gain control over the flood and the fire is the central focus of this book.

Basic Architecture of the Lung and its Circulation

Dual Circulation Systems

To gain an understanding of the causes and consequences of pulmonary edema, it is first necessary to understand the fundamental construction of the lung and its circulation. Simply stated, the lungs have two different circulatory systems: the pulmonary and the bronchial circulations (Harris and Heath, 1986). Pulmonary circulation is a low pressure, high flow system, which receives the total output of the right heart ventricle. Although the bloodless, lungs are only ~ 2% of the total body mass, remarkably this pulmonary flow (~ 5L/min., at rest) equals that of the entire remainder of the body, which receives the matched systemic output of the left ventricle, termed cardiac output. In other words, pulmonary blood flow equals systemic blood flow, as right ventricular output must match the output of the left ventricle. As basic as this flow balance is, this fundamental concept is widely unappreciated, especially in the context of development of pulmonary edema. Secondarily, the lungs also receive bronchial circulation, a small branch of the systemic system. In contrast to pulmonary circulation, bronchial circulation is a high pressure, low flow system.

Pulmonary Circulation and Gas Exchange

The high flow of the pulmonary circulation has one, well appreciated, primary purpose – the exchange of blood gasses with the air in the lung. Just as the lungs receive flow matched to the cumulative total of all other body organs, the pulmonary vascular/endothelial surface nearly matches that of the cumulative total endothelial surface of all the other organs. This lung endothelial surface area is estimated at 75-250 m^2 based on capillary radius and number (Ryan and Ryan, 1984; Muller and Griesmacher, 2000). The cell volume of pulmonary endothelium constitutes one fifth of the total lung parenchymal tissue volume. This significant portion is principally a function of the enormous capillary area, as the cells are exceedingly thin. The tortuous, interconnected capillary bed runs through the walls between adjacent alveolar air sacs (Figure 2). As adult humans lungs have ~ 300 million alveoli and each has 100 to 1000 interconnecting capillary segments, the enormity of the area is more easily understood. This large luminal area and minimal cell thickness provides for efficient oxygenation of blood and removal of carbon dioxide, with plenty of reserve. It is the redundant capillary network and accompanying arterioles and venules that allow the lungs to accept the high flow, while under a low propulsion force. It should be noted for clarity, that high flow does not mean that the blood volume in the lungs equals that of the rest of the body. Only 10-12% of total blood volume is contained in lung circulation at any instant in time and of this, only 10-15% is in the pulmonary capillary bed, or less than 100 ml. Moreover, at rest, many of the capillaries are not in use. Gravity partly dictates which capillaries are opened, but the pathway is constantly changing. When cardiac output increases to meet the demands of exercise, there is recruitment of reserve pulmonary capillaries to accommodate the increased flow (up to 40L/min).

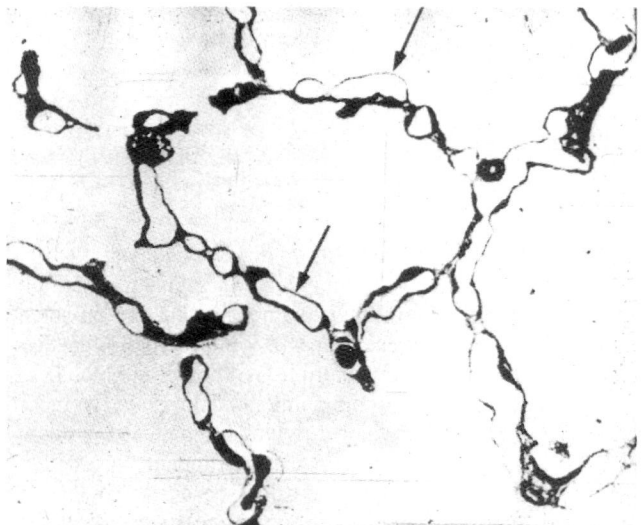

Figure 2: Pulmonary Capillaries. The large surface area of the multiple interconnected capillary segments in the walls between the alveoli provide for efficient gas exchange between blood and alveolar air.

In addition to gas exchange, this large pulmonary vascular bed serves other important purposes. The pulmonary microcirculation, with its large reserve, filters out blood clots; thereby, protecting the heart and brain, which don't have such redundant circulatory reserves. The large endothelial surface also serves as a clearinghouse for vasoactive and other bioactive mediators, before venous blood re-enters systemic circulation as arterial blood. Similarly, the pulmonary endothelium is the first major tissue to "see" iv-administered drugs. Because of this surface reserve, the lung also acts a major repository for leukocyte sequestration in cases of generalized activation. Finally, the pulmonary circulation does provide nutrients and waste exchange for the lung parenchymal cells, somewhat like flies at an elephant's trough.

Bronchial Circulation and Endothelial Heterogeneity

The bronchial circulation arises off the aortic arch and, thus, is a part of the systemic circulatory system and shares the high-pressure aspect of that system. It serves to provide nutrients and oxygenation to the airways, bronchial smooth muscle, and walls of the large vessels of the pulmonary circulation system. The total area and volume of the bronchial endothelium is much, much less than the endothelium of the extensive pulmonary circulation.

As will be discussed in Chapter 10 on endothelial heterogeneity, there are distinct differences in the endothelium of the bronchial and the pulmonary circulations. One such difference is that the bronchial endothelial cells are associated with a basement membrane containing collagen VI and VII, while the pulmonary endothelial cells are associated with collagens IV, V, and VIII (Kawanami, 1997). There is good evidence for anastamoses between the pulmonary and the bronchial system at the microcirculation level (Kawanami, 1997).

Not only do pulmonary and bronchial (or systemic) endothelium differ, but characteristics vary between organs, between arteries, veins, and capillaries, between large arteries and small arterioles, between areas of high and low shear stress, and between activated and quiescent endothelium, as will be discussed in Chapter 10. For instance, the blood-brain barrier is basically a characteristic of tight adhesion between adjacent endothelial cells in brain vasculature, which permit a very limited exchange between blood and cerebrospinal tissue; whereas, endothelium of the renal glomeruli are fenestrated and relatively leaky. The pulmonary endothelial dam is not absolute. It is intermediate between the extremes of the glomerular and the brain endothelium. Related to this intermediate selectivity, the vast area, and the high flow, there is a constant fluid exchange between the blood and the lung interstitium and small imbalances could spell big trouble as will be discussed in the following section. Thus, the lymphatic system in the lung is well developed to drain excess fluid from the interstitial spaces. The endothelial cells, which line the lymphatics, share some characteristics with vascular endothelium, but also have distinct differences as might be expected. Because the focus of this book is pulmonary edema, discussion of bronchial and lymphatic endothelium will be minimal as endothelium of the pulmonary circulation are the primary cells of import to dysregulation of lung fluid balance.

The Alveolar-Capillary Membrane and Lung Fluid Balance

Architecture, Area, and Leakage

The large pulmonary capillary surface area is generally matched by the equally large epithelial surface area of those 300 million alveoli. The interface providing for rapid gas exchange is known as the alveolar-capillary membrane, and as inferred from the name, consists of a thin type I alveolar epithelial cell laminated to an equally thin capillary endothelial cell (Figure 3). Thus, this fragile membrane, at its minimum, is little more than the cell membranes of the two facing cells, small amounts of cytoplasm, and a thin basal lamina between the cells to give strength. Over a great extent of the alveolar-capillary interface, this membrane is only ~200 nm thick, but other areas of the alveolar wall are thicker due to greater inclusion of cytoplasm, cell organelle, and the nuclei within the membranes, a thicker strength-imparting basal lamina between the two cells, and the intermittent presence of interstitial cells. The areas at the corners between the alveoli are more cellular, containing more interstitial fibroblasts and pericyte cells, the round type II epithelial cells – progenitors of the thin type I epithelial cells, the non-capillary vessels, and also more type I – structural collagen. These thicker areas are important sites for interstitial swelling, but are not important for gas exchange.

Resistance to fluid and solute movement from the blood to the air space is a combined function of the endothelial barrier, characteristics of the interstitial matrix, and the tight epithelial layer (Patterson and Lum, 2001). Actually, the alveolar epithelium is more water impermeable than the endothelium due to more numerous tight junctions (Doyle et al., 1999). Certainly, epithelial damage has profound consequences and will be discussed in the section on clinical edema, but the barrier properties of the endothelium are critical to normal maintenance of fluid balance and protection of the epithelium from excessive stress (Crandall et al., 1983; Pietra, 1984). Endothelium activation or injury is a major pathological factor in pulmonary edema (Ruben et al., 1990; Ketai and Godwin, 1998; Bannerman and Goldblum, 1999; Pararajasingam et al., 1999; Reinhart et al., 2002). Yet, even in the normal, intact, quiescent, lung endothelial layer, there is continuous movement of water in and out of the vasculature. Small neutral solutes like glucose move almost as freely as water, while the barrier is increasingly effective with regard to electrostatic charge and increasing solute size. Under basal conditions, fluid filtration in the lung has been estimated to occur ~ 20% in the arteries, ~ 40 % in the microvessels, and ~ 40% in the veins (Lin et al., 1998; Parker and Yoshikawa, 2002). Given the much, much larger area of the microvascular surface, the endothelium in the microvessels are considerably tighter than in the arteries and larger arterioles, which are tighter than endothelium in the larger venules and veins. Nevertheless, normally all the lung endothelium greatly restricts movement of large, charged proteins like the major plasma protein, albumin, with a molecular weight of about 70,000 (Effros et al., 1998). Under normal circumstances the endothelial barrier is nearly absolute with regard to undirected passage of blood cells out of the vasculature.

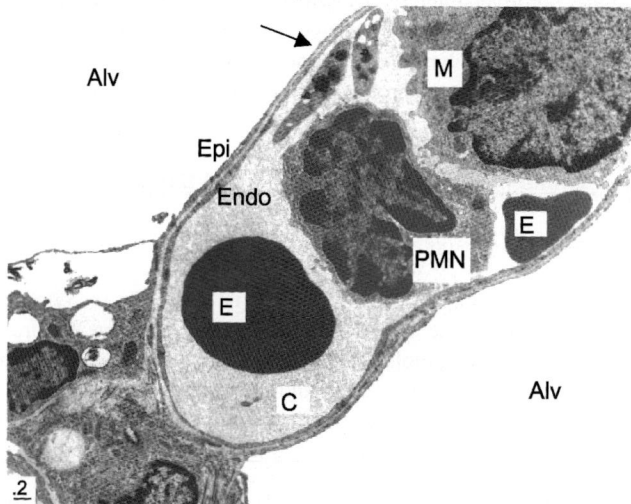

Figure 3. The Alveolar-Capillary Membrane. Electron micrograph of an alveolar wall capillary between two alveoli (Alv), showing the thin endothelial cell (Endo) lining of the vessel and the adjacent thin alveolar type I epithelium (Epi), facing the alveoli. The interstitial matrix is noted between the adjacent epithelial and endothelial cells over most of the alveolar-capillary membrane (~0.2 μm), while in some areas the membranes of the epithelium and endothelium appear fused. Type II epithelial cells, interstitial cells, microvessels, and larger vessels are in the corners of intersecting alveoli. Portions of formed elements are noted in the lumen of the capillary: E- erythrocytes, PMN – neutrophils, M – macrophage, P - platelets. The rat lung was fixed by tracheal instillation of glutaraldehyde. From *Patterson, C.E., Barnard, J.W., LaFuze, J.E., Hull, M.T., Baldwin, S.J., and Rhoades, R.A. (1989) Neutrophil activation and changes in microvascular pressure is acute pulmonary edema. Amer. Rev. Resp. Dis. 140:1052-1062.* By permission from The American Thoracic Society and The American Lung Association.

Forces Influence Leakage

Despite the relatively low pressures of pulmonary circulation, hydrostatic pressure in the vascular lumen provides filtration pressure to drive fluids from plasma across the endothelial layer. This pressure declines from the systolic pulmonary arterial pressure (P_a) of ~ 25 mmHg to ~ 10 mmHg – the mean capillary pressure (P_c) and to ~ 7 mmHg in the pulmonary veins. One might expect the hydrostatic pressure of the interstitial space (P_i) to directly counter this expulsive force, but it is often slightly negative (-5 mm Hg) in the lung, related to architecture and respiration. If there is a build up of matrix hydration with rapid leak or poor drainage, it can become positive and partially counter the hydrostatic filtration force. The greater the net hydrostatic driving pressure and the larger the filtration area and the more permeable the membrane is to water, the more total fluid will be filtered. The product of surface area and permeability is known as hydraulic conductance (K). In addition to hydrostatic pressures, there are protein osmotic pressures (sometimes referred to as oncotic pressures) with opposite vectors that influence fluid movement. This protein osmotic pressure is exerted due to the osmolarity of "impermeable" colloid-like proteins in plasma (again mainly albumin) of ~ 25 mmHg (π_{pl}), which in effect "pulls" water from the interstitium to the vessel lumen. Just how "impermeable" the membrane is to the macromolecules is described by σ, the macromolecule reflection coefficient. Similarly, large solutes in the interstitium exert a protein osmotic pressure of 5-15 mmHg (π_i) that partially counters the protein osmotic pressure of the plasma. The balance, or lack thereof, of these push-pull forces was described more than a century ago and is represented by the Starling-Landis equation (Figure 4). The balance of these factors in each segment determines the net fluid filtration (Q) of the vasculature.

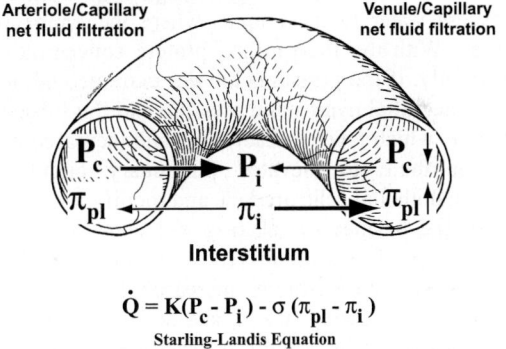

$$\dot{Q} = K(P_c - P_i) - \sigma(\pi_{pl} - \pi_i)$$
Starling-Landis Equation

Figure 4: The Starling-Landis Equation. The balance of hydrostatic and protein osmotic pressures determine filtration of fluid across the endothelium. Here P_c indicates the mean capillary blood pressure, but the pressure of any vessel segment could be used to calculate net transmural flow in that segment. P_i is the hydrostatic pressure of the interstitium, which may be positive or negative depending on the tissue and condition of the interstitial matrix. The pressure gradient ($P_c > P_i$) is multiplied by the hydraulic conductivity factor (K), which takes into account surface filtration area and the resistance characteristics of the barrier to water movement and the arrow indicates the direction of water movement driven by the usual hydrostatic pressure gradient. Protein osmotic (also termed by some as colloidal oncotic) pressure is represented by π. The plasma protein osmotic pressure is greater than the interstitial protein osmotic pressure, but the direction of water movement has an opposite vector than the gradient. This gradient is multiplied by σ, the reflection coefficient, that describes the characteristic impermeability of the endothelium to movement of charged macromolecules. The sum of these forces and resistances determine net fluid flux, Q.

On the arterial side of the pulmonary circulation the net force for filtration is normally slightly positive. Fortunately, these forces shift on the venous side to favor net reabsorption of the filtered fluid. This occurs because the loss of fluid and retention of protein drives up the protein osmotic pressure in the venous plasma, while the hydrostatic pressure falls along the vascular bed. In fact, if water is introduced into the alveoli at levels not interfering with gas exchange, it is actively transported by the epithelium to the interstitium and is also rapidly absorbed into the pulmonary circulation due to the favorable balance for reabsorption in the venous bed (Matthay et al., 1996; Song et al., 2000). Small imbalances are normally corrected by the flow of lymph from the lung interstitium. Thus, despite the large surface area, the delicate nature of the alveolar-capillary membrane, and the continuous dynamic fluid turnover, the alveoli normally remain dry and the interstitium at a steady state of hydration.

Obviously, if pulmonary vascular pressures increase, while endothelial permeability properties are unchanged, greater amounts of fluid are filtered. This occurs in common situations producing lung edema, known as "hydrostatic edema," and the clinical instances and implications will be further discussed below. When pressures rise, resembling the rising waters in the reservoir behind the dam and filtration increases, the interstitial space swells and perivascular and peribronchial fluids accumulate, and lung parenchyma begin to resemble the sodden, dam of Lake Conemaugh. The initial site of fluid accumulation in the interstitial space is partially dependent upon the site of leakage, but interstitial spaces are more or less continuous and the excess water and pressures can redistribute. Often the thicker alveolar-capillary spaces swell before there is evidence of widening of the thinner regions, which are so important for continued gas exchange. The alveolar wall interstitium is also continuous with the more loosely organized, compliant matrix in the interlobular, perivascular, and peribronchiolar spaces, which serve as fluid storage reservoirs. Pressure gradients in the interstitium drive excess water to these areas where the lymphatic drainage begins; thus, lymph flow increases. With the fluid influx, protein concentration in the interstitium and lymph declines. Normally, if the pressures are not extreme and decline, this is self-limiting due to the increase in interstitial hydrostatic pressure and the subsequent decline in interstitial protein osmotic pressure, which favor reabsorption, and increased lymph flow.

Many factors contribute to elevated pulmonary circulatory pressures, including endothelial products. Endothelial cells produce and modify a variety of vasoactive molecules, that significantly alter the balance of relaxing and constrictive influences on the vessels, contributing to development of pressure driven or "hydrostatic edema" (Crandall et al. 1983; Ketai and Godwin, 1998). Furthermore, increased luminal pressure causes mechanical distension of endothelium, increasing $[Ca^{++}]_i$ and activating the cells (Kuebler et al., 1999). When pulmonary hypertension is severe or sustained, vessel distension can result in mechanical stress and signaled loss of endothelial integrity, resulting in a second type of edema - "permeability edema" (Patterson and Lum, 2001). In studies of the tensile strength provided to the alveolar-capillary membrane by the interstitial proteins, increased capillary pressures (39 mm Hg) resulted in stress-rupture of the normal endothelial layer prior to disruption of collagen and laminin fibers or epithelium (West and Mathieu-Costello, 1999).

Barrier Permeability Influences Leakage

Thus, "permeability edema" can occur secondary to hydrostatic stress when there is mechanically induced compromise of endothelial properties. But permeability edema involving barrier disruption can even occur during pulmonary hypotension, if endothelia are activated or damaged by non-mechanical sabotage, that impairs selectivity and allows leakage of large molecules into the interstitium (Fein *et al.*, 2000; Ware and Matthay, 2000). Under these conditions, protein leak drives up the interstitial protein and the protein osmotic gradient favoring reabsorption is dissipated. Initially, lymph protein content rises with the rise in interstitial protein and lymph flow increases. If interstitial pressure exceeds 40 mmHg, excess water swells the alveolar-capillary membrane and the epithelial barrier may rupture, resulting in a severe form of permeability edema - alveolar edema (Crandall *et al.*, 1983; Ketai and Godwin, 1998). If this occurs, proteinacious fluid pours into the alveoli, blocking gas exchange. The combinations of hydrostatic, hydrostatic stress-induced permeability, primary endothelial permeability, and primary or secondary epithelial permeability in clinical edema will be discussed below. Thus lung edema cannot be simply classed as hydrostatic vs. permeability, as it can involve simple hydrostatic edema with no damage to the alveolar-capillary membrane, permeability edema related to endothelial activation but without membrane damage, interstitial or alveolar edema with damaged to the alveolar-capillary membrane, and mixed permeability and hydrostatic edema (Ketai and Godwin, 1998).

Potentially fluids and solutes can move across an endothelial layer by either going through or around the cells. There is evidence for solute exchange through the cells without an increase in the hydraulic conductance. Specialized substances such as immunoglobins are transported to tissues by formation of vesicles and transcytotic movement through the endothelial cell. Albumin, particularly when coupled to fatty acids, is regularly exchanged between plasma and interstitium by specific binding to gp-60 protein and transcytotic movement (Galis *et al.*, 1988; Schnitzer *et al.*, 1992; Tiruppathi *et al.*, 1997). Yet, the net active transport of albumin favors an abluminal to luminal direction by >10 fold, consistent with the low interstitial protein content (Shasby and Shasby, 1985). Although transcytosis is a major physiologic route for nutrient, hormone, and albumin flux, the most important mode by far for macromolecular movement from vessel lumen to the subendothelial abluminal space in conditions of abnormal permeability and edema development is the convective, pressure driven, bulk movement of fluid and solvents through gaps between the borders of cells, which is known as paracellular transport (Albelda *et al.*, 1988; Patterson *et al.*, 1994; Lum and Malik, 1994; Baldwin and Thurston, 2001; Patterson and Lum, 2001). Thus, permeability edema is largely identified as the compromise of the normal endothelial barrier integrity due to altered border characteristics between adjacent endothelial cells.

Endothelial Barrier Dysfunction: Activation & Injury

Weakening and Collapse of the Dam

Apart from mechanical stress-induced injury, insults to the pulmonary endothelium resulting in primary barrier dysfunction arise from broadly diverse etiologies. These include events originating at remote sites, as well as events originating within the alveoli or within the pulmonary circulation *per se*, as will be discussed in the next section on clinical lung edema. Depending on the severity, duration, extent, and the nature of the insult, the response of the endothelium can range from transient, mild, localized activation to widespread cell injury, death, and denudation of the vascular intima as illustrated in Figure 5. In concert with this spectrum of responses, the permeability that results may be readily reversible, self-limited interstitial edema or alveolar flooding, organ failure, and death of the organism. Low-level insult and a variety of biologic mediators induce barrier dysfunction by activation of cellular biochemical responses that return to the quiescent steady state as soon as the stimulus is removed. Such cellular responses, shown in Figure 5, include: membrane and second messenger signaling, kinase-mediated protein phosphorylation, reorganization of the endothelial cytoskeleton, and altered expression and function of luminal-, lateral-, and basal-adhesion molecules (Patterson and Lum, 2001). These responses may be almost immediate or require hours and cause retraction and gap formation between endothelium opening up pathways for paracellular fluid and abnormal protein movement without loss of the endothelium from the luminal surface. Such endothelial activation results not only in barrier dysfunction, but also in alteration of other important endothelial regulatory functions governing the balance of bioactive mediators in the vasculature, hemostasis, and leukocyte function. The consequences of these altered sequeli have a secondary exacerbating effect on the endothelium and its barrier integrity (Figure 5). For instance, endothelial activation is accompanied by apical surface expression of adhesion molecules that triggers recruitment and adherence of leukocytes, resulting in direct presentation of leukocyte-derived mediators and oxidants at the endothelial surface, in addition to those originally borne in blood.

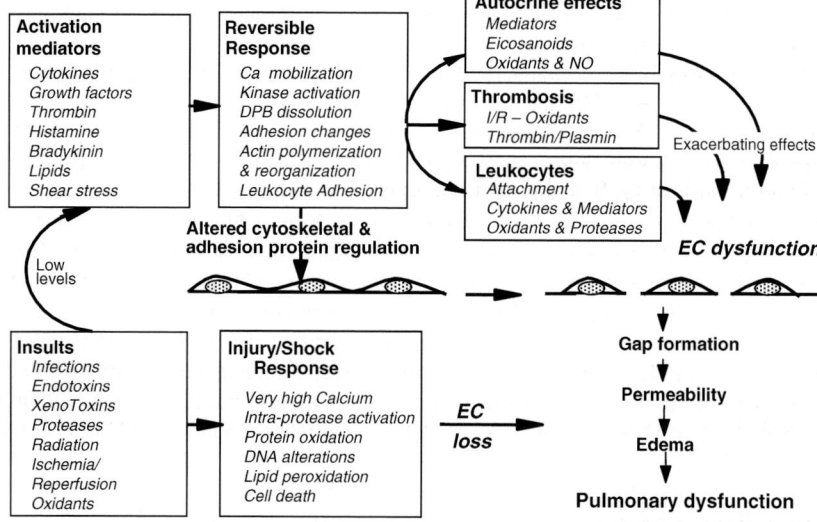

Figure 5: The role of endothelial activation and injury in pulmonary dysfunction. Injurious conditions can lead to extreme biochemical derangement in endothelium, culminating in cell sloughing and cell death; thus, removing the effective barrier that endothelial cells (EC) provide to fluid, protein, and cell loss from vascular to tissue and alveolar spaces. The ensuing proteinaceous alveolar edema compromises pulmonary circulation and gas exchange. At lower exposures, effects of injurious agents may be limited to activation of the EC. A variety of activating mediators lead to intracellular signaling of reversible alterations in EC that do not result in cytotoxicity, but do cause altered barrier function. These direct effects on regulation of the EC cytoskeleton and baso-lateral adhesions, resulting in increased monolayer permeability, are exacerbated by activation of other EC functions such as expression of apical adhesion molecules for leukocytes, release of certain autocrine factors, and enhanced thrombosis. The degree of edema formation is also influenced by altered endothelial control of vascular smooth muscle contraction, resulting in elevated filtration pressures. Figure modified from Patterson and Lum, 2001. Coyright (2001) from *Update on pulmonary edema: the role and regulation of endothelial barrier function.* by C.E. Patterson and H. Lum. Reproduced by permission of Taylor & Francis, Inc., http://www.taylorandfrancis.com

When the sabotage is severe and sustained, additional biochemical signaling may result in cell shock effects with cytotoxic cell death or activation of apoptosis (Figure 5). If the endothelia die or are otherwise lost from the intimal surface, repair will involve proliferation, migration, and re-endothelialization of the vessel lumen from the wound edges or from circulating progenitor cells, if the organism is to survive (Park *et al.*, 2004). Normal turnover of pulmonary endothelium is less than 1% of the cells/day, but this can increase dramatically in response to injury. Thus, the ability of endothelia to recover, regrow, and restore the barrier will influence the extent and outcome of pulmonary edema. Obviously, the continued presence of the underlying pathology with continued contribution of the insults, the extent of the denudation, the resultant damage to the interstitial components and the epithelium, and the extent of inflammatory activation will also affect whether there is adequate time for repair and recovery before alveolar edema results in respiratory failure. The mediators, mechanisms, and responses will be described in detail in the chapters of this volume.

Clinical Occurrence and Importance of Pulmonary Edema

Victims of the Flood

Shortness of breath, distress, and suffocation are the descriptions of dyspnea reported by patients at the first serious stages of edema. A dry cough and labored breathing may be noted, but the crackling rales present on physical exam raise the suspicion that alveolar edema is the actual cause of the respiratory problems. If the edema progresses, the cough produces pink foamy sputum, confirming that the dam has burst, that fluid, protein, and blood cells have now "thundered down the valley" and are filling the alveoli, and that cyanosis and even death may soon follow. As mentioned above, the precipitating factors in pulmonary edema are manifold and they may initially arise within the lung or at remote sites. The edema may be mild and limited to hydrostatic, interstitial edema, producing only a slight change in lung compliance and go undetected, especially if the history of other conditions that accompany hydrostatic edema doesn't motivate more careful probing. Or the edema may appear out of nowhere and claim the victim, before anything can be done to stop the flood. Regardless of origin, the acute or chronic nature of the edema, and the severity of manifestations, the pathological mechanisms converge upon disruption of the basic physiological principals outlined above - balance of hydrostatic pressures, balance of protein osmotic pressures, permeability of the alveolar-capillary membrane, or lymph flow. In the clinical setting, these factors are interactive, such that perturbation in one may alter another. Detailed discussion of lung injury in disease is found in Chapters 12 and 13, but here we will briefly introduce the relationship of pulmonary edema development to the physiologic factors described above.

Pathology Alters the Forces on the Dam

One of the most common causes of lung edema is cardiogenic edema due to left heart failure. While heart failure (generally meaning left heart failure) is defined as the inability of the heart to deliver sufficient systemic circulation to meet the body's needs for oxygen and nutrients, there are important pulmonary consequences (Edoute *et al.*, 2000; Gehlbach and Geppert, 2004). The volume of blood and the pressures build up in the pulmonary veins, which converge on the left atrium, returning blood to the ailing left heart. This increase in pulmonary venous pressure translates to a chronic rise upstream in lung capillary pressure and, thus, an increased hydrostatic drive for fluid filtration, tipping the balance for venous reabsorption of filtered lung water. Similarly, resistance due to mitral valve stenosis increases the pulmonary venous and, hence, capillary pressures. In both of these chronic conditions, there is a partial adaptation and thickening of the alveolar capillary membrane, especially following several incidents of pulmonary edema, that allow it to withstand somewhat higher pressures (40 mm Hg) before there is break-through alveolar edema (Gehlbach and Geppert, 2004). Elevated hydrostatic pressures also occur from neurogenic injury, which induces pulmonary vasoconstriction *via* direct neural and/or secondary hormonal output (Kerr, 1998). Volume overload can cause hydrostatic edema and, conversely, volume minimalization is often used to treat lung edema, whether hydrostatic or permeability. An altered balance of vasoactive mediators, released in many disease states, which bring about either dilation of the arterioles or constriction of the venules, cause elevated lung capillary pressure and hydrostatic edema (Barnard *et al.*, 1989; Lindenschmidt et al, 1983).

Primary pulmonary hypertension, involving vessel remodeling, produces chronic hydrostatic stress. In contrast to vasculature of systemic organs, which dilate in response to hypoxia to improve circulation, pulmonary vessels constrict to shift flow to aerated areas and improve perfusion/ventilation matching. But, global hypoxia, such as occurs at high altitudes, causes widespread rather than local corrective pulmonary vasoconstriction and creates hydrostatic, interstitial edema. If fluid accumulation is not extreme, edema appears as fluid cuffs in peribronchiolar and perivascular spaces, and blood gasses are relatively unchanged.

However, when hydrostatic edema overwhelms the ability of the lung to compensate, the mechanical stress on the endothelium and the epithelium can lead to secondary permeability edema, as explained above. Left heart failure, mitral valve stenosis, and even extremes of exercise may result in hydrostatic, cardiogenic stress-induced permeability edema (West and Mathieu-Costello, 1999). Difficulties in the differential diagnosis in cases of heart failure and pulmonary complications arise from the complex interactive nature of the maladies and other underlying pulmonary diseases (Gehlbach and Geppert, 2004). When high altitude-vasoconstrictive hypertension is not dealt with immediately, impaired endothelial function may cause severe respiratory failure due to the mechanical stress on the alveolar-capillary membrane (Koizumi *et al.*, 1999). Neurogenic edema is not uncommonly advanced to a permeability edema, although hydrostatic mechanisms seem to predominate in clinical neurogenic edema (Smith and Matthay, 1997). Moreover, experimental elevation of pulmonary pressure induced endothelial apoptosis (Gotoh *et al.*, 2000). Over-ventilation in clinical settings can cause direct mechanical stress failure to both the alveolar epithelium and the endothelium resulting in and increased permeability (West and Mathieu-Costello, 1999).

Disease conditions that decreased plasma protein osmotic pressure or increased tissue protein osmotic pressure instigate both increased loss of plasma fluid to the interstitium and also failure to return the fluids on to the venous circulation, due to the shift in Starling forces (Kramer *et al.*, 1983). Liver disease and under nutrition can reduce normal colloidal proteins in blood, while kidney disease leads to loss of protein due to abnormal renal filtration and failure of reabsorption mechanisms. On the other hand, tissue damage in lung could cause a release of protein into interstitial space, driving up tissue protein osmotic pressures, but this cause of edema is more common in other tissues, such as muscle. When hydration of the interstitial space increases, lymph flow increases to compensate and drain excess fluids, but injury or infection which blocks operation of the lymph system, would result in edema. Lung cancers, pulmonary fibrotic diseases, emphysema, and post-lung transplant abnormalities are conditions with effects on lymphatic integrity and function. As with hydrostatic edema, when the imbalance in protein osmotic pressures or lymph drainage or combinations are severe enough, tissue hydrostatic pressure can build to the level causing mechanical failure of the epithelial barrier and interstitial edema spills over into alveolar edema.

Pathology Alters Barrier Permeability

Edema resulting from primary altered permeability of the alveolar-capillary membrane is one of the most difficult situations for the pulmonary or critical care physician to manage. In general, pressures and plasma protein levels are more amenable to manipulation than metabolic and structural failure at a cellular level. When precipitating events, regardless of site or nature of origin, converge to produce acute life-threatening lung edema with proteinaceous exudates into the alveoli and concomitant arterial hypoxemia, the condition is now termed Acute Lung Injury (ALI) or Acute Respiratory Distress Syndrome (ARDS), depending on severity (Bernard *et al.*, 1994). The correlation between development of hemorrhagic broncho-pulmonary fluid and the high incidence of mortality among casualties due to a variety of traumas was recorded, analyzed, and reported in World War II. This syndrome was termed "wet lung of trauma" (Brewer, 1981). By the Vietnam War, there was improved ability to deal with the primary traumas, minimize shock, provide cardiovascular support, and sustain renal circulation and function; thus "the lung emerged as the limiting organ system" (Baue, 1974). A high-incidence of post-trauma mortality associated with pulmonary edema was evident. This non-cardiogenic respiratory failure from edema subsequent to initial treatment and stabilization was termed "Da Nang lung" and was later shown to be partly related to the battlefield use of packed red cells and crystalloids in lieu of whole blood, which provides more normal plasma protein osmotic pressure (Pearce and Lyons, 1999). At this same time, Ashbaugh, Bigelow, Petty, and Levine described a clinical syndrome of sudden, severe bilateral pulmonary infiltrates following seemingly unrelated traumas, shock, viral pneumonia, and hemorrhagic pancreatitis. Importantly, they discovered that this small sample of patients responded to positive end expiratory pressure ventilation with improved blood gas and improved incidence of survival (Ashbaugh *et al.*, 1967). They noted that the clinical course resembled that of babies with Infantile Respiratory Distress Syndrome and determined that the foamy fluid expressed from two patients who died had deficient surfactant activity. They termed this non-cardiogenic respiratory failure, "Adult Respiratory Distress Syndrome." Because negative pressure ventilation was normal procedure and it was generally thought that positive pressure ventilation would have untoward effects on venous return and cardiac function, it took four tries to get acceptance of their paper. But, it was then immediately noticed by the military physicians who were dealing with the dire consequences of Da Nang lung (Petty, 2001).

Compromised alveolar-capillary membrane permeability is a cardinal feature of ARDS and the degree of endothelial activation and injury is an important determinant of outcome (Ware *et al.*, 2001; Groeneveld, 2003; Matthay *et al.*, 2003). The term, syndrome, affirms the diverse nature of the underlying or initiating causes, the multifactoral processes involved in the lung damage, and the common result of lung failure. ARDS was recognized in children after trauma, as well as adults, and "Acute" was substituted for "Adult" (Murray *et al.*, 1988). Radiologic imaging has long been a standard method to determine whether dyspnea is due to pulmonary edema, but newer techniques improve precision in differentiating endothelial permeability and hydrostatic edema in patients. Increased pulmonary leak index, assessed with mobile probes at the bedside for ^{67}Gallium, assumed to bind to circulating transferrin, and ^{99m}technetium-labeled red blood cells to account for pulmonary blood volume, was shown to be a specific marker of pulmonary microvascular permeability and ARDS (Groeneveld et al., 1996). More sophisticated resolution of lung edema into hydrostatic edema – engorgement of vessels and interstitial edema, permeability edema without alveolar damage – interstitial edema, permeability edema caused by diffuse alveolar damage – ARDS, and mixed hydrostatic and permeability edema is possible with computed axial tomography and positron emission tomography, using ^{68}Ga-transferrin or ^{11}C-methylalbumin, (Schuster, 1998, Ketai and Godwin, 1998). Such resolution will be aid improved direction of therapy. Interestingly, such studies indicate that the extent of endothelial injury is often underestimated, because blood flow is diverted from injured to non-injured areas (Ehrhart et al., 1999).

There are two main sets of defining criteria used today, the expanded definition of ARDS (Murray et al., 1988) and the North American-European Conference definition (Bernard et al., 1994). These have different advantages and disadvantages, reflecting the complexity of the syndrome (Ware and Matthay, 2000). The expanded definition includes useful specific radiologic criteria, a lung function scoring mechanism, assessment of acute vs. chronic nature of the syndrome, and consideration of underlying pathologies. The conference definition scores pulmonary dysfunction and uses those criteria to stratify the cases into two levels of severity: ALI and ARDS. The conference definition does not include originating pathology, but does include criteria to rule out cardiogenic edema. These definitions have been helpful in standardization for both comparisons of clinical studies and for the determination of treatment. Applying these definitions, the incidence of ALI has been estimated at 6 in 10,000 and ARDS at roughly 1 in 10,000 per year with wide variability depending on the population selection and the strictness of the criteria used, but the incidence among patients with the known predisposing factors is 20-40% (Atabai and Matthay, 2002; Goss et al., 2003).

Unfortunately, it has not been possible using these criteria to predict which patients with trauma, sepsis, or other serious conditions, will progress to ARDS or how severe the alveolar-capillary permeability and the lung function will become in a pro-active fashion. Neither the injury scores nor the ratio of arterial oxygen to fraction of ventilation oxygen (Pa_{O2}/Fi_{O2}) correlates well with mortality. On the other hand, the appearance of procollagen III in the exudates, indicative of combined epithelial and interstitial matrix damage, has shown benefit in predicting the prolongation of the syndrome and the chances for development of interstitial fibrosis. And the plasma levels of von Willebrand factor, indicative of endothelial activation and damage, were predictive for ARDS occurrence in sepsis patients and for mortality in ARDS patients (Rubin et al., 1990; Ware et al., 2001; Atabai and Matthay, 2002). Once appropriate supportive positive pressure ventilation was indicated (Ashbaugh et al., 1967) and implemented, mortality was reduced to around 50–60 %, where it remained for several decades. Mortality is most closely associated with the underlying sepsis and multi-organ failure, but recent inroads in treatment of sepsis and reduced tidal volume ventilation for acute lung injury has brought this down to nearer 30% (Bernard et al., 2001, Eisner et al., 2001,Ware and Matthay, 2000, Groeneveld, 2003; Vincent et al., 2003a; Vincent et al., 2003b; Brower et al., 2004). Among the survivors, recovery can be fairly complete and by 6 months pulmonary function test may be nearly normal, with the exception of reduced carbon monoxide diffusion capacity and mild oxygen desaturation upon exercise (Herridge et al., 2003). Some non-pulmonary sequelae may persist, such as muscle weakness, which may be related to ARDS, to corticosteroid treatment, or to the underlying severe illness regardless of the development of ARDS (Herridge et al., 2003).

Saboteurs

Some cases of edema result from primary insults to the epithelium, particularly the type I pneumocyte, such paraquat poisoning due to rather specific accumulation of the oxidant producing paraquat by those cells (Patterson and Rhodes, 1982; Foth, 1995). Direct infection of the epithelium in viral, bacterial, and parasitic pneumonias and inhalation of certain toxins can also induce primary damage the epithelium (Matthay *et al.*, 2003). Aspiration of gastric contents (acid) and organic solvents cause direct epithelial injury. The thin type I epithelial cells are more susceptible to injury than the type II pneumocytes, but damage to the type II cells is important because of reduced surfactant production to provided reduction of surface tension at the alveolar air-liquid interface. Moreover, type II damage would impair the ability to regenerate type I cells to repair the alveolar-capillary membrane (Matthay *et al.*, 2003). Often it is difficult to sort out whether epithelium or endothelium is the primary target as damage may occur simultaneously or injury to one results in rapid dysfunction in the other as with inhaled toxic gasses, such as phosgene, carbon monoxide, chlorine, nitrogen dioxide, and the complex components in smoke. Similarly the oxidants produced in radiation injury are non-selective. Whether air-borne or blood-borne, numerous xenobiotic compounds are concentrated in the lung and all intravenous drugs encounter the large capillary bed of the lung first. Thus compounds like amphetamine, chlorphentermine, amiodarone, imipramine, chlorpromazine, 5-fluoro-2'-deoxyuridine, nicotine, benzo-α–pyrene, ipomeanol, butylhydroxytoluene, 4-methylnitrosamino-1-(3-pyridyl)-1-butanone, bleomycin, 3-methylindole, and nitrofurantoin may cause pneumotoxicity (Foth, 1995).

Epithelial and endothelial injury may occur concomitantly, but the endothelium is most often a primary target and the first failure in the alveolar-capillary barrier. Markers of endothelial activation and injury, such as plasma release of von Willebrand factor, ICAM-1, and thrombomodulin, are highly correlated with a variety of serious inflammatory disease states (Reinhart *et al.*, 2002). In some conditions, such as with IL-2 administration or Hantavirus pulmonary syndrome, endothelial permeability develops without clear evidence of increased epithelial injury (Ketai and Godwin, 1998; Peters *et al.*, 1999). In most conditions, primary injury to the endothelium results in secondary injury of the epithelium. First, endothelial barrier dysfunction and edema overwhelms the ability of the lymph system to clear interstitial edema fluid, leading to stress injury to the epithelium and alveolar flooding. Second, endothelial barrier dysfunction exposes the interstitium and epithelium to blood borne mediators. Third, endothelial activation results in altered mediator and oxidant production by the endothelium, which may directly affect the epithelium. Finally, with endothelial activation, the secondary influx of activated leukocytes can also damage epithelium. Thus, whether merely activated or injured to the point of cytotoxic or apoptotic death, endothelial dysfunction sets up a series of events that expand beyond the initial insult. As noted above these consequences include: increased leak of protein from plasma; decreased plasma protein osmotic pressure; increased tissue protein pressure; release of vasoactive mediators that alter hydrostatic gradients; altered hemostasis; increased adhesion and sequestration of leukocytes; and inflammation that compounds the damage; secondary injury to the epithelium and the consequent exacerbation of edema. Likewise, whether injury to the lung epithelium is primary or secondary to endothelial injury, epithelial dysfunction sets up a series of events that amplify the direct effects of fluid loss to the alveoli including: failure of the epithelial pump that actively removes fluid from the alveoli, reduction of amount and effectiveness of the surfactant produced by the type II epithelium, transmission of infection from the alveoli to the blood and vice versa, delayed repair of the alveolar-capillary barrier, and in some patients, fibrosing alveolitis (Ware and Matthay, 2000).

Infection *per se* and released bacterial products, such as endotoxin and β–hemolysin, are toxic to endothelium, whether the infection is present in the lung air space, carried to the lung by the circulation, or the bacterial products only are circulated to the pulmonary endothelium (Gibson *et al.*, 1999; Gao *et al.*, 2004). Although the lung involvement and edema may appear acutely, the underlying disease may have been present for some time. For instance, with direct infusion of *E. coli* endotoxin into sheep the permeability edema required 4-8 hours to develop (Brigham *et al.*, 1979; Gao *et al.*, 2004). Thus, infection in another part of the body would have delayed effects on the lungs, as it would take time for the infection to develop at the remote site, the endotoxin levels to build in the blood, and for the bacteria or the endotoxin borne from that site to the pulmonary endothelium to affect lung function. Often endotoxin, or other bacterial toxin, release is increased with effective antibiotic therapy. The vast endothelial surface, the position of the lung such that it first receives the cumulative venous blood flow, and the delicate nature of the alveolar-capillary membrane make it particularly vulnerable to any blood-borne toxin or activator as explained above. There is a high incidence of ARDS in patience with sepsis and the endothelium is a major target of sepsis damage and shock (Rubin *et al.*, 1990).

Because lung edema leads to arterial hypoxia, patients may require increasing levels of inspired oxygen (FiO_2) to maintain passable blood gas values. High levels of oxygen may be independently damaging to the lung (Patterson *et al.*, 1985) and thus oxygen levels are usually reduced to the minimum necessary (Artigas *et al.*, 1998). Moreover, in ARDS the low blood oxygenation is generally unresponsive to elevated oxygen alone due to the high shunt fraction and requires positive pressure ventilation to correct severe hypoxia. In fact, this is a central defining criteria and a major advance in the treatment of ARDS, as described above. Yet, uneven distribution of ventilation in the edematous lungs, alveolar over-distension, and the repeated opening and closing of atelectatic (collapsing) alveoli can directly exacerbate damage. As mentioned above, mechanical stress from over-ventilation can induce both epithelial and endothelial barrier failure even in non-diseased lungs. In addition to mechanical damage, imperfect ventilation can propagate and aggravate a cytokine, proinflammatory cascade (Slutsky and Tremblay, 1998). Thus, reduction of tidal volume ventilation has improved survival in ARDS (ARDS Network, 2000, Eisner *et al.*, 2001).

Pro-inflammatory cytokines are produced by a variety of cells, including lung endothelium, other lung cells, infiltrating leukocytes, and cells at remote injury sites. As will be discussed in greater detail in Chapters 3 and 11, these mediators can have direct effects on the lung cells and the balance of pro-inflammatory and anti-inflammatory factors is important in the sequestration and activation of leukocytes in the lung (Bhatia and Moochhala, 2004; Wiedermann *et al.*, 2004). Initial activation of endothelium by many stimuli in addition to certain cytokines, increases their adhesiveness for circulating leukocytes. Some definitions of ARDS include the increased presence of neutrophils in the lung as an essential marker (Lesur *et al.*, 1999). The activated neutrophils expose the lung tissue to further insult due to their production of proteases and reactive oxidants (Patterson *et al.*, 1989). Regardless of the source, oxidant stress activates and may injure endothelial cells. Experimental infusion of complement activated plasma or activation of complement, as occurs with pancreatitis, burn injury, bacteremia, and earlier with some renal dialysis systems results in permeability edema dependent upon neutrophil sequestration in the lungs (Heller *et al.*, 2000). Moreover, serum from burn patients was shown to directly induce expression of leukocyte adhesion factor by isolated endothelium (Chen *et al.*, 2004). On the other hand, acute respiratory failure has been identified in both patients and animals models in the absence of neutrophil involvement (Ware and Matthay, 2000). Since sepsis is highly correlated with increased mortality the presence of neutrophils may actually be an important mechanism of disease defense. Thus, the role of leukocytes in ARDS is complex and their ultimate benefit or detriment is dependent upon numerous other factors.

Remote events such as disseminated intravascular coagulation, systemic organ emboli, air emboli, blood transfusion, burn injury, and pancreatitis are common causes of ARDS in patients with no previous lung disease (Clark and Fick, 1984; Fein *et al.,* 2000). Even bone fractures may result in activation of lung endothelium *via* fat emboli or release of fatty acids, particularly oleic acid, which react with receptors and by direct effects on the plasma membrane, with secondary release of vasoactive agents from the lung (Selig *et al.,* 1987). Acute increases in free fatty acid released with both bone fracture and pancreatitis may also induce endothelial cell apoptosis (Artwohl *et al.*, 2004). In some conditions of neurogenic trauma, such as acute elevation of cranial pressure, a permeability lung edema results even when pulmonary pressures are not elevated (Beckman *et al.*, 1987). In all of these pathologic conditions, alteration of the mediators presented to the endothelium is an important mechanism for instigating the injury. Cytokines were mentioned above, but other circulating bioactive mediators and vasoactive mediators may directly activate the endothelial cells causing initial permeability and may also concomitantly alter the hydrostatic pressures aggravating the rate of edema formation. The activated endothelium themselves, produce numerous factors that can further alter the vascular characteristics and pressures to further enhance the rates of fluid and solute movement. In addition to altered balance of mediators, altered lung compliance, focal or widespread hypoxia (vasoconstrictive in lung), and swelling of the interstitium can secondarily lead to increased vascular pressures in the edematous lung. Currently, the 1994 consensus criteria of ARDS includes pulmonary wedge pressures < 18 mm Hg as an exclusion criterion in order to rule out cardiogenic edema (Bernard *et al.*, 1994). But, because of these other possible secondary association of elevated capillary pressure unrelated to heart failure, this is to be viewed critically, keeping cardiac function as well as secondary elevation of pulmonary pressures in mind.

Endothelial Dysfunction and Gas Exchange

Beyond the initiating pathology or trauma, the principal factor in the distress noted by the patients and of immediate concern to the clinician, as noted above, is the impaired arterial oxygenation due to insufficient pulmonary gas exchange. Low arterial oxygen and high arterial carbon dioxide may occur by a combination of several factors related to the compromised endothelial function. First, endothelial dysfunction can alter the normal fine-tuning that matches local blood flow to those areas, which are well ventilated. As introduced above, in systemic circulation hypoxia usually results in vasodilation in order to ensure adequate circulation to the dependent tissues; but in the pulmonary vascular bed, hypoxia triggers vasoconstriction to shift flow to alveoli that can best participate in gas exchange. This unique pulmonary hypoxic vasoconstriction, likely results from synergistic actions of both endothelium and the vascular smooth muscle, but it can proceed in the absence of endothelium, suggesting that the primary oxygen sensor is the smooth muscle cell. Nevertheless, endothelial-derived agents permit and modulate the muscle response and perhaps the unique nature of the pulmonary vascular smooth muscle phenotype (Aaronson *et al.*, 2002; Gurney, 2002). Thus, activation and injury of the endothelium may impair ventilation perfusion matching and gas exchange. Additionally, many of the very factors involved in initiating endothelial dysfunction have direct vasoactive or cytotoxic effects on the vascular smooth muscle. Often there is general vasoconstriction and pulmonary hypertension, but poor matching. Secondly, endothelial barrier dysfunction can result in widening of the alveolar-capillary membrane due to the interstitial edema. The non-polar gases readily traverse the lipid membranes of the alvoelar epithelium and the capillary endothelium, but a thick over-hydrated interstitium provides considerable resistance and a longer diffusion distance. Thirdly, when edema is severe and alveoli are flooded or become atelectatic (collapsed), there is less and less surface available for gas exchange. As indicated, efforts to force the lungs to ventilate and the pulmonary vasculature to dilate to improve gas exchange can result in untoward worsening the shunting of blood to unventilated areas or cause mechanical stress-injury to the alveolar capillary membrane.

Multifunctional Endothelium

Interactivity of Barrier and Other Vital Endothelial Functions

It should be clear from the preceding discussions that endothelial barrier function and the multiple other critical functions performed by the endothelium are interactive and are often altered concurrently (Cines *et al.*, 1998). The key role the endothelium plays in separation of blood fluids and elements from the fragile lung tissue will be dealt with extensively in this book and certain other functions of high importance in edema formation will be covered, while descriptions of other functions will necessarily be minimal. However, they will be briefly outlined here so that the reader can keep in mind that activation of the endothelium induces a complex, interactive, and variable set of acute and delayed responses that may relate to continuation of the barrier dysfunction and severity of the edema development.

Regulation of leukocyte adhesion and migration is a highly important function of endothelium with understandable significance in both health and disease. Inflammatory mediators increase leukocyte-endothelial interactions primarily *via* upregulation of increased expression of endothelial apical adhesion molecules. Mediators involved with leukocyte recruitment will be introduced in Chapter 3, mechanism of endothelial-leukocyte interaction will be dealt with more fully in Chapter 11, and injury related to leukocyte sequestration in lung will be discussed in Chapter 12. Here we will simply reiterate that leukocyte activation/adhesion is a formidable component in the pathologic sequelae during inflammation and in ARDS. Leukocytes protect us from foreign substances and organisms *via* activation, and release of oxidants, proteases, and cytokines and *via* critical participation in immunity, but these same properties can turn against us causing untoward collateral damage to the host tissue. For instance, complement activation with blood transfusion or as occurred earlier with renal dialysis triggers neutrophil sequestration in lung capillaries, resulting in acute lung injury (Dry *et al.*, 1999). Furthermore, endothelial-derived oxidants can initiate and exacerbate the oxidant stress produced by recruited and activated leukocytes.

Another critical function of endothelium with prominent pathologic implications is the regulation of vascular blood flow and tone *via* modulation of smooth muscle function and phenotype. Endothelial based paracrine mediators and gap junctions send signals to the underlying smooth muscle that have a powerful influence on their contractile tone. In both larger arteries and veins, but even more important in small pre-capillary arterioles and post-capillary venules, this endothelial function is a critical determinant of blood circulation, resistance, and pressures (Wolin et al., 1998). The term "endothelial dysfunction" in some settings, particularly in the systemic circulation in diseases such as diabetes, is in fact narrowly translated to mean "the failure of endothelium to relax vessels." More broadly, early endothelial dysfunction involving regulation of both leukocytes and smooth muscle is a primary step in the development of atherosclerosis and late disruption of the endothelial covering of plaques can be a precipitating event in plaque rupture, which is again more commonly associated with the high flow stress in the systemic circulatory systems. But, the endothelium also produces a number of factors that either keep smooth muscle in its non-proliferative "contraction able" state or cause it to shift to a proliferative phenotype, resulting in hypertrophy and thickening of the medial layer of the vessel wall. This altered endothelium-smooth muscle interaction results in chronic resistive and remodeling changes in systemic circulation and in persistent pulmonary hypertension. Similarly, shifts in endothelial-derived growth factors, inflammatory mediators, bioactive molecules, and oxidants can locally influence other tissues in intimate contact with the capillary endothelium, such as cardiomyocytes. Moreover, modulation of the circulating vasoactive and bioactive milieu can have downstream hormonal effects in addition to their local paracrine effects.

Endothelial cells participate in hemostasis in a number of ways. In normal endothelium, the tight barrier and non-reactive surface prevents activation of platelet adhesion and activation of the intrinsic clotting cascade. Moreover, several mediators produced by the quiescent endothelium also signal platelet quiescence; whereas activated or damaged endothelium releases pre-formed and newly synthesized mediators that activate the platelets to become adhesive and activate the extrinsic clotting cascade. Quiescent endothelium stores von Willebrand factor, which is released with activation and binds the anti-hemophilic factor to form Factor VIII-vWF. Quiescent endothelium binds and buffers activated thrombin, but thrombin binding to other endothelial receptors can activate endothelium and cause both barrier dysfunction and release of vWF. Moreover, thrombin-thrombomodulin binding by endothelium is a primary regulator of Protein C release. Several factors of the intrinsic and extrinsic clotting cascades, which are synthesized by the liver, are bound to the endothelial surface. Fibrin formation and fibrin dissolution is partly governed by endothelial-derived and by endothelial-bound factors, primarily tissue-type plasminogen activator and plasminogen activator inhibitor-1. As mentioned, the pulmonary microcirculation functions to filter out blood clots, thereby protecting the heart and brain; however, trapped thrombi can enhance presentation of activators, such as thrombin and plasmin, to the protease activated receptors on the endothelial surface (Garcia et al., 1995). Local, controlled activation of endothelial cells and clotting activity and activation of endothelial proliferation to close wounds and repair the vascular wall is a vital function, but widespread and uncontrolled thrombosis and formation of emboli can be devastating.

Importance of the Endothelium in Multiple Pathologies

Endothelial cell participation in pathophysiologic processes, such as inflammation, organ transplant failure, metastasis, and tumor growth, is now better understood. Vasculogenesis and angiogenesis are naturally critical functions in organ development and fetal growth, and limited activation of angiogenesis is necessary to repair vascular damage and maintain vascular health, but vessel expansion and the fragility of the tumor vascular beds is also a critical component of the growth and disease-effects of neoplasms (Cines *et al.*, 1998). Moreover, endothelial binding and endothelial-failure to prevent binding of cancer cells to tissue is a critical step in metastasis. Even with low-dose radiation treatment for cancer, endothelial activation and damage can actually enhance metastasis of escaping tumor cells by exposure of additional tumor cell sub-stratum adherence sites (Onoda *et al.*, 1999). Endothelial failure is now understood as a critical point for rejection and failure in organ transplantation (Faulk and Labarrere, 1994, Paik *et al.*, 2003). Although the focus of this volume is on the underlying principles and regulation lung endothelial barrier integrity, this information has important implications in all conditions of endothelial dysfunction and pathology. Multiorgan failure correlates with increased mortality in patients with ARDS (Bhatia and Moochhala, 2004). A recent report from the Conference on ARDS affirmed that, "acute lung injury and ARDS is a systemic syndrome in virtually all cases" (Matthay *et al.*, 2003). It is not unlikely that events causing injury to pulmonary endothelium, which can lead to respiratory failure, also injure endothelium of systemic organs, contributing to the failure in those organs. Consequently improved understanding of architecture, biochemical responsiveness and regulation of endothelial function, and the role of endothelial dysfunction in lung pathology will benefit the study of numerous acute and chronic human diseases.

Building Better Dams

Thus, the study of lung endothelium evolved from an early focus on barrier function to an intensive interest in the many interactive and regulatory functions of endothelium. However, the significance of the relationship between endothelial integrity and the clinically important problem of pulmonary edema is undiminished. To advance treatment and prevention of alveolar flooding and inflammation, we need an improved and up-to-date mastery of the construction and maintenance of the dam. This book is dedicated to that purpose. We will begin with fundamentals of the structural elements of the endothelium. Next, the forces that challenge barrier integrity and the mechanisms of response will be covered. Detailed descriptions of the individual structural elements, their regulation, their contribution to barrier integrity or dysfunction, and the heterogeneity of endothelium will follow. Next, the interactions of endothelium with blood-borne cells, factors, infectious agents, and the injurious consequences will be defined. Finally, the internal defenses and repair of the barrier by endothelium and the advances in treatment to aid in resistance to activation and injury and repair will be reviewed.

ACKNOWLEDGEMENTS:
This work was supported by a Merit Award from the Veteran's Administration Medical Research Service (CEP), and by RO1 HL51856 (MAM), RO1 HL51854 (MAM), and P50 HL740005 (MAM).

REFERENCES:

Aaronson, P.I., Robertson, T.P., and Ward, J.P. (2002) Endothelium-derived mediators and hypoxic pulmonary vasoconstriction. Respir. Physiol. Neurobiol. 132:107-120.

Albelda, S.M., Sampson, P., Haselton, F., McNiff J., Muellaer, S., Williams, S., Fishman, A., and Levine, E.M. (1988) Permeability characteristics of cultured endothelial cell monolayers. J. Appl. Physiol. 64:308-322.

ARDS Network. (2000) Ventilation with lower tidal volumes as compared with traditional tidal volumes for acute lung injury and ARDS. N. Eng. J. Med. 342:1301-1308.

Artigas, A., Bernard, G.R., Carlet, J., Dreyfuss, D., Gattinoni, L., Hudson, L., Lamy, M., Marini, J., Matthay M.A., Pinsky, M., Spragg, R., Suter, P.M. (1998) The American-European Consensus Conference on ARDS, part 2: Ventilatory, pharmacologic, supportive therapy, study design strategies, and issues related to recovery and remodeling. Acute respiratory distress syndrome. Amer. J. Respir. Crit. Care Med. 157:1332-47.

Artwohl, M., Roden, M., Waldhausl, W., Freudenthaler, A., and Baumgartner-Parzer, S. (2004) Free fatty acids trigger apoptosis and inhibit cell cycle progression in human vascular endothelial cells. FASEB J. 18:146-148.

Ashbaugh, D, Bigelow, D., Petty, T., and Levine, B. (1967) Acute respiratory distress in adults. Lancet 2:319-232.

Atabai, K. and Matthay M.A. (2002) The pulmonary physician in critical care: Acute lung injury and the acute respiratory distress syndrome: definitions and epidemiology. Thorax 57:452-458.

Baldwin, A.L. and Thurston, G. (2001) Mechanics of endothelial cell architecture and vascular permeability. Crit. Rev. Biomed. Eng. 29:247-278.

Bannerman, D.D. and Goldblum, S.E. (1999) Direct effects of endotoxin on the endothelium: barrier function and injury. Laboratory Investigation 79:1181-1199.

Barnard, J.W., Patterson, C.E., Hull, M., Wagner, W., and Rhoades, R.A. (1989) Oxygen radical-induced lung injury: Contribution of changes in microvascular pressure and permeability. J. Appl. Physiol. 66:1486-1493.

Baue, A.E. (1974) Mitochondrial function in shock. from Proceedings of a Symposium on Recent Research Developments and Current Clinic Practice in Shock. Upjohn, pp. 11-15.

Beckman, D.L., Ginty, D.D., and Gaither, A.C. (1987) Neurogenic pulmonary edema in a pulmonary normotensive model. Proc. Soc. Exp. Biol. 186:170-173.

Bernard, G.R., Artigas, A., Brigham, K., Carlet, J., Falke, K., Hudson, L., Lamy, M., LeGall, J., Morris, A., Spragg, R. (1994) Report of the American-European consensus conference on ARDS: definitions, mechanisms, relevant outcomes and clinical trial coordination. Intensive Care Med. 20:225-32.

Bernard, G.R., Vincent, J., Laterre, P., LaRosa, S., Dhainaut, J., Lopez-Rodriguez, A., Steingrub, J., Garber, G., Helterbrand, J., Ely, E., and Fisher, C. (2001) Efficacy and safety of recombinant human activated protein C for severe sepsis. N. Eng. J. Med. 344:699-709.

Bhatia M. and Moochhlala, S. (2004) Role of inflammatory mediators in the pathophysiology of ARDS. J. Pathol. 202:145-156.

Brewer, L.A. (1981) A historical account of "wet lung of trauma" and the introduction of intermittent positive pressure oxygen therapy in WWII. Ann. Thoracic Surg. 31:386-393.

Brigham, K.L., Bowers, R., and Haynes, J. (1979) Increased sheep lung vascular permeability caused by *Escherichi Coli* endotoxin. Circ. Res. 45:292-297.

Brower, R.G., Lanken, P., MacIntyre, N., Matthay, M., Morris, A., Ancukiewicz, M., Schoenfeld, D., Thompson, B.T.: NHLBI ARDS Clinical Trials Network. (2004) Higher versus lower PEEP in patients with ARDS. N. Eng. J. Med. 351:327-336.

Chen, X., Xia, Z., Wei, D., Liao, H., Ben, D., and Wang, G. (2004) Expression and regulation of vascular cell adhesion molecule-1 in human umbilical vein endothelial cells induced by sera from severely burned patients. Crit. Care Med. 32:77-82.

Cines, D., Pollak, E., Buck, C., Loscalzo, J., Zimmerman, G., McEver, R., Pober, J., Wick, T., Konkle, B., Schwartz, B., Barnathan, E., McCrae, B., Hug, B., Schmidt, A., and Stern, D. (1998) Endothelial cells in physiology and pathophysiology of vascular disorders. Blood 91:3527-3561.

Clark, M.C. and Fick M.R. (1984) Permeability pulmonary edema caused by venous air embolism. Amer. Rev. Respir. Dis. 129:633-635.

Crandall, E.D., Staub, N.C., Goldberg, H.S., and Effros, R.M. (1983) Recent developments in pulmonary edema. Ann. Internal Med. 99:808-822.

Doyle I.R., Nicholas, T.E., Bersten, A.D. (1999) Partitioning lung and plasma proteins: circulating surfactant proteins as biomarkers of alveolar-capillary permeability. Clin. Exper. Pharm. Physiol. 26:185-197.

Dry, S.M., Bechard, K., Milford, E., Churchill, W., and Benjamin, R.J. (1999) The pathology of transfusion-related acute lung injury. Amer. J. Clin. Pathol. 112:216-221.

Edoute, Y., Roguin, A., Behar, D., and Reisner, S.A. (2000) Prospective evaluation of pulmonary edema. Crit. Care Med. 28:330-335.

Effros, R.M., Schapira, R., Presberg, K., Ozker, K., and Jacobs, E.R. (1998) Stop-flow studies of solute uptake in rat lung. J. Appl. Physiol. 85:986-992.

Eisner, M.D., Thompson, T., Hudson, L., Luce, J., Hayden, D., Schoenfeld, D., Matthay, M.A. (2001) ARDS Network. Efficacy of low tidal volume ventilation in patients with different clinical risk factors for acute lung injury and the acute respiratory distress syndrome. Amer. J. Respir. Crit. Care Med. 164:231-236.

Ehrhart, I., Orfanos, S., McCloud, L., Sickles, D., Hofman, W., and Catravas, J. (1999) Vascular recruitment increases evidence of lung injury. Crit. Care Med. 27:120-129.

Faulk, W.P. and Labarrere, C.A. (1994) Antithrombin III in normal and transplanted human hearts. Sem. in Hematol. 31:26-33.

Fein, A.M. and Calalang-Colucci, M.G. (2000) Acute lung injury and acute respiratory distress syndrome in sepsis and septic shock. Crit. Care Clinics. 16:289-317.

Foth, H. (1995) Role of the lung in accumulation and metabolism of xenobiotic compounds. Crit. Rev. Tox. 25:165-205.

Galis Z. Ghitescu L. Simionescu M. Fatty acids binding to albumin increases its transcytosis by lung capillary endothelium. European Journal of Cell Biology. 47(2):358-65, 1988

Gao, J., Zeng, B., Zhou, L., and Yuan, S. (2004) Protective effects of early treatment with propofol on endotoxin-induced acute lung injury in rats. Brit. J. Anesth. 92:277-279.

Garcia, J.G.N., Pavalko, F.M., and Patterson, C.E. (1995) Vascular endothelial cell permeability responses to thrombin. Blood Coag. Fibrinol. 6:609-626.

Gehlbach, B.K. and Geppert, E. (2004) Pulmonary manifestations of left heart failure. Chest 125:669-682.

Gibson, R.L., Nizet, V., and Rubens, C.E. (1999) Group B streptococcal beta-hemolysin promotes injury of lung microvascular endothelial cells. Ped. Res. 45:626-634.

Goss, C.H., Brower, R., Hudson, L., and Rubenfeld, G.D. ARDS Network. (2003) Incidence of acute lung injury in the United States. Crit. Care Med. 31:1607-1611.

Gotoh, N., Kambara, K., Jiang, X.W., Ohno, M., Emura, S., Fujiwara, T., and Fujiwara, H. (2000) Apoptosis in microvascular endothelial cells of perfused rabbit lungs with acute hydrostatic edema. J. Appl. Physiol. 88:518-526.

Groeneveld, A.B.J. (2003) Vascular pharmacology of acute lung injury and acute respiratory distress syndrome. Vasc. Pharmacol. 39:247-256.

Groeneveld, A.B.J., Raijmakers, P., Teule, G., and Thijs, L.G. (1996) The ^{67}gallium pulmonary leak index in assessing severity and course of the adult respiratory distress syndrome. Crit. Care Med. 24:1467-1472.

Gurney, A.M. (2002) Multiple sites of oxygen sensing and their contributions to hypoxic pulmonary vasoconstriction. Respir. Physiol. Neurobiol. 132:43-53.

Harris, P. and Heath, D. (1986) Chapter 5: The Pulmonary Endothelial Cell. In: <u>The Human Pulmonary Circulation</u>. Churchill Livingstone.

Heller, A., Kunz, M., Samakas, A., Haase, M., Kirschfink, M., Koch, T. (2000) The complement regulators C1 inhibitor and soluble complement receptor 1 attenuate acute lung injury in rabbits. Shock. 13:285-90.

Herridge, M.S., Cheung, A., Tansey, C., Matte-Martyn, A., Diaz-Granados, N., Al-Saidi, F., Cooper, A., Guest, C., Mazer, C., Mehta, S., Stewart, T., Barr, A., Cook, D., Slutsky, A. Canadian

Critical Care Trials Group. (2003) One-year outcomes in survivors of the acute respiratory distress syndrome. New Eng. J. Med. 348:683-693.
Kawanami, O. (1997) The endothelium of the pulmonary microvessels. J. Nippon Med. School 64:495-511.
Kerr, G.W. (1998) Neurogenic pulmonary oedema. J. Accid. Emerg. Med. 15:275-276.
Ketai, L.H. and Godwin, J.D. (1998) A new view of pulmonary edema and acute respiratory distress syndrome. J. Thor. Imag. 13:147-171.
Koizumi, T., Kubo, K., Hanaoka, M., Yamamoto, H., Yamaguchi, S., Fujii, T., and Kobayashi, T. (1999) Serial scintigraphic assessment of iodine-123 metaiodobenzylguanidine lung uptake in a patient with high-altitude pulmonary edema. Chest 116:1129-1131.
Kramer, G.C., Harms, B.A., Bodai, B.I., Renkin, E.M., and Demling, R.H. (1983) Effects of hypoproteinemia and increase vascular pressure on lung fluid balance in sheep. J. Appl. Physiol. 55:1514-1522.
Kuebler, W.M., Ying, X., Singh, B., Issekutz, A.C., and Bhattacharya, J. (1999) Pressure is proinflammatory in lung venular capillaries. Journal of Clinical Investigation 104:495-502.
Lesur, O., Berthiaume, Y., Blaise, G., Damas, P., Deland E., Guimond, J.G., and Michel, R.P. (1999) ARDS: 30 years later. Can. Resp. J. 6:71-86.
Lin, W., Jacobs, E., Schapira, R., Presberg, K., and Effros, R.M. (1998) Stop-flow studies of distribution of filtration in rat lungs. J. Appl. Physiol. 84:47-52.
Lindenschmidt, R.C., Patterson, C.E., Forney, R.B., Rhoades, R.A. (1983) Selective action of $PGF_{2\alpha}$ during paraquat-induced pulmonary edema in perfused lung. Tox. Appl. Pharm. 70:105-114.
Lum, H. and Malik, A.B. (1994) Regulation of vascular endothelial barrier function. Amer. J. Physiol. 267:L223-L241.
Matthay, M.A., Folkesson, H., and Verkman, A.S. (1996) Salt and water transport across alveolar and distal airway epithelia in adult lung. Amer. J. Physiol. 270:L487-L503.
Matthay, M.A., Zimmerman, G., Esmon, C., Bhattacharya, J., Coller, B., Doerschuk, C., Floros, J., Gimbrone, M., Hoffman, E., Hubmayr, R., Leppert, M., Matalon, S., Munford, R., Parsons, P., Slutsky, A., Tracey, K., Ward, P., Gail, D., and Harabin, A. (2003) Future research directions in acute lung injury. Amer. J. Respir. Crit. Care Med. 167:1027-1035.
Muller, M.M. and Griesmacher, A. (2000) Markers of endothelial dysfunction. Clin.Chem.Lab Med. 38:77-85.
Murray, J.F., Matthay, M.A., Luce, J., Flick, M.R. (1988) An expanded definition of the adult respiratory distress syndrome. Amer. Rev. Respir. Dis. 138:720-723.
Onoda, J.M., Kantak, S.S., and Diglio, C.A. (1999) Radiation induced endothelial cell retraction in vitro: correlation with acute pulmonary edema. Pathol. Oncol. Res. 5:49-55.
Paik, H.C., Hoffmann, S.C., and Egan, T.M. (2003) Pulmonary preservation studies:effects on endothelial function and pulmonary adenine nucelotides. Transplant. 75:439-444.
Pararajasingam, R., Nicholson, M., Bell, P., and Sayers, R. (1999) Non-cardiogenic pulmonary oedema in vascular surgery. Europ. J. Vasc. Surg. 17:93-105.
Park, S., Tepper, O., Galiano, R., Capla, J., Baharestani, S., Kleinman, M., Pelo, C., Levine, J., and Gurtner, G. (2004) Selective recruitment of endothelial progenitor cells to ischemic tissues with increased neovascularization. Plast. Reconstr. Surg. 113:284-293.
Parker, J.C. and Yoshikawa, S. (2002) Vascular segmental permeabilities at high peak inflation pressure in isolated rat lungs. Amer. J. Physiol. 283:L1203-1209.
Patterson, C.E., Barnard, J.W., LaFuze, J.E., Hull, M.T., Baldwin, S.J., and Rhoades, R.A. (1989) Neutrophil activation and changes in microvascular pressure is acute pulmonary edema. Amer. Rev. Resp. Dis. 140:1052-1062.
Patterson, C.E., Davis, H., Schaphorst, K., and Garcia, J. (1994) Mechanisms of cholera toxin prevention of thrombin-and PMA-induced endothelial cell barrier function. Microvacs. Res. 48:212-235, 1994.
Patterson, C.E. and Lum, H. (2001) Update on pulmonary edema: the role and regulation of endothelial barrier function. Endothelium 8:75-105.
Patterson, C.E. and Rhodes, M.L. (1982) The effect of superoxide dismutase on paraquat mortality in mice and rats. Toxicol. App. Pharmacol., 62:65-72.

Patterson, C.E., Rhodes, M.L., Butler, J.A. and Byrne, F. (1985) Protection from normobaric hyperoxia by cysteamine and n-acetylcysteine. Lung 163:23-32.
Pearce, F.J. and Lyons, W.S. (1999) Logistics of parenteral fluids in battlefield resuscitation. Military Med. 164:653-655.
Peters, C.J., Simpson, G.L., and Levy, H. (1999) Spectrum of hantavirus infection: hemorrhagic fever with renal syndrome and pulmonary syndrome. Annual Review of Medicine 50:531-545.
Petty, T.L. In the cards was ARDS. (2001) Amer. J. Respir Crit. Care Med. 163:602-603.
Pietra, G. (1984) Insights into mechanisms of pulmonary edema. Lab. Invest. 51:489-494.
Reinhart, K., Bayer, O., Brunkhorst, F., Meisner, M. (2002) Markers of endothelial damage in organ dysfunction and sepsis. Crit. Care Med. 30:S302-S312.
Rubin, D.B., Wiener-Kronish, J., Murray, J., Green, D., Turner, J., Luce, J., Montgomery, A., Marks, J., Matthay, M.A. (1990) Elevated von Willebrand factor antigen is an early plasma predictor of acute lung injury in nonpulmonary sepsis syndrome. J. Clin. Invest. 86:474-80.
Ryan, U.S. and Ryan, J.W. (1984) Cell biology of the pulmonary endothelium. Circ. 70:46-62.
Schnitzer, J.E. (1992) Gp60 is an albumin-binding glycoprotein expressed by continuous endothelium involved in albumin transcytosis. Amer. J. Physiol. 262:H246-H254.
Schuster, D. (1998) Evaluation of lung function with PET. Seminars in Nuclear Medicine 28:341-351.
Selig, W.M., Patterson, C.E., and Rhoades, R.A. (1987) Cyclooxygenase metabolites contribute to oleic acid-induced lung edema by a pressure effect. Exper. Lung Res. 13:69-82.
Shasby, D.M. and Shasby, S.S. (1985) Active transendothelial transport of albumin. Circ. Res. 57:903-908.
Slutsky, A.S. and Tremblay, L.N. (1998) Multiple system organ failure: is mechanical ventilation a contributing factor? Amer. J. Respir. Crit. Care Med. 157:1721-1725.
Song, Y., Fukuda, N., Bai, C., Ma, T., Matthay, M.A., and Verkman, A.S. (2000) Role of aquaporins in alveolar fluid clearance in neonatal and adult lung, and in edema following acute lung injury: studies in transgenic aquaporin null mice. J. Physiol. 525:771-779.
Smith, W.S. and Matthay, M.A. (1997) Evidence for a hydrostatic mechanism in human neurogenic pulmonary edema. Chest. 111:1326-1333.
Tiruppathi, C., Song, W., Bergenfeldt M., Sass, P., and Malik, A.B. (1997) Gp60 activation mediates albumin transcytosis in endothelial cells by tyrosine-kinase-dependent pathways. J. Biol. Chem. 272:25968-25975.
Vincent, J.L., Angus, D., Artigas, A., Kalil, A., Basson, B., Jamal, H., Johnson, G., Bernard, G.R. (2003a) Recombinant Human Activated Protein C in Severe Sepsis. Effects of drotrecogin alfa (activated) on organ dysfunction in the PROWESS trial. Crit. Care Med. 31:834-840.
Vincent, J.L., Sakr, M.B., and Ranieri, V.M. (2003b) Epidemiology and outcome of acute respiratory failure in ICU patients. Crit. Care Med. 31:S296-S299.
Ware, L.B., Conner, E., and Matthay, M.A. (2001) von Willebrand factor antigen is an independent marker of poor outcome in patients with early acute lung injury. Crit. Care Med. 29:2325-2331.
Ware, L.B., Fang, X., and Matthay, M.A. (2003) Protein C and thrombomodulin in human acute lung injury. Amer. J. Physiol. 285:L514-L521.
Ware, L.B. and Matthay, M.A. (2000) The acute respiratory distress syndrome. New Eng. J. Med. 342:1334-1349.
Wiedermann, F.J., Mayr, A., Kaneider, N., Fuchs, D., Mutz, N., Schobersberger, W. (2004) Alveolar granulocyte colony stimulating factor and α-chemokines in relation to serum levels, pulmonary neutrophilia, and severity of lung injury in ARDS. Chest 125:212-219.
West, J. B. and Mathieu-Costello, O. (1999) Structure, strength, failure, and remodeling of the pulmonary blood-gas barrier. Ann. Rev. Physiol. 61:543-572.
Wolin, M.S., Davidson, C., Kaminski, P., Fayngersh, R., and Mohazzab, K., (1998) Oxidant-nitric oxide signaling mechanisms in vascular tissue. Biochem. (Moscow). 63:810-816.

Chapter 2. Molecular Architecture of the Endothelium

Carolyn E. Patterson, Ph.D.[1]* & Dimitrije Stamenović, Ph.D.[2]

[1]Departments of Medicine & Physiology, Indiana University School of Medicine & Roudebush VA Medical Center, Indianapolis IN., 46202, USA; [2] Dept. Biomechanical Engineering, Boston University, Boston, MA 02215 USA

CONTENTS:

Introduction

Determinants of the Cellular Structure-Function Relationship
Architectural Elements
Quiescent and Activation Endothelial Phenotypes and Function

The Endothelial Actin Cytoskeleton
Actin Organization in the Quiescent State
Actin Organization in the Activated State
Actin-associated Proteins

Intermediate Filaments
Function, Composition, and Structure
Vimentin and Activation

The Microtubule System
Function, Composition, and Structure
Activation and Microtubules

Mechanisms and Mechanics of Cell Shape Change with Activation
The contractile and Tensegrity Models
Activation Modeling: Controversies, Cooperation, and Resolution

Endothelial Attachment to the Basement Matrix
Cell/Matrix Complexes
Quiescence, Activation, and Signaling

Endothelial Cell-Cell Junctions
Cell-Cell Adhesion and Barrier Function
Adherens Junctions
Tight Junctions
Activation and Signaling
Integration of Adhesions in Quiescence, Activation, and Restoration

Structural Integrity of the Dam

Introduction

The integrity of the endothelial cell layer is a primary factor in the vascular barrier between underlying tissues and blood fluids, mediators, and formed elements as described in Chapter 1. The permeability characteristics of this vast, but very thin endothelial membrane essentially rests on the architectural characteristics of the individual endothelial cells, their links to each other, and their adhesion to the acellular basement membrane. As in all of nature, structure and function in the endothelium is intimately linked. The anatomical and physiologic characteristics of the endothelium vary somewhat between organ systems, with location in the vascular bed, and with physiologic conditions faced by the cells. Intrinsic differences in the structural components, signal transducers, and regulators of the cytoskeletal or adhesion complexes could account for differences in basal barrier function between endothelium from different vascular sites or variable responses to activating or toxic agents, as will be explored in Chapter 10. Nevertheless, there are fundamental structural design characteristics shared by endothelial cells and, certainly by endothelial cells in the pulmonary circulation that will be introduced in this chapter.

Determinants of the Cellular Structure-Function Relationship

Architectural Elements

The polar disk-like structure of the normal, quiescent endothelial cell is critically dependent upon the cytoskeletal architecture. This three dimensional, intracellular cytoskeletal network that determines and supports maintenance of the unique cell shape includes microfilaments (actin and myosin polymers), intermediate filaments (mainly, vimentin), and the microtubule assembly (tubulin) (Shasby *et al.*, 1982; Alexander *et al.*, 1988; Gotlieb *et al.*, 1991; Ettenson and Gotlieb, 1992; Stamenović and Wang, 2000). The cytoskeleton not only determines cell shape, but also anchors receptor complexes, enzyme complexes, and intracellular organelle. The microtubule system also functions as an intracellular transport system. The cytoskeleton provides the cellular anchorage for the different membrane spanning junctional complexes at the basal surface, the lateral borders, and the luminal aspect with it associated gylcocalyx. Thus, the cytoskeleton is important in organization of cellular biophysical and biochemical processes and in determining and maintaining the polarity of the cell. Coordinated regulation of the actin-cytoskeleton and the microtubule system is critical in directional migration and wound repair. Basic principles of the molecular organization of the cytoskeleton will be introduced in this chapter, but greater depth will be provided in Chapter 7, after signaling mechanisms have been discussed.

Quiescent and Activated Endothelial Phenotypes and Function

The importance of the thinness of the endothelial cell to efficient gas exchange and of the tightly joined pavement of the vascular wall to maintenance of fluid balance in the pulmonary circulation is clearly illustrated in Chapter 1. Thus, the architectural phenotype of the normal, quiescent endothelial cell is basic to normal physiologic function of the endothelium. As will be described in Chapter 3, a variety of mediators or stresses result in activation of endothelium expressed by alterations of many functional aspects, including reversible alteration of endothelial cell shape concomitant with alteration in barrier function. In this chapter we will limit discussion of activation to alterations in architectural features. The strong association between modification of the morphologic features and physiologic properties implies that in reversible barrier dysfunction, dynamic reorganization of the actin network due to altered regulation of actin and associated proteins is an important determinant of impaired endothelial barrier function. Moreover, reversible alteration of barrier function is generally associated with impaired adhesion between neighboring endothelial cells and altered adhesion to the underlying basement membrane matrix. Importantly, the connections between adjacent endothelial cells and between the cells and basal matrix are physically linked to the cytoskeleton and biochemically linked to regulation of the cytoskeleton. Components of these junctional structures and their relationship to cell shape and function will be introduced here and greater depth will be provided in Chapter 8 on endothelial-matrix connections and Chapter 9 on inter-endothelial junctions.

In addition to readily reversible changes in endothelium, pathologic and physical cell injury can open gaps between cells of a confluent endothelial monolayer, resulting in severe dysfunction of matrix adhesions with detachment of the cells from vessel walls, and may lead to cell death. Subsequent repair requires cell growth, probably proliferation, and cell motility to re-cover the denuded surface. These repair processes occur over longer time periods and require coordinated reorganization of the actin- and microtubule-cytoskeleton, cyclical formation and release of focal adhesions, and final re-formation of the cell-cell contacts in order to ultimately restore the quiescent phenotype. During this complex process, however, the endothelial cells are akin to activated endothelium. Thus, for simplicity, we will initially consider the endothelium to have two basic states or shapes: the quiescent and the activated.

The Endothelial Actin Cytoskeleton

Actin Organization in the Quiescent State

It is surprising to many scientists that the proteins, actin and myosin, generally associated with muscle contraction actually comprise 10-20% of the protein mass of endothelium (and other non-muscle cells), although these cells express thousands of different proteins. Moreover, there are about 8 actin molecules to each myosin molecule similar to the molecular ratio in muscle (Schnittler *et al.*, 1990). Thus, actin is the primary component of the microfilament cytoskeletal backbone of non-muscle cells and its interactions with the myosin microfilament, actin-binding proteins, and intermediate filaments are essential determinants of the form and function of the actin-cytoskeleton. Biomechanically, actin filaments are major force bearing components of the cytoskeleton and carry tensile mechanical stress that confers shape stability to the cell (Sato *et al.*, 1990; Satcher and Dewey, 1996; Pourati *et al.*, 1998; Wang *et al.*, 2001; Wang *et al.*, 2002). Several hundred of the actin-binding proteins have been identified, plus more secondary actin-associated proteins, and their direct and indirect relationships to the cytoskeleton, cell structure, and cell function is an area of intense current investigation (Dos Remedios *et al.*, 2003).

In quiescent endothelium there is a near equal balance of 42 kD monomeric, globular (G) actin and polymerized, filamentous (F) actin (Stossel *et al*, 1985). Formation/polymerization of F-actin begins by nucleation of three monomers and then elongation to form a polar (directional) doubled helical chain that is about 7-9 nm wide of indeterminate length depending on the amount of elongation. These single microfilaments, thin filaments in muscle terminology, are usually bundled or cross-linked to form thicker actin filaments. In stable vascular endothelium, part of the actin microfilaments are organized into a dense peripheral, cortical band (DPB) of thicker filaments and part exist as a fine, highly branched, network of short microfilaments throughout the cell body, but connected to the DPB (Fig. 1, Panel A). In this state, the central actin cytoskeleton can be viewed as an interconnected structural lattice and the DPB can be viewed as the hoop in the petticoat of the cell. The stability and disk-like shape resulting from connection of such structures mirrors the relatively flat, spread shape of quiescent endothelial cells, but other cellular elements and biomechanical factors are also important in determining endothelial structure, force generation, and force resistance, as will be discussed. Although, referred to as the "quiescent" or "resting" state, its maintenance is dynamic and the cells are highly metabolically active. In endothelium, this normal, quiescent state is associated with good barrier function (Patterson and Lum, 2001).

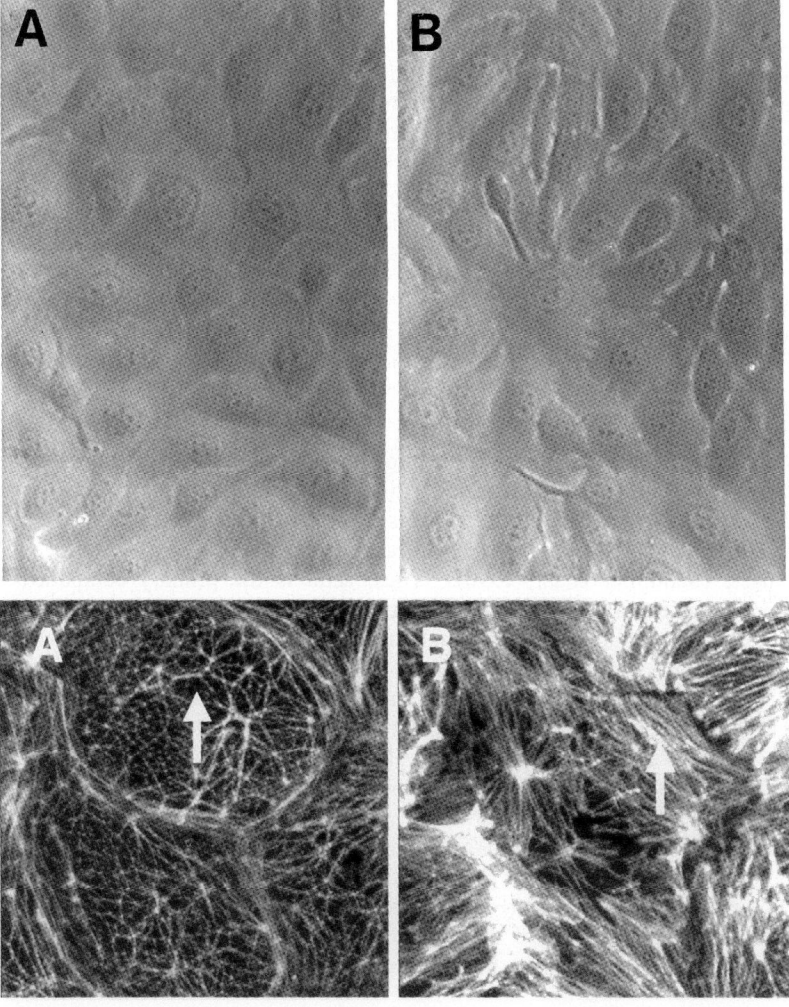

Figure 1: Endothelial cell activation and cytoskeleton.
Top panel: A. Confluent bovine pulmonary artery endothelium, viewed with Nemarski optics, display a cobblestone appearance and close apposition of cells, consistent with selective permeability. *B.* After 10 min. exposure to 100 nM bovine thrombin, the borders of some cells show uneven thinning and other show visible gaps, particularly the elongated cells, consistent with barrier dysfunction.
Lower panel: A. Rhodamine phalloidin staining of F-actin in bovine pulmonary artery endothelium illustrates the presence of dense peripheral bands (DPB) around the borders of the quiescent cells. The interconnected, highly branched web of central actin microfilaments and connections to the DPB is also apparent. *B.* After 10 min. of thrombin, the DPB has disappeared, there is more total F-actin, and stress fibers have replaced the web, some are parallel and others radiating from a central focus. The lower panel is reproduced from Patterson, C.E., *et al.*, 2000. Copyright (2000) from *Regulation of endothelial barrier function by the cAMP-dependent protein kinase. by C.E. Patterson, H. Lum, A.D. Verin, K.L. Schaphorst, and J.G.N. Garcia.* Reproduced by permission of Taylor & Francis, Inc., http://www.taylorandfrancis.com

Actin Organization in the Activated State

Chemically, actin thermodynamic equilibrium drives toward almost total polymerization. Controls exist in the cell to prevent this from occurring, as extreme polymerization is associated with barrier dysfunction. With activation of the endothelium by a variety of permeability-inducing mediators, there is usually a net increase in polymerized F-actin. Phallacidin stabilizes actin filaments and attenuates increased permeability due to mediator activation in polymerization (Alexander *et al.*, 1988; Phillips *et al.*, 1989). On the other hand, unlimited depolymerization results in barrier dysfunction. Cytochalasin prevents elongation of actin filaments and, thus, promotes net depolymerization. Cytochalasin increases endothelial layer permeability (Shasby *et al.*, 1982; Kielbassa *et al.*, 1998), but phallacidin stabilization of the filaments attenuates this kind of dysfunction (Alexander *et al.*, 1988).

In physiologic states and during transitions, neither extreme generally occurs; but, instead, there is remodeling of the cytoskeleton that involves depolymerization of some actin microfilaments and then repolymerization of the produced G-actin into new filaments with different organizations by both nucleation of new fibers and elongation. Most mediator-driven reversible increases in endothelial permeability are accompanied by a loss of peripheral DPB, requiring depolymerization, followed by increased polymerization and organization of actin filaments into long, parallel, bundles known as stress fibers as seen in Figure 1b (Shasby *et al.,* 1982; Garcia *et al.,* 1986; Gotlieb *et al.,* 1991; Ettenson and Gotlieb, 1992; Partridge *et al.,* 1992; Wu and Baldwin, 1992; Lum and Malik, 1994 & 1996; Patterson *et al.,* 1994; Ochoa *et al.,* 1997). The increased abundance of total polymerized actin is also shown in Fig. 1b by increased intensity of total rhodamine phalloidin staining, since phalloidin only binds F-actin, but not the G monomers. With this kind of activation the total amount of the actin pool is not changed, just the proportion in the polymerized state (Gotlieb *et al.*, 1991). Myosin filaments are bundled polymers of the long 500 kD myosin molecules with regulated actin binding sites at the head end. Although myosin is found in the DPB of resting cells, its distribution is highly diffuse; with activation it is strongly co-localized with actin along stress fibers (Hormia, 1985; Wong and Gottlieb, 1990; Gottlieb *et al.,* 1991).

Thus, there is complex, dynamic re-organization of actin microfilaments in both *in vivo* vessel endothelium and in *in vitro* cell cultures in response to physiologic mediators, pathological stress, shear stress, pressure stress, and the wound healing processes. The resulting shape change to an activated state is paralleled by increased in endothelial permeability (Brett *et al.*, 1989; Johnson *et al.*, 1989; Lum *et al.*, 1992; Patterson *et al.*, 1994; Ehringer *et al.*, 1999; Patterson *et al.*, 2000). The cell shape changes from flat (viewed along the wall of the vessels or in the plane of the monolayer in culture) and round or polygonal (viewed from inside of the vessel or looking down on the monolayer in culture) to elongated or ellipsoidal along the axis of the stress fibers. The central role of actin metabolism and organization in the transition from the quiescent to the activated state is illustrates in Figure 2. This schema serves merely as a highly simplified beginning for discussion and introduces some of the proteins and regulatory enzymes that interact with actin to form the cytoskeleton.

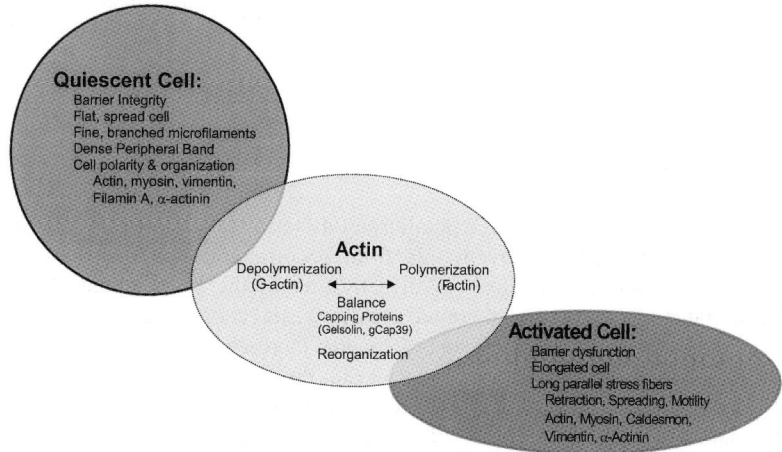

Figure 2: The central role of actin in endothelial form and function. Although endothelia exist in different phenotypes and variations, we will consider two basic states for the sake of simplicity: the quiescent state, which appear as flat, polygonal endothelium with excellent barrier function, and the activated state, which appear as elongated cells with stress fibers and poor barrier function. Actin plays a critical role in the characterization and determination of these physiologic states and the transitions from one to the other. Regulation of actin polymerization and organization is accomplished by a multitude of signaling pathways and numerous actin-binding proteins. The resulting actin cytoskeleton constructed provides the framework for determination of cell polarity, anchoring of functional and enzymatic complexes, and a basis for the mechanical properties of the cell.

Actin-associated Proteins

Actin-binding proteins control assembly of actin filaments into three-dimensional structures and many have been studied in endothelium (Stossel *et al.*, 1985; Dudek and Garcia, 2001; Do Remedios *et al.*, 2003; Winder, 2003). Specific functions and interactions of these proteins with the cytoskeleton will be discussed in Chapter 7, but their importance to barrier function and regulation will be highlighted here to lay the groundwork for the intermediate chapters. One of the most abundant actin-binding proteins in endothelium is α-actinin, which cross links or bundles actin microfilaments into thicker filaments and therefore is present along stress fibers, but also is important in binding the actin cytoskeleton to adhesion complexes at the membrane. Filamin is important in cross-linking actin microfilaments and attaching them to membrane glycoproteins (Hastie *et al.*, 1997b). Filamin A (ABP280) and, to a lesser extent, Arp2/3 are involved in elongation and nucleation/branching of the actin network (Flanagan *et al.*, 2001). The actin in the body of quiescent endothelium is highly branched and filamin A has promotes the orthogonal branching that gives stability to the network in endothelium (Gorlin *et al.*, 1990; Wang *et al.*, 1997; Shojaee *et al.*, 1998). Some of the actin-binding proteins that have been implicated in regulation of actin polymerization/depolymerization in endothelium are: cortactin (mainly present at the cortical matrix adhesion sites), profilin (binds actin monomers and promotes polymerization), cofilin (severs actin polymers), VASP (nucleation promoter), and a small gelsolin-like protein, gCap39 (capping, severing, nucleation). Numerous other actin-binding proteins (such as vinculin, paxillin, talin, zyxin, and fodrin) are primarily located in the matrix-adhesion or cell-cell adhesion sites and serve to link these structures with the cortical DPB in resting cells or with the stress fibers in activated cells. The ezrin-radixin-moesin (ERM) family connects membrane proteins to the cortical cytoskeleton and are involved in microvilli formation, cell-cell adhesion, cell shape and motility, but just what membrane complexes they associate with in endothelium and their role is not resolved (Bretscher *et al.*, 2000; Gruenheid and Finlay, 2003). They anchor ICAM and VCAM at the apical membrane, are a target for invasion by some bacteria, and interplay with Rho signaling.

Since, the actin network is reorganized with activation and these proteins bind actin, the re-organization/translocation of these proteins in endothelium is not surprising. For instance, histamine altered actin binding with gelsolin, which promotes both actin severing and filament assembly - depending on other conditions, suggesting that this might be an important regulatory molecule (Carson *et al.*, 1992). The difficulty is in defining the pathway sequence. A possible causal relation between actin-binding proteins, filamin, and barrier impairment is indicated by filamin translocation from the membrane to the cytosol prior to actin microfilament reorganization and intercellular gap formation in endothelial cells treated with H_2O_2 (Hastie *et al.*, 1997b). Many of the actin-binding proteins are regulated by a multitude of signaling molecules, including phosphoinositides (Fukami *et al.*, 1992; Sohn *et al.*, 1995; Gilmore and Burridge, 1996; De Corte *et al.*, 1997), PKA (Hastie *et al.*, 1997a; Jay and Stracher, 1997), PKC (Stasek *et al.*, 1992; Perez-Moreno *et al.*, 1998), Ras-related small GTPases (Aspenstrom, 1999; Ohta *et al.*, 1999), and tyrosine kinases (Garcia *et al.*, 1999).

Intermediate Filaments

Function, Composition, and Structure

Intermediate filaments provide mechanical stability and connection to membrane-associated complexes (Wang and Stamenović, 2000; Wang and Stamenović, 2002). In endothelium, vimentin and certain cytokeratins are the primary identified intermediate filaments (Alexander *et al.*, 1991; Miettinen and Fetsch, 2000). These structurally related proteins form long, non-polarized polymers as the individual units may orient in either direction. These long polymers generally form double strand α-helical coiled coils, and are called intermediate filaments because at ~10-12 nm they are between the width of actin thin filaments and the thicker microtubules. These double coils cluster into long cables that radiate from a centralized area toward the cell periphery, like a curly pom pom (Strelkov *et al.*, 2003). These filaments are able to resist tensile stress and the more they are stretched, the stiffer or resistant to deformation they become (Janmey *et al.*, 1991). Desmin is found in the vascular smooth muscle, but not usually the endothelium. In addition to their structural role, the cytokeratins are associated with binding of kininogen and kinin activation at the cell membrane (Shariat-Madar and Schmaier, 1999). Aortic endothelium did not stain for cytokeratins, but in pulmonary microvascular endothelium there was increased content and progressive organization of cytokeratin 8 and 19 into membrane-plaques with establishment of confluency at day 3 and then into the radial pom pom with quiescence at day seven (Alexander *et al.*, 1991). In contrast, vimentin content was not changed; but organization into filaments decreased slightly with confluency.

Vimentin and Activation

Vimentin knockout mice were grossly normal (Colucci *et al.*, 1994; Evans, 1998), but fibroblasts from these mice were 40% less stiff, less stable, and less motile (Eckes *et al.*, 1998). Moreover, focal contact protein organization and actin microfilaments were disturbed. Similarly, vimentin disruption by a peptide caused alteration in both the microfilament and microtubule systems (Goldman *et al.*, 1996). More recently, fibroblast and endothelial cells from vimentin knockout mice or cells with acrylamide disruption of vimentin filaments were shown to less stiff and respond to applied stress with less stiffening than cells from wild-type mice (Wang and Stamenović, 2000). Furthermore, these vimentin-deficient cells had slowed proliferation further supporting an important, though not critical role for vimentin in the mechanical properties of cells. The significance is not known, but vimentin moves bi-directionally along microtubules using the dynein motor (Clarke and Allan, 2002). Subjection of confluent endothelium to shear stress flow resulted in actin stress fiber organization and also in rapid reorganization of intermediate filament protein, vimentin (Davies *et al.*, 1994; Helmke *et al.*, 2000; Helmke *et al.*, 2001). Likewise, exposure of endothelial cells to agonist-stimulation resulted in rapid stress fiber formation, association of vimentin with the Triton-insoluble cytoskeleton fraction, and PKC-mediated vimentin phosphorylation (Stasek *et al.*, 1992). Additionally, PKC inhibition prevented vimentin filament formation in brain endothelial cells (Chen *et al.*, 2000).

The Microtubule System

Function, Composition, and Structure

Microtubules extend throughout the cytoplasm and serve to transport molecules within the cell, aid in secretion of molecules from the cell, organize the location of cellular organelle, help maintain the shape of cells, participate in cell division, and facilitate directed cell migration. In resting endothelium some microtubules radiate from the centrosomes located around the nucleus while others are non-centrosomal (Vinogradova *et al.*, 2000; Lee and Gotlieb, 2003). The inner diameters are 12-15 nm and the outer diameters are 24-30 nm. They are formed by helical arrays of bundles of tubulin heterodimers composed of α and β tubulin and the polymers are polar like actin. Regulation of microtubule growth and assembly and the involvement of microtubules in nuclear signaling in endothelium were recently reviewed (Lee and Gotlieb, 2003).

Activation and Microtubules

In their elegant study of endothelial wound repair, Ettenson and Gotlieb demonstrated the importance and interaction of the microtubule cytoskeleton with the actin microfilament cytoskeleton (Ettenson and Gotlieb, 1992). When a band of cells was scraped from the monolayer, cells at the wound edge responded, but not immediately, with activation. Beginning at 30 min. and almost complete at 3 hrs., microtubule centrosome reorganized from around the nucleus to the leading edge of the cells. By 3 hrs., the DPB began to disassemble; later, stress fibers formed, the cells elongated, and extended lamellipodia toward the wound edge. Cytochalasin B pre-treatment to inhibit actin filament elongation retarded, but did not prevent, centrosome relocation and retarded wound closure. Although the wound closed, there were gaps between cells that never gained their typical polygonal shape. However, colchicine, at a dose that disrupted microtubules, prevented centrosome reorientation, cell migration, and wound repair. Moreover, colchicine resulted in a dose dependent, but indirect, disruption of the microfilaments, indicating interaction between these cytoskeletal systems. Similarly, inhibition of microtubule polymerization caused *via* phosphatase inhibition, resulting in myosin-actin interaction and barrier dysfunction (Verin *et al.*, 2001). Recently, PKA activation was shown to attenuate barrier dysfunction induced by microtubule disassembly, indicating a new mechanism for cAMP barrier promotion (Birukova *et al.*, 2004). Biomechanical studies of endothelial cells established the elasticity (restoration following deformation) of the microtubules and the mechanical interaction between the microfilament and microtubule cytoskeleton such that the microtubules oppose the contractile effects of stress fiber activation (Wang, 1998; Wang *et al.*, 2001). These data indicate the importance of the microtubule system in support of the cell shape and the directed migration and its interaction with the actin cytoskeleton.

Mechanisms and Mechanics of Cell Shape Change with Activation

The Contractile and Tensegrity Models

Rearrangement of cytoskeletal elements accompanies, or rather directs, the shape change from quiescent, flattened, spread cells with good barrier properties to the activated, elongated, retracted cells with poor barrier function (Figure 3a). To help conceptualize the biochemical, morphologic, and the mechanical characteristics of the endothelial cell in these two states, the transition between states, and the response to external stimuli, two basic models have developed: the contractile/tethered model (Garcia *et al.* 1995b; Goeckeler and Wysolmerski, 1995; Moy *et al.*, 1996) and the tensegrity model (Ingber and Jamieson, 1985; Stamenović *et al.*, 1996; Ingber, 2003). The contractile model arose from drawing analogies to biochemical events in smooth muscle contraction, when the abundance of actin and myosin and the presence of stress fibers were recognized in endothelium (Hammersen, 1980; Herman *et al.*, 1982) and, particularly, when the correlation of myosin light chain (MLC) phosphorylation and endothelial retraction was made (Wysolmerski and Lagunoff, 1990). Recognizing the importance of cell-matrix and cell-cell attachments to barrier function, this idea was incorporated into the contractile model and termed the contractile/tethered model (Garcia *et al.*, 1995b). The tensegrity model arose from biomechanical considerations of a variety of attached cells to describe the manner by which cells maintain their shape stability using the stress-strain responses, or cell stiffness, and the relationship of these properties to the architecture of the cytoskeletal system (Ingber and Jameison, 1985). These models are not mutually exclusive and are based on similar principles. They differ in approach and are just now converging to give a more unified understanding of interdependency of biomechanical properties and cellular biochemistry and how these relate to endothelial function.

a.
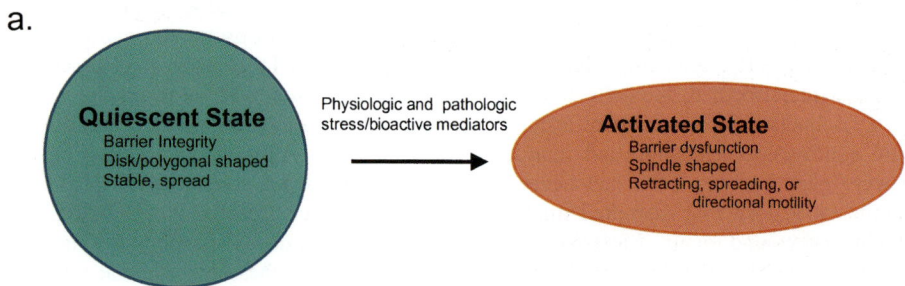

b. CONTRACTILE/TETHERED MODEL

Tethered and relaxed

Contracted and poorly tethered

c. TENSEGRITY PRESTRESSED MODEL

Interconnected microtubule struts, tensed actin cables, and tethers

Main forces along long axis of stress fibers, tightly anchored to matrix adhesions

d. TENSEGRITY GEODESIC MODEL

Interconnected, multivectoral struts, dense peripheral bands, and primarily lateral tethering.

Disassembly of geodesic structure. Main forces along long axis of new stress fibers, tightly anchored to matrix adhesions

Figure 3: Models of endothelial mechanical function.
A. Inflammatory and bioactive mediators, oxidant stress, and physical stress can trigger the transition of endothelium from the normal, quiescent state to the activated state. Reorganization of the cytoskeleton and the cellular attachments is reflected by a change in cell shape and mechanical properties, which can be modeled in several ways.
B. Contractile/tethered model: This model is directed at biochemical events that alter barrier function, but has mechanical implications. In the relaxed, quiescent mode, inward directed actomyosin contractile forces (red arrows) are small and outward directed cell-to-cell (green arrows) and cell-to-extracellular matrix tethering (blue arrows) is strong. The balance favors tethering and, thus, the cell is polygonal and spread. In the activated state, actomyosin contraction results in a larger centripetally directed force, while the tethering is weakened, leading to a shift in the balance resulting in cellular contraction and barrier dysfunction.
C. Prestressed tensegrity model: This model views the cell cytoskeleton, as analogous to compression bearing struts (pink) attached to tensed cables, which give shape and stability to the cell. It incorporates all elements of the cytoskeleton: the microtubules (struts), actin microfilaments, and intermediate filaments (tensile structures), as well as actomyosin units (cables developing the stabilizing tension). Endothelial activation results in an increase in the pre-existing tension, resulting in stiffening of the cell. Tethering provides anchoring for the tensed actomyosin cables.
D. Geodesic tensegrity model: This model views the actin cytoskeleton as a scaffolding structure in addition to the microtubules and intermediated filaments. Although not strictly geodesic, this term is used due to the resemblance of the central actin network in the quiescent state to triangulated geodesic structures. It is anchored in place by matrix tethering. It also emphasizes the structural importance of the prominent peripheral thick bands of actin (the larger curved red arrows) in stabilizing the disk-like form of quiescent endothelium and in providing anchoring for the lateral, barrier-imparting adhesions. Structural members of the geodesic dome bear both compression and tension as opposed to the prestressed tensegrity structure where tension bearing cables and compression bearing struts are distinct. With activation, collapse of the interconnected actin framework accounts for cell retraction and re-assembly into long stress fibers accounts for elongation of the cell. Actomyosin interactions are viewed to cross-link and stabilize the actin fibers and create tension for directed movements.

Once endothelial cells were known to contain contractile proteins and respond to activation with phosphorylation of MLC, strong evidence from many investigators soon followed strengthening the parallels between activation of endothelium and activation of the smooth muscle contractile apparatus. There are similarities in ligand/receptor activation of membrane signaling, with release of inositol phosphates that trigger release of stored calcium into the cytoplasm, calcium/calmodulin activation of MLC kinase, MLC phosphorylation, activation of myosin ATPase activity, and association of the myosin with the actin filament assembly (Moy *et al.*, 1993; Garcia *et al.* 1995b; Stevens *et al.*, 2000; Baldwin and Thurston, 2001; Dudek and Garcia, 2001). Endothelia grown on an elastic support show wrinkling of this artificial matrix and tension development consistent with contraction (Morel *et al.*, 1989; Kolodney and Wysolmerski, 1992). This evidence indicated that endothelial cells develop and inward directed force, often referred to as a centripetal force, although this term has a different technical meaning in physics. Such a radially directed force would logically be opposed by the tethering of the cells at their lateral and basal surfaces. These reports and others concluded that the inward force of contraction resulted in disruption of cell to matrix and cell-to-cell adhesions during contraction and that this disruption, as well as primary signaling of disassembly of adhesions, resulted in the observed barrier dysfunction. Tethering adhesions will be introduced in the following sections and discussed in detail in Chapters 8 and 9. During quiescence, the polygonal morphology and flattened cells was primarily attributed to the lateral and basal tethering, which was relatively unopposed by the relaxed state of the actin/myosin system. Thus, this contraction/tethered model represents the cytoskeletal and shape changes in transition from quiescence to the activated state as a perturbation of the balance of the inward contractile forces and the tethering forces that anchor the cell outwards as seen in Figure 3.b.

Coupling of the actin cytoskeleton to cell shape changes has also been explained by the "cellular tensegrity" concept (Ingber *et al.*, 1985; Stamenović *et al.*, 1996; Ingber, 1997; Ingber, 2003). This conceptual model provides additional understanding of the mechanical forces undergirding the basic cell shape, mechanical responses of the cell to external perturbation, bases for motility of spreading and crawling cells, and forces driving the cell shape changes associated with increases in endothelial permeability. It helps explain how the integrins, which span the cell membrane anchoring to the basal matrix externally and to the cytoskeleton on their cytoplasmic end, allow the cell to organize its cytoskeleton and to move when directed. Moreover, the association of numerous signaling molecules with this integrin-linked site makes clear how mechanical signals can be transmitted into the cells from external stimuli. Thus, in the tensegrity model the cytoskeleton not only supports the cell shape by exerting, resisting, and directing mechanical forces, but also senses and responds to mechanical stress applied to the cell. The tensegrity model includes two structural subclasses: prestressed structure and geodesic dome (Ingber, 2003). Mechanics of both of these structures are based upon synergy between continuous tension balanced by discrete compression. The former maintains its structural integrity *via* pretension (prestress) in cable-like structural elements attached to struts, which bear the compression generated by the tensed cables. The latter is comprised exclusively of struts and maintains its structural stability *via* triangulation of its struts oriented along geodesic paths, i.e., the minimal paths between 2 junctions, under continuous compression. The geodesic dome needs no prestress as the struts can carry both tension and compression.

Most work in cellular mechanics has focused on the prestressed tensegrity model (Figure 3.c.), rather than on the geodesic dome model, because it best describes the observed responses to application of external forces and the changes observed during cell contraction. Sometimes, the prestressed tensegrity model is likened to a camp tent, with internal poles (struts) that bear compression, and internal or external guide wires that place tension on the poles at various angles to give stability to the structure. The tension on the canvas imparts further structural stability and stiffness. The anchoring of the cables to the ground or nearby trees permits development of the tension in the guide wires, but also imparts location to the structure. If one of the guide wires is then pulled or pushed the force is transmitted to the other elements of the structure, which help provide resistance to movement and elastic recoil when the external force is removed. If several of the pegs anchoring the guide wires are suddenly pulled out of the ground or if the tensed cables are cut, the structure may collapse in whole or part. This image correctly explains the experimental biomechanical observations, which have been made on cells.

The tensile elements in the cell are thought to be actin filaments and intermediate filaments, which carry the pre-existing mechanical tensile stress (termed "prestress") prior to application of an external load. This prestress is partly generated actively, by low-level myosin contraction (latch state-like contraction), and partly passively, by swelling pressure of the cytoplasm and by cell distention over the substrate. The role of prestress is to confer stability to the entire structure; the greater the prestress, the more stable (i.e. less deformable) the structure is. In the absence of prestress, the structure loses its stability and rigidity and collapses as the compression elements alone are not stable, like the poles of the tent without the guide wires. In the tensegrity model, elastic restoring forces arise mainly from changes in the orientation and spacing of the tensed filaments as the network is deformed (Stamenović et al., 1996; Stamenović and Wang, 2000). Experimental biomechanical measurements of cultured smooth muscle cells, fibroblasts, and endothelial cells are consistent with the notion that cell shape stability is closely associated with cytoskeletal prestress (Wang and Ingber, 1994; Pourati et al., 1998; Wang et al., 2001, Wang et al., 2002). This model predicts that cell stiffness will increase in direct proportion to the level of cytoskeletal prestress. Such correlation was demonstrated in attached smooth muscle cells subjected to relaxing and contractile agonist (Wang et al., 2002). In skinned, attached fibroblast, ATP-initiated acto-myosin contraction also resulted in increased tension on the matrix while addition of cytochalasin prevented the increase in tension; further supporting the idea that actin-myosin contraction provides the tension on the cables (Thoumine and Ott, 1996).

The prestress is balanced internally by compression of microtubules and externally by the forces at the cell anchoring points to the matrix substrate (Wang et al., 2001; Stamenović et al. 2002). When external stress was applied directly to discrete matrix adhesion structures, internal movement was seen in mitochondria associated with the microtubules, but not the cortical DPB. With end-on deformation the microtubules buckled and sprung back when the load was removed (Wang et al., 2001). Thrombin activation of the cells also caused microtubules to buckle. This was prevented by cytochalasin disruption of the actin fibers, indicating that microfilaments tension caused the microtubule deformation. These data indicate that microtubules resist the stress conferred by the actin fiber tension. If microtubules are the main compression bearing elements in the tensegrity model, then removing them should increase traction on the matrix connections. In attached smooth muscle cells, the prestress tensegrity prediction was confirmed in that colchicine significantly increased traction, indicating that the microtubules carried a substantial compression in these cells (Wang et al., 2001; Stamenović et al., 2002). Nocodazol disruption of microtubules likewise resulted in a significant increase in force in intact fibroblast and smooth muscle cells, however this effects was attributed to increased MLC phosphorylation rather than removal of the compression struts (Kolodney and Elson, 1995; Polte et al., 2004). In skinned fibroblast, neither nocodazol-disassemble nor taxol-stabilization of the tubules had an affect, indicating that the microtubules were not resisting the traction (Thoumine and Ott, 1996). Interestingly, in mechanically or trypsin detached endothelial cells with no matrix connections, when colchicine was used to disrupt the microtubules or cytochalasin was used to disrupt the actin cytoskeleton, the actin microfilaments, rather than the microtubules, were indicated as the main component determining the viscoelastic response measured by the micropipette technique (Sato et al., 1990). Lastly, the prestressed tensegrity model illustrates many principles of how forces are directed in the forward and aft ends of migrating cells to provide propulsion and retraction necessary to movement of an attached cell. Thus, the prestressed tensegrity model fits many observations made in the endothelial and other attached cells.

While the prestressed tensegrity model was likened to a traditional camp tent, the geodesic model is somewhat like a child's playhouse constructed from PVC pipes with 3-way, right angled joiners at each corner and a colorful cloth loosely draped over for effect, but providing little structural contribution. It is internally complete and can even be picked up, moved to a new space, or rotated. It does not rely on gravity for continuous compression.

Additional triangulating crossbeams provide further stability and anchoring to the ground can provide locational stability. The morphologic similarities of the interconnected actin web in the quiescent cells and rigid, self-supporting frameworks suggest that this three-dimensional actin lattice of quiescent endothelial cells could act as a main structural element of a geodesic tensegrity model as shown in Figure 3.d. Under some circumstances, such as exposure to large, short duration compression stress, this model is consistent with rheologic measurements made in living cells; however, the shear modulus calculations for this model are an order of magnitude too high compared to that obtained by actual measurement in living cells with either the micropipette aspiration or the magnetic twisting technique (Sato et al., 1990; Wang and Ingber, 1994). Actually, such determinations on polymers of purified actin microfilaments or purified microtubules are even higher, emphasizing the importance of the geometrical arrangement of the cytoskeletal structures to the overall mechanical property (Stamenović and Wang, 2000). Moreover, the lack of cross linking and various other factors affecting the quality of the purified actin in these artificially produced actin polymers compromise the applicability of calculation on these viscoelastic gels to living cells (Xu et al., 1998). Structural similarities between the F-actin network and microstructures that resist deformation also led to examination of the biomechanical properties of the actin cytoskeleton by analogy to an open-cell foam model, with good agreement between the predictions and experimental properties of endothelial cells (Satcher and Dewey, 1996). In such a model, bending of the actin network is the principal mode of resistance to applied stress; whereas in the geodesic dome model the members are struts in tension and compression. Regardless of how the complex cellular architecture of the quiescent cell is modeled, qualitatively it is apparent that actin filaments *per se* are able to both store and dissipate mechanical energy. Finally, the ring-like structure of prominent circumferential DPB composed of thick actin fibers, characteristic of quiescent endothelium with the best barrier function, would impart shape and stiffness to the cell for mechanical stimuli acting more or less in parallel with the plane of the ring. Importantly, the DPB is interconnected at multiple points with the finer actin network, providing further structural support as suggested in Figure 3.d. In these quiescent cells myosin is mainly associated with the circumferential ring and contraction in this ring preceded the disassembly and remodeling to stress fiber, explaining rapid contraction (Kolodny and Wysolmerski, 1992). Moreover, direct association of lateral cell-cell adhesion with the DPB accounts for association of intercellular gap formation with contractile activation.

Activation Modeling: Controversies, Cooperation, and Resolution

Thus, each model helps to explain certain characteristics and behaviors of endothelium; yet isolated, each falls short of giving us a complete picture. While explaining the biochemical relevance of actin-myosin interactions to endothelial function and acknowledging the importance of tethering to barrier function, the contractile/tethered model does not fully address many observations on endothelial cells. First, this model does not explain mechanical properties of the cell or the responses to mechanical stimuli. Second, the contractile model does not generally take into account the dramatic, fundamental geometric alterations in the actin cytoskeletal structure, the simple effects of disassembly of the DPB and fine network on cell shape, or the effects of the cytoskeletal alterations on integrity of the cell-cell connections beyond increasing tension on these junctions. Although, contraction of the DPB with its associated myosin could adequately explain very early tension development, the rapid disassembly of the DPB argues against this for more than a few seconds, while tension peaks at 5 minutes. Third, the contractile model does not adequately explain the alterations in endothelial shape and barrier function, accompanied by dissolution of the DPB and altered actin polymerization, induced with other modes of activation, such as PMA and pertussis toxin, none of which involve changes in MLC phosphorylation or activation of actomyosin contraction (Shasby et al., 1982; Wong and Gottlieb, 1990; Garcia et al., 1995a; Patterson et al., 1995; Moy et al., 2004). Fourth, in endothelial activation that does correlate with and depend upon MLC phosphorylation (via either enhanced phosphorylation by MLC kinase or decreased dephosphorylation by rho kinase inhibition of phosphatase), if the sliding filament model of actomyosin contraction were the primary effect of the increased interaction of myosin with actin, then this should occur along the axis of the stress fibers, yet stress fiber formation is a later event and the cells actually elongate in this direction and retract horizontally. While data supports increased tension at the point of focal adhesion upon activation of actin-myosin interaction indicative of contraction (Thoumine and Ott, 1996), it does not explain how those long stress fibers formed and the cell elongated in the first place. Fifth, observed changes in silicone substrate wrinkling and tension development can be attributed to altered architecture and cell-cell and cell-matrix adhesion, as well as to muscle-like contraction. Sixth, contractile model does not take into account the contribution of the microtubule system or intermediate filaments to cell shape or shape change. Whereas, there are reports that inhibitors of microtubule polymerization enhanced MLC phosphorylation (and thus activated myosin-actin interaction and contraction) in fibroblast (Kolodney and Elson, 1995) and in endothelium (Verin et al., 2001), microtubules disrupted has also been shown to increase tractional forces on the substrate in the absence of MLC phosphorylation (Wang et al., 2000; Wang et al., 2001; Stamenović et al., 2002). Seventh, the model as usually presented treats cell-cell and cell-matrix tethering together as oppositional to inward contractile force. However, during agonist-activation with increased MLC phosphorylation, while there is a dramatic decrease in cell-to cell tethering, there is actually a major increase in the true focal adhesions to the matrix, as will be discussed in following sections. Moreover, studies of cAMP regulation of barrier function indicate that PKA maintenance of cell-cell adhesion maintains barrier function despite activation of contraction (Cioffi et al., 2002).

The tethered/contractile model predicts that the main contribution to the resting shape is the tethering of cells to each other and to the matrix. Yet, when a single endothelial cell within a truly cobblestoned monolayer of cultured endothelium is mechanically lifted from its position in the monolayer, it consistently maintains a disk-like shape for at least several minutes, despite the sudden rupture of the cell-cell and cell-matrix tethering. It behaves initially like a tinker toy structure lifted into the air, indicating that internal structure is a prime determinant of shape. Likewise, its former neighbors in the monolayer also maintain their polygonal, spread shape for a minute or two (unpublished observation, Patterson). The same appears to be true of endothelium left behind in the wound model, which did not begin to change shape until 3 or more hours after the scrape removal of their neighbors (Ettenson and Gotlieb, 1992). On the other hand, a similar study with wounding either along the side of the cells or beneath the cells did show elongation, indicative of rapid retraction of the remaining cells in the monolayer (Thoumine et al., 1995). The cells and culture conditions were similar in the two studies and both displayed three panels of control-actin stained cells of approximately the same magnification and number; however, in the second study it was apparent in the controls that the cells were somewhat activated. In the first study, the DPB were prominent in 100% of the control cells and stress fibers appeared in on 6% of cells. In the second study, DPB were visible in only 50% of the cells and they were thin and single rather than a multiple fiber band; prominent parallel thick stress fibers were seen in 68% of the control cells. Moreover, in the second study the elongated cells were all angled in the direction of the razor cut implying physical dragging and external compression of the monolayer and its matrix. Finally, in the undercut wound of the second study, the cells were fixed within two minutes, but at that time there was very dramatic dissolution of the actin filaments, DPB and stress fibers, with obvious clumping through the field. Similarly, there was disassembly of the vimentin network with in the short period. These data indicate significant biochemical events were occurring due to the applied stress. Together these three observations suggests that in quiescent barrier-competent endothelium shape depends more on multivectorial forces from a geodesic tensegrity structure of the actin cytoskeleton, with complementary strutting by the microtubules and the vimentin structure, than on the tethering to basement or adjacent cells. While on the other hand, in somewhat activated cells, without strong DPB and orthogonal actin network support, but with formation of anchored actin stress fibers, tethering is much more important for maintenance of the polygonal shape.

The prestressed tensegrity model accounts well for the observed correlation of cell stiffness and activation of the contractile apparatus and it explains the observed perturbations of internal structures with discrete or flow-stress force application to the cells. Furthermore it takes into account contributions of the microtubule and intermediate filament arrays and the contribution of both cell-cell and cell-matrix tethering to the mechanical properties of the cell, yet the prestressed tensegrity model alone does not fully address the behavior of the endothelium. Originally, it was naturally focused on the mechanical model *per se* and did not include the biochemical mechanisms involved in interactions of myosin and actin, but as this was identified in endothelium and could readily explain the mechanism responsible for the tension of the cables and the increased tension and stiffness with activation, it was rapidly incorporated. A recent paper embarks on establishing experimental relationships of MLC phosphorylation to tensegrity in smooth muscle cells (Wang *et al.*, 2001). The understanding of the role of mechanical stress in biochemical signaling is growing, particularly as it relates to the prestressed tensegrity model (Chicurel *et al.*, 1998; Stamenović and Wang, 2000).

The primary difficulty with the prestressed tensegrity model is that while the data make an excellent fit with properties observed in living cells, including endothelium, the endothelial cells are nearly always in an activated state, even when described as confluent. Although endothelial cells may often appear to be spread out, somewhat flat, and confluent, they are not necessarily cobblestoned, quiescent, and barrier intact. In this state they may have weak early contacts between adjacent cells but not good sealing junctions, lack the DPB, exhibit a preponderance of stress fibers, have minimal fine web-like actin network, and are either still in a state of transition or are sustained in the activated state. The immunofluorescent images reveal that the cells are actually isolated or form only a few tenuous contacts. In most of the studies, the cells are replated sparsely and allowed only 6 or so hours until experiments are performed. This is done because very spread cells are very stiff cells and the measurements become problematic to make and interpret. At such a time the cells will, invariably, have formed strong focal adhesions, marked stress fibers attached at the focal points, and a partially spread morphology, but neither the fine actin network, nor dense peripheral bands. For instance, to determine rapid, biomechanical response to mechanical severing of the actin cables, endothelial cells were plated "sparsely in serum-free medium ---for 4 hrs" then cells were cut with a microneedle and observed. The remaining halves or the nicked cells, indeed, retracted is less than 1 second. Importantly, this retraction did not occur when the cells were pre-treated with cytochalasin, confirming the actin cable-based pre-existing tension (Pourati *et al.*, 1998). Nevertheless, under these conditions the attached endothelial cells are not representative of quiescent cells.

Thus, the measurements provide strong support for the prestressed tensegrity model in these cells, but do not well describe the normal intact monolayer of connected cells with good barrier function and the cytoskeletal arrangement of quiescent cells. For instance, the prestressed tensegrity model predicts that a cell spread on the substrate will immediately round up when detached from the substrate (Coughlin and Stamenović, 1998; Ingber 2003). As discussed in the last paragraph, this is exactly what is seen in the partially activated cells exhibiting stress fibers – tension generating cables; while the above observation on temporary structural maintenance upon lifting single cell from very quiescent monolayers or with scrape wounding in monolayers is not consistent with prestressed tensegrity behavior and is more consistent with the geodesic tensegrity model. Importantly, in quiescent endothelium, the myosin is very prominently located with the DPB actin with almost none visible with the central web-like actin lattice (Schnittler *et al.*, 1990). Likely the transition from quiescence to activation and vice versa, involves not only an altered biochemical, and altered barrier state, and an altered cytoskeletal state, but an altered biomechanical state, as well. One study of confluent mammary tumor epithelial cells makes several conclusions that support a different mechanical state in confluent cells (Potard *et al.*, 1997). In this study, stress was applied by magnetic twisting at either E-cadherins or at integrins. They found that: at low stress microfilaments contributed little to stiffness since cytochalasin had no effect; at high stress, stiffness coupled through integrins was greater than stiffness coupled through the cadherins; at high stress the effect of cytochalasin on stress reduction coupled to the integrins was greater than for colchicine indicating microfilament were more important than microtubules but that both conducted force; at high stress only cytochalasin reduced stress conducted through cadherin indicating that these adhesion only connect to the microfilaments; the progressive relationship of stiffness to stress was much lower than that found in prestressed, sub-confluent cells; and that the cells exhibited strain dependent mechanical recovery after stress. They concluded that the microfilaments are not under the same kind of tension as in the non-confluent cells and could not be well described by the prestressed tensegrity model and surmised that in such confluent cells, which may be more like quiescent cells in tissues with low turnover rates and low shear stress, the cytoskeleton might be a stable entity.

When cytochalasin was added to disassemble actin fibers in smooth muscle, the cells retracted as predicted by the prestress model; however, when an inhibitor of myosin ATPase was added to dissipate the cell prestress without disrupting integrity of the cytoskeleton, cell shape was maintained. This suggests that actin fiber stiffness or that the microtubules and matrix adhesion was adequate to prevent a prominent collapse from loss of the prestress (Polte *et al.*, 2004). Probably, the effect of the inhibition of myosin ATPase was more analogous to loosening tension in a tent structure, than to total detachment of the pegs from the ground, i.e., without disrupting the connectedness of the structure. The tent would become much more compliant and "rickety" but would not lose its shape. Similarly, when airway smooth muscle cells are highly relaxed by membrane permeable dibutyryl cAMP, they become an order of magnitude more compliant than the control cells, but they do not lose their shape. Thus, basic shape of the cell may be influenced and stabilized by the prestress, but is not solely dependent on the prestress. Moreover, in most physical models of prestressed tensegrity structure, the geometry is one of apparent random vectorial arrangements of struts and cables, but in activated cells, stress fibers are markedly parallel with three to four areas in the cell of slightly different orientation. The stress fibers generally run parallel to the microtubule and the vimentin arrays as well. Thus, it is not yet known how geometry of the cytoskeletal elements correlates with the model, especially when there is a dramatic, rapid remodeling of the actin cytoskeleton with activation. While it is clear that prestress is present in these cells, it is not clear how the actin is associated with the compression elements. The prestressed tensegrity model does not address the specific role of the dense peripheral actin band, a prominent feature of quiescent endothelium. Nor does it contemplate the differential roles that would be played by the fine, highly branched actin network of the quiescent endothelium vs. the role of thick parallel stress fibers of the activated cell. Lastly, the model does not account for the recognized role of actin filament formation in the forward projections of lamellipodia that aid in spreading, crawling, and formation of initial cell-cell contacts (Elson *et al.*, 1999). When spherical, isolated cells are first plated, they form attachments between the cell and the "ground" within 10-30 minutes. These activated cells then send out lamellipodia laterally in all directions to form new attachments and spread out the cell in star-like patterns. Similarly, when a monolayer of cultured cells in artificially wounded by scraping off a band of cells, the cells at the wound edge become activated in about 3 hours and at 6 hours they begin to send out lamellipodia into the denuded area and initiate cell migration (Ettenson and Gotlieb, 1992). The lamellipodia extend by elongation of actin microfilaments from a stationary internal position and generation of an outward force, distorting and protruding the membrane. Possibly, similar actin elongation accounts for the initial elongation of endothelial cell that are activated by contractile agonists.

The geodesic tensegrity model could explain the morphology and a low-energy consumptive stability of the quiescent cell. It also is consistent with the internal structural integrity of truly quiescent cells observed when suddenly removed from matrix and cell-cell tethering. Assuming that geodesic tensegrity describes the quiescent endothelium, activation resulting in disassembly of actin filaments and concomitant loss of both the DPB and the internal lattice network would reduce the force that spreads the cell into the thin disk accounting for apparent retraction. Additionally, this disassembly would disturb cytoskeletal connections with the lateral adhesions. Re-polymerization of actin and bundling into long stress fibers could conceivably drive cell elongation. Phosphorylated myosin, which moves rapidly from a diffuse cytosolic pool to the actin, could act to cross-link actin filaments into stress fibers, rather than participate as sliding filament contraction, which requires firm anchoring of both thin and thick filaments at appropriate intervals for shortening to occur. But, beyond gross morphologic observations this is purely speculative and not tested by studies in cobblestoned monolayers. In studies with smooth muscle cells plated so that the cells were generally individual and measured 6 hr post plating, approximately 14% of the compression was due to the microtubules and the remainder of the resistance to deformation was attributed to matrix adhesions (Stamenović *et al.*, 2002). Other recent studies in smooth muscle have ascribed up to 80% of the balancing of the acto-myosin prestress to microtubule compression in moderately to little spread cells; whereas, in highly spread cells this contribution is only several percent (Hu *et al.*, 2004). It would be very interesting to know what this ratio is in quiescent, cobblestoned endothelium and if either or both the DPB and the actin network bears external load. Finally, actin filament disassembly by cytochalasin might be expected to have just as much effect in the geodesic tensegrity model of quiescent endothelium due to removal of the supporting structure as in the prestress model, where the aim is to cancel the actin prestress. As quiescent cells are generally in normal monolayer, tethering might be expected to counteract some of the immediate effects of actin disassembly, even though not as important a factor for shape as the actin network as suggested by the lifting and wounded monolayer studies. In different studies of confluent endothelial cells subjected to cytochalasin, the more confluent, quiescent, and laterally allied the endothelium the less the effect of the cytochalasin, but the cells generally showed retraction or shape alteration and gaps sometimes formed (Shasby et al., 1982; Alexander et al., 1991; Vischer et al., 2000; Sawyer et al., 2001; DeMaio et al., 2004). Interesting, but not surprising, when cells were then further treated with thrombin, no typical elongation occurred, suggesting that stress fiber formation participates in elongation of the cells with activation (Vischer et al., 2000). Recent studies on endothelial response to high shear stress demonstrating that cell elongation is the result of actin polymerization, shear stress fiber assembly, and fiber driven protrusion of the cell membrane supports the concept of cytoskeletal organization (as opposed to contraction) as the fundamental basis for cell shape and shape change (Noria et al., 2004).

Regardless of which model is discussed, with the drastic reorganization of the actin cytoskeleton and the cell-cell and cell-matrix adhesions as summarized in Figure 4, there is obviously much more going on in endothelial activation than mere development of centripetal contraction. When endothelial cells were suddenly subjected to stress flow, there was rapid displacement of vimentin intermediate filaments, indicating their response to external force and a significant contribution to deformation and resistance to deformation (Helmke et al., 2001). Directionality and nature of the deforming stress must also be considered. It is not surprising that push or pull force applied directly to matrix integrin connections is transmitted along the attached stress fibers to the microtubules, but not to the dense peripheral band, which is oriented behind stress fibers when they appear together in a cell. This might suggest that DPB have little mechanical import, but in cells with prominent DPBs subjected to lateral forces, such as stress flow, the DPB could be highly important. Similarly, when blunt force is applied from above, the biomechanical stiffness of the nucleus, which reaches from the top to bottom of the spread endothelial cell comes into play (Caille et al., 2002). Much work is needed to explain the molecular and vectorial interactions of actin and myosin before and during agonists-directed reorganization of various cytoskeletal molecules. We are beginning to gather information regarding interactions between the cytoskeleton and the proteins of adhesion complexes and their associated regulatory enzymes, but this information must be resolved with the mechanical models of contraction and tensegrity. In addition to understanding the relationship of signaling and integrin-cytoskeletal connections, how the cytoskeletal structure is coupled to other signaling is a wide-open field. For instance, disruption of the actin cytoskeleton in human umbilical vein endothelial cells resulted in a 3 to 6 fold increase in synthesis and in release of prostanoids into the medium, while disruption of the microtubules with nocodazol or colchicine increased synthesis, but not released (Sawyer et al., 2001). What is the mechanistic connection between this observation and cytoskeletal structure is not explained. Finally, there is evidence of cross-talk between the microfilament and the microtubule systems in endothelium as described above, but the nature and number of these interactions is not resolved. Models are necessary to understand and test the properties of highly complex biologic systems and no model reproduces the reality; thus, it is not expected that any one model is completely accurate. Likely, aspects of these models presented occur simultaneously within the cell or alternately as cell undergo transition from quiescence to activation. Because of the dynamic changes in the endothelium we have yet to arrive at a unified model that adequately describes the endothelium in either extreme of quietude or agitation, but because of the models and the testing of the defined elements, we have progressed greatly in our effort at understanding the nature of the endothelium.

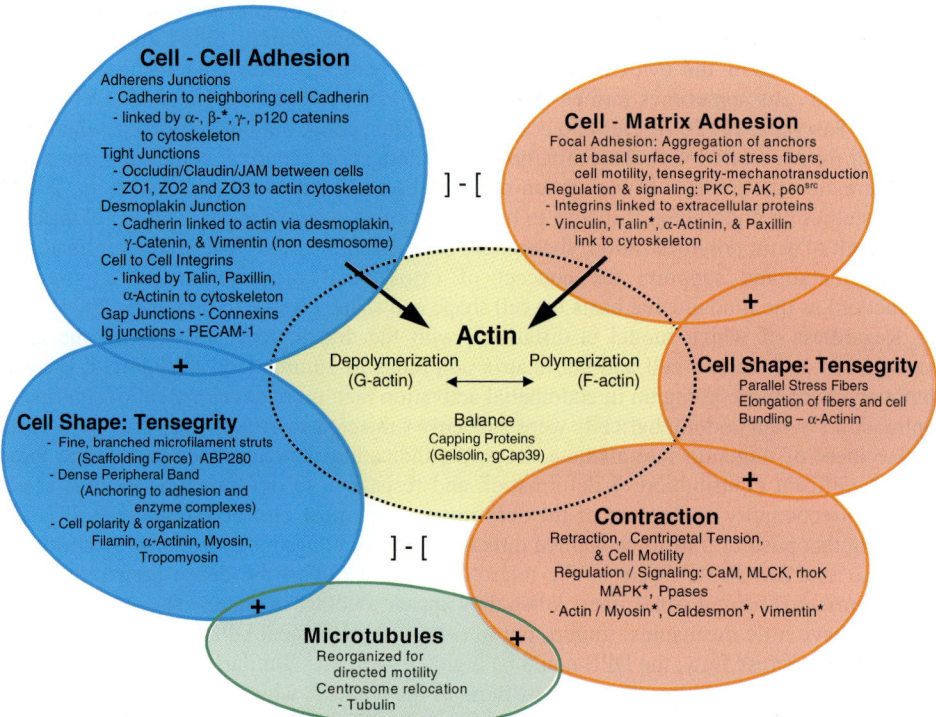

Figure 4. Relationship of endothelial cell cytoskeletal and associated proteins to barrier function/dysfunction. Yellow signifies the pool of the cytoskeletal actin. Blue indicates attributes associated with barrier integrity. Pink indicates attributes associated with barrier dysfunction and/or cell motility. * indicates proteins with well-known function-associated phosphorylation. + indicates parallel changes and general cooperativity in the reorganization of cytoskeletal elements.]-[indicates structural/functional changes that usually occur in opposition. **Actin** constitutes ~ 10% of total endothelial protein and, in normal confluent cells, the actin pool is distributed ~ 50/50 between monomeric G-actin and filamentous F-actin. Either extreme, depolymerization or polymerization, is associated with cellular and barrier dysfunction. Regulatory proteins associated with actin microfilament capping and severing, such as gelsolin or gCap39, agents that prevent actin filament elongation, such as cytochalasin, and monomer-linking proteins, such as profilin, influence the balanced formation of microfilaments. In normal confluent endothelium, which presents a selective barrier to fluid and protein movement, the central actin cytoskeleton is a fine network of highly branched microfilaments and at the lateral cell edges the actin forms a dense peripheral band. This cytoskeletal organization forms a tensegrity scaffolding that partially accounts for the flat, spread morphology of the endothelial cell and/or tensegrity tensile elements that account for cell stability and stiffness. Microtubules and intermediate filaments may also support the cell structure. **Adhesion complexes** at the lateral borders link externally to an adjacent endothelial cell and internally to the dense peripheral actin band and actin network. The primary cell-to-cell junction in endothelium appears to be the adherens junction, but tight junctions, gap junctions, PECAM, and integrins also contribute to formation and maintenance of the cell-to-cell seal that forms the barrier to paracellular solute flux. The cell-to-cell integrin complexes may be similar to the fine punctate focal adhesions to the extracellular matrix at the basolateral edges. Disassembly of the dense peripheral band,

disassociation of the junctional complexes from the cytoskeleton, and disassembly of the junctional connections lead to gap formation and barrier dysfunction. When endothelial cells are subjected to shear stress, activating agonists, oxidant stress, or mechanical injury, the disassembly of the pro-barrier cytoskeleton results in opposing assembly/activation of actin **stress fibers, contractile elements, focal adhesion** aggregates, and **microtubule** reorganization. Generally there is an increase in overall actin polymerization, elongation of filaments, elimination of branching, and bundling of filaments into thicker fibers, likely involving rho kinase activation. Reorganization of the tensegrity scaffolding could account for the elongated, spindle shape of the activated EC as the actin fibers are reorganized into semi-parallel arrays. The fibers converge and anchor at the focal adhesion sites that coalesce away from the cell edge and net focal adhesion increases. With directed cell movement, as in wound healing, the complexes may disassemble/reassemble at the leading edges and the microtubules reorganize with movement of centrosomes from a sub-nuclear location to a peripheral location. Finally, these changes are accompanied by activation of classic components of contraction and tension development by the cells, but the structural basis and the role of sliding filament, actomyosin shortening are likely very different from that in muscle and are consistent with a prestressed tensegrity structure. Figure modified from Patterson and Lum, 2001. Coyright (2001) from *Update on pulmonary edema: the role and regulation of endothelial barrier function. by C.E. Patterson and H. Lum.* Reproduced by permission of Taylor & Francis, Inc., http://www.taylorandfrancis.com

---

Endothelial Attachment to the Basement Matrix

Cell/Matrix Complexes

As might be expected for cells that face blood on one surface and tissue-adjacent matrix on the other surface, attached endothelial cells are highly polar. The asymmetric arrangement of the cytoskeletal structure and the distinct differences in the apical, lateral, and basal outward attachments molecules reflect this apical/basal directionality of environment. As depicted in Figure 5, the cell-matrix attachments are intimately linked to the cytoskeleton and involved in altered cell state and function. The most important cell-to-substrate adhesions are integrin-based focal adhesion complexes (Burridge *et al.*, 1990; Brakebusch and Fassler, 2003). Heparan protoglycans on the basal cell surface also form adhesions to the substrate proteins (Iivanainen *et al.*, 2003). Figure 5.a. illustrates these cell-matrix adhesions and shows some of the components of the integrin-based focal adhesion complex. True focal adhesions are defined microscopically as sites of very close, 10-15 nm, apposition of basal membrane and the cover slip and are identified as focal adhesion plaques with different interference reflectance patterns compared to the remainder of the basal membrane surface. But even when these plaques are not present, the focal adhesions proteins are usually complexed together at the membrane and looser adhesions are formed to matrix proteins (Schaphorst *et al.*, 1997). Although similar in structure and function, these complexes should be referred to as such and not termed "focal adhesions" as they do not conform to the original definition. This is especially true as integrin adhesions are also present on the lateral surface to link with other endothelial cells or substrate proteins between the cells and on the apical surface to bind leukocytes, but the particular integrins expressed at the different site show distinct site-specific distribution (Lampugnani *et al.*, 1995).

A.

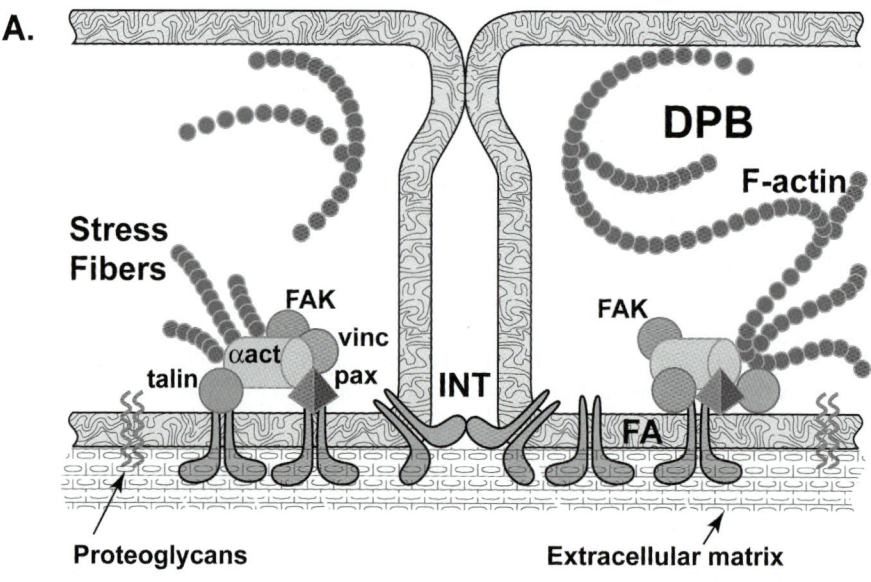

B.

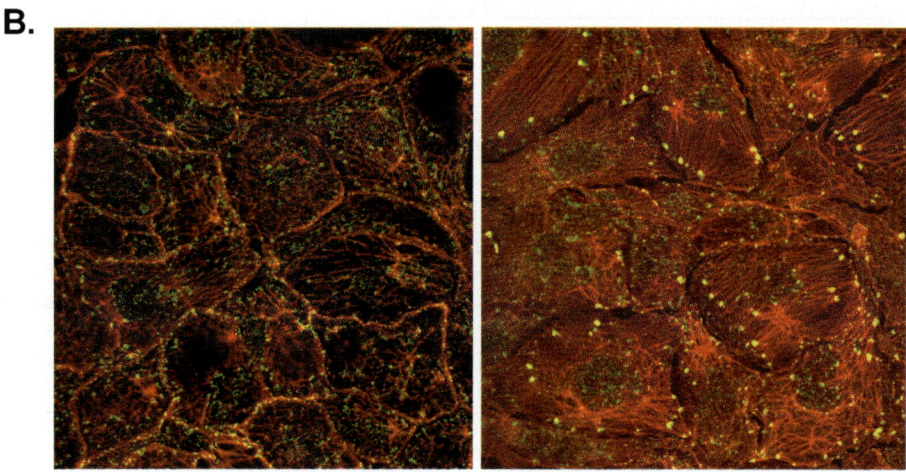

C.

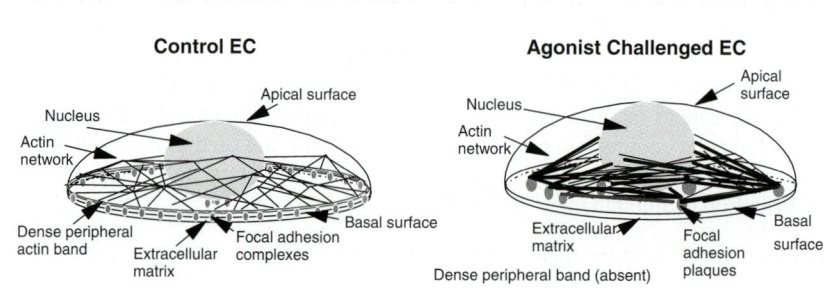

Figure 5. Endothelial-Matrix Adhesion.
A. <u>Endothelial cell-to-basal lamina adhesion</u>: Basolateral adhesions anchor the endothelium and contribute importantly to cell shape and barrier function. The primary connections between the base of the endothelial cells and the protein matrix of the basal lamina are formed by membrane-spanning integrins (**INT**), composed of one α and one β protein, from a selection of such proteins that render ligand specificity for the matrix substrate binding. Specific integrins and associated proteins also participate in lateral cell-to-cell binding, though the arrangements and ligand are less clear. Other integrins are differentially expressed on the apical surface. When integrins cluster at the basal surface to form very close connections between the basal membrane and the basal lamina, focal adhesion plaques (focal adhesions - **FA**) are formed, with unique microscopic appearances and are characteristic of activated endothelium. The cytosolic ends of the integrins are anchored to the actin cytoskeleton by complexes of several intermediate molecules, which include talin, paxillin, vinculin, α-actinin, and other proteins. These focal adhesion complexes also are closely associated with a number of signaling molecules, such as PKC, p60^{src}, and p125 focal adhesion kinase (FAK), which are involved in assembly, in actin polymerization, and in signal transduction of external condition and mechanical stress to the cell. These structures are continuously remodeling and are important in cell motility. Quiescent endothelia have few true focal adhesion plaques, but the integrins and associated proteins form complexes that similarly anchor the cell, but with a different distribution and wider membrane-matrix apposition. Additionally, heparan proteoglycans extending from the basal membrane bind matrix components and contribute to cell attachment.
B. <u>Confocal immunofluorescence of FAK</u>: In quiescent endothelium focal adhesion kinase, a marker of the focal adhesion complex, is distributed around the basal surface, but concentrated at the cell periphery and directly under the nucleus as seen by the green fluorescence or yellow where there is overlap with the red fluorescent F-actin (left panel). In activated cells, FAK staining appears in fewer, larger plaques at the ends of stress fibers or under the nucleus (right panel).
C. <u>Three-dimensional reconstruction</u> depicting endothelial FAK and F-actin determined by stack analysis of immunofluorescent confocal imaging. <u>Control EC</u> - In quiescent, confluent endothelial cells F-actin is organized into a highly branched, web-like network of fine filaments in the body of the cell and into a dense peripheral band of roughly parallel actin fibers that circle inside and just above the basolateral edge of the cell. FAK, a regulatory tyrosine kinase enzymes co-localized with integrin-focal adhesions, is distributed at discrete sites throughout the basal surface but with marked abundance at the lowest edge of the basolateral circumference and in a small area under the nucleus (left panel). <u>Agonist-challenged EC</u> – Subsequent to treatment with a variety of activating, edemagenic mediators, there is a marked increase in total actin staining, representative of actin polymerization; yet, the dense peripheral band disappears. There is transition from the filament network pattern to one of thicker, nearly parallel arrays of fibers and cell elongation along the axis of these parallel fibers. Groups of these fibers converge at the basal surface at areas where the FAK staining seems to have shifted from the edges of the cell. These basal sites are fewer in number and somewhat larger, suggestive of integrin/focal adhesion plaque clustering (right panel). Panel B from Patterson and Lum, 2001. Coyright (2001) from *Update on pulmonary edema: the role and regulation of endothelial barrier function. by C.E. Patterson and H. Lum.* Reproduced by permission of Taylor & Francis, Inc., http://www.taylorandfrancis.com

---

Integrins are bimolecular complexes with different α and β proteins that render localization and ligand specificity. In endothelial cells, αvβ3 integrins bind vitronectin, fibronectin, and von Willebrand factor; α4β1 also binds matrix fibronectin; and α1β1 binds collagen in microvascular but not in macrovascular endothelial cells. The αvβ3, α2β1, α3β1, α5β1, α6β1 integrins are also expressed on the apical surface of endothelium (Conforti et al., 1992; Bhattacharya et al., 2000). While α5β1 and α2β1 bind laminin and fibronectin, they are also involved in cell-to-cell adhesion. RGD peptides, that mimic ligand-binding sites of αv-, α5-, and α3- integrins, disrupt normal integrin-ligand binding. The RGD peptides cause permeability, but do not decrease matrix adherence, indicating a primary effect on cell-cell integrin junctions (Curtis *et al.*, 1995). Similarly, mice deficient in α5 form leaky vessels (Yang *et al.*, 1993). Besides, locational and binding specificities, different integrins are regulated differently. TNFα increased endothelial expression of laminin-binding-α1β1 integrin, but the vitronectin/fibronectin-binding-αvβ3 integrin expression was decreased, and other integrins were unaffected (Defilippi *et al.*, 1992). Associated proteins, such as α-actinin, paxillin, talin, and vinculin, link the cytoplasmic end of integrin to the cytoskeleton as shown in Figure 5.a. Regulatory enzymes, such as PKC, $p60^{src}$, and p125 focal adhesion kinase (FAK), localized to focal adhesion complexes, may phosphorylate and alter function of the focal adhesion complex proteins (Burridge *et al.*, 1990; Schaphorst *et al.*, 1997). These kinases also form a basis for matrix to cell signaling through the integrins, by tyrosine phosphorylations, which signal numerous cellular changes. Just as the integrin complex is involved in transmitting signals, they are also subject to regulation by cell signaling.

Quiescence, Activation, and Signaling

In confluent, quiescent endothelial cells, true focal adhesion plaques are sparse, but focal adhesion complexes, including integrins, associated structural proteins, and kinases are present at multiple, small foci scattered around the basement membrane but concentrated at basolateral edges and under nuclei (Figure 5.b). Such integrin dispersal from focal adhesions actually increased total cell adhesion (Lampugnani *et al.*, 1990 & 1991), yet activation and formation of focal adhesion plaques correlates with tighter substrate adhesion at the plaques (Hormia *et al.*, 1985; Alexander and Elrod, 2002). When endothelia are activated by agents, such as thrombin, the complexes coalesce to form fewer, larger, tightly bound focal adhesion plaques seen in Figure 5.b. (Garcia *et al.*, 1995b; Schaphorst *et al.*, 1997). Similarly, high shear stress or cyclic-strain caused rapid aggregation of focal adhesion proteins at the abluminal surface, tight adhesion, and alignment of plaques in the direction of flow (Yano *et al.*, 1997; Barakat and Davies, 1998; Helmke *et al.*, 2000). *In vivo*, most endothelia have few stress fibers, but endothelia in areas of high shear stress show these cytoskeletal structures. These agonists- and stress-mediated reorganizations of focal adhesion complexes are associated with parallel reorganization of F-actin. It is generally understood that formation of stress fibers drives the positioning of these plaques and the stress fibers are anchored to the substratum *via* these complexes, as also seen in Figure 5.b. Focal adhesions also provide the foci for actin polymerization and development of stress fiber formation and organization.

A schematic illustration of the 3-dimension reorganization of F-actin and focal adhesions upon activation is shown in Figure 5.c. Focal adhesions also participate in cell motility with coordinated formation at the leading edge to anchor both forward projection of lamellipodia and the traction exercised by the stress fibers to pull the cell forward; while at the tailing edge focal plaques are dissolved (Brakebusch and Fassler, 2003). When cell-cell junctions are established and the cell returns to quiescence, the plaques break up to form multiple loose basal junctions; yet, mechanisms of coordinated regulation in endothelium are only partially known. While cAMP elevation favors quiescence and impairs focal adhesion, it is not known how this is actually mediated. Cell activation correlates with small GTPase Rho activation resulting in paxillin and focal adhesion kinase phosphorylation and actin polymerization, but there are many missing steps (Flinn and Ridley, 1996; Lim *et al.*, 1996; Schulze *et al.*, 1997). Different regulatory pathways are indicated in endotoxin-mediated activation of endothelium for the early permeability effects and for the late caspase-induced cleavage of focal adhesion kinase and paxillin dephosphorylation involved in cell detachment (Bannerman and Goldblum, 1999). Importantly, external mechanical stimuli or ligand-induced integrin clustering affect signaling through activation of kinases associated with the focal adhesion complex (Bhattacharya *et al.*, 2001; Quadri *et al.*, 2003). Apparent coordinate regulation of matrix adhesion with cell-cell adhesion will be discussed at the end of the next section.

Endothelial Cell-Cell Junctions

Cell-Cell Adhesion and Barrier Function

Polarity of normal endothelium is also reflected and determined by the confinement of cell-cell attachments at the lateral borders and the different pattern of cell-to-matrix attachments at the basal surface and cell adhesion molecules and receptors on the apical surface. Intercellular junctions are interactive with the cytoskeleton and altered in the transition from quiescence to activation as summarized in Figure 4 and fully discussed in Chapter 9. Indeed, impairment of these junctions is a primary determinant of loss of barrier selectivity and the consequent increases in paracellular flux of fluid and solutes, prominent features in lung edema. Similar to squamous epithelial cells, lateral cell junctions include adherens and tight junctions that are critical for restrictive selectivity of the barrier. Inter-endothelial cell attachments also include $\alpha 5\beta 1$- and $\alpha 2\beta 1$-integrin linkages (Figure 6.a.) (Lampugnani *et al.*, 1991; Lampugnani and Dejana, 1997; Dejana *et al.*, 2001), gap junctions (Curtis *et al.*, 1995; DePaola *et al.*, 1999), PECAM-1 of the Ig superfamily (Ayalon *et al.*, 1994; Newman, 1997; Aurrand-Lions *et al.*, 2003; Jackson, 2003), CD34, and endoglin (Dejana *et al.*, 2001). The fine, punctate, peripheral location of the integrin linked proteins, paxillin and FAK, in confluent pulmonary endothelium (Fig. 5.b) supports a potential role for integrins in cell-to-cell adhesion in quiescent cells, but whether these are actually cell-cell, cell-lateral matrix, or cell-basal matrix at the edge is not known (Schaphorst *et al.*, 1997, Patterson *et al.*, 2000). Formation of these junctions, particularly VE-cadherin adherens junctions signal contact inhibition typical of normal endothelium.

Figure 6. Endothelial-Endothelial Adhesions: **A.** Lateral attachments between endothelia are formed by several complexes and adhesion molecules: adherens junctions, tight junctions, syndesmos junctions, gap junctions, PECAM, endoglin, S-endo-1, CD34, and integrins. PECAM is distributed over the surface of newly plated cells, but localizes to lateral borders early after contact, forming homophilic bridges. It is also involved in leukocyte transmigration. The primary cell-cell junction, the adherens junction, forms via transmembrane cadherin homophilic association with homologous cadherin of adjacent cells through Ca^{++}-dependent binding. The adherens junction is strengthened by its attachment to the cytoskeleton via β and γ-catenin binding to α-catenin, which binds actin. In addition, cadherin may also link to the actin cytoskeleton via desmoplakin, γ-catenin, and vimentin complexes (syndesmos junctions, not shown). Gap junctions form intercellular linkages, which permit exchange of some cytosolic material between adjacent cells. Tight junctions are more abundant in epithelial cells, but contribute to tightness of endothelial junctions and, thus, important to permeability characteristics. Tight junctions, generally located more toward the apical border, involve cell-to-cell occludin, claudin, and/or JAM binding, which bring the membranes closely together. They are linked to the cytoskeleton via ZO-1, 2, and 3. **B.** Catenin immunofluorescence: A continuous border of adherens junctions (α-catenin, green) shows is seen around the cell-to-cell borders of quiescent cells (left), sealing these borders against paracellular leaking. These junctional complexes are closely associated with the dense peripheral band as seen by the yellow blend of green-catenin and red F-actin staining. When the endothelium is activated 10 min with 100 μM thrombin (right), there is a net loss of catenin from the borders, and the remaining catenin forms discontinuous strands between retracting cells. Areas of frank gaps are devoid of catenin.

A.

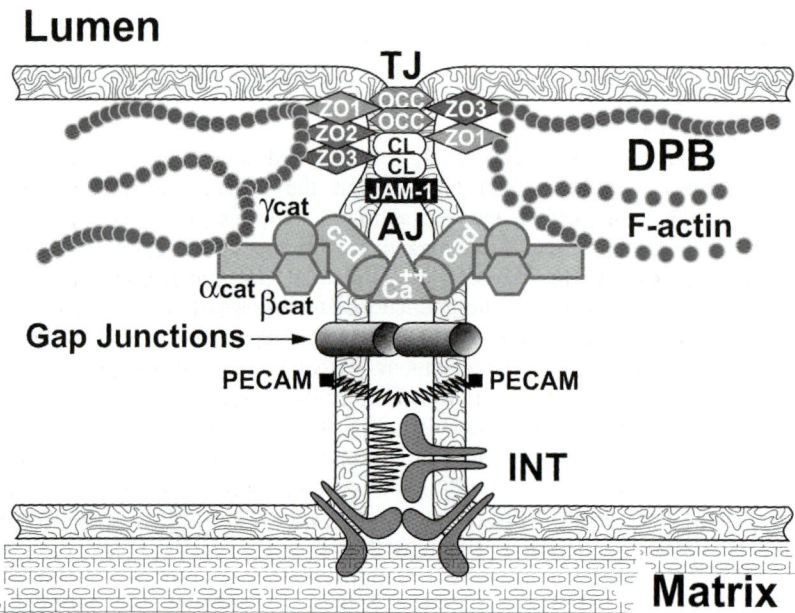

B.

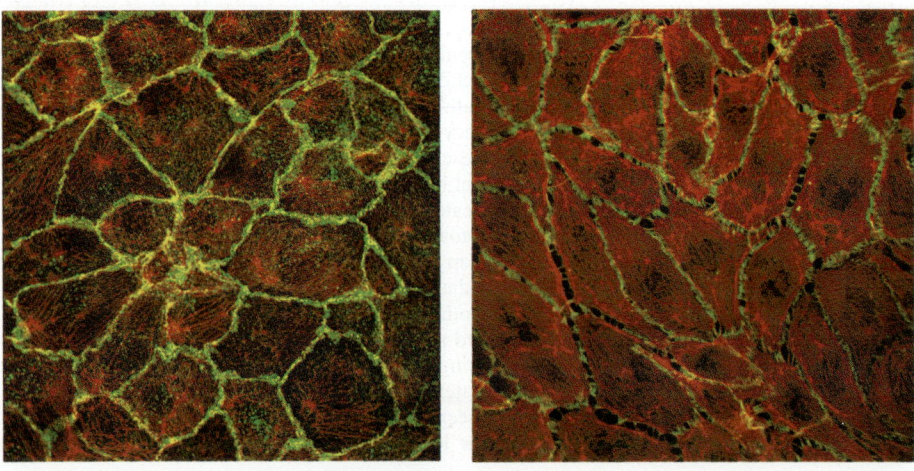

Adherens Junctions

Adherens junctions are formed by homotypic adhesion of the extracellular amino termini of cadherin of one cell to cadherin of the neighboring cell in a Ca^{++}-dependent manner (Figure 6.a) (Lutz *et al.*, 1995). Cadherins, glycoprotein adhesion molecules, are important in cell sorting and organ development and, thus, have cell-type specificity. Endothelial-exclusive, VE-cadherin, has a unique amino end (Hinck *et al.*, 1994; Gumbiner, 1996; Huber *et al.*, 1996; Lampugnani and Dejana, 1997; Nieset *et al.*, 1997). Endothelial cells also express N- and P-cadherins (Salomon *et al.*, 1992; Alexander *et al.*, 1993). Exposure of endothelium to certain anti-cadherin antibodies disrupts the adherens junctions and increases permeability, underscoring the importance of these junctions in barrier integrity (Alexander *et al.*, 1993; Haselton and Heimark, 1997; Corada *et al.*, 2002).

Internal anchorage of adherens junctions to the cytoskeleton is an important determinant of junctional stability and barrier integrity (Takeichi, 1990; Navarro *et al.*, 1995). Proteins, such as α-, β-, γ-catenin (plakoglobin), and p120, form the critical linkages (Figure 6.a.) (Hinck *et al.*, 1994; Gumbiner, 1996; Lampugnani and Dejana, 1997; Nieset *et al.*, 1997; Iyer *et al.*, 2004). The cadherin cytoplasmic domain contains the binding site for β-catenin, which then connects actin *via* α- and γ-catenin (Jou *et al.*, 1995). Transfection of endothelium with truncated cytoplasmic domain-VE-cadherin caused permeability, although cell-cell adhesion was maintained (Navarro *et al.*, 1995). As the cells grow in culture, cadherin recognition recruits other components to the junction and other lateral adhesions, such as PECAM, are established, but formation of complete adherens junctions may depend on linkage to the DPB to locate and stabilize the complex. In epithelial cells, cytochalasin- or constitutively activated V12rac-disrupted the DPB and redistributed β-catenin to perinuclear Golgi vesicles; whereas, DPB reformation returned catenins to the border (Quinlan and Hyatt, 1999). Similarly, in endothelial cells sphingosine-1-phosphate signaled increased DPB and localization of VE-cadherin, α-, β- and γ-catenins at cellular junctions (Lee *et al.*, 1999), further indicating the importance of the DPB to assembly and function of adherens junctions. P120 participates in VE-cadherin-adherens junctions and either p120 phosphorylation or excess p120 expression decreases junctional strength (Iyer *et al.*, 2004). Alternatively, signaling or external proteolysis, resulting in junction disintegration may signal changes in the DPB (Carden *et al.*, 1998; Alexander and Elrod, 2002). Figure 6.b shows the near complete encircling of quiescent endothelial cells with α-catenin, indicating intact adherens junctions. Although endothelial cells lack critical components of desmosomes, a semi-desmosome or syndesmos junction may be formed in confluent endothelium from cadherins, desmoplakin, γ-catenin, and vimentin (Valiron *et al.*, 1996, Kowalczyk *et al.*, 1998).

Tight Junctions

Tight junctions are more abundant in epithelial cells, which have tighter barriers, but contribute to barrier function in endothelium. They are formed last in establishment of the cobblestoned monolayer and appear as intramembranous strands that join adjacent membranes and effectively seal the border toward the apical aspect. They consist of transmembrane proteins: occludin, claudin (endothelium, claudin-5), and JAMs (small Ig family glycoprotein), which engage in homophilic binding with the adjacent cell (Dejana *et al.*, 2001). These integral membrane proteins associate with cytoplasmic proteins, such as ZO-1, 2, and 3 and cingulin that link them to the actin cytoskeleton (Fanning *et al.*, 1998;

Furuse et al., 1998; Matter and Balda, 1999). Greater permeability correlates with decreased occludin in endothelium (Hirase et al., 1997; Kevil et al., 1998; Jiang et al., 1999; Balda et al., 2000). Claudin knockout mice formed defective junctions indicating their importance in sealing the cell-to-cell border (Leach et al., 2000; Mitic et al., 2000). Differences in type and amounts of claudin between epithelium and endothelium of different tissues may account for differences in basal permeability (Morita et al., 1999; Lippoldt et al., 2000).

Activation and Signaling

Inflammatory mediator-induced permeability, described in Chapter 3, disrupts adherens junctions (Rabiet et al., 1996; Wong et al., 1999; Patterson et al., 2000) and tight junctions (Gardner et al., 1996; Blum et al., 1997; Antonetti et al., 1998). This may involve decreased expression or destruction of the proteins. For instance, decreased occludin and adherens junction proteins were found in increased vascular permeability in diabetic rats (Antonetti et al., 1998; Lee et al., 2004). Alternately, protein levels may not change, but serine/threonine or tyrosine phosphorylation may alter function (Volberg et al., 1992; Gloor et al., 1997; Staddon et al., 1997; Esser et al., 1998; Knaus et al., 1998; Andriopoulou et al., 1999; Braga et al., 1999; Cowell and Garrod, 1999; Ratcliffe et al., 1999). Damage of intercellular junctions also results from direct injury. For example, excessive release of neutrophil elastases results in cadherin proteolysis and barrier disruption (Carden et al., 1998). As mentioned, dismantling of the DPB and concomitant increase in stress fibers is accompanied by adherens junction loss. Agonist-activation induces such changes in the cytoskeleton and adherens junctions, as seen by loss of catenin from the cell borders (Fig. 6.b) (Patterson et al., 2000). Sphingosine-1-phosphate promotion of actin filament and junction formation was inhibited by expression of Rho and Rac mutants, indicating these are important in control of junctional integrity (Lee et al., 1999). Rac was shown to be important for maintenance of the DPB and cell-cell junctions (Waschke et al., 2004). Thus, mediators, which increase paracellular permeability, likely lead to DPB disruption, resulting in junction disassembly, but cytoskeletal-independent signaling may also be important, as will be further discussed in Chapter 9.

Integration of Adhesions in Quiescence, Activation, and Restoration

Increased cAMP enhances basal barrier function in cultured endothelium and attenuates permeability increases due to a number of activators, indicating the potential for cAMP-dependent kinase, PKA, regulation of cell-matrix and/or cell-to-cell tethering, either directly or *via* alterations in the cytoskeleton, as will be discussed in Chapters 5 and 15. Moreover, PKA activation protects from lung edema in models of endothelial barrier dysfunction. Just how PKA exerts its regulatory influence is unresolved, but cAMP elevation attenuates MLC phosphorylation induced by several contractile agonists (Moy et al., 1993; Patterson et al., 1994a; Garcia et al., 1995a). Yet, PKA activation also prevents permeability induced by other agonists, which do not cause MLC phosphorylation (Patterson et al., 1994a; Patterson et al., 1994b; Garcia et al., 1995; Patterson et al., 1995; Moy et al., 1998). Possible PKA targets include Rho proteins, which regulate actin polymerization and assembly (Flinn and Ridley, 1996; Arimura et al., 1997; Schulze et al., 1997). Conversely, PKA activity may be altered by Rho activation (Tigyi et al., 1996). Decreased PKA activity by transfection of endothelium with PKI, an inhibitor of PKA, or with pharmacologic inhibitors decreased basal barrier function and blocked the protective effects of agents that elevated cAMP (Lum et al., 1991;

Patterson et al., 2000). Immunofluorescent studies of cell-matrix and cell-cell junctions indicate disparate effects on junctions, whether changes were induced by PKA promotion or agonists-activation (Patterson et al., 2000). These facts further suggest the inappropriateness of grouping responses of cell-matrix and cell-cell adhesions. One means to examine junction behavior during endothelial transition from quiescence to activation and during restoration of barrier integrity is the use of electrical cell impedance, which can sort cell-cell resistance from cell-matrix resistance. Endothelium, pre-treated to elevate PKA, have higher total basal resistance, consistent with their higher reflectance coefficient for macromolecular proteins and better sealing of the paracellular border (Fig. 7). Agonist-activation causes rapid decrease in total resistance followed by a slow recovery, consistent with other measures of transendothelial protein flux. This drop is attenuated with PKA activation and recovery is more rapid. When total resistance is separated into resistance due to between cell and beneath cell resistances, interesting differences are revealed. First, while PKA elevation alone increased total resistance, there was actually a decrease in beneath cell resistance indicating that focal adhesions were decreased while the increased resistance due to adherens junctions and/or tight junctions was even greater than expected from measure of total resistance alone. Second, in endothelium challenge with permeability-inducing agonists, the drop in resistance was actually greater in between cell resistance, while the beneath cell resistance actually increased; again showing divergence in the break-up and formation of the two classes of junctional adhesions. This pattern is reversed during recovery. That is – while cell-to-cell junctions were reforming, cell-to-matrix junctions were diminishing. Third, in PKA elevated cells then challenged with the permeability agonist, the recovery of the cell-to-cell junctions was more rapid and more exaggerated than revealed by the change in total resistance because dissolution of the cell-to-matrix adhesions was exaggerated. Temporal-studies of junction proteins immunofluorescent confirmed this interpretation (unpublished data, Patterson). Thus, under conditions of reversible activation, it is apparent that regulation drives cell-cell junctions and cell-matrix junctions in opposite directions. Similar conclusions were reached with lung microvascular cells challenged with H_2O_2, bradykinin, histamine, and thrombin (Alexander et al., 2001). This is not surprising when viewed with the divergent regulation of the DPB and the stress fibers, which are important for each class of attachment.

Figure 7. Reciprocal changes in cell-to-cell and cell-to-matrix adhesions. Bovine pulmonary arterial endothelial cells were grown to optimal post-confluent barrier function determined by electrical cell impedance. Media was replaced with serum-free, 4% albumin media and cells were equilibrated 1 hr, during which 1 μg/ml cholera toxin was added to some wells to activate PKA. Low dose trypsin (no EDTA) was then added to some wells to activate PAR-receptors and induce barrier dysfunction. The permeability with agonist-activation and the PKA protection mirrors that shown with other methods. A. Total resistance (k ohms - calculated from impedance) is shown in the top panel. B. Total endothelial resistance (total with subtraction of system constant resistance), the resistance portion due to cell-to-cell junctions, and the resistance portion due to current impedance beneath the cells (related to the area of true focal adhesion) is shown for cells challenged with trypsin. C. Total cell resistance and distribution between cell-cell and cell-matrix resistance is shown for cells challenged with trypsin after PKA activation. The loss of cell-cell adhesion is clearly greater and recovers faster than revealed by total resistance measure only. Analysis of the resistance changes with regard to the two classes of junctions demonstrates that they changes in opposite directions during PKA-activation, agonist-activation, recovery from agonist-activation, and PKA-reversal of agonist activation. That is, cell-to-cell junctions are decreased when cell-to-matrix junctions increase and cell-to-cell junctions increase when cell-to-matrix junctions decrease in each of these cases.

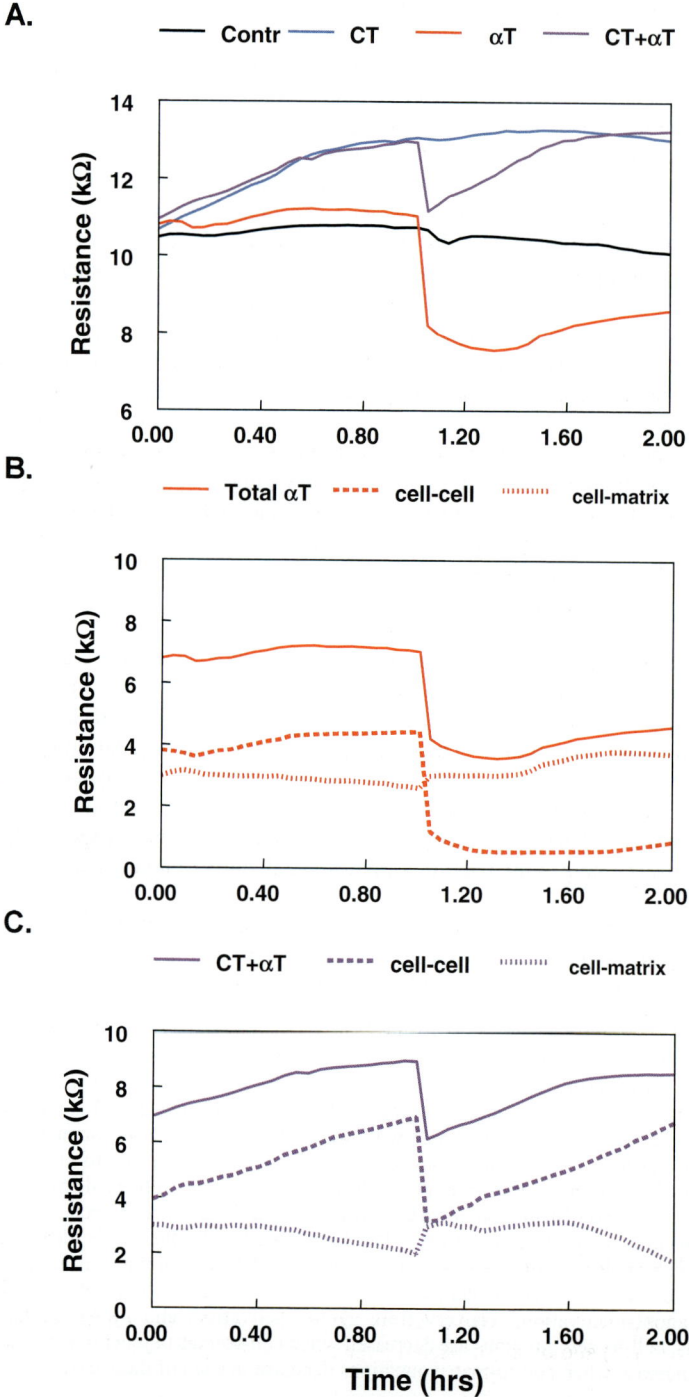

Structural Integrity of the Dam

Great progress in mechanical engineering has been made from that early construction of the faulty earthen dam at Lake Conemaugh to the modern day marvels such as the Itaipú dam and hydroelectric power plant on the Paraná River, with a reliable output of 75 million megawatts per year, enough to supply 25% of the total electrical power need of Brazil and 78% of electrical power in Paraguay. Likewise, we have made great progress in understanding the amazing architecture and important function of our large, fragile endothelial dam. Advancements in imaging, identification of the components of the construction, and in measurement of mechanical properties of live cells have strengthened the important relationship of structure and function as will be more fully explained in Chapters 3 on activation, 4 - 6 on signaling, and 7 - 10 on architecture. Much has been learned regarding coupling of biochemical and mechanical signaling. The formulation of models to describe barrier properties, tethering, contraction, and cell mechanics has been highly important in this progress, but many questions remain. Obviously, the presentation of the complex endothelium as existing in merely two states in this introductory chapter is an extreme simplification, when endothelial cells can be totally quiescent or under continuous or intermittent flow- and pressure-stress, temporarily or chronically activated by a wide variety of mediators and conditions, attached or isolated, motile or stationary, dividing or dying, injured or recovering, and arising or existing in different tissues and in different vascular segments, as will be described in Chapters 11-16.

Nevertheless, it has become apparent that past and present studies do not always address or define even these two simplified states of quiescence and activation. Some very basic questions remain unanswered. Does the actin microfilament cytoskeleton contribute supporting scaffolding in quiescent endothelium? Or in other words, does actin act like a skeleton or like a muscle and under what conditions? Is rapidly induced barrier dysfunction (15 sec) related to DPB contraction or disassembly? What are the mechanical and biochemical links between microfilaments, microtubules, and intermediate filaments in the different conditions and transitions of the cell state? What is the temporal relationship and cause-effect interdependence of actin reorganization, focal adhesion alteration, and adherens/tight junction disassembly in activation? Despite progress, many gaps remain in our understanding of inward and outward signaling, mechanical to biochemical signaling, biochemical to mechanical signaling, and the relationship of molecular structure to monolayer barrier function. Thus, there is a critical need to define the cellular and molecular mechanisms controlling cell shape, cytoskeletal organization, cytoskeletal interaction with adhesion complexes, and organization and contribution of adhesion complexes to barrier function in pulmonary endothelial cells.

ACKNOWLEDGEMENTS

This work was supported by a Merit Award from the Veteran's Administration Medical Research Service (CEP) and by a NIH Grant from NHLBI, #HL-33009 (DS).

REFERENCES:

Alexander, J.S., Blaschuk, O.W., and Haselton, F.R. (1993) N-cadherin-like protein contributes to solute barrier maintenance in cultured endothelium. J. Cell. Physiol. 156:610-618.
Alexander, J.S. and Elrod, J.W. (2002) Extracellular matrix, junctional integrity, and matrix metalloproteinase interactions in endothelial permeability regulation. J. Anat. 200:561-574.
Alexander, J.S., Hechtman, H.B., and Shepro, D. (1988) Phalloidin enhances endothelial barrier function and reduces inflammatory permeability *in vitro*. Microvas. Res. 35:308-315.
Alexander, J.S., Patton, W.F., Yoon, M.U., and Shepro, D. (1991) Cytokeratin filament modulation n pulmonary microvessel endothelial cells by vasoactive agent and culture confuency. Tissue and Cell 23:141-150.
Alexander, J.S., Zhu, Y., Elrod, J., Alexander, B., Coe, L., Kalogeris, T., and Fuseler, J. (2001) Reciprocal regulation of endothelial substrate adhesion and barrier function. Microcirc. 8:389-401.
Andriopoulou, P., Navarro, P., Zanetti, A., Lampugnani, M.G., and Dejana, E. (1999) Histamine induces tyrosine phosphorylation of endothelial cell-to-cell adherens junctions. Arterioscl. Thromb. Vasc. Biol. 19:2286-2297.
Antonetti, D., Barber, A., Khin, S., Lieth, E., Tarbell, J., and Gardner, T. (1998) Vascular permeability in experimental diabetes is associated with reduced endothelial occludin content. Diabetes 47:1953-1959.
Arimura, S., Nakata, H., Tomiyama, K., and Watanabe, Y. (1997) Phosphorylation of H-ras proteins by protein kinase A. Cell. Signalling 9:37-40.
Aspenstrom, P. (1999) Effectors for the Rho GTPases. Curr. Opin. Cell Biol. 11:95-102.
Aurrand-Lions, M., Leger-Johnson, C., and Imhof, B.A. (2003) Role of interendothelial adhesion molecules in the control of vascular function. Vasc. Pharm. 39:239-246.
Ayalon, O., Sabanai, H., Lampugnani, M., and Dejana, E. (1994) Spatial and temporal relationships between cadherins and PECAM-1 in cell-cell junctions of human enodthelial cells. J. Cell Biol. 126:247-258.
Balda, M.S., Maldonado, C., Cereijido, M., and Matter, K. (2000) Multiple domains of occludin are involved in regulation of paracellular permeability. J. Cell. Biochem. 78:85-96.
Baldwin, A.L., Thurston, G. (2001) Mechanics of endothelial cell architecture and vascular permeability. Crit. Rev. Biomed. Engin. 29:247-78.
Bannerman, D.D. and Goldblum, S.E. (1999) Direct effects of endotoxin on the endothelium: barrier function and injury. Lab. Invest. 79:1181-1199.
Barakat, A.I. and Davies, P.F. (1998) Mechanisms of shear stress transmission and transduction in endothelial cells. Chest 114:58S-63S.
Bhattacharya, S., Patel, R., Sen, N., Quadri, S., Parthasarathi, K., and Bhattacharya, J. (2001) Dual signaling by the $\alpha v\beta 3$-integrin activates cytosolic PLA(2) in bovine pulmonary artery endothelial cells. Amer. J. Physiol. 280:L1049-L1056.
Bhattacharya, S., Ying, X., Fu, C., Patel, R., Kuebler, W., Greenberg, S., and Bhattacharya, J. (2000) $\alpha v\beta 3$-integrin induces tyrosine phosphorylation-dependent Ca(2+) influx in pulmonary endothelial cells. Circ. Res. 86:456-462.
Birukova, A.A., Liu, F., Garcia, J., and Verin, A.D. Protein kinase A attenuates endothelial cell barrier dysfunction induced by microtubule disassembly. Amer. J. Physiol. 287:L86-L93.
Blum, M.S., Toninelli, E., Anderson, J.M., Balda, M.S., Zhou, J.Y., ODonnell, L., Pardi, R., and Bender, J.R. (1997) Cytoskeletal rearrangement mediates human microvascular endothelial tight junction modulation by cytokines. Amer. J. Physiol. 42:H286-H294.
Braga, V.M., Del Maschio, A., Machesky, L., and Dejana, E. (1999) Regulation of cadherin function by Rho and Rac: modulation by junction maturation and cellular context. Molec. Biol. Cell 10:9-22.
Brakebusch, C. and Fassler, R. (2003) The integrin-actin connection, an eternal love affair. EMBO J. 22:2324-2333.

Bretscher, A., Chambers, D., Nguyen, R., and Reczek, D. (2000) ERM-Merlin and EBP50 protein families in plasma membrane organization and function. Ann. Rev. Cell Devel. Biol. 16:113-143.
Brett, J., Gerlach, H., Nawroth, P., Steinberg, S., Godman, G., and Stern, D. (1989) Tumor necrosis factor/cachectin increases permeability of endothelial cell monolayers by a mechanism involving regulatory G proteins. J. Exper. Med. 169:1977-1991.
Burridge, K., Nuckolls, G., Otey, C., Pavalko, F., Simon, K., and Turner, C. (1990) Actin-membrane interaction in focal adhesions. Cell Diff. & Devel. 32:337-342.
Caille, N., Thoumine, O., Tardy, Y., and Meister, J.J. (2002) Contribution of the nucleus to the mechanical properties of endothelial cells. J. Biomech. 35:177-187.
Carden, D., Xiao, F., Moak, C., Willis, B.H., Robinson-Jackson, S., and Alesander, S. (1998) Neutrophil elastase promotes lung microvascular injury and proteolysis of endothelial cadherins. Amer. J. Physiol. 275:H385-H392.
Carson M.R., Shasby, S., Lind, S., and Shasby, D.M. (1992) Histamine, actin-gelsolin binding, and polyphosphoinositides in human umbilical vein endothelial cells. Amer. J. Physiol. 263:L664-669.
Chen, Y., McCarron, R., Ohara, Y., Bembry, J., Azzam, N., Lenz, F., Shohami, E., Mechoulam, R., and Spatz. M. (2000) Human brain capillary endothelium: 2-arachidonoglycerol (endocannabinoid) interacts with endothelin-1. Circ. Res. 87:323-327.
Chicurel, M.E., Chen, C.S., and Ingber, D.E. (1998) Cellular control lies in the balance of forces. Curr. Opin. Cell Biol. 10:232-239.
Chrzanowskawodnicka, M. and Burridge, K. (1996) Rho-stimulated contractility drives formation of stress fibers and focal adhesions. J. Cell Biol. 133:1403-1415.
Cioffi, D.L., Moore, T., Schaack, J., Creighton, J., Cooper, D., and Stevens, T. (2002) Dominant regulation of interendothelial cell gap formation by Ca^{++}-inhibited type 6 adenylyl cyclase. J. Cell. Biol. 157:1267-1278.
Clarke, E.J. and Allan, V. (2002) Intermediate filaments: vimentin moves in. Curr. Biol. 12:R596-R598.
Colucci-Guyon, E., Portier, M.M., Dunia, I., Paulin, D., Pournin, S., and Babinet, C. (1994) Mice lacking vimentin develop and reproduce without an obvious phenotypic change. Cell 79:679-694.
Conforti, G., Dominguez-Jiminez, C., Zanetti, A., Gimbrone, M., Cremona, O., Marchisio, P., and Dejana, E. (1992) Human endothelial cells express integrin receptors on the luminal aspect of their membrane. Blood 80:437-446.
Corada, M., Zanetta, L., Orsenigo, F., Breviario, F., Lampugnani, M., Bernasconi, S., Liao, F., Hicklin, D., Bohlen, P., and Dejana, E. (2002) A monoclonal antibody to vascular endothelial-cadherin inhibits tumor angiogenesis without side effects on endothelial permeability. Blood 100:905-911.
Coughlin, M.F., and Stamenović, D. (1998) A tensegrity model of the cytoskeleton in spread and round cells. ASM J. Biomech. Eng. 120:770-777.
Cowell, H.E. and Garrod, D.R. (1999) Activation of protein kinase C modulates cell-cell and cell-substratum adhesion of a human colorectal carcinoma cell line and restores 'normal' epithelial morphology. Internat. J. Cancer 80:455-464.
Curtis, T., McKeown, P., Vincent, P., Homan, S., Wheatley, E., and Saba, T. (1995) Fibronectin attenuates endothelial monolayer permeability after RGD peptide, anti-$\alpha 5\beta 1$, or TNF-α exposure. Amer. J. Physiol. 269:L248-L260.
Davies, P.F., Robotewskyj, A., and Griem, M.L. (1994) Quantitative studies of endothelial cell adhesion. Directional remodeling of focal adhesion sites in response to flow forces. J. Clin. Invest. 93:2031-2038.
De Corte, V., Gettemans, J., and Vandekerckhove, J. (1997) Phosphatidylinositol 4,5-bisphosphate specifically stimulates PP60(c-src) catalyzed phosphorylation of gelsolin and related actin-binding proteins. FEBS Letters 401:191-196.

Defilippi, P., Bozzo, C., Geuna, M., Rossino, P., Silngo, L., and Tarone, G. (1992) Modulation of extracellular matrix receptors on human endothelial cells by cytokines. in Angiogenesis, ed. R. Steiner, B. Weisz, and R. Langer. Berkhauser Verlag, Basil, Switzerland.

Dejana, E., Spagnuolo, R., and Bazzoni, G. (2001) Interendothelial junctions and their role in the control of angiogenesis, vascular permeability, and leukocyte transmigration. Thromb. Heamost. 86:308-315.

DeMaio, L., Tarbell, J., Scaduto, R., Gardner, T., and Antonetti, D.A. (2004) A transmural pressure gradient induces mecahnical and biological adaptive responses in endothelial cells. Amer . J. Physiol. 286:H731-H741.

DePaola, N., Davies, P.F., Pritchard, W., Florez, L, Harbeck, N., and Polacek, D.C. (1999) Spacial and temporal regulation of gap junction connexin43 in vascular endothelial cells exposed to disturbed flows. PNAS USA 96:3154-3159.

Dos Remedios, C.G., Chhabra, D., Kekic, M., Dedova, I.V., Tsubakihara, M., Berry, D.A., and Nosworthy, N.J. (2003) Actin Binding Proteins:regulation of cytoskeletal microfilaments. Physiol. Rev. 83:433-473.

Dudek, S.M. and Garcia, J.G.N. (2001) Cytoskeletal regulation of pulmonary vascular permeability. J. Appl. Physiol. 91:1487-1500.

Eckes, B., Dogic, D., Colucci-Guyon, E., Wang, N., Maniotis, A., Ingber, D., Merckling, A., Langa, F., Aumailley, M., Delouvee, A., Koteliansky, V., Babinet, C., and Krieg, T. (1998) Impaired mechanical stability, migration and contractile capacity in vimentin-deficient fibroblasts. J. Cell Sci. 111:1897-1907.

Ehringer, W.D., Yamany, S., Steier, K., Farag, A., Roisen, F.J., Dozier, A., and Miller, F.N. (1999) Quantitative image analysis of F-actin in endothelial cells. Microcirc. 6:291-303.

Elson, E.L., Felder, S., Jay, P., Kolodney, M., and Pasternak, C. (1999) Forces in cell locomotion. Biochem. Soc. Symp. 65:299-314.

Esser, S., Lampugnani, M.G., Corada, M., Dejana, E., and Risau, W. (1998) Vascular endothelial growth factor induces VE-cadherin tyrosine phosphorylation in endothelial cells. J. Cell Sci. 111:1853-1865.

Ettenson, D.S. and Gotlieb, A.I. (1992) Centrosomes, microtubules, and microfilaments in the reendothelialization and remodeling of *in vitro* wounds. Lab. Invest. 66:722-733.

Evans, R.M. (1998) Vimentin: the conundrum of the intermediate filament gene family. Bioessays 20:79-86.

Fanning, A.S., Jameson, B., Jesaitis, L., and Anderson, J.M. (1998) The tight junction protein ZO-1 establishes a link between the transmembrane protein occludin and the actin cytoskeleton. J. Biol. Chem. 273:29745-29753.

Flanagan, L.A., Chou, J., Falet, H., Neujahr, R., Hartwig, J., and Stossel, T.P. (2001) Filamin A, the Arp2/3 complex, and the morphology and function of cortical actin filaments in human melanoma cells. J. Cell Biol. 155:511-517.

Flinn, H.M. and Ridley, A.J. (1996) Rho stimulates tyrosine phosphorylation of focal adhesion kinase, p130 and paxillin. J. Cell. Sci. 109:1133-1141.

Fukami, K., Furuhashi, K., Inagaki, M., Endo, T., Hatano, S., and Takenawa, T. (1992) Requirement of PI 4,5-bisphosphate for alpha-actinin function. Nature 359:150-152.

Furuse, M., Fujita, K., Hiiragi, T., Fujimoto, K., and Tsukita, S. (1998) Claudin-1 and -2: novel integral membrane proteins localizing at tight junctions with no sequence similarity to occludin. J. Cell Biol. 141:1539-1550.

Garcia, J.G.N., Davis, H.W., and Patterson, C.E. (1995a) Regulation of endothelial cell gap formation and barrier dysfunction: role of myosin light chain phosphorylation. J. Cell. Physiol. 163:510-522.

Garcia, J.G.N., Pavalko, F.M., and Patterson, C.E. (1995b) Vascular endothelial cell permeability responses to thrombin. Blood Coag. Fibrin. 6:609-626.

Garcia, J.G.N., Siflinger-Birnboim, A., Bizios, R., DelVecchio, P., Fenton, J., and Malik, A.B. (1986) Thrombin-induced increase in albumin permeability across the endothelium. J. Cell. Physiol. 128:96-104.

Garcia, J.G., Verin, A.D. Schaphorst, K., Siddiqui, R., Patterson, C.E., Csortos, C., and Natarajan, V. (1999) Regulation of endothelial cell myosin light chain kinase by Rho, cortactin, and p60(src). Amer. J. Physiol. 276:L989-L998.

Gardner, T.W., Lesher, T., Khin, S., Vu, C., Barber, A.J., and Brennan, W.A. (1996) Histamine reduces ZO-1 tight-junction protein expression in cultured microvascular endothelial cells. Biochem. J. 320:717-721.

Gilmore, A.P. and Burridge, K. (1996) Regulation of vinculin binding to talin and actin by phosphatidyl-inositol-4-5-bisphosphate. Nature 381:531-535.

Gloor, S.M., Weber, A., Adachi, N., and Frei, K. (1997) Interleukin-1 modulates protein tyrosine phosphatase activity and permeability of brain endothelial cells. Biochem. Biophys. Res. Comm. 239:804-809.

Goeckeler, Z.M. and Wysolmerski, R.B. (1995) Myosin light chain kinase-regulated endothelial cell contraction: the relationship between isometric tension, actin polymerization, and myosin phosphorylation. J. Cell Biol. 130:613-627.

Goldman, R.D., Khuon, S., Chou, Y.H., Opal, P., and Steinert, P.M. (1996) The function of intermediate filaments in cell shape and cytoskeletal integrity. J. Cell. Biol. 134:971-983.

Gorlin, J.B., Yamin, R., Egan, S., Stewart, M., Stossel, T.P., Kwiatkowski, D.J., and Hartwig, J.H. (1990) Human endothelial actin-binding protein (ABP-280, nonmuscle filamin): a molecular leaf spring. J. Cell Biol. 111:1089-1105.

Gotlieb, A.I., Langille, B.L., Wong, M.K., and Kim, D.W. (1991) Structure and function of the endothelial cytoskeleton. Lab. Invest. 65:123-137.

Gruenheid, S. and Finlay, B.B. (2003) Microbial patholgenesis and cytoskeletal function. Nature 422:775-781.

Gumbiner, B.M. (1996) Cell adhesion: The molecular basis of tissue architecture and morphogenesis. Cell 84:345-357.

Hammersen, F. (1980) Endothelial contractility – does it exist? Adv. Microcirc. 9:95-134.

Haselton, F.R. and Heimark, R.L. (1997) Role of cadherins 5 and 13 in the aortic endothelial barrier. J. Cell Physiol. 171:243-251.

Hastie, L.E., Patton, W.F., Hechtman, H.B., and Shepro, D. (1997a) Filamin redistribution in an endothelial cell reoxygenation injury model. Free Rad. Biol. Med. 22:955-966.

Hastie, L.E., Patton, W.F., Hechtman, H.B., and Shepro, D. (1997b) H2O2-induced filamin redistribution in endothelial cells is modulated by the cyclic AMP-dependent protein kinase pathway. J. Cell Physiol. 172:373-381.

Helmke, B.P., Goldman, R.D., and Davies, P.F. (2000) Rapid displacement of vimentin intermediate filaments in living endothelial cells exposed to flow. Circ. Res. 86:745-752.

Helmke, B.P., Thakker, D., Goldman, R., and Davies, P.F. (2001) Spaciotemporal analysis of flow-induced intermediate filament displacement in living endothelial cells. Biophys. J. 80:184-194.

Herman, I.M., Pollard, T.D., and Wong, A.J. (1982) Contractile proteins in endothelial cells. Ann. N.Y. Acad. Sci. 401:50-60.

Hinck, L., Nathke, I., Papkoff, J., and Nelson, W.J. (1994) Dynamics of cadherin/catenin complex formation: novel protein interactions and pathways. J. Cell Biol. 125:1327-1340.

Hirase, T., Staddon, J., Saitou, M., Ando-Akatsuka, Y., Itoh, M., Furuse, M., Fujimoto, K., Tsukita, S., and Rubin, L. (1997) Occludin as a possible determinant of tight junction permeability in endothelial cells. J. Cell Sci. 110:1603-1613.

Hormia, M., Bradley, R.A., Lehto, V. and Virtanen, I. (1985) Actomyosin organization in stationary and migrating sheets of cultured endothelial cells. Exper. Cell Res. 157:116-126.

Hu, S., Chen, J., and Wang, N. (2004) Cell spreading controls balance of prestress by microtubules and extracellular matrix. Frontiers in Biosci. 9:2177-2182.

Huber, P., Dalmon, J., Engiles, J., Breviario, F., Gory, S., Siracusa, L.D., Buchberg, A.M., and Dejana, E. (1996) Genomic structure and chromosomal mapping of the mouse VE- cadherin gene (Cdh5). Genomics 32:21-28.

Iivanainen, E., Kahari, V., Heino, J., and Elenius, K. (2003) Endothelial cell-matrix interactions. Micros. Res. Techniq. 60:13-22.

Ingber, D.E. (2003) Tensegrity I. Cell structure and hierarchical systems biology. J. Cell Sci. 116:1157-1173.
Ingber, D.E. (1997) Tensegrity: The architectural basis of cellular mechanotransduction. Ann. Rev. Physiol. 59:575-599.
Ingber, D.E. and Jameison, J.D. (1985) Cells as tensegrity structures: architectural regulation of histodifferentiation by physical forces transduced over basement membrane. In: *Gene Expression during Normal and Malignant Differentiation.* (eds. Anderson, L.C., Gahmberg, G.C., and Ekblom, P.), Academic Press, Orlando, FL, pp. 13-32.
Iyer, S., Ferreri, D., DeCocco, N., Minnear, F., and Vincent, P.A. (2004) VE-cadherin-p120 interaction is required for maintenance of endothelial barrier function. Amer. J. Physiol. 286:L1143-L1153.
Jackson, D.E. (2003) The unfolding tale of PECAM-1. FEBS Lett. 540:7-14.
Janmey, P.A. Euteneuer, U., Traub, P., and Schliwa, M. (1991) Viscoelastic properties of vimentin compared to other filamentous biopolymer networks. J. Cell Biol. 113: 155-160.
Jay, D. and Stracher, A. (1997) Expression in E. coli, phosphorylation with cAMP-dependent protein kinase and proteolysis by calpain of a 71-kDa domain of human endothelial actin binding protein. Biochem. Biophy. Res. Comm. 232:555-558.
Jiang, W.G., Martin, T.A., Matsumoto, K., Nakamura, T., and Mansel, R.E. (1999) Hepatocyte growth factor decreases expression of occludin and transendothelial resistance and increases paracellular permeability in human vascular endothelial cells. J. Cell. Physiol. 181:319-329.
Johnson, A., Phillips, P., Hocking, D., Tsan, M., and Ferro, T. (1989) Protein kinase C inhibitor prevents pulmonary edema in response to H_2O_2. Amer. J. Physiol. 256:H1012-H1022.
Jou, T., Stewart, D., Stappert, J., Nelson, W., and Marrs, J. (1995) Genetic and biochemical dissection of protein links in the cadherin-catenin complex. PNAS USA 92:5067-5071.
Kevil, C.G., Okayama, N., Trocha, S.D., Kalogeris, T.J., Coe, L.L., Specian, R.D., Davis, C.P., and Alexander, J.S. (1998) Expression of zonula occludens and adherens junctional proteins in human venous and arterial endothelial cells: role of occludin in endothelial solute barriers. Microcirc. 5:197-210.
Kielbassa, K., Schmitz, C., and Gerke, V. (1998) Disruption of endothelial microfilaments reduces the transendothelial migration of monocytes. Exp. Cell Res. 243:129-141.
Knaus, U.G., Wang, Y., Reilly, A.M., Warnock, D., and Jackson, J.H. (1998) Structural requirements for PAK activation by Rac GTPases. J. Biol. Chem. 273:21512-21518.
Kolodny, M.S. and Elson, E.L. (1995) Contraction due to microtublule disruption is associated with increased phosphorylation of myosin regulatory light chain. PNAS 92:10252-10256.
Kolodny, M.S. and Wysolmerski, R.B. (1992) Isometric contraction by fibroblast and endothelial cells in culture. J. Cell Biol. 117:73-82.
Kowalczyk, A.P., Navarro, P., Dejana, E., Bornslaeger, E., Green, K., Kopp, D., Borgwardt, J.E. (1998) VE-cadherin and desmoplakin are assembled into dermal microvascular endothelial intercellular junctions: a pivotal role for plakogloin. J. Cell Sci. 111:3045-3057.
Lampugnani, M. G., Corada, M., Caveda, L., Breviario, F., Ayalon, O., Geiger, B., and Dejana, E. (1995) The molecular organization of endothelial cell to cell junctions:differential association of plakoglobin, β-catenin, and γ-catenin with vascular endothelial cadherin (VE-cadherin). J. Cell Biol. 129:203-217.
Lampugnani, M. G. and Dejana, E. (1997) Interendothelial junctions: structure, signaling and functional roles. Curr. Opin. Cell Biol. 9:674-682.
Lampugnani, M.G., Giorgi, M., Gaboli, M., Dejana, E., and Marchisio, P. (1990) Endothelial cell motility, integrin receptor clustering, and microfilament organization are inhibited by agents that increase cAMP. Lab. Invest. 63:521-531.
Lampugnani, M.G., Resnati, M., Dejana, E., and Marchisio, P. (1991) Role of integrins in maintenance of endothelial monolayer integrity. J. Cell Biol. 112:479 - 490.
Leach, L., Lammiman, M., Babawale, M., Hobson, S., Bromilou, B., Lovat, S., and Simmonds, M.J.R. (2000) Molecular organization of tight and adherens junctions in the human placental vascular tree. Placenta 21:547-557.

Lee, H.Z., Yeh, F.T., and Wu, C.H. (2004) The effect of elevated extracellular glucose on adherens junction proteins in cultured rat heart endothelial cells. Life Sci. 74:2085-2096.
Lee, M.J., Thangada, S., Claffey, K.P., Ancellin, N., Liu, C.H., Kluk, M., Volpi, M., Sha'afi, R.I., and Hla, T. (1999) Vascular endothelial cell adherens junction assembly and morphogenesis induced by sphingosine-1-phosphate. Cell 99:301-312.
Lee, T.J. and Gotlieb, A.I. (2003) Microfilaments and microtubules maintain endothelial integrity. Micros. Res. Tech. 60:115-127.
Lim, L., Manser, E., Leung, T., and Hall, C. (1996) Regulation of phosphorylation pathways by p21 GTPases - The p21 Ras- Rho subfamily and its role in phosphor-ylation signalling pathways. Eur. J. Biochem. 242:171-185.
Lippoldt, A., Liebner, S., Andbjer, B., Kalbacher, H., Wolburg, H., Haller, H., and Fuxe, K. (2000) Organization of choroid plexus epithelial and endothelial cell tight junctions and regulation by claudin-1, -2, and -5 expression by PKC. Molec. Neurosci. 11:1427-1431.
Lum, H., Aschner, J.L., Phillips, P.G., Fletcher, P.W., and Malik, A.B. (1992) Time course of thrombin-induced increase in endothelial permeability: relationship to Ca^{2+}_i and inositol polyphosphates. Amer. J. Physiol. 263:L219-L225.
Lum, H. and Malik, A.B. (1994) Regulation of vascular endothelial barrier function. Amer. J. Physiol. 267:L223-L241.
Lum, H. and Malik, A.B. (1996) Mechanisms of increased endothelial permeability. Can. J. Physiol. Pharmacol. 74:787-800.
Lum, H., Siflinger-Birnboim, A., Blumenstock, F., and Malik, AB. (1991) Serum albumin decreases transendothelial permeability to macromolecules. Microvasc. Res. 42:91-102.
Lutz, K.L., Jois, S.D.S., and Siahaan, T.J. (1995) Secondary structure of the HAV peptide which regulates cadherin- cadherin interaction. J. Biomolec. Struct. Dyn. 13:447-455.
Matter, K. and Balda, M.S., 1999. Occludin and functions of tight junctions. Intern'l. Rev. Cyt. 186:117-146.
Miettinen, M., Fetsch, J.F. (2000) Distribution of keratins in normal endothelial cells and a spectrum of vascular tumors: implications in tumor diagnosis. Human Pathol. 31:1062-1067.
Mitic, L.M., Van Itallie, C.M., and Anderson, J.M. (2000) Tight junction structure and function: lessons from mutant animals and proteins. Amer. J. Physiol. 279:G250-G254.
Morel, N.M., Dodge, A.B., Patton, W.F., Herman, I.M., Hechtman, H.B., and Shepro, D. (1989) Pulmonary microvascular endothelial cell contractility on silicone rubber substrate. J Cell. Physiol. 141:653-659.
Morita, K., Sasaki, H., Furuse, M., and Tsukita, S. (1999) Endothelial claudin: Claudin-5/TMVCF constitutes tight junction strands in endothelial cells. J. Cell Biol. 147:185-194.
Moy, A.B., Blackwell, K., Wang, N., Haxhinasto, K., Kasiske, M., Bodmer, J., Reyes, G., and English, A. (2004) Phorbol ester-mediated pulmonary artery endothelial barrier dysfunction through regulation of actin cytoskeletal mechanics. Amer. J. Physiol. 287: L153-L167.
Moy, A.B., Bodmer, J.E., Blackwell, K., Shasby, S., and Shasby, D.M. (1998) cAMP protects endothelial barrier function independent of inhibiting MLC20-dependent tension development. Am. J. Physiol. 274:L1024-L1029.
Moy, A., Shasby, S., Scott, B., and Shasby, D. (1993) The effect of histamine and cyclic adenosine monophosphate on myosin light chain phosphorylation in human umbilical vein endothelial cells. J. Clin. Invest. 92:1198-1206.
Moy, A., VanEngelenhoven, J., Bodmer, J., Kamath, J., Keese, C., Giaever, I., Shasby, S., and Shasby, D. (1996) Histamine and thrombin modulate endothelial focal adhesion through centripetal and centrifugal forces. J. Clin. Invest. 97:020-1027.
Navarro, P., Caveda, L., Breviario, F., Mandoteanu, I., Lampugnani, M.G., and Dejana, E. (1995) Catenin-dependent and -independent functions of vascular endothelial cadherin. J. Biol. Chem. 270:30965-30972.
Newman, P.J. (1997) The biology of PECAM-1. J. Clin. Invest. 99:3-7.
Nieset, J.E., Redfield, A.R., Jin, F., Knudsen, K.A., Johnson, K.R., and Wheelock, M.J. (1997) Characterization of the interactions of alpha-catenin with alpha- actinin and beta-catenin/plakoglobin. J. Cell Sci. 110:1013-1022.

Noria, S., Xu, F., McCue, S., Jones, M., Gotlieb, A.I., and Langille, B.L. (2004) Assembly and reorientation of stress fibers drives morphological changes to endothelial cells exposed to shear stress. Amer. J. Pathol. 164:1211-1223.

Ochoa, L., Waypa, G., Mahoney, J.R., Rodriguez, L., and Minnear, F.L. (1997) Contrasting effects of hypochlorous acid and hydrogen peroxide on endothelial permeability: Prevention with cAMP drugs. Amer. J. Respir. Crit. Care Med. 156:1247-1255.

Ohta, Y., Suzuki, N., Nakamura, S., Hartwig, J., and Stossel, T.P. (1999) The small GTPase RalA targets filamin to induce filopodia. PNAS USA 96:2122-2128.

Partridge, C.A., Horvath, C.J., Delvecchio, P.J., Phillips, P.G., and Malik, A.B. (1992) Influence of extracellular matrix in tumor necrosis factor-induced increase in endothelial permeability. Amer. J. Physiol. 263:L627-L633.

Patterson, C.E., Davis, H., Schaphorst, K., and Garcia, J.G. (1994a) Mechanisms of cholera toxin prevention of thrombin- and PMA-induced endothelial cell barrier dysfunction. Microvasc. Res. 48:212-235.

Patterson, C.E. and Garcia, J.G.N. (1994b) Regulation of thrombin-induced endothelial cell activation by bacterial toxins. Blood Coag. Fibrinol. 5:63-72.

Patterson, C.E. and Lum, H. (2001) Update on pulmonary edema: the role and regulation of endothelial barrier function. Endothelium 8:75-105.

Patterson, C.E., Lum, H., Verin, A., Schaphorst, K., and Garcia, J.G.N. (2000) Regulation of endothelial barrier function by the cAMP-dependent protein kinase. Endothel. 7:287-308.

Patterson, C.E., Stasek, J., Schaphorst, K., Davis, H., and Garcia, J.G.N. (1995) Mechanisms of pertussis toxin-induced barrier dysfunction in bovine pulmonary artery endothelial cell layers. Amer. J. Physiol. 268:L926-L934.

Perez-Moreno, M., Avila, A., Islas, S., Sanchez, S., and Gonzalez-Mariscal, L. (1998) Vinculin but not alpha-actinin is a target of PKC phosphorylation during junctional assembly induced by calcium. J. Cell Sci. 111:3563-3571.

Phillips, P., Lum, H., Malik, A., and Tsan, M.F. (1989) Phallacidin prevents thrombin-induced increases in endothelial permeability to albumin. Amer. J. Physiol. 257:C562-C567.

Polte T.R., Eicher, G., Wang, N., and Ingber, D.E. (2004) Extracellular matrix controls MLC phosphorylation and cell contractility through modulation of cell shape and cytoskeletal prestress. Amer. J. Physiol. 286:C518-C528.

Potard, U.S.B., Butler, J.P., and Wang, N. (1997) Cytoskeletal mechanics in confulent epithelial cells probed through integrins and E-cadherins. Amer. J. Physiol. 272:C1654-C1663.

Pourati, J., Maniotis, A., Spiegel, D., Schaffer, J., Butler, J., Fredberg, J., Ingber, D., Stamenović, D., and Wang, N. (1998) Is cytoskeletal tension a major determinant of cell deformability in adherent endothelial cells. Am. J. Physiol. 274;C1283-C1289.

Quadri, S.K., Bhattacharjee, M., Parthasarathi, K., Tanita, T., and Bhattacharya J. (2003) Endothelial barrier strengthening by activation of focal adhesion kinase. J. Biol. Chem. 278:13342-13349.

Quinlan, M.P. and Hyatt, J.L. (1999) Establishment of the circumferential actin filament network is a prerequisite for localization of the cadherin-catenin complex in epithelial cells. Cell Growth & Diff. 10:839-854.

Rabiet, M., Plantier, J., Rival, Y., Genoux, Y., Lampugnani, M.G., and Dejana, E. (1996) Thrombin-induced increase in endothelial permeability is associated with changes in cell-to-cell junction organization. Arterioscl. Thromb. Vasc. Biol. 6:488-496.

Ratcliffe, M.J., Smales, C., and Staddon, J.M. (1999) Dephosphorylation of catenins p120 and p100 in endothelial cells in response to inflammatory stimuli. Biochem. J. 338:471-478.

Salomon, D., Ayalon, O., Patel-King, R., Hynes, R.O., and Geiger, B. (1992) Extrajunctional distribution of N-cadherin in cultures human endothelial cells. J. Cell Sci. 102:7-17.

Satcher, R.L. and Dewey, C.F. (1996) Theoretical estimates of mechanical properties of the endothelial cell cytoskeleton. Biophys. J. 71:109-118.

Sato, M., Theret, D., Wheeler, L., Ohshima, N., and Nerem, R. (1990) Application of the micropipette technique to the measurement of cultured porcine aortic endothelial cell viscoelastic properties. ASME J. Biomech. Eng. 112:263-268.

Sawyer, S.J., Norvell, S., Ponik, S., and Pavalko F.M. (2001) Regulation of PGE_2 and PDI_2 release from human umbilical vein endothelial cells by actin cytoskeleton. Amer. J. Physiol. 281:C1038-C1045.
Schaphorst, K.L., Pavalko, F., Patterson, C.E., and Garcia, J.G.N. (1997) Thrombin-mediated focal adhesion plaque reorganization in endothelium: Role of protein phosphor-ylation. Amer. J. Resp. Cell Molec. Biol. 17:443-455.
Schnittler, H.J., Wilke, A., Gress, T., Suttorp, N., and Drenckhahn, D. (1990) Role of actin and myosin in the control of paracellular permeability in pig, rat and human vascular endothelium. J. Physiol. 431:379-401.
Schulze, C., Smales, C., Rubin, L., and Staddon, J.M. (1997) Lysophosphatidic acid increases tight junction permeability in cultured brain endothelial cells. J. Neurochem. 68:991-1000.
Shariat-Madar, Z. and Schmaier, A.H. (1999) Kininogen-cytokeratin 1 interactions in endothelial cell biology. Trends Cardiovasc. Med. 9:238-244.
Shasby, D., Shasby, S., Sullivan, J., and Peach, M. (1982) Role of endothelial cytoskeleton in control of endothelial permeability. Circ. Res. 51:657-661.
Shojaee, N., Patton, W.F., Chung-Welch, N., Su, Q., Hechtman, H.B., and Shepro, D. (1998) Expression and subcellular distribution of filamin isotypes in endothelial cells and pericytes. Electrophoresis. 19:323-32.
Sohn, R.H., Chen, J., Koblan, K., Bray, P., and Goldschmidt-Clermont, P.J. (1995) Localization of a binding site for phosphatidylinositol 4, 5- bisphosphate on human profilin. J. Biol. Chem. 270:21114-21120.
Staddon, J., Ratcliffe, M., Morgan, L., Hirase, T., Smales, C., and Rubin, L. (1997) Protein phosphorylation and the regulation of cell-cell junctions in brain endothelial cells. Heart & Vessels 12:106-109.
Stamenović, D., Fredberg, J., Wang, N., Butler, J., and Ingber, D.E. (1996) A microstructural approach to cytoskeletal mechanics based on tensegrity. J. Theoret. Biol. 181:125-136.
Stamenović, D., Mijailovich, S.M., Tolić-Nørrelykke, I.M., Chen, J., and Wang, N. (2002). Cell prestress. II. Contribution of microtubules. Am. J. Physiol. 282:C617-C624.
Stamenović, D. and Wang, N. (2000) Cellular Responses to Mechanical Stress: Engineering approaches to cytoskeletal mechanics. J. Appl. Physiol. 89:2085-2090.
Stasek, J.E., Patterson, C.E., and Garcia, J.G.N. (1992) Protein kinase C phosphorylates caldesmon77 and vimentin and enhances albumin permeability across cultured bovine pulmonary artery endothelial cell monolayers. J. Cell. Physiol. 153:62-75.
Stevens, T., Garcia, J.G., Shasby, D.M., Bhattacharya, J., Malik, A.B. (2000) Mechanisms regulating endothelial cell barrier function. Amer. J. Physiol. 279:L419-L422.
Stossel, T., Chaponnier, C., Ezzell, R., Hartwig, J., Janmey, P., Kwiatkowski, D., Lind, S., Smith, D., Southwick, F., and Yin, H. (1985) Nonmuscle actin-binding proteins. Ann. Rev. Cell Biol. 1:353-402.
Strelkov, S.V., Herrmann, H., and Aebi, U. (2003) Molecular architecture of intermediate filaments. Bioessays 25:243-251.
Takeichi, M. (1990) Cadherins: a molecular family important in selective cell-cell adhesion. Ann. Rev. Biochem. 59:237-252.
Thoumine, O. and Ott, A. (1996) Influence of adhesion and cytoskeletal integrity on fibroblast traction. Cell Motil. Cytosk. 35:269-280.
Thoumine, O., Ziegler, T., Girard, P., and Nerem, R.M. (1995) Elongation of confluent endothelial cells in culture: the importance of fields of force in the associated alterations of the cytoskeletal structure. Exp. Cell Res. 219:427-441.
Tigyi, G., Fischer, D.J., Sebok, A., Yang, C., Dyer, D.L., and Miledi, R. (1996) Lysophosphatidic acid-induced neurite retraction in PC12 cells: control by phosphoinositide-Ca^{2+} signaling and Rho. J. Neurochem. 66:537-548.
Valiron, O., Chevrier, V., Usson, Y., Breviario, F., Job, D., and Dejana, E. (1996) Desmoplakin expression and organization at hyman umbilical vein endothelial cell-to-cell junctions. J. Cell Sci. 109:2141-2149.

Verin, A.D., Birulova, A., Wang, P., Liu, F., Becker, P., Birukov, K., and Garcia, J.G.N. (2001) Microtubule disassembly increases endothelial cell barrier function: role of MLC phosphorylation. Amer. J. Physiol. 281:L565-L574.
Vinogradova, T.M., Roudnik, V., Bystrevskaya, V., and Smirnov, V.N. (2000) Centrosome-directed translocation of Weibel-Palade bodies is rapidly induced by thrombin, calyculin A, or cytochalasin B in human aortic endothelial cells. Cell Mot. Cytoskel. 47:141-153.
Vischer, U.M., Barth, H., and Wollheim, C.B. (2000) Regulated vWF secretion is associated with agonist-specific patterns of cytoskeletal remodeling in cultured endothelial cells. Arterioscl. Thromb. Vasc. Biol. 20:883-891.
Volberg, T., Zick, Y., Dror, R., Sabanay, L., Gilon, C., Levitzki, A., and Geiger, B. (1992) Effect of tyrosine-specific protein phosphorylation on the assembly of adherens-type junctions. EMBO J. 11:1733-1742.
Wang, N. (1998) Mechanical interactions among cytoskeletal filaments. Hyperten. 32:162-165.
Wang, N. and Ingber, D.E. (1994) Control of cytoskeletal mechanics by extracellular matrix, cell tension, and mechanical stress. Biophys J. 66:2181-2189.
Wang, N., Naruse, K., Stamenović, D., Fredberg, J.J., Mijailovich, S.M., Tolić-Nørrelykke, I.M., Polte, T., Mannix, R., and Ingber, D.E. (2001) Mechanical behavior in living cells consistent with the tensegrity model. PNAS USA 98:7765-7770.
Wang, N., Tolić-Nørrelykke, I.M., Chen, J., Mijailovich, S.M., Butler, J. P., Fredberg, J. J., and Stamenović, D. (2002) Cell prestress. I. Stiffness and prestress are closely associated in adherent contractile cells. Amer. J. Physiol. 282:C606-C616.
Wang, N. and Stamenović, D. (2000) Contribution of intermediate filaments to cell stiffness, stiffening, and growth. Amer. J. Physiol. 279:C188-C194.
Wang, N. and Stamenović, D. (2002) Mechanics of vimentin intermediate filaments. J. Musc. Res. Cell Motil. 23:535-540.
Wang, Q., Patton, W.F., Hechtman, H., and Shepro, D. (1997) A novel anti-inflammatory peptide inhibits endothelial cell cytoskeletal rearrangement, nitric oxide synthase translocation, and paracellular permeability increases. J. Cell. Physiol. 172:171-182.
Waschke, J., Baumgartner, W., Adamson, R., Zeng, M., Aktories, K., Barth, H., Wilde, C., Curry, F., and Drenckhahn, D. (2004) Requirement of Rac activity for maintenance of capillary endothelial barrier properties. Am. J. Physiol. 286:H394-H401.
Winder, S.J. (2003) Structural insights into actin-binding, branching, and bundling proteins. Curr. Opinion 15:14-22.
Wong, M. and Gotlieb, A. (1990) Endothelial monolayer integrity: Perturbation of F-actin filaments and the dense peripheral band- network. Arterioscl. 10:76-84.
Wong, R.K., Baldwin, A.L., and Heimark, R.L. (1999) Cadherin-5 redistribution at sites of TNF-α and IFN-γ -induced permeability. Amer. J. Physiol. 276:H736-H748.
Wu, N.Z. and Baldwin, A. (1992) Transient venular permeability increas and endothelial gap formation induced by histamine. Amer. J. Physiol. 262:H1238-H1247.
Wysolmerski, R.B. and Lagunoff, D. (1990) Involvement of myosin light chain kinase in endothelial cell retraction. PNAS USA 87:16-20.
Xu, J., Schwartz, W., Kas, J., Stossel, T., Janmey, P., and Pollard, T. (1998) Mechanical properties of actin filament networks depend on preparation, polymerization conditions, and storage of actin momomers. Biophys. J. 74:2731-2740.
Yang, J.T., Rayburn, H., and Hynes, R.O. (1993) Embryonic mesodermal defects in α5 integrin deficient mice. Devel. 119:1093-1105.
Yano, Y., Geibel, J., and Sumpio, B.E. (1997) Cyclic strain induces reorganization of integrin α5β1 and α2β1 in human umbilical vein endothelial cells. J. Cell. Biochem. 64:505-513.

Chapter 3
The Activated Endothelial Cell Phenotype

Hazel Lum, PhD*

Department of Pharmacology, Rush University Medical Center, Chicago, IL

CONTENTS

What Is Activation?

Activated Responses of Endothelium
The Pro-adhesive Phenotype
Loss of Endothelial Barrier Function
The Proliferative (Motile) Phenotype
The Pro-coagulant Phenotype
The Secretory Phenotype

Cell-derived Mediators in Endothelial Activation
Vasoactive Mediators
Cytokines and Chemokines
Growth Factors
Lysophospholipids
Oxidants

Activation by Microorganisms
Bacterial Infection
Viral Infection

Activation by Mechanical Stimuli
Stress and Response
Response Mechanisms and Chronic Disease

Provocation and Perspective

What Is Activation?

The vascular endothelium of the lung, as in other tissues, is normally a quiescent contact-inhibited monolayer of cells, which critically maintains anti-coagulant properties on the cell surface. Equally important, it functions as a gate to regulate fluid and solute traffic, actively directs leukocytic transmigration, controls vascular smooth muscle function and phenotype, transports nutrients and bioactive mediators to target organs, and contributes towards polarity of the tissue. Under conditions of various physiological (e.g., humoral agonists), infectious (e.g., bacteria), or non-infectious stimuli (e.g., hemodynamic perturbations), the quiescent endothelium shifts to an "activated" phenotype. Endothelial cells are activated by a wide range of stimuli, and the ensuing activation response profile varies with the type, duration, and amount of stimulus, as well as the immediate cellular milieu, and the cellular context.

The activation response may be rapid in onset and not dependent on protein synthesis, or the response delayed involving transcriptional activation of genes such as responses stimulated by growth factors and cytokines. The activation may be reversible, particularly if the stimulus is limited in scope. For instance, activated endothelium can revert to the quiescent state if a localized bacterial pneumonia resolves or a physiological stimulus (i.e., growth factor) becomes inactivated or removed with time. However, chronic or potent stimuli may induce prolonged endothelial cell activation, developing into phenotypes not readily reversible, which if beneficial, is adaptive, or if detrimental, pathological. Thus the activated endothelium has multiple guises; its regulatory mechanisms are complex; and it engages multiple factors. It is apparent that endothelial activation is not distinguished by one phenotype, but rather it encompasses a spectrum of responses that range from physiological activation to a dysregulated pathological activation, the net outcome dependent on integration of existing cellular regulatory factors.

The activated endothelium presents with heightened activity, particularly increased adhesion for leukocytes, cellular turnover (from 0.1% replications per day to 1-10% replications per day), cell motility, secretory activity, permeability, and metabolism, as summarized in Figure 1. These heightened activities are often accompanied by distinct changes in cell morphology and cell cytoskeletal architecture, as described in Chapter 2. Often, endothelial activation is thought to be synonymous with increased expression of adhesion molecules, a notion that underscores the central role of activated endothelium in the inflammatory response. Indeed, it is widely held that only activated endothelium and not quiescent endothelium participates in inflammatory responses. Increased expression of adhesion molecules on the luminal plasma membrane attended by increased leukocyte adhesion is a dominant and significant feature of activated endothelium, and is discussed below. However, it is clear that increased adhesion is not always presented by the activated endothelium, or more often, increased adhesion is one of several activation responses occurring. Increased permeability is a hallmark of inflammation and often, the activated endothelium presents with impaired barrier function. This is characterized by increased extravasation of solutes out of the circulation without barrier selectivity to large molecules, promoting tissue edema and ultimately organ dysfunction or failure.

In the lung, this condition is of serious concern since increased extravasation can result in alveolar flooding (water lung), causing impairment of gas exchange, which can be life threatening. In the brain cerebrovascular edema, which occurs with a wide range of disorders, can lead to serious brain functional deficits and mortality. A severe non-reversible increased permeability can be the result of an overt breach of the barrier caused by direct injury or cell death (such as with bacterial infection), or a reversible increased permeability can occurred in which endothelial cellular mechanisms tightly regulate its progression and resolution (Patterson and Lum, 2001). Some mediators activate angiogenic activities such as cell proliferation, these responses occurring in wound repair as part of a regulated physiological response or in tumorgenesis. Other mediators induce endothelial cells to exhibit heightened adhesion for leukocytes. Moreover, endothelial cells produce, activate, or modify a wide range of bioactive mediators that regulate other biological systems or autoregulate endothelial function. The overall pattern of this endothelial-derived regulatory milieu is dramatically altered with activation. Thus, the activated endothelium does indeed actively participate in fundamental processes regulating inflammation, angiogenesis, immunity, cell growth and differentiation.

In this chapter, we will examine current information on endothelial cell activation, providing overviews of some of the typical endothelial responses and activation stimuli. We will first present activation responses, and out of necessity, keeping it general because, again, the activation response profile is characteristic of a stimulus. Subsequent sections will focus on activators of vascular endothelium, providing current understanding of their activating mechanisms. The multitude of stimuli is broadly grouped by similarity of structure and function, ranging from cell-derived mediators (e.g., cytokines, vasoactive mediators), to microbial organisms, and to mechanical stimuli (e.g., shear stress). Tables 1, 2, and 3 tabulate some of the more studied activators, but are not meant to be comprehensive. There are excellent and extensive reviews on these topics and references for some are included.

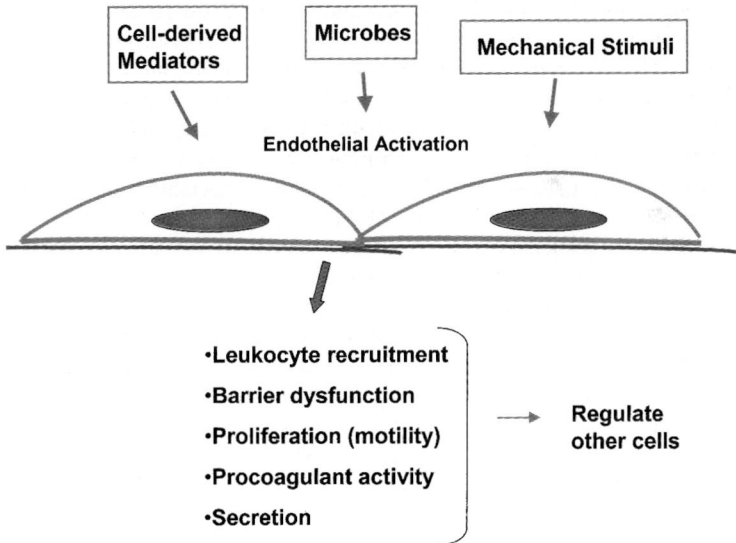

Figure 1. Endothelial activation and response. Because of the vast surface area at the interface of tissue and blood, the endothelium is subject to many potential activators, such as pathogens and mediators, derived from other cells or derived from the endothelial cells themselves. Due to the central role of the endothelium in regulation of vascular integrity, blood flow, blood clotting, smooth muscle function, leukocyte trafficking, and inflammation, activation of endothelium has many important consequences.

Activated Responses of Endothelium

Endothelial cells respond to activating stimuli by undergoing phenotypic changes, with increased metabolic activity and increases in a) adhesion for leukocytes, b) permeability, c) proliferation, d) pro-coagulant activity, and e) secretion. A particular stimulus can activate one or a combination of these endothelial phenotypes. More commonly, a stimulus triggers a combination of these phenotypes, the predominance of phenotype(s) being characteristic of the particular stimulus. While some activators produce rapid effects, others such as TNFα and growth factors may involve gene transcription, requiring hours or days to reach full effect.

The Pro-adhesive Endothelial Phenotype

A wide range of vasoactive mediators, cytokines, oxidants, microbes, and shear stress induce this hyperadhesive state of the endothelium. This increased adhesiveness is a vital host defense response to recruit immune cells for extravasation to tissue sites of microbial infection and injury. However, under conditions of sustained and/or severe activation, this normally protective action of host defense becomes pathological and is associated with inflammatory injury instead. Whether host defense or inflammatory injury, the activated endothelium is central in the initiation and successful completion of the passage of leukocytes from the circulation to the extravascular tissue spaces. It is now recognized that leukocyte extravasation is regulated by three sequential coordinated phases of leukocyte-endothelial cell interactions: i) activation of transient low-affinity adhesion (rolling), followed by ii) activation of firm shear-resistant adhesion, and finally iii) transmigration across the endothelium (diapedesis). There is evidence that the rolling phase is less important in the lung microcirculation.

Induction of leukocyte rolling on the endothelial surface is attributed to expression of selectins, the superfamily of adhesion molecules whose ligands are the sulfated, sialylated, fucosylated glycoproteins on the leukocyte surface (Vestweber and Blanks, 1999). Quiescent, nonactivated endothelial cells do not express (or at barely detectable levels) the two endothelial selectins (P- and E-selectins) on the surface. A third family member, L-selectin, is constitutively expressed exclusively by leukocytes. Activated endothelia exhibit increased expression of selectins on the plasma membrane surface. Inflammatory mediators, such as thrombin, oxidants, and lysophosphatidylcholine, are known to rapidly increase endothelial surface expression of P-selectin, which is stored within Weibel-Palade bodies (Patel *et al.*, 1991; Sugama *et al.*, 1992; Collins *et al.*, 1993; Wagner, 1993; Murohara *et al.*, 1996). The P-selectin expressed on the cell membrane can be cleared relatively rapidly (within hours) by endocytosis; where it is re-stored in Weibel-Palade bodies (Subramaniam *et al.*, 1993) or delivered to lysosomes for degradation (Green *et al.*, 1994). An activating stimulus can also induce gene transcription of both E- and P-selectins (as is the case for TNFα) and selective for one or the other selectins (Weller *et al.*, 1992; Khewgoodal *et al.*, 1996; Wong and Dorovinizis, 1996; Wyble *et al.*, 1996; Yao, Pan *et al.*, 1996). In several models of inflammation, redundancy of function by E- and P-selectins is evident (Labow *et al.*, 1994; Bullard *et al.*, 1996), indicating that the two selectins likely have some degree of overlapping function. However, the precise determinants dictating the preferred expression for a particular selectin or both selectins are not clear.

Activation stimuli also induce firm shear-resistant adhesion of leukocytes on the endothelial surface. This type of adhesion is attributed to the endothelium's ability to arrest rolling leukocytes onto its surface and upregulation of the IgG class of adhesion molecules, ICAM-1 (intercellular adhesion molecule-1) and VCAM-1 (vascular adhesion molecule-1). Induction of firm adhesion for leukocytes on the activated endothelial surface is an essential step for leukocyte transmigration. The arrest of rolling leukocytes occurs through surface-immobilized chemokines on the endothelial surface. Chemokines released by cells are bound to heparan sulfate and related glycosaminoglycans of proteoglycans on the endothelial surface, providing a haptotactic gradient, which attracts leukocytes (Tanaka *et al.*, 1993; Lortat-Jacob *et al.*, 2002; Middleton *et al.*, 2002; Carter *et al.*, 2003). The endothelial-displayed chemokines capture the rolling leukocytes by binding to specific G protein-coupled receptors on leukocytes, triggering signal transduction events that rapidly (milliseconds) activate β_1 and β_2 integrins to a state of high-affinity and increased avidity (Weber *et al.*, 1999a; Constantin *et al.*, 2000). The activated β integrins in turn engages the counter-receptors, ICAM-1, ICAM-2, and VCAM-1 express by the endothelium, inducing firm shear-resistant adhesion of leukocytes to the endothelial surface (Smith *et al.*, 1988). Quiescent endothelial cells express low levels of ICAM-1, ICAM-2, and VCAM-1; but ICAM-1 (but not ICAM-2) and VCAM-1 are upregulated transcriptionally by inflammatory stimuli (Gerritsen *et al.*, 1993; Abe *et al.*, 1996; Rahman *et al.*, 1996; Amberger *et al.*, 1997; Derhaag *et al.*, 1997; Shahgasempour *et al.*, 1997; Roebuck and Finnegan, 1999). ICAM-1 and VCAM-1 can undergo post-transcriptional modification such as dimerization or clustering, which may increase binding efficiency for β integrins (Miller *et al.*, 1995; Reilly *et al.*, 1996; Woska, Jr. *et al.*, 1996; Wojciak-Stothard *et al.*, 1999). In-depth discussion of mechanisms of regulation of leukocyte-endothelial communications in health and disease is provided in Chapters 11 and 12.

Loss of Endothelial Barrier Function

Another phenotypic change in response to activating stimuli is increase in vascular endothelial permeability, a hallmark of inflammation. Typically the increases in permeability are accompanied by morphological changes of the endothelial cell shape, appearance of gaps between adjacent endothelial cells, and remodeling of the cell cytoskeleton, particularly the actin filamentous network. These morphological features implicate impaired cell-cell adhesion (i.e. disruption of intercellular junctions) as a primary determinant of increased permeability. Endothelium, like the epithelium, has adherens and tight junctions, both of which are critical for maintaining a restrictive barrier. Although definitive mechanisms of regulation are under intense investigation, among the more favored hypotheses include a) actin-myosin mediated endothelial contraction, and b) regulation of adherens and tight junctional proteins by phosphorylation signals. An additional determinant that is critically important in maintenance of barrier restrictiveness is regulation of adhesion to extracellular matrix proteins, which contribute to the characteristic permeability response induced by the activating stimulus.

Activation of increased permeability can be relatively rapid (in minutes) or delayed (hours) in onset. The vasoactive mediators, such as thrombin and histamine, have been the most studied with respect to regulation of endothelial permeability and most, if not all, of these activators appear to elicit rapid and reversible increases in permeability. This rapid onset can be appreciated with real-time measurements of transendothelial resistance, which has been documented by numerous investigators. Studies of the time-course of endothelial contraction and phosphorylation of myosin light chain (Goeckeler and Wysolmerski, 1995; Moy *et al.*, 1996; Shasby *et al.*, 1997) as well as phosphorylation of junctional proteins (Shasby *et al.*, 2002; Konstantoulaki *et al.*, 2003) positively correlate with rapid onset of the increased permeability activated by these mediators. However, other mediators such as growth factors (i.e., TGFβ, VEGF) and cytokines (i.e., TNFα, IL-1α) usually required hours before the increased permeability is detected. The mechanisms of VEGF-induced barrier dysfunction are receiving increasing attention due to the recognized importance of VEGF in vasculogenesis and angiogenesis. In *in vitro* studies, VEGF (Chang *et al.*, 2000; Kurimoto *et al.*, 2004), as well as another growth factor, TGFβ1 (Hurst *et al.*, 1999), induce decreases in endothelial barrier function within 1-3 hours. Yet, *in vivo* animal studies, VEGF stimulated a much more rapid onset (5-15 min) of increased permeability (Sirois and Edelman, 1997; Murohara *et al.*, 1998). Explanations for these variances in VEGF responses remain unclear, but possibly in the *in vivo* models VEGF activates or interacts with other factors (blood borne?) to trigger the more rapid permeability responses. Similarly, *in vitro* cell culture studies report that TNFα-induced barrier dysfunction has a long onset time, and multiple mechanisms are implicated, including *de novo* protein synthesis of PKCα (Ferro *et al.*, 2000), extracellular matrix remodeling (Partridge *et al.*, 1993), and fragmentation of tight junctions (Blum *et al.*, 1997).

Whether the increased permeability has a rapid or delayed onset, the apparent morphological changes or the frequency and severity of these changes vary in degree with the particular stimulus (Schaeffer *et al.*, 1993; Ehringer *et al.*, 1996). These differences are attributed to inherent, divergent mechanisms in regulation of the barrier dysfunction. For example, several reports indicate that vasoactive mediators thrombin, histamine, and bradykinin elicit distinct differences in the signaling cascades such as Ca^{2+} and MLC phosphorylation activated in association with the increase in permeability (Ehringer *et al.*, 1996; Moy *et al.*, 1996; Aschner *et al.*, 1997). Furthermore, endothelial cells from different regions of the vascular bed (i.e., microvascular versus macrovascular) are known to be heterogeneous in structure and function, and their activated permeability responses are predictably varied (Kelly *et al.*, 1998; Stevens *et al.*, 2001). Although substantial information has been generated, our understanding of endothelial barrier dysfunction regulation is fragmented and much remains to be discovered regarding the diversity of the signaling cascades and their regulatory mechanisms.

Proliferative (and Motile) Phenotype

Activation of vascular endothelial proliferation is part of the normal physiological function in the generation of new capillary blood vessels in typical growth and wound repair, and is of particular interest in tumor growth and metastasis. Enhanced endothelial proliferation coupled with dissociation of homotypic cell-cell adhesion, heightened motility, break-down of extracellular matrix proteins, and reformation of cell-cell junctions comprise events producing the overall "angiogenic phenotype", forming neo-blood vessels with patent lumen. Often, activating stimuli which increase endothelial proliferation also regulate other angiogenic activities such as disruption of junctional organization and remodeling of extracellular matrix proteins, events tightly coupled with enhanced endothelial motility.

VEGF is a critical activator of endothelial cell-specific mitogenesis and angiogenesis. Usually, it is synthesized and secreted by normal or neoplastic cells that are in close proximity to the endothelial cells. VEGF induces a pleotropic response allowing endothelial cells to proliferate, migrate, assemble into tubes, to survive, and to increase permeability (Carmeliet and Collen, 2000). The importance of VEGF is underscored by the finding that targeted inactivation of the gene causes massive endothelial cell apoptosis, resulting in severe hemorrhage and lethality (Ferrara *et al.*, 1996). Other growth factors with direct angiogenic actions on endothelial cells are HGF (hepatocyte growth factor) (Koch *et al.*, 1996; Sengupta *et al.*, 2003) and FGF (fibroblast growth factor) (Landgren *et al.*, 1998; Tanaka *et al.*, 1999). Growth factors such as VEGF and FGF additionally elicit pro-survival signals in endothelial cells, providing an additional level of regulation of angiogenesis (Dimmeler and Zeiher, 2000).

Vasoactive mediators, bradykinin (Morbidelli *et al.*, 1998), endothelin-1 (ET-1) (Morbidelli *et al.*, 1995), and thrombin (Herbert *et al.*, 1994) increased endothelial cell proliferation. Thrombin-induced mitogenesis was related to autocrine release of bFGF (Herbert *et al.*, 1994). Effects of NO on proliferation are less clear. Reports indicate that chemical donors of NO inhibited bFGF-stimulated endothelial cell proliferation (RayChaudhury *et al.*, 1996). Interestingly, using *in vitro* models of endothelial cells and NO-producing lung cancer cell lines, NO inhibited capillary formation (Phillips *et al.*, 2001). However, other investigators report that inhibition of endogenous NO generation inhibited endothelial cell sprouting (Murohara *et al.*, 1999) and proliferation (Morbidelli *et al.*, 1996).

Other mediators also induce proliferation and angiogenesis. LPS directly promoted endothelial cell angiogenesis through toll-like receptors, engaging TNF receptor-associated factor 6 (TRAF6) (Pollet *et al.*, 2003). However, anti-IL-8 or anti-VEGF antibody blocked TNFα-induced neovascularization in rabbit cornea (Yoshida *et al.*, 1997), suggesting indirect regulation by paracrine and autocrine factors. Some chemokines are mitogenic. For example, stimulation of endothelial cells with IL-8 induces proliferation and chemotaxis through the IL-8 receptor, CXCR2 (Heidemann *et al.*, 2003; Szekanecz *et al.*, 1994). MCP-1A, a CC chemokine, is also angiogenic, mediating its effects through the receptor, CCR2 (Salcedo *et al.*, 2000).

Pro-coagulant Phenotype

An essential function of the vascular endothelium is to facilitate blood flow by maintaining an anti-thrombotic surface, which prevents platelet adhesion and clotting. Under conditions of vascular damage such as with overt trauma, infection, and inflammation, the endothelium is activated and in turn, becomes pro-coagulant as a protective response against the vascular injury. Yet, sustained or intense pro-coagulant activity can lead to thrombotic disorders. The activated endothelium regulates three key activities that cause the endothelial surface to lose its anti-coagulant characteristics: i) decreases thrombomodulin expression on the cell surface, ii) increases tissue factor production, and iii) decreases fibrinolytic activity of endothelial cells (Cines *et al.*, 1998). Endothelial cells express the transmembrane protein, thrombomodulin, which forms a 1:1 complex with thrombin. Formation of the complex activates the anti-coagulant protein C on the endothelial surface and helps maintain blood fluidity (Cines *et al.*, 1998). Activation of endothelial cells by inflammatory cytokines, particularly TNF, decreases surface expression of thrombomodulin by suppression of transcriptional and translational regulation of the protein and also increases internalization with subsequent degradation (Dittman and Majerus, 1990). However, tissue factor production is induced by pro-inflammatory and injurious stimuli, including shear stress (Lin *et al.*, 1997), oxidized phosphatidylcholine (Birukov *et al.*, 2004), and bacteria (Courtney *et al.*, 1996; Fryer *et al.*, 1997; Cines *et al.*, 1998). Elevation of tissue factor levels dramatically accelerates activation of coagulation factors X and IX. Regulation of endothelial fibrinolytic activity is control by the balance of tissue plasminogen activator (tPA) and the inhibitor (PAI-1), in which quiescent endothelial cells express minimal PAI-1. Inflammatory stimuli such as IL-8, TNF, TGFβ (Klagsbrun and Edelman, 1989; Hamaguchi *et al.*, 2003), native and oxidized LDL (Allison *et al.*, 1999), and cyclic strain (Cheng *et al.*, 1996) increase PAI-1, leading to inhibition of fibrinolytic activity and vessel engulfment with fibrin.

Secretory Phenotype

The endothelium is no longer perceived as a passive sheet of cells, but rather a highly responsive organ capable of producing and releasing a host of biochemically and biologically diverse molecules. Indeed, it would be more an exception than the rule that an activating stimulus does not induce secretion of some mediators or reduce secretion of others. These endothelial-secreted mediators have autocrine functions and or paracrine functions, regulating activities of nearby cells such as smooth muscle cells and leukocytes. Other released products may not act on cells, but remodel extracellular matrix proteins and carbohydrates.

A variety of activating stimuli, including shear stress (Kuchan and Frangos, 1993; Boo *et al.*, 2002; Krizanac-Bengez *et al.*, 2003), thrombin (Eto *et al.*, 2001), TNFα (Bove *et al.*, 2001), and LPS (Kan *et al.*, 2002) induce endothelial cells to release the vasoactive mediators NO and ET-1, both potent regulators of vasomotor activity as well as endothelial function. Inflammatory stimuli, particularly TNFα and bacterial products, including lysophosphatidylcholine, the bioactive lipid product of PLA_2, and oxidants are effective inducers of cytokine and chemokine generation from endothelial cells (Burns *et al.*, 1997; Lakshminarayanan *et al.*, 1998; Goebeler *et al.*, 1999; Weber *et al.*, 1999b; Garcia *et al.*, 2000; Murugesan al., 2003). Activated endothelial cells synthesize and secrete several chemokines (IL-8, MCP-1, GROα, RANTES), most of which are directed to and bind to the luminal cell membrane (Wolff *et al.*, 1998; Weber *et al.*, 1999b). The increased secretion of IL-8, is also attributed to a preformed stored pool in Weibel-Palade Bodies in cultured endothelial cells and in *in situ* vessels, and stimulation by histamine or thrombin results in increased IL-8 concentration in the medium and a concomitant disappearance of IL-8-containing granules (Utgaard *et al.*, 1998), suggesting that secretion of preformed IL-8 provides a rapid means for activation of leukocyte adhesion. A recent study also reported that eotaxin-3, GROα, and MCP-1, but not T-cell chemokines RANTES and IP-10, could be rapidly released from endothelial cells stimulated with histamine (Oynebraten *et al.*, 2004).

Another endothelial product is oxidants. Endothelial activation by several mediators increases oxidant generation, which are secreted and therefore may be a significant contributor in maintaining the oxidant-rich environment at an inflammatory locus. For instance, treatment of human umbilical vein endothelial cells with cytokines IL-1 and IFNγ induced dose- and time-dependent increases in O_2^- (Matsubara and Ziff, 1986). Also, vasoactive peptides, such as bradykinin, induce production of O_2^- (40 nmol/2×10^6 cells) within 5 min of ligand binding to its receptor on endothelial cells (Holland *et al.*, 1990). Further, pathological conditions, such as hypoxia followed by ischemia exposure, has also been reported to produce and release of 10-50 nmol O_2/hr/1.5 million endothelial cells (Zweier *et al.*, 1988; Lum *et al.*, 1992b).

Pro-inflammatory lipid mediators are also secreted by activated endothelial cells. For example, VEGF (Bernatchez *et al.*, 2001a,b), oxidants (Lewis *et al.*, 1988), and thrombin (Prescott *et al.*, 1984; McHowat *et al.*, 2001) induce PAF secretion (platelet activating factor). The released PAF has been shown to bind to the endothelial surface and mediate arrest of rolling leukocytes (Lewis *et al.*, 1988; Lorant *et al.*, 1991; Kuijpers *et al.*, 1992; Rainger *et al.*, 1997). Analogous to chemokines, endothelial cell-bound PAF is also a ligand for leukocyte G protein-coupled receptors (LeVan *et al.*, 1997; Honda *et al.*, 2002), and its activation, in turn, activates β_2 integrins on leukocytes and subsequent binding to counter-receptors on endothelium for firm adhesion (Lorant *et al.*, 1991; Rainger *et al.*, 1997).

Not all endothelial-released products are targeted to cells. Endothelial cells activated by ET-1 (Salani et al., 2000), TNFα (Partridge et al., 1993), or thrombin (DuhamelClerin et al., 1997) upregulate and release metalloproteinases, which subsequently bind to and remodel extracellular matrix proteins. And finally, the expression of glycosaminoglycans (GAGs) by the endothelium is regulated by inflammatory stimuli. Interleukins can induce cultured endothelial cells to produce a highly organized, GAGs-rich pericellular matrix, (Montesano et al., 1984). Further, stimulation of microvascular endothelial cells with the cytokines IFNγ and TNFα causes a transient decrease and a subsequent prolonged increase in the expression of N-deacetylase/N-sulphotransferase, the enzyme that catalyzes the sulfation of heparan sulphate and promoted binding with the chemokine RANTES (Carter et al., 2003). This mechanism is believed to be important in establishing the haptotactic gradient of chemokines to present to and to arrest rolling leukocytes onto the endothelial surface.

Cell-derived Mediators in Endothelial Activation

A vast array of bioactive molecules (e.g., vasoactive mediators, cytokines, growth factors, lysophospholipids- described in Table 1) are released by cells which activate endothelium and, as mentioned earlier, endothelial cells themselves are prolific secretory cells, releasing paracrine and autocrine factors. The mediators in this category activate endothelial cells through mostly ligand-receptor interactions, subsequently transducing and integrating cascades of signaling molecules, which mediate the activated endothelial responses. These activating stimuli are biochemically diverse and the transduction mechanisms are characteristically diverse as well. Below, we discuss in details some selected mediators and their mechanisms of activation in endothelial cells.

Table 1. Cell-Derived Mediators

Mediators	Transduction[2]	Activated Responses[3]	References
Vasoactive Mediators:			
Angiotensin II	GPCRs AT1, AT2) coupled to Gi/o or Gq/11/14	Secretion, Adhesion	Morgan-Boyd et al., 1987; Arai and Escobedo, 1996; Chua et al., 1996; Kim et al., 1996; Balmforth et al., 1997; Conchon et al., 1997; Grafe et al., 1997; Li and Shah, 2003;
Bradykinin	GPCRs (B1, B2); coupled to Gi/o or Gq/11/14	Barrier dysfunction, Mitogenesis, Secretion	Mcintyre et al., 1985; Holland et al., 1990; Liao and Homcy, 1993; Smith et al., 1995; Ehringer et al., 1996; Gooch and Frangos, 1996; Aschner et al., 1997; Morbidelli et al., 1998; Prado et al., 2002
Endothelin-1	GPCRs (ETA, ETB); coupled to Gi/o, Gq/11/14, or Gs	Barrier dysfunction, Adhesion, Mitogenesis, Motility, Secretion	Aramori and Nakanishi, 1992; Eguchi et al., 1993; Kurose et al., 1993; Stanimirovic et al., 1994; Morbidelli et al., 1995; Hayasaki et al., 1996; Kawai et al., 1997; Noiri et al., 1997; Zidovetzki et al., 1999; Salani et al., 2000; Albertini et al., 2003

Histamine	GPCRs (H1, H2); coupled Gq/11/14 or Gs	Barrier dysfunction, Adhesion, Secretion	Mcintyre et al., 1985; Lorant et al., 1991; Hekimian et al., 1992; Asako et al., 1994; Ehringer et al., 1996; Jilma et al., 1998; Andriopoulou al., 1999; Moy et al., 2000
Nitric Oxide (NO)	Guanylyl cyclase; Other mechanisms?	Mitogenesis, Motility, Barrier dysfunction, (Adhesion), Barrier enhacencement	Kubes et al., 1991; Ziche et al., 1994; Biffl et al., 1996; Morbidelli et al., 1996; Takahashi et al., 1996; Armstead et al., 1997; Liu and Sundqvist, 1997; LopezFarre et al., 1997; Terada et al., 1997; Volk et al., 1997; Murohara et al., 1998; Murohara et al., 1999; Albrecht et al., 2003; Krizanac-Bengez et al., 2003; Mark et al., 2004
Thrombin	GPCRs (PAR1, 3, 4); coupled to Gi/o, Gq/11/14, or G12/1	Barrier dysfunction, Adhesion, Secretion, Mitogenesis, Motility	Kitazumi and Tasaka, 1992; Sugama and Malik, 1992; McHowat and Corr, 1993; Herbert et al., 1994; Ehringer et al., 1996; Garcia et al., 1996; Rabiet et al., 1996; Duhamel-Clerin et al., 1997; Papadimitriou et al., 1997; Wang et al., 1997; Holland et al., 1998; Gilchrist et al., 2001; Marin et al., 2001; McHowat et al., 2001; Rahman et al., 2002; Vanhauwe et al., 2002; Kim et al., 2004
Cytokines:			
IFNγ	Cytoplasmic tyrosine kinases	Barrier dysfunction, Adhesion, Secretion	Marumo, et al. 1993; Matsubara et al., 1986; Matsubara & Ziff 1986; Kawai et al., 1999; Ranta et al., 1999; Wong et al., 1999; Balyasnikova et al., 2000; Kotenko and Pestka, 2000; Shukaliak and Dorovini-Zis, 2000; Staykova et al., 2000; Oshima et al., 2001; Shaw et al., 2001
InterleukinS (IL-1, IL-2, IL-3, IL-4, IL-6, IL-10)	Cytoplasmic tyrosine kinases	Barrier dysfunction, Adhesion, Secretion	Bevilacqua et al., 1985; Matsubara and Ziff, 1986; Bucana et al., 1988; Downie et al., 1992; Wogensen et al., 1993; Khewgoodall et al., 1996; Vora et al., 1996; Yao et al., 1996; Burns and Furie, 1998; Patterson et al., 1998; Ranta et al., 1999; Tamaru and Narumi, 1999; Garcia et al., 2000; Desai et al., 2002
Tumor Necrosis Factor: (TNFα, TNFβ)	TNFR1 (p55), TNFR2 (p75)	Barrier dysfunction, Adhesion, Secretion, Motility, Cell death	Szekanecz et al., 1994; Wyble et al., 1997; Yoshida et al., 1997; Wojciak-Stothard et al., 1998; Goebeler et al., 1999; Marumo et al., 1999; Ferro et al., 2000; Garcia et al., 2000; Rahman et al., 2000; Petrache et al., 2001; Frey et al., 2002; Paysant et al., 2002
Chemokines:			
IL-8, Monocyte chemoattractant protein-1 (MCP-1)	GPCRs (CXCR, CCR)	Barrier dysfunction,	Berger et al., 1999; Murdoch et al., 1999; Heidemann et al., 2003
Eotaxin		Mitogenesis, Motility	Szekanecz et al., 1994; Salcedo et al., 2000; Schraufstatter et al., 2001

Growth Factors:			
Fibroblast Growth Factor (FGF)	FGFR-1	Barrier dysfunction, Mitogenesis, Motility	Landgren et al., 1998; Tanaka et al., 1999; Kanda et al., 2000; Dye et al., 2001; Halama et al., 2001
Hepatocyte Growth Factor (HGF)	HGFR	Motility, Secretion, Barrier enhancement	Koch et al., 1996; Wang and Keiser, 2000; Liu et al., 2002; Sengupta et al., 2003;
Transforming Growth Factor (TFG-β)	TGFR	Barrier dysfunction	Hurst et al., 1999
Vascular Endothelial Growth Factor (VEGF)	VEGFR-1 (Flt1), VEGFR-2 (Flk1; KDR), VEGFR-3 (Flt-4)	Barrier dysfunction, Mitogenesis, Motility, Secretion	Hippenstiel et al., 1998; Wu et al., 1999; Montrucchio et al., 2000; Bernatchez et al., 2001; Robinson and Stringer, 2001; Wary et al., 2003
Bioactive Phospholipids			
Lysophosphatidic Acid (LPA)	GPCRs (LPA_1, LPA_2, LPA_3); coupled to Gi/o, Gq/11/14	Barrier dysfunction, Barrier enhancement, Mitogenesis, Motility, Secretion	Panetti et al., 1997; Schulze et al., 1997; Alexander et al., 1998; Chua et al., 1998a; English et al., 1999; Palmetshofer et al., 1999; Rizza et al., 1999; Lee et al., 2000; Panetti et al., 2000; van Nieuw Amerongen et al., 2000
Lyso-phosphatidyl choline (LPC)	GPCRs (G2A, GPR4); coupled to Gi/o, others?	Adhesion, Cell death, Secretion, (motility)	Kume et al., 1992; Kume and Gimbrone, 1994; Murohara et al., 1996; Takahara et al., 1996; Zhu et al., 1997; Rikitake et al., 2000; Takeshita et al., 2000; Inoue et al., 2001; Rikitake et al., 2001; Zhu et al., 2001; Chaudhuri et al., 2003; Lum et al., 2003; Murugesan et al., 2003
Sphingosine 1-phosphate (S1P)	GPCRs ($S1P_1$, $S1P_2$, $S1P_3$, $S1P_4$, $S1P_5$); coupled to Gi/o, Gq/11/14, and G12/13	Barrier enhancement, Motility	Panetti et al., 2000; Garcia et al., 2001; Liu et al., 2001
SPC	GPCRs (OGR1, GPR4, GPR12, others?); coupled to Gi/o, Gq, others?	Motility, Secretion	Mogami et al., 1999; Boguslawski et al., 2000; Mijares et al., 2000; Zhu et al., 2001
Oxidants[1]:			
(i.e., H_2O_2, O_2^-, OH^-, $OONO^-$)	?	Barrier dysfunction, Adhesion, Cell death, Mitogenesis, Motility, Secretion	Lewis et al., 1988; Pietra and Johns, 1990; Siflinger-Birnboim et al., 1992; Gaboury et al., 1994; Morita et al., 1995; Rahman et al., 1996; Ochoa et al., 1997; Chua et al., 1998b; Gow et al., 1998; Lakshminarayanan et al., 1998; Okayama et al., 1999; Zadeh et al., 2000; Kevil et al., 2001; Cai et al., 2002; Stone and Collins, 2002; Chen et al., 2003

[1] NO = Nitric Oxide, H_2O_2 = hydrogen peroxide, O_2^- = superoxide anion, OH^- = hydroxy radical, $OONO^-$ = peroxynitrite; [2] GPCR = G protein coupled receptor; [3] Parenthesis indicates evidence for both negative and positive regulation

Vasoactive Mediators

Mediators in this category not only activate endothelium but cause smooth muscle contraction and, thus, are regulate peripheral vasomotor tone. Functionally, they are ligands that specifically bind G protein-coupled receptors (GPCRs), triggering the activation of coupled G protein subtypes and consequent signaling cascades as detailed in Chapter 4. However, there are exceptions, such as NO, which directly binds and activate the enzyme, guanylyl cyclase.

The procoagulant α-thrombin is a multifunctional serine proteinase generated by coagulation pathways and plays an important role in blood hemostasis by catalyzing conversion of fibrinogen to fibrin (Fenton, II, 1988). Thrombin also regulates several fundamental physiological and pathological activities during vascular injury and recovery, including barrier function, leukocyte adhesion, and mitogenesis, and it activates a variety of pro-inflammatory activities. Endothelial cells express thrombin receptor(s), which belong to a relatively novel subset of GPCRs, the proteinase-activated receptors (PAR1, 2, 3, and 4) that are activated by enzymatic proteolysis of the receptor (Vu *et al*., 1991; Hamilton *et al*., 2001; Kim *et al*., 2004). Only PAR1, 3, and 4 are activated by thrombin (as well as by trypsin); whereas PAR2 is activated by trypsin and tryptase (Hollenberg and Compton, 2002). Evidence indicates that PAR1 expressed by the endothelium is coupled to multiple members of the four G protein families, Gi, Gs, Gq, and G12, such that thrombin engagement of PAR1 elicits the proteolysis and subsequent receptor activation, leading to regulation of the appropriate G protein(s) (Manolopoulos *et al*., 1997; Gilchrist *et al*., 2001; Rahman *et al*., 2002; Vanhauwe *et al*., 2002). The net outcome of the signaling cascades activated by thrombin is believed to mediate activated endothelial responses, such as hyper-adhesion for leukocytes (Sugama and Malik, 1992; Murphy and Mcgregor, 1994; Kameda *et al*., 1997; Rahman *et al*., 2002), increased permeability (Lum *et al*., 1992a; Patterson and Garcia, 1994; Garcia *et al*., 1996; Manolopoulos *et al*., 1997; Shasby *et al*., 1997; Konstantoulaki *et al*., 2003; Birukova *et al*., 2004), increased secretory activity (Prescott *et al*., 1984; Ngaiza and Jaffe, 1991; Emori *et al*., 1992; Ueno *et al*., 1996; Holland *et al*., 1998), and mitogenesis (Herbert *et al*., 1994). Although evidence indicates that thrombin also activates PAR3 and PAR4, the specific endothelial responses activated remain to be determined (Macfarlane *et al*., 2001).

Bradykinin is a nine amino acid peptide belonging to the kinin family, a class of molecules, which mediate pain, inflammation, and hyperalgesia. Tissue injury activates a family of proteolytic enzymes, kallikrein, resulting in the proteolytic cleavage of the precursor protein, kininogen, which generates the naturally occurring kinins, Met-Lys-bradykinin (kallidin), Lys-bradykinin, and bradykinin itself (Regoli and Barabe, 1980). Endothelial cells express two receptor subtypes for bradykinin (B1 and B2) (Sung *et al*., 1988; Smith *et al*., 1995). The B2 receptor is generally constitutively expressed and mediates the actions of kallidin, Lys-bradykinin, and bradykinin; whereas the B1 receptor is induced following tissue injury or inflammatory challenge (i.e., endotoxins or cytokines) and mediates the actions of the metabolites of kallidin and bradykinin, desArg(10)-kallidin, desArg(9)-bradykinin (Prado *et al*., 2002). The B1 receptor has been reported to be constitutively active; whereas B2 exhibits little if any constitutive activity (Kang and Leeb-Lundberg, 2002).

Despite that both bradykinin receptor subtypes are coupled to Gi and Gq, there is evidence that they generate distinct signaling pathways (Smith *et al*., 1995; Austin *et al*., 1997; Prado *et al*., 2002). Stimulation of endothelial cells by bradykinin initiates an array of intracellular signaling responses predominantly through the B2 receptor, which activates increased endothelial permeability (Regoli and Barabe, 1980; Ehringer *et al*., 1996; Wahl *et al*., 1996; Aschner *et al*., 1997), as well as increased release of secretory products such as NO (Liao and Homcy, 1993;

Gooch and Frangos, 1996), O_2^- (Holland et al., 1990), PAF (Mcintyre et al., 1985), and prostacylin (Mcintyre et al., 1985). Interestingly, B1 but not B2, mediates endothelial proliferative responses (Morbidelli et al., 1998).

Vasoactive histamine disrupts vascular endothelial barrier function (Majno et al., 1961; Majno and Palade, 1961). Numerous reports based on the use of *in vitro* cell cultures (Rotrosan and Gallin, 1986; Niimi et al., 1992; Ikeda et al., 1999; Moy et al., 2000) and *in vivo* animal models, such as isolated pig coronary venules (Yuan et al., 1993), hamster cheek pouch (Sun and Mayhan, 2001), and rat blood brain barrier (Mayhan, 1996) have corroborated these early findings. Controversy still exist, however, regarding whether histamine increases pulmonary vascular permeability (Nakahara et al., 1979), or the increases in permeability is limited to the bronchial circulation (Pietra et al., 1971), or no effect at all on permeability (Pietra et al., 1979). Yet, histamine has been reported to increase permeability of *in vitro* cultures of pulmonary artery and microvessel endothelial cells (Mineau-Hanschke et al., 1990) as well as pulmonary artery intimal explants (Meyrick and Brigham, 1984). Histamine-activated endothelium also shows enhanced expression of the adhesion molecule, P-selectin, and the promotion of adhesion and rolling of leukocytes (Lorant et al., 1991; Asako et al., 1994). Further, histamine induces endothelial release of products such as prostacyclin (Mcintyre et al., 1985), PAF (Lorant et al., 1991), and von Willebrand factor (Jilma et al., 1998). To-date, three histamine receptor subtypes (H1, H2, and H3), which belong to the family of GPCRs, provide the mechanism of transduction (Bull et al., 1992; Arrang et al., 1995). Competition binding studies on intact blood vessels indicate that the relative expression of these receptor types is likely tissue (or regional) selective which varied with species as well (Heltianu et al., 1982; Carman-Krzan, 1989). Functional studies comparing bovine aortic, human aortic, and bovine vein endothelial cells provide further evidence of tissue and species heterogeneity of the histamine-induced responses (Ikeda et al., 1999). Evidence indicates that activation of the histamine receptor triggers multiple cell signaling cascades, presumably through coupling to multiple G proteins. The H1 subtype is primarily coupled to Gq since its activation results in increases in PLCβ-induced signaling molecules (i.e., $[Ca^{2+}]_i$), inositol phosphates (Arrang et al., 1995; Kuhn et al., 1996; Ikeda et al., 1999), and is likely the receptor subtype responsible for increases in permeability (Rotrosan and Gallin, 1986; Niimi et al., 1992) and leukocytic transmigration (Asako et al., 1994). However, the H2 receptor is coupled to both Gq and Gs, which transduce increases in both inositol phosphates and cAMP following activation (Delvalle et al., 1992; Arrang et al., 1995; Kuhn et al., 1996).

The endothelins are a family of potent vasoactive peptides (ET-1, ET-2, and ET-3) with a wide range of biological effects, mediated through the GPCRs, ETA and ETB. In vascular endothelium, both ETA and ETB are functionally coupled to Gq with activation of the effector, PLCβ, and to Gi with its effector, adenylate cyclase (Eguchi et al., 1993). Receptor activation promotes enhanced adhesion for neutrophils (Hayasaki et al., 1996), and cell motility (Noiri et al., 1997), impairs endothelial function (Kurose et al., 1993; Albertini et al., 2003), and releases products such as prostaglandin F2α (Stanimirovic et al., 1994), IL-8 (Zidovetzki et al., 1999), and metalloproteinase-2 (Salani et al., 2000). Further ET-1 is a mitogen for endothelial cells, inducing angiogenic responses in cultured endothelial cells through ETB and neovascularization *in vivo* in concert with VEGF (Morbidelli et al., 1995; Salani et al., 2000).

Nitric oxide (NO), a potent vasodilator, is a free radical gas produced by NO synthase in a variety of cells, including leukocytes, neurons, myocytes, and endothelial cells (Albrecht et al., 2003). NO transduces its endothelial responses through at least in part, by activation of the receptor kinase, guanylyl cyclase by direct binding to the heme moiety, generating cGMP (Morbidelli et al., 1996; Liu and Sundqvist, 1997; Volk et al., 1997; Martin et al., 2003). Tonic amounts of NO are vital for maintenance of homeostatic regulation of endothelial function, as well as for endothelial interaction with surrounding cells and tissues. However, perturbation of this balance of NO (i.e., increased or decreased levels) induces endothelial activation responses. Interruption of endogenous NO generation by cessation of flow or pharmacological inhibitors [N-G-nitro-L-arginine methyl ester (L- NAME)] impair endothelial barrier function (Liu and Sundqvist, 1997; Krizanac-Bengez *et al.*, 2003) and promote increased adhesion for leukocytes through enhanced expression of adhesion molecules such as P-selectin (Kubes *et al.*, 1991; Armstead *et al.*, 1997). On the other hand, NO production induced by vasoactive agents or injury, such as VEGF or wound edge, appears necessary for endothelial cell proliferation, migration, and barrier dysfunction, events associated with neovascularization and angiogenesis (Ziche *et al.*, 1994; Morbidelli *et al.*, 1996; Murohara *et al.*, 1998; Murohara *et al.*, 1999; Mark *et al.*, 2004). However, NO has also been shown to be a growth inhibitor under conditions of subconfluency (LopezFarre *et al.*, 1997). See Chapter 15 for extensive discussion.

Cytokines and Chemokines

Cytokines and chemokines are secreted proteins integral in regulation of the immune response and as such, are integral in regulation of cell growth, differentiation, and immune cellular trafficking. The interleukin cytokines transduce signaling cascades through two cytokine receptor families, Class I and Class II, characterized by their homology of extracellular binding domains (Bazan, 1990). Several interleukins transduce their signaling through binding with Class I receptors; whereas, interferons, IL-10, and coagulation factor VIIa mediate through Class II receptors (Kotenko and Pestka, 2000). Cytokine receptors generally do not have intrinsic tyrosine kinases, but rather form stable complexes with cytoplasmic tyrosine kinases [i.e., Janus Kinases (JAKs), Signal Transducers and Activators of Transcription (STATs)] upon ligand binding. Much evidence indicates that several interleukin cytokines (i.e., IL-1, IL-3, IL-4, IL-10) upregulate adhesion molecule expression on endothelial cells (Bevilacqua *et al.*, 1985; Khewgoodall *et al.*, 1996; Vora, Romero *et al.*, 1996;Yao *et al.*, 1996; Tamaru and Narumi, 1999), and induce release of other cytokines (Burns and Furie, 1998; Ranta *et al.*, 1999; Garcia *et al.*, 2000) and oxidants (Matsubara and Ziff, 1986). Further, IL-1, IL-2, and IL-6 impair endothelial barrier function (Bucana *et al.*, 1988; Downie *et al.*, 1992; Desai *et al.*, 2002).

TNF is a potent pro-inflammatory cytokine eliciting multiple biological effects, including immunomodulation, proliferation, differentiation, apoptosis, and septic shock. It is comprised of two homologous proteins. There are two forms, of which TNFα is secreted primarily by monocytic phagocytes and TNFβ by lymphocytes. However, other cell types also secrete TNFα, including neutrophils, mast cells, microglia, and endothelial cells. These cytokines form homotrimers which bind to TNF receptors, TNFR1 (p55) or TNFR2 (p75) (Tartaglia *et al.*, 1991), inducing trimerization of the receptors, recruitment of several proteins including death domain proteins, and subsequent activation of transcription factor NFκB and c-Jun N-terminal kinase (Hsu *et al.*, 1996). It appears that TNFR1 mediates the majority of the actions of TNF. The activation of endothelial cells by TNF increases adhesion for and transmigration of leukocytes through upregulation of adhesion molecules such as ICAM-1 and selectins (Wyble *et al.*, 1997; Kelly *et al.*, 1998; Rahman *et al.*, 2000). TNF is a highly effective inducer of the synthesis and release of cytokines, chemokines, and oxidants by the endothelium (Marumo *et al.*, 1993; Yoshida *et al.*, 1997; Goebeler *et al.*, 1999; Garcia *et al.*, 2000; Frey *et al.*, 2002). TNF is also angiogenic. One report indicates that administration of anti-IL-8, anti-VEGF, and anti-bFGF antibodies inhibited TNF-induced endothelial tubular morphogenesis, suggesting that the angiogenesis is indirectly controlled through paracrine or autocrine mechanisms (Yoshida et al., 1997). Not surprisingly, TNF directly impairs endothelial barrier function (Wojciak-Stothard et al., 1998; Wong et al., 1999; Ferro et al., 2000; Petrache et al., 2001).

Chemokines (or chemoattractant cytokines) are members of a large superfamily of secreted proteins, which were originally discovered through their ability to recruit leucocytes to the inflammatory sites. The over 50 known types of chemokines are low molecular weight (8-15 kD) polypeptides classified into four major groups by the number and spacing of interchain disulfide bonds: CC (e.g., MCP-1, RANTES), CXC (e.g., 1L-8, GRO, NAP-2), C (Lymphotactin), and CX3XC (Fractalkine). These chemokines attract leukocytes by binding to and activating GPCRs specific for each major group of chemokines expressed on the leukocyte surface (Ono et al., 2003). However, chemokines can also directly activate endothelial cell activities such as migration, proliferation, angiogenesis (Szekanecz et al., 1994; Heidemann et al., 2003) and impair barrier function (Biffl et al., 1995; Conti et al., 1997; Stamatovic et al., 2003). Receptors for both CXC and CC chemokines are expressed by several endothelial cell types (Berger et al., 1999; Murdoch et al., 1999; Salcedo et al., 2000; Salcedo et al., 2001; Heidemann et al., 2003). It is generally believed that the chemokine receptors transduce signals through predominantly pertussis toxin-sensitive GPCRs (Ono et al., 2003).

Growth Factors

VEGF is an endothelial cell-specific angiogenic and vasculogenic mediator, activating proliferation, increases in permeability, cell motility, and synthesis and secretion of proteases and other regulatory molecules (Hippenstiel et al., 1998; Wu et al., 1999; Montrucchio et al., 2000; Bernatchez et al., 2001a,b; Robinson and Stringer, 2001). Six isoforms of human VEGF ranging from 121-206 amino acids have been identified which bind to VEGF receptors (VEGFR-1, VEGFR-2, VEGFR-3) almost exclusively expressed on endothelial cells (Robinson and Stringer, 2001). Targeted inactivation of a single VEGF allele or loss of its receptor results in embryonic lethality caused by abnormal blood vessel development (Robinson and Stringer, 2001). The receptors are characterized by seven extracellular Ig-like domains, a membrane-spanning region, and a conserved cytoplasmic tyrosine kinase (Robinson and Stringer, 2001).

Other growth factors such as Fibroblast Growth Factor (FGF), Hepatocyte Growth Factor (HGF), and Transforming Growth Factor-β (TGF-β) also activate endothelial cells. HGF, a recognized angiogenic factor and endothelial chemoattractant (Koch et al., 1996; Sengupta et al., 2003), also paradoxically enhances endothelial barrier function (Liu et al., 2002). TGF-β regulates a plethora of cellular function and is critical for numerous developmental and homeostatic controls. Evidence indicates that TGF-β has anti-inflammatory action in that it inhibits cytokine production and neutrophil transmigration (Smith et al., 1996; Flanders et al., 1997), but promotes endothelial barrier dysfunction (Hurst et al., 1999). The activation of endothelial cells by FGF elicits primarily endothelial barrier dysfunction (Dye et al., 2001; Halama et al., 2001), cell motility, (Landgren et al. 1998; Tanaka et al., 1999; Kanda et al., 2000), and proliferation (Landgren et al., 1998; Tanaka et al., 1999).

Lysophospholipids

These highly bioactive lipids have come to be recognized as mediators of numerous fundamental processes including angiogenesis, immunity, atherosclerosis, tumorigenisis, and neuronal survival. The lysophospholipids (LPs) are glycerol- or sphingosine-based lipids generated from membrane phospholipids as part of normal physiological activities or disease processes (Hla *et al.*, 2001). The LPs include lysophosphatic acid (LPA), lysophosphatidylcholine (LPC), sphingosylphosphorylcholine (SPC), and sphingosine-1-phosphate (S1P). In recent years, a subfamily of eight GPCRs has been identified for LPA and S1P, namely LPA_{1-3} and $S1P_{1-5}$, respectively (also known as endothelial differentiation gene or EDGs) (Chun *et al.*, 2002). In endothelial cells, LPA and S1P activate the transcription factor, nuclear factor kappa B (NFκB), generate production of cytokines MCP-1 and IL-8, upregulate expression of adhesion molecules VCAM-1, E-selectin, and ICAM-1 as well as increase leukocyte adhesion to the endothelial cell surface (Palmetshofer *et al.*, 1999; Rizza *et al.*, 1999). Further, they are mitogenic, and in particular, S1P is a potent endothelial cell chemotactic agent, and therefore, these lipids are proposed to be critical regulators in wound healing processes (Lee *et al.*, 2000; Panetti *et al.*, 2000). LPA and S1P appear to promote endothelial barrier restrictiveness (Alexander *et al.*, 1998; English *et al.*, 1999; Lee *et al.*, 1999), but LPA can also impair endothelial barrier function as well (Schulze *et al.*, 1997; van Nieuw Amerongen *et al.*, 1998). These seemingly opposing actions of LPA underscore our incomplete understanding of the function, regulation, and physiological role of these LPs vascular biology.

Another subfamily of four GPCRs (GPR4, G2A, OGR1, and TDAG8) has been proposed to mediate the multitude of actions of SPC and LPC (Xu, 2002). To-date, GPR4 has been shown to be expressed by vascular endothelial cells (Lum *et al.*, 2003). The predominant actions of LPC on the vascular endothelium are pro-inflammatory, inducing upregulation of adhesion molecules (Kume *et al.*, 1992; Murohara *et al.*, 1996; Zhu *et al.*, 1997), and production and release of cytokines, oxidants, and other inflammatory factors (Takahara *et al.*, 1996; Takeshita *et al.*, 2000; Inoue *et al.*, 2001; Rikitake *et al.*, 2001; Murugesan *et al.*, 2003). Although LPC is a normal circulating plasma lipid, it accumulates in inflammatory tissues such as ischemic myocardium, atherosclerotic aortas, and endometriosis (Sobel *et al.*, 1978; Katz and Messineo, 1981; Murphy *et al.*, 1998). SPC, however, appears to regulate endothelial responses predominantly associated with proliferation and angiogensis (Boguslawski *et al.*, 2000). Little information is currently available regarding the physiological and pathological functions, regulation, and transduction mechanisms of SPC and LPC in vascular endothelial cell function.

Oxidants

Oxidative stress (increased generation of reactive oxygen species such as O_2^-, ·OH, and H_2O_2) is a serious causative factor in the development of numerous vascular endothelial disorders and plays an important role in the pathophysiology of vascular diseases including atherosclerosis, diabetes, neuronal disorders, and lung injury. Oxidants are well documented to function as signaling molecules, stimulating cellular activities ranging from cytokine secretion to cell proliferation, and at higher concentrations can induce cell injury and death by oxidant modification of proteins and carbohydrates, lipid peroxidation, and DNA strand nicks. Such diverse responses are related to multiple factors, such as the oxidant species prevailing at the inflammatory locus, concentration and turnover of the oxidants, and the antioxidant capacity of the local environment as well as target cells.

There is substantial evidence that oxidants promote increased endothelial adhesiveness for leukocytes and vascular permeability (Lum and Roebuck, 2001). A major oxidant source is from blood leukocytes, which become activated and adherent to the endothelial cell surface. Further, endothelial cells activated by various mediators (i.e., bradykinin, TNF, IL-1) can produce intracellular oxidants through oxidant-generating systems such as the NADPH oxidase complex

(Matsubara and Ziff, 1986; Holland et al., 1990; Frey et al., 2002; Gertzberg et al., 2004)

Despite much information regarding the actions of oxidants, the actual transduction pathways, which lead to activation of signaling cascades and the subsequent cellular activity remain somewhat obscure. There is convincing evidence that low sub-cytotoxic levels of oxidants function as important physiologic signaling molecules mediating basic cellular activities such as cell growth and differentiation (Natarajan, 1995; Kamata and Hirata, 1999). Oxidants are ubiquitous and likely regulate signaling cascades at multiple points. For example, oxidants activate phospholipases PLA_2, PLC, and PLD, which in turn generate a host of cellular messengers and cofactors that regulate downstream cellular activities, including protein kinases and phosphatases (Natarajan, 1995). Additionally, oxidants may directly modulate activity of the downstream molecules. There is also evidence suggesting that sublethal levels of oxidant stress inactivate tyrosine and serine/threonine protein phosphatases, contributing to increased activities of protein kinases in several signaling pathways (Whisler et al., 1995).

Evidence indicates that endothelium exposure to oxidants produces oxidized phospholipids, which are potent pro-inflammatory mediators, and may in turn activate cellular signals in an autocrine fashion. Endothelial cells exposed to H_2O_2 were produced PAF, which is released extracellularly and binds to the endothelial surface to mediate arrest of rolling leukocytes (Lewis et al., 1988; Lorant et al., 1991; Kuijpers et al., 1992; Rainger et al., 1997). Accumulating evidence indicates that oxidants induce fragmentation of cell membrane phosphatidylcholine species, generating bioactive phospholipids in greater abundance and with greater potency than PAF (Tokumura et al., 1992; Leitinger et al., 1999; Marathe et al., 1999). Species of these oxidized phosphatidylcholines upregulated E-selectin and VCAM-1, increasing neutrophil and monocyte adhesion to endothelia (Leitinger et al., 1999).

Activation by Microorganisms

The average human vascular endothelial surface area has been estimated to exceed $1000\ m^2$, ~600-times that of the epidermis (Muller and Griesmacher, 2000). In blood-borne infections, this vast surface is exposed to a large number and variety of microorganisms (i.e., protozoa, fungi, bacteria, and viruses). Exposure of the endothelium to microorganisms activates profound changes in gene expression and function, which determine the state of immunity, inflammation, thrombosis, and pathogenic outcome. We will introduce some of the pathogens and specific responses and Chapter 13 will describe pathogenic mechanisms and consequences in more detail.

Bacterial Infection

Bacterial infection and the consequent activation of the endothelium involve multiple microbial components, with the mechanisms leading to pathogenesis likely microorganism-specific. Table 2 lists some of the bacteria and bacterial products, shown to activate endothelial cell responses. Microbial antigens such as Gram-negative bacterial lipopolysaccharide (endotoxin/LPS component of the outer envelope), Gram-positive bacterial cell wall components, such as soluble peptidoglycan, lipoteichoic acid, and bacterial lipoproteins, initiate a network of host-derived pro-inflammatory mediators, and if uncontrolled ultimately can lead to development of sepsis, cardiovascular shock and death. These bacterial components are now recognized to specifically bind to and activate a family of at least eight pattern recognition receptors called toll-like receptors (TLR 1-8) involved in innate immune recognition and cellular activation (Henneke

Table 2. Bacteria-Induced Endothelial Activation Responses

Bacterium/Bacterial Components	Activation Response[1]	References
Borrelia crocidurae	+ ICAM-1, E-selectin	Shamaei-Tousi et al., 2000
C3-transferase toxin from *Clostridium botulinum*	RhoA, RhoB, and RhoC inactivation, abolished actin stress fibers/focal adhesions	Aepfelbacher et al., 1997
Cytotoxic necrotizing factor 1 toxin	Rho GTPase activation, actin reorganization	Chung et al., 2003
EDIN toxin from *Staphylococcus aureus*	RhoA, RhoB, and RhoC inactivation, abolished actin stress fibers/focal adhesions	Aepfelbacher et al., 1997
Escherichia coli	CD11/CD18-dependent neutrophil emigration in lungs	Tasaka et al., 2003
Francisella tularensis	+ VCAM-1 and ICAM-1, + Chemokine CXCL8	Forestal et al., 2003
Listeria monocytogenes	+ VCAM-1, E-selectin and ICAM-1 + Neutrophil adhesion	Drevets, 1997
LPS – Lipopolysaccharide = Gram negative bacterial endotoxin	Angiogenesis + cJun N-terminal kinase + p38 MAPK + NF-κB + E-selectin + ICAM-1 + Chemokines: MCP-1, IL-8, & RANTES MIP-1β, Fractalkine + Cytokines: TNFα + Barrier dysfunction Actin reorganization + Matrix metalloproteinase-2 Neutrophil adhesion + Endothelin-1 + iNOS	Pollet et al., 2003 Bannerman et al., 1998; Faure et al., 2000 Hippenstiel et al., 2000; Kan et al., 2002 Huang et al., 1995; Wong and Dorovinizis, 1996 Yan et al., 2002 Beck, Yard et al., 1999; Garcia, Xia et al., 2000 Harrison et al., 1999; Shukaliak and Dorovini-Zis, 2000; Zhao et al., 2001 Ranta et al., 1999; Imaizumi et al., 2000 Bannerman et al., 1998; Gaillard, et al., 2003 Goldblum et al., 1993 Kim and Koh, 2000 Essler et al., 2000 Douthwaite et al., 2003 Kan et al, 2002
Neisseria meningitidis	+ Endothelial NOS	Constantin et al., 2002
Neisseria meningitidis, LPS-deficient strain	+ E-selectin	Dixon et al., 2004
Staphylococcus aureus	+ VCAM-1, ICAM-1, and MHC I + adhesiveness for monocytes and granulocytes	Beekhuizen et al., 1997
Streptococcus pneumoniae	CD11/CD18-independent neutrophil emigration in lungs	Tasaka et al., 2003
Toxin A from Clostridium *difficile*	RhoA, RhoB, and RhoC inactivation, abolished actin stress fibers/focal adhesions	Aepfelbacher et al., 1997

[1] + indicates increase, TNFα = Tumor Necrosis Factor α, MCP-1 = Monocyte Chemoattractant Protein-1, NOS = Nitric oxide synthase, tPA = Tissue Plasminogen Activator

and Golenbock, 2002). Endothelial cells express several of these TLRs (Faure et al., 2000). The highly pro-inflammatory actions of LPS have been most studied and evidence indicates that TLR4 is the primary receptor for LPS, whereas TLR2 has been implicated as the receptor for Gram-positive and fungal cell wall components and for bacterial, mycobacterial, and spirochetal lipoproteins (Faure et al., 2001). Direct activation of endothelium by LPS requires expression of TLR4 and recruitment of several cytoplasmic proteins (MyD88, Toll/IL-1 receptor associated protein, Toll-interacting protein, and Rac1/phosphatidylinositol 3-kinase), resulting in the activation of the transcription factor, NFκB, and transcription of pro-inflammatory genes such as cytokines (i.e., IL-6, IL-8, IL-1β), adhesion molecules (i.e., E-selectin, ICAM-1, VCAM-1), and tissue factor (Faure et al., 2000; Henneke and Golenbock, 2002). The importance of endothelial TLR4 is underscored by the finding that in endothelium of TLR4(-/-) mice, little increase in neutrophil sequestration in lungs occurred in response to LPS, in contrast to the rapid and selective accumulation of neutrophils into control lungs, a condition found in pulmonary failure such as that associated with sepsis (Andonegui et al., 2003). Recent evidence suggests that TLR4 may not be exclusive for LPS, but also oxidized phosphatidylcholine induce synthesis of IL-8 in endothelial cells through this receptor (Walton et al., 2003).

Bacterial protein toxins can also regulate endothelial activities by direct modification of members of the family of Rho GTPases, which are critical determinants of actin cytoskeleton function and therefore, in fundamental cellular activities. Several toxins inhibit Rho function by ADP-ribosylation or glucosylation and activate them by deamidation and transglutamination (Lerm et al., 2000). These toxins have been widely used as research tools to for study of the Rho signaling pathways. In cultured endothelial cells, the inactivation of Rho GTPases by the toxins C3-transferase from *Clostridium botulinum*, EDIN from *Staphylococcus aureus*, and toxin A from *Clostridium difficile* blocked migration of human umbilical vein endothelial cells in an *in vitro* wound repair assay (Aepfelbacher et al., 1997). Accumulating evidence indicates that these bacterial toxins may either exacerbate or protect against endothelial barrier dysfunction, dependent on which Rho GTPases are activated (or inhibited) (Essler et al., 1998a,b; Vouret-Craviari et al., 1999; Adamson et al., 2002).

Viral Infection

There is clear evidence that infection by viruses activates the vascular endothelium and some of these known viruses are listed in Table 3. In human immunodeficiency virus-1 (HIV-1)-infected patients, the viral load correlated significantly with vWF and tissue-type plasminogen activator (t-PA) (de Larranaga et al., 2003), two recognized endothelial cell products associated with activation. Direct infection of *in vitro* cultures of endothelial cells with viruses such as human herpesvirus 6 (Caruso et al., 2003) and human cytomegalovirus (Shahgasempour et al., 1997) is reported to induce activation responses, which include production and release of cytokines and upregulation of adhesion molecules. While specific pathogenetic mechanisms of viral-mediated diseases remain to be fully elucidated, it is evident that the vascular endothelium contributes to the direct and indirect regulation of the inflammatory, immune, and coagulation systems that determine the outcome of the disease.

Table 3. Virus-Induced Endothelial Activation Responses

Virus	Activation Response[1]	References
Bluetongue virus	IL-1, IL-6, IL-8; Cyclooxygenase-2, and iNOS	DeMaula et al., 2002
Caprine arthritis-encephalitis virus	Leukocyte transmigration	Milhau et al., 2003
Coxsackievirus	ICAM-1, VCAM-1; IL-6, IL-8, and TNF-alpha	Zanone et al., 2003
Cytomegalovirus, human	ICAM-1, ELAM-1, VCAM-1 Barrier dysfunction IL-8, leukocyte transmigration CD40	Altannavch et al., 2002; Shahgasempour et al., 1997 Scholz et al., 1999 Craigen et al., 1997 Maisch et al., 2002
Dengue virus	tPA	Huang et al., 2003; Warke et al., 2003
Hantaviruse	IL-8, IL-6, GRO-β, ICAM-1	Geimonen et al., 2002
Herpes virus 6, human	RANTES	Caruso et al., 2003
Herpes simplex virus type 1	Receptor for complement C3b	Smiley et al., 1985
Human immunodeficiency virus-1 protein, Tat	interleukin-6 mRNA	Zidovetzki et al., 1998
Human immunodeficiency virus-1	tPA, vWF	de Larranaga et al., 2003
PVC-211 murine leukemia virus	iNOS	Jinno-Oue et al., 2003

[1] iNOS=inducible nitric oxide synthase, vWF=von Willebrand factor, tPA=Tissue Plasminogen Activator

Activation by Mechanical Stimuli

Stress and Response

  Endothelial cells are located at a unique interface between the vessel wall and the flowing blood, constantly exposed to fluid mechanical forces. These forces, occurring in parallel (laminar shear stress) and tangential (pressure or distension) to the vessel wall can affect endothelium, and excess or disturbed force can cause activation with remodeling of the endothelial structure and alteration in its function. In this context, endothelial cells may behave as hemodynamic sensors, translating mechanical information from the blood flow into intracellular biochemical signals. Activated endothelial cells change cell shape, alignment of microfilament network, and remodeling of membrane surface topography in the direction of flow (Levesque and Nerem, 1985; Wechezak et al., 1985; Barbee et al., 1994; Birukov et al., 2002). This cellular alignment is associated with regulated disassembly/assembly of cell-cell junctions (Schnittler et al., 1997; Noria et al., 1999) and cell-matrix adhesion protein complexes (Thoumine et al., 1995; Tzima et al., 2001). Physiological or low levels of shear stress appear to be important in maintenance of the endothelial barrier. Direct application of low shear stress to cultured endothelial monolayers enhanced barrier function as determined by transendothelial resistance (Seebach et al., 2000;

Krizanac-Bengez et al., 2003). However, increased shear stress (>10 dynes/cm^2) increased the hydraulic conductivity of the bovine endothelial-retinal barrier (Lakshminarayanan et al., 2000; Seebach et al., 2000). Interestingly, shear stress as low as 1 dyne/cm^2 was sufficient to increase albumin permeability across bovine aortic endothelial cells (Jo et al., 1991). These studies suggest differential sensitivity to fluid shear stress by different vascular beds.

Shear stress of the endothelium also induces secretion of several biologically important mediators such as NO (Knudsen and Frangos, 1997; Malek et al., 1999; Boo et al., 2002; Krizanac-Bengez et al., 2003), von Willebrand factor, reactive oxygen species (Chiu et al., 1997), and ET-1 (Kuchan and Frangos, 1993). There is clear evidence that shear stress upregulates many endothelial genes (Resnick et al., 1997), including Epidermal Growth Factor-Like Growth Factor (Morita et al., 1993), endothelial NO synthase (Uematsu et al., 1995), Cu/Zn SOD (Inoue et al., 1996) and tissue factor (Lin et al., 1997). Further, pressure distension of human saphenous veins induces upregulation of ICAM-1, VCAM-1, and P-selectin, with subsequent increase in adhesion of both unstimulated and stimulated neutrophils to the endothelium (Chello et al., 2003). In another study, the degree of mechanical stress (product of duration and degree of pressure) correlated with increased levels of coronary artery ET-1 (Hasdai et al., 1997), indicating that pressure induced secretion of ET-1 in these cells.

Response Mechanisms and Chronic Disease

Definitive transduction mechanisms by which the mechanical stimuli mediate endothelial activation responses are unknown. Stretch-activated cation channels are proposed to act as mechanosensors for changes in hemodynamic forces. For instance, an 18 pS cation channel responded with a multi-fold increase in channel activity when positive pressure was applied to the luminal endothelial cell surface in intact arteries (Hoyer et al., 1996). Alternatively, mechanotransduction may also involve a conformational activation of integrin $\alpha v \beta 3$, resulting in increased binding to extracellular cell matrix proteins (Tzima et al., 2001). The transduction of mechanical stresses, in turn, triggers several signaling pathways (Ishida et al., 1997). These signaling cascades include activation of PLA$_2$ (Nollert et al., 1990; Berk et al., 1995; Pearce et al., 1996), Ca^{2+} transients (Schwarz et al., 1992; Ando et al., 1993), protein kinases (Hu and Chien, 1997; Jalali et al., 1998), and the family of Rho GTPases (Li et al., 1999).

Altered mechanical stresses play a particularly influential role in chronic lung diseases (such as pulmonary hypertension), contributing to remodeling pulmonary arteries, veins and even the bronchial circulation (Johnson et al., 1997). Pulmonary hypertension *per se* augmented NO-dependent arterial dilation and upregulation of arterial eNOS in lungs (Resta et al., 1997). Increased pulmonary blood flow in an experimental model of congenital heart disease indicated upregulation of VEGF receptors in small pulmonary arteries, which was accompanied by increased vessel number (Mata-Greenwood et al., 2003). Regulation of other angiogenesis-related genes and apoptosis-related genes have been implicated as well. Recently, pulmonary hypertension is shown to be associated with loss of peroxisome proliferator-activated receptor gamma (PPARγ) expression in endothelial cells, which was characterized by an abnormal, proliferating, apoptosis-resistant phenotype (Ameshima et al., 2003).

Provocation and Perspective

It is quite apparent that the vascular endothelium has vitally important physiological functions such as normal maintenance of host defense, hemostasis, and fluid balance, in addition to its primary function as a nutrient delivery system. Its physical location is ideal, that is at the blood-cell interphase, where it responds exquisitely to a host of activation stimuli in the circulation, ranging from cell-derived mediators to cells to mechanical changes of the blood vessels. As will be discussed in the later chapters, the ability of the endothelium to respond to and to elicit appropriate activation responses is attributed to its extensive array of receptors and transduction cascades, providing acute and chronic regulatory signals. Equally apparent is that activation of the endothelium can be inappropriate or unregulated, leading to serious pathological consequences, such as inflammatory injury, tissue edema, or thrombosis. Thus, the health of an organ is intimately dependent on the vascular endothelium functioning properly. Inappropriate activation responses can be manifested by events at cell apical membrane, junctional complexes, or focal adhesions, and mechanisms of regulation at the genomic and protein levels at these cellular sites remains to be fully elucidated. At present, there are no reliable markers of endothelial activation. Some markers of endothelial activation have been proposed (such as Factor VIII) to predict potential adverse endothelial events, but it is realized that any one marker of endothelial activation would not be sufficient to predict the varied activation responses elicited.

Anf finally, of great interest are known endogenous ligands (i.e., angiopoietin-1, S1P) as well as signaling systems (i.e, cAMP), which can either limit or prevent endothelial activation responses such as barrier dysfunction. These pathways may provide the braking (or reversal?) systems to maintain proper physiological control. Indeed, questions remain as to whether these braking systems are also regulated as part of the normal activation responses.

ACKNOWLEDGEMENT
This work was supported by NIH HL-071081.

REFERENCES

Abe, Y., Sugisaki, K., and Dannenberg, A.M. (1996) Rabbit vascular endothelial adhesion molecules: ELAM-1 is most elevated in acute inflammation, whereas VCAM-1 and ICAM-1 predominate in chronic inflammation. J. Leu. Biol. 60:692-703.

Adamson, R.H., Curry, F., Adamson, G., Liu, B., Jiang, Y., Aktories, K., Barth, H., Daigeler, A., Golenhofen, N., Ness, W., and Drenckhahn, D. (2002) Rho and rho kinase modulation of barrier properties: cultured endothelial cells and intact microvessels. J. Physiol. 539:295-308.

Aepfelbacher, M., Essler, M., Huber, E., Sugai, M., and Weber, S.C. (1997) Bacterial toxins block endothelial wound repair - Evidence that Rho GTPases control cytoskeletal rearrangements in migrating endothelial cells. Arterioscler. Thromb. Vasc. Biol. 17:1623-1629.

Albertini, M., Clement, M.G., and Hussain, S.N. (2003) Role of endothelin ETA receptors in sepsis-induced mortality, vascular leakage, and tissue injury in rats. Eur. J. Pharmacol. 474:129-135.

Albrecht, E.W., Stegeman, C., Heeringa, P., Henning, R.H., van Goor, H. (2003) Protective role of endothelial nitric oxide synthase. J. Pathol. 199:8-17.

Alexander, J.S., Patton, W., Christman, B., Cuiper, L., and Haselton, F.R. (1998) Platelet-derived lysophosphatidic acid decreases endothelial permeability in vitro. Amer. J. Physiol. 274:H115-H122.

Allison, B.A., Nilsson, L., Karpe, F., Hamsten, A., and Eriksson, P. (1999) Effects of native, triglyceride-enriched, and oxidatively modified LDL on plasminogen activator inhibitor-1 expression in human endothelial cells. Arterioscler. Thromb. Vasc. Biol. 19:1354-1360.

Altannavch, T.S., Roubalova, K., Kucera, P., Juzova, O., and Andel, M. (2002) Effect of human cytomegalovirus and glucose on adhesion molecules expression in cultured human endothelial cells. Acta Virol. 46:183-186.

Amberger, A., Maczek, C., Jurgens, G., Michaelis, D., Schett, G., Trieb, K., Eberl, T., Jindal, S., Xu, Q., and Wick, G. (1997) Co-expression of ICAM-1, VCAM-1, ELAM-1 and Hsp60 in human arterial and venous endothelial cells in response to cytokines and oxidized LDL. Cell Stress Chaperones 2:94-103.

Ameshima, S., Golpon, H., Cool, C., Chan, D., Vandivier, R., Gardai, S., Wick, M., Nemenoff, R., Geraci, M., and Voelkel, N.F. (2003) PPARγ expression is decreased in pulmonary hypertension and affects endothelial cell growth. Circ. Res. 92:1162-1169.

Ando, J., Ohtsuka, A., Korenaga, R., Kawamura, T., and Kamiya, A. (1993) Wall shear stress rather than shear rate regulates cytoplasmic Ca^{++} responses to flow in vascular endothelial cells. Biochem. Biophys. Res. Commun. 190:716-723.

Andonegui, G., Bonder, C., Green, F., Mullaly, S., Zbytnuik, L., Raharjo, E., Kubes, P. (2003) Endothelium-derived Toll-like receptor-4 is the key molecule in LPS-induced neutrophil sequestration into lungs. J. Clin. Invest. 111:1011-1020.

Andriopoulou, P., Navarro, P., Zanetti, A., Lampugnani, M.G., Dejana, E. (1999) Histamine induces tyrosine phosphorylation of endothelial cell-to-cell adherens junctions. Arterioscl. Thromb. Vasc. Biol. 19:2286-2297.

Arai, H. and Escobedo, J.A. (1996) Angiotensin II type 1 receptor signals through Raf-1 by a protein kinase C-dependent, Ras-independent mechanism. Molec. Pharmacol. 50:522-528.

Aramori, I. and Nakanishi, S. (1992) Coupling of two endothelin receptor subtypes to differing signal transduction in transfected Cho cells. J. Biol. Chem. 267:12468-12474.

Armstead, V.E., Minchenko, A., Schuhl, R., Hayward, R., Nossuli, T., and Lefer, A.M. (1997) Regulation of P-selectin expression in human endothelial cells by nitric oxide. Amer. J. Physiol. 42:H740-H746.

Arrang, J.M., Drutel, G., Garbarg, M., Ruat, M., Traiffort, E., and Schwartz, J.C. (1995) Molecular and functional diversity of histamine receptor subtypes. Ann. N.Y. Acad. Sci. 757:314-323.

Asako, H., Kurose, I., Wolf, R., Defrees, S., Zheng, Z., Phillips, M., Paulson, J., and Granger, D. (1994) Role of H1 receptors and P-selectin in histamine-induced leukocyte rolling and adhesion in postcapillary venules. J. Clin. Invest. 93:1508-1515.

Aschner, J.L., Lum, H., Fletcher, P.W., Malik, A.B. (1997) Bradykinin- and thrombin-induced increases in endothelial permeability occur independently of phospholipase C but require protein kinase C activation. J. Cell. Physiol. 173:387-396.

Austin, C.E., Faussner, A., Robinson, H., Chakravarty, S., Kyle, D., Bathon, J., and Proud, D. (1997) Stable expression of the human kinin B1 receptor in Chinese hamster ovary cells. Characterization of ligand binding and effector pathways. J. Biol. Chem. 272:11420-11425.

Balmforth, A.J., Lee, A., Shepherd, F., Warburton, P., Donnelly, D., and Ball, S.G. (1997) G-protein-coupled receptors for peptide hormones: angiotensin II. Biochem. Soc. Trans. 25:1041-1046.

Balyasnikova, I.V., Pelligrino, D., Greenwood, J., Adamson, P., Dragon, S., Raza, H., and Galea, E. (2000) Cyclic adenosine monophosphate regulates the expression of ICAM and iNOS in brain endothelial cells. J. Cereb. Blood Flow Metab. 20:688-699.

Bannerman, D.D., Sathyamoorthy, M., and Goldblum, S.E. (1998) Bacterial lipopolysaccharide disrupts endothelial monolayer integrity and survival signaling events through caspase cleavage of adherens junction proteins. J. Biol. Chem. 273:35371-35380.

Barbee, K.A., Davies, P.F., and Lal, R. (1994) Shear stress-induced reorganization of the surface topography of living endothelial cells imaged by atomic force microscopy. Circ. Res. 74:163-171.

Bazan, J.F. (1990) Structural design and molecular evolution of a cytokine receptor superfamily. PNS USA 87:6934-6938.

Beck, G.C., Yard, B., Breedijk, A., van Ackern, K., and van der Woude, F.J. (1999) Release of CXC-chemokines by human lung microvascular endothelial cells (LMVEC) compared with macrovascular umbilical vein endothelial cells. Clin. Exp. Immunol. 118:298-303.

Beekhuizen, H., vandeGevel, J., Olsson, B., vanBenten, I., and Vanfurth, R. (1997) Infection of human vascular endothelial cells with Staphylococcus aureus induces hyperadhesiveness for human monocytes and granulocytes. J. Immunol. 158:774-782.
Berger, O., Gan, X., Gujuluva, C., Burns, A., Sulur, G., Stins, M., Way, D., Witte, M., Weinand, M., Said, J., Kim, K.S., Taub, D., Graves, M.C., and Fiala, M. (1999) CXC and CC chemokine receptors on coronary and brain endothelia. Mol. Med. 5:795-805.
Berk, B.C., Corson, M., Peterson, T., Tseng, H. (1995) Protein kinases as mediators of fluid shear stress stimulated signal transduction in endothelial cells: a hypothesis for calcium-dependent and calcium-independent events activated by flow. J. Biomech. 28:1439-1450.
Bernatchez, P.N., Allen, B., Gelinas, D., Guillemette, G., and Sirois, M.G. (2001a) Regulation of VEGF-induced endothelial cell PAF synthesis: role of p42/44 MAPK, p38 MAPK and PI3K pathways. Brit. J. Pharmacol. 134:1253-1262.
Bernatchez, P.N., Winstead, M., Dennis, E., and Sirois, M.G. (2001b) VEGF stimulation of endothelial cell PAF synthesis is mediated by group V 14 kDa secretory phospholipase A2. Brit. J Pharmacol. 134:197-205.
Bevilacqua, M.P., Pober, J., Wheeler, M.E., Cotran, R.S., and Gimbrone, M.A., Jr. (1985) Interleukin 1 acts on cultured human vascular endothelium to increase the adhesion of polymorphonuclear leukocytes, monocytes, and related leukocyte cell lines. J. Clin. Invest. 76:2003-2011.
Biffl, W.L., Moore, E., Moore, F., and Barnett, C.C. (1996) Nitric oxide reduces endothelial expression of intercellular adhesion molecule (ICAM)-1. J. Surg. Res. 63:328-332.
Biffl, W.L., Moore, E., Moore, F., Carl, V., Franciose, R., Banerjee, A. (1995) Interleukin-8 increases endothelial permeability independent of neutrophils. J. Trauma 39:98-102.
Birukov, K.G., Birukova, A., Dudek, S., Verin, A., Crow, M., Zhan, X., DePaola, N., and Garcia, J.G. (2002) Shear stress-mediated cytoskeletal remodeling and cortactin translocation in pulmonary endothelial cells. Amer. J. Respir. Cell Mol. Biol. 26:453-464.
Birukov, K.G., Leitinger, N., Bochkov, V., Garcia, J.G. (2004) Signal transduction pathways activated in human pulmonary endothelium by OxPAPC, a bioactive component of ox-lipoproteins. Microvasc. Res. 67:18-28.
Birukova, A.A., Smurova, K., Birukov, K., Kaibuchi, K., Garcia, J.G., and Verin,A. (2004) Role of Rho GTPases in thrombin-induced lung vascular endothelial cells barrier dysfunction. Microvasc. Res. 67:64-77.
Blum, M.S., Toninelli, E., Anderson, J., Balda, M., Zhou, J., ODonnell, L., Pardi, R., and Bender, J.R. (1997) Cytoskeletal rearrangement mediates human microvascular endothelial tight junction modulation by cytokines. Amer. J. Physiol. 42:H286-H294.
Boguslawski, G., Lyons, D., Harvey, K., Kovala, A., and English, D. (2000) Sphingosyl-phosphorylcholine induces endothelial cell migration and morphogenesis. Biochem. Biophys. Res. Commun. 272:603-609.
Boo, Y.C., Hwang, J., Sykes, M., Michell, B., Kemp, B., Lum, H., and Jo, H. (2002) Shear stress stimulates phosphorylation of eNOS at Ser(635) by a protein kinase A-dependent mechanism. Amer. J. Physiol. 283:H1819-H1828.
Bove, K., Neumann, P., Gertzberg, N., and Johnson, A. (2001) Role of eNOS-derived NO in mediating TNF-induced endothelial barrier dysfunction. Amer. J. Physiol. 280:L914-L922.
Bucana, C.D., Trial, J., Papp, A., and Wu, K.K. (1988) Bovine aorta endothelial cell incubation with interleukin 2: morphological changes correlate with enhanced vascular permeability. Scanning Microsc. 2:1559-1566.
Bull, H.A., Courtney, P., Rustin, M., and Dowd, P. (1992) Characterization of histamine receptor subtypes regulating prostacyclin release from human endothelial cells. Brit. J. Pharmacol. 107:276-281.
Bullard, D.C., Kunkel, E., Kubo, H., Hicks, M., Lorenzo, I., Doyle,N., Doerschuk, C., Ley, K., and Beaudet, A.L. (1996) Infectious susceptibility and severe deficiency of leukocyte rolling and recruitment in E-selectin and P-selectin double mutant mice. J. Exp. Med. 183:2329-2336.
Burns, M.J. and Furie, M.B. (1998) Borrelia burgdorferi and IL-1 promote the transendothelial migration of monocytes in vitro by different mechanisms. Infect. Immun. 66:4875-4883.
Burns, M.J., Sellati, T., Teng, E., and Furie, M.B. (1997) Production of IL-8 by cultured endothelial cells in response to Borrelia burgdorferi occurs independently of secreted IL-1 and TNFα and is required

for subsequent transendothelial migration of neutrophils. Infect. Immun. 65:1217-1222.
Cai, H., Li, Z., Dikalov, S., Holland, S., Hwang, J., Jo, H., Dudley, S., and Harrison, D.G. (2002) NAD(P)H oxidase-derived hydrogen peroxide mediates endothelial nitric oxide production in response to angiotensin II. J. Biol. Chem. 277:48311-48317.
Carman-Krzan, M. (1989) Histaminergic H_1-receptors in smooth muscle and endothelium of bovine thoracic aorta. Agents Actions 27:198-201.
Carmeliet, P., Collen, D. (2000) Molecular basis of angiogenesis. Role of VEGF and VE-cadherin. Ann. N.Y. Acad. Sci. 902:249-262.
Carter, N.M., Ali, S., and Kirby, J.A. (2003) Endothelial inflammation: the role of differential expression of N-deacetylase/N-sulphotransferase enzymes in alteration of the immunological properties of heparan sulphate. J. Cell Sci. 116:3591-3600.
Caruso, A., Favilli, F., Rotola, A., Comar, M., Horejsh, D., Alessandri, G., Grassi, M., Di Luca,D., Fiorentini, S. (2003) Human herpesvirus-6 modulates RANTES production in primary human endothelial cell cultures. J. Med. Virol. 70:451-458.
Chang, Y.S., Munn, L., Hillsley, M., Dull, R., Yuan, J., Lakshminarayanan, S., Gardner, T., Jain,R., and Tarbell, J.M. (2000) Effect of vascular endothelial growth factor on cultured endothelial cell monolayer transport properties. Microvasc. Res. 59:265-277.
Chaudhuri, P., Colles, S., Damron, D., and Graham, L.M. (2003) Lysophosphatidylcholine inhibits endothelial cell migration by increasing intracellular calcium and activating calpain. Arterioscler. Thromb. Vasc. Biol. 23:218-223.
Chello, M., Mastroroberto, P., Frati, G., Patti, G., D'Ambrosio, A., Di Sciascio, G., and Covino, E. (2003) Pressure distension stimulates the expression of endothelial adhesion molecules in the human saphenous vein graft. Ann. Thorac. Surg. 76:453-458.
Cheng, J.J., Chao, Y., Wung, B., and Wang,D. L. (1996) Cyclic strain-induced PAI-1 release from endothelial cells involves ROS. Biochem. Biophys. Res. Commun. 225:100-105.
Chen, X.L., Zhang, Q., Zhao, R., Ding, X., Tummala, P., and Medford, R.M. (2003) Rac1 and superoxide are required for the expression of cell adhesion molecules induced by tumor necrosis factor-alpha in endothelial cells. J. Pharmacol. Exp. Ther. 305:573-580.
Chiu, J.J., Wung, B., Shyy, J., Hsieh, H., and Wang, D.L. (1997) Reactive oxygen species are involved in shear stress- induced ICAM-1 expression in endothelial cells. Arterioscler. Thromb. Vasc. Biol. 17:3570-3577.
Chua, C.C., Hamdy, R.C., and Chua, B.H. (1996) Angiotensin II induces TIMP-1 production in rat heart endothelial cells. Biochim. Biophys. Acta 1311:175-180.
Chua, C.C., Hamdy, R.C., and Chua, B.H. (1998a) Upregulation of endothelin-1 production by lysophosphatidic acid in rat aortic endothelial cells. Biochim. Biophys. Acta 1405:29-34.
Chua, C.C., Hamdy, R.C., and Chua, B.H. (1998b) Upregulation of vascular endothelial growth factor by H_2O_2 in rat heart endothelial cells. Free Rad. Biol. Med. 25:891-897.
Chun, J., Goetzl, E., Hla, T., Igarashi, Y., Lynch, K., Moolenaar, W., Pyne, S., and Tigyi, G. (2002) Lysophospholipid receptor nomenclature. Pharmacol. Rev. 54:265-269.
Chung, J.W., Hong, S., Kim, K., Goti, D., Stins, M., Shin, S., Dawson, V., Dawson, T., and Kim, K.S. (2003) 37-kDa laminin receptor precursor modulates cytotoxic necrotizing factor 1-mediated RhoA activation and bacterial uptake. J. Biol. Chem. 278:16857-16862.
Cines, D.B., Pollak, E., Buck, C., Loscalzo, J., Zimmerman, G., McEver, R., Pober, J., Wick, T., Konkle, B., Schwartz, B., Barnathan, E., McCrae, K., Hug, B., Schmidt, A., Stern, D.M. (1998) Endothelial cells in physiology and in the pathophysiology of vascular disorders. Blood 91:3527-3561.
Collins, P.W., Macey, M., Cahill, M., and Newland, A.C. (1993) vWF release and P-selectin expression is stimulated by thrombin and trypsin but not IL-1 in cultured endothelial cells. Thromb. Haemost. 70:346-350.
Conchon, S., Barrault, M., Miserey, S., Corvol, P., and Clauser, E. (1997) The C-terminal third intracellular loop of the rat AT(1A) angiotensin receptor plays a key role in G protein coupling specificity and transduction of the mitogenic signal. J. Biol. Chem. 272:25566-25572.
Constantin, D., Ala'Aldeent, D., and Murphy, S. (2002) Transcriptional activation of nitric oxide synthase-2, and NO-induced cell death, in mouse cerebrovascular endothelium exposed to Neisseria meningitidis. J. Neurochem. 81:270-276.

Constantin, G., Majeed, M., Giagulli, C., Piccio, L., Kim, J., Butcher, E., and Laudanna, C. (2000) Chemokines trigger immediate β2 integrin affinity and mobility changes: differential regulation and roles in lymphocyte arrest under flow. Immunity 13:759-769.

Conti, P., Pang, X., Boucher, W., Letourneau, R., Reale, M., Barbacane, R.C., Thibault, J., and Theoharides, T.C. (1997) Monocyte chemotactic protein-1 is a proinflammatory chemokine in rat skin injection sites and chemoattracts basophilic granular cells. Int. Immunol. 9:1563-1570.

Courtney, M.A., Haidaris, P., Marder, V., and Sporn, L.A. (1996) Tissue factor mRNA expression in the endothelium of an intact umbilical vein. Blood 87:174-179.

Craigen, J.L., Yong, K.L., Jordan, N.J., MacCormac, L.P., Westwick, J., Akbar, A.N., and Grundy, J.E. (1997) Human cytomegalovirus infection up-regulates interleukin-8 gene expression and stimulates neutrophil transendothelial migration. Immunol. 2:138-145.

de Larranaga, G.F., Petroni, A., Deluchi, G., Alonso, B., and Benetucci, J.A. (2003) Viral load and disease progression as responsible for endothelial activation and/or injury in human immunodeficiency virus-1-infected patients. Blood Coag. Fibrinol. 14:15-18.

Delvalle, J., Wang, L., Gantz, I., and Yamada, T. (1992) Characterization of H2 histamine receptor: linkage to both adenylate cyclase and Ca^{2+}_i signaling systems. Amer. J. Physiol. 263:G967-G972.

Derhaag, J.G., Duijvestijn, and A.M., Van Breda, V. (1997) Heart EC respond heterogeneous on cytokine stimulation in ICAM-1 and VCAM-1, but not in MHC expression. Endoth. 5:307-319.

Desai, T.R., Leeper, N., Hynes, K., and Gewertz, B.L. (2002) Interleukin-6 causes endothelial barrier dysfunction via the protein kinase C pathway. J. Surg. Res. 104:118-123.

Dimmeler, S. and Zeiher, A.M. (2000) Endothelial cell apoptosis in angiogenesis and vessel regression. Circ. Res. 87:434-439.

Dittman, W.A. and Majerus, P.W. (1990) Structure and function of thrombomodulin: a natural anticoagulant. Blood 75:329-336.

Dixon, G.L., Heyderman, R., Van Der, L., and Klein, N.J. (2004) High-level endothelial E-selectin (CD62E) expression by a lipopolysaccharide-deficient strain of *Neisseria meningitidis* despite poor activation of NF-κB. Clin. Exp. Immunol. 135:85-93.

Douthwaite, J.A., Lees, D.M., and Corder, R. (2003) A role for increased mRNA stability in the induction of ET-1 synthesis by lipopolysaccharide. Biochem. Pharmacol. 66:589-594.

Downie, G.H., Ryan, U., Hayes, B., and Friedman, M. (1992) IL-2 directly increases albumin permeability of bovine and human vascular endothelium in vitro. Amer. J. Respir. Cell Mol. Biol. 7:58-65.

Drevets, D.A. (1997) *Listeria monocytogenes* infection of cultured endothelial cells stimulates neutrophil adhesion and adhesion molecule expression. J. Immunol. 158:5305-5313.

DuhamelClerin, E., Orvain, C., Lanza, F., Cazenave, J., and KleinSoyer, C. (1997) Thrombin receptor-mediated increase of two matrix metalloproteinases, MMP-1 and MMP-3, in human endothelial cells. Arterioscler. Thromb. Vasc. Biol. 17:1931-1938.

Dye, J.F., Leach, L., Clark, P., and Firth, J.A. (2001) Cyclic AMP and acidic FGF have opposing effects on tight and adherens junctions in microvascular endothelial cells in vitro. Microvasc. Res. 62:94-113.

Eguchi, S., Hirata, Y., Imai, T., and Marumo, F. (1993) Endothelin receptor subtypes are coupled to adenylate cyclase via different guanyl nucleotide-binding proteins in vasculature. Endocrinol. 132:524-529.

Ehringer, W.D., Edwards, M.J., and Miller, F.N. (1996) Mechanisms of α-thrombin, histamine, and bradykinin induced endothelial permeability. J. Cell. Physiol. 167:562-569.

Emori, T., Hirata, Y., Imai, T., Ohta, K., Kanno, K., Eguchi, S., and Marumo, F. (1992) Cellular mechanism of thrombin on endothelin-1 biosynthesis and release in endothelial cells. Biochem. Pharmacol. 44:2409-2411.

English, D., Kovala, A., Welch, Z., Harvey, K., Siddiqui, R., Brindley, D., and Garcia, J.G. (1999) Induction of endothelial cell chemotaxis by sphingosine 1-phosphate and stabilization of endothelial monolayer barrier function by lysophosphatidic acid, potential mediators of hematopoietic angiogenesis. J. Hematother. Stem Cell Res. 8:627-634.

Essler, M., Amano, M., Kruse, H., Kaibuchi, K., Weber, P., and Aepfelbacher, M. (1998a) Thrombin inactivates myosin light chain phosphatase via Rho and its target Rho kinase in human endothelial

cells. J. Biol. Chem. 273:21867-21874.
Essler, M., Hermann, K., Amano, M., Kaibuchi, K., Heesemann, J., Weber, P., Aepfelbacher, M. (1998b) Pasteurella multocida toxin increases endothelial permeability via Rho kinase and myosin light chain phosphatase. J. Immunol. 161:5640-5646.
Eto, M., Barandier, C., Rathgeb, L., Kozai, T., Joch, H., Yang, Z., and Luscher, T.F. (2001) Thrombin suppresses endothelial NOS and upregulates endothelin-converting enzyme-1 expression by distinct pathways: role of Rho/ROCK and MAPK. Circ. Res. 89:583-590.
Faure, E., Equils, O., Sieling, P., Thomas, L., Zhang, F., Kirschning, C., Polentarutti, N., Muzio, M., and Arditi, M. (2000) Bacterial lipopolysaccharide activates NF-κB through toll-like receptor 4 (TLR-4) in cultured human dermal endothelial cells. J. Biol. Chem. 275:11058-11063.
Faure, E., Thomas, L., Xu, H., Medvedev, A., Equils, O., and Arditi, M. (2001) Bacterial LPS and IFN-gamma induce Toll-like receptor 2 and Toll-like receptor 4 expression in human endothelial cells: role of NF-κB activation. J. Immunol. 166:2018-2024.
Fenton, J.W., II (1988) Thrombin bioregulatory functions. Adv. Clin. Enzymol. 6:186-193.
Ferrara, N., Carver-Moore, K., Chen, H., Dowd, M., Lu, L., O'Shea, K., Powell-Braxton, L., Hillan, K., and Moore, M.W. (1996) Heterozygous embryonic lethality induced by targeted inactivation of the VEGF gene. Nature 380:439-442.
Ferro, T., Neumann, P., Gertzberg, N., Clements, R., and Johnson, A, (2000) PKC-α mediates endothelial barrier dysfunction induced by TNF-a. Amer. J. Physiol. 278:L1107-L1117.
Flanders, K.C., Bhandiwad, A.R., and Winokur, T.S. (1997) TGF-βs block cytokine induction of catalase and XO mRNA levels in cultured rat cardiac cells. J. Mol. Cell Cardiol. 29:273-280.
Forestal, C.A., Benach, J., Carbonara, C., Italo, J., Lisinski, T., and Furie, M.B. (2003) *Francisella tularensis* selectively induces proinflammatory changes in endothelial cells. J. Immunol. 171:2563-2570.
Frey, R.S., Rahman, A., Kefer, J., Minshall, R., and Malik, A.B. (2002) PKCζ regulates TNF-α-induced activation of NADPH oxidase in endothelial cells. Circ. Res. 90:1012-1019.
Fryer, R.H., Schwobe, E., Woods, M.L., and Rodgers, G.M. (1997) Chlamydia species infect human vascular endothelial cells and induce procoagulant activity. J. Invest. Med. 45:168-174.
Gaboury, J.P., Anderson, D.C., and Kubes, P. (1994) Molecular mechanisms involved in superoxide-induced leukocyte endothelial cell interactions in vivo. Amer. J. Physiol. 266:H637-H642.
Gaillard, P.J., de Boer, A.B., and Breimer, D.D. (2003) Pharmacological investigations on LPS-induced permeability changes in the blood-brain barrier in vitro. Microvasc. Res. 65:24-31.
Garcia, J.G., Liu, F., Verin, A., Birukova, A., Dechert, M., Gerthoffer, W., Bamberg, J., English, D. (2001) Sphingosine 1-phosphate promotes endothelial cell barrier integrity by Edg-dependent cytoskeletal rearrangement. J. Clin. Invest. 108:689-701.
Garcia, J.G.N., Verin, A., Schaphorst, K.L. (1996) Regulation of thrombin-mediated endothelial cell contraction and permeability. Semin. Thromb. Hemost. 22:309-315.
Garcia, G.E., Xia, Y., Chen, S., Wang, Y., Ye, R., Harrison, J., Bacon, K., Zerwes, H., Feng, L. (2000) NF-κB-dependent fractalkine induction in rat aortic endothelial cells stimulated by IL-1β, TNF-α, and LPS. J. Leuk. Biol. 67:577-584.
Geimonen, E., Neff, S., Raymond, T., Kocer, S., Gavrilovskaya, I., and Mackow, E.R. (2002) Pathogenic and nonpathogenic hantaviruses differentially regulate endothelial cell responses. PNAS USA 99:13837-13842.
Gerritsen, M.E., Kelley, K., Ligon, G., Perry, C., Shen, C., Szczepanski, A., andCarley, W.W. (1993) Regulation of the expression of ICAM-1 in cultured human endothelial cells derived from rheumatoid synovium. Arthr. Rheum. 36:593-602.
Gertzberg, N., Neumann, P., Rizzo, V., and Johnson, A. (2004) NAD(P)H oxidase mediates the endothelial barrier dysfunction induced by TNF-α. Amer. J. Physiol. 286:L37-L48.
Gilchrist, A., Vanhauwe, J., Li, A., Thomas, T., Voyno-Yasenetskaya, T., and Hamm, H.E. (2001) Gα minigenes expressing C-terminal peptides serve as specific inhibitors of thrombin-mediated endothelial activation. J. Biol. Chem. 276:25672-25679.
Goebeler, M., Kilian, K., Gillitzer, R., Kunz, M., Yoshimura, T., Brocker, E., Rapp, U., and Ludwig, S. (1999) The MKK6/p38 stress kinase cascade is critical for TNF-α-induced expression of monocyte-chemoattractant protein-1 in endothelial cells. Blood 93:857-865.

Goeckeler, Z.M. and Wysolmerski, R.B. (1995) MLCK-regulated endothelial cell contraction: the relationship between isometric tension, actin polymerization, and myosin phosphorylation. J. Cell. Biol. 130:613-627.
Goldblum, S.E., Ding, X., Brann, T., and Campbell-Washington, J. (1993) Bacterial lipopolysaccharide induces actin reorganization, intercellular gap formation, and endothelial barrier dysfunction in pulmonary vascular endothelial cells. J. Cell. Physiol. 157:13-23.
Gooch, K.J. and Frangos, J.A. (1996) Flow- and bradykinin-induced nitric oxide production by endothelial cells is independent of membrane potential. Amer. J. Physiol. 39:C546-C551.
Gow, A.J., Thom, S.R., and Ischiropoulos, H. (1998) Nitric oxide and peroxynitrite-mediated pulmonary cell death. Amer. J. Physiol. 274:L112-L118.
Grafe, M., Auchschwelk, W., Zakrzewicz, A., RegitzZagrosek, V., Bartsch, P., Graf, K., Loebe, M., Gaehtgens, P., and Fleck, E. (1997) Angiotensin II-induced leukocyte adhesion on human coronary endothelial cells is mediated by E-selectin. Circ. Res. 81:804-811.
Green, S.A., Setiadi, H., McEver, R., and Kelly, R.B. (1994) The cytoplasmic domain of P-selectin contains sorting determinant that mediates rapid degradation in lysosomes. J. Cell Biol. 124:435-448.
Halama, T., Groger, M., Pillinger, M., Staffler, G., Prager, E., Stockinger, H., Holnthoner, W., Lechleitner, S., Wolff, K., and Petzelbauer, P. (2001) PECAM-1 and VE-cadherin cooperatively regulate FGF-induced modulations of adherens junction functions. J. Invest. Dermatol. 116:110-117.
Hamaguchi, E., Takamura, T., Shimizu, A., and Nagai, Y. (2003) TNF-α and troglitazone regulate PAI-1 production through ERK- and NF-κB-dependent pathways in cultured human umbilical vein endothelial cells. J. Pharmacol. Exp. Ther. 307:987-994.
Hamilton, J.R., Frauman, A.G., and Cocks, T.M. (2001) Increased expression of protease-activated receptor-2 and PAR4 in human coronary artery by inflammatory stimuli unveils endothelium-dependent relaxations to PAR2 and PAR4 agonists. Circ.Res. 89:92-98.
Harrison, J.K., Jiang, Y., Wees, E., Salafranca, M., Liang, H., Feng, L., and Belardinelli, L. (1999) Inflammatory agents regulate in vivo expression of fractalkine in endothelial cells of the rat heart. J. Leu. Biol. 66:937-944.
Hasdai, D., Holmes, D., Garratt, K., Edwards, W., and Lerman, A. (1997) Mechanical pressure and stretch release ET-1 from human atherosclerotic coronary arteries in vivo. Circ. 95:357-362.
Hayasaki, Y., Nakajima, M., Kitano, Y., Iwasaki, T., Shimamura, T., and Iwaki, K. (1996) ICAM-1 expression on cardiac myocytes and aortic endothelial cells via their specific endothelin receptor subtype. Biochem. Biophys. Res. Commun. 229:817-824.
Heidemann, J., Ogawa, H., Dwinell, M., Rafiee, P., Maaser, C., Gockel, H., Otterson, M., Ota, D., Lugering, N., Domschke, W., and Binion, D.G. (2003) Angiogenic effects of IL-8 (CXCL8) in human microvascular endothelial cells are mediated by CXCR2. J. Biol. Chem. 278:8508-8515.
Hekimian, G., Cote, S., van Sande, J., and Boeynaems, J.M. (1992) H2 receptor-mediated responses of aortic endothelial cells to histamine. Amer. J. Physiol. 262:H220-H224.
Heltianu, C., Simionescu, M., and Simionescu, N. (1982) Histamine receptors of the microvascular endothelium revealed in situ with a histamine-ferritin conjugate: Characteristic high-affinity binding sites in venules. J. Cell Biol. 93:357-364.
Henneke, P. and Golenbock, D.T. (2002) Innate immune recognition of lipopolysaccharide by endothelial cells. Crit. Care Med. 30:S207-S213.
Herbert, J.M., Dupuy, E., Laplace, M., Zini, J., Barshavit, R., and Tobelem, G. (1994) Thrombin induces endothelial cell growth via both a proteolytic and a non-proteolytic pathway. Biochem. J. 303:227-231.
Hippenstiel, S., Krull, M., Ikemann, A., Risau, W., Clauss, M., and Suttorp, N. (1998) VEGF induces hyperpermeability by a direct action on endothelial cells. Amer. J. Physiol. 274:L678-L684.
Hippenstiel, S., Soeth, S., Kellas, B., Fuhrmann, O., Seybold, J., Krull, M., Eichel-Streiber, C., Goebeler, M., Ludwig, S., and Suttorp, N. (2000) Rho proteins and the p38-MAPK pathway are important mediators for LPS-induced IL-8 expression in human endothelial cells. Blood 95:3044-3051.
Hla, T., Lee, M., Ancellin, N., Paik, J., and Kluk, M.J. (2001) Lysophospholipids--receptor revelations. Science 294:1875-1878.
Holland, J.A., Meyer, J., Chang, M., O'Donnell, R., Johnson, D., and Ziegler, L.M. (1998) Thrombin

stimulated ROS production in cultured human endothelial cells. Endothel. 6:113-121.
Holland, J.A., Pritchard, K., Pappolla, M., Wolin, M., Rogers, N., and Stemerman, M.B. (1990) Bradykinin induces O_2^- release from human endothelial cells. J. Cell. Physiol. 143:21-25.
Hollenberg, M.D. and Compton, S.J. (2002) International Union of Pharmacology. XXVIII. Proteinase-activated receptors. Pharmacol. Rev. 54:203-217.
Honda, Z., Ishii, S., Shimizu, T. (2002) PAF receptor. J. Biochem. (Tokyo) 131:773-779.
Hoyer, J., Kohler, R., Haase, W., and Distler, A. (1996) Upregulation of pressure-activated Ca^{2+}- cation channel in intact vascular endothelium of hypertensive rats. PNAS.USA 93:11253-11258.
Hsu, H., Shu, H., Pan, M., and Goeddel, D.V. (1996) TRADD-TRAF2 and TRADD-FADD interactions define two distinct TNFR1 signal transduction pathways. Cell 84:299-308.
Hu, Y.L. and Chien, S. (1997) Effects of shear stress on PKC distribution in endothelial cells. J. Histochem. Cytochem. 45:237-249.
Huang, K., Fishwild, D., Wu, H., and Dedrick, R.L. (1995) Lipopolysaccharide-induced E-selectin expression requires continuous presence of LPS and is inhibited by bactericidal/permeability-increasing protein. Inflammation 19:389-404.
Huang, Y.H., Lei, H., Liu, H., Lin, Y., Chen, S., Liu, C., and Yeh, T.M. (2003) Tissue plasminogen activator induced by dengue virus infection of human endothelial cells. J. Med. Virol. 70:610-616.
Hurst, V.I., Goldberg, P., Minnear, F., Heimark, R., and Vincent, P.A. (1999) Rearrangement of adherens junctions by TGF-β1: role of contraction. Amer. J. Physiol. 276:L582-L595.
Ikeda, K., Utoguchi, N., Makimoto, H., Mizuguchi, H., Nakagawa, S., and Mayumi, T. (1999) Different reactions of aortic and venular endothelial cell monolayers to histamine on macromolecular permeability: role of cAMP, Ca^{2+} and F-actin. Inflamm. 23:87-97.
Imaizumi, T., Itaya, H., Fujita, K., Kudoh, D., Kudoh, S., Mori, K., Fujimoto, K., Matsumiya, T., Yoshida, H., and Satoh, K. (2000) Expression of TNFα in cultured human endothelial cells stimulated with LPS or IL-1α. Arterioscler.Thromb.Vasc.Biol. 20:410-415.
Inoue, N., Ramasamy, S., Fukai, T., Nerem, R., and Harrison, D.G. (1996) Shear stress modulates expression of Cu/Zn superoxide dismutase in human aortic endothelial gels. Circ. Res. 79:32-37.
Inoue, N., Takeshita, S., Gao, D., Ishida, T., Kawashima, S., Akita, H., Tawa, R., Sakurai, H., and Yokoyama, M. (2001) Lysophosphatidylcholine increases the secretion of MMP2 through the activation of NADH/NADPH oxidase in cultured aortic endothelial cells. Atheroscl. 155:45-52.
Ishida, T., Takahashi, M., Corson, M., and Berk, B.C. (1997) Fluid shear stress-mediated signal transduction: how do endothelial cells transduce mechanical force into biological responses? Ann. N.Y. Acad. Sci. 811:12-23.
Jalali, S., Li, Y., Sotoudeh, M., Yuan, S., Li, S., Chien, S., Shyy, J.Y.J. (1998) Shear stress activates p60src-Ras-MAPK signaling pathways in vascular endothelial cells. Arterioscler. Thromb. Vasc. Biol. 18:227-234.
Jilma, B., Pernerstorfer, T., Dirnberger, E., Stohlawetz, P., Schmetterer, L., Singer, E., Grasseli, U., Eichler, H., and Kapiotis, S. (1998) Effects of histamine and nitric oxide synthase inhibition on plasma levels of von Willebrand factor antigen. J. Lab. Clin. Med. 131:151-156.
Jinno-Oue, A., Wilt, S., Hanson, C., Dugger, N., Hoffman, P., Masuda, M., and Ruscetti, S.K. (2003) Expression of iNOS and elevation of tyrosine nitration of a 32-kD cellular protein in brain capillary endothelial cells from rats infected with a neuropathogenic murine leukemia virus. J. Virol. 77:5145-5151.
Jo, H., Dull, R., Hollis, T., and Tarbell, J.M. (1991) Endothelial albumin permeability is shear dependent, time dependent, and reversible. Amer. J. Physiol. 260:H1992-H1996.
Johnson, J.E., Perkett, E., and Meyrick, B. (1997) Pulmonary veins and bronchial vessels undergo remodeling in sustained pulmonary hypertension induced by continuous air embolization. Exp. Lung Res. 23:459-473.
Kamata, H. and Hirata, H. (1999) Redox regulation of cellular signalling. Cell. Sig. 11:1-14.
Kameda, H., Morita, I., Handa, M., Kaburaki, J., Yoshida, T., Mimori, T., Murota, S., Ikeda, Y. (1997) Re-expression of functional P-selectin molecules on the endothelial cell surface by repeated stimulation with thrombin. Brit. J. Haematol. 97:348-355.
Kan, W.H., Yan, W., Jiang, Y., Wang, J., Qin, Q., and Zhao, K.S. (2002) Role of p38 mitogen-activated protein kinase in lipopolysaccharide-induced expression of inducible nitric oxide synthase in human

endothelial cells. Di Yi. Jun. Yi. Da. Xue. Xue. Bao. 22:388-392.

Kanda, S., Lerner, E., Tsuda, S., Shono, T., Kanetake, H., and Smithgall, T.E. (2000) The nonreceptor protein-tyrosine kinase c-Fes is involved in fibroblast growth factor-2-induced chemotaxis of murine brain capillary endothelial cells. J. Biol. Chem. 275:10105-10111.

Kang, D.S., Leeb-Lundberg, L.M. (2002) Negative and positive regulatory epitopes in the C-terminal domains of the human B1 and B2 bradykinin receptor subtypes determine receptor coupling efficacy to G(q/11)-mediated - PLCβ activity. Molec. Pharmacol. 62:281-288.

Katz, A.M. and Messineo, F.C. (1981) Lipid-membrane interactions and the pathogenesis of ischemic damage in the myocardium. Circ. Res. 48:1-16.

Kawai, N., Yamamoto, T., Yamamoto, H., McCarron, R., and Spatz, M. (1997) Functional characterization of endothelin receptors on cultured brain capillary endothelial cells of the rat. Neurochem. Int. 31:597-605.

Kawai, T., Seki, M., Hiromatsu, K., Eastcott, J.W., Watts, G., Sugai, M., Smith, D., Porcelli, S., and Taubman, M.A. (1999) Selective diapedesis of Th1 cells induced by endothelial cell RANTES. J. Immunol. 163:3269-3278.

Kelly, J.J., Moore, T., Babal, P., Diwan, A., Stevens, T., and Thompson, W.J. (1998) Pulmonary microvascular and macrovascular endothelial cells: differential regulation of Ca^{2+} and permeability. Amer. J. Physiol. 274:L810-L819.

Kelly, S.A., Goldschmidtclermont, P., Milliken, E., Arai, T., Smith, E., and Bulkley, G.B. (1998) Protein tyrosine phosphorylation mediates TNF-induced endothelial-neutrophil adhesion. Amer. J. Physiol. 43:H513-H519.

Kevil, C.G., Oshima, T., and Alexander, J.S. (2001) The role of p38 MAP kinase in hydrogen peroxide mediated endothelial solute permeability. Endothel. 8:107-116.

Khewgoodall, Y., Butcher, C., Litwin, M., Newlands, S., Korpelainen, E., Noack, L., Berndt, M., Lopez, A., Gamble, J., and Vadas, M.A. (1996) Chronic expression of P-selectin on endothelial cells stimulated by the T-cell cytokine, interleukin-3. Blood 87:1432-1438.

Kim, H. and Koh, G. (2000) Lipopolysaccharide activates MMP-2 in endothelial cells through an NF-κB-dependent pathway. Biochem. Biophys. Res. Comm. 269:401-405.

Kim, J.A., Berliner, J.A., and Nadler, J.L. (1996) Angiotensin II increases monocyte binding to endothelial cells. Biochem. Biophys. Res. Comm. 226:862-868.

Kim, Y.V., Di Cello, F., Hillaire, C., and Kim, K.S. (2004) Differential Ca^{++} signaling by thrombin and PAR-1-activating peptide in human brain microvascular endothelial cells. Amer. J. Physiol. 286:C31-C42.

Kitazumi, K. and Tasaka, K. (1992) Thrombin-stimulated phosphorylation of MLC and its possible involvement in ET-1 secretion from porcine aortic endothelial cells. Biochem. Pharm. 43:1701-1709.

Klagsbrun, M. and Edelman, E.R. (1989) Biological and biochemical properties of fibroblast growth factors. Implications for the pathogenesis of atherosclerosis. Arterioscl. 9:269-278.

Knudsen, H.L. and Frangos, J.A. (1997) Role of cytoskeleton in shear stress-induced endothelial nitric oxide production. Amer. J. Physiol. 42:H347-H355.

Koch, A.E., Halloran, M., Hosaka, S., Shah, M., Haskell, C., Baker, S., Panos, R., Haines, G., Bennett, G., Pope, R., and Ferrara, N. (1996) Hepatocyte growth factor. A cytokine mediating endothelial migration in inflammatory arthritis. Arthr. Rheum. 39:1566-1575.

Konstantoulaki, M., Kouklis, P., and Malik, A.B. (2003) PKC modifications of VE-cadherin, p120, and β-catenin contribute to endothelial barrier dysregulation induced by thrombin. Amer. J. Physiol. 285:L434-L442.

Kotenko, S.V. and Pestka, S. (2000) Jak-Stat signal transduction pathway through the eyes of cytokine class II receptor complexes. Oncogene 19:2557-2565.

Krizanac-Bengez, L., Kapural, M., Parkinson, F., Cucullo, L., Hossain, M., Mayberg, M., and Janigro, D. (2003) Effects of transient loss of shear stress on blood-brain barrier endothelium: role of nitric oxide and IL-6. Brain Res. 977:239-246.

Kubes, P., Suzuki, M., Granger, D.N. (1991) Nitric oxide: an endogenous modulator of leukocyte adhesion. PNAS USA 88:4651-4655.

Kuchan, M.J. and Frangos, J.A. (1993) Shear stress regulates endothelin-1 release via PKC and cGMP in cultured endothelial cells. Amer. J. Physiol. 264:H150-H156.

Kuhn, B., Schmid, A., Harteneck, C., Gudermann, T., and Schultz, G. (1996) G proteins of the Gq family couple the H2 histamine receptor to phospholipase C. Molec. Endocrinol. 10:1697-1707.

Kuijpers, T.W., Hakkert, B., Hart, M., and Roos, D. (1992) Neutrophil migration across monolayers of cytokine-prestimulated endothelial cells: a role for PAF and IL-8. J. Cell Biol. 117:565-572.

Kume, N., Cybulsky, M., and Gimbrone, M.A., Jr. (1992) Lysophosphatidylcholine, a component of atherogenic lipoproteins, induces mononuclear leukocyte adhesion molecules in cultured human and rabbit arterial endothelial cells. J. Clin. Invest. 90:1138-1144.

Kume, N. and Gimbrone, M.A., Jr. (1994) Lysophosphatidylcholine induces growth factor gene expression in cultured human endothelial cells. J. Clin. Invest. 93:907-911.

Kurimoto, N., Nan, Y., Chen, Z., Feng, G., Komatsu, T., Kandatsu, N., Ko, J., Kawai, N., and Ishikawa, N. (2004) Effects of specific signal transduction inhibitors on increased permeability across rat endothelial monolayers induced by neuropeptide Y or VEGF. Amer. J. Physiol. 287:H100-H106.

Kurose, I., Miura, S., Fukumura, D., and Tsuchiya, M. (1993) Mechanisms of endothelin-induced macromolecular leakage in microvascular beds of rat mesentery. Eur. J. Pharmacol. 250:85-94.

Labow, M.A., Norton, C., Rumberger, J., Lombardgillooly, K., Shuster, D., Hubbard, J., Bertko, R., Knaack, P., Terry, R., Harbison, M., Kontgen, F., Stewart, C., Mcintyre, K., Will, P., Burns, D., and Wolitzky, B.A. (1994) Characterization of E-selectin-deficient mice. Immunity 1:709-720.

Lakshminarayanan, S., Gardner, T.W., and Tarbell, J.M. (2000) Effect of shear stress on hydraulic conductivity of cultured bovine retinal microvascular endothelial cell monolayers. Curr. Eye Res. 21:944-951.

Lakshminarayanan, V., Drab-Weiss, E., and Roebuck, K.A. (1998) H_2O_2 and TNFα induce differential binding of the redox-responsive transcription factors AP-1 and NF-κB to the IL-8 promoter in endothelial and epithelial cells. J. Biol. Chem. 273:32670-32678.

Landgren, E., Klint, P., Yokote, K., and Claesson-Welsh, L. (1998) Fibroblast growth factor receptor-1 mediates chemotaxis independently of direct SH2-domain protein binding. Oncogene 17:283-291.

Lee, H., Goetzl, E.J., and An, S. (2000) Lysophosphatidic acid and sphingosine 1-phosphate stimulate endothelial cell wound healing. Amer. J. Physiol. 278:C612-C618.

Lee, M.J., Thangada, S., Claffey, K., Ancellin, N., Liu, C., Kluk, M., Volpi, M., Sha'afi, R., and Hla, T. (1999) Vascular endothelial cell adherens junction assembly and morphogenesis induced by sphingosine-1-phosphate. Cell 99:301-312.

Leitinger, N., Tyner, T., Oslund, L., Rizza, C., Subbanagounder, G., Lee, H., Shih, P., Mackman, N., Tigyi, G., Territo, M., Berliner, J., and Vora, D.K. (1999) Structurally similar oxidized phospholipids differentially regulate endothelial binding of monocytes and neutrophils. PNAS USA 96:12010-12015.

Lerm, M., Schmidt, G., and Aktories, K. (2000) Bacterial protein toxins targeting rho GTPases. FEMS Microbiol. Lett. 188:1-6.

LeVan, T.D., Dow, S., Chase, P., Bloom, J., Regan, J., Cunningham, E., and Halonen, M. (1997) Evidence for PAF receptor subtypes on human PMN membranes. Biochem. Pharm. 54:1007-1012.

Levesque, M.J. and Nerem, R.M. (1985) The elongation and orientation of cultured endothelial cells in response to shear stress. J. Biomech. Eng. 107:341-347.

Lewis, M.S., Whatley, R., Cain, P., Mcintyre, T., Prescott, S., and Zimmerman, G.A. (1988) Hydrogen peroxide stimulates the synthesis of PAF by endothelium and induces endothelial cell-dependent neutrophil adhesion. J. Clin. Invest. 82:2045-2055.

Li, J.M. and Shah, A.M. (2003) Mechanism of endothelial cell NADPH oxidase activation by angiotensin II. Role of the p47phox subunit. J. Biol. Chem. 278:12094-12100.

Li, S., Chen, B., Azuma, N., Hu,Y., Wu, S., Sumpio, B., Shyy, J., and Chien, S. (1999) Distinct roles for the small GTPases Cdc42 and Rho in endothelial responses to shear stress. J. Clin. Invest. 103:1141-1150.

Liao, J.K. and Homcy, C.J. (1993) The G proteins of the G αi and G αq family couple the bradykinin receptor to the release of endothelium-derived relaxing factor. J. Clin. Invest. 92:2168-2172.

Lin, M.C., AlmusJacobs, F., Chen, H., Parry, G., Mackman, N., Shyy, J., and Chien, S. (1997) Shear stress induction of the tissue factor gene. J. Clin. Invest. 99:737-744.

Liu, F., Schaphorst, K., Verin, A., Jacobs, K., Birukova, A., Day, R., Bogatcheva, N., Bottaro, D., and Garcia, J.G. (2002) HGF enhances endothelial cell barrier function and cortical cytoskeletal

rearrangement: potential role of GSK-3β. FASEB J. 16:950-962.
Liu, F., Verin, A., Wang, P., Day, R., Wersto, R., Chrest, F., English, D., and Garcia, J.G. (2001) Differential regulation of sphingosine-1-phosphate- and VEGF-induced endothelial cell chemotaxis. Involvement of Giα2-linked Rho kinase activity. Amer. J. Respir. Cell Molec. Biol. 24:711-719.
Liu, S.M. and Sundqvist, T. (1997) Nitric oxide and cGMP regulate endothelial permeability and F- actin distribution in hydrogen peroxide-treated endothelial cells. Exp. Cell Res. 235:238-244.
LopezFarre, A., DeMiguel, L., Caramelo, C., GomezMacias, J., Garcia, R., Mosquera, L., DeFrutos, T., Millas, I., Rivas, F., Echezarreta, G., and Casado, S. (1997) Role of nitric oxide in autocrine control of growth and apoptosis of endothelial cells. Amer. J. Physiol. 41:H760-H768.
Lorant, D.E., Patel, K., Mcintyre, T., McEver, R., Prescott, S., and Zimmerman, G.A. (1991) Coexpression of GMP-140 and PAF by endothelium stimulated by histamine or thrombin: a juxtacrine system for adhesion and activation of neutrophils. J. Cell Biol. 115:223-234.
Lortat-Jacob, H., Grosdidier, A., and Imberty, A. (2002) Structural diversity of heparan sulfate binding domains in chemokines. PNAS.USA 99:1229-1234.
Lum, H., Aschner, J., Phillips, P., Fletcher, P., and Malik, A.B. (1992a). Time course of thrombin-induced increase in endothelial permeability: relationship to $Ca^{2+}{}_i$ and inositol polyphosphates. Amer. J. Physiol. 263:L219-L225.
Lum, H., Barr, D., Shaffer, J., Gordon, R., Ezrin, A., and Malik, A.B. (1992b) Reoxygenation of endothelial cells increases permeability by oxidant-dependent mechanisms. Circ.Res. 70:991-998.
Lum, H., Qiao, J., Walter, R.J., Huang, F., Subbaiah, P.V., Kim, K.S., and Holian, O. (2003) Inflammatory stress increases receptor for lysophosphatidylcholine in human microvascular endothelial cells. Amer. J. Physiol. 285:H1786-H1789.
Lum, H. and Roebuck, K.A. (2001) Oxidant stress and endothelial dysfunction. Amer. J. Physiol. 280:C719-C741.
Macfarlane, S.R., Seatter, M., Kanke, T., Hunter, G., and Plevin, R. (2001) Proteinase-activated receptors. Pharmacol. Rev. 53:245-282.
Maisch, T., Kropff, B., Sinzger, C., and Mach, M. (2002) Upregulation of CD40 expression on endothelial cells infected with human cytomegalovirus. J. Virol. 76:12803-12812.
Majno, G. and Palade, G.E. (1961) Studies on inflammation. I. The effect of histamine and serotonin on vascular permeability: an electron microscopic study. J. Biophys. Biochem. Cytol. 11:571-605.
Majno, G., Palade, G.E., and Schefl, G.I. (1961) Studies on inflammation. II. The site of action of histamine and serotonin along the vascular tree: a topographic study. J. Biophys. Biochem. Cytol. 11:607-626.
Malek, A.M., Izumo, S., and Alper, S.L. (1999) Modulation by pathophysiological stimuli of shear stress-induced up-regulation of endothelial nitric oxide synthase expression in endothelial cells. Neurosurg. 45:334-344.
Manolopoulos, V.G., Fenton, J.W., and Lelkes, P.I. (1997) The thrombin receptor in adrenal medullary microvascular endothelial cells is negatively coupled to adenylyl cyclase through a Gi protein. Biochim. Biophys.Acta 1356:321-332.
Marathe, G.K., Davies, S., Harrison, K., Silva, A., Murphy, R., Castro-Faria-Neto, H., Prescott, S., Zimmerman, G., and Mcintyre, T.M. (1999) Inflammatory PAF-like phospholipids in oxidized LDL are fragmented alkyl phosphatidylcholines. J. Biol. Chem. 274:28395-28404.
Marin, V., Farnarier, C., Gres, S., Kaplanski, S., Su, M., Dinarello, C., and Kaplanski, G. (2001) The p38 MAPK pathway plays a critical role in thrombin-induced endothelial chemokine production and leukocyte recruitment. Blood 98:667-673.
Mark, K.S., Burroughs, A., Brown, R., Huber, J., and Davis, T.P. (2004) Nitric oxide mediates hypoxia-induced changes in paracellular permeability of cerebral microvasculature. Amer. J. Physiol. 286:H174-H180.
Martin, E., Sharina, I., Kots, A., and Murad, F. (2003) A constitutively activated mutant of human soluble guanylyl cyclase (sGC): implication for the mechanism of sGC activation. PNAS.USA 100:9208-9213.
Marumo, T., Nakaki, T., Adachi, H., Esumi, H., Suzuki, H., Saruta, T., and Kato, R. (1993) Nitric oxide synthase mRNA in endothelial cells - Synergistic induction by IFN-γ, TNF-α and LPS and inhibition by dexamethasone. Jpn. J. Pharmacol. 63:327-334.

Mata-Greenwood, E., Meyrick, B., Soifer, S., Fineman, J., and Black, S.M. (2003) Expression of VEGF and its receptors Flt-1 and Flk-1/KDR is altered in lambs with increased pulmonary blood flow and pulmonary hypertension. Amer. J. Physiol. 285:L222-L231.

Matsubara, T. and Ziff, M. (1986) Increased superoxide anion release from human endothelial cells in response to cytokines. J. Immunol. 137:3295-3298.

Mayhan, W.G. (1996) Role of nitric oxide in histamine-induced increases in permeability of the blood-brain barrier. Brain Res. 743:70-76.

McHowat, J., Kell, P., O'Neill, H., and Creer, M.H. (2001) Endothelial cell PAF synthesis following thrombin stimulation utilizes Ca^{2+}-independent phospholipase A_2. Biochem. 40:14921-14931.

Mcintyre, T.M., Zimmerman, G., Satoh, K., and Prescott, S.M. (1985) Cultured endothelial cells synthesize both platelet-activating factor and prostacyclin in response to histamine, bradykinin, and adenosine triphosphate. J. Clin. Invest. 76:271-280.

Meyrick, B. and Brigham, K.L. (1984) Increased permeability associated with dilatation of endothelial cell junctions caused by histamine in intimal explants from bovine pulmonary artery. Exper. Lung Res. 6:11-25.

Middleton, J., Patterson, A., Gardner, L., Schmutz, C., and Ashton, B.A. (2002) Leukocyte extravasation: chemokine transport and presentation by the endothelium. Blood 100:3853-3860.

Milhau, N., Bellaton, C., Balleydier, S., Gaonach, M., and Le Jan, C. (2003) In vitro infection of aortic endothelial cells by caprine arthritis encephalitis virus enhances transmigration of pmn and modulates their phenotypic expression. Vet.Res. 34:273-284.

Miller, J., Knorr, R., Ferrone, M., Houdei, R., Carron, C., and Dustin, M.L. (1995) Intercellular adhesion molecule-1 dimerization and its consequences for adhesion mediated by lymphocyte function associated-1. J. Exp. Med. 182:1231-1241.

Mineau-Hanschke, R., Wiles, M., Morel, N., Hechtman, H., and Shepro, D. (1990) Modulation of cultured pulmonary microvessel and arterial endothelial cell barrier structure and function by serotonin. Microvasc. Res. 39:140-155.

Mogami, K., Mizukami, Y., Todoroki-Ikeda, N., Ohmura, M., Yoshida, K., Miwa, S., Matsuzaki, M., Matsuda, M., and Kobayashi, S. (1999) Sphingosylphosphorylcholine induces Ca^{2+} elevation in endothelial cells in situ and causes endothelium-dependent relaxation through NO production in bovine coronary artery. FEBS Lett. 457:375-380.

Montesano, R., Mossaz, A., Ryser, J., Orci, L., and Vassalli, P. (1984) Leukocyte interleukins induce cultured endothelial cells to produce a highly organized, glycosaminoglycan-rich pericellular matrix. J. Cell Biol. 99:1706-1715.

Montrucchio, G., Lupia, E., Battaglia, E., Del Sorbo, L., Boccellino, M., Biancone, L., Emanuelli, G., and Camussi, G. (2000) PAF enhances VEGF-induced endothelial cell motility and neoangiogenesis in a murine matrigel model. Arterioscler. Thromb. Vasc. Biol. 20:80-88.

Morbidelli, L., Chang, C., Douglas, J., Granger, H., Ledda, F., and Ziche, M. (1996) NO mediates mitogenic effect of VEGF on coronary venular endothelium. Amer. J. Physiol. 39:H411-H415.

Morbidelli, L., Orlando, C., Maggi, C., Ledda, F., and Ziche, M. (1995) Proliferation and migration of endothelial cells is promoted by endothelins via activation of ETB receptors. Amer. J. Physiol. 269:H686-H695.

Morbidelli, L., Parenti, A., Giovannelli, L., Granger, H., Ledda, F., and Ziche, M. (1998) B1 receptor involvement in the effect of bradykinin on venular endothelial cell proliferation and potentiation of FGF-2 effects. Brit. J. Pharmacol. 124:1286-1292.

Morgan-Boyd, R., Stewart, J., Vavrek, R., Hassidd, A. (1987) Effects of bradykinin and angiotensin II on intracellular Ca^{2+} dynamics in endothelial cells. Amer. J. Physiol. 253:C588-C598.

Morita, Y., Clemens, M., Miller, L., Rangan, U., Kondo, S., Miyasaka, M., Yoshikawa, T., and Bulkley, G.B. (1995) Reactive oxidants mediate TNF-α-induced leukocyte adhesion to rat mesenteric venular endothelium. Amer. J. Physiol. 38:H1833-H1842.

Morita, T., Yoshizumi, M., Kurihara, H., Maemura, K., Nagai, R., Yazaki, Y. (1993) Shear stress increases heparin-binding EGF-like growth factor mRNA levels in human vascular endothelial cells. Biochem. Biophys. Res. Comm. 197:256-262.

Moy, A.B., VanEngelenhoven, J., Bodmer, J., Kamath, J., Keese, C., Giaever, I., Shasby, S., Shasby, D.M. (1996) Histamine and thrombin modulate endothelial focal adhesion through centripetal and

centrifugal forces. J. Clin. Invest. 97:1020-1027.
Moy, A.B., Winter, M., Kamath, A., Blackwell, K., Reyes, G., Giaever, I., Keese, C., and Shasby, D.M. (2000) Histamine alters endothelial barrier function at cell-cell and cell-matrix sites. Amer. J. Physiol. 278:L888-L898.
Muller, M.M. and Griesmacher, A. (2000) Markers of endothelial dysfunction. Clin. Chem. Lab. Med. 38:77-85.
Murdoch, C., Monk, P.N., and Finn, A. (1999) Cxc chemokine receptor expression on human endothelial cells. Cytokine 11:704-712.
Murohara, T., Horowitz, J., Silver, M., Tsurumi, Y., Chen, D., Sullivan, A., and Isner, J.M. (1998) VEGF/VPF enhances vascular permeability via nitric oxide and prostacyclin. Circ. 97:99-107.
Murohara, T., Scalia, R., Lefer, A.M. (1996) Lysophosphatidylcholine promotes P-selectin expression in platelets and endothelial cells. Possible involvement of PKC and its inhibition by NO donors. Circ. Res. 78:780-789.
Murohara, T., Witzenbichler, B., Spyridopoulos, I., Asahara, T., Ding, B., Sullivan, A., Losordo, D., and Isner, J.M. (1999) Role of endothelial nitric oxide synthase in endothelial cell migration. Arterioscler. Thromb. Vasc. Biol. 19:1156-1161.
Murphy, A.A., Santanam, N., Morales, A.J., Parthasarathy, S. (1998) Lysophosphatidyl choline, a chemotactic factor for monocytes/T-lymphocytes is elevated in endometriosis. J. Clin. Endocrinol. Metab. 832110-2113.
Murphy, J.F. and Mcgregor, J.L. (1994) Two sites on P-selectin (lectin and EGF- like domains) are involved in adhesion of monocytes to thrombin-activated endothelial cells. Biochem. J. 303:619-624.
Murugesan, G., Sandhya Rani, M., Gerber, C., Mukhopadhyay, C., Ransohoff, R., Chisolm, G., and Kottke-Marchant, K. (2003) Lysophosphatidylcholine regulates human microvascular endothelial cell expression of chemokines. J. Molec. Cell Cardiol. 35:1375-1384.
Nakahara, K., Ohkuda, K., and Staub, N.C. (1979) Effect of infusing histamine into pulmonary or bronchial artery on sheep pulmonary fluid balance. Amer. Rev. Respir. Dis. 120:875-882.
Natarajan, V. (1995) Oxidants and signal transduction in vascular endothelium. J. Lab. Clin. Med. 125:26-37.
Ngaiza, J.R. and Jaffe, E.A. (1991) A 14 amino acid peptide derived from the amino terminus of the cleaved thrombin receptor elevates intracellular calcium and stimulates prostacyclin production in human endothelial cells. Biochem. Biophys. Res. Commun. 179:1656-1661.
Niimi, N., Noso, N., and Yamamoto, S. (1992) The effect of histamine on cultured endothelial cells. A study of the mechanism of increased vascular permeability. Eur. J. Pharmacol. 221:325-331.
Noiri, E., Hu, Y., Bahou, W., Keese, C., Giaever, I., and Goligorsky, M.S. (1997) Permissive role of nitric oxide in endothelin-induced migration of endothelial cells. J. Biol. Chem. 272:1747-1752.
Nollert, M.U., Eskin, S.G., and Mcintire, L.V. (1990) Shear stress increases inositol trisphosphate levels in human endothelial cells. Biochem. Biophys. Res. Comm. 170:281-287.
Noria, S., Cowan, D., Gotlieb, A., and Langille, B.L. (1999) Transient and steady-state effects of shear stress on endothelial cell adherens junctions. Circ. Res. 85:504-514.
Ochoa, L., Waypa, G., Mahoney, J., Rodriguez, L., and Minnear, F.L. (1997) Contrasting effects of hypochlorous acid and hydrogen peroxide on endothelial permeability: Prevention with cAMP drugs. Amer. J. Respir. Crit. Care Med. 156:1247-1255.
Okayama, N., Coe, L., Oshima, T., Itoh, M., and Alexander, J.S. (1999) Intracellular mechanisms of hydrogen peroxide-mediated neutrophil adherence to cultured human endothelial cells. Microvasc. Res. 57:63-74.
Ono, S.J., Nakamura, T., Miyazaki, D., Ohbayashi, M., Dawson, M., and Toda, M. (2003) Chemokines: roles in leukocyte development, trafficking, and effector function. J. Allergy Clin. Immunol. 111:1185-1199.
Oshima, T., Laroux, F., Coe, L., Morise, Z., Kawachi, S., Bauer, P., Grisham, M., Specian, R., Carter, P., Jennings, S., Granger, D., Joh,T., and Alexander, J.S. (2001) IFγ and IL-10 reciprocally regulate endothelial junction integrity and barrier function. Microvasc. Res. 61:130-143.
Oynebraten, I., Bakke, O., Brandtzaeg, P., Johansen, F.E., Haraldsen, G. (2004) Rapid chemokine secretion from endothelial cells originates from 2 distinct compartments. Blood 104:314-320.
Palmetshofer, A., Robson, S.C., Nehls, V. (1999) Lysophosphatidic acid activates NFκB and induces

proinflammatory gene expression in endothelial cells. Thromb.Haemost. 82:1532-1537.
Panetti, T.S., Chen, H., Misenheimer, T., Getzler, S., and Mosher, D.F. (1997) Endothelial cell mitogenesis induced by LPA: Inhibition by thrombospondin-1 and thrombospondin-2. J. Lab. Clin. Med. 129:208-216.
Panetti, T.S., Nowlen, J., and Mosher, D.F. (2000) Sphingosine-1-phosphate and LPA stimulate endothelial cell migration. Arterioscler. Thromb. Vasc. Biol. 20:1013-1019.
Papadimitriou, E., Manolopoulos, V., Hayman, G., Maragoudakis, M., Unsworth, B., Fenton, J., and Lelkes, P.I. (1997) Thrombin modulates vectorial secretion of extracellular matrix proteins in cultured endothelial cells. Amer. J. Physiol. 272:C1112-C1122.
Partridge, C.A., Jeffrey, J., and Malik, A.B. (1993) 96-kDa gelatinase induced by TNF-α contributes to increased microvascular endothelial permeability. Amer. J. Physiol. 265:L438-L447.
Patel, K.D., Zimmerman, G., Prescott, S., McEver, R., and Mcintyre, T.M. (1991) Oxygen radicals induce human endothelial cells to express GMP-140 and bind neutrophils. J. Cell Biol. 112:749-759.
Patterson, C.E. and Garcia, J.G.N. (1994) Regulation of thrombin-induced endothelial cell activation by bacterial toxins. Blood Coagul. Fibrinol. 5:63-72.
Patterson, C.E. and Lum, H. (2001) Update on pulmonary edema: the role and regulation of endothelial barrier function. Endothel. 8:75-105.
Patterson, C.E., Stasek, J., Bahler, C., Verin, A., Harrington, M., Garcia, J.G. (1998) Regulation of IL-1-stimulated GMCSF mRNA levels in human endothelium. Endothel. 6:45-59.
Paysant, J.R., Rupin, A., and Verbeuren, T.J. (2002) Effect of NADPH oxidase inhibition on E-selectin expression induced by concomitant anoxia/reoxygenation and TNF-α. Endothel. 9:263-271.
Pearce, M.J., Mcintyre, T., Prescott, S., Zimmerman, G., and Whatley, R.E. (1996) Shear stress activates cytosolic PLA_2 and MAPK in human endothelial cells. Biochem. Biophys. Res. Comm. 218:500-504.
Petrache, I., Verin, A.D., Crow, M.T., Birukova, A., Liu, F., Garcia, J.G.N. (2001) Differential effect of MLCK in TNF-a-induced endothelial cell apoptosis and barrier dysfunction. Amer. J. Physiol. 280:L1168-L1178.
Phillips, P.G., Birnby, L., Narendran, A., and Milonovich, W.L. (2001) NO modulates capillary formation at the endothelial cell-tumor cell interface. Amer. J. Physiol. 281:L278-L290.
Pietra, G.G. and Johns, L. (1990) Leaky intra-acinar arteries in rat lungs perfused with hydrogen peroxide. J. Appl. Physiol. 69:1110-1116.
Pietra, G.G., Magno, M., and Johns, L. (1979) Morphological and physiological study of the effect of histamine on the isolated perfused rabbit lung. Lymphology 12:165-176.
Pietra, G.G., Szidon, J., Leventhal, M., and Fishman, A.P. (1971) Histamine and interstitial pulmonary edema in the dog. Circ. Res. 29:323-337.
Pollet, I., Opina, C., Zimmerman, C., Leong, K., Wong, F., and Karsan, A. (2003) Bacterial lipopolysaccharide directly induces angiogenesis through TRAF6-mediated activation of NF-κB and c-Jun N-terminal kinase. Blood 102:1740-1742.
Prado, G.N., Taylor, L., Zhou, X., Ricupero, D., Mierke, D., Polgar, P. (2002) Mechanisms regulating expression, self-maintenance, and signaling of the bradykinin B2 and B1 receptors. J. Cell. Physiol. 193:275-286.
Prescott, S.M., Zimmerman, G.A., Mcintyre, T.M. (1984) Human endothelial cells in culture produce PAF when stimulated with thrombin. PNAS USA 81:3534-3538.
Rabiet, M.J.P., Plantier, J., Rival, Y., Genoux, Y., Lampugnani, M., and Dejana, E. (1996) Thrombin-induced increase in endothelial permeability is associated with changes in cell-to-cell junction organization. Arterioscl. Thromb. Vasc. Biol. 16:488-496.
Rahman, A., Anwar, K.N., and Malik, A.B. (2000) PKCζ mediates TNF-α-induced ICAM-1 gene transcription in endothelial cells. Amer. J. Physiol. 279:C906-C914.
Rahman, A., Roebuck, K.A., and Malik, A.B. (1996) Transcriptional regulation of endothelial adhesion molecule gene expression by oxidants and cytokines. In: Weir, E.K., Archer, S.L., Reeves, J.T. (Eds.), Nitric Oxide and Radicals in the Pulmonary Vasculature. Futura Publishing Company, Inc., Armonk, NY, pp. 63-85.
Rahman, A., True, A., Anwar, K., Ye, R., Voyno-Yasenetskaya, T., and Malik, A.B. (2002) Gαq and Gβγ regulate PAR-1 signaling of thrombin-induced NF-κB activation and ICAM-1 transcription in endothelial cells. Circ. Res. 91:398-405.

Rainger, G.E., Fisher, A.C., and Nash, G.B. (1997) Endothelial-borne platelet-activating factor and interleukin-8 rapidly immobilize rolling neutrophils. Amer. J. Physiol. 41:H114-H122.

Ranta, V., Orpana, A., Carpen, O., Turpeinen, U., Ylikorkala, O., and Viinikka, L. (1999) Human vascular endothelial cells produce tumor necrosis factor-alpha in response to proinflammatory cytokine stimulation. Crit. Care Med. 27:2184-2187.

RayChaudhury, A., Frischer, H., and Malik, A.B. (1996) Inhibition of endothelial cell proliferation and bFGF- induced phenotypic modulation by nitric oxide. J. Cell Biochem. 63:125-134.

Regoli, D. and Barabe, J. (1980) Pharmacology of bradykinin and related kinins. Pharm. Rev. 32:1-46.

Reilly, P.L.Woska, J., Jeanfavre, D. McNally, E., Rothlein, R., Bormann, B.J. (1996) The native structure of intercellular adhesion molecule-1 (ICAM- 1) is a dimer: Correlation with binding to LFA-1. J. Immunol. 155:529-532.

Resnick, N., Yahav, H., Khachigian, L., Collins, T., Anderson, K., Dewey, F., and Gimbrone, M.A., Jr. (1997) Endothelial gene regulation by laminar shear stress. Adv. Exp. Med. Biol. 430:155-164.

Resta, T.C., Gonzales, R., Dail, W., Sanders, T., and Walker, B.R. (1997) Selective upregulation of arterial endothelial NOS in pulmonary hypertension. Amer. J. Physiol. 41:H806-H813.

Rikitake, Y., Hirata, K., Kawashima, S., Takeuchi, S., Shimokawa, Y., Kojima, Y., and Inoue, N., and Yokoyama, M. (2001) Signaling mechanism underlying COX-2 induction by lysophosphatidylcholine. Biochem. Biophys. Res. Commun. 281:1291-1297.

Rizza, C., Leitinger, N., Yue, J., Fischer, D., Wang, D., Shih, P., Lee, H., Tigyi, G., and Berliner, J.A. (1999) Lysophosphatidic acid as a regulator of endothelial/leukocyte interaction. Lab. Invest. 79:1227-1235.

Robinson, C.J. and Stringer, S.E. (2001) The splice variants of vascular endothelial growth factor (VEGF) and their receptors. J. Cell Sci. 114:853-865.

Roebuck, K.A. and Finnegan, A. (1999) Regulation of intercellular adhesion molecule-1 (CD54) gene expression. J. Leuk. Biol. 66:876-888.

Rotrosan, D. and Gallin, J.I. (1986) Histamine type I receptor occupancy increases endothelial cytosolic calcium, reduces F-actin, and promotes albumin diffusion across cultured endothelial monolayers. J. Cell Biol. 103:2379-2387.

Salani, D., Taraboletti, G., Rosano, L., Di, C., Borsotti, P., Giavazzi, R., Bagnato, A. (2000) Endothelin-1 induces an angiogenic phenotype in cultured endothelial cells and stimulates neovascularization in vivo. Amer. J. Pathol. 157:1703-1711.

Salcedo, R., Ponce, M., Young, H., Wasserman, K., Ward, J., Kleinman, H., Oppenheim, J., and Murphy, W.J. (2000) Human endothelial cells express CCR2 and respond to MCP-1: direct role of MCP-1 in angiogenesis and tumor progression. Blood 96:34-40.

Salcedo, R., Young, H., Ponce, M., Ward, J., Kleinman, H., Murphy, W., and Oppenheim, J.J. (2001) Eotaxin (CCL11) induces angiogenic responses by human CCR3+ endothelial cells. J. Immunol. 166:7571-7578.

Schaeffer, R.C., Gong, F., Bitrick, M., and Smith,T.L. (1993) Thrombin and Bradykinin Initiate Discrete Endothelial Solute Permeability Mechanisms. Amer. J. Physiol. 264:H1798-H1809.

Schneeberger, P.M., Vanlangevelde, P., Vankessel, K., Vandenbrouckegrauls, C., and Verhoef, J. (1994) LPS induces hyperadhesion of endothelial cells for neutrophils leading to damage. Shock 2:296-300.

Schnittler, H.J., Puschel, B., and Drenckhahn, D. (1997) Role of cadherins and plakoglobin in interendothelial adhesion under resting conditions and shear stress. Amer. J. Physiol. 42:H2396-H2405.

Schraufstatter, I.U., Chung, J., and Burger, M. (2001) IL-8 activates endothelial cell CXCR1 and CXCR2 through Rho and Rac signaling pathways. Amer. J. Physiol. 280:L1094-L1103.

Scholz, M., Blaheta, R., Vogel, J., Doerr, H., and Cinatl, J.J. (1999) Cytomegalovirus-induced transendothelial cell migration. Intervirology 42:350-356.

Schulze, C., Smales, C., Rubin, L., and Staddon, J.M. (1997) Lysophosphatidic acid increases tight junction permeability in cultured brain endothelial cells. J. Neurochem. 68:991-1000.

Schwarz, G., Callewaert, G., Droogmans, G., and Nilius, B. (1992) Shear stress-induced Ca^{2+} transients in endothelial cells from human umbilical cord veins. J. Physiol-London 458:527-538.

Seebach, J., Dieterich, P., Luo, F., Schillers, H., Vestweber, D., Oberleithner, H., Galla, H., and Schnittler, H.J. (2000) Endothelial barrier function under laminar fluid shear stress. Lab. Invest. 80:1819-1831.

Sengupta, S., Gherardi, E., Sellers, L., Wood, J., Sasisekharan, R., and Fan, T.P. (2003) Hepatocyte growth factor/scatter factor can induce angiogenesis independently of vascular endothelial growth factor. Arterioscl. Thromb. Vasc. Biol. 23:69-75.
Shahgasempour, S., Woodroffe, S.B., and Garnett, H.M. (1997) Alterations in the expression of ELAM-1, ICAM-1 and VCAM-1 after in vitro infection of endothelial cells with a clinical isolate of human cytomegalovirus. Microbiol. Immunol. 41:121-129.
Shamaei-Tousi, A., Burns, M., Benach, J., Furie, M., Gergel, E., and Bergstrom, S. (2000) The relapsing fever spirochaete, Borrelia crocidurae, activates human endothelial cells and promotes the transendothelial migration of neutrophils. Cell Microbiol. 2:591-599.
Shasby, D.M., Ries, D., Shasby, S., and Winter, M.C. (2002) Histamine stimulates phosphorylation of adherens junction proteins and alters their link to vimentin. Amer. J. Physiol. 282:L1330-L1338.
Shasby, D.M., Stevens, T., Ries, D., Moy, A., Kamath, J., Kamath, A., and Shasby, S.S. (1997) Thrombin inhibits myosin light chain dephosphorylation in endothelial cells. Amer. J. Physiol. 16:L311-L319.
Shaw, S.K., Perkins, B., Lim, Y., Liu, Y., Nusrat, A., Schnell, F., Parkos, C., and Luscinskas, F.W. (2001) Reduced expression of junctional adhesion molecule and PECAM-1 (CD31) at human vascular endothelial junctions by cytokines TNF-α plus INF-γ does not reduce leukocyte transmigration under flow. Amer. J. Pathol. 159:2281-2291.
Shukaliak, J.A. and Dorovini-Zis, K. (2000) Expression of the β-chemokines RANTES and MIP-1β by human brain microvessel endothelial cells in primary culture. J. Neuropath. Exp. Neurol. 59:339-352.
Siflinger-Birnboim, A., Goligorsky, M., del Vecchio, P., and Malik, A.B. (1992) Activation of PKC pathway contributes to H_2O_2-induced endothelial permeability. Lab. Invest. 67:24-30.
Sirois, M.G. and Edelman, E.R. (1997) VEGF effect on vascular permeability is mediated by synthesis of platelet-activating factor. Amer. J. Physiol. 41:H2746-H2756.
Smiley, M.L., Hoxie, J.A., and Friedman, H.M. (1985) Herpes simplex virus type 1 infection of endothelial, epithelial, and fibroblast cells induces a receptor for C3b. J. Immunol. 134:2673-2678.
Smith, C.W., Rothlein, R., Hughes, B., Mariscalco, M., Rudloff, H., Schmalstieg, F., and Anderson, D.C. 91988) Recognition of an endothelial determinant for CD 18-dependent human neutrophil adherence and transendothelial migration. J. Clin. Invest. 82:1746-1756.
Smith, J.A., Webb, C., Holford, J., and Burgess, G.M. (1995) Signal transduction pathways for B1 and B2 bradykinin receptors in bovine pulmonary artery endothelium. Molec. Pharmacol. 47:525-534.
Smith, W.B., Noack, L., Khewgoodall, Y., Isenmann, S., Vadas, M., and Gamble, J.R., (1996) TGFβ1 inhibits the production of IL- 8 and transmigration of neutrophils through endothelium. J. Immunol. 157:360-368.
Sobel, B.E., Corr, P., Robison, A., Goldstein, R., Witkowski, F., and Klein, M.S. (1978) Accumulation of lysophosphoglycerides with arrhythmogenic properties in ischemic myocardium. J. Clin. Invest. 62:546-553.
Stamatovic, S.M., Keep, R., Kunkel, S., and Andjelkovic, A.V. (2003) Potential role of MCP-1 in endothelial cell tight junction 'opening': signaling via Rho and ROCK. J. Cell Sci. 116:4615-4628.
Stanimirovic, D.B., Yamamoto, T., Uematsu, S. and Spatz, M. (1994) ET-1 receptor binding and cellular signal transduction in cultured human brain endothelial cells. J. Neurochem. 62:592-601.
Staykova, M., Maxwell, L., and Willenborg, D. (2000) Kinetics and polarization of the membrane expression of cytokine-induced ICAM-1 on rat brain endothelial cells. J. Neuropathol. Exper. Neurol. 59:120-128.
Stevens, T., Rosenberg, R., Aird, W., Quertermous, T., Johnson, F., Garcia, J., Hebbel, R., Tuder, R., and Garfinkel, S. (2001). NHLBI workshop report: endothelial cell phenotypes in heart, lung, and blood diseases. Amer. J. Physiol. 281:C1422-C1433.
Stone, J.R. and Collins, T. (2002) The role of H_2O_2 in endothelial proliferative responses. Endothel. 9:231-238.
Subramaniam, M., Koedam, J.A., Wagner, D.D. (1993) Divergent fates of P- and E-selectins after their expression on the plasma membrane. Moecl. Biol. Cell 4:791-801.
Sugama, Y. and Malik, A.B. (1992) Thrombin receptor peptide mediates endothelial hyperadhesivity and neutrophil adhesion by P-selectin-dependent mechanism. Circ. Res. 71:1015-1019.
Sugama, Y., Tiruppathi, C., Janakidevi, K., Andersen, T., Fenton, J., and Malik, A.B. (1992) Thrombin-

induced expression of endothelial P-selectin and ICAM-1: stabilizing neutrophil adhesion. J. Cell Biol. 119:935-944.

Sun, H. and Mayhan, W.G. (2001) Temporal effect of alcohol consumption on reactivity of pial arterioles: role of oxygen radicals. Amer. J. Physiol. 280:H992-H1001.

Sung, C.P., Arleth, A., Shikano, K., and Berkowitz, B.A. (1988) Characterization and function of bradykinin receptors in vascular endothelial cells. J. Pharmacol. Exp. Ther. 247:8-13.

Szekanecz, Z., Shah, M., Harlow, L., Pearce, W., and Koch, A.E. (1994) Interleukin-8 and tumor necrosis factor-alpha are involved in human aortic endothelial cell migration. The possible role of these cytokines in human aortic aneurysmal blood vessel growth. Pathobiol. 62:134-139.

Takahara, N., Kashiwagi, A., Maegawa, H., and Shigeta, Y. (1996) Lysophosphatidylcholine stimulates the expression and production of MCP-1 by human vascular endothelial cells. Metabol. Clin. Exper. 45:559-564.

Takahashi, M., Ikeda, U., Masuyama, J., Funayama, H., Kano, S., and Shimada, K. (1996) NO attenuates adhesion molecule expression in human endothelial cells. Cytokine 8:817-821.

Takeshita, S., Inoue, N., Gao, D., Rikitake, Y., Kawashima, S., Tawa, R., Sakurai, H., Yokoyama, M. (2000) Lysophosphatidylcholine enhances superoxide anions production via endothelial NADH/NADPH oxidase. J. Atheroscler. Thromb. 7:238-246.

Tamaru, M. and Narumi, S. (1999) E-selectin gene expression is induced synergistically with the coexistence of activated classic PKC and signals elicited by IL-1β but not TNF-α. J. Biol. Chem. 274:3753-3763.

Tanaka, K., Abe, M., and Sato, Y. (1999) Roles of ERK1/2 and p38 MAPK in the signal transduction of bFGF in endothelial cells during angiogenesis. Jpn. J. Cancer Res. 90:647-654.

Tanaka, Y., Adams, D.H., and Shaw, S. (1993) Proteoglycans on endothelial cells present adhesion-inducing cytokines to leukocytes. Immunol. Today 14:111-115.

Tartaglia, L.A., Weber, R., Figari, I., Reynolds, C., Palladino, M., and Goeddel, D.V. (1991) The two different receptors for TNF mediate distinct cellular responses. PNAS USA 88:9292-9296.

Terada, L.S., Repine, J., Piermattei, D., and Hybertson, B.M. (1997) Endogenous NO decreases XO-mediated neutrophil adherence: Role of P-selectin. J. Appl. Physiol. 82:913-917.

Thoumine, O., Nerem, R.M., and Girard, P.R. (1995) Oscillatory shear stress and hydrostatic pressure modulate cell- matrix attachment proteins in cultured endothelial cells. In Vitro Cell Dev. Biol-Animal. 31:45-54.

Tokumura, A., Tanaka, T., Yotsumoto, T., and Tsukatani, H. (1992) Formation of PAF-like compounds by peroxidation of phospholipids from bovine brain. J. Lipid Mediat. 5:127-130.

Tzima, E., del Pozo, M., Shattil, S., Chien, S., and Schwartz,M.A. (2001) Activation of integrins in endothelial cells by fluid shear stress mediates Rho-dependent cytoskeletal alignment. EMBO J. 20:4639-4647.

Uematsu, M., Ohara, Y., Navas, J., Nishida, K., Murphy, T., Alexander, R., Nerem, R., and Harrison, D.G. (1995) Regulation of endothelial cell NOS expression by shear stress. Amer. J. Physiol. 38:C1371-C1378.

Ueno, A., Murakami, K., Yamanouchi, K., Watanabe, M., and Kondo, T. (1996) Thrombin stimulates production of interleukin-8 in human umbilical vein endothelial cells. Immunol. 88:76-81.

Utgaard, J.O., Jahnsen, F., Bakka, A., Brandtzaeg, P., and Haraldsen, G. (1998) Rapid secretion of prestored IL-8 from Weibel-Palade bodies of microvascular endothelial cells. J. Exp. Med. 188:1751-1756.

van Nieuw Amerongen, G.P., Draijer, R., Vermeer, M., and van Hinsbergh, V.W. (1998) Transient and prolonged increase in endothelial permeability induced by histamine and thrombin. Circ. Res. 83:1115-1123.

Vanhauwe, J.F., Thomas, T., Minshall, R., Tiruppathi, C., Li, A., Gilchrist, A., Yoon, E., Malik,A., and Hamm, H.E. (2002) Thrombin receptors activate G(o) proteins in endothelial cells to regulate intracellular calcium and cell shape changes. J. Biol. Chem. 277:34143-34149.

Vestweber, D. and Blanks, J.E. (1999) Mechanisms that regulate the function of the selectins and their ligands. Physiol. Rev. 79:181-213.

Volk, T., Mading, K., Hensel, M., and Kox, W.J. (1997) Nitric oxide induces transient Ca^{2+} changes in endothelial cells independent of cGMP. J. Cell. Physiol. 172:296-305.

Vora, M., Romero, L.I., Karasek, M.A. (1996) Interleukin-10 induces E-selectin on small and large blood vessel endothelial cells. J. Exp. Med. 184:821-829.
Vouret-Craviari, V., Grall, D., Flatau, G., Pouyssegur, J., Boquet, P., and Obberghen-Schilling, E. (1999) Effects of cytotoxic necrotizing factor 1 and lethal toxin on actin cytoskeleton and VE-cadherin localization in human endothelial cell monolayers. Infect. Immun. 67:3002-3008.
Vu, T.K.H., Hung, D., Wheaton, V., and Coughlin, S.R. (1991) Molecular cloning of a functional thrombin receptor reveals a novel proteolytic mechanism of receptor activation. Cell 64:1-20.
Wagner, D.D. (1993) The Weibel-Palade Body - The Storage Granule for von Willebrand Factor and P-Selectin. Thromb. Haemost. 70:105-110.
Wahl, M., Whalley, E., Unterberg, A., Schilling, L., Parsons, A., Baethmann, A., and Young, A.R. (1996) Vasomotor and permeability effects of bradykinin in cerebral microcirculation. Immunopharm. 33:257-263.
Walton, K.A., Hsieh, X., Gharavi, N., Wang, S., Wang, G., Yeh, M., Cole, A., and Berliner, J.A. (2003) Receptors involved in the ox-phosphorylcholine-mediated synthesis of IL-8. J. Biol. Chem. 278:29661-29666.
Wang, H. and Keiser, J.A. (2000) Hepatocyte growth factor enhances MMP activity in human endothelial cells. Biochem. Biophys. Res. Comm. 272:900-905.
Wang, H.S., Li, F., Runge, M., and Chaikof, E.L. (1997) Endothelial cells exhibit differential chemokinetic and mitogenic responsiveness to α-thrombin. J. Surg. Res. 68:139-144.
Warke, R.V., Xhaja, K., Martin, K., Fournier, M., Shaw, S., Brizuela, N., De Bosch, N., Lapointe, D., Ennis, F., Rothman, A., and Bosch, I. (2003) Dengue virus induces novel changes in gene expression of human umbilical vein endothelial cells. J. Virol. 77:11822-11832.
Wary, K.K., Thakker, G., Humtsoe, J., and Yang, J. (2003) Analysis of VEGF-responsive Genes Involved in the activation of endothelial cells. Molec. Cancer 2:25.
Weber, K.S., Klickstein, L.B., and Weber, C. (1999a) Specific activation of leukocyte beta2 integrins lymphocyte function-associated antigen-1 and Mac-1 by chemokines mediated by distinct pathways via the alpha subunit cytoplasmic domains. Molec. Biol. Cell 10:861-873.
Weber, K.S., von Hundelshausen, P., Clark-Lewis, I., Weber, P., and Weber, C. (1999b) Differential immobilization and hierarchical involvement of chemokines in monocyte arrest and transmigration on inflamed endothelium in shear flow. Eur. J. Immunol. 29:700-712.
Wechezak, A.R., Viggers, R.F., and Sauvage, L.R. (1985) Fibronectin and F-actin redistribution in cultured endothelial cells exposed to shear stress. Lab. Invest. 53:639-647.
Weller, A., Isenmann, S., and Vestweber, D. (1992) Cloning of the mouse endothelial selectins. Expression of both E- and P-selectin is inducible by TNFα. J. Biol. Chem. 267:15176-15183.
Whisler, R.L., Goyette, M., Grants, I., and Newhouse, Y.G. (1995) Sublethal oxidant stress stimulate multiple serine/threonine kinases and suppress phosphatases in Jurkat T cells. Arch. Biochem. Biophy. 319:23-35.
Wogensen, L., Huang, X.J., and Sarvetnick, N. (1993) Leukocyte extravasation into pancreatic tissue in transgenic mice expressing IL-10 in the islets of langerhans. J. Exp. Med. 178:175-185.
Wojciak-Stothard, B., Entwistle, A., Garg, R., and Ridley, A.J. (1998) Regulation of TNF-α-induced reorganization of the actin cytoskeleton and cell-cell junctions by Rho, Rac, and Cdc42 in human endothelial cells. J. Cell. Physiol. 176:150-165.
Wojciak-Stothard, B., Williams, L., and Ridley, A.J. (1999) Monocyte adhesion and spreading on human endothelial cells is dependent on Rho-regulated receptor clustering. J. Cell Biol. 145:1293-1307.
Wolff, B., Burns, A., Middleton, J., Rot, A. (1998) Endothelial cell "memory" of inflammatory stimulation: human venular endothelial cells store interleukin 8 in Weibel-Palade bodies. J. Exp. Med. 188:1757-1762.
Wong, D. and Dorovinizis, K. (1996) Regulation by cytokines and lipopolysaccharide of E- selectin expression by human brain microvessel endothelial cells in primary culture. J. Neuropath. Exp. Neurol. 55:225-235.
Wong, R.K., Baldwin, A.L., and Heimark, R.L. (1999) Cadherin-5 redistribution at sites of TNF-α and IFN-γ-induced permeability in mesenteric venules. Amer. J. Physiol. 276:H736-H748.
Woska, J.R., Jr., Morelock, M., Jeanfavre, D., and Bormann, B.J. (1996) Characterization of molecular interactions between ICAM-1 and leukocyte function- associated antigen-1. J. Immunol. 1564680-

4685.

Wu, H.M., Yuan, Y., Zawieja, D., Tinsley, J., and Granger, H.J. (1999) Role of PLC, PKC, and calcium in VEGF-induced venular hyperpermeability. Amer. J. Physiol. 276:H535-H542.

Wyble, C.W., Desai, T., Clark, E., Hynes, K., and Gewertz, B.L. (1996) Physiologic concentrations of TNFα and IL-1β released from reperfused human intestine upregulate E-selectin and ICAM-1. J. Surg. Res. 63:333-338.

Wyble, C.W., Hynes, K., Kuchibhotla, J., Marcus, B., Hallahan, D., Gewertz, B.L. (1997) TNF-α and IL-1 upregulate membrane-bound and soluble E- selectin through a common pathway. J. Surg. Res. 73:107-112.

Xu, Y. (2002) Sphingosylphosphorylcholine and lysophosphatidylcholine: G protein-coupled receptors and receptor-mediated signal transduction. Biochim. Biophys. Acta 1582:81-88.

Yan, W., Zhao, K., Jiang, Y., Huang, Q., Wang, J., Kan, W., and Wang, S. (2002) Role of p38 MAPK in ICAM-1 expression of vascular endothelial cells induced by lipopolysaccharide. Shock 17:433-438.

Yao, L., Pan, J., Setiadi, H., Patel, K., and McEver, R.P. (1996) IL-4 or oncostatin M induces prolonged increase in P-selectin mRNA and protein in human endothelial cells. J. Exp. Med. 184:81-92.

Yoshida, S., Ono, M., Shono, T., Izumi, H., Ishibashi, T., Suzuki, H., and Kuwano, M. (1997) Involvement of IL-8, VEGF, and bFGF in TNFα-dependent angiogenesis. Molec. Cell Biol. 17:4015-4023.

Yuan, Y.A., Granger, H., Zawieja, D., Defily, D., and Chilian, W.M. (1993) Histamine increases venular permeability via PLC- NO synthase-guanylate cyclase. Amer. J. Physiol. 264:H1734-H1739.

Zhu, Y., Lin, J., Liao, H., Verna, L., and Stemerman, M.B. (1997) Activation of ICAM-1 promoter by LPC: involvement of protein tyrosine kinases. Biochim. Biophys. Acta 1345:93-98.

Ziche, M., Morbidelli, L., Masini, E., Amerini, S., Granger, H., Maggi, C., Geppetti, P., and Ledda, F. (1994) Nitric oxide mediates angiogenesis in vivo and endothelial cell growth and migration in vitro promoted by substance P. J. Clin. Invest. 94:2036-2044.

Zidovetzki, R., Chen, P., Chen, M., Hofman, F.M. (1999) ET-1-induced IL-8 production in human brain-derived endothelial cells is mediated by the PKC and protein tyrosine kinase pathways. Blood 94:1291-1299.

Zweier, J.L., Kuppusamy, P., and Lutty, G.A. (1988) Measurement of endothelial cell free radical generation: evidence for a central mechanism of free radical injury in postischemic tissues. PNAS USA 85:4046-4050.

Chapter 4

Membrane and Cellular Signaling of Integrity and Acute Activation

Viswanathan Natarajan, Ph.D.,[1]* Peter V. Usatyuk, Ph.D.,[1] and Carolyn E. Patterson, Ph.D.[2,3]

[1]Johns Hopkins University School of Medicine, Baltimore, MD, USA 21224,
[2]Indiana University School of Medicine, Indianapolis, IN, USA 46202, and
[3]The Roudebush VA Medical Center, Indianapolis, IN 46202

CONTENTS:

Introduction

Principles of Signaling
Acute Signaling vs. Transcription and Translation
Signaling of State Transition and Maintenance

Signaling of Quiescence and Barrier Integrity
Mechanisms and Mediators for Making and Maintaining the Barrier
cAMP- the First Second Messenger
Sphingosine 1- Phosphate
Angiopoietin

Signaling of Activation and Motility
Membrane Receptors and Events
Ca^{++} - Second Messenger Signaling
Lipid – Second Messenger Signaling
Signaling by Tyrosine Kinases and Phosphatases
Small GTPase Signaling
Oxidant Signaling

Integration and Interactions of Signaling Mechanisms
Cooperation, Coordination, and Contradiction
Integration with Long-term Signaling
Autocrine and Paracrine Signaling

Signaling Synopsis

Introduction

The importance of endothelial barrier function to vascular integrity and to pulmonary health is clear from the discussion in Chapter 1 and the structural basis for this vital function was described in Chapter 2. Chapter 3 revealed that the endothelia respond to a plethora of stimuli and is in a unique position at the cross roads of key processes, such as coagulation and inflammation, to influence those processes. Additionally, as a director of vascular smooth muscle function, the endothelium exercises some control over organ perfusion, particularly at the microcirculatory level. In order to perform these various functions as responder and regulator, endothelial cells must be able to communicate with their environment and put this information into action. The basis of this physiologic functionality is the biochemical signaling in response to the external environment and the genetically driven signaling toward cellular integrity. The purpose of this chapter is to introduce some of the primary players and pathways for the molecular integration and coordination of these functions.

Principles of Signaling

Acute Signaling vs. Transcription and Translation

As we simplified the very complex and dynamic existence of endothelium by focusing on two states, quiescence and activation, we will make a similar simplistic division of signaling into "acute" and "long term" mechanisms. By acute, we mean processes that can be altered in a matter of seconds or perhaps minutes, involving release of intracellular second messengers and rapid responses, such as altered phosphorylation, redox, and conformational states of proteins. Such signals are often triggered by the binding of external mediators to membrane receptors and the consequent changes in cellular proteins often result in their translocation to membranes or to complexes with other proteins. By long term, we mean processes that may require a number of minutes to days for full expression and may involve transcription of new protein and translation of genetic information into new mRNA. These long term signaling mechanisms will be discussed in Chapter 6. Of course the cell doesn't make this artificial division and there is temporal and spatial integration of signaling, which we will attempt to describe as we proceed. Furthermore, certain processes with fairly short-term effects that are usually associated with pathways leading to transcription will be primarily discussed in Chapter 6, especially when initiated by growth factors.

Signaling of State Transition and Maintenance

Furthermore, we will divide signaling according to the "two states" of the endothelium. Some pathways are primarily associated with establishment of the architecture of quiescent endothelial cells, formation of cell-cell junctions, and maintenance of barrier integrity. Other signals lead to cell activation resulting in altered cell architecture, temporary or extended barrier dysfunction, altered release of paracrine factors, processes involved with cell motility, and processes resulting in either proliferation or apoptosis. Generally, endothelial cells have a low turnover rate and are in a stable state of functional homeostasis, yet even at these times biochemical processes proceed to maintain their viability and functional integrity. At other times, the cell is in transition from either quiescence to activation or activation to quiescence. These times of dramatic changes in signaling have been the focus of most experimental work. There is also heterogeneity of the vascular endothelial population in their ability to respond to particular mediators and in the responses, which they make. Some of this heterogeneity is determined genetically and some is determined by the environmental history of the cell.

Signaling of Quiescence and Barrier Integrity

Mechanisms and Mediators for Making and Maintaining the Barrier

In vasculogenesis of embryologic development, in angiogenesis of development, repair, and tumor growth, in repair of small wounds that occurs routinely in the vascular bed, and in cultured endothelium plated on new medium, endothelial cells reach a point where contacts are made between cells and a barrier is established. Similarly, after activation without further stimulation, the endothelial cells return to a state of confluency and quiescence. These signals also result in biochemical resistance to activation and injury. The signaling associated with these processes and with maintenance of the intact endothelial barrier will be briefly introduced here, as the main focus of this chapter will be on signals of activation. Further details are contained in the following chapters on cytoskeleton elements, cell-to-cell junctions, and cell-to-matrix junctions and particularly in Chapter 15 on protective mechanisms and in Chapter 5 on regulation of cAMP, a key signal for quiescence in many cells.

Formation of an endothelial monolayer depends on recognition of endothelial cells for each other and a sequential lacing together at their mutual lateral borders by connecting a number of junctional molecules and complexes as shown in Chapter 2. Details of this process will be given in Chapter 15. Signaling molecules are associated with these cell to cell and also to cell-to-matrix junctions. These kinases and phosphatases are important is translating the initial contact into assembly of a stable multimolecular complexes with anchoring attachments to the cytoskeleton. Further, they signal changes in the cytoskeleton. Moreover, certain growth factor receptors, such as VEGFR2 (KDR, *flk1*), and GTPases, such as rac, are localized at these cell-cell junctions (Braga *et al.*, 1999; Vestweber, 2000; Shay-Salit *et al.*, 2002; Lampugnani *et al.*, 2003). In all very little is actually known about the temporal aspects and specific signals guiding this border sealing process and the relationship of these mechanisms to biochemical regulation of the cytoskeletal architecture.

Maintenance of the barrier intact state is the normal genetically predetermined phenotype and thus the normal ongoing signals participate in the direction of the structural elements and metabolic processes consistent with this state. Hormone and autocrine mediators that support quiescence, e.g. β-agonists and prostacyclin, interact with specific receptors on the endothelial surface that trigger membrane events resulting in synthesis and intracellular release of the second messenger, cAMP. Additionally, albumin, thrombocytes, and erythrocytes, which are in normal contact with the endothelial apical surface, signal endothelial quiescence. Albumin has several beneficial effects, by binding at discrete sites in or near the invaginated coated pits on the luminal surface results in lowering basal $[Ca^{++}]_i$ and reducing Ca^{++}-mediated activation signals (He and Curry, 1993a). Quiescent platelets release S1P (S1P), which signals endothelial quiescence (Haselton and Alexander, 1992). Red cells protect endothelium, primarily by scavenging oxidant radicals (Tsan and White, 1988). Finally the growth factor, angiopoietin, promotes endothelial barrier function (Gamble *et al.*, 2000; Jones, 2003).

cAMP- the First Second Messenger

Numerous membrane receptors are coupled to specific α,β,γ-trimeric GTP-binding proteins and certain ones of these receptors are coupled to adenylyl cyclase. Ligation of these receptors leads to an exchange of GDP for GTP on the Gsα subunit and dissociation from the βγ subunit. The α and βγ subunits bind different moieties on the integral membrane adenylate cyclases within the receptor-G protein-cyclase complex. There are 9 adenylyl cyclase isoforms with different tissue expressions and different interactions with βγ and Ca^{++}/calmodulin (Hanoune and Defer, 2001). Isoform expression in macrovascular and microvascular endothelium and differential regulation is covered in the following chapter. Whereas the Gsα subunit is stimulatory, βγ may be stimulatory or inhibitory depending on the isotype. Cyclase activation results in conversion of ATP to cAMP (adenosine 3', 5'-cyclic monophosphate). Metabolism of cAMP by several different phosphodiesterases also influences the cellular cAMP level (Lugnier and Schini, 1990).

Pharmacologic elevation of cAMP by various means, such as forskolin stimulation of the catalytic core of adenylyl cyclase or use of membrane-permeable modified cAMPs, results in improved basal barrier function in cultured endothelial cells (Farrukh *et al.*, 1987; Minnear *et al.*, 1989; Stelzner *et al.*, 1989; He and Curry, 1993b; Barnard *et al.*, 1994; Patterson *et al.*, 1994a; Patterson *et al.*, 1994b; Lum *et al.*, 1999). Increased cAMP also effectively opposes activation by many agonists and accelerates reversal of established dysfunction (Patterson *et al.*, 1994a). This will be explained in greater detail in Chapter 15. Thus, cAMP is a key intracellular second messenger signaling endothelial quiescence. This molecule conveys its messages within the cell primarily by binding and activating the cAMP-dependent protein kinase (PKA) as depicted in Figure 1. PKA is known in many cell types to elicit a large cascade, involving phosphorylation and dephosphorylation of many proteins. In endothelium some of the PKA targets result in inhibition of components in the activation pathway, such as myosin light chain kinase (MLCK) (Moy *et al.*, 1993; Patterson *et al.*, 1994a; Garcia *et al.*, 1995a), but other targets may directly promote barrier function. A few other possible identified targets include: IP_3 receptors (Moore *et al.*, 1998; Wojcikiewicz and Luo, 1998), filamin (Hastie *et al.*, 1997), actin binding proteins (Hastie *et al.*, 1997; Jay and Stracher, 1997), serine/threonine protein phosphatases (Usui *et al.*, 1998), and ras and RhoA (Lang *et al.*, 1996; Arimura *et al.*, 1997).

Whereas PKA activation promotes endothelial barrier integrity, antagonistic inhibition or overexpression of PKI, an endogenous PKA inhibitor, results in cytoskeletal reorganization to the activated pattern, disassembly of adhesion complexes, and barrier dysfunction (Stevens *et al.*, 1995; Lum *et al.*, 1999; Patterson *et al.*, 2000). Moreover, some conditions or mediators that cause endothelial permeability, such as hypoxia, TNFα, and H_2O_2, do so by decreasing cAMP (Ogawa *et al.*, 1992; Suttorp *et al.*, 1993; Koga *et al.*, 1995). This may be mediated *via* other receptors coupled to G proteins inhibitory for adenylate cyclase (e.g. Giα-coupled receptors), by Ca^{++}- mediated inhibition of adenylate cyclase, by phosphodiesterase activation, or possibly *via* PKI (Stevens *et al.*, 1995; Chetham *et al.*, 1997; Manopoulos *et al.*, 1997; Stevens *et al.*, 1997).

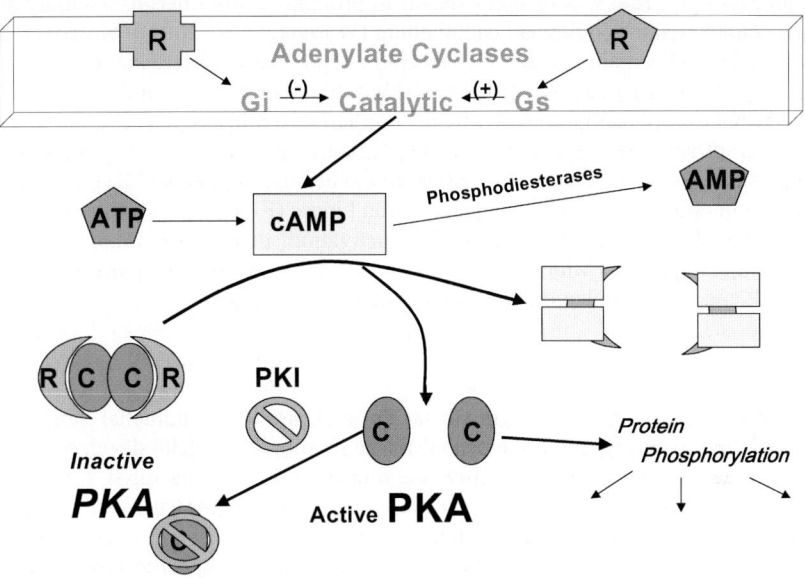

Figure 1. Activation of PKA. A variety of receptors are linked to adenylyl cyclase via coupling to Gi and Gs, which respectively inhibit or stimulate adenylyl cyclase, upon receptor occupation and activation of nucleotide exchange on the G-protein. Whereupon, adenylyl cyclase catalyzes less or more cAMP synthesis, altering cAMP levels in the particular cellular locality of the receptor-cyclase complex. The concentration of cAMP also depends on catabolism by a variety of phosphodiesterases. This is not usually an ON/OFF regulation, but synthesis and degradation are in dynamic balance that is modified by the respective coupled G-protein activation and regulation of the phosphodiesterases. Four cAMP molecules cooperatively bind to inhibitor moieties of the cAMP-dependent protein kinase (PKA), freeing up the catalytic subunit (C). PKA-mediated phosphorylation activates a large number of downstream substrates, including kinases and phosphatases, depending on tissue and compartmentation. In endothelium, intrinsic activity and receptor-activation are prime mechanisms for maintaining barrier integrity. When the cAMP level drops, the inhibitor units re-bind the PKA and terminate the catalysis. Alternatively, there is a separate endogenous inhibitor of the catalytic subunit (PKI), which can inactivate PKA, but little is known about its regulation and physiologic significance.

Sphingosine 1- Phosphate

Platelets were long known to express a factor that promoted endothelial barrier function. This factor was narrowed down to a lipid, then to lysophosphatidic acid, and finally to S1P (Haselton and Alexander, 1992; Alexander *et al.*, 1998; Lee *et al.*, 1999; Yatomi *et al.*, 2000; Schaphorst *et al.*, 2003). The signaling was found to be unrelated to cAMP elevation (Gainor *et al.*, 2001; Minnear *et al.*, 2001), but rather to activation of a small GTPase, Rac (Lee *et al.*, 1999). Interestingly, Rac was recently shown to promote stress fiber disassembly and DPB assembly, characteristic of quiescent endothelium (Waschke *et al.*, 2004). Moreover, S1P was shown to bind the G protein coupled Edg receptors (endothelial differentiation gene receptor) (Fischer *et al.*, 1998; Spiegel and Milstein, 2000), which is important in the barrier effects of S1P in endothelium (Schaphorst *et al.*, 2003). Furthermore it behaves as both an extracellular signal and an intracellular signal, with many effects in various cells, including opposition to apoptosis (Watterson *et al.*, 2003). In contrast to the beneficial effects of S1P in endothelium, it was shown in one study to induce focal adhesion kinase (FAK) phosphorylation, stress fiber formation, and MLC phosphorylation *via* Rho activation; thus, there is much more to learn about this lipid regulator (Miura *et al.*, 2000). For more on the receptors and differences between S1P and lysophosphatidic acid see lipid mediators below.

Angiopoietin

Angiopoietins are a family of growth factors that promote endothelial barrier integrity, cell survival, angiogenesis, and are essential for embryologic and fetal development and will mainly be discussed in Chapter 6. But angiotensin also exhibits rapid signaling effects (under 10 minutes) with the ability to reduce basal permeability (measured after 30 minutes of angiopoietin exposure) and block activation by a several mediators, including thrombin, TNF, and VEGF (Gamble *et al.*, 2000). While the mechanisms and specific targets for its rapid promotion of barrier function are unclear, they are not due to the same signaling mechanisms that are important in promoting cell survival (Gamble *et al.*, 2000; Ward and Dumont, 2002; Jones, 2003). One possible mechanism is suggested by the demonstration that angiopoietin-1 decreased phosphorylation of the intercellular junction proteins, PECAM-1 and VE-cadherin (Gamble *et al.*, 2000). Long-term effect will be considered in Chapter 6.

Signaling of Activation and Motility

Membrane Receptors and Events

Numerous extracellular mediators, both proteins and phospholipids, exert their effects by engaging specific receptors on the endothelial surface, which generally span the membrane 7 times and are coupled to integral membrane heterotrimeric GTP-proteins (Lum and Malik, 1994; Garcia *et al.*, 1995b; Stevens, *et al.*, 2000; Yen *et al.*, 2003). Receptor activation leads to exchange of GDP in the GTP-ase protein (G-protein) for GTP, which activates the G-protein. Notably, many signaling systems and receptors are concentrated in plasmalemma invaginations on the endothelial luminal surface, known as caveolae. The G-proteins, in turn, activate or inhibit associated enzymes, such as phospholipase C (PLC), phospholipase A_2 (PLA_2), adenylate cyclase, and associated ion channels, as shown in Figure 2. For instance, thrombin activates its receptor (mainly PAR-1, protease activated receptor), which is coupled to Gq, resulting in activation of PLC (Garcia *et al.*, 1993; Yan *et al.*, 1998). Histamine activates H1- G-protein coupled receptors to induce Ca^{++} release, degradation of the tight junction protein ZO1, and permeability (Rotrosen and Gallin, 1986; Gardner *et al.*, 1996; MacGlashan, 2003); whereas, H2 engagement leads to increased cAMP and attenuation of permeability. The net effect depends on receptor expression, dose, duration of activation, and

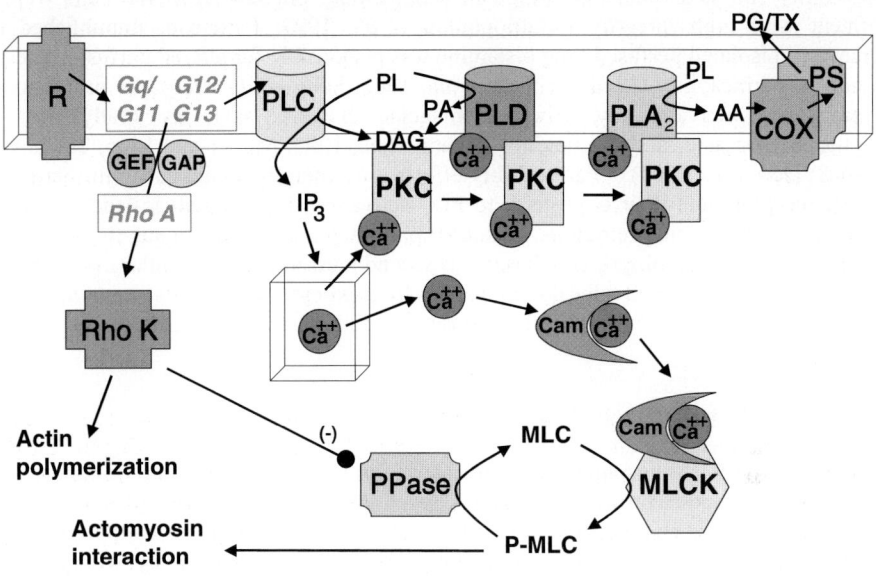

Figure 2. Signaling of endothelial cell activation. Receptors for activating agonists, such as thrombin, histamine, and bradykinin, (R) are coupled to a variety of GTPase-proteins, which propagate the signal to various phospholipase Cs (PLC), that catalyze the hydrolysis of particular membrane phospholipids producing diacylglycerol (DAG) and possibly inositol 1,4,5 trisphosphate (IP$_3$), depending on the particular PLC, the phospholipid, and the duration of activation. The DAG is a potent activator of certain protein kinase Cs (PCK), such as PKCα. This classic PKC also requires Ca^{++} for activation, which is initially derived by release from the endoplasmic reticular stores, under the command from the IP$_3$, and subsequently derived by activation of membrane ion channels. The activated PKC can further activate phospholipase D (PLD), which catalyzes hydrolysis of other membrane lipids and produces phosphatidic acid, which is then converted to DAG. The active PKC and the increased Ca^{++} also activates phospholipase A$_2$ (PLA$_2$) releasing arachidonic acid, that is metabolized by cyclooxygenase (COX) and prostaglandin synthase (PS) to form released prostanoids, such as prostacyclin and thromboxane. At the same time, the increased Ca^{++} binds calmodulin and the Cam/Ca^{++} complex activates myosin light chain kinase (MLCK), which phosphorylates the light chains, rendering them capable of binding actin. Another consequence of receptor activation of particular membrane G-proteins, such as G13, is activation of nucleotide exchange on a small GTPase, such as RhoA, by activation of GEF and GAP, protein modifyers of the small G-protein. RhoA, can then in turn activate an effector kinase, Rho kinase. The Rho K can inhibit MLC phophatase and independently increase MLC phosphorylation. Rho K also phosphorylates several actin-binding proteins, initiating actin polymerization and reorganization.

on the state and type of the endothelium. For instance, histamine was shown to induced permeability and junctional dissolution in endothelium, but not when the cells were long confluent with high integrity (Andriopoulou *et al.*, 1999; Patterson, unpublished data). Moreover, in isolated perfused lung histamine was responsible for altered perfusion pressures, but not for permeability (Lindenschmidt *et al.*, 1986; Selig *et al.*, 1986). G protein-coupled receptors are activated by other edemagenic ligands, such as bradykinin, PAF, lysophosphatidic acid, and chemokines (Kuang *et al.*, 1996; Muller and Lipp, 2001; Vogt *et al.*, 2003; Deo *et al.*, 2004). Permeability effects are generally mediated by further coupling of these receptors and their G-proteins to PLC activation, which catalyzes the hydrolysis of membrane phospholipids releasing, inositol-triphosphates and diacylglycerols, depending in part on the lipids metabolized, which serve as second messengers for cellular events described below. Engagement of cell adhesion molecules by leukocytes also elicits signaling that alters junctional properties and cytoskeletal responses as will be described in Chapter 11.

Ca^{++} - Second Messenger Signaling

Inositol 1,4,5-trisphosphate (IP$_3$) released from particular membrane phospholipids interacts with receptors on the endoplasmic reticulum triggering release of stored Ca^{++} into the cytoplasm (Ferris and Snyder, 1992). One effect of depletion of the endoplasmic store is activation of an ion-channel leading to a slower, more sustained influx of extracellular Ca^{++} (Wu *et al.*, 2000; Cioffi *et al.*, 2003). Lipids released after membrane G-protein activation trigger opening of other ion-channels for additional Ca^{++} entry (Bird *et al.*, 2004; Nilius and Voets, 2004). The voltage gated and non-voltage gated Ca^{++} channels and the interactive regulation of Ca^{++} and cAMP will be considered in detail in the following chapter. Although, most work has been done with G-protein coupled receptors for bioactive mediators, such as thrombin and histamine, VEGF activation of VEGF-R2 and angiotensin activation of the AT-1 receptor activates PLC, increases intracellular Ca^{2+}, leads to DAG generation, and subsequent PKC-activation consistent with its effects on endothelial permeability (Takahashi and Shibuya, 1997; Wu *et al.*, 1999; Yin *et al.*, 2003). Integrin induced tyrosine phosphorylation was shown to induce a Ca^{2+} influx in pulmonary endothelial cells (Bhattacharya *et al.*, 2000) and the opening of store-operated calcium channels has also been linked to activation of pp60c-src (Babnigg *et al.*, 1997).

Mechanical disturbance of the endothelium can also trigger Ca^{++} release (Kuebler *et al.*, 1999; Lehoux and Tedgui, 2003; Nilius and Voets, 2004). Increased $[Ca^{++}]_i$ results in multiple effects that contribute to permeability. Although Ca^{++} alone is not sufficient to induce barrier dysfunction, blockers and chelators of Ca^{++} attenuate permeability induced by a number of mediators, indicating its importance (Chetham *et al.*, 1997; Garcia *et al.*, 1997; Kelly *et al.*, 1998; Siflinger-Birnboim *et al.*, 1996; Gao *et al.*, 2000; Usatyuk *et al.*, 2003a).

One such effect is the activation of MLCK when four Ca^{++} molecules bind calmodulin and the complex binds MLCK, similar to Ca^{++} activation of contraction in smooth muscle cells (Lum *et al.*, 1992; Garcia *et al.*, 1993; Sheldon *et al.*, 1993; Garcia *et al.*, 1995b; Moy *et al.*, 1996; Shasby *et al.*, 1997). Although not *via* G-protein coupled receptors, hydrogen peroxide also triggers an increase in cellular Ca^{++} and MLCK activation (Zhao and Davis, 1998; Lopez-Ongil *et al.*, 1999). Activated MLCK directly phosphorylates myosin light chains at Thr-18 and Ser-19, which alters their conformation and increases myosin binding for actin. Consistent with the importance of actomyosin interaction, ML-7 and KT5926, inhibitors of MLCK, attenuate MLC phosphorylation and permeability induced by thrombin and histamine (Garcia et al., 1995a; Garcia et al., 1995b; Moy et al., 1996). Endothelial cells normally maintain a low rate of MLCK activation and MLC-phosphorylation so inhibition of phosphatases, which alter the steady state level, is another mechanism for mediator- and oxidant-induced increases in actomyosin interaction (Essler et al., 1998; Shasby et al., 1997). Thus, inhibition of protein phosphatase-1 increased MLC phosphorylation and endothelial permeability and phosphatase 2B inhibition sustained agonist-induced permeability (Verin et al., 1995; Verin et al., 1998). Actomyosin shortening may occur in the DPB within the first few seconds of agonist-activation and actomyosin contraction contributes to tension exerted by stress fibers on the focal adhesion complexes, but the massive reorganization of the actin cytoskeleton as described in Chapter 2 suggests much more is occurring than muscle-pattern actomyosin shortening in activation of endothelial cells. Further, endothelial permeability can occur by mechanisms independent of MLC phosphorylation (Patterson et al., 1994; Patterson et al., 1995; Shasby et al., 1995; Moy et al., 1996; Garcia et al., 1997; Patterson, et al., 2000), implicating other regulatory mechanisms, such as PKC and tyrosine kinases, and other targets such as actin binding proteins, that regulate actin polymerization, depolymerization, and organization, and junctional proteins, that regulate the intercellular barrier tightness.

Increased $[Ca^{++}]_i$ activates numerous other cellular enzymes. Membrane PLA_2, which hydrolyzes membrane phospholipids at the second position, releases arachidonic acid, which is rate limiting for synthesis of prostanoids. Thus, agonists, that increase endothelial Ca^{++}, increase release of PGI_2, and perhaps thromboxane A_2 and $PGF_{2\alpha}$ depending on further metabolism, as also seen in Figure 2. Moreover, released lysophospholipid may be further metabolized to PAF, which reacts with a G-protein receptor to stimulate permeability itself. Ca^{++} increase is also an activator of membrane phospholipase D (PLD). PLD lysis of membrane lipids producing phosphatidic acid is more sustained than PLC lysis of phospholipids (see next section). Classic protein kinase C isoforms α and β are also activated by Ca^{++} (Stasek et al., 1992; Bussolino et al., 1994; Vuong et al., 1998). Certain types of adenylyl cyclase isoforms are inhibited by Ca^{++} resulting in mediator-induced decreases in cAMP (Stevens et al., 1995), but cAMP increases do not decrease basal or mediator-induced Ca^{++} increases (Patterson et al., 2000).

Lipid – Second Messenger Signaling

Diacylglycerol (DAG) is initially released from membrane phospholipids by the G-protein coupled PLC activation as IP_3 is released, but the net amount is low and the release is transient. Subsequent activation of PLD releases much more phosphatidic acid, which is then metabolized to form the more significant pool of DAG. DAG is a physiologically important activator of the serine/threonine protein kinases, PKC. There are at least 12 known PKC isoforms that are classified by their activating cofactor requirements, distributed to different intracellular sites, and sensitive to different inhibitors and conditions of down-regulation. Importantly, they exhibit different substrate selectivity, indicating that the isoforms have functional specificity (Knopf et al., 1986; Nishizuka, 1992; Yamamura et al., 1996; Zhou et al., 1996). PKC α and β are classic, Ca^{++}-dependent isoforms and are abundant in endothelial cells (Bussolino et al., 1994; Vuong et al. 1998). In contrast, ϵ, δ, ζ, and λ are Ca^{++}-independent isoforms (Haller et al., 1996; Yamamura et al., 1996; Zhou et al., 1996). In addition to mediators activating G protein-coupled receptors, growth factors, cytokines, and oxidative stress activate PKC (Lynch *et al.*, 1990; Ferro *et al.*, 1993; Taher *et al.*, 1993; Haller *et al.*, 1996; Konishi *et al.*, 1997; Ross and Joyner, 1997).

Partially due to the multiple PKC isoforms and varied effects at different times, signaling of PKCs is complex and not fully resolved, but it is clear that PKC activation contributes to barrier dysfunction. Various inhibitors acting at different sites and on different PKC isoforms attenuate agonist-induced permeability (Johnson, *et al.*, 1989; Gescher, 1992; Siflinger-Birnboim *et al.*, 1992; Stasek *et al.*, 1992; Ferro *et al.*, 1993). Ca^{++} and DAG release, with agonist-activation of endothelium, co-coordinately increase classic isoform PKC translocation to the membrane fraction, a requirement for activation (Lynch *et al.*, 1990; Stasek *et al.*, 1992; Patterson *et al.*, 1995; Ferro *et al.*, 2000). Transfection of endothelium with antisense PKC β1 inhibited PMA-mediated endothelial permeability (Vuong *et al.*, 1998). In a cyclical fashion, DAG activates PKC and PKC activates PLD, which produces DAG. VEGF-induced DAG release resulted in PKCβ and PKCα translocation and phosphorylation of intercellular junctions molecules, β-catenin and PECAM-1 (Gamble *et al.*, 2000; Wang *et al.*, 2004).

Both intracellular lipid messengers and PKC regulate various cytoskeletal proteins. Phosphatidylinositol 4,5-bisphosphate (PIP_2) binds and regulates several actin-binding proteins such as profilin (Sohn *et al.*, 1995), α-actinin (Fukami *et al.*, 1992), vinculin (Gilmore and Burridge, 1996), and gelsolin (De Corte *et al.*, 1997). PKC mediated phosphorylation of the intermediate filament, vimentin (Perez-Moreno *et al.*, 1998; Stasek *et al.*, 1992).

Lysophosphatidic acid and S1P are closely related lysophospholipids, which act as both intracellular signaling molecules and as extracellular mediators *via* binding specific G-protein coupled receptors, as noted above. They affect various functions in many cell types (Moolenaar, 1999). The surface receptors are termed Edg receptors for their family of encoding genes - endothelial differentiation genes (An *et al.*, 1998). The receptors for the two lipids differ. Lysophosphatidic acid receptors include: Edg2 (LPA1), Edg4 (LPA2), and Edg7 (LPA3). S1P receptors include: Edg1 (S1P1), Edg5 (S1P2), Edg3 (S1P3), Edg6 (S1P4), and Edg8 (S1P5). Lysophosphatidic acid and S1P are intermediates in intracellular lipid metabolism, but can be released by cells and notably by platelets; such as their serum concentration can reach the micromolar range (An *et al.*, 1998). Phospholipase secretion in inflammation can significantly boost this level further (Moolenaar, 1999). Their effects on endothelial cells are divergent and include both short term-mediated and long-term mediated effects. Whereas lysophosphatidic acid induces permeability (English *et al.*, 1999), S1P promotes barrier function, as noted above (Schaphorst *et al.*, 2003). Consistent with these differences, lysophosphatidic acid receptor (Edg2 & 4) activation results in Gi – inhibition of adenylate cyclase and Gq-PLCγ-stimulated Ca^{++} increase and subsequent activation of various PKC enzymes, similar to other permeability inducing agents (Cunningham *et al.*, 1997). But S1P (Edg 1 & 3 receptors) does not (An *et al.*, 1998; Panetti, 2002). Lysophosphatidic acid also directly activates PLD to produce phosphatidic acid and causes activation of PI3K and tyrosine phosphorylation of FAK and p43MAPK (Kumagai *et al.*, 1993; An *et al.*, 1998; Moolenaar, 1999). PI3K activation, resulting from both coupling to tyrosine phosphorylated receptors and G-proteins, results in synthesis of multiple intracellular signaling lipids, that interact with other signaling pathways. Thus, PI3K activation is both downstream and upstream from receptor tyrosine phosphorylations (Siddiqui and English, 2000). In endothelium, PI3K is recruited to and activated by assembled adherens junctions, resulting in activation of p38 MAPK and Akt, which will be discussed in Chapter 6 (Laprise *et al.*, 2002). As with other G-protein coupled receptors, Edg-signaling involves tyrosine kinases activation, phosphatases, and small G-proteins, to be further discussed in the following sections.

Another important sphingolipid is lactosyl ceramide, which is a specific second messenger involved in Rac1-mediated assembly/activation of NADPH oxidase (Chatterjee. 1998).

Signaling by Tyrosine Kinases and Phosphatases

Endothelial activation by a variety of agents is accompanied by increased tyrosine phosphorylation, implicating tyrosine kinases and phosphatases as signaling factors in barrier dysfunction (Esser *et al.*, 1998; Shi *et al.*, 1998; Andriopoulou *et al.*, 1999; Cohen *et al.*, 1999). Inhibition of tyrosine kinase blocked both tyrosine phosphorylation of proteins and attenuates increased albumin permeability, implicating tyrosine kinases in the permeability response (Carbajal and Schaeffer, 1998; Shi *et al.*, 1998). As might be expected, inhibition of tyrosine phosphatases increased tyrosine phosphorylation and permeability (Yuan *et al.*, 1998). Tyrosine kinase inhibitors attenuated mediator–induced rises in Ca^{++} (Sharma and Davis, 1996; Suzuki *et al.*, 1997); while vanadate inhibition of tyrosine phosphatases increased endothelial Ca^{2+} (Sharma and Davis, 1996; Usatyuk *et al.*, 2003a). There are numerous proteins that can undergo tyrosine phosphorylation. A few of the important tyrosine kinase targets related to endothelial cell activation and barrier dysfunction include: PLD, MLCK (Garcia *et al.*, 1999), and focal adhesion kinase (Burridge *et al.*, 1990; Schaphorst *et al.*, 1997); VE-cadherin and β-catenin (McLees *et al.*, 1995; Esser *et al.*, 1998; Andriopoulou *et al.*, 1999; Cohen *et al.*, 1999; Exton, 1999). Tyrosine phosphatases associated with adherens junctions have also been shown to be of importance, as their inhibition resulted in net tyrosine phosphorylation of VE-cadherin, β-catenin, γ-catenin, and p120-catenin, which increase transendothelial permeability but did not cause disassembly of the junctions or their dissociation from the DPB (Young *et al.*, 2003). Interestingly, mice lacking either pp60-src or pp62c-yes tyrosine kinases, showed normal angiogenic responses to VEGF, but no acute permeability effects, indicating the importance of this particular signaling pathway to the VEGF-induced acute barrier dysfunction (Elicieri *et al.*, 1999). Shear stress resulted in loss of the protein tyrosine phosphatase, SHP-2, from adherens junctions and also resulted in tyrosine phosphorylation of β-catenin, but this did result in displacement of α-catenin from the complex (Ukropec *et al.*, 2002). Shear stress, osmotic shock, and direct application of mechanical stress to PECAM-1 homotypic lateral adhesions results in src-mediated tyrosine phosphorylation (Newman and Newman, 2003). In resting cells, PECAM is associated with SHP-2 tyrosine phosphatase and is in the unphosphorylated state. Notably, focal adhesion integrin complexes contain several non-receptor type tyrosine kinases, including src family kinases, p125 FAK, and proline rich tyrosine kinase-2 (Pyk2), which may cause tyrosine phosphorylation and altered function of other focal adhesion complex proteins or other cellular proteins (Ruegg and Mariotti, 2003; Orr *et al.*, 2004).

Besides these junction-linked tyrosine kinases, receptors, such as the VEGF receptors, have intrinsic tyrosine kinase capacity, which is activated by ligand-induced dimerization and autophosphorylation (Clauss, 2000). The PLC activation with VEGF, noted above, is related to tyrosine phosphorylation of PLCγ. Thus, some of the early signaling downstream from VEGFR2 is similar to that of the G-protein coupled receptors, including the PLC-mediated hydrolysis of membrane lipids, IP_3 and DAG release, and PKC activation. Activation of VEGFR2, but not VEGFR1, also resulted in PGI_2 release and increased NO production (Hippenstiel *et al.*, 1998) and others have attributed the permeability effect to this combined release (Murohara *et al.*, 1998). PLA_2 activation and subsequent PAF release has also been implicated in endothelial permeability as PAF antagonist blocked the permeability (Sirois and

Edelman, 1997). However, angiopoietin-1, which acutely blocks VEGF-permeability, does not block these pathways (Wang et al., 2004). Similarly, cAMP elevation was shown to inhibit VEGF-induced permeability and not all endothelial cells respond to VEGF with barrier dysfunction, partly due to differences in basal cAMP levels and also differences in VEGFR2 expression (Hippenstiel et al., 1998). But possibly also due to attenuation of the VEGFR2 tyrosine phosphorylation signal, by cadherin-associated phosphatase, which prevents VEGF-proliferative signaling (Lampugnani et al., 2003). VEGF-induced permeability can occur rapidly, especially in vivo, in association with other factors (Fu and Shen, 2004), or it may be delayed and occur over a much longer period (Hippenstiel et al., 1998). Undoubtedly this relates to the various different pathways activated through the tyrosine kinase receptor and also to synergy with other factors and antagonism with innate inhibitors. Further discussion of VEGF-permeability and other long-term responses will be found in Chapter 6.

Other growth factors, such as TGFβ1, and cytokines, such as TNFα and IFNγ, also function *via* their specific tyrosine kinase receptors. Note, the distinction between cytokines and growth factors is rather blurry, as growth factors can exhibit pro-inflammatory effects, such as upregulation of chemokines in endothelium, and cytokines can lead to proliferation in some cells, and downstream signaling can be rather similar. A very basic generalized mechanism of signaling involves ligand binding, dimerization of the receptor, tyrosine phosphorylation, and downstream signaling via subsequent activation of other tyrosine and serine/threonine kinases. As many downstream kinases, such as the MAP kinases, have known effects on transcription, detailed discussion of this cascade will be found in Chapter 6.

Small GTPase Signaling

In addition to the GTPases coupled to membrane-bound receptors, there is a large family of small G-proteins, which may or may not be membrane associated, depending on cell activation. These include the Ras family with 19 known mammalian members, the Rho family (19 members), the Arf family (16 members), the Rab family (42 members), and Ran (Takai et al., 2001). Besides their close association with the usual G-protein coupled receptors, integrin complexes also contain these Rho and Ras family GTPases (Ruegg and Mariotti, 2003). Of particular importance in endothelial barrier function is the Rho family, including Rac1-3, Cdc42, TC10, RhoA-E, G, H and Rnd1 and 2 (Aspenstrom, 1999). The main mechanism identified for activation of these small G-proteins is through G-protein coupled membrane receptor-induced activation of guanine nucleotide exchange factors (GEFs), as also seen in Figure 2. The GEFs are associated with membrane receptor Gα12/13, and when the membrane receptor is activated, the GEF is released to the small G-protein, which increases nucleotide exchange of GDP for GTP on the small GTPase proteins (Hart et al., 1998; Vogt et al., 2003). GAPs also modulate the activation level by regulation of the hydrolysis rate for GTP to GDP. Alternatively, tyrosine kinase activity of the Edg receptors has been implicated in activation of G-proteins (Gohla et al., 1998). The cytoplasmic tail of VE-cadherin is also capable of activating Cdc42 and Rac, dependent upon association with p120 catenin and in reverse, Rac1 is important in junctional formation (Braga et al., 1999; Vincent et al., 2004). Most G-protein receptors activate the MAPK cascade (see Chapter 6).

Generally speaking, RhoA contributes to barrier dysfunction, while Rac1 contributes to integrity. Figure 3, from a recent review of Rho-GTPases in endothelium, summarizes some of the signaling features and the opposing effects of Rho and Rac on endothelial barrier function (Wojciak-Stothard and Ridley, 2003). RhoA, as in other cell types, regulates several

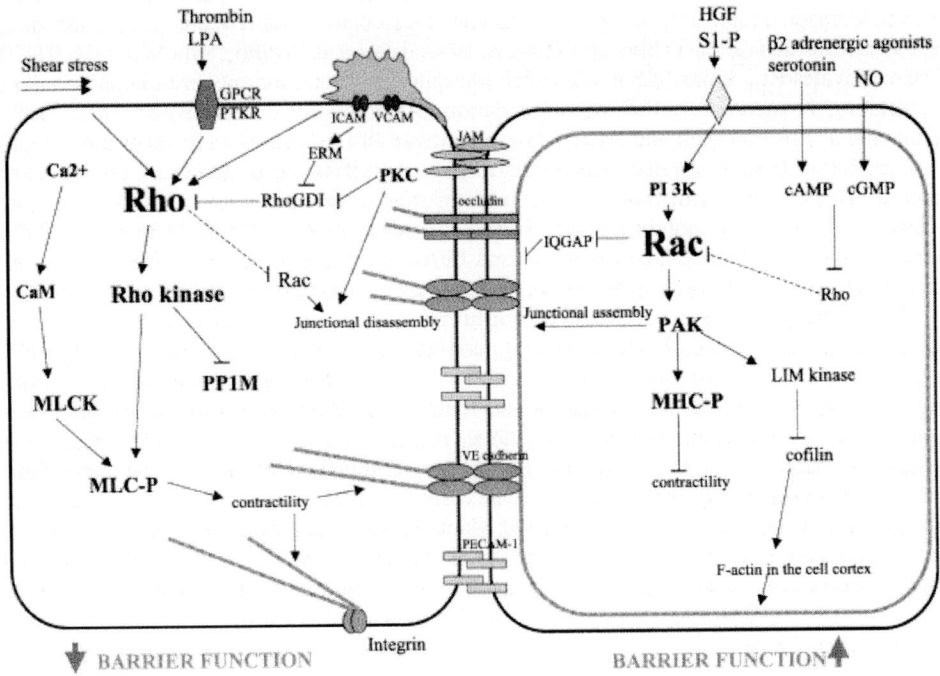

Figure 3. **Opposing effects of Rac and Rho on endothelial barrier function.** Actomyosin contractility induced by Rho activation contributes to breakdown of intercellular junctions (left). Rac activation increases barrier function by stabilizing intercellular junctions and decreasing actomyosin contractility (right). Abbreviations: LPA =lysophosphatidic acid; GPCR = G-protein-coupled receptor; PTKR = protein tyrosine kinase receptor; ERM = ezrin, radixin and moezin; GDI = guanine nucleotide dissociation inhibitor; CaM = calcium-calmodulin; PP1M = MLC phosphatase; PAK = p21-activated kinase; MHC =myosin heavy chain; HGF =hepatocyte growth factor; S1-P = sphingosine 1-phosphate. Reproduced from *Wojciak-Stothard, B. and Ridley, A.J. (2002) Rho GTPases and the regulation of endothelial permeability. Vasc. Pharmacol. 39:187-199.* by permission of Elsevier, Amsterdam.

---

cytoskeletal regulatory proteins, such as profilin and cofilin (Aspenstrom, 1999; Ohta *et al.*, 1999; Takai *et al.*, 2001). Another study confirmed stress fiber formation with RhoA activation, but found barrier dysfunction to be independent of stress fibers and RhoA, while dependent on Rac (Vouret-Craviari *et al.*, 1998). This interpretation was supported by a study with constitutively active Rac and Rho in endothelial cells, which showed that Rho-mediated stress fiber and focal adhesion formation did not decrease cell-cell adhesion or cause VE-cadherin redistribution, whereas the active Rac resulted in an oxidant-dependent dissolution of adherens junctions (van Wetering *et al.*, 2002). Thus, leaving the question of the roles of Rho and Rac and cytoskeletal formation open for further resolution. In endothelium, these and other effects of RhoA activation on cytoskeletal proteins result in stress fiber formation, focal adhesion formation, and redistribution of VE cadherin associated with barrier dysfunction (Chrzanowskawodnicka and Burridge, 1996; Hippenstiel et al., 1997; Braga et al., 1999; Adamson et al., 2002). Both FAK and it associated paxillin are targets for small GTPases Rho activation (Flinn and Ridley, 1996; Lim et al., 1996; Schulze et al., 1997). RhoA activates Rho kinase (ROCK), which phosphorylates and inhibits protein phosphatase-1, resulting in increased MLC phosphorylation (Amano et al., 1996; Essler et al., 1998; Kimura et al., 1996). Rho and ARF GTPases activate PLD1, with the attendant downstream PLD signaling (Exton, 1999). RhoA-kinase phosphorylates ezrin/radixin/moesin proteins, interfering with their actin filament/plasma membrane cross-linking function (Matsui et al., 1996). C3 exotoxin inhibition of RhoA abolished diperoxovanadate-mediated MLC phosphorylation and attenuated thrombin- and LPA-induced permeability (Essler et al., 1998; van Nieuw Amerongen et al., 1998; Garcia et al., 1999; van Nieuw Amerongen et al., 2000).

Rac regulates LIM kinase-mediated cofilin phosphorylation, resulting in inactivation of cofilin and F-actin accumulation at the cell periphery rather than as stress fibers (Yang et al., 1998). Cdc42 regulates formation of filopodia, whereas lamellipodia formation is under Rac1 control. Rac-1 is crucial for regulation and function of adherens junctions associated with improved barrier function (Braga et al., 1999). Yet, it is not as simple as barrier destroying-RhoA and barrier promoting- Rac-1. Rac-1 is also involved in assembly of endothelial NADPH oxidase and as will be seen in the next section, oxidants have their untoward barrier effects (Abid et al., 2001; Sohn et al., 2000; Cook-Mills et al., 2004). And above we related that constitutively active Rac caused an oxidant-dependent decrease in adherens junctions (van Wetering et al., 2000). Rac was also shown to be essential for the vascular permeability effects of VEGF (Eriksson et al., 2003). On the other hand, an effector of Rac1, IQGAP1, caused dissociation of the cadherin-catenin complex, while Rac-1 counteracted this dissolution (Kaibuchi et al., 1999). Likely, timing and compartmentation contribute to these seemingly contradictory effects of Rac on cell-cell adhesion. Further, details on this subject are discussed in the review and will also be found in Chapter 9, including interactions of the G-proteins with the adherens junction proteins (Wojciak-Stothard and Ridley, 2003).

Oxidant Signaling

Reactive oxygen species (ROS) and reactive nitrogen intermediates generated in the vasculature by activated neutrophils or vascular cells themselves have been implicated in the pathobiology of barrier dysfunction. The exact mechanisms that regulate ROS-induced endothelial barrier dysfunction are unclear, but they appear to affect virtually all of the signaling mechanisms described above. ROS-induced permeability alterations and signaling for cell growth and differentiation in the endothelium are modulated by transduction pathways involving calcium, protein kinases, and phosphatases (Natarajan, 1995; McQuaid and Keenan, 1997; Wolin et al., 1998; Kunsch et al., 1999; Shi et al., 2000; Usatyuk, et al., 2003a). Indices of barrier dysfunction, i.e. albumin flux and transendothelial electrical resistance (TER) are respectively increased and decreased in the presence of ROS (Parinandi et al., 2001).

Oxidants activate endothelial phospholipases, PLC, PLA_2, and PLD, and downstream signals consistent with their effects on permeability (Shasby et al., 1985; Shasby et al., 1988; Chakraborti et al., 1989; Duane et al., 1991; Natarajan et al., 1993; Dreher and Junod, 1995; Siflinger-Birnboim et al., 1996; Volk et al., 1997; Exton, 1999). ROS activation of PLD results in generation of phosphatidic acid and its conversion to lysophosphatidic acid or DAG, catalyzed by PLA1/PLA2 or phosphatidic acid phosphatase, respectively (Billah et al., 1981; Brindley and Waggoner, 1996; Cummings et al., 2002). Endothelial exposure to exogenous phosphatidic acid enhanced albumin flux across the monolayer, suggesting that intracellularly generated phosphatidic acid may exhibit a similar response (English et al., 1999). Moreover, ROS-induced permeability in pulmonary artery endothelial cells is attenuated by 1-butanol, a PLD inhibitor, but not by 3-butanol, a non-inhibitor. Bovine pulmonary artery endothelial cells transfected with catalytically inactive PLD1 and PLD2 mutants attenuated ROS-induced TER alteration; whereas, wild type PLD1 and 2 overexpression enhanced H_2O_2 - and phorbol ester-mediated PLD activation and TER changes (Parinandi et al., 2001). These studies clearly define a role for phosphatidic acid generated from PLD1 and PLD2 in endothelial barrier dysfunction. The signaling pathways downstream of PLD leading to permeability changes have not been clearly defined, but activation of several pathways will be discussed.

A number of studies have described the ability of oxidants such as H_2O_2 and oxidized lipoproteins, to elevate $[Ca^{++}]_i$ in endothelial cells, while chelation of extracellular Ca^{++} partly prevents the oxidant effect, indicating that Ca^{++} influx and mobilization of intracellular Ca^{++} are involved in regulation of barrier function (Sharma and Davis, 1996; Mikalsen and Kaalhus, 1996; Siflinger-Birnboim et al., 1996; Natarajan et al., 1998; Norwood et al., 2000; Shi et al., 2000: Lum and Roebuck, 2001; Natarajan et al., 2001; Mehta et al., 2002). One mechanism for increased $[Ca^{++}]_i$ by oxidants is by PLC generation of DAG and IP_3, and endoplasmic reticulum release of Ca^{++}, as described above (Takahashi, et al. 1997). Oxidants also affect cell Ca^{++} by triggering the opening of Ca^{++} ion channels (Az-ma et al., 1999) and by inhibiting re-uptake of cytoplasmic Ca^{++} by the endoplasmic reticulum (Grover et al. 1992; Lee and Okabe 1995). In contrast to H_2O_2, 4-hydroxynonenol, generated by cellular lipid peroxidation/oxidative stress, did not increase $[Ca^{2+}]_i$, though it triggered actin rearrangement and decreased TER (Usatyuk et al., 2003b; Usatyuk and Natarajan, 2004). This suggests Ca^{++}-independent mechanisms for barrier dysfunction in lung microvascular endothelia, although earlier studies demonstrated that 4-hydroxynonenol increased $[Ca^{2+}]_i$ in hepatocytes (Carini et al., 1996; Retta et al., 1996) and neurons (Romer et al., 1994; Mark et al., 1997; Sandoval et al., 2001; Lu et al., 2002). Although mediator- and ROS-induced Ca^{++} increases are implicated in permeability, as described above, the mechanisms remain incompletely defined. Possible mechanisms include activation of PKC, PLD, MLCK, and tyrosine kinases.

ROS regulate PKC activity, as might be expected from oxidant-induced increases in DAG and Ca^{++} (Natarajan et al., 1996; English et al., 1999). Moreover, phosphatidic acid, from ROS-activated PLD, directly activate PKC ζ (Cross et al., 1996). PIP_2 and PI-3,4,5-trisphosphate are important regulators of the actin cytoskeleton. Phosphatidic acid activates PI-4-kinase in vitro and type I PIP-5-kinase in vivo (Jenkins et al., 1994). Subsequently, phosphatidic acid-mediated kinase activation alters cell PIP2, in turn modulating interactions between actin and actin binding proteins such as vinculin and filamin (Wakelam et al., 1997). PIP_2 can also stimulate PLD activity (Hammond et al., 1997), thereby amplifying production of phosphatidic acid and the dependent kinases. Furthermore, PIP_2 enhances interaction between proteins that contain PH and PX domains (Xu et al., 2001; Wishart et al., 2001).

Oxidants regulate the activity of tyrosine kinases and tyrosine phosphatases. In endothelial cells, ROS activated the non-receptor tyrosine kinases, Src and FAK; while, ROS-mediated barrier dysfunction was blocked by tyrosine kinase inhibitors (Natarajan et al., 1996; English et al., 1999; Shi et al., 2000). Recent studies in lung endothelial cells show that diperoxovanadate, a potent activator of tyrosine kinases and inhibitor of tyrosine phosphatases, induced increases in $[Ca^{2+}]_i$ (Usatyuk et al., 2003a). Inhibitors of store operated Ca^{++} channels partially blocked the rise in intracellular Ca^{++}. While BAPTA-chelation of intracellular Ca^{++} blocked DPV-activation of src, EGTA-chelation of extracellular Ca^{++} did not attenuate src activation. Both BAPTA-and EGTA-calcium chelation attenuated the cytoskeleton remodeling and the decrease in TER (Mehta et al., 2002; Usatyuk et al., 2003a). Thus, the DPV-induced increase in intracellular Ca^{++} likely resulted in the src activation. Oxidant-mediated inactivation of tyrosine and serine/threonine protein phosphatases, also contributes to increased effects of protein kinases (Whisler et al., 1995). Notably, tyrosine phosphorylation of various PKC isotypes by ROS enhanced their activity (Konishi et al. 1997). Oxidant also activated small G-proteins and the MAPK cascade as will be discussed in Chapter 6. Notably, H2O2-induced permeability in isolated perfused lungs was attenuated by blocking the Rho effector kinase, p160 ROCK (Chiba et al., 2001).

Reactive oxidants may be derived from external sources, such as activated neutrophils, or they may be produced in endothelium by mitochondrial respiration, eicosanoid metabolism, NADPH oxidase, and uncoupled xanthine oxidase. Activation of components of endothelial NADPH oxidase ($p22^{phox}$, $p47^{phox}$, $p67^{phox}$, $gp91^{phox}$, and Nox4) in pathologic states is a major contributor of superoxide, which can then oxidize the tetrahydrobiopterin cofactor of NOS and convert it into an oxidase (Jones et al., 1996; Bayraktutan et al., 1998; Ago et al., 2004; Verhaar, 2004). Inhibition of HSP90 protein also resulted in eNOS uncoupling and superoxide production (Ou et al., 2004). Endothelial NADPH oxidase was the primary oxidant source in perfused lung and in cultured cells during oxygenated ischemia, then when oxidant levels build and ATP levels fall, xanthine oxidase becomes a major contributor (Zhao et al., 1997; Al-Mehdi et al., 1998; Wei et al., 1999).

Integration and Interactions of Signaling Mechanisms

Cooperation, Coordination, and Contradiction

Oxidant-mediated signals promotes growth, differentiation, apoptosis, inflammation, and gene expression, to be discussed in Chapter 6. Acute activation by oxidants, introduced above, also causes transactivation of agonists-signals. Figure 4 summarizes some of the important interactions between oxidants and agonists, between different signaling systems, and between short and long-term signaling. ROS, such as H_2O_2, activate growth factor receptor (e.g. EGF, PDGF and VEGF) tyrosine kinases (Frank and Eguchi, 2003). This is true whether ROS are added exogenously or produced endogenously. Moreover, oxidants activate non-receptor tyrosine kinases in both vascular endothelium and smooth muscle; thus, contributing to the pathophysiology of vascular disorders. In smooth muscle cells, H_2O_2 induced tyrosine phosphorylation of the EGF-R, thereby enhancing association of Shc-Grb2-SOS adapter protein complex (Frank et al., 2003; Rao, 1996). Angiotensin II transactivation of EGF receptors was blocked by N-acetylcysteine, supporting a role for angiotensin-induced ROS generation in the cross activation (Ushio-Fukai et al., 2001; Frank et al., 2001). Similarly, angiotensin II, cyclic stretch, oxidized-LDL, and ceramide have all been shown to transactivate the PDGF-Rβ in a ligand-independent manner (Frank and Eguchi, 2003). Such oxidant-mediated transactivation likely involves inhibition of tyrosine phosphatases and/or activation of metalloproteases. Activation of several G-protein coupled receptors by cognate ligands results in metalloprotease-dependent release of heparin-binding EGF-like growth factor (Prenzel et al., 1999). The released factor then binds the EGF receptor, activating downstream signaling. Thus, ROS-dependent transactivation of growth factor receptors may play a critical role in the pathogenesis of atherosclerosis and vascular remodeling.

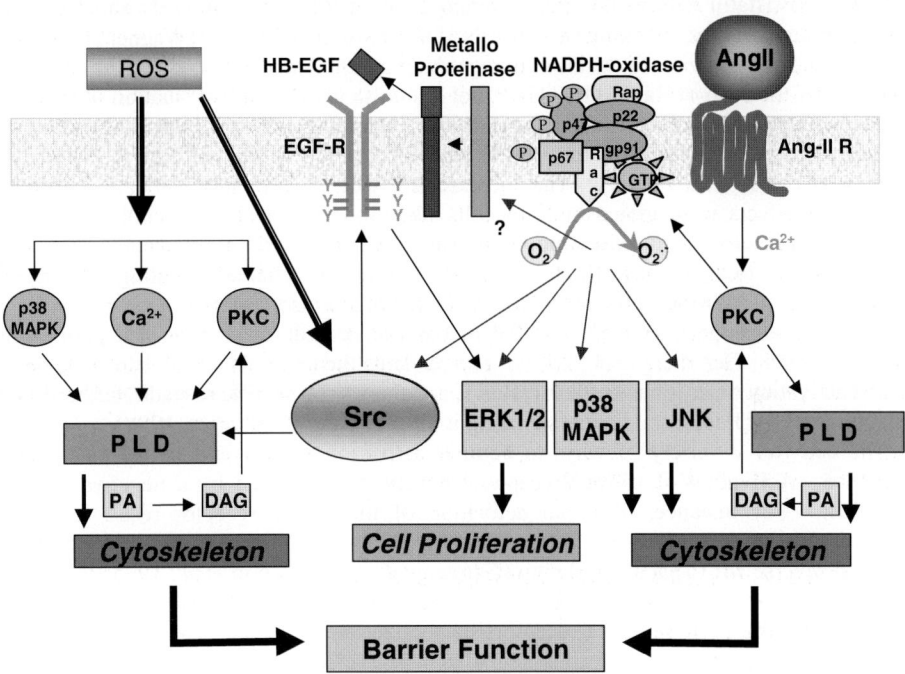

Figure 4. Schema of signal transduction pathways activated by ROS and angiotensin II in vascular cells. Reactive oxygen species (ROS) generated by activated PMNs activates MAPKs, PKC, Src and increases Ca^{2+}, which in turn activates PLD. Activation of PLD results in hydrolysis of membrane phospholipids such as phosphatidylcholine and/or phosphatidylethanolamine and generation of phosphatidic acid (PA). PA acts as a second messenger and alters endothelial cell barrier function via PA-dependent reorganization of cytoskeleton. PA can be acted upon by lipid phosphate phosphatases to generate diacylglycerol (DAG), which can activate PKC. Similarly, Angiotensin II (Ang II) binds to its receptor (AT-1R) leading to activation of NADPH oxidase via PKC and Ca^{2+} dependent pathways. Ang II mediated of superoxide/hydrogen peroxide production activates Src, PLD and MAPKs, which in turn modulates cell proliferation, cytoskeleton and barrier function. Furthermore, Ang II-dependent intracellular ROS generation leads to EGF-R transactivation via metalloproteinase-dependent release of heparin-binding epidermal growth factor (HB-EGF) which binds to EGF-R, activates extracellular-signal regulated kinase (ERK) and causes cell proliferation.

There is significant interaction of oxidants with small GTPases. Rac1 activation is responsible for assembly of the cytosolic components of NADPH oxidase with membrane components to activate the enzyme and produce superoxide. Interestingly, pre-treatment of cells with simvastatin reduces the geranylgeranylation of Rac, attenuating its translocation to the membrane and, thus, blocking sustained NADPH oxidase activation (Wagner et al., 2000). Delayed effects of simvastatin result in upregulation of RhoA and Rac and activation of Rac1 (measured by the ratio of Rac GTP to GDP, not by translocation) and attenuation of thrombin-induced barrier dysfunction (Jacobson et al., 2004). Oxidants, whether from NADPH oxidase or other sources, cross-activate the PDGF receptor leading to further Ras-mediated ERK signaling (Accorsi et al., 2001). The increase in oxidants, probably H_2O_2, is critical for proliferative effects of angiotensin II in cells like smooth muscle, but endothelia do not respond to angiotensin with migration or proliferation. Yet, endothelial cells have a strong oxidant response to angiotensin II and many of the other AP1-elicited signaling pathways are similar (Hsu et al., 2004). TNF activation of endothelium also results in lactosyl ceramide-lipid second messenger-mediated NADPH oxidase superoxide production and permeability (Chatterjee, 1998; Gertzberg et al., 2004). The oxidants then stimulate GTP exchange on p21 Ras and activation of its kinase effector, Raf-1, and thence the MAPK cascade, further linking oxidants, small GTPases, and kinase signaling in endothelium (Chatterjee, 1998).

The cell has a variety of ways to counter activation signals and return to a state of quiescence. Activation of various receptors may be accompanied by a negative feedback component. For instance, thrombin activation of the PAR1 receptor result in a rapid activation of MLCK, but it also activates both phosphatase 1 and phosphatase 2b in a delayed fashion to reverse MLC phosphorylation (Garcia et al., 1995a; Verin et al., 1995; Verin et al., 1998). Many the activators themselves are metabolized at the luminal surface by bound peptidases and the PAR series of receptors catabolize their ligands. The receptors become phosphorylated, internalized, and ether degraded or re-cycled (Kozasa and Gilman, 1996; Yan et al., 1998). Some activating receptors have counterparts, which bind the same ligands, but have opposing, although often weaker, signaling mechanisms. From a signaling point of view, activation of other pathways, which respond to different ligands are perhaps the most interesting. Many of these ligands may be derived from either an autocrine negative feedback loop or a paracrine loop. For instance, most activating agonists increase endothelial production of PGI_2, which binds a Gαs-protein coupled endothelial receptor linked to adenylate cyclase, resulting in increased cAMP to promote barrier restoration. When endothelial activation results in platelet activation, platelets release S1P, which can ligate its Edg receptors on the endothelium. Finally, angiopoietin promotes barrier function by TIE receptor signaling. Less is known about downstream targets of these three attenuating modulators than in the activating pathways, but some interactions have been revealed.

PKA prevention and reversal of barrier dysfunction does not depend upon altered Ca^{++} signaling. Although, this is an important means for cGMP and cAMP relaxation in smooth muscle, various means to elevate cAMP in endothelial cells failed to show any effect on basal or agonist-induced Ca^{++} mobilization (Patterson et al., 1994a; Stevens et al., 1997; Patterson et al., 2000), yet PKA activation does reduce MLC phosphorylation (Moy et al., 1993; Patterson et al., 1994a; Garcia et al., 1995a). Whether this is mainly due to direct inhibitory phosphorylation of MLCK or to activation of phosphatases or is unresolved (Lamb et al., 1988; Giembycz and Raeburn, 1991; Essler et al., 2000; Patterson et al., 2000; Carr et al., 2001). PKA does inhibit RhoA-inhibition of the MLC phosphatases, but this effect was not attributed to direct PKA-mediated RhoA phosphorylation, but by action at some point upstream from the phosphatase (Essler *et al.*, 2000). In contrast, others have shown that small GTPase-proteins are subject to PKA phosphorylation (Flinn and Ridley, 1996; Lang *et al.*, 1996; Tigyi *et al.*, 1996; Arimura *et al.*, 1997; Schulze *et al.*, 1997). Conversely, PKA activity may be altered by Rho-GTPases (Tigyi *et al.*, 1996). Notably, PKA inhibition did not alter basal or mediator-induced MLC phosphorylation but produced the cytoskeletal changes and barrier function similar to that seen with agonist activation; thus, other PKA targets are likely to be more important than MLC or MLCK (Patterson *et al.*, 2000; Liu *et al.*, 2001). Interactions of PKA with prolonged signals and MAPK will be discussed in the next chapter.

The endothelial protective lipid, S1P, activation of small GTPase proteins was introduced above. It is becoming clear that Rac1 and RhoA have different and somewhat oppositional signaling in endothelium, as had been shown in Figure 3. S1P actually stimulated sequential Rho and Rac activations, resulting in very early MLC phosphatase inactivation, followed by activation in the 1-15 minute range, and then by low level sustained inactivation (Essler *et al.*, 2002). The differential Rac/Rho regulation would be even more interesting to put into a compartmental context, since this is the timeframe of dynamic reorganization of the cytoskeleton. S1P activation of Edg-1 tyrosine kinase signaling also indicates early links between lipid signaling, receptor GTPases, small GTPases, and tyrosine kinase signaling in regulation of endothelial barrier properties by barrier promoters in opposition to activators of the endothelial cells. This opposition was shown in a study comparing S1P and thrombin signaling (Vouret-Craviari *et al.*, 2002). Whereas thrombin increased RhoA and decreased Rac1 levels, S1P resulted in only a slow, weak increase in RhoA, but a robust increase in Rac1 and src kinase signaling. While Rac1 activation is associated with promotion of the DPB and disassembly of stress fibers consistent with barrier integrity, it is also critical for assembly of NADPH oxidase, which contributes to barrier dysfunction. Other contradictory reports on Rac effects on adherens junction assembly/disassembly and barrier function were noted above. Likely, the answer to these contradictory effects lies in very specific compartmentation, contingent on the receptor activated and on the temporal pattern of activation, as untoward effects of Rac were found with constitutive overexpression.

Mediator-activation and permeability is opposed by angiopoietin 1 (Gambel *et al.*, 2000; Jones, 2003; Pizurki *et al.*, 2003). As mentioned above, PECAM-1 and VE-cadherin phosphorylation was reduced by angiopoietin-1 (Gamble *et al.*, 2000). The temporal pattern in this study indicates that short-term mechanisms accounted for this reduction in permeability, yet the phosphorylation effect on VE-cadherin and its co-precipitation with β-catenin was maximal at 10 minutes, but by 30 min the phosphorylation and complex of VE-cadherin with β-catenin had returned to basal levels, and permeability was only measured after 30 minutes. Whether there is interactive signaling or precisely what signaling accounted for these short-term benefits are not resolved. There are a number of potential clues from the studies on interactions of angiopoietin and VEGF, which will be explored in Chapter 6.

Finally, quiescent cells exert a stubborn resistance to activation. For instance, there is evidence for the cadherin complex to bind the VEGFR2 receptor, bringing it into union with adherens junction associated tyrosine phosphatases, so that the activation signal is quenched (Lampugnani *et al.*, 2003). Moreover, the catenin, p120, was recently found to mitigate thrombin-induced MLC phosphorylation (Iyer *et al.*, 2004). As will be explored in the ensuing chapter, under usual circumstances endothelial cells also possess intrinsic adenylyl cyclase activity to maintain a "steady state" PKA activation to ensure normal barrier function.

Integration with Long-term Signaling

Signaling involving increased or decreased expression of proteins will be explained in Chapter 6. Importantly, it is not possible to actually make this clear distinction as many of the pathways discussed, which have prompt consequences on endothelial function, are the same early signals that result in long-term mediation of upregulation and downregulation of gene expression. For instance, there are links between PKC activation and the MAPK cascade, which activates a number of transcription factors. Tyrosine phosphorylation signaling and small G-protein signaling also lead toward transcritption effects. Oxidants have early and intermediate effects on barrier function and signaling as described, but are important players in the upregulation of leukocyte adhesion molecule expression and in signaling transcription of other genes. Some of these links are illustrated in Figure 4. Other than the nuclear hormone receptors, virtually all transcription factors are activated in some way by short-term phosphorylation events. Importantly expression of such molecules, like ICAM, on the endothelial surface and their ligation by white cells, leads to a whole new realm of signaling in the endothelium as will be discussed in Chapters 11 and 12.

Autocrine and Paracrine Signaling

We will only touch on this briefly as our aim was to discuss intracellular signaling, but the particular signaling elicited by activating or calming mediators has important consequences for the types of mediators released by the endothelium, which affect and regulate other cells and processes. Autocrine mediators released can also have either a positive or negative feedback effect on the endothelium that may be important to the summation of signaling and physiologic response. Moreover, the endothelial cells receive feedback from the other cells with which it interacts. Before, we mentioned two such examples. Activation of membrane G-protein coupled receptors and also tyrosine kinase receptors increases PLA_2 activity, cleavage of arachidonic acid from membrane phospholipids, and increased PGI_2 release into the luminal space/blood. This prostacyclin affects blood elements such as platelets, but also acts upon the endothelium to trigger signaling and increase cAMP. Likewise platelet activation causes release of stored S1P to induce signaling in the endothelium as described.

One mediator release locally by endothelium to signal vascular smooth muscle relaxation is NO. This function is so basic that when many cardiovascular researchers refer to endothelial function, they actually mean endothelial NO production for control of blood flow and pressure. NO may or may not exhibit an autocrine effect. Some endothelial cells possess guanylate cyclase receptors and some do not, explaining why cGMP, unlike cAMP, has had variable effects on endothelial permeability from protection (Lofton *et al.*, 1991) to total lack of effect (Boulanger *et al.*, 1990; Buchan and Martin, 1992; Patterson *et al.*, 1994a; Schnittler *et al.*, 1990). Regardless, NO influence the endothelium itself by influencing the balance of superoxide, hydrogen peroxide, and peroxynitrite species, having either an anti- or a pro-oxidant effect, depending on its level of production and on the immediate cellular environment (Gow *et al.*, 1998; Okayama *et al.*, 1998). NO has been reported to both protect (Schulze-Neick *et al.*, 1999; Sprague *et al.*, 1998) and disrupt barrier function (Fischer *et al.*, 1999; Murohara *et al.*, 1998). Shear stress and stretch are important physiologic regulators of eNOS activity. These mechanical stimuli are thought to be mediated by PI3K activation and subsequent Akt kinase and by PKA activation, resulting in activating serine/threonine phosphorylation of eNOS (Fleming and Busse, 2003). Ca^{++} variations have less effect than phosphorylation state on activity level, but Ca^{++} is necessary. Caveolin-1 binding is generally taken to be inhibitory and in confluent endothelium, they are not generally associated (Minshall *et al.*, 2003). Tyrosine phosphorylation of HSP90 increases its binding to eNOS, which increases eNOS affinity for calmodulin, and results in displacement of caveolin (Fleming and Busse, 2003). As mentioned above HSP90 helps maintain eNOS in the coupled state, whereas superoxide-mediated oxidation of the cofactor causes uncoupling and superoxide production rather than the desirable NO production.

Signaling Synopsis

Because, the endothelium was considered as nothing but a passive barrier for many years, signaling mechanisms were worked out in other tissues considered to be metabolically important, like hepatocytes, and physiologically important, like smooth muscle cells. With the recognition of the highly active biochemistry conducted by the endothelial cells and their central role in physiologic and pathologic phenomena, there is fervent effort to apply the information from other investigations and to understand the unique integration of the signaling and functional aspects of endothelium. Publications in this area have expanded exponentially in the past decade, but there is still a critical need to understand the cellular and molecular mechanisms governing cell shape related to organization of the cytoskeletal and its interactions with adhesion complexes, as these complexes are vital to barrier function of the pulmonary endothelium. Chapters 7, 8, and 9 will give more depth regarding the current insights in regulation of cytoskeletal, matrix adhesion, and endothelial-endothelial adhesion. Equally important is the urgent need to define the signaling mechanisms involved in endothelial regulation of vascular function and inflammation, which play a key role in pathophysiology of many diseases. Chapters 11 through 14 will build upon the basic description of signaling in this chapter and in Chapters 5 and 6, to explain the signaling involved in the interaction of the endothelium with blood elements and pathogens and with the regulation of coagulation, complement, and inflammatory systems and in chronic changes in the vasculature. Chapter 15 will provide details on the significance and mechanisms related to protection of endothelial stability and barrier function. At last, Chapter 16 will put this into context in human respiratory distress and explore how improved understanding of endothelial biochemistry and physiology will be important to translation into clinic benefits.

ACKNOWLEDGEMENTS:

This review was supported by HL 69909 and HL 58064 for (VN) and a Merit Grant from the Veteran's Administration Medical Research Service (CEP)

REFERENCES:

Abid, M.R., Tsai, J., Spokes, K., Deshpande, S., Irani, K., and Aird, W.C. (2001) Vascular endothelial growth factor induces manganese-superoxide dismutase expression in endothelial cells by a Rac1-regulated NADPH oxidase-dependent mechanism. FASEB J. 15:2548-2550.

Accorsi, K., Giglione, C., Vanoni, M., Parmeggiani, A. (2001) The Ras GDP/GTP cycle is regulated by oxidizing agents at the level of Ras regulators and effectors. FEBS Lett. 492:139-145.

Adamson, R.H., Curry, F., Adamson, G., Liu, B., Jiang, Y., Aktories, K., Barth, H., Daigeler, A., Golenhofen, N., Ness, W., and Drenckhahn, D. (2002) Rho and ROCK modulation of barrier properties: cultured endothelial cells and intact microvessels. J. Physiol. 539:295-308.

Ago, T., Kitazono, T., Ooboshi, H., Iyama, T., Han, Y., Takada, J., Wakisaka, M., Ibayashi, S., Utsumi, H., and Iida, M. (2004) Nox4 as the major catalytic component of an endothelial NAD(P)H oxidase. Circ. 109:227-233.

Alexander, J.S., Patton, W., Christman, B., Cuiper, L., and Haselton, F.R. (1998) Platelet-derived LPA decreases endothelial permeability in vitro. Amer. J. Physiol. 274:H115-H122.

Al-Mehdi, A., Zhao, G., Dodia, C., Tozawa, K., Costa, K., Muzykantov, V., Ross, C., Blecha, F., Dinauer, M., and Fisher, A. (1998) Endothelial NADPH oxidase as the source of oxidants in lungs exposed to ischemia or high K^+. Circ. Res. 83:730-737.

Amano, M., Ito, M., Kimura, K., Fukata, Y., Chihara, K., Nakano, T., Matsuura, Y., and Kaibuchi, K. (1996) Phosphorylation and activation of myosin by Rho-associated kinase (Rho-kinase). J. Biol. Chem. 271:20246-20249.

An, S., Goetzl, E.J., and Lee, H. (1998) Signaling mechanisms and molecular characteristics of G protein-coupled receptors for lysophosphatidic acid and S1P. J. Cell. Biochem. 30S:147-157.

Andriopoulou, P., Navarro, P., Zanetti, A., Lampugnani, M., and Dejana, E. (1999) Histamine induces tyrosine phosphorylation of endothelial cell-to-cell adherens junctions. Arterioscler. Thromb. Vasc. Biol. 19:2286-2297.

Arimura, S., Nakata, H., Tomiyama, K., and Watanabe, Y. (1997) Phosphorylation of H-ras proteins by protein kinase A. Cell. Signal. 9:37-40.

Aspenstrom, P. (1999) Effectors for the Rho GTPases. Curr. Opin. Cell Biol. 11:95-102.

Az-ma, T., Saeki, N., and Yuge, O. (1999) Cytosolic Ca^{++} movements of endothelial cells exposed to reactive oxygen intermediates: role of hydroxyl radical-mediated redox alteration of cell-membrane Ca2+ channels. Brit. J. Pharmacol. 126:1462-1470.

Babnigg, G., Bowersox, S.R., and Villereal, M.L. (1997) The role of pp60c-src in the regulation of calcium entry via store-operated calcium channels. J. Biol. Chem. 272:29434-29437.

Barnard, J.W., Seibert, A.F., Prasad, V.R., Smart, D.A., Strada, S.J., Taylor, A.E., and Thompson, W.J. (1994) Reversal of pulmonary capillary ischemia-reperfusion injury by rolipram, a cAMP phosphodiesterase inhibitor. J. Appl.Physiol. 77:774-781.

Bayraktutan, U., Draper, N., Lang, D., and Shah, A. (1998) Expression of functional neutrophil-type NADPH oxidase in rat cornonary microvascular endothelial cells. Cardiovasc. Res. 38:256-262.

Bhattacharya, S., Ying, X., Fu, C., Patel, R., Kuebler, W., Greenberg, S., and Bhattacharya, J. (2000) $\alpha v \beta_3$ integrin induces tyrosine phosphorylation-dependent Ca^{2+} influx in pulmonary endothelial cells. Circ. Res. 86:456-462.

Billah, M.M., Lapetina, E.G., and Cuatrecasas, P. (1981) PLA2 activity specific for phosphatidic acid. A possible mechanism for the production of arachidonic acid in platelets. J. Biol. Chem. 256:5399-5403.

Bird, G.S., Aziz, O., Lievremont, J., Wedel, B., Trebak, M., Vazquez, G., and Putney, J.W. (2004) Mechanisms of PLC-regulated calcium entry. Curr. Mol. Med. 4:291-301.

Boulanger, C., Schini, V.B., Moncada, S., and Vanhoutte, P.M. (1990) Stimulation of cyclic GMP production in cultured endothelial cells of the pig by bradykinin, adenosine diphosphate, calcium ionophore A23187 and nitric oxide. Brit. J. Pharmacol. 101:152-156.

Braga, V.M., Del Maschio, A., Machesky, L., and Dejana, E. (1999) Regulation of cadherin function by Rho and Rac: modulation by junction maturation and cellular context. Molec. Biol. Cell 10:9-22.
Brindley, D.N. and Waggoner, D.W. (1996) Phosphatidate phosphohydrolase and signal transduction. Chem. Phys. Lipids 80:45-57.
Buchan, K.W. and Martin, W. (1992) Modulation of barrier function of novine aortic and pulmonary artery endothelial cells - dissociation from cytosolic calcium content. Brit. J. Pharmacol. 107:932-938.
Burridge, K., Nuckolls, G., Otey, C., Pavalko, F., Simon, K., and Turner, C. (1990) Actin-membrane interaction in focal adhesions. Cell Diff. Devel. 32:337-342.
Bussolino, F., Silvagno, F., Garbarino, G., Costamagna, C., Sanavio, F., Arese, M., Soldi, R., Aglietta, M., Pescarmona, G., Camussi, G., and Bosia, A. (1994) Human endothelial cells are targets for platelet-activating factor (PAF). Activation of α and β PKC isozymes in endothelial cells stimulated by PAF. J. Biol. Chem. 269:2877-2886.
Carbajal, J.M. and Schaeffer, R.C., Jr. (1998) H_2O_2 and genistein differentially modulate protein tyrosine phosphorylation, endothelial morphology, and monolayer barrier function. Biochem. Biophys. Res. Comm. 249:461-466.
Carini, R., Bellomo, G., Paradisi, L., Dianzani, M., and Albano, E. (1996) 4-OH-nonenal triggers Ca^{2+} influx in isolated hepatocytes. Biochem. Biophys. Res. Comm. 218:772-776.
Carr, A.N., Sutliff, R., Weber, C., Allen, P., Greengard, P., de Lanerolle, P., Kranias, E., Paul, R.J. (2001) Is myosin phosphatase regulated in vivo by inhibitor-1? Evidence from inhibitor-1 knockout mice. J. Physiol. 534:357-366.
Chakraborti, S., Burtner, G.H., and Michael, J.R. (1989) Oxidant-mediated activation of PLA_2 in pulmonary endothelium. Amer. J. Physiol. 257:L430-L437.
Chatterjee, S. (1998) Sphingolipids in atherosclerosis and vascular biology. Arterioscl. Thromb. Vasc. Biol. 18:1523-1533.
Chetham, P.M., Guldemeester, H., Mons, N., Brough, G., Bridges, J., Thompson, W., and Stevens, T. (1997) Ca^{2+}-inhibitable adenylyl cyclase and pulmonary microvascular permeability. Amer. J. Physiol. 273:L22-L30.
Chiba, Y., Ishii, Y., Kitamura, S., and Sugiyama, Y. (2001) Activation of rho is involved in hydrogen-peroxide-induced lung edema in isolated perfused rabbit lung. Microvasc. Res. 62:164-171.
Chrzanowskawodnicka, M. and Burridge, K. (1996) Rho-stimulated contractility drives the formation of stress fibers and focal adhesions. J. Cell Biol. 133:1403-1415.
Cioffi, D.L., Wu, S., and Stevens, T. (2003) On the endothelial cell I(SOC). Cell Calcium 33:323-336.
Clauss, M. (2000). Molecular biology of the VEGF and the VEGF receptor family. Sem. Thromb. Hemost. 26:561-569.
Cohen, A.W., Carbajal, J.M., and Schaeffer, R.C.J. (1999) VEGF stimulates tyrosine phosphorylation of β-catenin and small-pore endothelial barrier dysfunction. Amer. J. Physiol. 277:H2038-H2049.
Cook-Mills, J.M., Johnson, J., Deem, T., Ochi, A., Wang, L., and Zheng, Y. (2004) Calcium mobilization and Rac1 activation are required for VCAM-1 stimulation of NADPH oxidase activity. Biochem. J. 378:539-547.
Cross, M.J., Roberts, S., Ridley, A., Hodgkin, M., Stewart, A., Claesson-Welsh, L., and Wakelam, M.J. (1996) Stimulation of actin stress fibre formation by activation of PLD. Curr. Biol. 6:588–597.
Cummings, R., Parinandi, N., Wang, L., Usatyuk, P., Natarajan, V. (2002) PLD/phosphatidic acid signal transduction: role and physiological significance in lung. Molec. Cell. Biochem. 234:99-109.
Cunningham, S.A., Arrate, M., Brock, T., and Waxham, M.N. (1997) Interactions of FLT-1 and KDR with PLCγ: identification of phosphotyrosine binding sites. Biochem. Biophys. Res. Comm. 240:635-639.

De Corte, V., Gettemans, J., and Vandekerckhove, J. (1997) Phosphatidylinositol 4,5-bisphosphate specifically stimulates PP60(c-src) catalyzed phosphorylation of gelsolin and related actin-binding proteins. FEBS Lett. 401:191-196.

Deo, D.D., Bazan, N., and Hunt, J.D. (2004) Activation of PAF receptor-coupled Gαq leads to stimulation of Src and FAK via two separate pathways in human umbilical vein endothelial cells. J. Biol. Chem. 279:3497-3508.

Dreher, D. and Junod, A.F. (1995) Differential effects of superoxide, hydrogen peroxide, and hydroxyl radical on Ca^{2+} in human endothelial cells. J. Cell Physiol. 162:147-153.

Duane, P.G., Rice, K.L., Charboneau, D.E., King, M.B., Gilboe, D.P., and Niewoehner, D.E. (1991) Relationship of oxidant-mediated cytotoxity to phospholipid metabolism in endothelial cells. Amer. J. Respir. Cell Molec. Biol. 4:408-416.

Eliceiri, B.P., Paul, R., Schwartzberg, P.L., Hood, J.D., Leng, J., Cheresh, D.A. (1999) Selective requirement for Src kinases during VEGF-induced angiogenesis and vascular permeability. Molec. Cell 4:915-924.

English, D., Cui, Y., Siddiqui, R., Patterson, C., Natarajan, V., Brindley, D., and Garcia, J.G. (1999) Induction of endothelial monolayer permeability by phosphatidate. J. Cell. Biochem. 75:105-117.

Eriksson, A., Cao, R., Roy, J., Tritsaris, K., Wahlestedt, C., Dissing, S., Thyberg, J., and Cao, Y (2003) Small GTP-binding protein Rac is an essential mediator of VEGF-induced endothelial fenestrations and vascular permeabiity. Circ.107:1532-1538.

Esser, S., Lampugnani, M.G., Corada, M., Dejana, E., and Risau, W. (1998) Vascular endothelial growth factor induces VE-cadherin tyrosine phosphorylation in endothelial cells. J. Cell Sci. 111:1853-1865.

Essler, M., Amano, M., Kruse, H.J., Kaibuchi, K., Weber, P.C., and Aepfelbacher, M. (1998) Thrombin inactivates myosin light chain phosphatase via Rho and its target Rho kinase in human endothelial cells. J. Biol. Chem. 273:21867-21874.

Essler, M., Retzer, M., Ilchmann, H., Linder, S., and Weber, P.C. (2002) S1P dynamically regulates myosin light chain phosphatase activity in human endothelial cells. Cell. Sig. 14:607-613.

Essler, M., Staddon, J., Weber, P., and Aepfelbacher, M. (2000) Cyclic AMP blocks bacterial LPS-induced MLC phosphorylation in endothelial cells through inhibition of Rho/Rho kinase signaling. J. Immunol. 164:6543-6549.

Exton, J.H. (1999) Regulation of phospholipase D. Biochim. Biophys. Acta 1439:121-133.

Farrukh, I.S., Gurtner, G.H., and Michael, J.R. (1987) Pharmacological modification of pulmonary vascular injury: possible role of cAMP. J. Appl. Physiol. 62:47-54.

Ferris, C.D. and Snyder, S.H. (1992) Inositol 1,4,5-trisphosphate-activated calcium channels. Ann. Rev. Physiol. 54;469-488.

Ferro, T., Neumann, P., Gertzberg, N., Clements, R., and Johnson, A. (2000) PKC-α mediates endothelial dysfunction induced by TNF-α. Amer. J. Physiol. 278:L1107-L1117.

Ferro, T.J., Parker, D., Commins, L., Phillips, P., and Johnson, A. (1993) TNFα activates pulmonary artery endothelial protein kinase C. Amer. J. Physiol. 264:L7-L14.

Fischer, D., Liliom, K., Guo, Z., Nusser, N., Virag, T., Murakami-Murofushi, K., Kobayashi, S., Erickson, J., Sun, G., Miller, D., and Tigyi, G. (1998) Naturally occurring analogs of lysophosphatidic acid elicit different cellular responses through selective activation of multiple receptor subtypes. Molec. Pharm. 54:979-988.

Fischer, S., Clauss, M., Wiesnet, M., Renz, D., Schaper, W., and Karliczek, G.F. (1999) Hypoxia induces permeability in brain microvessel endothelium via VEGF and NO. Amer. J. Physiol. 276:C812-C820.

Fleming, I. and Busse, R. (2003) Molecular mechanisms involved in the regulation of the endothelial nitric oxide synthase. Amer. J. Physiol. 284:R1-R12.

Flinn, H.M. and Ridley, A.J. (1996) Rho stimulates tyrosine phosphorylation of focal adhesion kinase, p130 and paxillin. J. Cell. Sci. 109:1133-1141.

Frank, G.D., Eguchi, S., Inagami, T., and Motley, E.D. (2001) N-Acetylcysteine inhibits angiotensin II-mediated activation of extracellular signal-regulated kinase and epidermal growth factor receptor. Biochem. Biophys. Res. Comm. 286;692-696.

Frank, G.D. and Eguchi, S. (2003) Activation of tyrosine kinases by reactive oxygen species in vascular smooth cells: Significance and involvement of EGF receptor transactivation by angiotensin II. Antiox. Redox Sig. 5:771-780.

Frank, G.D., Mifune, M., Inagami, T., Ohba, M., Sasaki, T., Higashiyama, S., Dempsey, P., and Eguchi, S. (2003) Distinct mechanisms of receptor and non-receptor tyrosine kinase activation by reactive oxygen species in vascular smooth muscle cells; role of metalloproteases and protein kinase C-delta. Molec. Cell Biol. 23:1581-1589.

Fu, B.M. and Shen S. (2004) Acute VEGF effect on solute permeability of mammalian microvessels in vivo. Microvasc. Res. 68:51-62.

Fukami, K., Furuhashi, K., Inagaki, M., Endo, T., Hatano, S., and Takenawa, T. (1992) Requirement of phosphatidylinositol 4,5-bisphosphate for α-actinin function. Nature 359:150-152.

Gainor, J.P., Morton, C., Roberts, J., Vincent, P., and Minnear, F.L. (2001) Platelet-conditioned medium increases endothelial electrical resistance independently of cAMP/PKA and cGMP/PKG. Amer. J. Physiol. 281:H1992-2001.

Gamble, J.R., Drew, J., Trezise, L., Underwood, A., Parsons, M., Kasminkas, L., Rudge, J., Yancopoulos, G., and Vadas, M.A. (2000) Angiopoietin-1 is an antipermeability and anti-inflammatory agent in vitro and targets cell junctions. Circ. Res. 87:603-607.

Gao, X., Kouklis, P., Xu, N., Minshall, R., Sandoval, R., Vogel, S., and Malik, A.B. (2000) Reversibility of increased microvessel permeability in response to VE-cadherin disassembly. Am J Physiol Lung Cell Mol Physiol 279:L1218-L1225.

Garcia, J.G.N., Davis, H., and Patterson, C.E. (1995a) Regulation of endothelial cell gap formation and barrier dysfunction: role of myosin light chain phosphorylation. J. Cell. Physiol. 163:510-522.

Garcia, J.G.N., Patterson, C.E., Bahler, C., Dukes, R., Aschner, J., and D. English. (1993) Receptor activating peptides induce Ca^{2+} mobilization, barrier dysfunction, PG synthesis, and PDGF mRNA expression in cultured endothelium. J. Cell. Physiol. 156:541-549.

Garcia, J.G.N., Pavalko, F.M., and Patterson, C.E. (1995b) Vascular endothelial cell permeability responses to thrombin. Blood Coag. Fibrin. 6:609-626.

Garcia, J.G.N., Schaphorst, K.L., Shi, S., Verin, A.D., Hart, C.M., Callahan, K.S., and Patterson, C.E. (1997) Mechanisms of ionomycin-induced endothelial cell barrier dysfunction. Amer. J. Physiol. 273;L172-L184.

Garcia, J.G.N., Verin, A., Schaphorst, K., Siddiqui, R., Patterson, C.E., Csortos, C., and Natarajan, V. (1999) Regulation of endothelial cell MLCK by Rho, cortactin, and $p60^{src}$. Amer. J. Physiol. 276:L989-L998.

Gardner, T.W., Lesher, T., Khin, S., Vu, C., Barber, A.J., and Brennan, W.A. (1996) Histamine reduces ZO-1 tight-junction protein expression in cultured retinal microvascular endothelial cells. Biochem. J. 320:717-721.

Gertzberg, N., Newmann, P., Rizzo, V., and Johnson, A. (2004) NADPH oxidase mediates the endothelial barrier dysfucntion induced by TNFα. Amer. J. Physiol. 286:L37-L48.

Gescher, A. (1992) Towards selective pharmacological modulation of protein kinase C - opportunities for the development of novel antineoplastic agents. Brit. J. Cancer 66:10-19.

Giembycz, M.A. and Raeburn, D. (1991) Putative substrates for cyclic nucleotide dependent protein kinases and the control of airway smooth muscle tone. J. Auton. Pharmacol. 11:365-398.

Gilmore, A.P. and Burridge, K. (1996) Regulation of vinculin binding to talin and actin by phosphatidyl-inositol-4-5-bisphosphate. Nature 381:531-535.

Gohla, A., Harhammer, R., and Schultz, G. (1998) The G-protein G13 but not G12 mediates signaling from lysophosphatidic acid receptor via epidermal growth factor receptor to Rho. J. Biol. Chem. 273:4653-4659.

Goldberg, P.L., MacNaughton, D., Clements, R., Minnear, F., and Vincent, P.A. (2002) p38 MAPK activation by TGF-β1 increases MLC phosphorylation and endothelial monolayer permeability. Amer. J. Physiol. 282:L146-L154.

Gow, A.J., Thom, S R., and Ischiropoulos, H. (1998) Nitric oxide and peroxynitrite-mediated pulmonary cell death. Amer. J. Physiol. 274:L112-L118.

Grover, A.K., Samson, S.E., and Fomin, V.P. (1992) Peroxide inactivates calcium pumps in pig coronary artery. Amer. J. Physiol. 263:H537-H543.

Haller, H., Ziegler, W., Lindschau, C., and Luft, F.C. (1996) Endothelial cell tyrosine kinase receptor and G protein- coupled receptor activation involves distinct protein kinase C isoforms. Arterioscler. Thromb. Vasc. Biol. 16:678-686.

Hammond, S.M., Jenco, J., Nakashima, S., Cadwallader, K., Gu, Q., Cook, S., Nozawa, Y., Prestwich, G., Frohman, M., and Morris, A.J. (1997) Characterization of two spliced forms of PLD1. Activation of the purified enzymes by PI 4,5-PP, ADP-ribosylation factor, and Rho family monomeric GTP-binding proteins and PKC-α. J. Biol. Chem. 272:3860–3868.

Hanoune, J. and Defer N. (2001) Regulation and role of adenylyl cyclase isoforms. Ann. Rev. Pharmaol. Toxicol. 41:145-174.

Hart, M.J., Jiang, X., Kozasa, T., Roscoe, W., Singer, W., Gilman, A., Sternweis, P., and Bollag, G. (1998) Direct stimulation of the guanine nucleotide exchange activity of p115 RhoGEF by Gα13. Science 280:2112-2114.

Haselton, F.R. and Alexander, J.S. (1992) Platelets and a platelet-released factor enhance endothelial barrier. Amer. J. Physiol. 263:L670-L678.

Hastie, L.E., Patton, W., Hechtman, H., and Shepro, D. (1997) Filamin redistribution in an endothelial cell reoxygenation injury model. Free Rad. Biol. Med. 22:955-966.

He, P. and Curry, F.E. (1993a) Albumin modulation of capillary permeability: role of endothelial cell $[Ca^{2+}]_i$. Amer. J. Physiol. 265:H74-H82.

He, P. and Curry, F.E. (1993b) Differential actions of cAMP on endothelial [Ca2+]i and permeability in microvessels exposed to ATP. Amer. J. Physiol. 265:H1019-H1023.

Hippenstiel, S., Krull, M., Ikemann, A., Risau, W., Clauss, M., and Suttorp, N. (1998) VEGF induces hyperpermeability by a direct action on endothelial cells. Amer. J. Physiol. 274:L678-L684.

Hippenstiel, S., Tannert-Otto, S., Vollrath, N., Krull, M., Just, I., Aktories, K., von Eichel-Streiber, C., and Suttorp, N. (1997) Glucosylation of small GTP-binding Rho proteins disrupts endothelial barrier function. Amer. J. Physiol. 272:L38-L43.

Hsu, Y.H., Chen, J., Chang, N., Chen, C., Liu, J., Chen, T., Jeng, C., Chao, H., and Cheng, T.H. (2004) Role of reactive oxygen species-sensitive extracellular signal-regulated kinase pathway in angiotensin II-induced endothelin-1 gene expression in vascular endothelial cells. J. Vasc. Res. 41:64-74.

Issbrucker, K., Marti, H., Hippenstiel, S., Springmann, G., Voswinckel, R., Gaumann, A., Breier, G., Drexler, H., Suttorp, N., and Clauss, M. (2003) p38 MAP kinase-a molecular switch between VEGF- angiogenesis and vascular hyperpermeability. FASEB J. 17:262-264.

Iyer, S., Ferreri, D., DeCocco, N., Minnear, F., and Vincent, P.A. (2004) VE-cadherin-p120 interaction is required for maintenance of endothelial barrier function. Amer. J. Physiol. 286:L1143-L1153.

Jay, D. and Stracher, A. (1997) Expression in Escherichia coli, phosphorylation with cAMP-dependent protein kinase and proteolysis by calpain of a 71-kDa domain of human endothelial actin binding protein. Biochem. Biophys. Res. Comm. 232:555-558.

Jenkins, G.H., Fisette, P.L., Anderson, R.A. (1994) Type I phosphatidylinositol 4-phosphate 5-kinase isoforms are specifically stimulated by phosphatidic acid. J. Biol. Chem. 269:11547–11554.

Johnson, A., Phillips, P., Hocking, D., Tsan, M., and Ferro, T. (1989) PKC inhibitor prevents pulmonary edema in response to H_2O_2. Amer. J. Physiol. 256:H1012-H1022.

Jones, P.F. (2003) Not just angiogenesis--wider roles for the angiopoietins. J. Path. 201:515-527.

Jones, S., O'Donnell, V., Wood, J., Broughton, J., Hughes, E., and Jones, O. (1996) Expression of phagocyte NADPH oxidase components in human endothelial cells. Amer. J. Physiol. 271:H1626-H1634.

Kaibuchi, K., Kuroda, S., Fukata, M., and Nakagawa, M. (1999) Regulation of cadherin-mediated cell-cell adhesion by the Rho family GTPases. Curr. Opin. Cell Biol. 11:591-596.

Kelly, J.J., Moore, T.M., Babal, P., Diwan, A.H., Stevens, T., and Thompson, W.J. (1998) Pulmonary microvascular and macrovascular endothelial cells: differential regulation of Ca^{++} and permeability. Amer. J. Physiol. 274;L810-L819.

Kimura, K., Ito, M., Amano, M., Chihara, K., Fukata, Y., Nakafuku, M., Yamamori, B., Feng, J., Nakano, T., Okawa, K., Iwamatsu, A., and Kaibuchi, K. (1996) Regulation of myosin phosphatase by Rho and Rho-associated kinase. Science 273:245-248.
Knopf, J.L., Lee, M., Sultzman, L.A., Kriz, R.W., Loomis, C.R., Hewick, R.M., and Bell, R.M. (1986) Cloning and expression of multiple protein kinase C cDNAs. Cell 46:491-502.
Koga, S., Morris, S., Ogawa, S., Liao, H., Bilezikian, J., Chen, G., Thompson, W., Ashikaga, T., Brett, J., and Stern, D.M. (1995) TNF modulates endothelial properties by decreasing cAMP. Amer. J. Physiol. 268:C1104-C1113.
Konishi,H., Tanaka,M., Takemura,Y., Matsuzaki,H., Ono,Y., Kikkawa,U., and Nishizuka,Y. (1997) Activation of protein kinase C by tyrosine phosphorylation in response to H2O2. PNAS USA 94:11233-11237.
Kozasa, T. and Gilman, A.G. (1996) Protein kinase C phosphorylates G(12 α) and inhibits its interaction with G(β and γ). J. Biol Chem 271:12562-12567.
Kuang, Y., Wu, Y., Jiang, H., and Wu, D. (1996) Selective G protein coupling by C-C chemokine receptors. J. Biol. Chem. 271:3975-3978.
Kuebler, W.M., Ying, X., Singh, B., Issekutz, A., and Bhattacharya, J. (1999) Pressure is proinflammatory in lung venular capillaries. J. Clin. Invest. 104:495-502.
Kunsch, C. and Medford, R.M. (1999) Oxidative stress as a regulator of gene expression in the vasculature. Circ. Res. 85:753-66.
Kumagai, N., Morii, N., Fujisawa, K., Nemoto, Y., and Narumiya, S. (1993) ADP-ribosylation of Rho p21 inhibits LPA-induced protein tyrosine phosphorylation and phosphatidylinositol 3-kinase activation. J. Biol. Chem. 268:24535-24538.
Lamb, N.J.C., Fernandez, A., Conti, M., Adelstein, R. Glass, D., Welch, W., Feramisco, J.R. (1988) Regulation of actin microfilament integrity in living non-muscle cells by PKA and MLCK. J. Cell Biol. 106:1955-1971.
Lampugnani, M., Zanetti, A., Corada, M., Takahashi, T., Balconi, G., Breviario, F., Orsenigo, F., Cattelino, A., Kemler, R., Daniel, T., and Dejana, E. (2003) Contact inhibition of VEGF-induced proliferation requires vascular endothelial cadherin, β-catenin, and the phosphatase DEP-1/CD148. J. Cell Biol. 161:793-804.
Lang, P., Gesbert, F., Delespine-Carmagnat, M., Stancou, R., Pouchelet, M., and Bertoglio, J. (1996) PKA phosphorylation of RhoA mediates the morphological and functional effects of cyclic AMP in cytotoxic lymphocytes. EMBO J. 15:510-519.
Lee, C.I. and Okabe, E. (1995) Hydroxyl radical-mediated reduction of Ca^{++}-ATPase activity of masseter muscle sarcoplasmic reticulum. Jap. J. Pharmacol. 67:21-28.
Lee, M.J., Thangada, S., Claffey, K., Ancellin, N., Liu, C., Kluk, M., Volpi, M., Sha'afi, R., and Hla, T. (1999) Vascular endothelial cell adherens junction assembly and morphogenesis induced by sphingosine-1-phosphate. Cell 99:301-312.
Lim, L., Manser, E., Leung, T., and Hall, C. (1996) Regulation of phosphorylation pathways by p21 GTPases - The p21 Ras-related Rho subfamily and its role in phosphorylation signalling pathways. Eur. J. Biochem. 242:171-185.
Lindenschmidt, R.C., Selig, W., Patterson, C.E., Verburg, K., Henry D., and Rhoades, R.A. (1986) Histamine release in paraquat-induced lung injury. Amer. Rev. Respir. Dis. 133:274-278.
Liu, F., Verin, A., Borbiev, T., and Garcia, J.G. (2001) Role of PKA activity in endothelial cell cytoskeleton rearrangement. Amer. J. Physiol. 280:L1309-L1317.
Lofton, C.E., Baron, D.A., Heffner, J.E., Currie, M.G., and Newman, W.H. (1991) Atrial natriuretic peptide inhibits oxidant-induced endothelial permeability. J. Molec. Cell. Cardiol. 23;919-927.
Lopez-Ongil, S., Torrecillas, G., Perez-Sala, D., Gonzalez-Santiago, L., Rodriguez-Puyol, M., and Rodriguez-Puyol, D. (1999) Mechanisms involved in the contraction of endothelial cells by hydrogen peroxide. Free Rad. Biol. Med. 26:501-510.
Lu, C., Chan, S., Fu, W., and Mattson, M.P. (2002) The lipid peroxidation product 4-hydroxynonenal facilitates opening of voltage-dependent Ca2+ channels in neurons by increasing protein tyrosine phosphorylation. J. Biol. Chem. 277:24368-24375.
Lugnier, C. and Schini, V.B. (1990) Characteriztion of cyclic nucleotide phosphodiesterases from cultured bovine aortic endothelial cells. Biochem. Pharmacol. 39:75-84.

Lum, H., Aschner, J.L., Phillips, P.G., Fletcher, P.W., and Malik, A.B. (1992) Time course of thrombin-induced increase in endothelial permeability: relationship to Ca^{2+}_i and inositol polyphosphates. Amer. J. Physiol. 263:L219-L225.

Lum, H., Jaffe, H., Schulz, I., Masood, A., RayChaudhury, A., and Green, R.D. (1999) Expression of PKA inhibitor (PKI) gene abolishes cAMP-mediated protection to endothelial barrier dysfunction. Amer. J. Physiol. 46:C580-C588.

Lum, H. and Malik, A.B. (1994) Regulation of vascular endothelial barrier function. Amer. J. Physiol. 267:L223-L241.

Lum, H. and Roebuck, K.A. (2001) Oxidant stress and endothelial cell dysfunction. Amer. J. Physiol. 280:C719-C741.

Lynch, J., Ferro, T., Blumenstock, A., and Malik, A.B. (1990) Increased endothelial albumin permeability mediated by protein kinase C activation. J. Clin. Invest. 85:1991-1998.

MacGlashan, D. (2003) Histamine: A mediator of inflammation. J. Allergy Clin. Immunol. 112:S53-S59.

Manolopoulos, V.G., Fenton, J., and Lelkes, P.I. (1997) The thrombin receptor in adrenal medullary microvascular endothelial cells is negatively coupled to adenylyl cyclase through a Gi protein. Biochim. Biophys. Acta 1356:321-332.

Mark, R.J., Lovell, M., Markesbery, W., Uchida, K., and Mattson, M.P. (1997) A role for 4-OH-2-nonenal, an aldehydic product of lipid peroxidation, in disruption of ion homeostasis and neuronal death induced by amyloid beta-peptide. J Neurochem (68): 255-264.

Matsui, T., Amano, M., Yamamoto, T., Chihara, K., Nakafuku, M., Ito, M., Nakano, T., Okawa, K., Iwamatsu, A., and Kaibuchi, K. (1996) Rho-associated kinase, a novel serine/threonine kinase, as a putative target for small GTP binding protein Rho. EMBO J. 15:2208-2216.

McLees, A., Graham, A., Malarkey, K., Gould, G., and Plevin, R. (1995) Regulation of lysophosphatidic acid-stimulated tyrosine phosphorylation of mitogen-activated protein kinase by PKC- and pertussis toxin-dependent pathways in the endothelial cell line EAhy 926. Biochem. J. 307:743-748.

McQuaid, K.E. and Keenan, A.K. (1997) Endothelial barrier dysfunction and oxidative stress: roles for nitric oxide. Exp. Physiol. 82:233-241.

Mehta, D., Tiruppathi, C., Sandoval, R., Minshall, R., Holinstat, M., and Malik, A.B. (2002) Modulatory role of focal adhesion kinase in regulating human pulmonary arterial endothelial barrier function. J. Physiol. 539:779-789.

Mikalsen, S.O. and Kaalhus, O. (1996) A characterization of pervanadate, an inducer of cellular tyrosine phosphorylation and inhibitor of gap junctional intercellular communication. Biochim. Biophys. Acta 1290:308-318.

Minnear, F.L., DeMichele, M., Moon, D., Rieder, C., and Fenton, J.W. (1989) Isoproterenol reduces thrombin-induced pulmonary endothelial permeability in vitro. Amer. J. Physiol. 257: H1613-H1623.

Minnear, F.L., Patil, S., Bell, D., Gainor, J., and Morton, C.A. (2001) Platelet lipid(s) bound to albumin increases endothelial electrical resistance: mimicked by LPA. Amer. J. Physiol. 281:L1337-L1344.

Minshall, R.D., Sessa, W., Stan, R., Anderson, R., and Malik AB. (2003) Caveolin regulation of endothelial function. Amer. J. Physiol. 285:L1179-L1183.

Miura, Y., Yatomi, Y., Rile, G., Ohmori, T., Satoh, K., and Ozaki, Y. (2000) Rho-mediated phosphorylation of FAK and MLCK in human endothelial cells stimulated with sphingosine 1-phosphate, a bioactive lysophospholipid released from activated platelets. J. Biochem. 127:909-914.

Moolenaar, W.H. (1999) Bioactive lysophospholipids and their G protein-coupled receptors. Exp. Cell Res. 253:230-238.

Moore, T.M., Chetham, P., Kelly, J., and Stevens, T. (1998) Signal transduction and regulation of lung endothelial permeability. Interaction between calcium and cAMP. Amer. J. Physiol. 275:L203-L222.

Moy, A.B., Shasby, S., Scott, B., and Shasby, D.M. (1993) The effect of histamine and cyclic adenosine monophosphate on myosin light chain phosphorylation in human umbilical vein endothelial cells. J. Clin. Invest. 92:1198-1206.

Moy, A.B., VanEngelenhoven, J., Bodmer, J., Kamath, J., Keese, C., Giaever, I., Shasby, S., and Shasby, D M. (1996) Histamine and thrombin modulate endothelial focal adhesion through centripetal and centrifugal forces. J. Clin. Invest. 97:1020-1027.

Muller, G. and Lipp, M. (2001) Signal transduction by the chemokine receptor CXCR5: structural requirements for G protein activation analyzed by chimeric CXCR1/CXCR5 molecules. Biol. Chem. 382:1387-1397.

Murohara, T., Horowitz, J., Silver, M., Tsurumi, Y., Chen, D., Sullivan, A., and Isner, J.M. (1998) Vascular endothelial growth factor/vascular permeability factor enhances vascular permeability via nitric oxide and prostacyclin. Circ. 97:99-107.

Natarajan, V. (1995) Oxidants: signal transduction in vascular endothelium. J. Lab. Clin. Med. 125:26-37.

Natarajan, V., Scribner, W., Morris, A., Roy, S., Vepa, S., Yang, J., Wandgaonkar, R., Reddy, S., Garcia, J., and Parinandi, N.L. (2001) Role of p38 MAP kinase in diperoxovanadate-induced PLD activation in endothelial cells. Amer. J. Physiol. 281:L435-L449.

Natarajan, V., Taher, M., Roehm, B., Parinandi, N., Schmid, H., Kiss, Z., and Garcia, J.G. (1993) Activation of endothelial cell PLD by H_2O_2 and fatty acid hydroperoxide. J. Biol. Chem. 268:930-937.

Natarajan, V., Vepa, S., Shamlal, R., Al-Hassani, M., Ramasarma, T., Ravishanker, H., and Scribner, W.M. (1998) Tyrosine kinases and Ca^{++} dependent activation of endothelial cell PLD by diperoxovanadate. Molec. Cell Biochem. 183:113-124.

Natarajan, V., Vepa, S., Verma, R., and Scribner, W. (1996) Role of protein tyrosine phosphorylation in H_2O_2-induced activation of endothelial cell PLD. Amer. J. Physiol. 271:L400-L408.

Newman, P.J. and Newman, D.K. (2003) Signal transduction pathways mediated by PECAM-1. Arterioscler. Thromb. Vasc. Biol. 23:953-964.

Nilius, B. and Voets, T. (2004) Diversity of TRP channel activation. Novartis Found. Sym. 258:140-149.

Nishizuka, Y. (1992) Intracellular signaling by hydrolysis of phospholipids and activation of PKC. Science 258:607-614.

Norwood, N., Moore, T., Dean, D., Bhattacharjee, R., Li, M., and Stevens, T. (2000) Store-operated calcium entry and increased endothelial cell permeability. Amer. J. Physiol. 279:L815-L824.

Ogawa, S., Koga, S., Kuwabara, K., Brett, J., Morrow, B., Morris, S., Bilezikian, J., Silverstein, S., and Stern, D. (1992) Hypoxia-induced increased permeability of endothelial monolayers occurs through lowering of cellular cAMP. Amer. J. Physiol. 262:C546-C554.

Ohta, Y., Suzuki, N., Nakamura, S., Hartwig, J., and Stossel, T.P. (1999) The small GTPase RalA targets filamin to induce filopodia. PNAS USA 96:2122-2128.

Okayama, N., Ichikawa, H., Coe, L., Itoh, M., and Alexander, J.S. (1998) Exogenous NO enhances hydrogen peroxide-mediated neutrophil adherence to cultured endothelial cells. Amer. J. Physiol. 274:L820-L826.

Orr, A.W. and Murphy-Ullrich, J.E. (2004) Regulation of endothelial cell function by FAK and PYK2. Front. Biosci. 9:1254-1266.

Ou, J., Fontana, J., Ou, Z., Jones, D., Ackerman, A., Oldham, K., Yu, J., Sessa, W., and Pritchard, K.A. Jr. (2004) Heat shock protein 90 and tyrosine kinase regulate eNOS NO* generation but not NO* bioactivity. Amer. J. Physiol. 286:H561-H569.

Panetti, T.S. (2002) Differential effects of S1P and lysophosphatidic acid on endothelial cells. Biochim. Biophys. Acta 1582:190-196.

Parinandi, N.L., Roy, S., Shi, S., Cummings, R., Morris, A., Garcia, J., and Natarajan, V. (2001) Role of Src kinase in diperoxovanadate-mediated activation of PLD in endothelial cells. Arch. Biochem. Biophys. 396:231-243.

Patterson, C.E., Davis, H., Schaphorst, K., and Garcia, J.G.N. (1994a) Mechanisms of cholera toxin prevention of thrombin- and PMA-induced endothelial cell barrier dysfunction. Microvasc. Res. 48:212-235.

Patterson, C.E. and Garcia, J.G.N. (1994b) Regulation of thrombin-induced endothelial cell activation by bacterial toxins. Blood Coag. Fibrinol. 5:63-72.
Patterson, C.E., Lum, H., Verin, A., Schaphorst, K., and Garcia, J.G.N. (2000) Regulation of endothelial barrier function by the cAMP-dependent protein kinase. Endoth. 7:287-308.
Patterson, C.E., Stasek, J., Schaphorst, K., Davis, H., and Garcia, J.G. (1995) Mechanisms of pertussis toxin-induced barrier dysfunction in bovine pulmonary artery endothelial cell monolayers. Amer. J. Physiol. 268:L926-L934.
Perez-Moreno, M., Avila, A., Islas, S., Sanchez, S., and Gonzalez-Mariscal, L. (1998) Vinculin but not alpha-actinin is a target of PKC phosphorylation during junctional assembly induced by calcium. J. Cell Sci. 111:3563-3571.
Pizurki, L., Zhou, Z., Glynos, K., Ro ussos, C., and Papapetropoulos, A. (2003) Angiopoietin-1 inhibits endothelial permeability, neutrophil adherence and IL-8 production. Brit. J. Pharmacol. 139:329-336.
Prenzel, N., Zwick, E., Daub, H., Leserer, M., Abraham, R., Wallasch, C., and Ullrich, A. (1999) EGF receptor transactivation by G-protein coupled receptors requires metalloproteinase cleavage of proHB-EGF. Nature 402:884-888.
Rao, G.N. (1996) H_2O_2 induces complex formation of SHC-Grb2-SOS with receptor kinase and activates Ras and ERK group of MAPK. Oncogene 13:713-719.
Retta, S.F., Barry, S., Critchley, D., Defilippi, P., Silengo, L., and Tarone, G. (1996) Focal adhesion and stress fiber formation is regulated by tyrosine phosphatase activity. Exp. Cell Res. 229:307-317.
Romer, L.H., McLean, N., Turner, C., and Burridge, K. (1994) Tyrosine kinase activity, cytoskeletal organization, and motility in human vascular endothelial cells. Molec. Biol. Cell 5:349-361.
Ross, D. and Joyner, W.L. (1997) Resting distribution and stimulated translocation of PKC isoforms α, ε, and ζ in response to bradykinin and TNF in human endothelial cells. Endothelium 5:321-332.
Rotrosen, D. and Gallin, J.I. (1986) Histamine type I receptor occupancy increases endothelial cytosolic calcium, reduces F-actin, and promotes albumin diffusion across cultured endothelial monolayers. J. Cell Biol. 103:2379-2387.
Ruegg, C. and Mariotti, A. (2003) Vascular integrins: pleiotropic adhesion and signaling molecules in vascular homeostasis and angiogenesis. Cell. Molec. Life Sci. 60:1135-1157.
Sandoval, R., Malik, A., Naqvi, T., Metha, D., Tirrupathi, C. (2001) Requirement for Ca^{2+} signaling in the mechanism of thrombin-induced endothelial permeability. Amer. J. Physiol. 280:L239-L247.
Schaphorst, K.L., Chiang, E., Jacobs, K., Zaiman, A., Natarajan, V., Wigley, F., and Garcia, J.G. (2003) Role of sphingosine 1-phosphate in enhancement of endothelial barrier integrity by platelet-released products. Amer. J. Physiol. 285:L258-L267.
Schaphorst, K.L., Pavalko, F., Patterson, C.E., and Garcia, J.G. (1997) Thrombin-mediated focal adhesion plaque reorganization in endothelium: Role of protein phosphorylation. Amer. J. Respir. Cell Molec. Biol. 17: 443-455.
Schnittler, H.J., Wilke, A., Gress, T., Suttorp, N., and Drenckhahn, D. (1990) Role of actin and myosin in the control of paracellular permeability in pig, rat and human vascular endothelium. J. Physiol. 431:379-401.
Schulze, C., Smales, C., Rubin, L.L. and Staddon, J.M. (1997) Lysophosphatidic acid increases tight junction permeability in cultured brain endothelial cells. J. Neurochem. 68:991-1000.
Schulze-Neick, I., Penny, D.J., Rigby, M.L., Morgan, C., Kelleher, A., Collins, P., Li, J., Bush, A., Shinebourne, E.A., and Redington, A.N. (1999) L-arginine and substance-P reverse the pulmonary endothelial dysfunction caused by congenital heart surgery. Circ. 100:749-755.
Selig, W.M., Patterson, C.E. and Rhoades, R.A. (1986) Role of histamine in acute oleic-acid induced lung injury. J. Appl. Phys. 61:233-239.
Sharma, N.R and Davis, M.J. (1996) Calcium entry activated by store depletion in coronary endothelium is promoted by tyrosine phosphorylation. Amer. J. Physiol. 270:H267-H274.

Shasby, D.M., Kamath, J.M., Moy, A.B., and Shasby, S.S. (1995) Ionomycin and PDBU increase MDCK monolayer permeability independently of MLC phosphorylation. Amer. J. Physiol. 269:L144-L150.
Shasby, D.M., Lind, S.E., Shasby, S S., Goldsmith, J.C., and Hunninghake, G.W. (1985) Reversible oxidant-induced increases in albumin transfer across cultured endothelium: alterations in cell shape and calcium homeostasis. Blood 65:605-614.
Shasby, D.M., Stevens, T., Ries, D., Moy, A., Kamath, J., Kamath, A., and Shasby, S.S. (1997) Thrombin inhibits myosin light chain dephosphorylation in endothelial cells. Amer. J. Physiol. 16:L311-L319.
Shasby, D.M., Yorek, M., and Shasby, S.S. (1988) Exogenous oxidants initiate hydrolysis of endothelial cell inositol phospholipds. Blood 72:491-499.
Shay-Salit, A., Shushy, M., Wolfovitz, E., Yahav, H., Breviario, F., Dejana, E., and Resnick, N. (2002) VEGF receptor 2 and the adherens junction as a mechanical transducer in vascular endothelial cells. PNAS USA 99:9462-9467.
Sheldon, R., Moy, A., Lindsley, K., Shasby, S., and Shasby, D.M. (1993) Role of myosin light-chain phosphorylation in endothelial cell retraction. Amer. J. Physiol. 265:L606-L612.
Shi, S., Garcia, J., Roy, S., Parinandi, N., and Natarajan, V. (2000) Involvement of c-Src in DPV-induced endothelial cell barrier dysfunction. Amer. J. Physiol. 279:L441-L451.
Shi, S., Verin, K.L., Gilbert-McClain, L., Paterson, C.E., Irwin, R., Natarajan, V., and Garcia, J.G. (1998) Role of tyrosine phosphorylation in thrombin-induced endothelial cell contraction and barrier function. Endothelium 6, 153-171.
Siddiqui, R.A. and English, D. (2000) PI3K-mediated calcium mobilization regulates chemotaxis in phosphatidic acid-stimulated human neutrophils. Biochim. Biophys. Acta 1483:161-173.
Siflinger-Birnboim, A., Goligorsky, M.S., Delvecchio, P.J., and Malik, A.B. (1992) Activation of protein kinase C pathway contributes to hydrogen peroxide-induced increase in endothelial permeability. Lab. Invest. 67:24-30.
Siflinger-Birnboim, A., Lum, H., del Vecchio, P., and Malik, A. (1996) Involvement of Ca^{2+} in the H_2O_2-induced increase in endothelial permeability. Amer. J. Physiol. 270:L973-L978.
Sirois, M.G. and Edelman, E.R. (1997) VEGF effect on vascular permeability is mediated by synthesis of PAF. Amer. J. Physiol. 272:H2746-H2756.
Sohn, H.Y., Keller, M., Gloe, T., Morawietz, H., Rueckschloss, U., and Pohl, U. (2000) The small G-protein Rac mediates depolarization-induced superoxide formation in human endothelial cells. J. Biol. Chem. 275:18745-18750.
Sohn, R.H., Chen, J., Koblan, K.S., Bray, P.F., and Goldschmidtclermont, P.J. (1995) Localization of a binding site for PI 1 4, 5- bisphosphate on human profilin. J. Biol. Chem. 270:21114-21120.
Spiegel, S. and Milstien, S. (2000) Functions of a new family of sphingosine-1-phosphate receptors. Biochim. Biophys. Acta 1484:107-116.
Sprague, R.S., Stephenson, A., McMurdo, L., and Lonigro, A.J. (1998) Nitric oxide opposes phorbol ester-induced pulmonary microvascular permeability in dogs. J. Pharm. Exp. Therap. 284:443-448.
Stasek, J.E., Patterson, C.E., and Garcia, J.G. (1992) PKC phosphorylates caldesmon77 and vimentin and enhances albumin permeability across cultured bovine pulmonary artery endothelial cell monolayers. J. Cell. Physiol. 153:62-75.
Stelzner, T.J., Weil, J.V., and O'Brien, R.F. (1989) Role of cyclic adenosine monophosphate in the induction of endothelial barrier properties. J. Cell. Physiol. 139;157-166.
Stevens, T., Fouty, B., Hepler, L., Richardson, D., Brough, G., McMurtry, I.F., and Rodman, D.M. (1997) Cytosolic Ca^{2+} and adenylyl cyclase responses in phenotypically distinct pulmonary endothelial cells. Amer. J. Physiol. 272:L51-L59.
Stevens, T., Nakahashi, Y., Cornfield, D., McMurtry, I., Cooper, D., and Rodman, D.M. (1995) Ca^{2+}-inhibitable adenylyl cyclase modulates pulmonary artery endothelial cell cAMP content and barrier function. PNAS USA 92:2696-2700.
Suttorp, N., Weber, U., Welsch, T., and Schudt, C. (1993) Role of phosphodiesterases in the regulation of endothelial permeability in vitro. J. Clin. Invest. 91:1421-1428.

Suzuki, Y.J., Forman, H.J., and Sevanian, A. (1997) Oxidants as stimulators of signal transduction. Free Rad. Biol. Med. 22:269-285.

Taher, M., Garcia, J.G., and Natarajan, V. (1993) Hydroperoxide-induced diacylglycerol formation and PKC activation in vascular endothelial cells. Arch. Biochem. Biophy. 303:260-266.

Takahashi, R., Watanabe, H., Zhang, X., and Kakizawa, H. (1997) Roles of inhibitors of MLCK and tyrosine kinase on cation influx in agonist-stimulated endothelial cells. Biochem. Biophys. Res. Comm. 235:657-662.

Takahashi, T. and Shibuya, M. (1997) The 230 kDa mature form of KDR/Flk-1 (VEGF receptor-2) activates the PLC-gamma pathway and partially induces mitotic signals in NIH3T3 fibroblasts. Oncogene 14:2079-2089.

Takai, Y., Sasaki, T., and Matozaki, T. (2001) Small GTP-binding proteins. Physiol. Rev. 81:153-208.

Tigyi, G., Fischer, D., Sebok, A., Yang, C., Dyer, D.L., and Miledi, R. (1996) Lysophosphatidic acid-induced neurite retraction in PC12 cells: control by phosphoinositide-Ca^{2+} signaling and Rho. J. Neurochem. 66:537-548.

Tsan, M. and White, J.E. (1988) Red blood cells protect endothelial cells against H_2O_2-mediated but not hyperoxia-induced damage. Proc. Soc. Exp. Biol. Med. 188:323-327.

Ukropec, J.A., Hollinger, M.K., and Woolkalis, M.J. (2002) Regulation of VE-cadherin linkage to the cytoskeleton in endothelial cells exposed to fluid shear stress. Exp. Cell Res. 273:240-247.

Usatyuk, P.V., Fomin, V., Shi, S., Garcia, J., Schaphorst, K., and Natarajan, V. (2003a) Role of Ca2+ in DPV-cytoskeletal remodeling and endothelial barrier function. Amer. J. Physiol. 285:L1006-L1017.

Usatyuk, P.V. and Natarajan, V. (2004) Role of MAPKs in 4-hydroxy-2-nonenal-induced actin remodeling and barrier function in endothelial cells. J. Biol. Chem. 279:11789-11797.

Usatyuk, P.V., Vepa, S., Watkins, T., He, D., Parinandi, N., and Natarajan, V. (2003b) Redox regulation of ROS-induced p38 MAPK activation and barrier dysfunction in lung microvascular endothelial cells. Antiox. Redox Sig. 5:723-730.

Ushio-Fukai, M., Griendling, K., Becker, P., Hilenski, L., Halleran, S., and Alexander, R.W. (2001) Epidermal growth factor receptor transactivation by angiotensin II requires ROS in vascular smooth muscle cells. Arterioscler.Thromb. Vasc.biol. 21, 489-495.

Usui, H., Inoue, R., Tanabe, O., Nishito, Y., Shimizu, M., Hayashi, H., Kagamiyama, H., and Takeda, M. (1998) Activation of protein PPase 2A by PKA-catalyzed phosphorylation of the 74-kDa B" (delta) regulatory subunit in vitro and identification of the phosphorylation sites. FEBS Lett. 430:312-316.

Van NieuwAmerongen, G.V., Draijer, R., Vermeer, M., and van Hinsbergh, V.M. (1998) Transient and prolonged increase in endothelial permeability induced by histamine and thrombin - Role of protein kinases, calcium, and RhoA. Circ. Res. 83:1115-1123.

Van Nieuw Amerongen, G.P., Vermeer, M.A., and van Hinsberg, V.W. (2000) Role of RhoA and ROCK in LPA-induced endothelial barrier dysfunction. Arterioscl. Thromb. Vasc. Biol. 20:E127-E133.

van Wetering, S., van Buul, J., Quik, S., Mul, F., Anthony, E., Ten Klooster, J., Collard, J., and Hordijk, P.L. (2002) Reactive oxygen species mediate Rac-induced loss of cell-cell adhesion in primary human endothelial cells. J. Cell. Sci. 115:1837-1846.

Verhaar, M.C., Westerweel, P., van Zonneveld, A., and Rabelink, T.J. (2004) Free radical production by dysfunctional eNOS. Heart (Brit. Card. Soc.) 90:494-495.

Verin, A.D., Cooke, C., Herenyiova, M., Patterson, C.E. and Garcia, J.G. (1998) Role of Ca^{2+}/calmodulin-dependent phosphatase 2B in thrombin-induced endothelial cell contractile responses. Amer. J. Physiol. 275:L788-L799.

Verin, A.D., Patterson, C.E., Day, M., and Garcia, J.G.N. (1995) Regulation of endothelial cell gaps and barrier function by myosin-associated phophatase activities. Amer. J. Physiol. 269:L99-L108.

Vestweber, D. (2000) Molecular mechanisms that control endothelial cell contacts. J. Pathol. 190:281-291.

Vincent, P.A., Xiao, K., Buckley, K., and Kowalczyk, A.P. (2004) VE-cadherin: adhesion at arm's length. Amer. J. Physiol. 286:C987-C997.
Vogt, S., Grosse, R., Schultz, G., and Offermanns, S. (2003) Receptor-dependent RhoA activation in G12/G13-deficient cells: evidence for an involvement of Gq/G11. J. Biol. Chem. 278:28743-28749.
Volk, T., Hensel, M., Kox, W.J. (1997) Transient Ca^{2+} changes in endothelial cells induced by low doses of ROS: role of hydrogen peroxide. Molec. Cell Biochem. 171:11-21.
Vouret-Craviari, V., Boquet, P., Pouyssegur, J., and Van Obberghen-Schilling, E. (1998) Regulation of the actin cytoskeleton by thrombin in human endothelial cells: role of Rho proteins in endothelial barrier function. Molec. Biol. Cell 9:2639-2653.
Vouret-Craviari, V., Bourcier, C., Boulter, E., and van Obberghen-Schilling, E. (2002) Distinct signals via Rho GTPases and Src drive shape changes by thrombin and sphingosine-1-phosphate in endothelial cells. J. Cell Sci. 115:2475-2484.
Vuong, P.T., Malik, A., Nagpala, P., and Lum, H. (1998) PKCβ modulates thrombin-induced Ca^{2+} signaling and endothelial permeability increase. J. Cell. Physiol. 175:379-387.
Wagner, A.H, Koehler, T., Rueckschloss, U., Just, I., and Heckler, M. (2000) Improvement of NO-dependent vasodilation by HMG-CoA reductase inhibitors through attenuation of EC superoxide formation. Arterio. Thromb. Vasc. Biol. 20:61-69.
Wakelam, M.J., Hodgkin, M., Martin, A., Saqib, K. (1997) PLD. Sem. Cell Dev. Biol. 8:305-310.
Wang, Y., Pampou, S., Fujikawa, K., and Varticovski, L. (2004) Opposing effect of angiopoietin-1 on VEGF-disruption of endothelial cell-cell interactions requires PKC β. J. Cell. Physiol. 198:53-61.
Ward, N.L. and Dumont, D.J. (2002) The angiopoietins and Tie2/Tek: adding to the complexity of cardiovascular development. Sem. Cell Dev. Biol. 13:19-27.
Waschke, J., Baumgartner, W., Adamson, R., Zeng, M., Aktories, K., Barth, H., Wilde, C., Curry, F., and Drenckhahn, D. (2004) Requirement of Rac activity for maintenance of capillary endothelial barrier properties. Amer. J. Physiol. 286:H394-H401.
Watterson, K., Sankala, H., Milstien, S., and Spiegel, S. (2003) Pleiotropic actions of sphingosine-1-phosphate. Prog. Lipid Res. 42:344-357.
Wei, Z., Costa, K., Al-Mehdi, A., Dodia, C., Muzykantov, V., Fisher, A. (1999) Simulated ischemia in flow-adapted endothelial cells leads to generation of ROS and cell signaling. Circ. Res. 85:682-689.
Whisler, R.L., Goyette, M., Grants, I., and Newhouse, Y.G. (1995) Sublethal oxidant stress stimulates multiple serine/threonine kinases and suppress phosphatases. Arch. Biochem. Biophys. 319:23-35.
Wishart, M.J., Taylor, G.S., and Dixon, J.E. (2001) Phoxy lipids: Revealing PX domains as phosphoinositide binding modules. Cell 105:817-820.
Wojciak-Stothard, B. and Ridley, A.J. (2002) Rho GTPases and the regulation of endothelial permeability. Vasc. Pharmacol. 39:187-199.
Wojcikiewicz, R. and Luo, S. (1998) Phosphorylation of inositol 1,4,5-trisphosphate receptors by PKA - Type I, II and III receptors are differentially susceptible to phosphorylation in intact cells. J. Biol. Chem. 273:5670-5677.
Wolin, M.S., Davidson, C., Kaminski, P., Fayngersh, R., Mohazzab-H, K.M. (1998) Oxidant-nitric oxide signalling mechanisms in vascular tissue. Biochem. (Moscow) 63:810-816.
Wu, H. M., Yuan, Y., Zawieja, D., Tinsley, J., and Granger, H. J. (1999) Role of phospholipase C, PKC, and calcium in VEGF-induced venular hyperpermeability. Amer. J. Physiol. 276:H535-H542.
Wu, S., Moore, T., Brough, G., Whitt, S., Chinkers, M., Li, M., and Stevens, T. (2000) Cyclic nucleotide-gated channels mediate membrane depolarization following activation of store-operated calcium entry in endothelial cells. J. Biol. Chem. 275: 8887-8896.
Xu, Y., Seet, L.F., Hanson, B., and Hong, W. (2001) The Phox homology (PX) domain, a new player in phosphoinositide signaling. Biochem. J. 360:513-530,
Yamamura, S., Nelson, P.R., and Kent, K.C. (1996) Role of PKC in attachment, spreading, and migration of human endothelial cells. J. Surg. Res. 63:349-354.

Yan, W., Tiruppathi, C., Lum, H., Qiao, R., and Malik, A.B. (1998) PKCβ regulates desensitization of thrombin receptor (PAR-1) in endothelial cells. Amer. J. Physiol. 274:C387-C395.

Yang, N., Higuchi, O., Ohashi, K., Nagata, K., Wada, A., Kangawa, K., Nishida, E., and Mizuno, K. (1998) Cofilin phosphorylation by LIM-kinase 1 and its role in Rac-mediated actin reorganization Nature 393:809-812.

Yatomi, Y., Ohmori, T., Rile, G., Kazama, F., Okamoto, H., Sano, T., Satoh, K., Kume, S., Tigyi. G., Igarashi, Y., and Ozaki, Y. (2000) S1P a major bioactive lysophospholipid that is released from platelets and interacts with endothelial cells. Blood 96:3431-3438.

Yin, G., Yan, C., and Berk, B.C. (2003) Angiotensin II signaling pathways mediated by tyrosine kinases. Internat'l. J. Biochem. Cell Biol. 35:780-783.

Yuan, Y., Meng, F., Huang, Q., Hawker, J., and Wu, H.M. (1998) Tyrosine phosphorylation of paxillin/pp125FAK and microvascular endothelial barrier function. Amer. J. Physiol. 275:H84-H93.

Young, B.A., Sui, X., Kiser, T., Hyun, S., Wang, P., Sakarya, S., Angelini, D., Schaphorst, K., Hasday, J., Cross, A., Romer, L., Passaniti, A., and Goldblum, S.E. (2003) Protein tyrosine phosphatase activity regulates endothelial cell-cell interactions, the paracellular pathway, and capillary tube stability. Amer. J. Physiol. 285:L63-L75.

Zhao, G., Al-mehdi, A., and Fisher, A. (1997) Anoxia-reoxygenation versus ischemia in isolated rat lungs. Amer. J. Physiol. 273:L1112-L1117.

Zhao, L., Zhang, M.M., and Ng, K.Y. (1998) Effects of VPF on the permeability of cultured endothelial cells from brain capillaries. J. Cardiovasc. Pharm. 32:1-4.

Zhou, L.Y., Disatnik, M., Herron, G., Mochly-Rosen, D., and Karasek, M.A. (1996) Differential activation of PKC isozymes by PMA and collagen in microvascular endothelium. J. Invest. Derm. 107:248-252.

Chapter 5

Adenylyl Cyclase and cAMP Regulation of the Endothelial Barrier

Sarah Sayner and Troy Stevens, Ph.D.*

Department of Pharmacology, Center for Lung Biology
The University of South Alabama College of Medicine, Mobile AL 36688

CONTENTS:

    Introduction

    A Historic Perspective on cAMP Signal Transduction
    Discovery of Second Messengers
    Adenylyl Cyclases
    Phosphodiesterases
    Achieving the cAMP-Signaling Threshold

    Molecular Diversity of cAMP Signaling
    Specificity and Fine Control of the cAMP level
    Downstream Signaling

    cAMP Signaling in Microdomains
    Spatial Resolution
    Compartmentalization

    cAMP and Calcium in Endothelial Cell Permeability
    Rises in cAMP Protect the Barrier
    Rises in Cytosolic Calcium Disrupt the Barrier
    SOC Entry and the Endothelial Cell Barrier
    SOC Entry Inhibits cAMP Synthesis
    Endothelial Cell Heterogeneity: cAMP Signal Transduction
    AC6 in Control of the Endothelial Cell Barrier

    Summary

Introduction

As illustrated in Chapters 1 and 2, the endothelium functions as a semi-permeable barrier, which regulates transudation of water, solutes, and molecules into the underlying tissue. Neurohumoral inflammatory mediators, described in Chapter 3, trigger a breach in the endothelial cell barrier that promotes water and protein transudation necessary for tissue repair. However, if inflammation is sustained or unregulated, such fluid accumulation causes tissue edema that compromises lung gas exchange and may culminate in acute respiratory failure, as depicted in Chapters 12, 13, and 16.

The principal transendothelial pathway for fluid and protein flux is between cells (Lum and Malik, 1996; Moore et al., 1998b). Majno and Palade (Majno et al., 1967) first described the appearance of gaps between adjacent endothelial cells after histamine exposure and proposed endothelial contraction was responsible for this action. Indeed, subsequent findings have shown elevated cytosolic Ca^{2+} activates cytoskeletal elements to induce actomyosin interaction and contraction (Lum and Malik, 1996; Schaphorst et al., 1997; Moore et al., 1998a; Curry and Adamson, 1999; Murphy et al., 2001). Increased actomyosin interaction promotes a change in inward tension that, along with decreased cell-cell and cell-matrix adhesion, induces intercellular gaps forming a paracellular pathway for increased permeability (Moore et al., 1998a). However, Ca^{2+}-induced barrier dysfunction is only observed when membrane adenosine 3',5'-cyclic monophosphate (cAMP) concentrations are not increased (Stevens et al., 2000). cAMP inhibits actomyosin interaction, stabilizes the cortical actin rim, and promotes cell-cell and cell-matrix adhesion (Moore et al., 1998a; Patterson et al., 2000). Discovery that endothelial cells express a Ca^{2+} inhibited, type 6 adenylyl cyclase (Stevens et al., 1995) illustrated how calcium agonists decrease cAMP necessary for gap formation. This chapter addresses cAMP signaling events, which control the endothelial cell barrier function.

A Historic Perspective on cAMP Signal Transduction

Discovery of Second Messengers

Studies seeking to determine the mechanism of epinephrine- and glucagon-induced hyperglycemia in dog livers led Sutherland and Rall to discover a heat stable factor characterized as an adenine ribonucleotide [("had a function, found a nucleotide" (Sutherland, 1972)], now known as cAMP (Sutherland and Rall, 1957). Subsequently, cAMP has been found in many tissues and cell types mediating diverse biological functions (Sutherland et al., 1968), including proliferation, cell shape, membrane excitability, gene expression, metabolism, differentiation, apoptosis, and control of the endothelial cell barrier.

Discovery of cAMP provided a means to understand how extracellular stimuli trigger an intracellular response, and greatly shaped our concept of signal transduction. For the first time it was possible to examine how extracellular stimuli (*e.g.* hormones, neurotransmitters, or growth factors) encode intracellular signals (Sutherland, et al., 1968). These extracellular ligands, or first-messengers, disseminate their responses through specific plasma membrane receptors to activate a second messenger cascade – *e.g.* cAMP. Intracellular transducers (*e.g.* trimeric GTP binding proteins) either positively ($G_{\alpha s}$) or negatively ($G_{\alpha i}$) convey receptor activation to adenylyl cyclases to alter cAMP synthesis, thereby dynamically regulating the cAMP signal (Krupinski, 1991; Iyengar, 1993; Cooper et al., 1995; Taussig & Gilman, 1995).

Temporal changes in intracellular cAMP concentrations do not solely reflect adenylyl cyclase-G protein activity, but also phosphodiesterase activity (Taussig et al., 1994; Cooper et al., 1995; Taussig and Gilman, 1995). Phosphodiesterases comprise a large family of enzymes that inactivate cAMP (or cGMP) by hydrolyzing it to 5'AMP (Butcher and Sutherland, 1962). cAMP signaling can also be terminated through its extrusion (Brunton and Mayer, 1979), however relatively little is known regarding this emerging field. Since its initial discovery, the complexity of cAMP signaling has continued to expand whereby many levels of regulation by multiple modulators instigate the subtle changes necessary to "fine tune" signaling and optimize the physiological response.

Adenylyl Cyclases

Adenylyl cyclases are the only enzymes known to catalyze ATP to cAMP conversion. These enzymes have been identified in a wide variety of phyla, ranging from prokaryotes to eukaryotes, including plants (Danchin, 1993; Barzu and Danchin, 1994; Kasahara et al., 1997; Tang et al., 1998). Diversity within the cyclase domain accounts for their organization into three classes: class I adenylyl cyclases from Enterobacteria (including *E. coli*), class II toxic adenylyl cyclases from bacteria, and class III enzymes (Danchin, 1993; Tang et al., 1998). The latter class is further subdivided into four groups based on phylogenic analysis, and includes mammalian transmembrane enzymes (Tang et al., 1998). Restricted membrane localization, low abundance (0.1%-0.001% of plasmalemma proteins) (Sunahara et al., 1996; Tang et al., 1998), and the labile nature of these adenylyl cyclases hindered their structural determination. The diterpene, forskolin, which activates most eukaryotic adenylyl cyclases to varying degrees, was then identified (Seamon and Daly, 1981; Seamon et al., 1981). Forskolin affinity chromatography (Pfeuffer et al., 1985; Pfeuffer and Pfeuffer, 1989) was used to augment protein purification techniques, which ultimately led to cloning of the first mammalian adenylyl cyclase cDNA (Krupinski et al., 1989).

Since cloning of the original mammalian adenylyl cyclase gene, this family has been expanded to include nine closely related, transmembrane glycoproteins (adenylyl cyclase 1-9) with numerous splice variants (Cooper, 2003b). These isoforms are all activated by forskolin, with the exception of adenylyl cyclase 9 (AC9) (Premont et al., 1996), and are non-competitively inhibited by purine nucleotides, otherwise known as P-site inhibitors (Londos and Wolff, 1977). In addition, a single soluble enzyme has been described, which is synergistically activated by bicarbonate and calcium (Buck et al., 1999; Chen et al., 2000; Litvin et al., 2003; Zippin et al., 2003). Surprisingly, the soluble adenylyl cyclase shows no sequence homology to transmembrane isoforms, and exhibits greater similarity to the catalytic domain of bacterial enzymes (Buck et al., 1999).

As the primary structure of adenylyl cyclase unraveled, a general topology for the enzyme was proposed (Krupinski et al., 1989; Tang et al., 1991). The structural complexity of adenylyl cyclases resembles that of ion channels and ABC transporter family proteins (Krupinski et al., 1989). AC1-9 comprises two six-transmembrane cassettes (M_1 and M_2), with their associated cytosolic loop (C_1 and C_2), seen in Figure 1. These loops are subdivided into "a" and "b" subdomains. High sequence homology exists between the C_{1a} and C_{2a} domains. Indeed, fusion of C_{1a} of AC1 and C_{2a} of AC2 with a 14 amino acid linker generates a soluble, $G_{\alpha s}$ and forskolin activated adenylyl cyclase (Tang and Gilman, 1995). Thus, this fusion protein contains catalytic machinery of the enzyme and residues required to bind $G_{\alpha s}$. Insights into the catalytic mechanism of mammalian enzymes have been revealed through this and other soluble adenylyl cyclase chimeras (Dessauer and Gilman, 1996; Whisnant et al., 1996; Scholich et al., 1997; Tesmer et al., 1997; Tesmer et al., 2002). Crystal structure evaluation established that the C_{1a} and C_{2a} domains form a wreath-like dimer (Zhang et al., 1997; Hurley, 1998) through their head to tail arrangement. The interface of these domains encodes a βαββαβ motif to form a single catalytic site resembling the palm domain of prokaryotic DNA polymerases, where both enzymes catalyze a 3'-5' diester bond (Artymiuk et al., 1997). A single ATP binds the catalytic site with the purine ring residing in a hydrophobic pocket (Sunahara et al., 1997). Magnesium coordinates the reaction to generate cAMP and pyrophosphate (Rall and Sutherland, 1962; Sutherland et al., 1962). A second pseudo-catalytic site is responsible for binding forskolin (Liu et al., 1997). Adenylyl cyclase nucleotide specificity resides within the hydrophobic pocket such that three amino acids are responsible for conferring binding of ATP over GTP. Exchanging these residues converts adenylyl cyclase into a non-selective purine nucleotide cyclase (Sunahara et al., 1998).

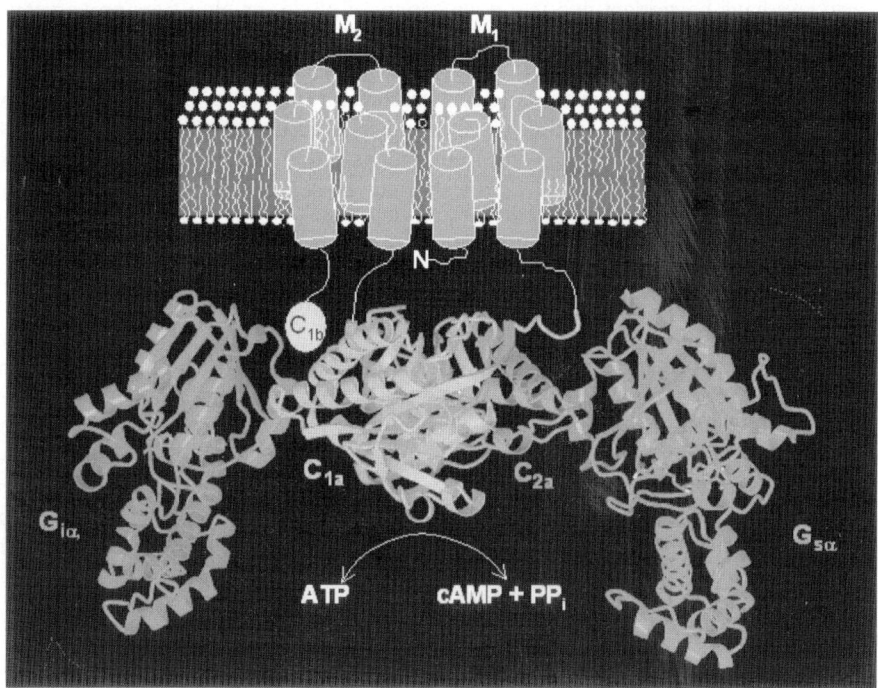

Figure 1. Structure of Adenylyl Cyclases and Catalytic Site. The proposed topology of mammalian membrane-bound adenylyl cyclase includes 12 hydrophobic regions arranged into 2 six-transmembrane spanning cassettes (M_1 and M_2). A short variable cytosolic amino terminus precedes M_1. The first cytosolic loop (C1) is flanked by M_1 and M_2 and further subdivided into region C_{1a} and C_{1b}. M_2 is followed by the C_{2a} and in some isoforms the C_{2b} region. C_1 domain and the C_{2a} domain form the catalytic moiety of the enzyme. $G_{s\alpha}$ forms hydrophobic and polar contracts with C_{2a} to enhance enzymatic activity while $G_{\alpha i}$ associates with C_{1a} to inhibit enzymatic activity. From http://ibp.med.uth.tmc.edu/faculty/cdessaue/dessauer_model.gif and modified from: *Dessauer et al., (1998) Identification of a $G_i\alpha$ Binding Site on Type V Adenylyl Cyclase. J. Biol. Chem. 273:25831-25839.* by permission of Dr. Carmen Dessauer and reproduced with permission from the Journal of Biological Chemistry and the American Association for Biochemistry and Molecular Biology.

While advances have been made in establishing catalytic mechanisms, functions of the two six transmembrane-spanning helices has yet to be defined and the significance of their topological homology to ABC transporters remains elusive. M_1 and M_2 are speculated to be responsible for creating a framework permitting optimal spatial orientation of catalytic loops, or they possibly play a role in enzyme dimerization (Gu *et al.*, 2001; Seebacher, *et al.*, 2001). As little sequence homology exists between transmembrane cassettes of different family members, these regions could confer isoform specificity (Gu *et al.*, 2001). Further, extracellular loops are glycosylated. All isoforms are N-glycosylated between transmembrane spans 9 and 10, while other glycosylation sites are isoform specific (Krupinski and Cali, 1998). The extracellular loop which contains this conserved glycosylation site is required for trafficking adenylyl cyclases to the plasma membrane (Gu *et al.*, 2001). Certain isoforms (AC6, and -8) are highly enriched in plasma membrane compartments like lipid rafts and caveolae (Cooper, 2003a; Cooper, 2003b). Transmembrane domains could be responsible for lateral positioning of isoforms within plasma membrane microdomains [*e.g.* "address domain" to a membrane compartment (Hanoune and Defer, 2001)]. In addition, they could confer isoform specific regulatory properties. Ultimately, determining the tertiary structure of this protein family with high resolution will be necessary to develop a comprehensive understanding of the adenylyl cyclase catalytic and regulatory mechanisms.

Phosphodiesterases

Phosphodiesterase expression and function has been reviewed recently (Bushnik and Conti, 1996; Houslay and Milligan, 1997; Francis *et al.*, 2001; Houslay and Adams, 2003; Maurice *et al.*, 2003), and will only be briefly introduced here and discussed in the following section on molecular diversity of signaling. Phosphodiesterases hydrolyze cAMP to 5' AMP necessary to terminate cAMP signaling. The action of these enzymes maintain cAMP levels within a relatively narrow range to fine tune the cellular response (Francis *et al.*, 2001). Members of this superfamily contain a conserved catalytic domain with 20-45% sequence identity between families (Maurice *et al.*, 2003). Sequences with greater variability reside near the phosphodiesterase catalytic domain and are responsible for isoform-selective regulatory properties. These regions modulate catalytic activity, and contain dimerization and targeting domains. Methylxanthines inhibit the enzymatic activity of most family members, while isoform specific inhibitors have also been developed that act as valuable pharmacological tools to study the cellular regulation of cAMP.

Achieving the cAMP-Signaling Threshold

Modulating adenylyl cyclase and/or phosphodiesterase activities can alter cAMP concentrations, although cells differ in their mechanisms. "Low cycling" is accomplished when both enzymes have low basal activity (Houslay and Milligan, 1997). Under these conditions, increasing adenylyl cyclase activity is the only mechanism capable of sufficiently increasing cAMP to affect downstream signaling. In contrast, high basal adenylyl cyclase activity coupled with high phosphodiesterase activity maintains cAMP levels below threshold for activation of downstream targets. In this case, either stimulating adenylyl cyclase or inhibiting phosphodiesterase activity rapidly elevates cAMP; this "high" or "futile cycling" allows cAMP concentrations to be modulated by its synthesis or degradation (Houslay and Milligan, 1997). This ability to regulate cAMP permits both a more subtle change and a more rapid response. Further, it enables integration of a greater number of signaling pathways to govern intracellular cAMP level. Ultimately, adenylyl cyclase and phosphodiesterase enzyme activities determine temporal and spatial cAMP-signaling networks.

Molecular Diversity of cAMP Signaling

Specificity and Fine Control of the cAMP Level

Specificity in cAMP signaling is achieved not only through unique expression of hormone receptor isotypes, but also through distinctive expression of multiple adenylyl cyclase and phosphodiesterase isoforms. While the large number of adenylyl cyclase and phosphodiesterase isoforms suggests redundancy, their unique tissue, cellular, and subcellular distribution profiles coupled with individual regulatory properties eliminates functional overlap and indeed provides a mechanism for specificity. Coupled with multiple downstream effectors, the diversity of cAMP signaling networks is extensive.

The nine mammalian transmembrane adenylyl cyclases are regulated by a variety of intracellular signals, as shown in Table 1. All isoforms are inhibited by mM Ca^{2+}, while submicromolar Ca^{2+} have isoform-specific direct or indirect effects on cyclase activity (Cooper *et al.*, 1994a). AC-5 and -6 are directly inhibited by the low Ca^{2+} concentrations (Ishikawa *et al.*, 1992; Yoshimura and Cooper, 1992). AC-1, - 3, and - 8 are all activated by calcium *via* its association with calmodulin. AC-2, -4, -7 are indirectly stimulated by Ca^{2+} through its activation of PKC. The fourth subfamily, AC9, is stimulated by the Ca^{2+}-activated phosphatase calcineurin. Such regulation highlights crosstalk between Ca^{2+} and cAMP signaling. All adenylyl cyclase isoforms are stimulated to varying degrees by $G_{s\alpha}$ and $G_{\beta\gamma}$ subunits. AC6 activity is preferentially inhibited by $G_{i\alpha}$ subunits, when compared with AC-2 and -4. Recent investigations suggest dimerization state (homo- and/or hetero-dimerization) influences enzymatic activity (Tang *et al.*, 1995; Cooper *et al.*, 1998; Gu *et al.*, 2001; Gu *et al.*, 2002). Therefore, each adenylyl cyclase isoform has its own idiosyncratic mode of regulation *via* multiple cellular signals. Integration of many signals to govern adenylyl cyclases establishes them as coincident detectors (Cooper *et al.*, 1995; Sunahara *et al.*, 1996).

Higher order cAMP signaling complexity is achieved through multiple phosphodiesterase isoforms. Eleven family memebers have been identified and classified according to their cyclic nucleotide substrate specificity, as itemized in Table 2. Phosphodiesterases 4, -7, and – 8 hydrolyze cAMP, phosphodiesterases 5 and -6 hydrolyze cGMP and phosphodiesterases 1, -2, -3, -9, -10, and -11 hydrolyze both. These phosphodiesterase families are differentially regulated by calcium-calmodulin (CaM), G-proteins, phosphorylation, and cyclic nucleotides (Francis *et al.*, 2001). In addition, multiple isoforms exist within families. For example, phosphodiesterase 4 has four separate genes (A-D) located on three chromosomes. Further diversity is achieved through different promoters, multiple transcription start sites, and splice variants to generate more than 15 isoenzymes of the phosphodiesterase 4 family (Houslay and Adams, 2003). Regulation of activity also occurs by altering protein expression – i.e. elevated phosphodiesterase 4 in response to chronic elevation of cAMP and by its subcellular localization. While some isoforms associate with the particulate fraction, others are found in cytosolic fraction. Emerging evidence indicates phosphodiesterases are targeted to subcellular compartments via association with anchoring proteins [i.e. anchoring *via* AKAPs; (Colledge and Scott, 1999)] or signaling complexes [*e.g.* β arrestin (Perry *et al.*, 2002)].

Table 1. Isoform specific regulation of adenylyl cyclases.

Isoform	$G_{\alpha s}$	$G_{\beta\gamma}$	G_i	Fsk	P-site analogues	Calcium	Protein kinases	Other regulators
AC1	↑	↓	↓ (CaM or Fsk ↑)	↑	↓	↑ (CaM)	↑ PKC (weak) ↓ CaM kinase IV	
AC2	↑	↑(with $G_{\alpha s}$)		↑	↓		↑ PKC	
AC3	↑		↓	↑	↓	↑ (CaM)	↑ PKC (weak) ↓ CaM kinase II	
AC4	↑	↑(with $G_{\alpha s}$)		↑	↓		↑ PKC	
AC5 (2)	↑	↓	↓	↑	↓	↓	↓ PKA ↑ PKC α/ζ	
AC6 (2)	↑	↓	↓	↑	↓	↓	↓ PKA ↓ PKC	Phorbol Esters
AC7	↑	↑		↑	↓		↑ PKC	
AC8	↑		↓ (Ca^{2+} rise)	↑	↓	↑ (CaM)		
AC9	↑		↓	↑ (weak)	↓	↓ (calcineurin)		
sAC								HCO_3^-

Each adenylyl cyclase isoform (splice variant) has a unique repertoire of regulators leading to unique isoform characteristics. (Hanoune and Defer, 2001) Adapted and reprinted, with permission from Ann. Review of Pharmacol. and Toxicol., Vol. 41 ©2001 by Annual Reviews www.annualreviews.org

Table 2. Isoform specific regulation of phosphodiesterases.

Isoform	Nucleotide specificity	Regulators	Inhibitors
PDE1	cAMP, cGMP	↑ Ca^{2+}/Calmodulin	IBMX Vinpocetine
PDE2	cAMP, cGMP	↑ cGMP	IBMX EHNA
PDE3	cAMP, cGMP	↓ CGMP	IBMX Milrinone
PDE4	cAMP	↓ Ca^{2+} (PDE4D) PKA	IBMX Rolipram
PDE5	cGMP	PKG	IBMX Sildenafil, Zaprinast
PDE6	cGMP	Light (Photoreceptor)	IBMX Zaprinast
PDE7	cAMP	Unknown	IBMX
PDE8	cAMP	Unknown	IBMX insensitive Dipyridamole
PDE9	cGMP	Unknown	IBMX
PDE10	cAMP, cGMP	Unknown	IBMX, Sildenafil, Zaprinast, Dipyridamole

Phosphodiesterases hydrolyze cAMP, cGMP or both nucleotides. Enzymatic activity is modified by nucleotides themselves or other cellular factors. (Lugnier and Schini, 1990; Beltman et al., 1993; Suttorp et al., 1993; Houslay and Milligan, 1997; Francis, Turko et al., 2001).

Downstream Signaling

Fidelity of cAMP signaling is achieved through a variety of downstream effectors. Protein kinase A (PKA) was the first downstream cAMP target to be identified (Walsh *et al.*, 1968). Cooperative binding of 4 molecules of cAMP to the regulatory subunit of the PKA dimer leads to release of two active catalytic subunits from the holoenzyme. These catalytic subunits differentially phosphorylate serine/threonine residues on hundreds of cellular substrates (Karpen and Rich, 2001). Expression of different PKA isoforms and anchoring of PKA to sites in proximity of cAMP generation also render specificity and control of the cAMP downstream effects, as defined in the following section. Both location of cAMP generation and the duration of the cAMP elevation also affect whether the catalytic subunits enter the nucleus in sufficient amount to phosphorylate cAMP responsive elements (CRE) and trigger altered gene expression. The list of other cAMP effectors has expanded to include cyclic nucleotide gated ion channels (Kaupp *et al.*, 1989; Yau 1994) and nucleotide exchange protein activated by cAMP (Epac) (de Rooij *et al.*, 1998; Kawasaki *et al.*, 1998). Expansion of cAMP effectors has implicated cAMP in more diverse, site-specific cellular functions.

cAMP Signaling in Microdomains

Spatial Resolution

The hydrophilic nature of cAMP led researchers of the time to propose that it was involved in rapid, long range cellular signaling. In this "one compartment" model, elevated cAMP diffuses throughout the cytoplasm to activate downstream targets (Rall, 1975). However, Rall proposed this one compartment model could not account for the biological diversity in cAMP responsiveness, stating it provided "the unsatisfying picture of the catalytic subunit of protein kinase swimming about, happily phosphorylating a variety of cellular constituents whether they need it or not." In time, Rall's concern was borne out. Indeed, stimulation of rat heart with either isoproterenol or PGE_1 caused a comparable increase in cAMP, yet lead to unique target phosphorylation patterns and physiological events (Hayes *et al.*, 1980). Similar findings were reported in lymphocytes and adipocytes (Hayes *et al.*, 1980), and activation of L-type calcium currents in cardiomyocytes was due to highly compartmentalized cAMP transitions (Jurevicius and Fischmeister, 1996). A model of compartmentation emerged as a means to explain how cAMP maintained signaling specificity despite its involvement in activating parallel signaling pathways (Brunton *et al.*, 1981).

Traditional techniques, such as radiochemical assays and later the sensitive and rapid immunological assays detect global (whole cell) cAMP concentrations (Gilman, 1970). While these techniques provide valuable insights regarding changes in whole cell cAMP, they do not permit resolution of compartmentalized cAMP or of intracellular cAMP gradients. Technical advances now permit resolution of spatial and temporal changes in free cAMP. For instance, fluorescence imaging technique are used to observe dynamic changes in cAMP in living cells (Adams *et al.*, 1991). The regulatory and catalytic subunits of PKA are labeled with different fluorescent dyes capable of fluorescence resonance energy transfer (FRET). Binding of the regulatory subunit by cAMP, leading to catalytic subunit dissociation eliminates the FRET signal. Change in fluorescence emission spectrum permits cAMP concentrations and PKA activation to be visualized within the cell. Using this technique, cAMP gradients were demonstrated in sensory neurons (Bacskai et al., 1993). Subsequent studies using a modification of this technique supported the cAMP compartmentalization model in cardiomyocytes (Zaccolo and Pozzan, 2002). While these approaches provided superior subcellular resolution of cAMP, they were limited by the nature of the probe; cAMP was being measured by its binding to PKA. Thus, a technique was developed which exploited exogenously expressed cyclic nucleotide gated channels to act as biosensors, permitting resolution of cAMP concentrations residing in plasma membrane microdomains (Rich et al., 2000). Using ionic currents through cyclic nucleotide gated channels as biosensors for cAMP, these investigators demonstrated cAMP concentrations in the vicinity of the plasma membrane of C6-2B glioma cells are at least 12-fold higher than concentrations present in the bulk cytosol. These data suggest that cAMP gradients exist within a single cell, with higher cAMP concentrations in microdomains at the plasmalemma compared with the bulk cytosol.

Compartmentalization

The idea is therefore emerging that cells achieve specificity and rapidity in cAMP signaling by organizing macromolecular complexes in the plasma membrane that contain G-protein coupled receptors, G-protein(s), adenylyl cyclase(s) and effector proteins (Davare et al., 2001; Laporte et al., 2001; Brunton, 2003; Bundey and Insel, 2004). Further, multivalent anchoring proteins (*e.g.* A kinase anchoring proteins, AKAPs) integrate signaling effectors, thereby enhancing signaling specificity and allowing compartmentalization of signaling cascades (Smith and Scott, 2002). Caveolae and lipid rafts are an example of such plasma membrane compartments that contain a variety of signal transduction molecules (Shaul and Anderson, 1998). The light buoyant density of these fractions reflects the elevated cholesterol, glycosphingolipids, sphingomyelin, and lipid-anchored membrane protein composition. This specialized localization and composition suggests caveolae compartmentalize and integrate signaling events at the plasma membrane (Shaul and Anderson, 1998). Selective activation of downstream targets can also be explained by cAMP synthesis in diffusion-restricted domains, creating a model where second messenger signaling is temporally and spatially confined by restricted domains. Several ideas are emerging that could account for the morphological basis of diffusion restriction. Firstly, most adenylyl cyclases are transmembrane proteins, and increase cAMP in plasma membrane microdomains. Second, organelles like the endoplasmic reticulum reside as lateral stacks immediately adjacent to the plasma membrane and, hence, physically obstruct free cAMP diffusion. In addition, phosphodiesterases critically localized to the peripheral compartments restrict the distance that cAMP diffuses from its site of synthesis, and therefore control the magnitude and duration of cAMP rise nearby adenylyl cyclase. It is noteworthy that without these diffusional constraints, cAMP concentrations juxtaposed to adenylyl cyclase would not be high enough to activate PKA unless the entire cell filled with cAMP (Rich et al., 2000).

Oscillations in the cAMP level have been described, and are proposed to play a critical role in transducing signals in excitable cells (Brooker, 1973; Cooper et al., 1995; DeBernardi and Brooker, 1996; Gorbunova and Spitzer, 2002; Zaccolo and Pozzan, 2003). While the idea of cAMP oscillations has been advanced, clear evidence for the physiological role of such elevations remains uncertain. Nonetheless, a collective theme has emerged which suggests that second messengers compartmentalized along with signaling molecules provides the microenvironment necessary to generate cAMP gradients that evoke site-specific control over cell function.

cAMP and Calcium in Endothelial Cell Permeability

Rises in cAMP Protect the Barrier

For many years the anti-inflammatory property of epinephrine has been recognized, in part for its important role in alleviating anaphylaxis, urticaria, and the tissue edema that accompanies inflammatory agonists (histamine, bradykinin), hypoxia, shock, and ischemia-reperfusion. While activation of α_1-mediated vasoconstriction limits tissue perfusion and plays a role in the anti-inflammatory actions of epinephrine, there is an increasing recognition that epinephrine activates β-receptors on endothelial cells important to strengthen barrier properties and limit fluid transudation. Catecholamines were first demonstrated to increase cAMP in all endothelial cells, including those isolated from the pulmonary artery (Makarski, 1981; Stevens et al., 1995) and microvasculature (Stevens et al., 1999), as seen in Figure 2.

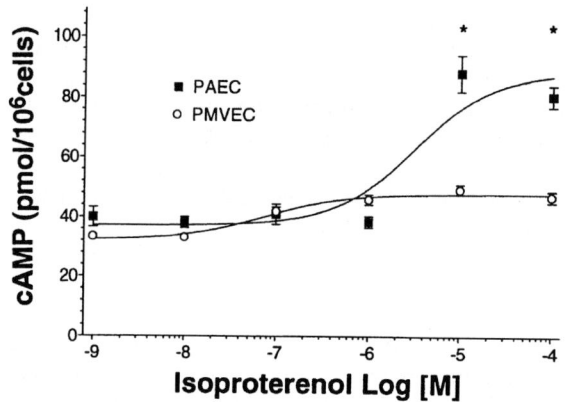

Figure 2. *Isoproternol elevates cAMP in endothelial cells isolated from the pulmonary artery and from the pulmonary microvasculature.* The maximal response to isoproternol is higher in pulmonary artery endothelial cells (PAECs) compared with the microvascular cells (PMVECs). From (Stevens, Creighton et al., 1999), by permission of The American Physiological Society.

Both *in vivo* and *in vitro* studies demonstrate that a cAMP rise attenuates pulmonary edema formation and limits trans-endothelial fluid flux (Farrukh et al., 1987; Kobayashi et al., 1987). Early studies seeking to resolve the mechanism of cAMP's barrier protective actions demonstrated it reorganizes the actin cytoskeleton into a DPB that stabilizes cell-cell apposition (Stelzner et al., 1989). These findings are consistent with evidence that cAMP increases the number of tight junctions in endothelial cells (Adamson et al., 1998) and, further, that it inactivates actomyosin interaction to decrease inwardly-directed tension in response to inflammatory agonists (Minnear et al., 1989; Langeler and van Hinsbergh, 1991; Moy et al., 1993; Patterson et al., 1994). Nonetheless, how cAMP acts to control cytoskeletal organization (*e.g.* through PKA, Epac, or cyclic nucleotide gated channels) and which cytoskeletal structures are phosphorylated by PKA are still incompletely understood.

Rises in Cytosolic Calcium Disrupt the Barrier

In contrast to barrier protective effects of cAMP, elevated intracellular Ca^{2+} is barrier disruptive (Lum and Malik, 1994; Moore et al., 1998b; Dudek and Garcia, 2001; Tiruppathi et al., 2002b). However, the specific ion channels activated to provide the necessary Ca^{2+} source are still incompletely understood. Inflammatory agonists (including bradykinin, ATP, thrombin, substance P, proteases, leukotrienes) bind G protein coupled receptors that activate

G_q proteins (Patterson and Lum, 2001; Bogatcheva et al., 2002). $G\alpha_q$ then activates phospholipase C that catalyzes the hydrolysis of phosphatidylinositol-4,5-bisphosphate, generating inositol 1,4,5-trisphosphate (IP_3) and 1,2-diacylglycerol. IP_3 rapidly diffuses through the cytosol to interact with IP_3 receptors on the endoplasmic reticulum that serve as ion channels for transient Ca^{2+} release into the surrounding cytosol. The resulting depletion of endoplasmic reticulum calcium stores triggers calcium entry across the plasma membrane through calcium selective and non-selective store operated calcium (SOC) entry channels (Wu et al., 2000; Cioffi et al., 2003). In addition, diacylglycerol, arachidonic acid and its metabolites, and potentially components of heterotrimeric G proteins activate a separate class of calcium non-selective receptor operated calcium (ROC) entry channels (Bird et al., 2004). Thus, both SOC and ROC entry channels contribute to the global rise in cytosolic calcium.

Global cytosolic calcium measurements reveal that activation of IP_3 generates biphasic cytosolic calcium transients – the initial calcium transient is due to calcium release from intracellular stores, while the sustained calcium rise is due to calcium entry across the plasma membrane (Hallam and Pearson, 1986; Putney, 1986; Rotrosen and Gallin, 1986; Schilling et al., 1992). In the absence of extracellular calcium, only the first transient due to store depletion is detected. Replenishing extracellular calcium results in entry that generates a sustained increase in cytosolic calcium. It is this latter, sustained cytosolic calcium rise that disrupts the endothelial cell barrier. Whereas calcium release is not sufficient to increase endothelial cell permeability, calcium entry across the cell membrane reorganizes the cytoskeleton, increases cell tension and disrupts cell-cell adhesions (Kelly et al., 1998; Moore et al., 1998a; Moore et al., 2000; Norwood et al., 2000).

SOC Entry and the Endothelial Cell Barrier

Because SOC and ROC entry pathways are activated by different arms of the G_q signaling cascade, their independent roles in control of endothelial cell barrier can be difficult to discern. SOC entry channels are activated by endoplasmic reticulum Ca^{2+} store depletion, and not by Ca^{2+} release. Agents like TPEN that chelate endoplasmic reticulum calcium, and the plant alkaloid thapsigargin (or CPA and BHQ) that inhibit(s) the sarcoplasmic, endoplasmic reticulum calcium ATPase and prevents calcium re-uptake, both trigger SOC entry, as shown in Figure 3a. TPEN and thapsigargin activate SOC without the involvement of α_q, $\beta\gamma$, or diacylglycerol, which allows resolution of the physiological role of just SOC entry channels (Thastrup et al., 1990; Liu and Ambudkar, 2001).

A considerable number of studies have demonstrated the physiological consequence of directly activating SOC channels, both in the intact pulmonary circulation and in cultured cells. Thapsigargin increases fluid filtration in the intact pulmonary circulation (Chetham et al., 1997), as seen in Figure 3b, and it increases macromolecular permeability across cultured pulmonary artery endothelial cell monolayers (Kelly et al., 1998; Moore et al., 2000). These effects of thapsigargin require the presence of extracellular calcium, indicating calcium entry through SOC entry channels is necessary to disrupt the barrier (Chetham et al., 1997; Kelly et al., 1998; Moore et al., 1998a; Moore et al., 2000). While thapsigargin-induced gap formation depends on extracellular calcium, thrombin-induced gap formation in low extracellular calcium is attenuated compared to normal extracellular calcium (Moore et al., 2000). Thus, agonist activation of the G-protein cascade recruits additional barrier disrupting signaling pathways – e.g. $G_{12/13}$ – that interact with SOC entry to alter barrier function (Bogatcheva et al., 2002).

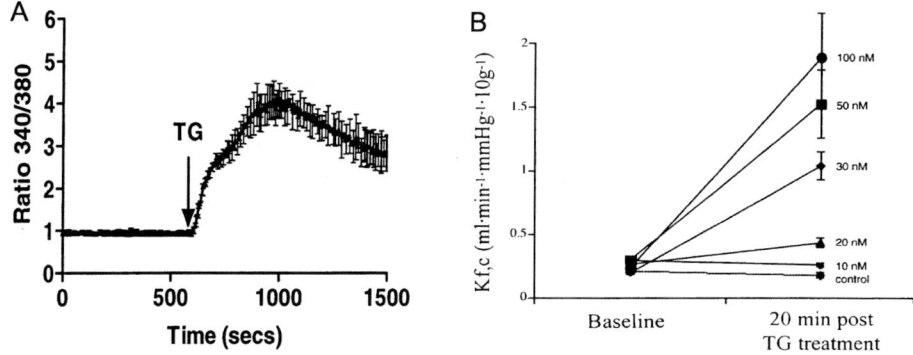

Figure 3. Thapsigargin induces calcium transients and increases isolated lung permeability. A) Thapsigargin (1µM) induces biphasic calcium transient in pulmonary artery endothelial cells representing calcium release from intracellular stores and calcium entry across the plasma membrane. From (Norwood, Moore et al., 2000) by permission of The American Physiological Society. **B)** Thapsigargin increases permeability in the isolated perfused lung. From (Chetham, Babal et al., 1999), by permission of The American Physiological Society.

The molecular identity of SOC entry channels in endothelial cells is incompletely resolved. However, discovery of the transient receptor potential (TRP) protein in *Drosophila*, which encodes a calcium entry pathway, provided a putative candidate. Seven mammalian homologues of TRP proteins have been identified (TRPC1-7) and proposed to form subunits of heterotetramer ion channels. TRPC1, 2, 4 and 5 subunits contribute to SOC entry channels, whereas TRPC 3, 6 and 7 are each ROC. Presently, the oligomeric state of endogenous channels is not described (Birnbaumer et al., 1996; Hofmann et al., 2000; Vennekens et al., 2002; Cioffi et al., 2003; Nilius, 2003).

Endothelial cells are generally thought to express (at least) TRPC1, 3 and 4; although, this expression pattern may vary among endothelial cell phenotypes. RT-PCR analysis of pulmonary artery endothelial cells revealed TRPC1 expression (Moore et al., 1998a). Antisense inhibition of TRPC1 decreased the thapsigargin-induced cytosolic calcium response 40-50%. Thapsigargin activates a calcium-selective store operated current, I_{SOC}, and inhibition of TRPC1 expression reduces this current more than 50%. These findings suggest that TRPC1 subunit(s) contribute to a calcium selective SOC entry channel in pulmonary endothelial cells (Brough et al., 2001; Cioffi et al., 2003). TRPC4 subunits also form a calcium selective, store operated channel (Philipp et al., 2000; Freichel et al., 2001). Indeed, heterologous expression of TRPC4 produces a channel that is activated by thapsigargin. Endothelial cells isolated from TRPC4 deficient mice do not possess I_{SOC} (Tiruppathi et al., 2002a). So while the oligomeric state of the channel responsible for I_{SOC} has not been resolved (Schilling and Goel, 2004), considerable functional evidence suggests both TRPC1 and TRPC4 subunits contribute to the heterotetramer. This idea is further supported by evidence that, among the TRPC proteins, TRPC1 and TRPC4 preferentially interact (Goel et al., 2002; Cioffi et al., 2003). It therefore appears that a TRPC1/4 containing channel importantly regulates a calcium transition in response to calcium store depletion.

Calcium entry through this TRPC1/4 channel increases endothelial cell permeability. TRPC4 deficient mice display a drastic reduction in thrombin-induced Ca^{2+} influx (Tiruppathi et al., 2002a). Such inhibition of Ca^{2+} influx protected endothelium from thrombin-induced stress-fiber formation and, in isolated perfused lungs, attenuated increased permeability. Therefore, the TRPC1/4 channel provides a Ca^{2+} entry pathway responsible for activating downstream signals that promote gap formation and increase paracellular transport.

SOC Entry Inhibits cAMP Synthesis

The barrier protective effects of cAMP supersede the barrier disrupting actions of increased $[Ca^{2+}]_i$ illustrated in Figure 4a (Moore et al., 1998b; Cioffi et al., 2002). Transitions in cytosolic calcium disrupt the endothelial cell barrier, but only when cAMP is not also increased. It is therefore important to understand whether calcium regulates cellular cAMP concentrations. Early studies in this area used pulmonary artery endothelial cells, and demonstrated inflammatory calcium agonists like thrombin, bradykinin and thapsigargin decrease whole cell cAMP. In permeabilized cells, increasing cytosolic calcium concentrations directly decreased cAMP content, as shown in Figure 4b; and, in intact cells, inhibiting constitutive calcium influx increased cAMP content. Calcium inhibition of cAMP was due to AC6 expression (Stevens et al., 1995; Chetham et al., 1997; Stevens et al., 1997). Thus, calcium influx inhibits AC6 activity and dynamically controls cAMP synthesis.

Calcium inhibition of cAMP in whole cells is typically no more than 30% of its basal concentration. Since AC6 is a transmembrane protein, endothelial membranes were isolated to enrich AC6 and better resolve the impact of Ca^{2+} on its activity. Isolated caveolin rich membranes demonstrated biphasic Ca^{2+} inhibition. As seen in Fig. 4c, microvascular cell membranes were particularly sensitive to high affinity Ca^{2+} inhibition (Stevens et al., 1995; Creighton et al., 2003). Indeed, the exquisite Ca^{2+} sensitivity suggests high levels of Ca^{2+}-inhibited adenylyl cyclase activity compared to tissue from the heart, anterior pituitary and striatum (Cooper, 2003a), and also compared with pulmonary artery endothelial cells (see below). Thus, AC6 is the predominant adenylyl cyclase isoform in pulmonary endothelium, and it provides a direct link between elevated intracellular calcium and the decrease in cAMP.

Whole cell studies suggest caveolae are sites of SOC entry. AMP-PNP is an adenylyl cyclase substrate that produces an electron dense product reflective of enzyme activity. Using this substrate, transmission electron micrographs, seen in Figure 5, reveal adenylyl cyclase activity is present in caveolae and at cell-cell borders (Sayner et al., 2004). SOC entry pathways provide the calcium source that inhibits AC6 activity (Cooper et al., 1994b; Chiono et al., 1995; Fagan et al., 1998), and calcium inhibition of AC6 only occurs when the enzyme is located in cholesterol-rich domains (Fagan et al., 2000). These data suggest an intimate spatial relationship between SOC entry channels and AC6. Notably, TRPC1 and TRPC4 are enriched in caveolae (Lockwich et al., 2000; Torihashi et al., 2002; Ambudkar et al., 2004), suggesting they may account for the SOC entry pathway that regulates AC6 activity in endothelial cells. However, studies firmly establishing the causal link between TRPC1/4 channels and calcium inhibition of AC6 activity have not been completed.

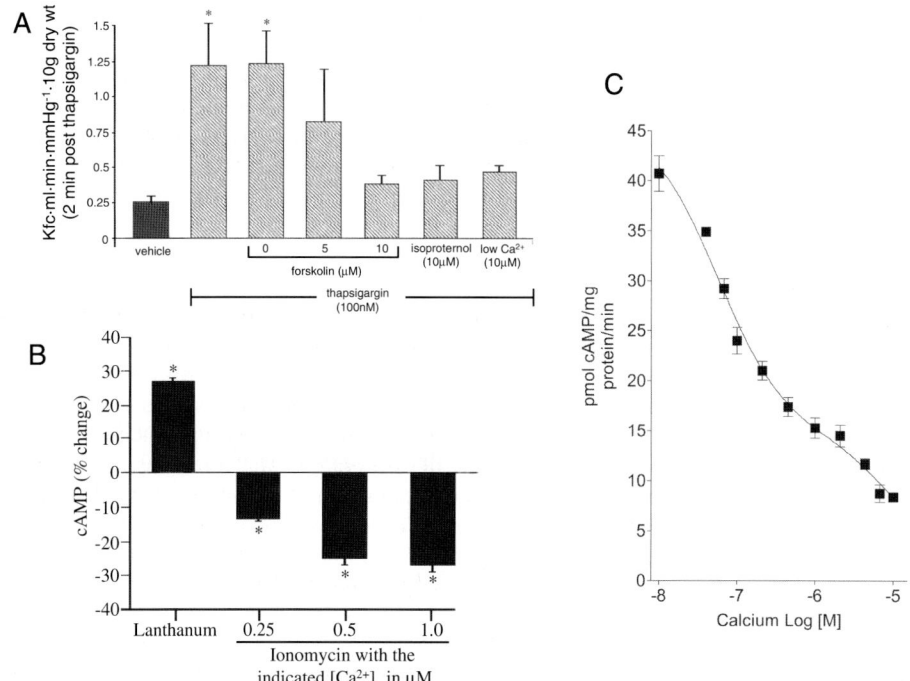

Figure 4. The barrier protective effects of cAMP supercede the barrier disruptive effects of increased cytosolic calcium. A) Forskolin dose dependently attenuates the elevated K_f induced by thapsigargin. From Chetham et al., 1997 by permission of The American Physiological Society. **B)** Ionomycin induced Ca^{2+} entry decreases intracellular cAMP in pulmonary artery endothelial cells demonstrating Ca^{2+} inhibition of AC6. From Stevens et al., 1995, author rights by policy of PNAS, USA. **C)** Calcium inhibits adenylyl cyclase activity in membrane preparations of pulmonary microvascular endothelial cells. Inhibition of enzymatic activity occurs at the submicromolar concentrations indicative of adenylyl cyclase isoform 6. From Creighton et al., 2003, permission of The American Physiological Society.

Figure 5. Adenylyl cyclase activity resides in caveolae and at cell-cell borders. (See next page) AC activity is localized to caveolae and along the luminal plasma membrane of PMVECs. (**A**) Uranyl acetate control experiments reveal that the buffer and fixation procedure does not produce an electron dense staining at cell membranes. In adenylylimidodiphosphate treated cells, forskolin (EC_{100}) and rolipram stimulates AC activity prominent in caveolae or caveolae like structures (**B**) and at sites of cell-cell tethering (**C**). From (Sayner, Frank et al., 2004) by premission of the Amer. Heart Association and the publishers: Lippincott, Williams, and Wilkins, London.

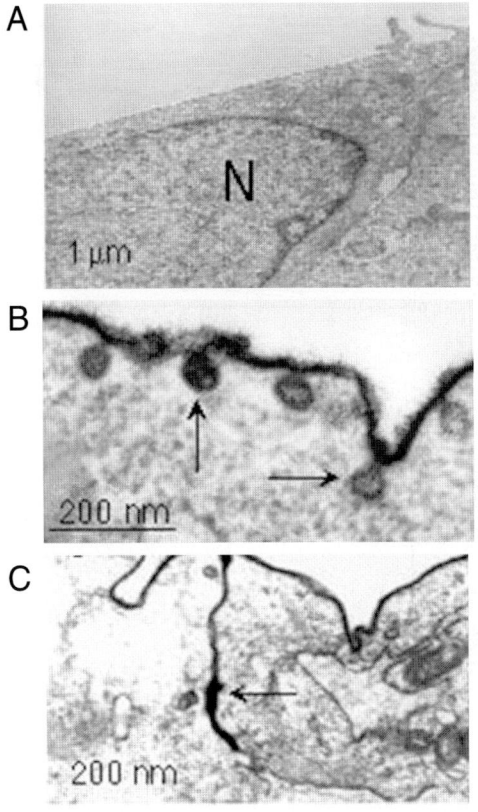

Endothelial Cell Heterogeneity: cAMP Signal Transduction

It is increasingly apparent that stable phenotypic differences exist between microvascular and conduit endothelium (King *et al.*, 2004). Endothelial cell biologists have long recognized that environmental stimuli alter cell behavior. Yet, only recently has it become evident that lung macrovascular and microvascular endothelium apparently arise from different embyrological origins, so that the respective cells are imprinted through development to ascertain a distinct phenotype, as explored in Chapter 10. Indeed, even when lung macro- and microvascular cells are studied in similar conditions, they exhibit distinct gene expression profiles and functional attributes reflective of their origin (Baldwin 1996; Stevens *et al.*, 2001; Chi *et al.*, 2003).

One of the many attributes that differ between the pulmonary artery and the microvascular endothelial cells is their control of cAMP concentrations (Stevens *et al.* 1999; Creighton *et al.*, 2003). Both of these cell types express AC6 (their calcium sensitivity is not equal, see below). However, pulmonary microvascular endothelial cells exhibit elevated global intracellular cAMP concentrations compared to pulmonary artery endothelial cells (Figure 6a). Microvascular endothelial cell cAMP concentrations appear to be governed by futile cycling, where high adenylyl cyclase and comparatively high phosphodiesterase activity rapidly synthesizes and then degrades cAMP. Thus, inhibiting phosphodiesterase activity increases cAMP content in microvascular cells (Figure 6b). In contrast, phosphodiesterase activity is much lower in pulmonary artery endothelial cells; consequently, inhibiting phosphodiesterase activity does not similarly increase cAMP, and activation of adenylyl cyclase (*e.g.* forskolin) induces a more profound cAMP rise in pulmonary artery than in pulmonary microvascular endothelial cells (Figure 6c). When adenylyl cyclase(s) are maximally stimulated and phosphodiesterase(s) are maximally inhibited, however, both cell types achieve similar cAMP concentrations (Figure 6d) (Stevens and Thompson, 1999). Thus, each cell type possesses different cAMP regulatory mechanisms germane to their function. Microvascular endothelial cells possess superior barrier properties when compared with pulmonary artery endothelial cells, both *in vivo* and *in vitro* (Chetham *et al.*, 1997; Kelly *et al.*, 1998; Moore *et al.*, 1998b). Increased global cAMP and "high" or "futile cycling" contributes to the enhanced barrier property in microvascular cells.

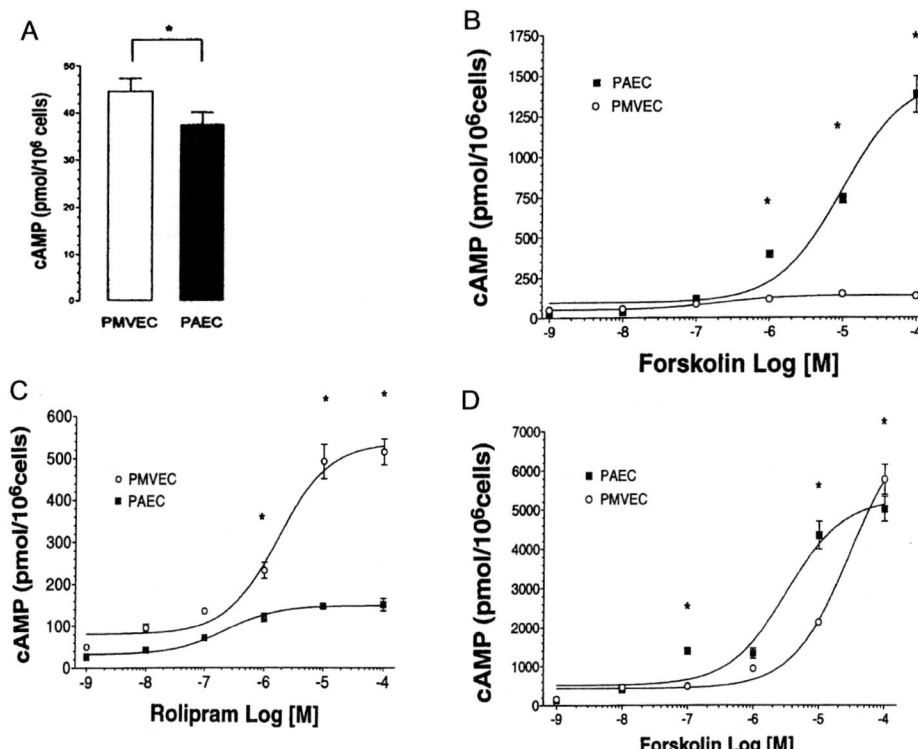

Figure 6. Pulmonary artery and microvascular endothelial cells exhibit unique cAMP characteristics. A) Basal $[cAMP]_i$ is higher in pulmonary microvascular endothelial cells (PMVECs) than pulmonary artery endothelial cells (PAECs). B) Direct stimulation of adenylyl cyclase produces a greater increase in cAMP in PAECs than PMVECs. C) PMVECs have greater phosphodiesterase 4 activity compared to PAECs. D) Maximal cell cAMP is achieved by adenylyl cyclase stimulation and phosphodiesterase 4 inhibition, and is similar in both cell types. Thus, basal adenylyl cyclase and phosphodiesterase activities of pulmonary microvascular endothelium are governed by futile cycling. From Stevens & Thompson, 1999 by permission of The Amer. Physiological Society.

Evidence that a cell possesses "increased" global cAMP and "futile" cycling at the same time may, at first glance, seem counterintuitive. However, as discussed in preceding sections, cAMP is highly compartmentalized. Membrane cAMP pools are generated by one of nine transmembrane adenylyl cyclases, which, in endothelial cells, is largely due to AC6. The extent to which cAMP produced by transmembrane cyclases contributes to the bulk cytosolic cAMP pool is still unclear, especially since soluble adenylyl cyclase (AC10) may be more ubiquitously expressed than previously thought (Sinclair et al., 2000; Wuttke et al., 2001).

To further understand how the membrane compartment may uniquely regulate cAMP production in pulmonary artery and microvascular endothelial cells, adenylyl cyclase activity in caveolae-rich fractions were directly compared. The findings revealed that total membrane cAMP synthesis is similar among the cells, but cAMP-specific phosphodiesterase activity is 30-fold greater in the microvascular cells. In addition, pulmonary microvascular endothelial cells exhibit a much greater sensitivity to calcium inhibition, suggesting a greater expression of, or reliance on, AC6 in membrane fractions in these cells (Creighton *et al.*, 2003).

It is interesting that microvascular endothelial cells exhibit greater calcium inhibition of cAMP in membrane fractions, because their global cytosolic calcium response to thapsigargin (*e.g.* activation of SOC entry) is reduced compared to pulmonary artery endothelial cells. Moreover, thapsigargin is sufficient to increase permeability in cultured pulmonary artery endothelial cells, but not in cultured microvascular endothelial cells (Kelly *et al.*, 1998). In the intact pulmonary circulation, thapsigargin induces intercellular gaps in extra-alveolar pulmonary artery and vein endothelial cells but does not induce gaps in intra-alveolar endothelial cells (*e.g.* capillaries) (Chetham *et al.*, 1999). In contrast to site-specific effects of thapsigargin, G_q-linked agonists that also reorganize cytoskeletal structures reveal gap formation in microvascular endothelial cells. Indeed, thrombin activates G_q, G_i and $G_{12/13}$ proteins. $G_{12/13}$ proteins activate Rho, which in concert with SOC entry, reorganizes F-actin (Dudek and Garcia, 2001; Bogatcheva *et al.*, 2002). Although microvascular endothelial cells are still less sensitive to thrombin-induced barrier disruption than are pulmonary artery endothelial cells, thrombin is capable of inducing inter-endothelial gaps in microvascular cells (Cioffi *et al.*, 2002). This relative insensitivity to calcium agonists in microvascular endothelial cells is consistent with their superior barrier function, as they express more adherens and tight junction proteins than do pulmonary artery endothelial cells.

AC6 in Control of the Endothelial Cell Barrier

Clearly demonstrating the physiological function of AC6 in endothelial cells has been difficult because the binding site that mediates Ca^{2+} inhibition is not resolved. In addition, both thapsigargin and thrombin decrease cAMP, yet these agonists do not readily induce gap formation in microvascular endothelial cells. Such disparate effects of thapsigargin and thrombin on pulmonary microvascular endothelial cell barrier function have brought into question exactly how important Ca^{2+} inhibition of AC6 is in controlling the barrier.

To address this issue, the type 8 adenylyl cyclase (AC8), which is activated by calcium-calmodulin, was heterologously expressed in microvascular endothelial cells (Cioffi *et al.*, 2002). Enzyme expression and function was restricted to the cell membrane, and appeared enriched at sites of cell-cell adhesion. The expression of AC8 did not change basal cAMP concentrations. Whereas thrombin normally decreased cAMP, in cells expressing AC8, thrombin modestly increased cAMP concentrations (*e.g.* 2-fold) when phosphodiesterase activity was inhibited. However, this small cAMP rise abolished thrombin-induced gap formation even without concurrent phosphodiesterase inhibition, revealing the dominant role that cAMP plays in control of cell-cell adhesion and the importance of AC6 in mediating the inverse relationship between calcium and membrane cAMP.

AC6 therefore generates a membrane delimited cAMP pool that is localized nearby its physiologically relevant downstream targets. Activating AC6 increases membrane cAMP that strengthens the barrier, whereas calcium inhibition of AC6 decreases membrane cAMP that weakens the barrier. Cooperativity between anatomical and biochemical constraints provides diffusion restricted domains that are necessary to establish cAMP gradients. Anatomically, restrictions to diffusion exist through the generation of three-dimensional compartments. Organelles, such as the endoplasmic reticulum lying in close apposition to the plasma membrane provide the boundaries to such compartments. Indeed, transmission electron micrographs demonstrate the endoplasmic reticulum in both pulmonary artery and microvascular endothelial cells comes in close proximity with the plasma membrane and caveolae (King et al., 2004). Visualizing sites of cAMP synthesis using ultrastructural approaches illustrate that cAMP is produced in caveolae microdomains and at cell-cell borders (Sayner et al., 2004). Localized phosphodiesterases represent biochemical constraints that hydrolyze cAMP and impede its diffusion away from membrane microdomains.

If it is correct that high cAMP concentrations are achieved specifically at the membrane necessary to activate localized targets that strengthen the endothelial cell barrier, then relocalizing cAMP synthesis to the cytosol should activate discrete cell targets with different physiological consequence(s). In this regard, the actions of certain pathogenic bacteria are interesting, because they have devised mechanisms to transfer soluble eukaryotic factor dependent adenylyl cyclases directly into the cytosol of target cells. One such adenylyl cyclase is edema factor of *Bacillus anthracis*, which is activated by calmodulin to generate high-level intracellular cAMP in CHO cells. This cAMP pool is not highly regulated by phosphodiesterase activity (Leppla, 1982). When edema factor is subcutaneously injected into the skin of whole animals, it induces edema (Smith et al., 1955; Stanley and Smith, 1961). While the direct actions of edema factor on the endothelial cell barrier have never been tested, these findings are interesting because they suggest that a soluble adenylyl cyclase could produce a cAMP pool that is barrier disruptive rather than barrier protective.

Like the *Bacillus anthracis* edema factor, *Pseudomonas aeruginosa* inserts an adenylyl cyclase toxin, ExoY, into eukaryotic cells (Yahr et al., 1998). When activated by an unknown eukaryotic factor it achieves high-level cAMP synthesis in pulmonary microvascular endothelial cells. Whereas endogenous AC6 is restricted to the plasma membrane and more specifically to lipid rafts and caveolae, immunocytochemistry studies demonstrate that ExoY resides only in the cytosolic compartment (Sayner et al., 2004). Indeed, ExoY does not co-localize with β-catenin, that resides at the plasma membrane in association with the adherens junction. As with edema factor, cAMP that is generated by ExoY is not highly regulated by phosphodiesterase activity. Moreover, cAMP produced by ExoY induces widespread gap formation and increases permeability. Thus, the location of cAMP synthesis plays a critical role in its ultimate biological activity, by regulating which effectors are activated. Whereas membrane cAMP activates targets that strengthen the endothelial cell barrier, soluble cAMP activates different cytosolic targets that disrupt the endothelial cell barrier.

Summary

cAMP was the first second messenger discovered. Since its original description, elements of the cAMP signal transduction cascade have grown in complexity. The idea that cAMP quickly and evenly disperses within the cytosolic compartment to achieve a uniform cell signal has slowly faded. The challenge before us is to understand how cAMP signaling is achieved with spatio-temporal fidelity. This fidelity is cell-type specific and in accordance with any cell's unique function. To fully accomplish such a high level of understanding, identifying the molecular anatomy of the cAMP signaling scaffold must be put in context with a cell's specialized phenotype. Pulmonary microvacular endothelial cells must maintain a tight barrier function to limit fluid, solute and macromolecular flux, and preserve alveolar gas exchange. Therefore, membrane cAMP concentrations are maintained at a high level, and cAMP spillover into the bulk cytosol is limited by high, membrane-associated phosphodiesterase activity. This arrangement allows cAMP to achieve concentrations high enough to activate localized effectors that promote barrier function, while maintaining bulk cytosolic cAMP concentrations below a threshold required to activate effectors that disrupt barrier function. Calcium inhibition of membrane cAMP allows transient gap formation. Generation of cAMP outside the membrane compartment produces a sustained increase in microvascular endothelial cell permeability. In this context, pulmonary microvascular endothelium has developed a highly ordered and specialized mechanism for regulating cAMP production to dynamically control alveolar-capillary stability on a moment-to-moment basis.

ACKNOWLEDGMENTS

This work was supported by HL66299 (TS) and HL60024 (TS) and by an AHA Southeast Consortium predoctoral fellowship award 0315134B (SS).

REFERENCES

Adams, S.R., Harootunian, A., Buechler, Y., Taylor, S., and Tsien, R.Y. (1991) Fluorescence ratio imaging of cyclic AMP in single cells. Nature 349:694-697.
Adamson, R.H., Liu, B., Fry, G., Rubin, L., and Curry, F.E. (1998) Microvascular permeability and number of tight junctions are modulated by cAMP. Amer. J. Physiol. 274:H1885-H1894.
Ambudkar, I.S., Brazer, S., Liu, X., Lockwich, T., and Singh, B. (2004) Plasma membrane localization of TRPC channels: role of caveolar lipid rafts. Novartis Found. Symp. 258:63-70; discussion 70-74, 98-102, 263-266.
Artymiuk, P.J., Poirrette, A., Rice, D., and Willett, P. (1997) A polymerase I palm in adenylyl cyclase? Nature 388:33-34.
Bacskai, B.J., Hochner, B., Mahaut-Smith, M., Adams, S., Kaang, B., Kandel, E., and Tsien, R.Y. (1993) Spatially resolved dynamics of cAMP and protein kinase A subunits in Aplysia sensory neurons. Science 260:222-226.
Baldwin, H.S. (1996) Early embryonic vascular development. Cardiovasc. Res. 31:E34-E45.
Barzu, O. and Danchin, A (1994) Adenylyl cyclases: a heterogeneous class of ATP-utilizing enzymes. Prog. Nucleic. Acid Res. Mol. Biol. 49:241-283.
Beltman, J., Sonnenburg, W.K., and Beavo, J.A. (1993) The role of protein phosphorylation in the regulation of cyclic nucleotide phosphodiesterases. 127:239-253.
Bird, G.S., Aziz, O., Lievremont, J., Wedel, B., Trebak, M., Vazquez, G., and Putney, J.W. (2004) Mechanisms of phospholipase C-regulated calcium entry. Curr. Mol. Med. 4:291-301.
Birnbaumer, L., Zhu, X., Jiang, M., Boulay, G., Peyton, M., Vannier, B., Brown, D., Platano, D., Sadeghi, H., Stefani, E., and Birnbaumer, M. (1996) On the molecular basis and regulation of cellular capacitative calcium entry: roles for Trp proteins. PNAS USA 93:15195-15202.
Bogatcheva, N.V., Garcia, J.G., and Verin, A.D. (2002) Molecular mechanisms of thrombin-induced endothelial cell permeability. Biochem. (Mosc.) 67:75-84.
Brooker, G. (1973) Oscillation of cyclic adenosine monophosphate concentration during the myocardial contraction cycle. Science 182:933-934.
Brough, G.H., Wu, S., Cioffi, D., Moore, T., Li, M., Dean, N., and Stevens, T. (2001) Contribution of endogenously expressed Trp1 to a Ca^{2+}-selective, store-operated Ca^{2+} entry pathway. FASEB J. 15:1727-1738.

Brunton, L.L. (2003) PDE4: arrested at the border. Sci. STKE 204:PE44.
Brunton, L.L., Hayes, J.S., and Mayer, S.E. (1981) Functional compartmentation of cyclic AMP and protein kinase in heart. Adv. Cyclic Nucl. Res. 14:391-397.
Brunton, L.L. and Mayer, S.E. (1979) Extrusion of cyclic AMP from pigeon erythrocytes. J. Biol. Chem. 254:9714-9720.
Buck, J., Sinclair, M., Schapal, L., Cann, M., and Levin, L.R. (1999) Cytosolic adenylyl cyclase defines a unique signaling molecule in mammals. PNAS USA 96:79-84.
Bundey, R.A. and Insel, P.A. (2004) Discrete intracellular signaling domains of soluble adenylyl cyclase: camps of cAMP? Sci. STKE 231:PE19.
Buonassisi, V. and Venter, J.C. (1976) Hormone and neurotransmitter receptors in an established vascular endothelial cell line. PNAS USA 73:1612-1616.
Bushnik, T. and Conti, M. (1996) Role of multiple cAMP-specific phosphodiesterase variants. Biochem. Soc. Trans. 24:1014-1019.
Butcher, R.W. and Sutherland, E.W. (1962) Adenosine 3',5'-phosphate in biological materials. I. Purification and properties of cyclic 3',5'-nucleotide phosphodiesterase and use of this enzyme to characterize adenosine 3',5'-phosphate. J. Biol. Chem. 237:1244-1250.
Chen, Y., Cann, M., Litvin, T., Iourgenko, V., Sinclair, M., Levin, L., and Buck, J. (2000) Soluble adenylyl cyclase as an evolutionarily conserved bicarbonate sensor. Science 289:625-628.
Chetham, P.M., Babal, P., Bridges, J., Moore, T., and Stevens, T. (1999) Segmental regulation of pulmonary vascular permeability by store-operated Ca^{2+} entry. Amer. J. Physiol. 276:L41-L50.
Chetham, P.M., Guldemeester, H., Mons, N., Brough, G., Bridges, J., Thompson, W., and Stevens, T. (1997) Ca^{2+}-inhibitable adenylyl cyclase and pulmonary microvascular permeability. Amer. J. Physiol. 273:L22-L30.
Chi, J.T., Chang, H., Haraldsen, G., Jahnsen, F., Troyanskaya, O., Chang, D., Wang, Z., Rockson, S., van de Rijn, M., Botstein, D., and Brown, P.O. (2003) Endothelial cell diversity revealed by global expression profiling. PNAS USA 100:10623-10628.
Chiono, M., Mahey, R., Tate, G., and Cooper, D.M. (1995) Capacitative Ca^{2+} entry exclusively inhibits cAMP synthesis in C6-2B glioma cells. Evidence that physiologically evoked Ca^{2+} entry regulates Ca^{2+} inhibitable adenylyl cyclase in non-excitable cells. J. Biol. Chem. 270:1149-1155.
Cioffi, D.L., Moore, T., Schaack, J., Creighton, J., Cooper, D.M., and Stevens, T. (2002) Dominant regulation of interendothelial cell gap formation by calcium-inhibited type 6 adenylyl cyclase. J. Cell. Biol. 157:1267-1278.
Cioffi, D.L., Wu, S., and Stevens, T. (2003) On the endothelial cell I(SOC). Cell Calcium 33:323-336.
Colledge, M. and Scott, J.D. (1999) AKAPs: from structure to function. Trends Cell Biol. 9:216-221.
Cooper, D.M. (2003a) Molecular and cellular requirements for the regulation of adenylate cyclases by calcium. Biochem. Soc. Trans. 31:912-915.
Cooper, D.M. (2003b) Regulation and organization of adenylyl cyclases and cAMP. Biochem. J. 375:517-529.
Cooper, D.M., Karpen, J., Fagan, K., and Mons, N.E. (1998) Ca^{2+} sensitive adenylyl cyclases. Adv. Second Mess. Phosphopr. Res. 32:23-51.
Cooper, D.M., Mons. N., and Fagan, K. (1994a) Ca^{2+}-sensitive adenylyl cyclases. Cell Sig. 6:823-840.
Cooper, D.M., Mons, N., and Karpen, J.W. (1995) Adenylyl cyclases and the interaction between calcium and cAMP signaling. Nature 374:421-424.
Cooper, D.M., Yoshimura, M., Zhang, Y., Chiono, M., and Mahey, R. (1994b) Capacitative Ca^{2+} entry regulates Ca^{2+}-sensitive adenylyl cyclases. Biochem. J. 297:437-440.
Creighton, J.R., Masada, N., Cooper, D.M., and Stevens, T. (2003) Coordinate regulation of membrane cAMP by Ca^{2+}-inhibited adenylyl cyclase and phosphodiesterase activities. Amer. J. Physiol. 284:L100-L107.
Curry, F.E. and Adamson, R.H. (1999) Transendothelial pathways in venular microvessels exposed to agents which increase permeability: gaps in our knowledge. Microcirc. 6:3-5.

Danchin, A. (1993) Phylogeny of adenylyl cyclases. Adv. Second Mess. Phosphopr. Res. 27:109-162.
Davare, M.A., Avdonin, V., Hall, D., Peden, E., Burette, A., Weinberg, R., Horne, M., Hoshi, T., and Hell, J.W. (2001) A β2 adrenergic receptor signaling complex assembled with the Ca^{2+} channel Cav1.2. Science 293:98-101.
de Rooij, J., Zwartkruis, F., Verheijen, M., Cool, R., Nijman, S., Wittinghofer, A., and Bos, J.L. (1998) Epac is a Rap1 guanine-nucleotide-exchange factor directly activated by cyclic AMP. Nature 396:474-477.
DeBernardi, M.A. and Brooker, G. (1996) Single cell Ca^{2+}/cAMP cross-talk monitored by simultaneous Ca^{2+}/cAMP fluorescence ratio imaging. PNAS USA 93:4577-4582.
Dessauer, C.W. and Gilman, A.G. (1996) Purification and characterization of a soluble form of mammalian adenylyl cyclase. J. Biol. Chem. 271:16967-16974.
Dudek, S.M. and Garcia, J.G. (2001) Cytoskeletal regulation of pulmonary vascular permeability. J. Appl. Physiol. 91:1487-1500.
Fagan, K.A., Mons, N., and Cooper, D.M. (1998) Dependence of the Ca^{2+}-inhibitable adenylyl cyclase of C6-2B glioma cells on capacitative Ca^{2+} entry. J. Biol. Chem. 273:9297-9305.
Fagan, K.A., Smith, K.E., and Cooper, D.M. (2000) Regulation of the Ca^{2+}-inhibitable adenylyl cyclase type VI by capacitative Ca^{2+} entry requires localization in cholesterol-rich domains. J. Biol. Chem. 275:26530-26537.
Farrukh, I.S., Gurtner, G.H., and Michael, J.R. (1987) Pharmacological modification of pulmonary vascular injury: possible role of cAMP. J. Appl. Physiol. 62:47-54.
Francis, S.H., Turko, I.V., and Corbin, J.D. (2001) Cyclic nucleotide phosphodiesterases: relating structure and function. Prog. Nucl. Acid Res. Mol. Biol. 65:1-52.
Freichel, M., Suh, S., Pfeifer, S., Schweig, U., Trost, C., Weissgerber, P., Biel, M., Philipp, S., Freise, D., Droogmans, G., Hofmann, F., Flockerzi, V., and Nilius, B. (2001) Lack of an endothelial store-operated Ca^{2+} current impairs agonist-dependent vasorelaxation in TRP4-/- mice. Nat. Cell Biol. 3:121-127.
Goel, M., Sinkins, W.G., and Schilling, W.P. (2002) Selective association of TRPC channel subunits in rat brain synaptosomes. J. Biol. Chem. 277:48303-48310.
Gorbunova, Y.V. and Spitzer, N.C. (2002) Dynamic interactions of cyclic AMP transients and spontaneous Ca^{2+} spikes. Nature 418:93-96.
Gu, C., Cali, J.J., and Cooper, D.M. (2002) Dimerization of mammalian adenylate cyclases. Eur. J. Biochem. 269:413-421.
Gu, C., Sorkin, A., and Cooper, D.M. (2001) Persistent interactions between the two transmembrane clusters dictate the targeting and functional assembly of adenylyl cyclase. Curr. Biol. 11:185-190.
Hallam, T.J. and Pearson, J.D. (1986) Exogenous ATP raises cytoplasmic free calcium in fura-2 loaded piglet aortic endothelial cells. FEBS Lett. 207:95-99.
Hanoune, J. and Defer, N. (2001) Regulation and role of adenylyl cyclase isoforms. Ann. Rev. Pharmacol. Toxicol. 41:145-174.
Hayes, J.S., Brunton, L.L., and Mayer, S.E. (1980) Selective activation of particulate cAMP-dependent protein kinase by isoproterenol and prostaglandin E1. J. Biol. Chem. 255:5113-5119.
Hofmann, T., Schaefer, M., Schultz, G., and Gudermann, T. (2000) Transient receptor potential channels as molecular substrates of receptor-mediated cation entry. J. Mol. Med. 78:14-25.
Houslay, M.D. and Adams, D.R. (2003) PDE4 cAMP phosphodiesterases: modular enzymes that orchestrate signaling cross-talk, desensitization and compartmentalization. Biochem. J. 370:1-18.
Houslay, M.D. and Milligan, G. (1997) Tailoring cAMP-signaling responses through isoform multiplicity. Trends Biochem. Sci. 22:217-224.
Hurley, J.H. (1998) The adenylyl and guanylyl cyclase superfamily. Curr. Opin. Struct. Biol. 8:770-777.
Ishikawa, Y., Katsushika, S., Chen, L., Halnon, N., Kawabe, J., and Homcy, C.J. (1992) Isolation and characterization of a novel cardiac adenylylcyclase cDNA. J. Biol. Chem. 267:13553-13557.
Iyengar, R. (1993) Molecular and functional diversity of mammalian Gs-stimulated adenylyl cyclases. FASEB J. 7:768-775.

Jurevicius, J. and Fischmeister, R. (1996) cAMP compartmentation is responsible for a local activation of cardiac Ca^{2+} channels by beta-adrenergic agonists. PNAS USA 93:295-299.
Karpen, J.W. and Rich, T.C. (2001) The fourth dimension in cellular signaling. Science 293:2204-2205.
Kasahara, M., Yashiro, K., Sakamoto, T., and Ohmori, M. (1997) The Spirulina platensis adenylate cyclase gene, cyaC, encodes a novel signal transduction protein. Plant Cell Physiol. 38:828-836.
Kaupp, U.B., Niidome, T., Tanabe, T., Terada, S., Bonigk, W., Stuhmer, W., Cook, N., Kangawa, K., Matsuo, H., and Hirose, T. (1989) Primary structure and functional expression from complementary DNA of the rod photoreceptor cyclic GMP-gated channel. Nature 342:762-766.
Kawasaki, H., Springett, G., Mochizuki, N., Toki, S., Nakaya, M., Matsuda, M., Housman, D., and Graybiel, A.M. (1998) A family of cAMP-binding proteins that directly activate Rap1. Science 282:2275-2279.
Kelly, J.J., Moore, T., Babal, P., Diwan, A., Stevens, T., and Thompson, W.J. (1998) Pulmonary microvascular and macrovascular endothelial cells: differential regulation of Ca^{2+} and permeability. Amer. J. Physiol. 274:L810-L819.
King, J., Hamil, T., Creighton, J., Wu, S., Bhat, P., McDonald, F., and Stevens, T. (2004) Structural and functional characteristics of lung macro- and microvascular endothelial cell phenotypes. Microvasc. Res. 67:139-151.
Kobayashi, H., Kobayashi, T., and Fukushima, M. (1987) Effects of dibutyryl cAMP on pulmonary air embolism-induced lung injury in awake sheep. J. Appl. Physiol. 63:2201-2207.
Krupinski, J. (1991) The adenylyl cyclase family. Mol. Cell Biochem. 104:73-79.
Krupinski, J. and Cali, J.J. (1998) Molecular diversity of the adenylyl cyclases. Adv. Second Mess. Phosphopr. Res. 32:53-79.
Krupinski, J., Coussen, F., Bakalyar, H., Tang, W., Feinstein, P., Orth, K., Slaughter, C., Reed, R., and Gilman, A.G. (1989) Adenylyl cyclase amino acid sequence: possible channel- or transporter-like structure. Science 244:1558-15564.
Langeler, E.G. and van Hinsbergh, V.W. (1991) Norepinephrine and iloprost improve barrier function of human endothelial cell monolayers: role of cAMP. Amer. J. Physiol. 260:C1052-C1059.
Laporte, S.A., Oakley, R.H., and Caron, M.G. (2001) Signal transduction. Bringing channels closer to the action! Science 293:62-63.
Leppla, S.H. (1982) Anthrax toxin edema factor: a bacterial adenylate cyclase that increases cyclic AMP concentrations of eukaryotic cells. PNAS USA 79:3162-3166.
Litvin, T.N., Kamenetsky, M., Zarifyan, A., Buck, J., and Levin, L.R. (2003) Kinetic properties of "soluble" adenylyl cyclase. Synergism between calcium and bicarbonate. J. Biol. Chem. 278:15922-15926.
Liu, X. and Ambudkar, I.S. (2001) Characteristics of a store-operated calcium-permeable channel: sarcoendoplasmic reticulum calcium pump function controls channel gating. J. Biol. Chem. 276:29891-29898.
Liu, Y., Ruoho, A., Rao, V., and Hurley, J.H. (1997) Catalytic mechanism of the adenylyl and guanylyl cyclases: modeling and mutational analysis. PNAS USA 94:13414-13419.
Lockwich, T.P., Liu, X., Singh, B., Jadlowiec, J., Weiland, S., and Ambudkar, I.S. (2000) Assembly of Trp1 in a signaling complex associated with caveolin-scaffolding lipid raft domains. J. Biol. Chem. 275:11934-11942.
Londos, C. and Wolff, J. (1977) Two distinct adenosine-sensitive sites on adenylate cyclase. PNAS USA 74:5482-5486.
Lugnier, C. and Schini, V.B. (1990) Characterization of cyclic nucleotide phosphodiesterases from cultured bovine aortic endothelial cells. Biochem. Pharm. 39:75-84.
Lum, H. and Malik, A.B. (1994) Regulation of vascular endothelial barrier function. Amer. J. Physiol. 267:L223-L241.
Lum, H. and Malik. A.B. (1996) Mechanisms of increased endothelial permeability. Can. J. Physiol. Pharmacol. 74:787-800.
Majno, G., Gilmore, V., and Leventhal, M. (1967) On the mechanism of vascular leakage caused by histaminetype mediators. A microscopic study in vivo. Circ. Res. 21:833-847.

Makarski, J.S. (1981) Stimulation of cyclic AMP production by vasoactive agents in cultured bovine aortic and pulmonary artery endothelial cells. In Vitro 17:450-458.

Maurice, D.H., Palmer, D., Tilley, D., Dunkerley, H., Netherton, S., Raymond, D., Elbatarny, H., and Jimmo, S.L. (2003) Cyclic nucleotide phosphodiesterase activity, expression, and targeting in cells of the cardiovascular system. Mol. Pharmacol. 64:533-546.

Minnear, F.L., DeMichele, M., Moon, D., Rieder, C., and Fenton, J.W. (1989) Isoproterenol reduces thrombin-induced pulmonary endothelial permeability. Amer. J. Physiol. 257:H1613-H1623.

Moore, T.M., Brough, G., Babal, P., Kelly, J., Li, M., and Stevens, T. (1998a) Store-operated calcium entry promotes shape change in pulmonary endothelial cells expressing Trp1. Amer. J. Physiol. 275:L574-L582.

Moore, T.M., Chetham, P., Kelly, J., and Stevens, T. (1998b) Signal transduction and regulation of lung endothelial cell permeability. Interaction between Ca^{2+} and cAMP. Amer. J. Physiol. 275:L203-L222.

Moore, T.M., Norwood, N., Creighton, J., Babal, P., Brough, G., Shasby, D.M., and Stevens, T. (2000) Receptor-dependent activation of store-operated calcium entry increases endothelial cell permeability. Amer. J. Physiol. 279:L691-L698.

Moy, A.B., Shasby, S., Scott, B., and Shasby, D.M. (1993) The effect of histamine and cyclic adenosine monophosphate on myosin light chain phosphorylation in human umbilical vein endothelial cells. J. Clin. Invest. 92:1198-1206.

Murphy, J.T., Duffy, S., Hybki, D., and Kamm, K. (2001) Thrombin-mediated permeability of human microvascular pulmonary endothelial cells is Ca dependent. J. Trauma 50:213-222.

Nilius, B. (2003) From TRPs to SOCs, CCEs, and CRACs: consensus and controversies. Cell Calcium 33:293-298.

Norwood, N., Moore, T., Dean, D., Bhattacharjee, R., Li, M., and Stevens, T. (2000) Store-operated calcium entry and increased endothelial cell permeability. Amer. J. Physiol. 279:L815-L824.

Patterson, C.E., Davis, H., Schaphorst, K., and Garcia, J.G. (1994) Mechanisms of cholera toxin prevention of thrombin- and PMA-induced endothelial cell barrier dysfunction. Microvasc. Res. 48:212-235.

Patterson, C.E. and Lum, H. (2001) Update on pulmonary edema: the role and regulation of endothelial barrier function. Endothelium 8:75-105.

Patterson, C.E., Lum, H., Verin, A., Schaphorst, K., and Garcia, J.G. (2000) Regulation of endothelial barrier function by the cAMP-dependent protein kinase. Endothel. 7:287-308.

Perry, S.J., Baillie, G. Kohout, T., McPhee, I., Magiera, M., Ang, K., Miller, W., McLean, A., Conti, M., Houslay, M., and Lefkowitz, R.J. (2002) Targeting of cyclic AMP degradation to beta 2-adrenergic receptors by beta-arrestins. Science 298:834-836.

Pfeuffer, E., Mollner, S., and Pfeuffer, T. (1985) Adenylate cyclase from bovine brain cortex: purification and characterization of the catalytic unit. Embo J. 4:3675-3679.

Pfeuffer, E. and Pfeuffer, T. (1989) Affinity labeling of forskolin-binding proteins. Comparison between glucose carrier and adenylate cyclase. FEBS Lett. 248:13-17.

Philipp, S., Trost, C., Warnat, J., Rautmann, J., Himmerkus, N., Schroth, G., Kretz, O., Nastainczyk, W., Cavalie, A., Hoth, M., and Flockerzi, V. (2000) TRP4 (CCE1) protein is part of native calcium release-activated Ca^{2+}-like channels in adrenal cells. J. Biol. Chem. 275:23965-23972.

Premont, R.T., Matsuoka, I., Mattei, M., Pouille, Y., Defer, N., and Hanoune, J. (1996) Identification and characterization of a widely expressed form of adenylyl cyclase. J. Biol. Chem. 271:13900-13907.

Putney, J. W., Jr. (1986) A model for receptor-regulated calcium entry. Cell Calcium 7:1-12.

Rall, T.W. (1975) Opening Remarks. Adv. Cyclic Nucleo. Res. 5:1-2.

Rall, T.W. and Sutherland, E.W. (1962) Adenyl cyclase. II. The enzymatically catalyzed formation of adenosine 3',5'-phosphate and inorganic pyrophosphate from adenosine triphosphate. J. Biol. Chem. 237:1228-1232.

Rich, T.C., Fagan, K., Nakata, H., Schaack, J., Cooper, D.M., and Karpen, J.W. (2000) Cyclic nucleotide-gated channels colocalize with adenylyl cyclase in regions of restricted cAMP diffusion. J. Gen. Physiol. 116:147-161.

Rotrosen, D. and Gallin, J.I. (1986) Histamine type I receptor occupancy increases endothelial cytosolic calcium, reduces F-actin, and promotes albumin diffusion across cultured endothelial monolayers. J. Cell Biol. 103:2379-2387.
Ryan, U.S., Clements, E., Habliston, D., and Ryan, J.W. (1978) Isolation and culture of pulmonary artery endothelial cells. Tissue Cell 10:535-554.
Sayner, S.L., Frank, D., King, J., Chen, H., VandeWa, J., and Stevens, T. (2004) Paradoxical cAMP-induced lung endothelial hyperpermeability revealed by *P. aeruginosa* ExoY. Circ. Res. 95:196 - 203.
Schaphorst, K.L., Pavalko, F., Patterson, C., and Garcia, J.G. (1997) Thrombin-mediated focal adhesion plaque reorganization in endothelium: role of protein phosphorylation. Amer. J. Respir. Cell Mol. Biol. 17:443-455.
Schilling, W.P., Cabello, O.A., and Rajan, L. (1992) Depletion of the inositol 1,4,5-trisphosphate-sensitive intracellular Ca^{2+} store in vascular endothelial cells activates the agonist-sensitive Ca^{2+}-influx pathway. Biochem. J. 284:521-530.
Schilling, W.P. and Goel, M. (2004) Mammalian TRPC channel subunit assembly. Novartis Found. Symp. 258:18-30; discussion 30-43, 98-102, 263-266.
Scholich, K., Wittpoth, C., Barbier, A., Mullenix, J., and Patel, T.B. (1997) Identification of an intramolecular interaction between small regions in type V adenylyl cyclase that influences stimulation of enzyme activity by Gsα. PNAS USA 94:9602-9607.
Seamon, K. and Daly, J.W. (1981) Activation of adenylate cyclase by the diterpene forskolin does not require the guanine nucleotide regulatory protein. J. Biol. Chem. 256:9799-9801.
Seamon, K.B., Padgett, W., and Daly, J.W. (1981) Forskolin: unique diterpene activator of adenylate cyclase in membranes and in intact cells. PNAS USA 78:3363-3367.
Seebacher, T., Linder, J.U., and Schultz, J.E. (2001) An isoform-specific interaction of the membrane anchors affects mammalian adenylyl cyclase type V activity. Eur. J. Biochem. 268:105-110.
Shaul, P.W. and Anderson, R.G. (1998) Role of plasmalemmal caveolae in signal transduction. Amer. J. Physiol. 275:L843-L851.
Sinclair, M.L., Wang, X., Mattia, M., Conti, M., Buck, J., Wolgemuth, D., and Levin, L.R. (2000) Specific expression of soluble adenylyl cyclase in male germ cells. Mol. Reprod. Dev. 56:6-11.
Smith, F.D. and Scott, J.D. (2002) Signaling complexes: junctions on the intracellular information super highway. Curr. Biol. 12:R32-R40.
Smith, H., Keppie, J., Stanley, J., and Harris-Smith, P.W. (1955) The chemical basis of the virulence of Bacillus anthracis. IV. Secondary shock as the major factor in death of guinea-pigs from anthrax. Br. J. Exp. Pathol. 36:323-335.
Stanley, J.L. and Smith, H. (1961) Purification of factor I and recognition of a third factor of the anthrax toxin. J. Gen. Microbiol. 26:49-63.
Stelzner, T.J., Weil, J.V., and O'Brien, R.F. (1989) Role of cyclic adenosine monophosphate in the induction of endothelial barrier properties. J. Cell Physiol. 139:157-166.
Stevens, T., Creighton, J., and Thompson, W.J. (1999) Control of cAMP in lung endothelial cell phenotypes. Implications for control of barrier function. Amer. J. Physiol. 277:L119-L126.
Stevens, T., Fouty, B., Hepler, L., Richardson, D., Brough, G., McMurtry, I.F., and Rodman, D.M. (1997) Cytosolic Ca^{2+} and adenylyl cyclase responses in phenotypically distinct pulmonary endothelial cells. Amer. J. Physiol. 272:L51-L59.
Stevens, T., Garcia, J., Shasby, D.M., Bhattacharya, J., and Malik, A.B. (2000) Mechanisms regulating endothelial cell barrier function. Amer. J. Physiol. 279:L419-L422.
Stevens, T., Nakahashi, Y., Cornfield, D., McMurtry, I.F., Cooper, D.M., and Rodman, D.M. (1995) Ca^{2+}-inhibitable adenylyl cyclase modulates pulmonary artery endothelial cell cAMP content and barrier function. PNAS USA 92:2696-2700.
Stevens, T., Rosenberg, R., Aird, W., Quertermous, T., Johnson, F., Garcia, J., Hebbel, R., Tuder, R., and Garfinkel, S. (2001) NHLBI workshop report: endothelial cell phenotypes in heart, lung, and blood diseases. Amer. J. Physiol. 281:C1422-C1433.
Stevens, T. and Thompson, W.J. (1999) Regulation of pulmonary microvascular endothelial cell cyclic adenosine monophosphate by adenylyl cyclase: implications for endothelial barrier function. Chest 116:32S-33S.

Sunahara, R.K., Beuve, A., Tesmer, J., Sprang, S., Garbers, D., and Gilman, A.G. (1998) Exchange of substrate and inhibitor specificities between adenylyl and guanylyl cyclases. J. Biol. Chem. 273:16332-16338.
Sunahara, R.K., Dessauer, C.W., and Gilman, A.G. (1996) Complexity and diversity of mammalian adenylyl cyclases. Ann. Rev. Pharmacol. Toxicol. 36:461-480.
Sunahara, R.K., Tesmer, J., Gilman, A.G., and Sprang, S.R. (1997) Crystal structure of the adenylyl cyclase activator Gsα. Science 278:1943-1947.
Sutherland, E.W. (1972) Studies on the mechanism of hormone action. Science 177:401-408.
Sutherland, E.W. and Rall, T.W. (1957) The properties of an adenine ribonucleotide produced with cellular particles, ATP, Mg^{++}, and epinephrine or glucagon. J. Amer. Chem. Soc. 79:3608.
Sutherland, E.W., Rall, T.w., and Menon, T. (1962) Adenyl cylase. I. Distribution, preparation, and properties. J. Biol. Chem. 237:1220-1227.
Sutherland, E.W., Robinson, G.W., and Butcher, R.W. (1968) Some aspects of the biological role of adenosine 3'5'-monophosphate. Circ. 37:279-306.
Suttorp, N., Weber, U., Welsch, T., and Schudt, C. (1993) Role of phosphodiesterases in the regulation of endothelial permeability in vitro. J. Clin. Invest. 91:1421-1428.
Tang, W.J. and Gilman, A.G. (1995) Construction of a soluble adenylyl cyclase activated by Gsα and forskolin. Science 268:1769-1772.
Tang, W.J., Krupinski, J., and Gilman, A.G. (1991) Expression and characterization of calmodulin-activated (type I) adenylylcyclase. J. Biol. Chem. 266:8595-8603.
Tang, W.J., Stanzel, M., and Gilman, A.G. (1995) Truncation and alanine-scanning mutants of type I adenylyl cyclase. Biochem. 34:14563-14572.
Tang, W.J., Yan, S., and Drum, C.L. (1998) Class III adenylyl cyclases: regulation and underlying mechanisms. Adv. Second Mess. Phosphopr. Res. 32:137-151.
Taussig, R. and Gilman, A.G. (1995) Mammalian membrane-bound adenylyl cyclases. J. Biol. Chem. 270:1-4.
Taussig, R., Tang, W., Hepler, J., and Gilman, A.G. (1994) Distinct patterns of bidirectional regulation of mammalian adenylyl cyclases. J. Biol. Chem. 269:6093-6100.
Tesmer, J.J., Sunahara, R., Fancy, D., Gilman, A.G., and Sprang, S.R. (2002) Crystallization of complex between soluble domains of adenylyl cyclase and activated Gsα. Meth. Enz. 345:198-206.
Tesmer, J.J., Sunahara, R., Gilman, A.G., and Sprang, S.R. (1997) Crystal structure of the catalytic domains of adenylyl cyclase in a complex with Gsalpha.GTPγS. Science 278:1907-1916.
Thastrup, O., Cullen, P., Drobak, B., Hanley, M., and Dawson, A.P. (1990) Thapsigargin, a tumor promoter, discharges intracellular Ca^{2+} stores by specific inhibition of the endoplasmic reticulum Ca^{2+}-ATPase. PNAS USA 87:466-470.
Tiruppathi, C., Freichel, M., Vogel, S., Paria, B., Mehta, D., Flockerzi, V., and Malik, A.B. (2002a) Impairment of store-operated Ca^{2+} entry in TRPC4(-/-) mice interferes with increase in lung microvascular permeability. Circ. Res. 91:70-76.
Tiruppathi, C., Minshall, R., Paria, B., Vogel, S., and Malik, A.B. (2002b) Role of Ca^{2+} signaling in the regulation of endothelial permeability. Vascul. Pharmacol. 39:73-85.
Torihashi, S., Fujimoto, T., Trost, T., and Nakayama, S. (2002) Calcium oscillation linked to pacemaking of interstitial cells of Cajal: requirement of calcium influx and localization of TRP4 in caveolae. J. Biol. Chem. 277:9191-9197.
Vennekens, R., Voets, T., Bindels, R., Droogmans, G., and Nilius, B. (2002) Current understanding of mammalian TRP homologues. Cell Calcium 31:53-64.
Walsh, D.A., Perkins, J.P., and Krebs, E.G. (1968) An adenosine 3',5'-monophosphate-dependant protein kinase from rabbit skeletal muscle. J. Biol. Chem. 243:763-765.
Whisnant, R.E., Gilman, A.G., and Dessauer, C.W. (1996) Interaction of the two cytosolic domains of mammalian adenylyl cyclase. PNAS USA 93:621-625.
Wu, S., Moore, T., Brough, G., Whitt, S., Chinkers, M., Li, M., and Stevens, T. (2000) Cyclic nucleotide-gated channels mediate membrane depolarization following activation of store-operated calcium entry in endothelial cells. J. Biol. Chem. 275: 8887-8896.

Wuttke, M.S., Buck, J., and Levin, L.R. (2001) Bicarbonate-regulated soluble adenylyl cyclase. J. Pancreas 2:154-158.

Yahr, T.L., Vallis, A., Hancock, M., Barbieri, J., and Frank, D.W. (1998) ExoY, an adenylate cyclase secreted by Pseudomonas aeruginosa type III system. PNAS USA 95:3899-3904.

Yau, K.W. (1994) Cyclic nucleotide-gated channels: an expanding new family of ion channels. PNAS USA 91:481-483.

Yoshimura, M. and Cooper, D.M. (1992) Cloning and expression of a Ca^{2+}-inhibitable adenylyl cyclase from NCB-20 cells. PNAS USA 89:716-720.

Zaccolo, M. and Pozzan, T. (2002) Discrete microdomains with high concentration of cAMP in stimulated rat neonatal cardiac myocytes. Science 295:711-715.

Zaccolo, M. and Pozzan, T. (2003) cAMP and Ca^{2+} interplay: a matter of oscillation patterns. Trends Neurosci. 26:3-5.

Zhang, G., Liu, Y., Ruoho, A., and Hurley, J.H. (1997) Structure of the adenylyl cyclase catalytic core. Nature 386:47-53.

Zippin, J.H., Chen, Y., Nahirney, P., Kamenetsky, M., Wuttke, M., Fischman, D., Levin, L.R., and Buck, J. (2003) Compartmentalization of bicarbonate-sensitive adenylyl cyclase in distinct signaling microdomains. FASEB J. 17:2-4.

Chapter 6

Signaling and Prolonged Endothelial Activation

Carolyn E. Patterson, Ph.D.[1,2]* and Matthias Clauss, Ph.D.[1]

[1]Indiana University School of Medicine, Indianapolis, IN, USA 46202, and
[2]The Roudebush VA Medical Center, Indianapolis, IN 46202, and

CONTENTS

Introduction

Pathways of Prolonged Signaling Effects
General Principles
The MAPK Cascade
The PI3K/Akt Cascade

Transcription
NFκB
AP-1
HIF-1
PPARs
β-catenin
Other Factors of Growth and Survival

Signaling Responses to Endothelial Effectors
Mechanical Forces
Bioactive Mediators
Bioactive Lipids
Cytokines
Growth Factors
Metabolic Stress
Hypoxia
Oxidants
cAMP/PKA

The Long and the Short of It

Introduction

Chapter 4 has highlighted key signaling pathways involved with acute activation signaling in endothelia, which primarily occur within a few seconds or a few minutes. Although, some downstream enzymes may show sustained activation for as long as 60 minutes, or is some circumstances even for 24 hours. Subsequent to this short-term or sustained signaling, these same signals initiate a chain of events that result in altered protein turnover and altered protein expression and function of cytoskeletal, junctional, regulatory proteins, contributing to ongoing permeability. Altered levels of enzymatic proteins and altered structural, scaffolding, and regulatory proteins thereby influence the function state of endothelia. This shift in expression occurs particularly with sustained activation by the agonist or condition. Simply put, protein expression is altered in two ways: a change in the rate of synthesis and a change in the rate of degradation. From there on out, there is nothing simple about it. Although we have deciphered some of the major pathways responsible for the acute or short-term signaling, less is known about signaling of the prolonged responses that account for endothelial dysfunction, especially in the lung. Moreover, these signals activate altered endothelial participation in coagulation and inflammation. When adhesion molecules are expressed on the luminal surface and activated leukocytes become involved, the initial endothelial activation is exacerbated and prolonged. Inflammatory mediators (TNFα, IL-1, IL-6, PAF, GM-CSF, C5a, ICAM, chemokines, VEGF, & ROS) are both produced by and act on the pulmonary endothelium in ARDS and, thus, understanding the signaling involved in this interaction is critical to gaining control over the pathophysiology of acute lung injury (Bhatia and Moochhala, 2004). This chapter will highlight what is know regarding some of the chief mediators and processes and the relationship between signaling and prolonged endothelial dysfunction. We will explore how endothelial participation in initiation and response to acute exacerbations of inflammation, which progress in minutes to hours to several days, is related to protein turnover, sustained activation, and even cell death.

Pathways of Prolonged Signaling Effects

General Principles

There are a number of levels at which signaling has prolonged effects on endothelial barrier properties: 1) Enzymes of pathways, normally associated with translation and transcription, have some effects that do not actually require altered protein synthesis and some of these signals may be sustained much longer than the transient activation of enzymes, such as MLCK. 2) Synthesis or degradation of cytoskeletal, cell-cell adhesion, cell-matrix adhesion, and regulatory proteins with direct participation in cell shape and barrier function have a prolonged influence over barrier integrity. 3) Altered levels of endothelial proteins, such as chemokines or leukocyte adhesion molecules, contribute to the inflammatory process and secondary, but important, effects on prolonged barrier dysfunction. 4) Some effects of these pathways of prolonged signaling are activation of cell propagation and angiogenesis. Although, angiogenesis does not normally play a major role in acute lung injury, these signals can have overlapping influence on barrier properties and endothelial proliferation can participate in needed repair possesses. 5) With certain activating mediators and conditions, the pathways can lead to cell death, especially when not prevented by survival signals and their imbalance can contribute to barrier dysfunction and acute lung injury. The contribution of endothelial cytotoxicity to ARDS is indicated by several lines of evidence as further described in Chapter 12, with up to 50% ablation of the pulmonary capillaries (Tomashefski *et al.*, 1983; Pittet *et al.*, 1997; Hamacher *et al.*, 2002). For example, BAL fluid from patients with high-risk, early, and late ARDS caused cytotoxicity in human lung microvascular endothelial cells, but not fibroblasts (Hamacher *et al.*, 2002). In these studies, the major agents of cell injury were TNFα and angiostatin, which is derived from plasminogen by cleavage with neutrophil elastase or MMP. Finally, trophic signaling of survival programs is important for maintenance of endothelial barrier integrity and restoration after activation.

In other conditions of severe dysfunction or in disease conditions lasting many days to years, chronic alterations in signaling may lead to chronic hyperpermeability and chronic pulmonary disease, involving interplay between endothelium and the other lung tissues. Such remodeling pathophysiology of chronic hyperpermeability, as in pulmonary hypertension, in both pulmonary and bronchial circulation will not be discussed here but will be considered in Chapter 14 in the pathophysiology section. Furthermore, we will not attempt to describe the prolonged signaling involved in processes of embryologic and fetal development, tumorogenesis, or atherosclerosis, although there is certainly some convergence of signaling. Rather, we will focus on those changes associated with diseases of relatively rapid onset, such as sepsis and acute lung injury.

As in acute signaling and issuing from acute signaling, signaling of prolonged responses involves phosphorylation/dephosphorylation, binding/release of co-activators, lipid/receptor interactions, all of which are highly organized by anchored complexes and scaffolding proteins. Scaffolding increases process catalytic-efficiency, decreases cross talk (or increases specificity of signal to target), and in other cases promotes signaling propagation along several pathways in parallel. In this respect cytoskeletal organization and anchoring of receptors and signaling-associated junction/adhesion proteins (whether luminal, abluminal, or lateral) to the cytoskeleton influences signaling. For example, focal adhesion-integrins couple synergistically with growth factor receptors to promote proliferation, while adherens junctions couple with growth factor receptors to inhibit their signaling to provide contact inhibition and stabilization of the confluent endothelium (Dejana, 2004). Receptor association with adhesion complexes provides both specificity of signaling and synergy. In prolonged signaling, targets generally include processes that increase protein synthesis or degradation and/or increase or suppress gene expression.

Protein synthesis may be altered acutely without any change in mRNA or translation may increase over time as new message is transcribed. Primarily this chapter will deal with altered transcription, as that is where most work in altered protein expression in response to endothelial activation is focused, however the controls over translation and protein degradation processes are tightly controlled and important to the balance of expression. Only a small percentage of proteins in eukaryotic cells have very short half-life of less than 10 minutes, whereas 99% of cellular proteins have an average half-life of around 2 days and are primarily catabolized after lysosomal uptake (Mortimore *et al.*, 1989). The rate of degradation of specific proteins can be increased as they are dislodged from their position in a complex by activation. Apoptotic activation releases a processes of highly active protein degradation, which contributes to cell shrinkage and death.

Particular small G-proteins (i.e. IF-2, EF-Tu, EF-G, and RF-3) are important regulators of activation, elongation, and protein release in synthesis. There are also activating proteins (e.g. IF4E), which assist in linking mRNA to the ribosome. These proteins are regulated by phosphorylation and inhibitory binding proteins that are subject to phosphorylation regulation themselves (Knauf *et al.*, 2001). Both MAPKs and PI3K catalyze these phosphorylations, resulting in stimulation or inhibition depending on the particular pathways initiated and followed (Wang *et al.*, 1998; Knauf *et al.*, 2001).

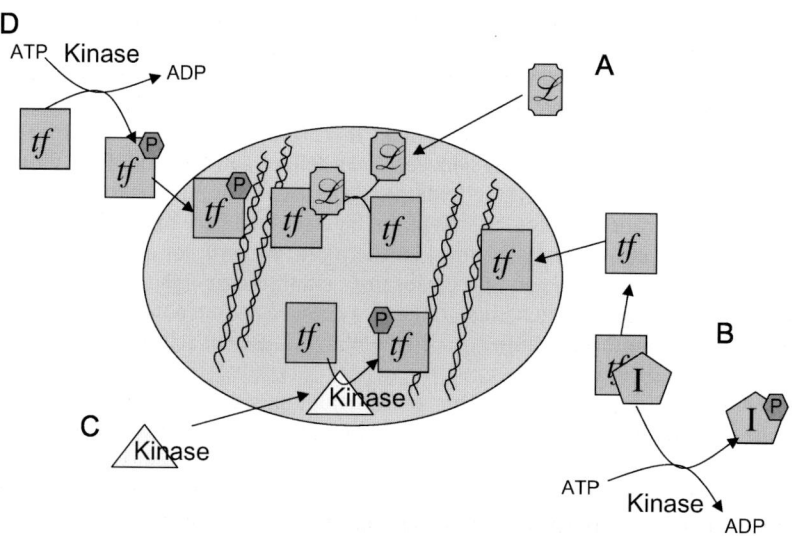

Figure 1. Modes of signaling transcriptional regulation. Regulation of transcription is understandably complex and there are many players; we will present four basic examples of signaling to activate transcription factors that regulate gene expression. Following clockwise:

A. The first mechanism discovered for signaling altered expression of mRNA from genes was the ability of steroid hormones to enter the cell, enter the nucleus, and bind to factors that resulted in transcription of certain genes. Other cell generated lipid second messengers share this mode of initiation, but target specific transcription factors.

B. The second mode involves the release of a transcription factor from the cytoplasm, permitting its translocation into the nucleus. Generally, the transcription factor is held in the cytoplasm by binding to inhibitory factors, whose binding affinities are altered by phosphorylation.

C. Alternatively, the activating kinase may enter the nucleus itself, where it phosphorylates the transcription factors causing activation.

D. Finally, a kinase may directly phosphorylate a transcription factor in the cytoplasm/plasma membrane, which then enters the nucleus and initiates transcription.

Increased mRNA does not guarantee increased expression of a specific protein, but more often than not, this does occur and is the main mechanism to influence the relative abundance of particular proteins or sets of proteins. Thus, initiation (or suppression) of transcription is a topic of great interest in endothelial function/dysfunction. Although transcription is a highly complex process with multiple interactions, there are four general ways that mediators signal altered transcription, as depicted in Figure 1. The first mechanism simply shows a lipid translocating to the nucleus, where it activates a transcription factor of the nuclear hormone receptor family. An example of this is PPARγ, which is activated by certain lipids, such as prostaglandin J2. The second mechanism is illustrated by the activation of a transcription factor that is commonly altered in inflammation, namely NFκB. Normally, the NFκB is retained in the cytoplasm, bound to IκB (inhibitor of NFκB). When IκB is phosphorylated, it is released from NFκB and is subject to degradation. NFκB is then is targeted to the nucleus where it activates gene transcription programs. The third basic scheme shows the kinase itself entering the nucleus, where it activates transcription factors by phosphorylation. This is illustrated by specific MAPKs, which translocate and phosphorylate factors, like c-jun. Lastly, some non-nuclear transcription factors, such as STAT, are phosphorylation targets and this allows them to be taken up into the nucleus. In some cases, mitogens also induce expression of genes, such as the immediate-early genes encoding transcription factors, without the need for de novo protein synthesis. Precisely how short-term signaling is related to these activation processes will be considered in the following section.

The MAPK Cascade

Mitogen activated protein kinases (MAPKs) as the name indicates are activated by mediators, which were originally recognized for their ability to induced mitosis and cell proliferation, and possess the ability to catalyze phosphorylation of other proteins of direct relevance to endothelial barrier function. The pathway signaling may be transient (seconds to a few minutes), intermediate (minutes to an hour), or sustained (up to 24 hours) and the resultant effects due to altered transcription and translation can be long sustained. Generally, these are serine/threonine phosphorylations, but some members have dual tyrosine phosphorylation activity. The cascade involves a sequence of kinases phosphorylating another kinase and yet another kinase, thereby amplifying the original signal. In its basic form this can be represented by a generalized series: MAPKKK ⇒ MAPKK⇒ MAPK, as shown in Figure 2. These kinases may be activated in several ways by various types of receptors as depicted in the figure. The final MAPK members may induce phosphorylations that alter transcription events. There are numerous specific members of the cascade, with only a few major ones shown in the figure.

G-protein coupled receptors, as discussed in Chapter 4, evoke several signaling effects that result in activation of one arm of the MAPK cascade. Agonists, like thrombin, activate membrane-receptor bound G-proteins (particularly G12/G13), which activated GEF and GAP regulatory proteins that in turn activate Ras. Ras activates Raf (a MAP kinase kinase kinase), which leads to phoshorylation of MEK (a MAP kinase kinase), which finally activates ERK 1/2. How this relates to transcription will be explained in the following section on transcription factors. Alternatively, receptor activation results in PKC activation, which can activate Raf. Finally, receptor activation releases the βγ dimer, which may stimulate Ras by a PI3K-dependent mechanism or by a Shc, GRB, SOS mechanism. Another type of receptor is the tyosine kinase receptor, common for growth factors, like VEGF. Ligand binding of these receptors results in receptor dimerization, auto-tyrosine-phosphorylation, recruitment of Shc and Grb, activation of SOS, which activates Ras and so forth leading down to ERK 1/2 (see Figure 2). Integrin complexes are similar to the tyrosine receptors in leading to tyrosine phsophorylation; but, since they have no catalytic capacity *per se,* this is due to associated tyrosine kinases, such as src. The integrin-associated kinases phosphorylate and activate PKCs, that are also located in the complex. PKC is then responsible for Raf activation as with the G-protein-coupled receptor. Thus far, all three receptor types have led to ERK activation, but there are two other primary arms of the MAPK cascade (and other minor specific arms as well, not discussed). These other two main arms are primarily activated by cytokines and the two pathways lead to either activation of JNK or activation of p38 MAPK, which are therefore known as stress-activated kinases (SAPKs). One effect of p38 MAPK activation is phosphorylation of another kinase, MAPK-AP-2, which activates heat shock protein-27 (HSP27), which then regulates uncapping proteins, resulting in cytoskeletal remodeling (Huot et al., 1997; Hoefen and Berk, 2002). While this is an acute MAPK effect, there is substantial evidence for activation of survival programs downstream from ERKs and BMK1 (big MAPK1= ERK5), which mediates activation of the MEF2 transcription factor that was recently found to be important for endothelial survival (Olson, 2004). Activated MAPKs phosphorylate and regulate both effector proteins and transcription factors, depending on the receptor, cell phenotype, co-activation of other receptors, and compartmentation.

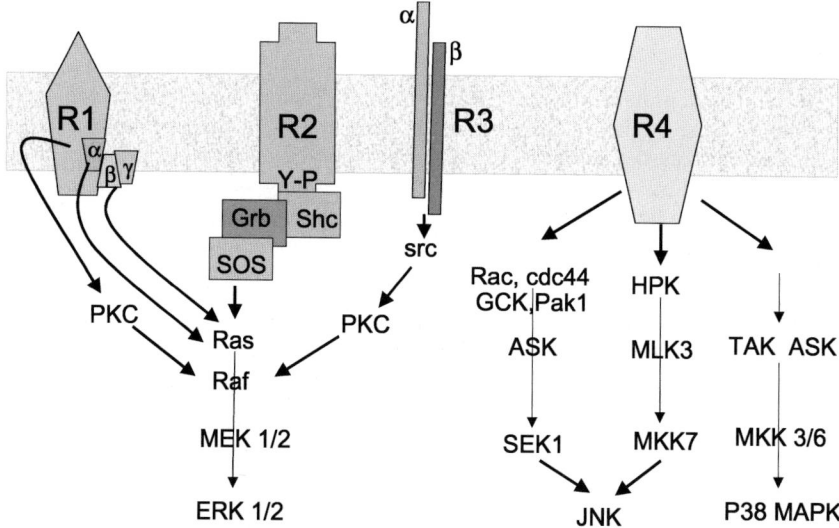

Figure 2. Activation of small GTPases and the MAPK cascade. The MAPK cascade is a primary signaling mechanism for both immediate or prolonged regulation of cell function by phosphorylation of target proteins and long-term regulation, as it activates a number of transcription factors resulting in production of certain new proteins and decreases in others. The cascade is highly complex with many members and interactions, but at its simplest is composed of a G-protein activator, kinases (MAPKKK), which phosphorylate yet other kinases (MAPKK), which phosphorylate the effector kinases (MAPK). There are three basic arms: the ERK path, the JNK path, and the p38 path. Excitation by different kinds of membrane signals is represented. **R1** represents G-protein coupled receptors. In addition to the other receptor-mediated signaling events, such as PLC activation, these receptors have two main means of activating the MAPK cascade: by activation of the small GTPase, Ras, by the α or the βγ G-proteins coupled to the receptor and by PKC activation of MAPKKKs, such as Raf. MEK 1 and 2 are major MAPKKs down from Raf and ERK 1 and 2 are major MAPKs downstream from MEK. **R2** represents tyrosine kinase receptors, which recruit cell components upon activation that signal down to Ras and activate the MAPK cascade. In addition to activating these MAPK pathways, they activate other systems, such as PLCγ and lipid hydrolysis or PI_3 kinase and Akt activation. The third membrane structure, **R3**, represents integrin complexes that serve as quasi-receptors for both external ligands, such as matrix proteins, and mechanical stimuli. They have no

intrinsic kinase activity, but are variably coupled to a number of kinase, such as src, that may be activated or inactivated by stress or by ligand binding. Ig superfamily family adhesion molecules and adhesion complexes, association with non-receptor tyrosine kinases or phosphatases, may participate in activation or deactivation of the MAPK cascade in similar ways. **R4** represents an assortment of other receptors for agonists, like cytokines, with varying actions, but also a common activation of the MAPK cascade. ERK1/2, JNK, and p38 MAPK are the major effector kinases, respectively, in the three basic arms of the cascade, with downstream effects on transcription factors.

The PI3K/Akt Cascade

Receptor-mediated activation of phosphatidyl inositol-3 kinase (PI3K) results in downstream activation of Akt. PI3K generation of PI-3,4P and PI-3,4,5,P phospholipids are primary intermediate messengers that can activate a number of downstream targets, such as tyrosine kinases, GTPase activation protein for the small GTPases, certain serine/threonine kinases like the atypical PKC isoforms, and Akt. The phospholipids assist in translocating Akt from the cytoplasm to the inner plasma membrane to association with kinases that phosphorylate and activate Akt. Three genes encode Akt, also termed PKB, the Akt1 form being expressed abundantly in lung endothelium (Shiojima and Walsh, 2002). PI3K/Akt has been shown to activate NFκB by three different pathways: 1) Akt activation of IKK (Romashkova and Makarov, 1999), 2) Akt activation of a specific MAPK, cot, which activates IKK (Kane *et al*, 2002), and IKK-independent Akt activation of calpain, which degrades IκBα (Pianetti *et al.*, 2001). Akt activation enhances cellular glucose uptake, protein synthesis, and NO production (Shiojima and Walsh, 2002; Minshall *et al.*, 2003). Interestingly, statins promote a protective Akt activation in a PI3K-dependent manner, but how this is brought about is not known (Kureishi *et al.*, 2000; Wolfram *et al.*, 2003).

Akt directs a major survival-signaling pathway, with multiple sites of action (Datta *et al.*, 1999). First, Akt activation of IKK and subsequent activation of NFκB contributes to transcription of survival genes. Second, Akt-mediated phosphorylation affects cytosolic sequestration of several transcription factors, such as the forkhead factors (FKHR/FOX01, FKHRL1, and AFX), which would decrease FKHR-directed transcription of apoptotic factors (Skurk *et al.*, 2004). Bad and Bax are members of the Bcl-2 protein family with apoptotic effects, whereas Bcl-2 and Bcl-X are anti-apoptotic. Suppression of Bad, Bax, and Fas ligand are major mechanisms for the endothelial protective effects of angiopoietin and VEGF (Abid *et al.*, 2004; Daly *et al.*, 2004). Third, Akt increases expression of the non-catalytic endogenous homologue of caspase 8, FLIP, which blocks caspase 8 activation (Skurk *et al.*, 2004). Interestingly, caspase 8 cleaves MLCK resulting in unregulated kinase active fragments (Petrache *et al.*, 2003b). Fourth, Akt promotes expression of Bcl-2 and Bcl-X. Trophic signaling of Bcl-2 and Bcl-X expression is important to prevent Bad and Bax activation. If survival signaling is removed, apoptosis proceeds without new activation by transcription-dependent and -independent mechanisms. Left unchecked, Bax induces mitochondrial cytC release, which nucleates with scaffolding/activator proteins and activates caspase 9. Fifth, Akt phosphorylates caspase 9, further blocking its activation. Sixth, Akt-phosphorylation of Bad causes it to complex with 14-3-3 proteins, rendering it inactive. PKA also suppresses Bad function by phosphorylation at a site different from the Akt site. Seventh, Akt phosphorylates and inactivates glycogen synthase kinase-3 (GSK-3), preventing its apoptotic signaling. Finally, Akt can translocate to the nucleus, where it influences transcription factors and expression of genes for pro- and anti- apoptotic proteins, including Bcl-2 members and caspases. For instance, Akt phosphorylation of CREB enhances its activity. Of course, the downside to uncontrolled Akt activation is oncogenesis.

Transcription

Some links between short-term signaling, targeted activation of transcription factors, and cellular responses identified in endothelial cells are summarized in Table 1. There are hundreds of transcription factors and many play a functional role in endothelium; thus, this list is not comprehensive as there is much yet to be learned.

NFκB

Many stress-related mediators activate NFκB with the major response being upregulation of genes producing cytokines, chemokines, and leukocyte adhesion molecules. Thus, inflammation is largely mediated by NFκB. But NFκB is also a survival transcription factor and has been shown to counter TNF-induced apoptosis. Pathway "B" shown in Figure 1 is highly applicable to activation of the transcription factor NFκB. As explained above, upstream activation of the PI3K/Akt pathway by tyrosine-kinase receptors results in activation of the kinase IKKα/β, which phosphorylates IκB (NFκB –inhibitory binding protein). This releases the IκB from NFκB freeing it for translocation into the nucleus to interact with its target genes. Alternatively, receptors coupled to Gαq or release of certain βγ units result in activation of NFκB. Most activators of the endothelial cell initiate activation of this transcription factor. Important downstream effects include upregulation of message for expression of ICAM. The retinoic acid receptor (RORα) is a nuclear receptor with inhibitory influence over NFκB translocation to the nucleus in human endothelium (Migita *et al.*, 2004).

AP-1

AP-1 comprises a collection of transcription factors, including Jun, Fos, and ATF, as they bind a common DNA site. Gene expression for TGFβ1, cytokines, and leukocyte adhesion molecules is signaled by AP-1 activation. Depending on other factors and conditions, AP-1 can either activate proliferation or apoptosis. AP-1 is activated by most endothelial activators downstream from the SAPKs, p38 and JNK and is pivotal for induction of genes associated with endothelial activation, such as adhesion molecules and tissue factor (Moll *et al.*, 1995). However, it is also required for vascular network formation during placenta development and junB gene deficient mice die from placental defects (Schorpp-Kistner *et al.*, 1999).

Table 1: This table summarizes some pathways in activation, prolonged barrier dysfunction, the inflammatory process, growth, and survival vs. apoptosis in endothelial cells as discussed and referenced in the text. Many gaps exist in knowledge of specific coupling. Unless specifically noted, listing indicates activation or upregulation. Abbreviations: ASK1 - apoptosis signal-related kinse; ATF2 – EDG-responsive, CAM – Leukocyte adhesion molecules (ICAM, VCAM, E-selectin); Cre activating transcription factor; CREB- cAMP response elemant; Egr - early growth response factor; FKHR –forkhead transcription factor (FOX01); JAK - Janus kinases-signal transducers; PPARg – peroxisome proliferator-activated receptor; SEK1 - stress-activated protein kinase ERK kinase; SRE- serum response element; STAT - signal transducers and activators of transcription; TCF- T cell transcription factor; TRAF2- TNF receptor associated factor 2; TRE- TPA response element.

Table 1.
Transcription Signals, Factors, and Endothelial Responses

Signals	Transcription Factor	Transcription Response
Gα13→MEKK1→ MEK1/2→ ERK 1/2 (= p42/p44 MAPK)	TCF/SRE	Egr-1 transcription factor, growth/survival
Grb2→SOS→Ras→Raf1→ MEK→ERK	TCF/SRE/Elk-1	Egr-1 transcription factor
β-catenin	LEF/TCF family/SRE	c-myc, cyclin D1 → cell cycle, growth and anti-apoptosis
Gα 12/13 → RhoGEF → MEK→ERK	SRF/SRE	"
"	Egr-1	PDGF, Tissue factor, VEGFR1, TGFβ, TNFα, MMPs
Raf1----ERK1/2	AP-2/Sp1, HIF1	VEGF, VEGFR2, iNOS, glycolysis enzymes
Gβγ →Akt → IKKβ → IκB release	P65 NFκB	ICAM-1, fractalkine, iNOS, cytokines
Gαq → Akt → IKKβ → IκB release	P65 NFκB	"
PKCδ → → → p38 MAPK	P65 NFκB	"
TRADD→TRAF2→NIK→ IKKβ → IκB release	P65 NFκB	"
PI3K → PKCδ → → → p38 MAPK	P65 NFκB	VCAM-1
FAK/ILK/shc → PI3K→ Akt → →	FKHR inhibition	survival
PI3K → PKCζ	GATA-2	VCAM-1
PI3K → PKCδ → → → p38 MAPK	ATF-2 (CRE-BP1)	E-selectin
dbpB		PDGF, EPCR, VEGFR2, thrombomodulin
PKA	GATA and Sp-1	ENOS, steroidogenesis, differentiation
→ Rac → GCK → MEKK1 → SEK1 → JNK/SAPK	TRE	Stress response, growth arrest, apoptosis
CrKy-P → C3G Ras like → MLK → DLK → JNK	c-Jun, AP-1	Proliferation and apoptosis depending on other factors
TRADD → TRAF2 → TAK1 → MKK4/SEK1 →	AP-1	Cytokine upregulation, CAM expression
ERK1/2 → BMK1 → decrease JNK	decreased cJun activation	Suppress apoptosis
GTPase? → Ras like → ASK1 → MKK4/SEK1 → decrease JNK	"	"
GTPase? → Rac/cdc42 → MEKK1 → MKK4/SEK1 → decrease JNK	"	"
TRADD → TRAF2 → TAK1 → MKK4/SEK1 → decrease JNK	"	"
JAKs	STATs	cell growth, suppression of apoptosis, and cell motility
oxidized-linoleic acid, PGJ2, ?	PPARγ	Counters upregulation of inflammatory responses such as ICAM expression
PKA	CREB	

HIF-1

HIF-1 is an important cell sensor for oxygenation level and is a heterodimer composed of HIF-1β, which is constitutively expressed, and HIF-1α, which is variably activated and expressed in relation to the oxygen level. With adequate oxygen, HIF-1α is hydroxylated by enzymes with a requirement for oxygen. The hydroxylation targets the HIF-1α for rapid proteosome degradation. Thus, the absence of oxygen (hypoxia) stabilizes the factor. The complex translocates to the nucleus and in combination with co-activators initiates gene programs to improve cell survival in conditions of hypoxia, such as increase uptake of glucose and glycolytic enzymes. HIF-1 increases VEGF expression (Berra *et al.*, 2000) and exerts an anti-apoptotic effect in anoxia-stressed endothelial cells (Yu *et al.*, 2004). Transgenic mice with overexpression of HIF-1α resulted in an increased number of vessels without leakage that is also seen in VEGF overexpressing mice (Elson *et al.*, 2001). Constiutive expression in cultured endothelial cells revealed that HIF-1α-induced angiopoietin-4 expression accounted for the increased proliferation and tube formation and for the barrier integrity of the new endothelium (Yamakawa *et al.*, 2003).

PPARs

The peroxisome proliferator-activated receptors, PPARγ and PPARα, are nuclear hormone family receptors, expressed in many cell types and with overall beneficial effects for endothelial cell function. Natural ligands for PPARγ are not fully known but may include oxidized linoleic acid and prostaglandin J_2 (Plutzky, 2001). PPARγ activation with a pharmacologic agonist, e.g. troglitazone, increased early repair and wound closure, but inhibited late neointima formation after balloon angioplasty (Hannan *et al.*, 2003). In cultured endothelial cells, troglitazone inhibited proliferation but prevented apoptosis (Hannan *et al.*, 2003). In endothelium PPARγ activation also decreased expression of endothelin-1 and cytokine-induced expression of chemokines (Plutzky, 2001). These effects not only benefit endothelia, but also the systems and cells regulated by endothelium. Of relevance to ARDS, activation of both PPARγ and PPARα decreased TNF- and IL-1-induced expression of ICAM-1, VCAM-1, and E-selectin, but results are variable (Jackson *et al.*, 1999; Plutzsky, 2001). Importantly, both PPARγ and PPARα increased expression of Cu^{2+}, Zn^{2+} SOD and decrease expression of components of NADPH oxidase, yielding an overall protection from excessive superoxide levels (Inoue *et al.*, 2001). Interestingly, activated PPARγ–upregulation of VEGF expression and subsequent Akt activation and eNOS phosphorylation is thought to be responsible for the PPARγ-induced increase in eNOS activation in the absence of increased eNOS expression (Cho *et al.*, 2004). While PPARγ activation enhanced VEGFR2 expression, mediated in part by physical interaction with Sp1 transcription factor (Sassa *et al.*, 2004), PPARα activation suppressed Sp-1 signaled VEGFR2 expression (Meissner *et al.*, 2004).

β-catenin

When β-catenin is released from its complex with cadherin into the cytoplasm, due to endothelial activation and dissolution of the adherens junction, it becomes available to translocate to the nucleus. Cytosolic processing is also under tight control of GSK-3, which enhances its stability and signaling in other cells and in endothelium (Vincent *et al.*, 2004). In the nucleus, it binds transcription factors of the lymphoid enhancing factor (LEF/TCF) family, promoters of the cyclin D1 and c-myc, thus augmenting progression of the endothelial cell into proliferation (Shtutman *et al.*, 1999). In complex with cadherin, β-catenin also functions to assist VEGF-survival signaling, as described below in the discussion of growth factors. Further details of β-catenin signaling are contained in Chapter 9. Additionally, other intercellular junction associated proteins, such as p120, γ-catenin, and several tight junction associated proteins have been reported to have transcription effects, but their roles in normal endothelium are unexplored (Dejana, 2004).

Other Factors of Growth and Survival

Recently, Akt inhibition of the forkhead transcription factor (FKHD/FOX01), which would decrease FKHD expression of apoptotic factors, was shown to be a major mechanism for angiopoietin protective effects (Daly *et al.*, 2004). Various genes associated with endothelial growth and survival are promoted by the SRE factor (An *et al.*, 1998). CREB is not only responsive to cAMP but is also activated by peptide hormones and growth factors via ERK, PKC, Ca^{++}/CaM kinases, p38MAPK, and as mentioned above Akt (e.g. TNF⇒p38⇒MSK1⇒CREB), with activation eliciting transcription of both survival and apoptosis program genes (Gustin *et al.*, 2004). A variety of stresses, including elevated temperature, ischemia, LPS, ox LDL, hypoxia, pressure overload, and decreased GSH can activate heat shock factors (HSFs), transcription factors that increase expression of a number of heat stress proteins (HSPs), cytokines, and SOD (Christians *et al.*, 2002). Both JNK and PKCδ can upregulate HSF1. Some HSPs are expressed constitutively. HSPs have multiple functions, which benefit the endothelial cell, such as their involvement in protein folding and degradation (chaperone functions) and their ability to help maintain metabolic and structural integrity of the cell during stress. For example, HSP90 participates in eNOS activation (Su *et al.*, 2002). Several HSPs prevent apoptosis by binding and inactivating various players (e.g. HSP60 binds Bax) (Christians *et al.*, 2002; Knowlton and Gupta, 2003). In extreme conditions, HSPs actually aids in necrosis and apoptosis (Pirkkala *et al.*, 2001).

Signaling Responses to Endothelial Effectors

Mechanical Forces

Endothelial cells respond to mild mechanical stimuli with improved survival, but high levels of stress can make the cells vulnerable to dysfunction and abnormal response to other stimuli. In systemic circulation areas of high or turbulent stress and areas of very low stress are both vulnerable sites for development of atherosclerosis. Although, not prone to atherosclerotic lesions, pulmonary endothelium also exhibits this response pattern. For instance, mild cyclic stretch conditioning rendered pulmonary endothelium more resistant to thrombin activation, even after removal of the stimuli and replating for 16 hrs compared to cells maintained in static culture; whereas, high stretch increased responsiveness (Birukov *et al.*, 2003). As might be expected this memory of conditioning was related to altered gene expression during the conditioning phase. Several junctional molecules are proven to complex with various signaling molecules and as such comprise mechanotransducers, capable of responding to force stimulation and to the presence of external ligands (Juliano, 2002). *In vivo*, there is a strong correlation between the degree of shear stress and the steady state presence of stress fibers coupled to focal adhesion/integrin plaques.

When PECAM-1 of adjacent endothelial cells are bound in a homotypic complex, their cytoplasmic tails with their associated tyrosine based inhibitory motifs are able to recruit and activate tyrosine phosphatases. These phosphatases exert anti-apoptotic effects via inhibition of tyrosine kinase-mediated transduction of Bax-apoptotic factors (Newman and Newman, 2003). Direct mechanical perturbation of PECAM, through magnetic bead-coupled antibodies, resulted in PECAM-1 tyrosine phosphorylation, SHP-2 binding, and downstream ERK activation (Osawa *et al.*, 2003). Likewise, physiologic fluid shear stress and hyperosmotic shock caused PECAM tyrosine phosphorylation and ERK signaling (Osawa *et al.*, 2003). These rapid signaling effects of mild stimulation are coupled to long-range survival by transcription dependent and independent means. Similarly, mechanical perturbation and ligand binding can elicit anti-apoptotic signaling through integrins. Integrins are usually coupled to non-receptor kinases, such as c-src, FAK, PKC, and PI3K, making them a quasi receptor kinase. Both mechanical disturbance and integrin ligand binding led to kinase activation and signaling (Katsumi *et al.*, 2004). The mechanosensor signaling can be transmitted in several ways: through a caveolin, fyn, Shc, SOS, Ras, ERK pathway, through a c-src, FAK, Ras, ERK pathway, through a FAK, PI3K, IKK, NFκB pathway, and via a PKC, ERK pathway dependent on the cellular context (Juliano, 2002; Ruegg and Mariotti, 2003). Conversely, mechanical stress can trigger apoptosis through caspase 8/Bax activation - the difference likely related to timing and degree of stress (Ruegg and Mariotti, 2003). Integrin-based activation of Rho, Rac, and CDC42 signals from the focal adhesion complex to cytoskeletal reorganization. Moreover, integrin anchoring affects functional signaling of other receptors, including tyrosine kinase type receptors, the link between G-protein-coupled receptors and the MAPK cascade, and cytokine receptors (Juliano, 2002). Further details of integrin-based signaling and function will be related in Chapter 8. Mechanical disturbance has been shown in numerous studies to alter membrane channel activities, which may contribute to activation of these intracellular pathways (Fisher *et al.*, 2001). And finally, signaling is bi-directional with the small GTPases signaling to the integrin/kinase complex, controlling assembly and maturation.

Physiologic shear stress activates and increases eNOS expression; and laminar flow prevents apoptosis by activation of Akt, preventing entry of the cell into cell cycling (Berk *et al.*, 2001; Shiojima *et al.*, 2002; Fleming and Busse, 2003). Shear stress also induced eNOS upregulation by a c-Src, ERK 1/2, IKK path (Davis *et al.*, 2004). Laminar shear stress in endothelial cells induced expression of iNOS by an oxidant, IKK, NFκB mechanism (Ozawa *et al.*, 2004). Importantly, physiologic laminar flow blocked endothelial activation by TNF (Berk *et al.*, 2001). Apparently, this is mediated via ERK 1/2 MAPK-mediated inhibition of TNF-JNK activation. This effect was independent from NO, cGMP, and cAMP, but due to BMK1. Similarly, shear stress activation of the JNK pathway was shown to be transient, while sustained laminar flow resulted in its downregulation (Fisher *et al.*, 2001). These reports are consistent with the association of laminar flow and improved survival, as JNK is generally an apoptotic pathway. Shear stress also induces coupling of VEGFR2 to adherens junctions, as will be further discussed in Chapter 9. Coupling of cadherin and VEGFR2 indicates they can coordinately act as mechanochemical transducers. This coupling response to physiologic laminar flow can actually substitute for VEGF activation of the VEGFR2-mediated Akt-survival signaling (Shay-Salit *et al.*, 2002). Similarly, the adherens junction-VEGR2 coupling modulates the VEGF proliferation signaling by attenuation of the tyrosine kinase activity via association of tyrosine phosphatases with the adherens junction complex, while enhancing VEGF survival signaling via PI3K-Akt activation (Carmeliet *et al*, 1999; Lamugnani *et al*, 2003). These interactions can partly account for the enhancement of endothelial survival by laminar flow. Conversely, using laser capture to compare expression in quiescent and wound edge motile endothelium under static conditions and shear stress, laminar flow stress enhanced wound closure, presumably related to enhanced endothelial proliferation and motility, and increase the relative expression of p120-catenin, VE-cadherin, and the transcription factor, Kaiso, at the wound edge (Kondapalli *et al.*, 2004).

Bioactive Mediators

As explained in Chapter 4, bioactive peptide/protein mediators evoke multiple signaling effects, including activation of MAPK-, Akt-, and other- pathways that consequently activated transcription. For example, thrombin induces endothelial proliferation related to induction of both VEGF and angiopoietin (Herbert *et al.*, 1994; Tsopanoglou and Maragoudakis, 1999; Wang *et al.*, 2002; Jin *et al.*, 2003). Furthermore, its receptors, PAR1 and PAR2, are upregulated in new tumor vessels (Jin *et al.*, 2003). One study analyzed the effects of 18 hrs thrombin exposure on expression of ~ 9,000 genes (Minami *et al.*, 2004). They found major consistent upregulation of 74 genes and downregulation of 20 others. Some of the pathologically important increases included: VEGFR2, tissue factor, IL-6, IL-8, ICAM-1, VCAM-1, MCP-1, COX-2, endothelin-1, the chemokine-fractalkine, the early growth response factor (EGF-1), and PDGF. One important gene downregulated was for eNOS.

How are early thrombin-signals coupled to these effects? In one mechanism, the receptor coupled -Gα13 leads to activation of MEK1/2, which leads to activation of the SRF transcription factor (Minami *et al.*, 2004). In another, Gαq and Gβ,γ lead to activation of NFκB, a key factor as discussed above. As shown in Table 1, activation of PI3K/Akt, in this case by thrombin, results in subsequent activation of PKCζ and PKCδ, that respectively activate and induce nuclear translocation of GATA-2 and NFκB. This pathway is thought to be specific for endothelial cells as thrombin-activation of smooth muscle results in a different pattern of transcription. Moreover, even in endothelium different activating mediators may yield similar, but not identical patterns of gene expression, related to the differences in their particular short term signaling patterns. Finally, thrombin also appears to have some protective effects on endothelial survival, possible via induction of the anti-apoptotic Bcl2 related proteins (Minami *et al.*, 2004).

There is considerable cross reactivity between various receptors and pathways. One example of this is signaling by angiotensin II, which will serve to illustrate similar crossovers with other receptors. The primary angiotensin II receptor, AT-1, is a usual G-coupled receptor with 7 membrane spans, a luminal ligand binding tail, and a cytoplasmic signaling tail (de Gasparo *et al.*, 2000). Through typical Gq coupling it activates PLC leading to Ca^{++} and PKC-β activation. Other membrane effector systems activated include PLD, PLA$_2$, adenylyl cyclase, and several ion channels. However signaling through AT1 is much more complex as shown in Figure 3 (Yin *et al.*, 2003). First, PKC can cross react with the MAPK cascade by activating ERK, so MAPK phosphorylations may result. Next Ca^{++}/PKC signaling can activate non-receptor based pp60^{src}- tyrosine kinases, resulting in autophosphorylation and in FAK tyrosine phosphorylation and activation. Receptor activation also results in nearly immediate tyrosine phosphorylation of PLCγ, similar to the growth factor receptors, but this is not due to intrinsic tyrosine kinase activity of the AT1 receptor *per se*, but due to cross activation of other non-receptor tyrosine kinases. Yet AT1 facilitates this by binding the effector on its cytoplasmic tail. Ca^{++}/PKC also is involved with direct cross talk with the epithelial growth factor receptor (EGFR) eliciting small GTPase Ras and Erk signaling. As well, Ca^{++}/PKC is involved in activation of Janus-activated kinase (JAK)2 in cooperation with docking on the cytoplasmic tail of the AT1 receptor in juxtaposition with STAT1. This kind of cross signaling illustrated in Figure 3 is common for many of the receptors. For instance, bradykinin activates the JAK/STAT pathway in vascular endothelial cells with localization of the JAK/STAT signaling in the plasmalemmal caveolae (Ju *et al.*, 2000). Ras activation of it effector kinase, Raf, which phosphorylates MEK and activates the MAPK cascade is a common link in many cell types between various receptors, small GTPases, and the MAPK signaling (Takai *et al.*, 2001, Lehoux and Tedgui, 2003). Importantly, these acute-signaling effects translate to altered transcription and cell migration.

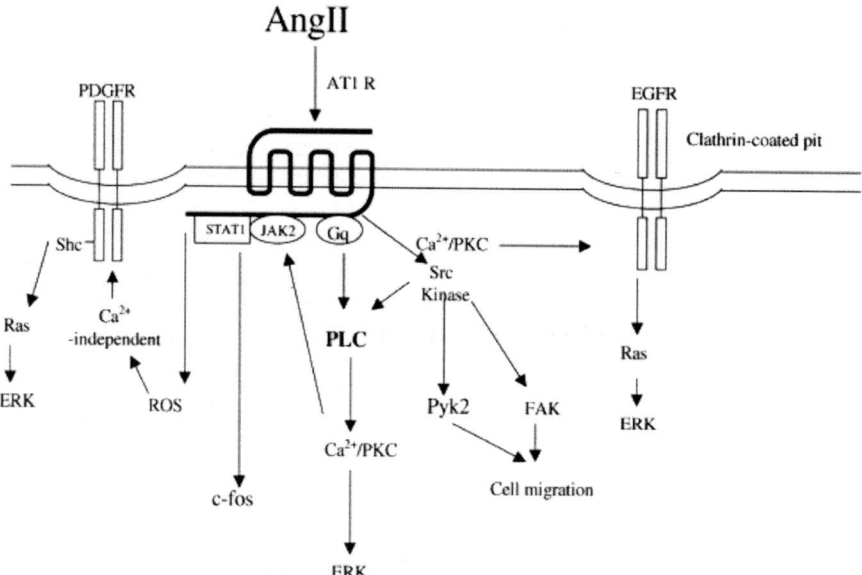

Figure 3. **Angiotensin II signaling interactions.** There is a staggering number of ways in which signaling systems interact with each other, both cooperatively and antagonistically. This schematic illustrates some simple interactions of the major angiotensin receptor – AT1 with other receptors and signaling systems in addition to coupling with G proteins to activate phospholipase C. First, the AT1 receptor is capable of activating non-receptor tyrosine kinases, such as Src, Fyn, Yes, Pyk2, FAK, and JAK2. Second, ligation of AT1 results in activation of oxidant production via activation of the small GTPase, Rac1. Third, the derived oxidants utilize the receptor tyrosine kinase scaffolding of the platelet derived growth factor receptor (PDGF-R). Fourth, PKC/Ca^{++}/src activation results in activation of the tyrosine kinase receptor epidermal growth factor receptor (EGF-R). Similarly, AT1 activation leads to activation of the insulin-like growth factor (IGF-R). Several of the pathways lead to altered transcription. Thus, this G-protein coupled receptor triggers activation of multiple receptor and non-receptor tyrosine kinases that are important in both short-term and prolonged physiologic and pathologic effects of angiotensin II. This figure is reproduced from: *Yin, G., Yan, C., and Berk, BC. (2003) Angiotensin II signaling pathways mediated by tyrosine kinases. International Journal of Biochemistry & Cell Biology 35:780-783*, by permission of Elsevier, Amsterdam.

We found that angiotensin II upregulated the p22^{phox} component of NADPH oxidase in pulmonary artery endothelium, as shown in Figure 4. Recently, angiotensin II-upregulation of the gp91^{phox} component of NADPH oxidase and also endothelin-1 was shown to occur via activation of Ras, Raf, and MEK1/2 (Hsu et al., 2004). Additionally, AP-1 activation contributed to endothelin-1 upregulation (Hsu et al., 2004). As will be seen in the discussion of oxidant signaling below, this primary oxidant effect is likely a central feature in the cross talk shown in Figure 3. For instance, angiotensin II-upregulation of endothelin-1 was dependent on oxidant production (Hsu et al, 2004) and reactive oxygen species are intermediary messengers for PDGF-, angiotensin II-, TNF, and thrombin-induced smooth muscle cell proliferation and hypertrophy (Flack et al., 1998; Patterson et al., 1999; Frank et al., 2000). Importantly, as seen in Figure 4.b. the upregulation of NADPH by angiotensin II was greatly attenuated by statins, which interfere with Ras prenylation and its membrane association. Furthermore, endothelin, itself, can upregulate critical components of NADPH oxidase (Duerrschmidt et al., 2000).

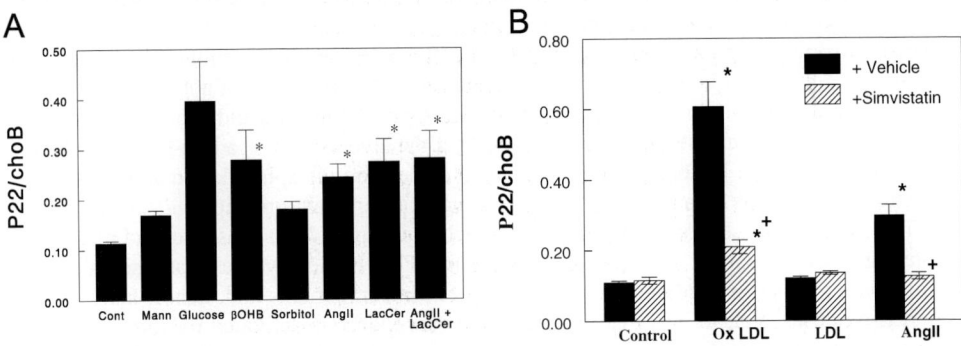

Figure 4. **Mediator-induced NAD(P)H oxidase expression in bovine pulmonary endothelium.** Bovine pulmonary artery endothelial cells were grown to 90% confluence and then serum starved 24 h in media with only 0.1% serum, then exposed to mediators in same media for an additional 18h for mRNA isolation. Mediators/substrates included 25 mM glucose (by supplementing the usual 5 mM media glucose), 5 mM β-hydroxybutyrate, 10 μM D-sorbitol, 1 μM angiotensin II, 10 μM, lactosyl ceramide, 250 μg/ml oxidized low density lipoprotein, and 250 μg/ml LDL. The ox LDL was prepared by incubation of fresh, human LDL for 24 h at room temperature with 10 μM CuSO$_4$. In some experiments, cells were pre-incubated with 10 μM simvastatin (Calbiochem) 30 min prior to addition of angiotensin II or LDL. Northern blots were prepared for both p22^{phox} probes (kindly provided by Mary Dinauer, Indiana Univ.) and then dehybridized and re-probed for constitutively expressed ChoB. **A.** Average p22^{phox}/choB mRNA ratios for 4-8 experiments are shown. Glucose (25 mM) was significantly different from the 20 mM mannitol (+5 mM glucose) osmotic control (+). Groups with β-OH butyrate, angiotensin II, lactosyl ceramide, and the combination of ang II and LacCer were sig. diff. from control (*). The group treated with sorbitol, which cause mild redox stress, was nearly significant (p<0.07). **B.** In other experiments the endothelial cells were exposed to oxLDL, fresh LDL, or angiotensin II in the presence or absence of simvastatin. Isolated mRNA was separated on gels and Northern blotted for p22^{phox}, reprobed for choB, and the relative densiotometry determined. Whereas, oxLDL resulted in a dramtic increase in p22^{phox} expression, simvastatin attenuated this effect and the increased expression due to angiotensin II. * indicates significant difference from control and + indicates significant difference between the simvastain and matched mediator ratios.

Bioactive Lipids

Sphingosine 1-phosphate (S1P) is a lipid mediator with diverse endothelial effects, including acute barrier promotion, as discussed in Chapter 4. The long-term protective effects of sphingosine 1-phosphate are mediated via its activation of the endothelial differentiation gene receptor proteins (EDG), a class of G protein coupled receptors that bind lipid ligands with high affinity and specificity (Fischer *et al.,* 1998; Spiegel and Milstein, 2000). This lipid can induce upregulation of ICAM, VCAM, and E-selectin by PKCα and p38-dependent mechanisms (Shimamura *et al.,* 2004). S1P promotes survival by activation of Akt (Morales-Ruiz *et al*, 2001). It is required for endothelial cell migration, proliferation, and morphologic differentiation, and assembly of adherens junctions (Lee *et al.,* 1999; Kimura *et al.,* 2000; Boguslawski *et al*, 2000). Similarly, estrogen and corticosteroids activate PI3K/Akt signaling (Shiojima and Walsh, 2002).

A similar lipid, lysophosphatidic acid bind different Edg receptors and activates Rho, but is barrier disruptive. Despite their different short-term effects on barrier function they both appear to activate the serum response element (SRE) transcription factor, which promotes growth related genes and endothelial survival (An *et al.,* 1998). Lysophosphatidic acid induced tyrosine phosphorylation activation of pp42 MAPK, within 5 minutes but activation was sustained at least 2 hours (McLees *et al.,* 1995). PKA activation did not alter early phase lysophosphatidic acid-induced MAPK tyrosine phosphorylation, but did block the sustained activation (McLees *et al.,* 1995). PKA interaction at the level of Raf-1, a Ras-GTPase effector kinase, upstream from MEK and ERK1/2, altered caldesmon phosphorylation and stress fiber formation, making this an attractive pathway for signaling interaction (Liu *et al.,* 2001).

The sphingolipid, lactosyl ceramide, was noted in Chapter 4 to act as a second messenger triggering Rac1 activation. Both angiotensin II and TNF induce synthesis and signaling via lactosylceramide. This mediated rapid assembly of the components and, hence, activation of NADPH oxidase. Yet, in every case where NADPH oxidase is activated its components are subsequently upregulated, as well, in an oxidant-dependent positive feed forward interaction. Lactosylceramide, but not other ceramides, was shown to be a key signal for oxLDL-induced upregulation of NADPH oxidase (Lang *et al.* 2000) and for TNFα-induced NFκB and ICAM expression in endothelial cells (Buhnia *et al.,* 1997; Chatterjee. 1998). This is important as these positive feed forward loops likely contribute to long-term activation. Thus, there are close ties between signaling by this lipid mediator and oxidant signaling discussed below. We have shown that exogenous lactosyl ceramide can induce upregulation of the important $p22^{phox}$ component of NADPH oxidase in lung endothelium (Figure 4).

Oxidized lipoproteins have important pathologic effects and interact directly with endothlial cells through a LOX receptor (Nagase *et al.,* 1998). In human coronary endothelium oxLDL was also shown to involve both TNF and FasL receptors, activation of JNK, and caspase activation by decreased Bcl-2 (Napoli *et al.,* 2000). These investigators reported oxidant-dependent activation of ATF-2, ELK-1, CREAB, and AP-1 transcription factors. Likewise, oxidized phospholipids have multiple effects on endothelium. For instance, ox-phosphatidylcholine induced upregulation of the early growth response factor 1 (EGR-1) and heme oxygenase (HO-1) within 3 hours in both isolated endothelial cells and in various tissues *in vivo*, including lung (Kadl *et al.,* 2002; Kronke *et al.,* 2003). Increased HO-1 expression was dependent on release and signaling by another lipid paracrine mediator, PAF (Kadl *et al.,*2003) and was signaled by PKA and PKC, ERK and p38, and CREB (Kronke *et al.,* 2003). This pattern differs from that elicted by LPS or TNF, which signal through NFκB and do not increase HO-1 expression (Kronke et al., 2003). For example, TNF, IL-1, and LPS cause endothelium to express receptors to bind monocytes and neutrophils, whereas oxLDL causes endothelium to only bind monocytes, which are particularly important in atherosclerosis associated with oxLDL elevation (Lusis, 2000). Similarly, 8-isoprostane, the oxidized lipid product from $PGF_{2\alpha}$ also induced upregulation of factors that only bound monocytes (Leitenjer et al., 2001). IL-8 upregulation due to oxPC was also independent of NFkB, C/EBPβ, and AP-1 and therefore different from the IL-8 upregulation due to TNF (Yeh et al., 2001). OxLDL induced activation of JAK2 and translocation to the nucleus of its downstream transcription factor, STAT (Maziere et al., 2001). Survival is also compromised, as oxLDL promotes dephosphorylation and inactivation of Akt (Shijima and Walsh, 2002). We have shown that oxLDL, as opposed to LDL, upregulated NADPH oxidase in pulmonary endothelium and that this effect was attenuated by statin (Figure 4). Several plasma enzymes rapidy metabolize such activating lipids and therefore endothelial activation was proposed to be a local event, yet levels of oxLDL rises significantly in a number of disease states. Ordinary LDL can also cause Ras-mediated activation and phenotypic changes in cultured vascular endothelial cells via JNK⇒AP-1 and p38⇒ATF-2 activation (Zhu et al., 2001). Finally, the lipid mediator, PAF, contributes to angiogenesis by STAT5B activation, but little else is known about long-term effects of PAF on endothelium (Brizzi et al., 1999).

LPS (endotoxin) has both lipid and poysaccharide domains, but LPS signaling will be considered here, because the lipid A portion in the main ligand for the TLRs. LPS signals activation of the p38, JNK, and ERK arms of the MAPK cascade and activates NFκB. Whereas NFκB promotes survival, JNK promotes upregulation of caspase 8 and cytotoxicity, but both effects apppear to be downstream from LPS-activation of TRAF (Hull et al., 2002). LPS-induced barrier dysfunction follows the same temporal pattern as apoptosis, suggesting a linkage. Indeed, the LPS-signaled apoptosis is via the usual activation of caspases, which were shown to catalyze direct cleavage of the adherens junction proteins, β- and γ-catenin, thus leading to prolonged barrier dysfunction (Bannermann et al., 1998). After either intraperitoneal or intranasal administration of LPS, the induced JAK and src activation throughout the lung resulted in activation of the transcription factor, STAT, which was downstream from upregulation and secretion of IL-6 (Severgnini et al., 2004). More information on LPS effects and signaling in endothelium can be found in Chapter 13.

Cytokines

Cytokines have various effects on endothelial cells, from upregulation of leukocyte adhesive factors and chemokines (i.e. exacerbation of inflammation), to permeability, to cell death. IL-1 and TNFα were the first cytokines described to lead to endothelial cell activation including the expression of vascular adhesion molecules (Bevilacqua et al., 1985). IL-1-induced permeability requires over 8 hours and altered protein synthesis (Figure 5). The signaling and effectors of the cytoskeletal reorganization and the permeability is unknown, but is independent from MLCK activation and is not prevented by cAMP elevation. IL-1 further elicits synthesis of GMCSF, IL-6, and IL-8 in endothelium by a PKC-NFκB and JNK-dependent manner, while p38 MAPK activation stabilizes IL-8 mRNA (Patterson et al., 1998; Holtmann et al., 2001). IL-1-induced increases in COX2 expression and MnSOD expression are also NFκB-dependent (Miralpeix et al., 1997; Rogers et al., 2001). IL-1-induced VCAM and E-selectin upregulation was prevented by antioxidant treatment (De Caterina et al., 1995). While TNF was not sufficient itself, it potentiated IL-1-induced upregulation of prostaglandin

H synthase-2 in human endothelial cells (Said *et al.*, 2002). Finally, IL-1 and LPS induced expression of TNF in microvascular endothelial cells, while INFγ failed to independently induce TNF expression but enhanced expression-induced by IL-1 or LPS (Hoffmann *et al.*, 2004). INFγ decreased occludin expression and cause endothelial barrier dysfunction (Minagar *et al.*, 2003). In vascular endothelial cells, IL-1 increased expression of cooperative transcription factors, ESE-1, and p300, but the significance of this is unknown (Wang *et al.*, 2004a). In mouse lungs, LPS caused IL-6 production, which then activated the JAK/STAT pathway prior to development of lung injury (Severgnini *et al.*, 2004).

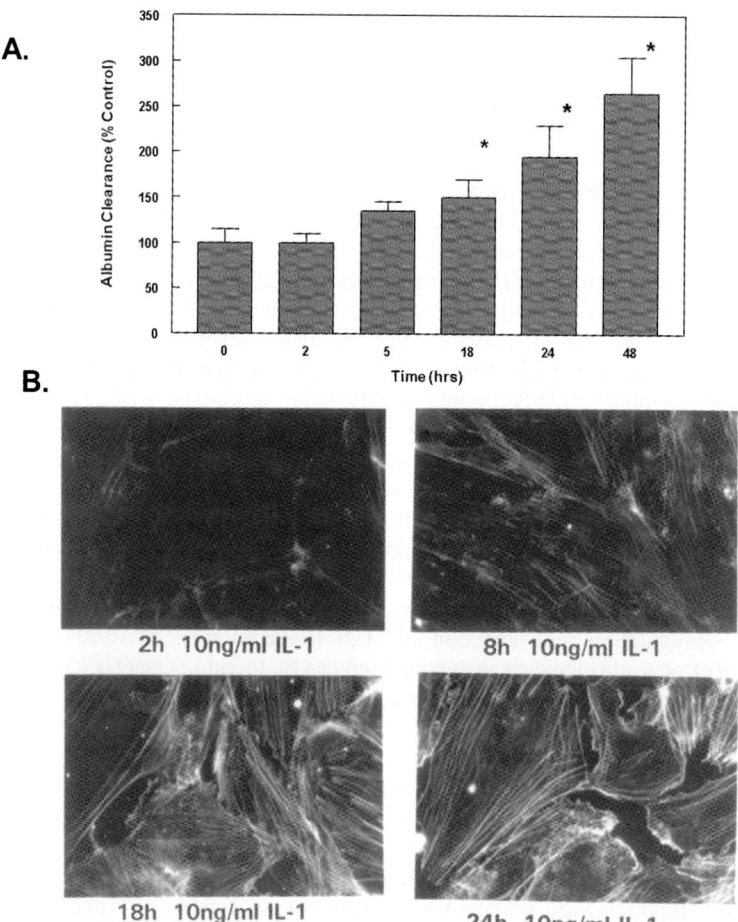

Figure 5. IL1 –induced permeability requires prolonged signaling and altered protein synthesis. Human umbilical vein endothelial cells grown to confluence on semipermeable membrane supports or glass cover slips were exposed to 10 ng/ml IL-1. **A.** Albumin permeability was determined by Evans blue dye-albumin clearance across the monolayers. Significant increases in permeability were noted at 18 hrs that was dependent on protein sysnthesis, but independent of MLC phosphorylation and not prevented by cAMP. **B.** There was a slow induction of net actin polymerization, stress fiber formation, cell elongation, and gap formation, revealed by rhodamine-phalloidin staining of F-actin.

A model of sustained signaling activation was produced by continuous expression of the TNF-precursor by transfection of endothelial cells with an uncleavable transmembrane TNF mutant (Clauss et al., 2001). This resulted in continuous activation of TNF pathways, such as p38 MAPK and NFκB activation. This activation was observed for at least 10 passages and led to persistent parameters of endothelial cell activation, including continuous IL-6 secretion, tissue factor production (Clauss, 2001), and adhesion molecule expression (ICAM-1 and VCAM-1 but not E-selectin) (Clauss, unpublished observation). Based on these results it can be hypothesized that short-term signaling can be converted into persistent chronic activation when the activation is appropriate, which is assumed to occur under pathological conditions.

In endothelium, TNF can induce an inflammatory response, an anti-apoptotic response, and a pro-apoptotic response, as shown in Figure 6. TNFα ligation of its receptor, TNFR1, activates binding of TNF-receptor associated death domain (TRADD) leading to activation of associated factor 2 (TRAF2) and RIP. TRAF then activates apoptosis signal-related kinase (ASK1). This kinase is upstream from activation of ERK kinase (SEK1), which is the upstream modulator of the SAPK, JNK, and activation of AP-1 (Pober, 2002; Yoshizumi et al., 2004). Many receptors utilize a PI3K/Akt pathway to activate NFκB, but TNF's use of this pathway in endothelium has been questioned (Zhang et al., 2001; Madge and Pober, 2000; Pober, 2002). Alternatively, direct interaction of IKK with TRAF2 has been shown to account for the sustained NFκB activation by TNF (Devin et al., 2001). Although PI3K activation of Akt does not appear to be important in endothelial survival pathways activated by TNF or IL-1 in endothelium, PI3K does activate survival pathways via inhibition of a cathepsin B-mediated apoptosis pathway (Madge and Pober, 2000; Madge et al., 2002). TNF-induced death is exacerbated by inhibition of PI3K, such that the cathepsin-mediated apoptosis is upstream from the caspase-mediated apoptosis. If other strong survival systems are present, they suppress activation of the TNF-death pathways. Most genes upregulated or suppressed by TNF in human endothelial cells are dependent upon the NFκB and AP-1 transcription factors, but a small number are related to the transient p38- signaling of different transcription factors (Gustin et al., 2004; Viemann et al., 2004). Downstream from TNF-induced p38MAPK, activation of the transcription factor CREB was due to MSK1 activation (Gustin et al., 2004). Recently, fractalkine upregulation by TNF was shown to be NFκB and Sp-1 mediated (Ahn et al., 2004). In endothelium of artery segments, TNF and INF-γ synergistically activate STAT-1 and NFκB and induce expression of iNOS, but not in smooth muscle, while constitutive eNOS was reduced (Wagner et al., 2002). In pulmonary microvascular endothelium, TNF decreased eNOS expression mediated by GATA-4 and Sp3 transcription factors (Neumann et al., 2004). However, TNF-induced decrease in eNOS was also reported to be due to increased degradation rather than altered gene expression (Fleming and Busse, 2003). When survival signals are supressed, TNF activation of ASK-1 can lead to JNK activation of FasL, caspase-8, and caspase-12 and, hence, apoptosis. TNF-induced apoptosis with nucleosomal fragmentation and caspase activation was also linked to RhoA activation and stress fiber formation (Petrache et al., 2003c). Recent studies demonstrated that the poly (ADP-ribose) polymerase (PARP-1), enhanced TNF-expression of some genes, while supressing expression of genes down from TNF-NFκB activation (Carrillo et al., 2004).

TNF-induced permeability requires 4-10 hrs for effect (Camussi et al., 1991; Petrache et al., 2003a). TNF activation of caspase resulted in generation of MLCK catalytic fragments, which were not governed by Ca^{++}/calmodulin and TNF also caused loss of tubulin content in a p38-dependent manner (Petrache et al., 2003a; Petrache et al., 2003b). Importantly, TNF-induced permeability was dependent upon NADPH-mediated oxidant production (Gertzberg *et al.*, 2004). TNFα- upregulation of the MCP-1 gene also depended on oxidant production (Chen *et al.*, 2004). And other showed that TNF-induced E-selectin expression was Rho-dependent and was inhibited by statin treatment (Nubel *et al.*, 2004), indicating this effect and possibly others might be oxidant-dependent in endothelium as shown in smooth muscle.

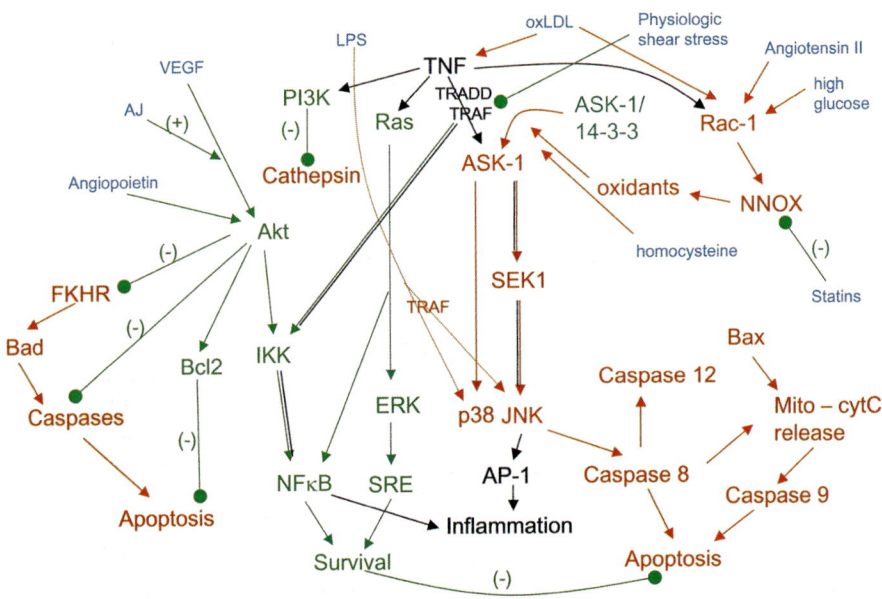

Figure 6. TNF-mediated signaling of inflammation, survival, and apoptosis in endothelium.
TNFα signaling in endothelial cells produces various effects on transcription, translation, and protein turnover. Here pro-apoptotic (red) and anti-apotoic (green) pathways are contrasted. Initial receptor signaling pathways and pathways contributing to inflammation are shown in black. Selected signaling interactions with other mediators and conditions (blue) are indicated. As TNF signals both pro- and anti-apoptotic pathways, the balance may depend upon the presence of these other aggravating or antagonizing co-mediators as well as on the degree and duration of the TNF ligation of its receptor and receptor expression. For instance, cell death does not occur when survival signaling blocks activation of the apoptotic pathway and, thus, co-exposure of endothelial cells to VEGF and TNF prevents apoptosis. This is due to enhanced Akt-NFkB activation and enhanced Akt-inhibition of both FKHR (a pro-apoptotic transcription factor) and Bad (a pro-apoptotic protein). For instance, Akt-mediated phosphorylation of FKHR prevents its ability to inhibit transcription of FLIP, which is necessary to prevent caspase 8 activation. TNF, emits anti-apoptotic signals through PI3K activation, but the downstream effects are not via Akt activation, but by inhibition of cathepsin B. In contrast, TNF-TRADD/oxidant-dependent ASK-1 activation leading to p38 and JNK activation is pro-apoptotic in the absence of survival (apoptosis blockade). This death effect is exacerbated by other factors contributing to oxidant stress, such as high glucose, angiotensin II and ox LDL, which activate NADPH oxidase (NNOX) like TNF. Statin treatment suppresses NNOX activation, endothelial dysfunction, and apoptosis. Once caspase 8 is activated, there seems to be no turning back, although the process can be delayed by various inhibitors; thus, survival factors work by blocking activation of apoptosis. TNF also signals an inflammatory response through an TRADD/oxidant-dependent mechanism via SEK1 and the SAPK, JNK-triggered upregulation of leukocyte adhesion molecules and chemokines. Finally, TNF signals both survival and inflammatory responses dependent on NFκB by a direct interaction of TRADD with IKK, rather than through Akt/IKK activation in endothelium.

Growth Factors

Numerous mediators and conditions lead to VEGF and VEGF receptor upregulation in endothelial cells and other tissues (Berra *et al.*, 2000). Hypoxia is a key stimulant. Other growth factors, such as EGF and TGF (α & β), and cytokines induce its expression. On the other hand, TGFa overexpression in mice leads to decreased VEGF expression and pulmonary vascular abnormalites (Le Cras *et al.*, 2003). VEGF and basic fibroblast growth factor (bFGF) activities require *de novo* protein sythesis and gene transcription and include expression of inflammatory cellular adhesion molecules and tissue factor and angiogenic direction of endothelial survival, proliferation, migration, and tube formation (Clauss, 2000; Cabrita and Christofori, 2003; Naik *et al.*, 2003). The two main receptors are VEGFR2 (KDR, flk-1) and VEGFR1 (flt-1), with dominant signaling through VEGFR2, as seen in Figure 7. VEGF increased expression of > 130 genes in human umbilical vein endotheliun within 24 hrs and 53 of these were induced within the first 2 hrs (Abe and Sato, 2001). Of these, Down syndrome candidate region 1 was most profoundly induced, followed by Mifl, COX2, EGR 2/3, CD1B antigen, and bactericidal/permeability-increasing protein. With a 9 day *in vivo* VEGF treatment to stimulate angiogenesis, proteins, such as tPA, Glut 1, VEGFR2, and Tie-2, were upregulated in pre-existing and new vessels, while in the new vessels there was a lack of tPA, Glut1, and Tie2 expression but a large increase in VEGFR2 (Witmer *et al.*, 2004).

Endothelial proliferation upon VEGFR2 ligation is mediated by Ras and PKC, which lead to Raf-ERK1/2- MAPK activation of intranuclear cyclin D1 that is important in shifting the cell from G1 to S-phase (D'Angelo *et al.*, 1999; Ilan *et al.*, 2003 ; Suhardja and Hoffman, 2003). Several different pathways activated *via* VEGFR2 are depicted in Figure 7. VEGF activates several transcription factors, including Ets1, Egr1, NFAT, Stat3 and 5 and NFκB, with long-range effects (Liu *et al.*, 2000; Zachary, 2001). VEGF utilization of the Ets-1 transcription factor was critical to its induction of angiogenesis (Watanabe *et al.*, 2004). The signaling itself may be short, intermediate, or sustained. As shown, one arm of VEGF-activation is by ERK1/2, which is temporally intermediate in pattern but generally disappears after 60 min. If however, p38 MAP kinase, which is also activated by VEGF, is inhibited, ERK1/2 phosphorylation can be observed for > 24 hrs (Issbrucker *et al.*, 2003). ERK is also activated in bFGF-induced migration and proliferation in endothelial cells (Naik *et al.*, 2003). PLD activation, downstream from PKCδ, was recently shown to be critical for VEGF-induced MEK/ERK signaling (Cho *et al.*, 2004). VEGFR2, other tyrosine kinase receptors, and also G-protein coupled receptors associate with and activate non-receptor tyrosine kinases, such as Src (Meyer *et al.*, 2002) and FAK (McLees *et al.*, 1995). The growth response to VEGF also depends on increased survival, mediated by PI3K/Akt, as is the angiopoietin-induced survival effect (Gerber *et al.*, 1998; Shiojima and Walsh, 2002). The role of VEGF as an essential survival factor may have underestimated by the focus on it effects on angiogenesis and vascular permeability. In fact, removal of VEGF in tumors causes endothelial apoptosis and vascular regression (Benjamin *et al.*, 1999) and blocking VEGF signal transduction in lungs causes vascular regression and emphysema formation (Le Cras *et al.*, 2002; Tuder *et al.*, 2003). VEGF-induced Akt activation has recently been linked with inhibition of FKHR in endothelial cells (Abid *et al.*, 2004) and increased expression of FLIP (Skurk *et al.*, 2004), which could partly explain the pro-survival effects of VEGF as discussed above. Both PI3K-Akt and p38 MAPK pathways are involved in migration (Rousseau *et al.*, 1997). VEGF-induced ICAM, VCAM, and E-selectin expression via NFkB activation, with some supression by the PI3K pathway, but little involvement of the ERK-MAPK pathway (Kim *et al.*, 2001).

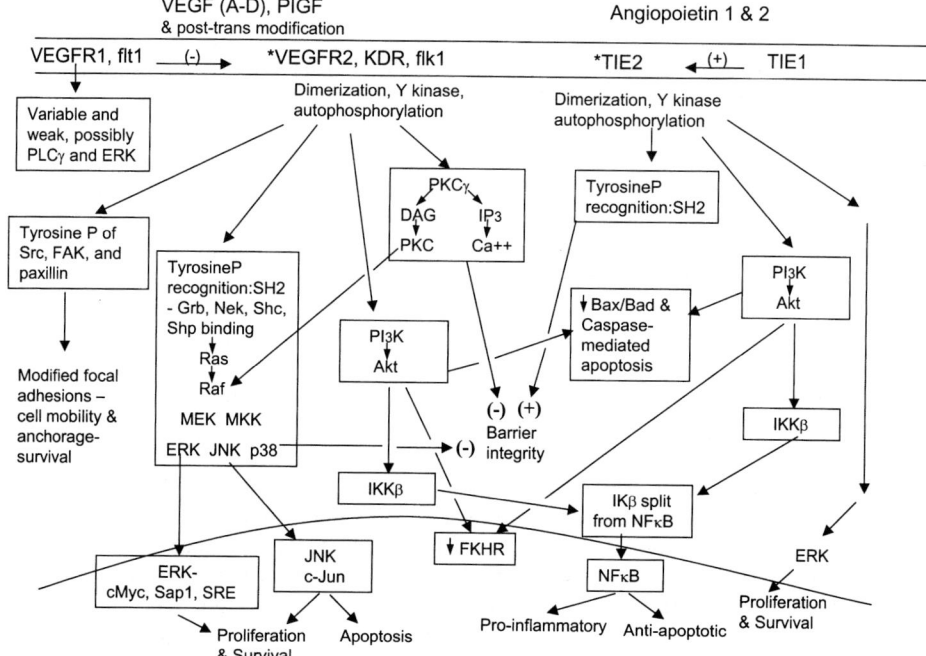

Figure 7. Growth factor signaling. This generalized schema shows the activation and resultant signaling of the two VEGF- and the two angiopoietin- receptors in endothelium. Ligand/receptor specificities for VEGF are reviewed by Zachary and Gliki, 2001 and for angiopoietins by Loughna and Sato, 2001 and are not shown. VEGFR3 is not shown, as it is primarily in lymphatic endothelium. The main VEGF receptor for acute and prolonged permeability effects and for the proliferative effects of VEGF is the tyrosine kinase-type receptor, VEGFR2. In addition to the normal pattern of SH2/Grb – activation of Ras, the receptor also activates non-receptor tyrosine kinases, which may participate in focal adhesion formation involved in actin polymerization, stress fiber formation, and cell mobility. Through the receptor tyrosine kinase activation of Ras, downstream MAPKs (ERK, JNK, and p38) exert their respective effects as shown. p38 MAPK is particularly important to the permeability effects of VEGF. The PI3K/Akt/NFκB pathway is also initiated, as is PKCγ pathway. The main angiopoietin receptor, Tie2, is also a tyrosine kinase receptor, but initiates pathways that antagonize the activating/barrier dysfunction effects of VEGF, mainly by countering VEGF activation of p38MAPK. Additionally Tie2 activates PI3K/Akt, which cooperates with VEGF signaling of cell survival and NFκB activation. These complementary Akt-survival effects are due in part to Akt-mediated phosphorylation of members of the forkhead (FKHR) transcription family, which prevents their translocation and inhibition of the transcription of FLIP (i.e. inhibition of inhibition leads to increased FLIP). FLIP then blocks activation of caspase 8, and thus blocks activation of apoptosis, enhancing cell survival. Inhibition of FKHR also decreases levels of Bad and Bax, pro-apoptotic proteins. VEGF and angiopoietin cooperate in suppression of Fas ligand expression and activation of Bcl2 and Bcl$_x$ expression (anti-apoptotic proteins). Akt also directly phosphorylates Bad, making it vulnerable to inhibitory binding. Other VEGF and angiopoietin factors may be upregulated or activated to block other players in the apoptotic pathway, but this has not been fully resolved. Importantly, VEGF and angiopoietin also cooperate in proliferative effects on the endothelium. Tie1 appears to lack direct signaling, but may cooperate with Tie2 activation. Importantly, Tie1 is crucial for endothelial integrity and is expressed at relatively high levels in the lung compared to other organs.

In some studies VEGFR2 was determined to be primarily located at the caveolae (Fisher et al., 2001; Shiojima and Walsh, 2001; Cho et al., 2004). Others assign it to the abluminal aspect in association with αvβ3 integrins in relation to initiation of the PI3K/Akt pathway (Vestweber, 2000; Eliceiri, 2001). And yet others reported VEGFR2 location in intercellular junctional complexes (Zhao et al., 1998; Shay-Salit et al., 2002; Lampugnani et al., 2003). The interesting interactions of VEGFR2 and the adherens junctions, which attenuate the proliferative ERK/JNK signaling, yet enhance Akt-survival signaling by a specific junction-associated phosphatase will be discussed in Chapter 9 (Carmeliet et al., 1999; Lamugnaini et al., 2003). Thus, it is not clear where this receptor is primarily located, as these sites are lateral, basal, and apical, respectively. Moreover, it is not clear, how this spacial distribution is related to activation state of the endothelium, temporal changes, or coupling to the various intracellular signaling pathways.

There is a strong association of VEGF with permeability and it was early identified separately as vascular permeability factor (VPF), as noted in Chapter 4. VEGF-adenovirus administered by an intratracheal route resulted in lung expression and permeability edema in mice compared to transfection with truncated VEGF (Kaner et al., 2000). *In vivo*, permeability can occur rapidly, but in cultured cells, induction generally requires about 2 hours. Actually, application of VEGF to human umbilical vein endothelial cells resuted in a rapid (5 min) increase in inositol phosphates and an intermediate (15 min) surface expression of P-selectin, but macromolecule permeability was not noted until 150 minutes (Hippenstiel et al., 1998). In contrast to VEGF, PDGF and GMCSF evoked the inositol phosphate response, but did not lead to permeability in this study. Numerous other laboratories have confirmed this direct effect on endothelial permeability, yet some endothelium do not respond to VEGF with permeability, especially highly confluent endothelium (Lampugnani et al, 2003). Paracellular permeability is mediated via the VEGFR2 (KDR, flk-1) receptor, but not the VEGFR1 (flt-1) receptor (Meyer et al., 1999). In VEGF-induced permeability, co-activation of membrane TNF and p38MAPK were shown to be critical for the barrier impairment (Clauss et al., 2001). In fact, p38 MAPK activation was not only confirmed as the mechanism of hypoxia- and VEGF- mediated edema formation in the brain, but it was also demonstrated to be a molecular switch between vascular permeability and angiogenesis (Issbrucker et al., 2003). Likewise, p38 MAPK was critically involved in TNF-induced endothelial permeability, with specific effects on the microtubule system (Petrache et al., 2003a). As introduced in Chapter 4, there is specificity of the particular src-family kinases with regard to differential direction of signaling toward permeability or angiogenesis in VEGF stimulation, in that fyn (-/-) mice responded with permeability and proliferation, whereas pp60c-src (-/-) and pp63c-yes (-/-) mice responded with only proliferation (Eliceiri et al., 1999). When VEGF overexpression in lung epithelial cells was induced in newborn mice, they developed capillary endothelial attenuation, vascular remodeling, and 50% mortality at two weeks due to pulmonary hemorrhage (Le Cras et al., 2004). Thus, ether deficiency or excess of VEGF results in pulmonary endothelial dysfunction.

TGFβ1 elicits rapid activation of Ras, ERKs, and JNK upstream from AP-1 activation (Mulder, 2000). TGF effects are mainly mediated by subsequent generation of a local inflammatory condition and its ability to increase VEGF generation, thus it behaves as a cytokine (Pepper, 1997). In lung endothelium, TGFβ1 induced the upregulation of endothelin-1 (Lee et al., 2000). *In vitro*, it induced upregulation of matrix components, integrin subunits, uPA, PAI-1, eNOS, PDGF, and TGFβ1 itself, while downregulating

VEGFR2, certain integrin subunits, TIMP-1, thrombomodulin, and vWF (Pepper, 1997; Mulder, 2000). While low levels induce migration and proliferation via induction of VEGF and bFGF, in co-culture expreiments TGFβ1 was a critical mediator of pericyte- or smooth muscle-paracrine signaling of endothelial quiescence by inhibition of migration and proliferation and downregulation of the VEGFR2 (Antonelli-Orlidge et al., 1989; Rifkin et al., 1991). This indicates that conditions and timing are important to the effects observed. TGF-signaled growth inhibition of vascular endothelial cells was opposed by experimental expression of the RUNX2 transcription factor, but TGF-signaled apoptosis was enhanced by experimental expression of the RUNX2 splice variant, RUNX2δ8, indicating the regulatory effects of these transcription factors over TGF-induced effect (Sun et al., 2004). While, TGFβ1 seems to be highly important for normal vessel development in gestation, its role in angiogenesis in adults appear to be related to the secondary inflammatory effects and VEGF. By itself, in vitro, it does not induce angiogenesis, but potentiates angiogenesis initiated with VEGF or bFGF at low levels and inhibits angiogenesis at high levels (Pepper et al., 1993).

In contrast to VEGF, hepatocyte growth factor and angiopoietin exert barrier protective effects, as depicted in Figure 7 (Gamble, 2000). Angiopoietin-1 was identified as an essential vascular promoter in embryogenesis less than 10 years ago (Suri et al., 1996). It belongs to a family of ligands (angiopoietin-1 to 5), which share the transmembrane receptor Tie-2. As seen in Figure 7, Tie-1 and Tie-2 are the two related members in vascular endothelium and they are tyrosine kinase type receptors, as are most growth factor receptors (Sato et al., 1995). Tie-2 deficient mice embryos failed to establish structural integrity of vascular endothelial cells, resulting in a "plethora of vascular abnormalities" (Ward and Dumont, 2002). The primary capillary plexus failed to undergo sprouting and remodeling, resulting in incomplete development of the heart and head regions with lethality between day 9 and 12; thus, it is essential. In contrast, Tie1 knock out mice develop rather normally in early gestation, but show vessel weakness ~d14, especially in the lung and they die shortly after birth from pulmonary hemorrhagic edema (Sato et al., 1995). Interestingly, Tie1 is increased specifically in the lung in late gestation and expressed at 10 or more fold greater levels than in any other organ (Taichman et al., 2003). The significance of this unique expression in pulmonary endothelium is not known, but Tie-1 signals endothelial survival via activation of PI3K/Akt (Kontos et al., 2002). Nevertheless, Tie2 is considered the main functional receptor for angiogenesis and is the most studied receptor. Several cells produce angiopoietin-1 including fibroblast, platelets, and endothelial cells (Jones, 2003).

Angiopoietin-1 and hepatic growth factor promotion of cell survival is mediated in part via tyrosine phosphorylation of PI3K and activation of Akt signaling (Kontos et al., 1998; Gamble et al., 2000; Kim et al., 2000b; Dixon, 2001; Shiojima and Walsh, 2002; Daly et al., 2004; Peters et al., 2004; Zhang et al., 2004). But angiopoietin-1 and hepatocyte growth factor also signal survival and growth transcription effects via ERK and several STATs (Dixon, 2001; Loughna and Sato, 2001; Nakagami et al., 2001a). Both angiopoietin-1 and VEGF oppose JNK-mediated apoptosis in endothelial cell through Akt as illustrated in Figures 6 and 7 (Harfouche et al., 2003). Additionally, angiopoietin promotes survival through upregulation of the IAP protein, survivin (Papapetropoulos et al., 2000) and heptocyte growth factor upregulated the protective Bcl2 (Nakagami et al., 2002). In contrast to many receptor tyrosine kinases, angiopoietin is not able to activate ERKs 1/2 and is not mitogenic for endothelial cells (Koblizek et al., 1998; Witzenbichler et al., 1998). The ability of angiopoietin to induce endothelial migration depended on PI3K-dependent increase in FAK tyrosine phosphorylation and induction of the proteolytic enzymes, plasmin and MMP-2 (Kim *et al.*, 2000a). Besides, survival effects, hepatocyte growth factor also produced upregulation of angiotensin converting enzyme (ACE) in pumonary artery endothelial cells, a characteristic of the quiescent pulmonary circulation (Day *et al.*, 2004).

In addition to prolonged benefits for barrier function related to survival, angiotensin exerts an acute protective effect on agonists- and VEGF-induced endothelial cell permeability as noted in Chapter 4 (Gamble *et al.*, 2000). Angiopoietin's potent inhibition of vascular leakage has been shown both *in vivo* and *in vitro* (Thurston *et al.*, 1999; Thurston *et al.*, 2000). Like angiopoietin, hepatocyte growth factor has short-term protective effects on endothelia. The responsible signaling and effectors are undetermined, but they are not due to the activation of PI3K/Akt that was responsible for the prolonged signaling, nor was it due to RhoA inhibition (Gamble *et al.*, 2000; Ward and Dumont, 2002; Jones, 2003). It did decrease PECAM-1 or VE-cadherin phosphorylation (Gamble *et al.*, 2000). Additional clues were recently obtained for the opposition of angiopoietin and VEGF in barrier function (Wang *et al.*, 2004b). Although, VEGF led to PLA_2 activation and subsequent PAF production, arachidonic acid release was not blocked when VEGF and angiopoietin were added together, indicating that this pathway makes little contribution to VEGF-permeability and is not a target for angiopoietin protection. Angiopoietin did block the VEGF-induced translocation and activation of PKCβ, but not PKCα, indicating that this might be an important target for angiopoietin-1. Unfortunately, neither PKCδ nor PKCγ were examined yet both have been implicated in VEGF-PKC signaling. Co-treatement also prevented the VEGF-induced dissociation of β-catenin from VE-cadherin, which could be downstream from the PKC activation. To test this, effects of both the non-specific PKC inhibitor and a specific PKCβ inhibitor were examined, and both prevented the VEGF dissociation of the junctional complex, supporting a role for PKCβ signaling in the VEGF permeability and the angiopoietin protection.

Similarly, angiopoietin blocked permeability induced by other edemagenic mediators. While, angiopoietin increased VE-cadherin and PECAM-1 phosphorylation and attenuated permeability induced by thrombin and neutrophil migration in response to TNF, the opposing signaling was not determined (Gamble *et al.*, 2000). In a recent study, addition of angiopoietin 30-60 min before thrombin blocked its permeability effects, but did not block thrombin activation of RhoA kinase, indicating this pathway was not a target for the angiopoietin protection (Pizurki *et al.*, 2003). Interestingly, thrombin-induced PKCζ activation is prevented by angiopoietin-1 (Li *et al.*, 2004). Thus, the temporal interactive effects of the protective and activating signals are complex and not resolved, but determining the specific points of opposition should reveal the most important of the multitude of signaling effects elicited by edemagenic agents. Unlike their oppositional roles on acute barrier function, angiopoietin and VEGF cooperate in angiogenesis as shown in Figure 7.

Metabolic Stress

Interestingly, we have shown that substrate reductive stress can lead to oxidative stress via upregulation of NADPH oxidase in pulmonary artery endothelium as seen in Figure 4. Both high glucose and β-OH butyrate cause a reductive redox stress that paradoxically leads to oxidative stress. This is interesting in light of the endothelial dysfunction, defined by NO underproduction and chronic hyperpermeability, seen in insulin-resistant and diabetic patients. These same levels of glucose, but not mannitol, induced both necrosis and apoptosis in endothelium related to increase expression of Bax (but not anti-apoptotic Bcl2) and Bax translocation to the mitochondria, where Bax induces cytC release and the apoptotic pathway (Nakagami et al., 2001b). They also found, high glucose induced caspase 3 and 9 activation, which led to sustained phosphorylation of p38 MAPK (Nakagami et al., 2002). Heptocyte growth factor attenuated this apoptosis. Moreover, there is strong correlation of microalbuminuria, indicative of endothelial barrier dysfunction, and severity of diabetes. In experimental diabetes, vascular permeability correlated with decreased expression of the tight junction protein occludin and VEGF (Antonetti et al., 1998). Thus, altered junctional protein content may contribute to chronic dysfunction. Finally, the role of chronic inflammation and oxidant stress is now recognized in diabetes. Consistent with these effects of high glucose on oxidant production, high glucose-induced apoptosis in endothelial was dependent on oxidant production, JNK activation, and caspase-3 induction (Ho et al., 2000). Moreover, in diabetic animal models, endothelial and lung structural abnormalities indicate the lung to be an organ affected by diabetes (Popov and Simonescu, 1997).

Hypoxia

A major system to sense hypoxia is status of the transcription factor, HIF1, described above (Wiesener et al., 2003). HIF-1α phosphorylation by p42/p44 MAP kinases indicates cooperation between hypoxic and growth factor signals, leading to increased HIF-1-dependent gene expression (Berra et al., 2000). Hypoxia-induced transcription increases endothelial expression of VEGF, PDGF, and endothelin-1 (Kourembanas et al., 1998; Berra et al., 2000; Semenza, 2000). Hypoxia-induced transcription and translation of angiopoietin-2 was independent of HIF-1 transcription, but rather due to intracellular COX-2- (but not COX-1 or exogenous prostanoid) linked PGE_2/PGI_2 synthesis (Pichiule et al., 2004). In lung epithelial cells, HIF-2α was similarly upregulated by hypoxia, but showed sustained activation and in extended hypoxia, whereas HIF-1α signal waned (Uchida et al., 2004). Similarly, bovine arterial endothelial cells expressed both HIF-1α and HIF-2α in response to hypoxia, but the HIF-2α displayed greater stability and was even expressed in the nucleus under conditions of normoxia (Takahashi et al., 2004).

Oxidants

Oxidants affect endothelial function by: 1) direct acute effects on cellular components, such as altered sulfhydryl-based tertiary and quaternary protein structure that impairs protein function; 2) participating as a signaling molecule, leading to altered transcription; and 3) contributing to endothelial necrosis and apoptosis. Thus, oxidants cause permeability by activation of acute signaling, induction of gene expression, and cytotoxicity at high levels (Kunsch and Medford, 1999; McQuaid and Keenan, 1997; Wolin et al., 1998). The bewildering array of interactions and involvement of oxidant signaling in multiple acute signaling pathways was described in Chapter 4. In lung microvascular endothelial cells, H_2O_2- or diamide-induced barrier dysfunction depends on decreased intracellular GSH and on p38 MAPK, but not ERK1/2 (Usatyuk et al., 2003b). Oxidants, such as H_2O_2 and diperoxovanadate, activate the MAPK cascade by direct effects and through activation of Ras (Natarajan et al., 2001; Takai et al., 2001; Lee and Esselman 2002; Usatyuk et al., 2003b; Yin et al., 2003). MAPK stimulation depended on decreased GSH/GSSG, indicating the importance of redox status in modulating MAPK activity (Kamata and Hirata, 1999; Thannickal and Fanburg, 2000; Usatyuk et al., 2003b). Consistent with this, addition of N-acetylcysteine prior to oxidant exposure partially restored GSH and attenuated the oxidant-induced MAPK phosphorylation and barrier effects (Usatuk et al. 2003b). Activation of p38 MAPK was critical to disruption of tight junctions and the permeability effects of H_2O_2 (Kevil et al., 2001). 4-hydroxynonenal triggered activation of ERK1/2, p38 MAPK, and JNK pathways (Usatyuk et al., 2003b; Usatyuk and Natarajan, 2004), while N-acetylcysteine inhibited JNK, p38 MAPK, AP-1, and NFκB activation (Zafarallah et al., 2003). Similarly, exposure of endothelium to toxic, polychlorinated biphenyls (PCB), resulted in oxidative stress and decreased GSH/GSSG ratios, activation of the JNK/SAPK pathway, and activation of caspase-3 (Slim et al., 2000). The importance of oxidant stress in PCB-induced cytotoxicity was demonstrated by its abrogation with N-acetylcysteine and exacerbation with BSO, an inhibitor of glutathione synthesis. At toxic H_2O_2 levels, cell death in pulmonary artery endothelial cells also depended on the activation of ASK-1 and downstream late phase (2h) activation of the SAPKs, p38 and JNK, and activation of caspase 3 at 18 to 48 hrs (Machino et al., 2003). In a study of lung ischemia/reperfusion injury, which is based in oxidant stress, inhibition of JNK attenuated TNFα upregulation, protein leakage, lung injury, and apoptosis (Ishii et al., 2004). PPARγ activation was also protective in I/R lung injury, determined by significant reductions in albumin extravasation, oxidant stress levels, neutrophil recruitment, TNFa production, and the chemokine - CINC-1 (Ito et al., 2004).

Hyperoxia affects both alveolar epithelial and endothelial cells by oxidant generation and secondary inflammation, resulting in both apoptosis and necrosis with profound effects on pulmonary edema formation (Pagano and Barazzone-Argiroffo, 2003). The oxidant role in TNF-induced effects, including activation of endothelial apoptosis were noted earlier. Members of the Bcl-2 familiy of proteins have both pro- and anti-apopototic effects. Bax is held in check in the cytoplasm in complex with HSP60, upon release under apoptotic conditions it plays an important role in the mitochondrial release of cytC, whereas Akt-activated Bcl-2 and Bcl-x_L suppresses activation of the Bax/caspase apoptotic pathway (Knowlton and Gupta, 2003). Fas null mice, anti- TNF-antibodies, and anti-CD40 antibodies did not protect mice from hyperoxic lung-injury, whereas overexpression of Bcl2 or Bcl-x_L rescued cells from hyperoxic death, indicating the cell receptor pathways were less important than the direct oxidant-mediated signaling (Katoh et al., 1997; Budinger et al., 2002; Pagano and Barazzone-Argiroffo, 2003). Consistent with this observation, N-acetylcysteine anti-oxidant protection significantly reduced hyperoxia-induced pulmonary edema and mortality in rats (Patterson et al., 1985).

Moreover, activation of endothelium with TNF, angiotensin II, thrombin, oxLDL, and high glucose occurs in an oxidant-dependent manner and all signaled upregulation of key components of the NADPH oxidase enzyme complex, through an oxidant-dependent mechanism. LPS upregulation of IL-6 in lung tissue was also oxidant-dependent (Severgnini et al., 2004). Indeed, upregulation of oxidant production is itself oxidant dependent (Chen et al., 2004; Gertzberg et al., 2004; Hsu et al., 2004; unpublished data, C. Patterson). While upregulation of components of the NADPH oxidase are NFκB-dependent, in a cyclical fashion, NFκB upregulation of ICAM-1 is oxidant sensitive, compounding the effects of the activating mediators and reductive substrates. TNF-induced endothelial permeability and TNF-induced gene expression depended on NADPH-oxidant production (Chen et al., 2004; Gertzberg et al., 2004). Endothelin-1 also signaled gp91^{phox} upregulation and ROS production in endothelium (Volk et al., 2000). Interestingly, ERK1/2, p38MAPK, and JNK activation due to angiotensin II was shown to be dependent upon functional NADPH oxidase (Li et al., 2004). Likewise, ox-LDL activation of endothelial cells and smooth muscle cells depends on upregulation and activation of NADPH oxidase, superoxide-mediated p21Ras activation, and Ras activation of the MAPK cascade (Chatterjee, 1998). Hyperoxia also increased endothelial superoxide production *via* activation and upregulation of NADPH oxidase dependent upon MEK, ERK1/2, and p38 MAPK (Parinandi *et al.*, 2003).

Possible immediate downstream targets of oxidants in permeability include cytoskeletal proteins (Garcia *et al.*, 2002; Goldberg *et al.*, 2002; Wang and Doerschuk, 2001), focal adhesion plaque proteins, adherens junction proteins (Vepa *et al.*, 1999; Wang and Doerschuk, 2001), and matrix metalloproteinases (Chen and Wang, 2004). Oxidant stress, shear stress, and advanced glycation end products (from high glucose) induce cadherin endocytosis and lysosomal degradation, which is opposed by stable binding with β-catenin and p120-catenin in pulmonary endothelial cells (Iyer *et al.*, 2004; Vincent *et al.*, 2004). Consistent with this, decreased GSH/GSSG with diamide or with diamide + BSO+BCNU resulted in a slow increase of lung and endothelial cell permeability and a significant decrease in cell VE-cadherin content, indicating the important of cell redox status to enzyme, protein, and physiologic function and expression (Zhao *et al.*, 2001). Downstream transcription factors activated by oxidants in endothelium include, ATF-2, AP-1, E2F, NFκB, Elk-1, and STAT1 (Griendling *et al.*, 2000). The second messenger effects of reactive oxidants activate multiple intracellular proteins and enzymes, including the EDG receptor, c-Src, p38 MAPK, Ras, and Akt, resulting in modulation of endothelial function. Through these pathways oxidants affect prolonged signaling and expression of a pro-inflammatory phenotype (Griendling *et al.*, 2000). Downstream phosphorylation of HSP27 may also play a role in oxidant effects on endothelium (Huot *et al.*, 1997).

Corticosteroids, which have anti-inflammatory effects, decreased expression of NADPH oxidase components (Chatterjee, 1998). In endothelium, cyotokine-induced MCP-1 and IL-6 expression was ROS dependent and suppressed by inhibitors of NADPH oxidase inhibitors (Marumo *et al.*, 1998). Statins, HMG CoA reductase inhibitors which have effects beyond cholesterol reduction, inhibited NADPH oxidase assembly via effects on Rac prenylation affecting its ability to membrane associate (Durrschmidt *et al.*, 2000; Wolfram *et al.*, 2003). Thus, statins decreased endothelial superoxide production (Wagner *et al.*, 2000), p22^{phox} mRNA, and p47 protein (Patterson, Figure 4). In summary, mediator activation and NADPH oxidase upregulation seems to be a central feature in propagation of the activating pathways that lead to both acute effects and to altered transcription.

cAMP/PKA

As introduced in Chapter 2 and reviewed in Chapter 15, cAMP and PKA are extremely important in sustaining barrier intergrity in the state of quiescence, conferring resistance to acute activation by a variety of mediators, and restoring integrity in cells that have been reversibly activated. And as revealed in Chapter 5, cAMP is highly regulated and highly compartmentalized in endothelium. The effectors of cAMP/PKA in sustained endothelial activation are largely undefined. Numerous studies in other cell types suggest possiblities. For instance, cAMP can either inhibit NFkB activation and exacerbate TNF-induced apoptosis (Satriano and Schlondorff, 1994; Yin *et al.*, 2000) or participate in LPS-induced NFkB activation, iNOS expression, and NO production (Chen *et al.*, 1999). In epithelium, hypoxia-induced NFκB activation and expression of TNF, IL-8, and MHC class II antigens (but not ICAM-1) depended on simultaneous suppression of cAMP and CREB activation, which was reversed by cAMP addition (Taylor *et al.*, 1999). Hypoxia-induced permeability in endothelium depended upon a similar reduction in cAMP (Ogawa *et al.*, 1992). Interestingly, in mesangial cells, superoxide generation, but not H_2O_2, activated NFκB, an effect that was prevented by cAMP (Satriano and Schlondorff, 1994). And in human aortic endothelial cells, cAMP elevation inhibited TNF-induced NFκB activation, but not the TNF-activation of MAPK and Akt apthways (Ouchi *et al.*, 2000).

The Long and the Short of It

Oxidant/redox signaling emerges as a critical point of intersection in the interactive signaling cascades. Oxidant have been shown to activate various signaling pathways (Ras proteins, MAPKs, Akt, tyrosine kinases) associated with both acute and prolonged endothelial dysfunction and pathways which result in altered transcription (Greindling *et al.*, 2000; Natarajan *et al.*, 2001; Takai *et al.*, 2001; Lee and Esselman 2002; Usatuk *et al.*, 2003b; Yin *et al.*, 2003; Li *et al.*, 2004). This is particularly interesting in that upregulation of NADPH oxidase and oxidant production is initiated by substrates, bioactive mediators, cytokines, growth factors, signaling lipids, oxidized LDL, oxidant stress itself, and hypoxia (Chatterjee, 1998; Flack *et al.*, 1998; Patterson *et al.*, 1999; Frank *et al.*, 2000; Chen *et al.*, 2004; Gertzberg *et al.*, 2004; Hsu *et al*, 2004) and attenuated by statins, with their proven benefits for endothelial function *in vivo*.

In the tangled web of interacting signals, pathways related to prolonged endothelial activation, p38MAPK is another key signaling factor that stands out at the intersection of many pathways converging on prolonged barrier dysfunction. Permeability induced by hypoxia, hyperoxia, H_2O_2, oxidized lipids, lipid mediators acting *via* Edg receptors, VEGF, TNFα, TGFβ1, and angiotensin II have all been shown to be critically dependent upon p38 MAPK activation (Clauss *et al.*, 2001; Niwa *et al.*, 2001; Goldberg *et al.*, 2002; Kayyali *et al.*, 2002; Issbrucker *et al.*, 2003; Parinandi *et al.*, 2003; Usatyuk *et al.*, 2003b; Li *et al.*, 2004; Usatyuk and Natarajan, 2004). Although it should be noted that p38 inhibition was reported not to block thrombin-induced acute (within 15 min) permeability (Vouret-Craviari *et al.*, 1998). Thus membrane GTPase-coupled receptors, membrane tyrosine kinase receptors, oxidants, small GTPases, and non-receptor tyrosine kinases cooperate in signaling immediate, intermediate, and long term-signaling.

Altered signaling of gene expression in the endothelium is central to its complex responses to activation and inflammation. Likewise, transcription plays a part in the active contribution of endothelium to amplification of the inflammatory response and to coagulopathy. Besides the short-term signaling of barrier dysfunction, identified in Chapter 4, and in addition to signaling involved with the inflammatory response, prolonged signaling influences endothelial health, phenotypic expression, and barrier function. The balance of contact-inhibition and proliferation signaling and the balance of survival and apoptosis signaling are critical. But it is also recognized that in acute lung injury, involvement of other types of lung cells, matrix alterations, differing inflammatory stimuli, differing levels involvement of leukocytes, and other blood elements contribute to the pathophysiology as well. When mechanical, toxic, infectious, or inflammatory injury leads to sloughing or death of endothelial cells, the disruption of the monolayer and the multiple paracrine and regulatory factors and likely the cytokine and growth factor milieu combine to restore the monolayer by increasing motility of remaining cells and by triggering proliferation. Within an hour, cells begin to crawl in to the spaces (temporarily covered by platelets) and within 6 hours they begin dividing. Normally large gaps are not observed and when cells die they remain in place until they are pushed off by the new cells from below. More generally, the cells only need to recover their barrier function, but that may itself involve synthesis of new functional proteins after injury. Likewise in the lung with epithelial injury, migration and phenotype conversion of dividing type II cells in thin type I cells must occur to restore the epithelial barrier. Motility, proliferation, and re-establishment of functional alveolar and vascular pavement require exquisite coordination of signaling. Thus, signaling is involved in the initial response or resistance to activating stimuli, amplification of inflammation, necrotic and apoptotic cells death or resistance to death, and recovery, repair, and restoration. Therefore, gaining an improved understanding of acute and prolonged signaling in the endothelium will help direct our efforts at effective intervention in both acute and chronic pulmonary diseases.

ACKNOWLEDGEMENTS

This review was supported by a Merit Grant from the Veteran's Administration Medical Research Service (CEP)

REFERENCES

Abe, M. and Sato, Y. (2001) cDNA microarray analysis of the gene expression profile of VEGF-activated human umbilical vein endothelial cells. Angiogen. 4:289-298.

Abid, M.R., Guo, S., Minami, T., Spokes, K., Ueki, K., Skurk, C., Walsh, K., and Aird, W.C. (2004) VEGF activates PI3K/Akt/forkhead signaling in endothelial cells. Arterioscl. Thromb. Vasc. Biol. 24:294-300.

Ahn, S.Y., Cho, C., Park, K., Lee, H., Lee, S., Park, S., Lee, I., and Koh, G.Y. (2004) TNFα induces fractalkine expression preferentially in arterial endothelial cells and mithramycin A suppresses TNFα-induced fractalkine expression. Amer. J. Path. 164:1663-1672.

An, S., Goetzl, E.J., and Lee, H. (1998) Signaling mechanisms and molecular characteristics of G-protein coupled receptors for LPA and S1P. J. Cell. Biochem. 30:147-157.

Antonelli-Orlidge, A., Smith, S., and D'Amore, P. (1989) Influence of pericytes on capillary endothelial growth. Amer. Rev. Respir. Dis. 140:1129-1131.

Antonetti, D.A., Barber, A., Khin, S., Lieth, E., Tarbell, J., and Gardner, T.W. (1998) Vascular permeability in experimental diabetes is associated with reduced endothelial occludin content: VEGF decreases occludin in retinal endothelial cells. Diabetes 47:1953-1959.

Bannerman, D.D., Sathyamoorthy, M., and Goldblum, S.E. (1998) Bacterial lipopolysaccharide disrupts endothelial monolayer integrity and survival signaling events through caspase cleavage of adherens junction proteins. J. Biol. Chem. 273:35371-35380.

Benjamin, L.E., Golijanin, D., Itin, A., Pode, D., and Keshet, E. (1999) Selective ablation of immature blood vessels in established human tumors follows VEGF withdrawal. J. Clin. Invest. 103:159-165.

Berk, B.C., Abe, J., Min, W., Surapisitchat, J., and Yan, C. (2001) Endothelial atheroprotective and anti-inflammatory mechanisms. Ann. NY Acad. Sci. 947:93-109.

Berra, E., Milanini, J., Richard, D., Le Gall, M., Vinals, F., Gothie, E., Roux, D., Pages, G., and Pouyssegur, J. (2000) Signaling angiogenesis via p42/p44 MAP kinase and hypoxia. Biochem. Pharm. 60:1171-1178.

Bevilacqua, M.P., Pober, J., Wheeler, M., Cotran, R., Gimbrone, M.A. (1985) IL1 acts on cultured human vascular endothelium to increase the adhesion of PMN, monocytes, and related leukocyte cell lines. J. Clin. Invest. 76:2003-2011.

Bhatia, M. and Moochhala, S. (2004) Role of inflammatory mediators in the pathophysiology of acute respiratory distress syndrome. J. Pathol. 202:145-156.
Birukov, K.G., Jacobson, J., Flores, A., Ye, S., Birukova, A., Verin, A., and Garcia, J.G.N. (2003) Magnitude-dependent regulation of pulmonary endothelial cells barrier function by cyclic stretch. Amer. J. Physiol. 285:L785-L797.
Boguslawski, G., Lyons, D., Harvey, K., Kovala, A., and English, D. (2000) Sphingosylphosphorylcholine induces endothelial cell migration and morphogenesis. Biochem. Biophys. Res. Comm. 272:603-609.
Brizzi, M.F., Battaglia, E., Montrucchio, G., Dentelli, P., Del Sorbo, L., Garbarino, G., Pegoraro, L., and Camussi, G. (1999) Thrombopoietin stimulates endothelial cell motility and neoangiogenesis by a PAF-dependent mechanism. Circ. Res. 84:785-796.
Budinger, G.R., Tso, M., McClintock, D., Dean, D., Sznajder, J., and Chandel, N.S. (2002) Hyperoxia-induced apoptosis does not require mitochondrial reactive oxygen species and is regulated by Bcl-2 proteins. J. Biol. Chem. 277:15654-15660.
Cabrita, M.A. and Christofori, G. (2003) Sprouty proteins: antagonists of endothelial cell signaling and more. Thromb. Haemost. 90:586-590.
Camussi, G., Turello, E., Bussalino, F., and Baglioni, C. (1991) TNF alters cytoskeletal organization and barrier function of endothelial cells. Int. Arch. Allerg. Appl. Immunol. 96:84-91.
Carmeliet, P., Lampugnani, M., Moons, L., Breviario, F., Compernolle, V., Bono, F., Balconi, G., Spagnuolo, R., Oostuyse, B., Dewerchin, M., Zanetti, A., Angellilo, A., Mattot, V., Nuyens, D., Lutgens, E., Clotman, F., deRuiter, M., Gittenberger, G., Poelmann, R., Lupu, F., Herbert, J., Collen, D., and Dejana, E. (1999) Targeted deficiency or cytosolic truncation of VE-cadherin impairs VEGF-mediated endothelial survival and angiogenesis. Cell 98:147-157.
Carrillo, A., Monreal, Y., Ramirez, P., Marin, L., Parrilla, P., Oliver, F., and Yelamos, J. (2004) Transcription regulation of TNFα-early response genes by poly(ADP-ribose) polymerase-1 in murine heart endothelial cells. Nucl. Acids Res. 32:757-766.
Chatterjee, S. (1998) Sphingolipids in vascular biology. Art Thromb Vasc Biol 18:1523-1533.
Chen, C.C., Chiu, K., Sun, Y., and Chen, W.C. (1999) Role of cAMP-PKA pathway in LPS-induced NOS expression in RAW 264.7 macrophages. J. Biol. Chem. 274:31559-31564.
Chen, H.H. and Wang, D.L. (2004) Nitric oxide inhibits MMP-2 expression via the induction of activating transcription factor 3 in endothelial cells. Molec. Pharm. 65:1130-1140.
Chen, X.L., Zhang, Q., Zhao, R., and Medford, R.M. (2004) Superoxide, H_2O_2, and iron are required for TNFα-induced MCP-1 gene expression in endothelial cells: role of Rac1 and NADPH oxidase. Amer. J. Physiol. 286:H1001-H1007.
Cho, D.H., Choi, Y., Jo, S., and Jo, I. (2004) NO production and regulation of eNOS phosphorylation by troglitazone: evidence for involvement of PPARγ-dependent and PPARγ-independent signaling pathways. J. Biol. Chem. 279:2499-2506.
Christians, E.S., Yan, L.J., and Benjamin, I.J. (2002) Heat shock factor 1 and heat shock proteins: critical partners in protection against acute cell injury. Crit. Care Med. 30:S43-S50.
Clauss, M. (2000) Molecular biology of the VEGF and the VEGF receptor family. Sem. Thromb. Hemost. 26:561-569.
Clauss, M., Sunderkotter, C., Sveinbjornsson, B., Hippenstiel, S., Willuweit, A., Marino, M., Haas, E., Seljelid, R., Scheurich, P., Suttorp, N., Grell, M., and Risau, W. (2001) A permissive role for TNF in VEGF-induced vascular permeability. Blood 97:1321-1329.
Daly, C., Wong, V., Burova, E., Wei, Y., Zabski, S., Griffiths, J., Lai, K., Lin, H., Ioffe, E., Yancopoulos, G., and Rudge, J.S. (2004) Angiopoietin-1 modulates endothelial cell function and gene expression via the transcription factor FKHR. Genes & Devel. 18:1060-1071.
D'Angelo, G., Martini, J., Iiri, T., Fantl, W., Martial, J., and Weiner, R.I. (1999) 16K human prolactin inhibits VEGF-induced activation of Ras in capillary endothelial cells. Molec. Endocrin. 13:692-704.
Datta, S.R., Brunet, A., and Greengerg, M.E. (1999) Cellular survival: a play in three Akts. Genes Devel. 13:2905-2927
Davis, M.E., Grumbach, I., Fukai, T., Cutchins, A., and Harrison, D.G. (2004) Shear stress regulates eNOS promoter activity through NFκB binding. J. Biol. Chem. 279:163-168.

Day, R.M., Thiel, G., Lum, J., Chevere, R., Yang, Y., Stevens, J., Sibert, L., and Fanburg, B.L. (2004) HGF regulates ACE expression. J. Biol. Chem. 279:8792-8801.
Dejana, E. (2004) Endothelial cell-cell junctions: happy together. Nat. Rev. Molec. Cell Biol. 5:261-270.
Devin, A., Lin, Y., Yamaoka, S., Li, Z., Karin, M., and Liu, Z. (2001) The α and β subunits of IκB kinase (IKK) mediate TRAF2-dependent IKK recruitment to TNF receptor 1 in response to TNF. Molec. Cell. Biol. 21:3986-3994.
Dixon, I.M.C. (2001) Help from within: cardioprotective properties of hepatocyte growth factor. Cardiovasc. Res. 51:4-6.
Duerrschmidt, N., Wippich, N., Goettsch, W., Broemme, H., and Morawietz, H. (2000) Endothelin-1 induces NAD(P)H oxidase in human endothelial cells. Biochem. Biophys. Res. Comm. 269:713-717.
Eliceiri, B.P., Paul, R., Schwartzberg, P., Hood, J., Leng, J., and Cheresh, D.A. (1999) Selective requirement for Src kinases during VEGF-induced angiogenesis and vascular permeability. Molec. Cell 4:915-924.
Eliceiri, B.P. (2001) Integrin and growth factor receptor crosstalk. Circ. Res. 89:1104-1110.
Elson, D.A., Thurston, G., Huang, L., Ginzinger, D., McDonald, D., Johnson, R., and Arbeit, J.M. (2001) Induction of hypervascularity without leakage or inflammation in transgenic mice overexpressing hypoxia-inducible factor-1α. Genes Dev. 15:2520–2532.
Fischer, D., Liliom, K., Guo, Z., Nusser, N., Virag, T., Murakami-Murofushi, K., Kobayashi, S., Erickson, J., Sun, G., Miller, D., and Tigyi, G. (1998) Naturally occurring analogs of lysophosphatidic acid elicit different cellular responses through selective activation of multiple receptor subtypes. Molec. Pharm. 54:979-988.
Fisher, A.B., Chien, S., Barakat, A., and Nerem, R.M. (2001) Endothelial cellular response to altered shear stress. Amer. J. Physiol. 281:L529-L533.
Flack, J., Hamaty, M., and Staffileno, B. (1998) Renin-angiotensin-aldosterone-kinin system influences on diabetic vascular disease. Miner. Electrolyte Metab. 24:412-422.
Fleming, I. and Busse, R. (2003) Molecular mechanisms involved in the regulation of eNOS. Amer. J. Physiol. 284:R1-R12.
Frank, G., Eguchi, S., Yamakawa, T., Tanaka, S., Inagami, T., and Motley, E. (2000) Involvement of ROS in activation of tyrosine kinase and ERK by angiotensin II. Endocrin 141:3120-3126.
Gamble, J.R., Drew, J., Trezise, L., Underwood, A., Parsons, M., Kasminkas, L., Rudge, J., Yancopoulos, G., and Vadas, M.A. (2000) Angiopoietin-1 is an antipermeability and anti-inflammatory agent in vitro and targets cell junctions. Circ. Res. 87:603-607.
Gerber, H.P., McMurtrey, A., Kowalski, J., Yan, M., Keyt, B., Dixit, V., and Ferrara, N. (1998) VEGF regulates endothelial cell survival through the PI3'-kinase/Akt signal transduction pathway. J. Biol. Chem. 273:30336-30343.
Gertzberg, N., Neumann, P., Rizzo, V., and Johnson, A. (2004) NAD(P)H oxidase mediates the endothelial barrier dysfunction induced by TNF-α. Amer. J. Physiol. 286:L37-L48.
Griendling, K.K., Sorescu, D., Lassegue, B., and Ushio-Fukai, M. (2000) Modulation of protein kinase activity and gene expression by ROS and their role in vascular physiology and pathophysiology. Arterioscler. Thromb. Vasc. Biol. 20:2175-2183.
Gustin, J.A., Pincheira, R., Mayo, L., Ozes, O., Kessler, K., Baerwald, M., Korgaonkar, C., and Donner, D.B. (2004) TNF activates CRE-binding protein through a p38 MAPK/MSK1 signaling pathway in endothelial cells. Amer. J. Physiol. 286:C547-C555.
Hamacher, J., Lucas, R., Lijnen, H., Buschke, S., Dunant, Y., Wendel, A., Grau, G., Suter, P., and Ricou, B. (2002) TNFα and angiostatin are mediators of endothelial cytotoxicity in BAL of patients with ARDS. Amer. J. Respir. Crit. Care Med. 166:651-656.
Hannan, K.M., Dilley, R., deDios, S., Little, P.J. (2003) Troglitazone stimulates repair of the endothelium and inhibits neointima formation in denuded rat aorta. Arterioscl. Thromb. Vasc. Biol. 23:762-768.
Harfouche, R., Gratton, J., Yancopoulos, G., Noseda, M., Karsan, A., and Hussain, S.N. (20030 Angiopoietin-1 activates both anti- and proapoptotic mitogen-activated protein kinases. FASEB J. 17:1523-1525.

Herbert, J.M., Dupuy, E., LaPlace, M., Zini, J., Bar Shavit, R., and Tobelem, G. (1994) Thrombin induces endothelial growth via both a proteolytic and non-proteolytic mechanism. Biochem. J. 303:227-231.
Hippenstiel, S., Krull, M., Ikemann, A., Risau, W., Clauss, M., and Suttorp, N. (1998) VEGF induces hyperpermeability by a direct action on endothelial cells. Amer. J. Physiol. 274:L678-L684.
Ho, F.M., Liu, S.H., Liau, C.S., Huang, P.J., and Lin-Shiau, S.Y. (2000) High glucose-induced apoptosis in human endothelial cells is mediated by sequential activations of c-Jun NH(2)-terminal kinase and caspase-3. Circ. 101:2618-2624.
Hoefen, R.J. and Berk, B.C. (2002) The role of MAP kinases in endothelial activation. Vasc. Pharm. 38:271-273.
Hoffmann, G., Schloesser, M., Czechowski, M., Schobersberger, W., Furhapter, C., and Sepp, N. (2004) TNFα gene expression and release in cultured human dermal microvascular endothelial cells. Exper. Derm. 13:113-119.
Holtmann, H., Enninga, J., Kalble, S., Thiefes, A., Dorrie, A., Broemer, M., Winzen, R., Wilhelm, A., Ninomiya-Tsuji, J., Matsumoto, K., Resch, K., and Kracht, M. (2001) The MAPK kinase kinase TAK1 plays a central role in coupling the IL-1 receptor to both transcriptional and RNA-targeted mechanisms of gene regulation. J. Biol. Chem. 276:3508-3516.
Hsu, Y.H., Chen, J., Chang, N., Chen, C., Liu, J., Chen, T., Jeng, C., Chao, H., and Cheng, T.H. (2004) Role of reactive oxygen species-sensitive extracellular signal-regulated kinase pathway in angiotensin II-induced endothelin-1 gene expression in vascular endothelial cells. J. Vasc. Res. 41:64-74.
Hull, C., McLean, G., Wong, F., Duriez, J., and Karsan, A. (2002) LPS signals an endothelial apoptosis pathway through TNFRAF-6-mediated activation of c-JNK. J. Immunol. 169:2611-2618.
Huot, J., Houl, F., Marceau, F., and Landry, J. (1997) Oxidant stress-induced actin reorganization mediated by the p38 MAPK/HSP27 pathway in vascular endothelial cells. Circ. Res. 80:383-392.
Ilan, N, Tucker, A. and Madri, J.A. (2003) VEGF expression, β-catenin tyrosine phosphorylation, and endothelial proliferation behavior. Lab. Invest. 83:1105-1115.
Inoue, I., Goto, S., Matsunaga, T., Nakajima, T., Awata, T., Hokari, S., Komoda, T., and Katayama, S. (2001) The ligands/activators for PPARγ and PPARα increase Cu2+,Zn2+-superoxide dismutase and decrease p22^{phox} message expressions in primary endothelial cells. Metab. Clin. Exp. 50:3-11.
Ishii, M., Suzuki, Y., Takeshita, K., Miyao, N., Kudo, H., Hiraoka, R., Nishio, K., Sato, N., Naoki, K., Aoki, T., and Yamaguchi, K. (2004) Inhibition of c-Jun NH2-terminal kinase activity improves ischemia/reperfusion injury in rat lungs. J. Immunol. 172:2569-2577.
Issbrucker, K., Marti, H., Hippenstiel, S., Springmann, G., Voswinckel, R., Gaumann, A., Breier, G., Drexler, H., Suttorp, N., and Clauss, M. (2003) p38 MAPK--a molecular switch between VEGF-angiogenesis and vascular hyperpermeability. FASEB J. 17:262-264.
Ito, K., Shimada, J., Kato, D., Toda, S., Takagi, T., Naito, Y., Yoshikawa, T., and Kitamura, N. (2004) Protective effects of preischemic treatment with pioglitazone, a PPARγ ligand, on lung ischemia-reperfusion injury in rats. Eur. J. Cardio-Thor. Surg. 25:530-536.
Iyer, S., Ferreri, D., DeCocco, N., Minnear, F., and Vincent, P.A. (2004) VE-cadherin-p120 interaction is required for maintenance of endothelial barrier function. Amer. J. Physiol. 286:L1143-L1153.
Jackson, S.M., Parhami, F., Xi, X., Berliner, J., Hsueh, W., Law, R., and Demer, L.L. (1999) PPAR activators target human endothelial cells to inhibit leukocyte-endothelial cell interaction. Arterioscl. Thromb Vasc. Biol. 19:2094-2104.
Jin, E., Fujiwara, M., Pan, X., Ghazizadeh, M., Arai, S., Ohaki, Y., Kajiwara, K., Takemura, T., and Kawanami, O. (2003) Protease-activated receptor (PAR)-1 and PAR-2 participate in the cell growth of alveolar capillary endothelium in primary lung adenocarcinomas. Cancer 97:703-713.
Juliano, R.L. (2002) Signal transduction by cell adhesion receptors and the cytoskeleton. Ann. Rev. Pharm. Tox. 42:283-323.

Kadl, A., Huber, J., Gruber, F., Bochkov, V., Binder, B., and Leitinger, N. (2002) Analysis of inflammatory gene induction by oxidized phospholipids in vivo by quantitative real-time RT-PCR in comparison with effects of LPS. Vasc. Pharm. 38:219-227.

Kamata, H. and Hirata, H. (1999) Redox regulation of cellular signaling. Cell Signal 11:1-14.

Kane, L.P., Mollenauer, M., Xu, Z., Turck, C., and Weiss, A. (2002) Akt-dependent phosphorylation specifically regulates Cot induction of NFκB-dependent transcription. Molec. Cell. Biol. 22:5962-5974.

Kaner, R.J., Ladetto, J., Singh, R., Fukuda, N., Matthay, M., and Crystal, R.G. (2000) Lung overexpression of VEGF gene induces pulmonary edema. Amer. J. Respir. Cell Molec. Biol. 22:657-664.

Katoh, S., Mitsui, Y., Kitani, K., and Suzuki, T. (1997) The rescuing effect of nerve growth factor is the result of up-regulation of Bcl-2 in hyperoxia-induced apoptosis of a subclone of pheochromocytoma cells, PC12h. Neurosci. Lett. 232:71-74.

Katsumi, A., Orr, A., Tzima, E., and Schwartz, M.A. (2004) Integrins in mechanotransduction. J. Biol. Chem. 279:12001-12004.

Kevil, C.G., Oshima, T., and Alexander, J.S. (2001) The role of p38 MAPK in H_2O_2 mediated endothelial solute permeability. Endothel. 8:107-116.

Kim, I., Kim, H., Moon, S., Chae, S., So, J., Koh, K., Ahn, B., and Koh, G.Y. (2000a) Angiopoietin-1 induces endothelial cell sprouting through the activation of focal adhesion kinase and plasmin secretion. Circ. Res. 86:952-959.

Kim, I., Kim, H., So, J., Kim, J., Kwak, H., and Koh, G.Y. (2000b) Angiopoietin-1 regulates endothelial cell survival through PI3K/Akt signal transduction pathway. Circ Res 86:24-29.

Kim, I., Moon, S., Kim, S., Kim, H., Koh, Y., and Koh, G.Y. (2001) VEGF expression of ICAM-1, VCAM-1, and E-selectin through NFκB activation in endothelial cells. J. Biol. Chem. 276:7614-7620.

Kimura, T., Watanabe, T., Sato, K., Kon, J., Tomura, H., Tamama, K., Kuwabara, A., Kanda, T., Kobayashi, I., Ohta, H., Ui, M., and Okajima, F. (2000) S1P stimulates proliferation and migration of endothelial cells through lipid receptors, Edg-1 and -3. Biochem. J. 348:71-76.

Knauf, U., Tschopp, C., and Gram, H. (2001) Negative regulation of protein translation by MAPK-Interacting Kinases 1 and 2. Molec. Cell Biol. 21:5500–5511.

Knowlton, A.A. and Gupta, S. (2003) HSP60, Bax, and cardiac apoptosis. Cardiovasc. Tox. 3:263-268.

Koblizek, T.I., Weiss, C., Yancopoulos, G., Deutsch, U., and Risau, W. (1998) Angiopoietin-1 induces sprouting angiogenesis in vitro. Curr. Biol. 8:529-532.

Kondapalli, J., Flozak, A., and Albuquerque, M.L. (2004) Laminar shear stress differentially modulates gene expression of p120 catenin, Kaiso transcription factor, and VE-cadherin in human coronary artery endothelial cells. J. Biol. Chem. 279:11417-11424.

Kontos, C.D., Stauffer, T., Yang, W., York, J., Huang, L., Blanar, M., Meyer, T., and Peters, K.G. (1998) Tyrosine 1101 of Tie2 is the major site of association of p85 and is required for activation of phosphatidylinositol 3-kinase and Akt. Molec. Cell Biol. 18:4131-4140.

Kontos, C.D., Cha, E., York, J., and Peters, K.G. (2002) The endothelial receptor tyrosine kinase Tie1 activates PI3K and Akt to inhibit apoptosis. Molec. Cell. Biol. 22:1704-1713.

Kourembanas, S., Morita, T., Christou, H., Liu, Y., Koike, H., Brodsky, D., Arthur, V., and Mitsial, S.A. (1998) Hypoxic responses of vascular cells. Chest 114:25S-28S.

Kronke, G., Bochkov, V., Huber, J., Gruber, F., Bluml, S., Furnkranz, A., Kadl, A., Binder, B., and Leitinger, N. (2003) Oxidized phospholipids induce human heme oxygenase-1 involving activation of cAMP-responsive element-binding protein. J. Biol. Chem 278:51006-51014.

Kunsch, C. and Medford, R.M. (1999) Oxidative stress as a regulator of gene expression in the vasculature. Circ. Res. 85:753-766.

Kureishi, Y., Luo, Z., Shiojima, I., Bialik, A., Fulton, D., Lefer, D., Sessa, W., and Walsh, K. (2000) The HMG-CoA reductase inhibitor simvastatin activates Akt and promotes angiogenesis in normocholesterolemic animals. Nat. Med. 6:1004-10010.

Lampugnani, M., Zanetti, A., Corada, M., Takahashi, T., Balconi, G., Breviario, F., Orsenigo, F., Cattelino, A., Kemler, R., Daniel, T., and Dejana, E. (2003) Contact inhibition of VEGF-

induced proliferation requires vascular endothelial cadherin, β-catenin, and the phosphatase DEP-1/CD148. J. Cell Biol. 161:793-804.
Lang, D., Mosfer, S., Shakesby, A., Donaldson, F., and Lewis, M.J. (2000) Coronary microvascular endothelial redox state in LV hypertrophy. Circ. Res. 86:463-469.
Le Cras, T.D., Hardie, W., Fagan, K., Whitsett, J., and Korfhagen, T.R. (2003) Disrupted pulmonary vascular development and pulmonary hypertension in transgenic mice overexpressing tGF-α. Amer. J. Physiol. 285:L1046-L1054.
Le Cras, T.D., Markham, N., Tuder, R., Voelkel, N., and Abman, S.H. (2002) Treatment of newborn rats with a VEGF receptor inhibitor causes pulmonary hypertension and abnormal lung structure. Amer. J. Physiol. 283:L555-L562.
Le Cras, T.D., Spitzmiller, R., Albertine, K., Greenberg, J., Whitsett, J., and Akeson, A.L. (2004) VEGF causes pulmonary hemorrhage, hemosiderosis, and air space enlargement in neonatal mice. Amer. J. Physiol. 287:L134-L142.
Lee, K. and Esselman, W.J. (2002) Inhibition of PTPs by H_2O_2 regulates the activation of distinct MAPK pathways. Free Rad. Biol. Med. 33:1121-1132.
Lee, M.J., Thangada, S., Claffey, K., Ancellin, N., Liu, C., Kluk, M., Volpi, M., Sha'afi, R., and Hla, T. (1999) Vascular endothelial cell adherens junction assembly and morphogenesis induced by S1P. Cell 99:301-312.
Lee, S.D., Lee, D., Chun, Y., Paik, S., Kim, W., Kim, D., Kim, W., Tuder, R., and Voelkel, N.F. (2000) Tgf-β1 induces endothelin-1 in a bovine pulmonary artery endothelial cell line and rat lungs via cAMP. Pulm. Pharm. Therap. 13:257-265.
Lehoux, S. and Tedgui, A. (2003) Cellular mechanics and gene expression in blood vessels. J. Biomech. 36:631-643.
Leitinger, N., Huber, J., Rizza, C., Mechtcheriakova, D., Bochkov, V., Koshelnick, Y., Berliner, J., Binder, B.R. (2001) The isoprostane 8-iso $PGF_{2\alpha}$ stimulates endothelial cells to bind monocytes: differences from thromboxane activation. FASEB J. 15:1254-1256.
Lerner-Marmarosh, N., Yoshizumi, M., Che, W., Surapisitchat, J., Kawakatsu, H., Akaike, M., Ding, B., Huang, Q., Yan, C., Berk, B.C., and Abe, J. (2003) Inhibition of TNFα-induced SHP-2 phosphatase by shear stress: reduced endothelial inflammation. Arterioscl. Thromb. Vasc. Biol. 23:1775-1781.
Li, X., Hahn, C., Parsons, M., Drew, J., Vadas, M., and Gamble, J.R. (2004) Role of PKCζ in thrombin-induced endothelial permeability changes: Inhibition by angiopoietin-1. Blood Published online.
Liu, L., Tsai, J.C., and Aird, W.C. (2000) Egr-1 gene is induced by the systemic administration of the VEGF and the epidermal growth factor. Blood 96:1772-1781.
Liu, F., Verin, A., Borbiev, T., and Garcia, J.G. (2001) Role of PKA activity in endothelial cell cytoskeleton rearrangement. Amer. J. Physiol. 280:L1309-L1317.
Loughna, S. and Sato, T.N. (2001) Angiopoietin and Tie signaling pathways in vascular development. Matrix Biol. 20:319-325.
Lusis, A.J. (2000) Atherosclerosis. Nature 407:233-241.
Machino, T., Hashimoto, S., Maruoka, S., Gon, Y., Hayashi, S., Mizumura, K., Nishitoh, H., Ichijo, H., and Horie, T. (2003) ASK1-mediated signaling regulates H_2O_2-induced apoptosis in human pulmonary vascular endothelial cells. Crit. Care Med. 31:2776-2781.
Madge, L.A., Li, J., Choi, J., and Pober, J.S. (2003) Inhibition of PI3K sensitizes vascular endothelial cells to cytokine-initiated cathepsin-dependent apoptosis. J. Biol. Chem. 278:21295-21306.
Madge, L.A. and Pober, J.S. (2000) PI3K/Akt pathway, activated by TNF or IL-1, inhibits apoptosis but does not activate NFκB in human endothelial cells. J. Biol. Chem. 275:15458-15465.
Marumo, T., Schini-Kerth, V., Brandes, R., and Busse, R. (1998) Glucocorticoids inhibit O_2^- and $p22^{phox}$ in SMC. Hypertens 32:1083-1088.
Maziere, C., Conte, M., and Maziere, J.C. (2001) Activation of JAK2 by the oxidative stress generated with oxidized low-density lipoprotein. Free Rad. Biol. Med. 31:1334-1340.
McQuaid, K.E. and Keenan, A.K. (1997) Endothelial barrier dysfunction and oxidative stress: roles for nitric oxide. Exp. Physiol. 82:233-241.

Meissner, M., Stein, M., Urbich, C., Reisinger, K., Suske, G., Staels, B., Kaufmann, R., and Gille, J. (2004) PPARα activators inhibit VEGFR2 expression by repressing Sp1-dependent DNA binding and transactivation. Circ. Res. 94:324-332.
Meyer, M., Clauss, M., Lepple-Wienhues, A., Waltenberger, J., Augustin, H. G., Ziche, M., Lanz, C., Buttner, M., Rziha, H. J., and Dehio, C. (1999) A novel VEGF encoded by Orf virus, VEGF-E, mediates angiogenesis via signalling through VEGFR-2 (KDR) but not VEGFR-1 (Flt-1) receptor tyrosine kinases. EMBO J. 18:363-374.
Meyer, R.D., Dayanir, V., Majnoun, F., and Rahimi, N. (2002) The presence of a single tyrosine residue at the carboxyl domain of VEGFR2/FLK-1 regulates its autophosphorylation and activation of signaling molecules. J. Biol. Chem. 277:27081-27087.
Migita, H., Satozawa, N., Lin, J., Morser, J., and Kawai, K. (2004) RORα1 and RORα4 suppress TNFα-induced VCAM-1 and ICAM-1 expression in endothelium. FEBS Lett. 557:269-274.
Minagar, A., Long, A., Ma, T., Jackson, T., Kelley, R., Ostanin, D., Sasaki, M., Warren, A., Jawahar, A., Cappell, B., and Alexander, J.S. (2003) IFN-β1a and IFN-β1b block IFN-γ-induced disintegration of endothelial junction integrity and barrier. Endoth. 10:299-307.
Minami, T., Sugiyama, A., Wu, S., Abid, R., Kodama, T., and Aird, W.C. (2004) Thrombin and phenotypic modulation of the endothelium. Arterioscler. Thromb. Vasc. Biol. 24:41-53.
Minshall, R.D., Sessa, W., Stan, R., Anderson, R., and Malik AB. (2003) Caveolin regulation of endothelial function. Amer. J. Physiol. 285:L1179-L1183.
Miralpeix, M., Camacho, M., Lopez-Belmonte, J., Canalias, F., Beleta, J., Palacios, J., and Vila, L. (1997) Selective induction of COX-2 activity in HUVEC-C. Brit. J. Pharm. 121:171-180.
Moll, T., Czyz, M., Holzmuller, H., Hofer-Warbinek, R., Wagner, E., Winkler, H., Bach, F., and Hofer, E. (1995) Regulation of the tissue factor promoter in endothelial cells. Binding of NF κB-, AP-1-, and Sp1-like transcription factors. J. Biol. Chem. 270:3849-3857.
Morales-Ruiz, M., Lee, M., Zollner, S., Gratton, J., Scotland, J., Shiojima, I., Walsh, K., Hla, T., and Sessa, W.C. (2001) S1P activates Akt, NO production, and chemotaxis through a Gi/PI3K pathway in endothelial cells. J. Biol. Chem. 276:19672-19677.
Mortimore, G.E., Poso, A.R., and Lardeux, B.R. (1989) Mechanism and regulation of protein degradation in liver. Diabetes-Metab. Rev. 5:49-70.
Mulder, K.M. (2000) Role of Ras and MAPKs in TGFβ signaling. Cytok. Growth Factor Rev. 11:23-35.
Nagase, M., Hirose, S., and Fujita, T. (1998) Unique repetitive sequence and unexpected regulation of expression of rat endothelial receptor for oxidized LDL (LOX-1). Biochem. J. 330:1417-1422.
Naik, M.U., Vuppalanchi, D., and Naik, U.P. (2003) Essential role of JAM-1 in bFGF-induced endothelial cell migration. Arterioscl. Thromb. Vasc. Biol. 23:2165-2171.
Nakagami, H., Morishita, R., Yamamoto, K., Taniyama, Y., Aoki, M., Matsumoto, K., Nakamura, T., Kaneda, Y., Horiuchi, M., and Ogihara, T. (2001a) Mitogenic and antiapoptotic actions HGF through ERK, Stat3, and Akt in endothelial cells. Hypert. 37:581-586.
Nakagami, H., Morishita, R., Yamamoto, K., Taniyama, Y., Aoki, M., Yamasaki, K., Matsumoto, K., Nakamura, T., Kaneda, Y., Ogihara, T. (2002) HGF prevents endothelial cell death through inhibition of bax translocation from cytosol to mitochondrial membrane. Diabetes 51:2604-2611.
Nakagami, H., Morishita, R., Yamamoto, K., Yoshimura, S., Taniyama, Y., Aoki, M., Matsubara, H., Kim, S., Kaneda, Y., and Ogihara, T. (2001b) Phosphorylation of p38 MAPK downstream of bax-caspase-3 leads to cell death induced by high D-glucose in human endothelial cells. Diabetes. 50:1472-1481.
Napoli, C., Quehenberger, O., De Nigris, F., Abete, P., Glass, C., and Palinski, W. (2000) Mildly oxLDL activates multiple apoptotic signaling pathways in human coronary cells. FASEB J. 14:1996-2007.
Neumann, P., Gertzberg, N., and Johnson, A. (2004) TNF-α induces a decrease in eNOS promoter activity. Amer. J. Physiol. 286:L452-L459.
Newman, P.J. and Newman, D.K. (2003) Signal transduction pathways mediated by PECAM-1. Arterioscl. Thromb. Vasc. Biol. 23:953-964.
Nubel, T., Dippold, W., Kleinert, H., Kaina, B., and Fritz, G. (2004) Lovastatin inhibits Rho-expression of E-selectin by TNFα and attenuates tumor cell adhesion. FASEB J. 18:140-142.

Ogawa, S., Koga, S., Kuwabara, K., Brett, J., Morrow, B., Morris, S.A., Bilezikian, J.P., Silverstein, S.C., and Stern, D. (1992) Hypoxia-increased permeability of endothelial monolayers occurs through lowering of cAMP levels. Amer. J. Physiol. 262:C546-C554.
Olson, E.N. (2004) Undermining the endothelium by ablation of MAPK-MEF2 signaling. J. Clin. Invest. 113:1110-1112.
Osawa, M., Masuda, M., Kusano, K., and Fujiwara, K. (2002) Evidence for a role of PECAM-1 in endothelial cell mechanosignal transduction. J. Cell Biol. 158:773-785.
Ouchi, N., Kihara, S., Arita, Y., Okamoto, Y., Maeda, K., Kuriyama, H., Hotta, K., Nishida, M., Takahashi, M., Muraguichi, M., Ohmoto, Y., Nakamura, T., Yamashita, S., Funahashi, T., and Matsuzawa, Y. (2000) Adiponectin inhibits endothelial NFκB signaling through a cAMP-dependent pathway. Circ. 102:1296-1301.
Ozawa, N., Shichiri, M., Iwashina, M., Fukai, N., Yoshimoto, T., and Hirata, Y. (2004) Laminar shear stress up-regulates iNOS in the endothelium. Hyperten. Res. Clin. Exp. 27:93-99.
Pagano, A. and Barazzone-Argiroffo, C. (2003) Alveolar cell death in hyperoxia-induced lung injury. Ann. NY Acad. Sci. 1010:405-416.
Papapetropoulos, A., Fulton, D., Mahboubi, K., Kalb, R., O'Connor, D., Li, F., Altieri, D., and Sessa, W.C. (2000) Angiopoietin-1 inhibits endothelial cell apoptosis via the Akt/survivin pathway. J. Biol. Chem. 275:9102-9105.
Parinandi, N.L., Kleinberg, M., Usatyuk, P., Cummings, R., Pennathur, A., Cardounel, A., Zweier, J., Garcia, J., and Natarajan, V. (2003) Hyperoxia-induced NAD(P)H oxidase activation and regulation by MAPK in human lung endothelial cells. Amer. J. Physiol. 284:L26-L38.
Patterson, C., Ruef, J., Madamanchi, N., Barry, P., Hu, Z., Horaist, C., Ballinger, C., Brasier, A., and Bode, C. (1999) Stimulation of vascular smooth muscle NAD(P)H oxidase by thrombin. J. Biol. Chem. 174:19814-19822.
Patterson, C.E., Rhodes, M., Butler, J. and Byrne, F. (1985) Protection from normobaric hyperoxia by cysteamine and n-acetylcysteine. Lung 163:23-32.
Patterson, C.E., Stasek, J., Bahler, C., Harrington, M., and Garcia, J.G.N. (1998) Regulation of IL-1-stimulated GMCSF mRNA levels in human endothelium. Endothel. 6:45-59.
Pepper, M.S. (1997) TGF-β: vasculogenesis, angiogenesis, and vessel wall integrity. Cytok. Growth Factor Rev. 8:21-43.
Pepper, M.S., Vassalli, J., Orci, L., and Montesano, R. (1993) Biphasic effects of transforming growth factor-β on in vitro angiogenesis. Exp. Cell Res. 204:356-363.
Peters, K.G., Kontos, C., Lin, P., Wong, A., Rao, P., Huang, L., Dewhirst, M., and Sankar, S. (2004) Functional significance of Tie2 signaling in the adult vasculature. Rec. Prog. Horm. Res. 59:51-71.
Petrache, I., Birukov, A., Ramirez, S., Garcia, J., and Verin, A.D. (2003a) The role of microtubules in TNFα-induced endothelial cell permeability. Amer. J. Respir. Cell Molec. Biol. 28:574-581.
Petrache, I., Birukov, a., Zaiman, A., Crow, M., Deng, H., Wadgaonkar, R., Romer, L., and Garcia, J.G. (2003b) Caspase-dependent cleavage of MLCK is involved in TNFα-mediated bovine pulmonary endothelial cell apoptosis. FASEB J. 17:407-416.
Petrache, I., Crow, M., Neuss, M., and Garcia, J.G. (2003c) Central involvement of Rho GTPases in TNFα-mediated pulmonary endothelial apoptosis. Biochem. Biophys. Res. Comm. 306:244-249.
Pianetti, S., Arsura, M., Romieu-Mourez, R., Coffey, R., and Sonenshein, G.E. (2001) Her-2/neu overexpression induces NFκB via a PI3K/Akt pathway involving calpain-mediated degradation of IκB-α that can be inhibited by the tumor suppressor PTEN. Oncogene 20:1287-1299.
Pichiule, P., Chavez, J.C., and LaManna, J.C. (2004) Hypoxic regulation of angiopoietin-2 expression in endothelial cells. J. Biol. Chem. 279:12171-12180.
Pirkkala, L., Nykanen, P., and Sistonen, L. (2001) Roles of the heat shock transcription factors in regulation of the heat shock response and beyond. FASEB J. 15:1118-1131.
Plutzky, J. (2001) PPAR in endothelial cell biology. Curr. Opin. in Lipidol. 12:511-518.
Pober, J.S. (2002) Endothelial activation: Intracellular signaling pathways. Arthritis Res. 4:S109-S116.
Popov, D. and Simonescu, M. (1997) Alterations of lung structure in experimental diabetes, and diabetes associated with hyperlipidemia in hamsters. Eur. Respir. J. 10:1850-1858.

Pittet, J.F., Mackersie, R., Martin, T., and Matthay, M.A. (1997) Biomarkers of acute lung injury: prognostic and pathogenetic significance. Amer. J. Respir. Crit. Care Med. 155:1187-1205.

Rifkin, D.B., Moscatelli, D., Flaumenhauft, R., Sato, Y., Saksela, O., and Tsuboi, R. (1991) Mechanisms controlling the extracellular activity of bFGF and TGFβ. Ann. NY Acad. Sci. 614:250-258.

Rogers, R.J., Monnier, J.M., and Nick, H.S. (2001) TNFα induces MnSOD expression via mitochondria-to-nucleus signaling, whereas IL-1β utilizes an alternative pathway. J. Biol. Chem. 276:20419-20427.

Romashkova, J.A. and Makarov, S.S. (1999) NFkB is a target of AKT in anti-apoptotic PDGF signalling. Nature 401:86-90.

Rousseau, S., Houle, F., Landry, J., and Huot, J. (1997) p38 MAPK activation by VEGF mediates actin reorganization and cell migration in human endothelial cells. Oncogene 15:2169-2177.

Ruegg, C. and Mariotti, A. (2003) Vascular integrins: pleiotropic adhesion and signaling molecules in vascular homeostasis and angiogenesis. Cell. Molec. Life Sci. 60:1135-1157.

Said, F.A., Werts, C., Elalamy, I., Couetil, J., Jacquemin, C., and Hatmi, M. (2002) TNFα, inefficient by itself, potentiates IL-1β-induced PGHS-2 expression in human pulmonary microvascular endothelial cells: requirement of NFκB and p38 MAPK pathways. Brit. J. Pharm. 136:1005-1014.

Sassa, Y., Hata, Y., Aiello, L., Taniguchi, Y., Kohno, K., and Ishibashi, T. (2004) Bifunctional properties of PPARγ1 in KDR gene regulation mediated via both Sp1 and Sp3. Diabetes 53:1222-1229.

Satriano, J. and Schlondorff, D. (1994) Activation and attenuation of NFκB in mouse mesangial cells in response to TNF, IgG, and cAMP: Evidence of ROS. J. Clin. Invest. 94:1629-1636.

Schorpp-Kistner, M., Wang, Z., Angel, P., and Wagner, E.F. (1999) JunB is essential for mammalian placentation. EMBO J. 18:934-948.

Semenza, G.L. (2000) Oxygen-regulated transcription factors in pulmonary disease. Resp. Res. 1:159-162.

Severgnini, M., Takahashi, S., Rozo, L., Homer, R., Kuhn, C., Jhung, J., Perides, G., Steer, M., Hassoun, P., Fanburg, B., Cochran, B., and Simon, A.R. (2004) Activation of the STAT pathway in acute lung injury. Amer. J. Physiol. 286:L1282-L1292.

Shimamura, K., Takashiro, Y., Akiyama, N., Hirabayashi, T., and Murayama, T. (2004) Expression of adhesion molecules by S1P and histamine in endothelial cells. Eur. J. Pharmacol. 486:141-150.

Shiojima, I. and Walsh, K. (2002) Role of Akt signaling in vascular homeostasis and angiogenesis. Circ. Res. 90:1243-1250.

Shtutman, M., Zhurinsky, J., Simcha, I., Albanese, C., D'Amico, M., Pestell, R., and Ben-Ze'ev, A. (1999) The cyclin D1 gene is a target of the β-catenin/LEF-1 pathway. PNAS USA 96:5522-5527.

Skurk, C., Maatz, H., Kim, H., Yang, J., Abid, M., Aird, W., and Walsh, K. (2004) The Akt-regulated forkhead transcription factor FOXO3a controls endothelial cell viability through modulation of the caspase-8 inhibitor FLIP. J. Biol. Chem. 279:1513-1525.

Slim, R., Toborek, M., Robertson, L., Lehmler, H., and Hennig, B. (2000) Cellular glutathione status modulates polychlorinated biphenyl-induced stress response and apoptosis in vascular endothelial cells. Tox. Appl. Pharm. 166:36-42.

Spiegel, S. and Milstien, S. (2000) Functions of a new family of sphingosine-1-phosphate receptors. Biochim. Biophys. Acta 1484:107-116.

Su, Y., Zharikov, S.I., and Block, E.R. (2002) Microtubule-active agents modify nitric oxide production in pulmonary artery endothelial cells. Amer. J. Physiol. 282:L1183-L1189.

Suhardja, A. and Hoffman, H. (2003) Role of growth factors and their receptors in roliferation and microvascular endothelial cells. Micros. Res. Tech. 60:70-75.

Sun, L., Vitolo, M., Qiao, M., Anglin, I., and Passaniti, A. (2004) Regulation of TGFβ1-mediated growth inhibition and apoptosis by RUNX2 in endothelial cells. Oncogene 23:4722-4734.

Takahashi, R., Kobayashi, C., Kondo, Y., Nakatani, Y., Kudo, I., Kunimoto, M., Imura, N., and Hara, S. (2004) Subcellular localization and regulation of HIF-2α in vascular endothelial cells. Biochem. Biophys. Res. Comm. 317:84-91.

Taylor, C.T., Fueki, N., Agah, A., Hershberg, R., and Colgan, S.P. (1999) Critical role of CREB expression in hypoxia-elicited epithelial TNF. J. Biol. Chem. 274:19447-19454.

Thurston, G., Rudge, J., Ioffe, E., Zhou, H., Ross, L., Croll, S., Glazer, N., Holash, J., McDonald, D., and Yancopoulos,G.D. (2000) Angiopoietin-1 protects the adult vasculature against plasma leakage. Nat. Med. 6:460-463.

Thurston, G., Suri, C., Smith, K., McClain, J., Sato, T., Yancopoulos, G., and McDonald, D.M. (1999) Leakage-resistant blood vessels in mice overexpressing angiopoietin-1. Science 286:2511-2514.

Tomashefski, J.F. Jr., Davies, P., Boggis, C., Greene, R., Zapol, W., and Reid, L.M. (1983) The pulmonary vascular lesions of ARDS. Amer. J. Path. 112:112-126.

Tsopanoglou, N.E. and Maragoudakis, M.E. (1999) On the mechanism of thrombin-induced angiogenesis. J. Biol. Chem. 274:23969-23976.

Tuder, R.M., Zhen, L., Cho, C., Taraseviciene-Stewart, L., Kasahara, Y., Salvemini, D., Voelkel, N., and and Flores, S.C. (2003) Oxidative stress and apoptosis interact and cause emphysema due to VEGF receptor blockade. Amer. J. Respir. Cell Molec. Biol. 29:88-97.

Uchida, T., Rossignol, F., Matthay, M.A., Mounier, R., Couette, S., Clottes, E., and Clerici, C. (2004) Prolonged hypoxia differentially regulates HIF-1α and HIF-2α expression in lung epithelial cells: implication of natural antisense HIF-1α. J. Biol. Chem. 279:14871-14878.

Vestweber, D. (2000) Molecular mechanisms that control endothelial cell contacts. J. Pathol. 190:281-291.

Viemann, D., Goebeler, M., Schmid, S., Klimmek, K., Sorg, C., Ludwig, S., and Roth, J. (2004) Transcriptional profiling of IKK2/NF-κB- and p38 MAP kinase-dependent gene expression in TNFα-stimulated primary human endothelial cells. Blood 103:3365-3373.

Vincent, P.A., Xiao, K., Buckley, K., and Kowalczyk, A.P. (2004) VE-cadherin: adhesion at arm's length. Amer. J. Physiol. 286:C987-C997.

Volk, T., Hensel, M., Schuster, H., and Kox, W. (2000) Secretion of MCP-1 and IL-6 by cytokine stimulated production of ROS in endothelial cells. Molec. Cell Biochem. 206:105-112.

Vouret-Craviari, V., Bourcier, C., Boulter, E., and van Obberghen-Schilling, E. (2002) Distinct signals via Rho GTPases and Src drive shape changes by thrombin and sphingosine-1-phosphate in endothelial cells. J. Cell Sci. 115:2475-2484.

Wagner, A.H., Kohler, T., Ruckschloss, U., Just, I., and Hecker, M. (2000) Improvement of NO-dependent vasodilation by HMG-CoA reductase inhibitors through attenuation of endothelial O_2^- production. Arterioscl. Throm. Vasc. Biol. 20:61-69.

Wagner, A.H., Schwabe, O., and Hecker, M. (2002) Atorvastatin inhibition of cytokine-inducible NOS expression in native endothelial cells in situ. Brit. J. Pharm. 136:143-149.

Wang, H., Fang, R., Cho, J., Libermann, T., and Oettgen, P. (2004a) Positive and negative modulation of the transcriptional activity of the ETS factor ESE-1 through interaction with p300, CREB-binding protein, and Ku 70/86. J. Biol. Chem. 279:25241-25250.

Wang, J., Morita, I., Onodera, M., and Murota, S.I. (2002) Induction of KDR expression in bovine arterial endothelial cells by thrombin. J. Cell Physiol. 190:238-250.

Wang, Q. and Doerschuk, C.M. (2001) p38 MAPK mediates cytoskeletal remodeling in pulmonary microvascular endothelial cells on ICAM-1 ligation. J. Immunol. 166:6877-6884.

Wang, Y., Pampou, S., Fujikawa, K., and Varticovski, L. (2004b) Opposing effect of angiopoietin-1 on VEGF-disruption of endothelial cell-cell interactions requires PKCβ. J. Cell. Physiol. 198:53-61.

Wang, X., Flynn, A., Waskiewicz, A., Webb, B., Vries, R., Baines, I., Cooper, J., and Proud, C.G. (1998) The phosphorylation of eukaryotic initiation factor eIF4E in response to phorbol esters, cell stresses, and cytokines is mediated by distinct MAP kinase pathways. J. Biol. Chem. 273:9373-9377.

Watanabe, D., Takagi, H., Suzuma, K., Suzuma, I., Oh, H., Ohashi, H., Kemmochi, S., Uemura, A., Ojima, T., Suganami, E., Miyamoto, N., Sato, Y., and Honda, Y. (2004) Transcription factor

Ets-1 mediates ischemia- and VEGF-dependent retinal neovascularization. Amer. J. Path. 164:1827-1835.

Wiesener, M.S., Jurgensen, J., Rosenberger, C., Scholze, C., Horstrup, J., Warnecke, C., Mandriota, S., Bechmann, I., Frei, U., Pugh, C., Ratcliffe, P., Bachmann, S., Maxwell, P., and Eckardt, K.U. (2003) Widespread hypoxia-expression of HIF-2α in distinct cells of different organs. FASEB J. 17:271-273.

Witmer, A.N., van Blijswijk, B., van Noorden, C., Vrensen, G., and Schlingemann, R.O. (2004) In vivo angiogenic phenotype of endothelial cells and pericytes induced by vascular endothelial growth factor-A. J. Histochem. Cytochem. 52:39-52.

Witzenbichler, B., Maisonpierre, P., Jones, P., Yancopoulos, G., and Isner, J.M. (1998) Chemotactic properties of angiopoietin-1 and -2, ligands for the endothelial-specific receptor tyrosine kinase Tie2. J. Biol. Chem. 273:18514-18521.

Wolfrum, S., Jensen, K.S., and Liao, J.K. (2003) Endothelium-dependent effects of statins. Arterioscler. Thromb. Vasc. Biol. 23:729-736.

Wolin, M.S., Davidson, C., Kaminski, P., Fayngersh, R., and Mohazzab-H, K.M. (1998) Oxidant-NO signaling mechanisms in vascular tissue. Biochem. (Moscow) 63:810-816.

Yamakawa, M., Liu, L., Date, T., Belanger, A., Vincent, K., Akita, G., Kuriyama, T., Cheng, S., Gregory, R., and Jiang, C. (2003) Hypoxia-inducible factor-1 mediates activation of cultured vascular endothelial cells by inducing multiple angiogenic factors. Circ. Res. 93:664-673.

Yamawaki, H., Lehoux, S., and Berk, B.C. (2003) Chronic physiological shear stress inhibits TNFα-induced proinflammatory responses in aorta perfused ex vivo. Circ. 108:1619-1625.

Yeh, M., Leitinger, N., de Martin, R., Onai, N., Matsushima, K., Vora, D., Berliner, J., and Reddy, S.T. (2001) Increased transcription of IL-8 in endothelial cells is differentially regulated by TNF-α and oxidized phospholipids. Arterioscl. Thromb. Vasc. Biol. 21:1585-1591.

Yin, G., Yan, C., and Berk, B.C. (2003) Angiotensin II signaling pathways mediated by tyrosine kinases. Int. J. Biochem. Cell Biol. 35:780-783.

Yoshizumi, M., Fujita, Y., Izawa, Y., Suzaki, Y., Kyaw, M., Ali, N., Tsuchiya, K., Kagami, S., Yano, S., Sone, S., and Tamaki, T. (2004) Ebselen inhibits TNFα-induced JNK activation and adhesion molecule expression in endothelial cells. Exp. Cell Res. 292:1-10.

Yu, E.Z., Li, Y., Liu, X., Kagan, E., and McCarron, R.M. (2004) Antiapoptotic action of hypoxia-inducible factor-1 α in human endothelial cells. Lab. Invest. 84:553-561.

Zachary, I. (2001) Signaling mechanisms mediating vascular protective actions of VEGF. Amer. J. Physiol. 280:C1375-C1386.

Zachary, I. and Gliki, G. (2001) Signaling transduction mechanisms mediating biological actions of the VEGF family. Cardiovasc. Res. 49:568-581.

Zafarullah, M., Li, W., Sylvester, J., and Ahmad, M. (2003) Molecular mechanisms of N-acetylcysteine actions. Cell. Molec. Life Sci. 60:6-20.

Zhang, F., Cheng, J., Hackett, N., Lam, G., Shindo, K., Pergolizzi, R., Jin, D., Crystal, R., and Rafii, S. (2004) Adenovirus E4 gene promotes selective endothelial cell survival and angiogenesis via activation of VE-cadherin/Akt signaling. J. Biol. Chem. 279:11760-11766.

Zhang, L., Himi, I., Morita, I., and Murota, S. (2001) Inhibition of PI3K/Akt or MAPK signaling sensitizes endothelial cells to TNFα cytotoxicity. Cell Death Diff. 8:528-536.

Zhao, X., Alexander, J., Zhang, S., Zhu, Y., Sieber, N., Aw, T., and Carden, D.L. (2001) Redox regulation of endothelial barrier integrity. Amer. J. Physiol. 281:L879-L886.

Zhao, L., Zhang, M.M., and Ng, K.Y. (1998) Effects of VPF on the permeability of cultured endothelial cells from brain capillaries. J. Cardiovasc. Pharm. 32:1-4.

Zhu, Y., Liao, H., Wang, N., Ma, K., Verna, L., Shyy, J., Chien, S., and Stemerman, M.B. (2001) LDL-activated p38 in endothelial cells is mediated by Ras. Arterioscl. Thromb. Vasc. Biol. 21:1159-1164.

Chapter 7. Dynamic Microfilaments and Microtubules Regulate Endothelial Function

Joanna Zurawska,[1] Mabel Sze, B.Sc.,[1] Joanne Lee, B.Sc.,[1] M.Sc. & Avrum I. Gotlieb, M.D., C.M.[1]*

[1]Department of Laboratory Medicine and Pathobiology, University of Toronto, and Department of Pathology, and Toronto General Research Institute, University Health Network, Toronto, ON M5G 1L5 Canada

CONTENTS:

Introduction

The Dynamic Role of Actin in Endothelial Cell Function
Endothelial Cell Migration
Actin in Structure and Cellular Processes

Structural Organization and Actin Assembly/Disassembly
Polymerization/Depolymerization
Actin-binding Proteins

Rho GTPases in Regulation of Actin Assembly
Rho GTPases Influence Actin Cytoskeletal Functions
Cellular Targets of Rac and Cdc42
Cellular Targets of Rho

The Endothelial Microtubule Cytoskeleton
Structure and Assembly/Disassembly of Microtubules
Microtubules and Cell Polarity
Crosstalk: Roles of Actin and Microtubules in Migrating Cells

Actin and Cell Adhesion
Cell-Cell Adhesion
Cell-Matrix Adhesion

Actin-Mediated Endothelial Integrity in Health & Disease
Permeability Barrier Dysfunction
Inflammation
Atherosclerosis

Perspective

Introduction

The structural integrity of the vascular endothelium is largely dependent on the function of its cytoskeleton, composed of microfilaments, microtubules, and intermediate filaments (Vyalov *et al.*, 1996; Kim *et al.*, 1989b; Helmke *et al.*, 2000; Lee and Gotlieb, 2003b). Microfilaments, which contain actin, are arranged centrally and peripherally in endothelial cells to form stress fibers (Colangelo *et al.*, 1994) and a dense peripheral band (DPB) (Wong and Gotlieb, 1986), respectively (Figure 1a). Stress fibers are long bundles of actin and myosin filaments that can contract and exert tension (Wong *et al.*, 1983). They are linked to the plasma membrane at sites called focal adhesions, where they promote strong attachment to the substratum (Lee and Gotlieb, 2002). The DPB, on the other hand, plays a role in cell-cell adhesion (Wong and Gotlieb, 1986; Noria *et al.*, 1999; Lee and Gotlieb, 2002). The distribution and organization of the actin and microtubular cytoskeleton can be altered in response to various external signals. For example, low hemodynamic shear stress favors peripheral microfilaments (Vyalov *et al.*, 1996), while both elevated hemodynamic shear stress (Langille *et al.*, 1991) and exposure to thrombin (Wong and Gotlieb, 1990) promote an increase in central microfilaments. In cells exposed to high shear conditions, the central microfilaments become both thicker and longer, while the peripheral band of microfilaments becomes disrupted (Kim *et al.*, 1989a). This is reversible if normal shear is reinstated (Langille *et al.*, 1991). Sustained exposure to flow also induces cytoskeletal remodeling that results in endothelial cell orientation in the direction of flow (Noria *et al.*, 2004). Activation sets the stage for a migration response, an essential process for maintaining and re-establishing integrity of the endothelium, associated with profound, dynamic changes in microfilaments and microtubules, which are complex in nature and which illustrate how the cytoskeleton acts to coordinate and control endothelial structure and function. In this chapter, we will utilize the process of cell migration as a framework to discuss the nature of the dynamic cytoskeleton in endothelial cells.

The Dynamic Role of Actin in Endothelial Cell Function

Endothelial Cell Migration

The ability of a cell to migrate is a milestone achievement of evolution. Migration not only provides a survival advantage by allowing movement away from undesirable environments, but the orchestration of a moving body of cells permits development of complex, multicellular organisms. The profoundly important mechanisms regulating cell migration have not been fully elucidated and are under intense study. Migration of individual cells has been characterized as a cycle of four steps in which microfilaments and microtubules are major players: extension of a leading edge protrusion (lamellipodium), adhesion to the substratum at the front, cell body contraction, and detachment of adhesions at the rear of the cell (Sheetz *et al.*, 1998; Giannone *et al.*, 2004; Raftopoulou and Hall, 2004). Lamellipodia are protrusive, sheet-like structures formed from a highly branched network of intermediate-length actin filaments. Through elongation and branching of these filaments, lamellipodia push the ruffling membrane forward (Small *et al.*, 2002b). The region behind the lamellipodia is called the lamella. Filopodia are microspike-structures that protrude from areas of the lamellipodia, and are constructed by radiating actin bundles arising from within the lamellipodia meshwork (Kaverina *et al.*, 2002b). Filopodia may function to sample the extracellular milieu and transmit signals that direct migration. In both filopodia and lamellipodia, F-actin is arranged with barbed ends subadjacent to the plasma membrane and pointed ends toward the cell interior (Rodriguez *et al.*, 2003). Both are also important as they form sites of adhesion, termed 'focal complexes', with the underlying substratum during migration. Behind the leading edge, other actin constructs exist. Stress fibers run from the cytosol towards the lamellipodia, and they form focal adhesions, which are specialized structures that contain a large number of cytoskeletal proteins and signaling molecules (Petit and Thiery, 2000b). Focal adhesions anchor the cell as lamellipodia project towards the direction of migration and the cell moves along the substratum. As the lamellipodium reaches forward, it grasps the underlying matrix via focal complexes. Meanwhile, contraction of stress fibers propels the cell body ahead. Retraction of the rear of the cell is also promoted by cell-substratum detachment (Lee and Gotlieb, 2003b).

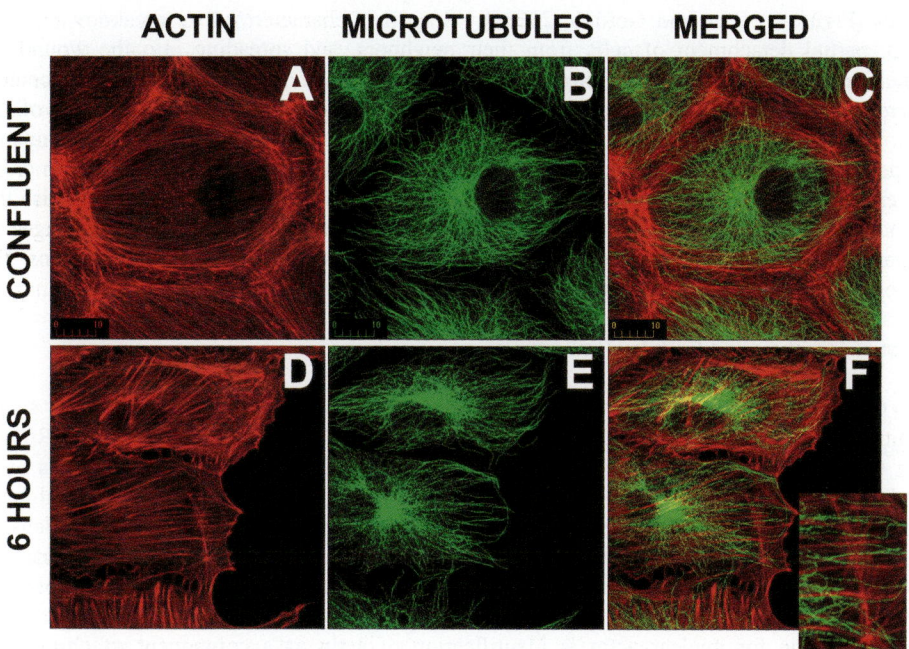

Figure 1. The redistribution of microtubules during wound repair. Porcine aortic endothelial cells were grown to confluency and a 1 mm wound was created using a scraper. Cells were fixed and stained for actin microfilaments and microtubules at confluency and 6 hours after wounding (w – wound edge). In a confluent monolayer, the microtubules emanate from the centrosome to the periphery of the cells. In a monolayer undergoing wound repair, the cells at the wound edge, at 6 hours, show reposition of the centrosome between the cell nucleus and the leading edge of the cell. Inset shows some alignment of some microtubules with microfilaments (see Figure 2 for details of actin redistribution at the wound edge) [1200x magnification].

The morphological changes that accompany cell migration have been studied extensively in many models, one example being *in vitro* observations of endothelial cells during wound healing, which appear to be similar to *in vivo* repair (Vyalov et al., 1996). Actin microfilaments, along with microtubules, are activated following endothelial wounding to mediate repair and re-establish integrity. Within the first few hours post wounding, endothelial cells at the wound edge become polarized perpendicular to the wound and show changes in morphology (Gotlieb et al., 1981; Gotlieb et al., 1983). Actin protrusions appear at the front edge. The microtubule-organizing centre, which is at or just adjacent to the slow growing ends of the microtubules (Gotlieb et al., 1981; Ettenson and Gotlieb, 1992; Ettenson and Gotlieb, 1993), becomes reoriented towards the wound, with microtubule plus-ends being captured at cortical sites near the leading edge of the cell; this is also characteristic of other cell types, including fibroblasts (Fukata et al., 2003). In addition, dynamic focal adhesions to the substratum are turned over as the cell moves (Davies et al., 1994). The actin cytoskeletal changes that quiescent endothelial cells in a confluent monolayer undergo in response to wounding to become activated migrating cells can be divided into three sequential stages (Figure 2) (Lee and Gotlieb, 2003a). Stage I is characterized by breakdown of the DPB, partial detachment of cells from their neighbors, and spreading into the wound by lamellipodia extension. The distinguishing feature of Stage II is the presence of central microfilaments oriented parallel to the wound edge, associated with lateral spreading of the cell. Finally, Stage III is characterized by organization of central microfilaments perpendicular to the wound edge and by initiation of cell migration into the wound. In order for endothelial cells and fibroblasts to migrate into the wound and eventually re-establish the monolayer, the processes described above must be well coordinated, especially between microtubules and microfilaments (Goode et al., 2000). The complex basic mechanisms that govern the functions described in endothelial cells are discussed below.

Actin in Structure and Cellular Processes

Cytoplasmic actin regulates numerous essential cellular processes including cell shape change, endocytosis, intracellular transport and cytokinesis (Papakonstanti et al., 2000; Rogers and Gelfand, 2000; Huckaba and Pon, 2002; Ascough, 2004). The role of actin has been studied predominantly in cytoplasmic actin, especially in stress fibers and the cortical actin network. However, mounting evidence points to the significance of actin in the nucleus as well (Bettinger et al., 2004). Studies have demonstrated β-actin in the nucleus(Clark and Merriam, 1978) and pointed to its role in diverse processes like transcription (Fomproix and Percipalle, 2004) and mRNA export (Hofmann et al., 2001). The strongest evidence for a functional role for nuclear actin is identification of actin as a component of chromatin-remodeling complexes (Galarneau et al., 2000; Shen et al., 2000). The role of nuclear actin in endothelial cells is a fruitful area for further investigation.

Cellular functions of actin can be classified into those carried out by more stable actin filaments and those that require rapid dynamic monomer-polymer transitions (Weber, 1999). In the cytoplasm, stable actin filaments play a number of structural roles, such as supporting cell membranes, anchoring cytoskeletal proteins, providing internal structure, and as a polarized actin bundle in microvilli (Louvet-Vallee, 2000; Weed and Parsons, 2001). Stable filaments, when associated with the motor protein myosin, function in force generation (Weber, 1999), for example, in the sliding filaments of skeletal and cardiac muscle cells (Cooke, 1997) and in stress fibers (Narumiya et al., 1997).

Actin association with myosin also regulates important cellular processes, such as cytokinesis, pinocytosis, vesicle transport, and chemotaxis. Dynamic interactions between actin monomers and polymers, on the other hand, enable actin to produce movement in the absence of motor proteins. Rapid actin assembly and disassembly produces vectoral force that drives a range of dynamic cellular processes, notably migration, in which microfilament assembly results in forward protrusion of the plasma membrane and retrograde flow of actin provides traction for cell movement (Pollard and Borisy, 2003), and shape change in stress-activated endothelium (Noria et al., 2004).

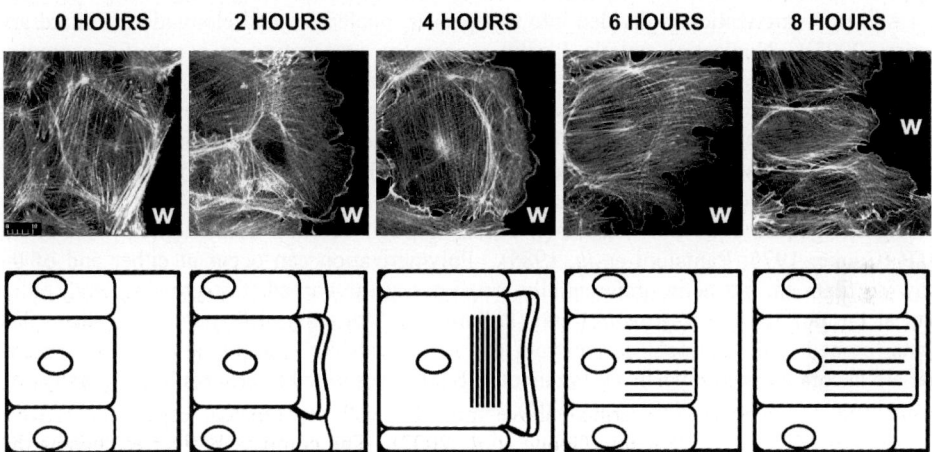

Figure 2. The redistribution of actin microfilaments during wound repair. Porcine aortic endothelial cells were grown to confluency and a 1 mm wound (w) created using a scraper. Cells were fixed at 0, 2, 4, 6, and 8 hrs after wounding, and actin was localized using rhodamine phalloidin. Immediately after wounding, the endothelial cells at the wound edge enter stage I of early repair. By 2 hours, there is disruption of DPB, partial detachment of cells from their neighbors at the wound edge as well as spreading by lamellipodia extension into the wound. The cells in stage II, by 4 hrs, undergo changes in microfilament orientation as microfilaments become more prominent and align parallel to the wound edge and the cells begin to spread. As the cells enter stage III 6 hours after wounding, the microfilaments rearrange to become perpendicular to the wound edge. By 8 hours, actin microfilaments are perpendicular to the wound edge, cells are elongated and begin moving into the wound. The wound closes at approximately 24-48 hours [1200x magnification].

Structural Organization and Actin Assembly/Disassembly

Polymerization/Depolymerization

In the cell, actin can exist in a monomeric globular form (G-actin; 43 kDa, 375 amino acids), or it can assemble into filaments (F-actin) (Sarmiere and Bamburg, 2004). Actin, which is only found in eukaryotes, is a highly conserved family of proteins that falls into three classes – α, β, and γ (dos Remedios et al., 2003). Alpha-actin is muscle-specific, and β and γ isoforms are ubiquitous (Bettinger et al., 2004). Monomeric actin has 40% α-helical content, and it has dimensions of 67 x 40 x 37 Å (dos Remedios et al., 2003). It is a spherical protein that separates into two lobes by a central, deep nucleotide-binding cleft (dos Remedios et al., 2003). ATP binds at this site in complex with either Ca^{2+} or Mg^{2+} (Guan et al., 2003). ATP-bound monomers assemble into microfilaments, which are 2 chains of G-actins twisted together to form a 7-nm double right-handed helix (Pollard, 1986; Bettinger et al., 2004). Assembly of G-actin into F-actin leads to ATP hydrolysis ADP and P_i (Pollard, 1986). G-actin spontaneously polymerizes to form F-actin *in vitro* under favourable solvent conditions, including elevated temperature, neutral or slightly acidic pH, high ionic strength, and high Mg^{2+} (Grazi and Trombetta, 1985; Zimmerle and Frieden, 1986; Wang et al., 1989).

Actin polymerization is divided into two phases: nucleation and elongation (Pollard and Borisy, 2003). Nucleation entails 3 actin monomers aggregating in a specific geometric configuration to provide a nucleus for further polymerization (Nishida and Sakai, 1983). The initial association to a dimer is unfavourable, as dimers are unstable, being more likely to rapidly dissociate than to assemble. The slow dimerization of actin accounts for the lag phase in actin polymerization (Sarmiere and Bamburg, 2004). Subsequent formation of a stable trimer leads to rapid assembly of F-actin (Grazi et al., 1983; Sept and McCammon, 2001). During elongation, actin polymerization occurs through addition of monomers with bound ATP (Cooke, 1975; Pantaloni et al., 1985). Polymerization can occur at either end of the growing filament, but actin preferentially associates at the barbed (fast-growing) end, rather than at the pointed (slow-growing) end (Pollard and Mooseker, 1981; Pollard 1986). The terms 'barbed' and 'pointed' derive from the arrowhead-like appearance of actin filaments that are decorated with myosin subfragment 1 (S1), as described in electron micrographs (dos Remedios et al., 2003; Millard et al., 2004).

An excellent review of the dynamics of actin assembly and disassembly is provided by Pollard and Borisy (Pollard and Borisy, 2003). The rate of F-actin elongation is directly proportional to the concentration of monomers in solution (Pollard, 1986). The ratio of the rate constants for dissociation and association (k^-/k^+) is the dissociation equilibrium constant for a subunit binding at the end of a polymer, also known as the critical concentration (C_c) (Pollard and Borisy, 2003). When the concentration of G-actin is below the C_c, actin disassembly occurs. Conversely, all actin above the C_c polymerizes, leaving the C_c exchanging with the end of the polymer. For Mg-ATP actin, the C_c for the barbed end (0.1 µM) is lower than that for the pointed end (0.7 µM). Consequently, steady-state concentration of ATP-G-actin is slightly above the C_c at the barbed end and below the C_c at the pointed end. Thus, actin associates at the barbed end and dissociates from the pointed end, leading to a slow 'treadmilling' of subunits from barbed to pointed end (Pollard and Borisy, 2003).

ADP-actin comprises the majority of F-actin subunits (ADP being nonexchangeable with solvent nucleotides), while ATP-actin makes up most of the monomeric G-actin pool (dos Remedios et al., 2003). Mg-ATP bound in actin's cleft stabilizes the molecule, but is not required for polymerization (DeLa Cruz et al., 2000). ATP hydrolysis is coupled to the polymerization process, and it is rapid, having a half time of about 2 seconds (Blanchoin and Pollard, 2002). Dissociation of P_i, however, is slower, with a half time of 350 seconds (Carlier and Pantaloni, 1986), meaning that ADP-P_i-actin is a relatively long-lived intermediate in newly formed filaments (Pollard and Borisy, 2003). As filament subunits progress from barbed to pointed end, they 'age', with monomers at the barbed end ATP-bound and those at the pointed end primarily ADP-bound (Pollard and Borisy, 2003).

Actin-binding Proteins

A host of actin-binding proteins is present in the cell to help coordinate actin's many functions (Figure 3). In fact, over 150 distinct proteins are reported to interact with actin; principal ones were comprehensively reviewed recently (dos Remedios *et al.*, 2003). One set of actin-binding proteins participates in stabilizing actin filaments, for example, by capping them, and in organizing them into networks by crosslinking or bundling them to each other (Winder, 2003). The myosin family of motor proteins interacts with actin filaments to move cargoes (e.g. organelles (Wu *et al.*, 2000) and mRNA (Stebbings, 2001)) along them or to generate contractile force (Narumiya *et al.*, 1997). The subset of actin-binding proteins that this review focuses on are proteins that regulate actin dynamics – the equilibrium between G-actin and F-actin (Weber, 1999).

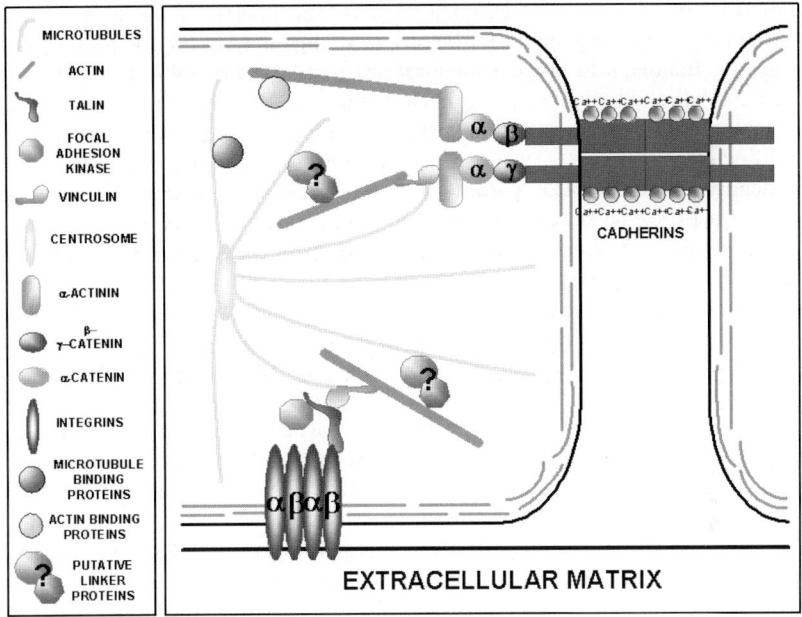

Figure 3. Diagram of cell-cell and cell-substratum adhesion. Clustering of integrin αβ heterodimers at the endothelial cell membrane is associated with binding of both cytoskeletal elements and catalytic signaling proteins to nucleate formation of focal adhesion complexes at cell-substratum sites. Actin cytoskeleton binds to focal adhesions via talin, vinculin, and FAK. Vinculin is composed of a globular head and a rod-like tail region. The globular head binds talin while the tail region binds actin. These focal adhesion proteins form a complex and link the integrins and the actin cytoskeleton. VE-cadherin at the cell-cell junctions (the major endothelial cadherin) interacts with catenins (α-, β-, & γ-catenin). β-catenin binds to the cadherin cytoplasmic domain, which in turn binds α-catenin. α-catenin binds either directly to actin or to α-actinin to link the cadherin/catenin complex with actin. α-catenin has substantial homology to vinculin, which binds α-actinin as well as talin. In some cases, plakoglobin or γ-catenin replaces β-catenin in mediating the cadherin-cytoskeletal complexes. Microtubules have also been shown to interact directly with vinculin. The diagram also shows actin binding proteins and microtubule binding proteins, which regulate microfilament and microtubule dynamics. Microtubules also interact with microfilaments either directly, or through linker proteins.

A significant fact, pointed out by Pollard and Borisy (2003), is that, although actin by itself undergoes polymerization and depolymerization, the rate of filament turnover *in vitro* is much slower than that *in vivo* - while migrating cells can advance quickly, pure actin filaments *in vitro* treadmill very slowly, filament growth being limited by subunit dissociation from the pointed ends. Furthermore, the cellular concentration of actin can exceed 100 µM but the C_c at the barbed end is only 0.1 to 1 µM; this creates a need for some mechanism to promote depolymerization at pointed ends and control the number of growing barbed ends in order to maintain a high concentration of polymerization-ready actin subunits at sites of assembly (Pollard and Borisy, 2003). Actin-binding proteins are employed in cells to coordinate nucleation, polymerization, and disassembly of actin to regulate processes, such as cell migration, that require rapid actin reorganization.

In vivo, actin-binding proteins perform the following functions: sequestering G-actin to maintain a pool of monomers (e.g. profilin -Goldschmidt-Clermont *et al.*, 1992; thymosin β4 - Goldschmidt-Clermont *et al.*, 1992), nucleating assembly (Arp2/3 complex - Mullins *et al.*, 1998), promoting branching and cross-linking of filaments (Arp2/3 complex - Mullins *et al.*, 1998), capping filaments to prevent monomer exchange at the pointed end (tropomodulin - Weber *et al.*, 1994) and barbed end (CapZ - Caldwell *et al.*, 1989; gelsolin - Harris and Weeds, 1984), stabilizing filaments (tropomyosin - Mudry *et al.*, 2003), severing filaments (ADF/cofilin - Carlier *et al.*, 1997); gelsolin - Harris and Weeds, 1984), and promoting filament depolymerization (ADF/cofilin - Carlier *et al.*, 1997). The minimum set of proteins sufficient for actin-based motility was defined by examining *Listeria* movement. *Listeria* recruits host proteins to assemble actin filaments on its surface, which in turn propels the bacteria through the cytoplasm of the infected cell. This system provides a model of actin polymerization that occurs in the leading edge of a moving cell (Cooper and Schafer, 2000). In this model, the combination of actin, Arp2/3 complex, an activator of Arp2/3 complex, ADF/cofilin, and capping protein was sufficient to reconstitute motility *in vitro*, with additional proteins (e.g. profilin) making the process more effective (Loisel *et al.*, 1999).

An overview of how actin-binding proteins participate at various stages of an actin filament's life cycle is described by Pollard and Borisy (2003). A pool of monomeric actin is maintained in cells through binding to sequestering proteins like thymosin β4. Filament growth is triggered by extracellular stimuli that activate signaling pathways, which lead to activation of the WASP/Scar family of nucleation-promoting factors. These proteins, in turn, stimulate Arp2/3 complex to initiate a new filament as a side-branch of a preexisting filament. As profilin shuttles ATP-actin to the assembly site, branches elongate rapidly and push the membrane forward (Pollard and Borisy, 2003). A mechanism by which a subunit can add on to a filament abutting the plasma membrane was conceived by Peskin *et al.*, (1993) and named the 'elastic Brownian ratchet' model. This model postulates that the actin filament behaves like a spring, constantly bending because of thermal energy. When bent away from the plasma membrane, an actin monomer can squeeze in and extend the filament. The restoring force of the filament straightening against the plasma membrane delivers the propulsive force. Before their elongation is terminated by capping proteins, filaments grow and push the plasma membrane transiently. Aging of the filaments through ATP hydrolysis and release of the γ-phosphate initiates debranching followed by the ADF/cofilin-executed filament severing and dissociation of ADP-actin from filament ends. Profilin then catalyzes the exchange of ADP for ATP, refilling the pool of assembly-ready ATP-G-actin (Pollard and Borisy, 2003).

The slow intrinsic rate of F-actin treadmilling is greatly increased through the synergistic action of ADF/cofilin and profilin. ADF/cofilin promotes subunit dissociation from the pointed end, and profilin-bound G-actin can only bind to the barbed end. Together, these two proteins cause a 75- to 125-fold increase in the rate of treadmilling (Didry et al., 1998).

Arp2/3 complex - There are three proposed mechanisms to explain how cells generate free barbed ends: severing existing filaments, uncapping existing filaments and *de novo* nucleation. *De novo* creation of new filament seeds is the principal mechanism operating in the leading edge (Pollard and Borisy, 2003). Since nucleation is kinetically unfavourable, accessory proteins are necessary to accelerate the formation of actin dimers and trimers and stabilize them. Arp2/3 (actin related proteins) complex has such nucleating properties(Millard et al., 2004). Arp2/3 complex is composed of Arp2, Arp3, and five smaller proteins (ARPC1 - ARPC5) (Robinson et al., 2001). The two Arps belong to a family of proteins whose sequence and structure are related to actin (dos Remedios et al., 2003). Arp2/3 complex binds to a preexisting filament, favoring newly polymerized ATP-bound filaments over older ADP-bound ones, and nucleates a new filament with a free barbed end (Weaver et al., 2003). Arp2 and Arp3 form the first subunits of the nascent daughter filament, and together with one incoming actin monomer, create a trimer that acts as a stable nucleus (Volkmann et al., 2001). Arp2/3 complex caps the pointed end of the daughter filament and binds it to the mother filament. This generates a branch having a characteristic 70° angle between the two filaments (Mullins et al., 1998). According to the widely accepted 'dendritic nucleation model', Arp2/3 complex binds the side of the mother filament and leads to formation of a Y-branched network of F-actin (Millard et al., 2004). There is also evidence (Pantaloni et al., 2000) to support a 'barbed-end branching model', in which Arp2/3 complex binds to the barbed end of the mother filament and initiates branch formation, while allowing growth of the mother filament to continue (Millard et al., 2004). In both models, a dense array of short-branched filaments is produced, like that seen in the quiescent endothelial cell.

Pure Arp2/3 complex is intrinsically inactive, because Arp2 and Arp3 are spaced too far apart to form the first two subunits of the new filament. Arp2/3 complex activators induce a conformational change that reorients the two Arps (Millard et al., 2004). Factors implicated in activation of Arp2/3 complex include ATP (Le Clainche et al., 2001), the Wiskott-Aldrich syndrome protein (WASP) family (Zalevsky et al., 2001), suppressor of cAMP receptor (Zalevsky et al., 2001), Scar, also known as WASP-family verprolin homologous (WAVE) proteins, and cortactin (Uruno et al., 2001).

WASP family of proteins - WASPs transduce extracellular signals into reorganization of the cytoskeleton, and are central to Arp2/3 regulation. Two WASP genes exist: WASP and neural-WASP (N-WASP). WASP is expressed exclusively in hematopoietic cells, while N-WASP is ubiquitously expressed (Millard et al., 2004). WASPs have a conserved domain arrangement (Millard et al., 2004). Within the N-terminus is a proline-rich region that can bind profilin and various SH3 (src homology 3)-containing proteins, a GTP-ase binding domain that can bind the Rho small GTPase Cdc42, and a basic stretch that can bind PIP_2 (Millard et al., 2004; Notarangelo and Ochs, 2003). The C-terminus contains one or two WASP homology 2 domains that bind monomeric actin and an acidic region that can bind to and activate Arp2/3 complex (Millard et al., 2004). WASP induces an activating conformational change in Arp2/3 complex, promotes nucleation by bringing an actin monomer to Arp2/3 via the WASP homology 2 domain, and stabilizes the newly formed Arp2-Arp3-actin trimer (Winder, 2003). Purified WASPs exist in an autoinhibited state, in which the GTP-ase binding domain in the N-terminus is folded onto the C-terminus, blocking the acidic region from interacting with Arp2/3 complex (Millard et al., 2004). This autoinhibition can be overcome by Cdc42 (Kim et al., 2000), SH3 domain-containing proteins

Nck and Grb2 (Benesch *et al.*, 2002) (adaptor proteins that are recruited to activated tyrosine kinase receptors through their SH2 domain), PIP_2 (Benesch *et al.*, 2002) or profilin (Yang *et al.*, 2000). These compete with the C-terminus to interact with their respective binding domains in the N-terminus and can act cooperatively to disrupt the autoinhibitory loop (Notarangelo and Ochs, 2003). Phosphorylation of a specific tyrosine residue in WASP also plays a role in regulation of WASP activity (Cory *et al.*, 2002).

Scar proteins - While WASPs have been clearly implicated in membrane-trafficking processes, such as endocytosis and vesicle motility, evidence for their involvement in lamellipodia formation is scarce. However, Scar proteins are believed to play a role in the latter process. Three Scar genes exist (Scar1, Scar2, and Scar3); Scar2 being the most widely expressed in adults (Millard *et al.*, 2004). Unlike WASPs, Scars are constitutively active, but can be inhibited by forming a multimeric complex with four other proteins (Eden *et al.*, 2002). GTP-bound Rac interacts with inhibited Scar via insulin receptor substrate (IRSp53) and induces dissociation of the inhibitory proteins. Consequently, Scar stimulates Arp2/3 complex, leading to lamellipodia formation (Miki *et al.*, 2000).

Cortactin - The vertebrate protein, cortactin, can stimulate Arp2/3 complex and stabilize branches; although it is weak in comparison to WASPs (Millard *et al.*, 2004). Like WASPs, cortactin binds Arp2/3 complex through an acidic domain (Weaver *et al.*, 2003). However, instead of binding monomeric actin, as do WASPs, cortactin-induced Arp2/3 complex activation requires filament binding by cortactin (Weaver *et al.*, 2001). It is proposed that cortactin activates Arp2/3 complex by bringing it into proximity with the actin filament (Weaver *et al.*, 2003).

Profilin - Profilin, a small (19 kDa), high-affinity actin monomer-binding protein, functions to maintain a high concentration of polymerization-ready actin (Pollard and Borisy, 2003). By modulating opening of the actin nucleotide cleft, profilin can increase the rate of exchange of ADP for ATP by 140-fold (Selden *et al.*, 1999). Profilin also suppresses spontaneous actin nucleation and inhibits hydrolysis of actin-bound ATP, thus keeping actin monomers in a state in which their addition to barbed ends is favorable (dos Remedios *et al.*, 2003; Pollard and Borisy, 2003). Profilin is recruited to regions of actin polymerization (e.g. lamellipodia), through its interactions with the vasodilator-stimulated phosphoprotein (VASP) family (Kang *et al.*, 1997) and formin homology domain proteins (Pring *et al.*, 2003), such as mDia1 and Bni1p. Once there, profilin-bound ATP-G-actin binds to the barbed ends of growing filaments (being blocked by profilin from binding to the pointed ends). Subsequently, profilin dissociates from actin (dos Remedios *et al.*, 2003).

Thymosin - A second mechanism by which cells maintain an unpolymerized actin pool is the association of ATP-actin monomers with thymosin (Pollard and Borisy, 2003). This small (5 kDa) protein, whose predominant mammalian isoform is thymosin β4, is exclusively an actin monomer-binding protein that acts as a buffer of G-ATP-actin (Sun *et al.*, 1995; dos Remedios *et al.*, 2003). While concentration of free actin monomers in cells exceeds that of profilin, thymosin β4 is present in sufficient concentration to sequester the reserve monomer pool (Sun *et al.*, 1995; Pollard and Borisy, 2003). Acting alone, thymosin β4 inhibits exchange of actin-bound nucleotide (in contrast to profilin) (Goldschmidt-Clermont *et al.*, 1992) and inhibits polymerization, including G-actin nucleation or association with the barbed or pointed end of filaments (Safer *et al.*, 1997). However, thymosin β4 and profilin cooperate to regulate actin assembly by competing for binding to monomers (Pollard and Borisy, 2003). Both proteins exchange on and off actin on a sub-second time scale, but profilin binds monomers tighter than thymosin β4. Consequently, profilin provides polymerization-ready G-ATP-actin to barbed ends, while thymosin β4 holds the rest of the monomers in reserve.

Capping Proteins - G-actin polymerization continues until the growing filaments are capped by capping proteins. Although capping may seem counterproductive since it opposes filament elongation, it increases the efficiency of actin-driven motility in two ways. First, capping, by limiting growth of branches, promotes short filaments, which are stiffer and thus more effective at pushing the membrane than long filaments. Second, capping controls where actin filaments push by preventing global consumption of actin subunits, thus funnelling the subunits to a select number of growing barbed ends (Pollard and Borisy, 2003). The principal pointed end capping protein is tropomodulin, while the primary barbed end capping proteins are CapZ and gelsolin (dos Remedios et al., 2003).

Tropomodulin - Tropomodulin is the only protein, with exception of Arp2/3 complex, known to bind to the pointed end of F-actin and inhibit elongation at this end (dos Remedios et al., 2003). The intrinsically weak binding of tropomodulin to F-actin is increased by three orders of magnitude in the presence of tropomyosin (Weber et al., 1994). In striated muscle, tropomodulin plays a key role in sarcomere assembly by capping pointed ends of actin-thin filaments (Borisy and Svitkina, 2000). In nonmuscle cells, tropomodulin function is not known (Cooper and Schafer, 2000).

CapZ - CapZ (also called 'capping protein' in nonmuscle cells) binds tightly to the barbed end of actin filaments. It facilitates actin nucleation by binding to and stabilizing actin monomers and oligomers, regulates actin assembly at barbed ends of filaments, and prevents annealing of filaments. It also mediates assembly of filaments at the Z-disk in skeletal muscle, but its role is less clear in non-muscle cells (dos Remedios et al., 2003).

Gelsolin superfamily - Gelsolin and its relatives fragmin, villin, severin, and adseverin/ scinderin sever preexisting filaments and cap them at the barbed end. These activities are regulated by Rac, Ca^{2+}, and PIP_2 (Sun et al., 1999). Rac is thought to be an upstream activator of gelsolin (Azuma et al., 1998). Ca^{2+} directly interacts with gelsolin and is required for its activation (Lin et al., 2000). PIP_2 displaces the otherwise stable barbed end cap formed by gelsolin (Janmey et al., 1987). The N-terminus of gelsolin contains two actin-binding sites, while the C-terminus is a regulatory domain. In absence of Ca^{2+}, gelsolin adopts a compact structure, in which a helix at the C-terminus binds to the N-terminus, blocking the actin-binding sites and inhibiting severing activity. Ca^{2+} activates gelsolin by binding to it and inducing a conformational change to accommodate actin binding (Sun et al., 1999). Gelsolin can then sever F-actin by inserting itself between two longitudinally associated actin subunits in a filament and weakening their non-covalent bonds (Sun et al., 1999). In apoptotic cells, gelsolin is cleaved by caspase-3, which abolishes its requirement for Ca^{2+} activation. Constitutively active gelsolin, by severing actin filaments, dismantles the membrane cytoskeleton to cause membrane blebbing, a hallmark of apoptosis (Kothakota et al., 1997). After severing, gelsolin remains attached to the barbed end of the filament as a cap. This prevents growth at the barbed end, while disassembly proceeds unchecked at the pointed end, resulting in disassembly of the actin network (dos Remedios et al., 2003).

Filament uncapping - Local factors that inhibit the capping can act to promote actin polymerization in specific regions in the cell (dos Remedios et al., 2003). Polyphosphoinositides, particularly PIP_2 (induced by thrombin or Rac), cause dissociation of both gelsolin and CapZ from actin filaments (Sun et al., 1999; Lee and Gotlieb, 2003b). VASP is also a candidate capping inhibitor (Bear et al., 2002). Uncapping can generate many free barbed ends, leading to rapid filament elongation (dos Remedios et al., 2003).

ADF/cofilin - Monomer-binding proteins (profilin and thymosin β4), together with capping proteins, reduce depletion of G-actin and its assembly into F-actin. ADF/cofilin complements the action of these proteins in maintaining a high cellular concentration of unpolymerized G-actin by promoting F-actin depolymerization, thus replenishing the monomer pool (Pollard and Borisy, 2003). ADF/cofilin enhances the rate of filament turnover by severing actin filaments (without capping their ends) and increasing the rate of depolymerization at the pointed end. It acts by changing the twist of the actin filament it binds to and weakening the lateral contacts between subunits (Ono, 2003). ADF/cofilin's depolymerization-inducing activity is essential in cells, since pure actin filaments are intrinsically stable, undergoing only slow monomer dissociation at their ends (Pollard and Borisy, 2003). Because it increases the rate of pointed end disassembly 25-fold, ADF/cofilin is essential for the high rate of treadmilling *in vivo* (Carlier *et al.*, 1997), although profilin contribution is also important (Didry *et al.*, 1998).

ADF/cofilin has the highest binding affinity for ADP-bound actin, binding only weakly to ATP or ADP-P_i actin filaments and not depolymerizing them. ATP hydrolysis and P_i dissociation thus act as a 'timer' for depolymerization (Pollard and Borisy, 2003). Following depolymerization, profilin competes with ADF/cofilin for binding to ADP-G-actin and promotes its conversion to the polymerization-ready ATP-G-actin form (Pollard and Borisy, 2003). ADF/cofilin is regulated through phosphorylation of its Serine-3 residue, which inhibits ADF/cofilin's actin-binding activity (Sarmiere and Bamburg, 2004). Phosphorylation is achieved by LIM-kinase proteins 1 and 2 (LIMK-1 and LIMK-2), whose activity is, in turn, regulated by Rho GTPases (dos Remedios *et al.*, 2003). The phosphatase Slingshot can remove the inhibitory phosphate and reactivate ADF/cofilin (Niwa *et al.*, 2002). Other negative regulators of ADF/cofilin include PIP_2, which can bind to ADF/cofilin and block its interaction with actin (Ojala *et al.*, 2001), and tropomyosin, which competes with ADF/cofilin for binding to actin (Ono and Ono 2002).

Rho GTPases in Regulation of Actin Assembly

Rho GTPases Influence Actin Cytoskeletal Functions

Rho GTPases, members of the Ras super-family of GTPases, cycle between an active GTP-bound state and an inactive GDP-state (Raftopoulou and Hall, 2004) and are important regulators of actin microfilaments. As such, Rho GTPases are critical players during cell migration when the actin cytoskeleton changes to accommodate and enable cell movement. In addition, Rho GTPases play a central role in regulation of other actin-directed processes such as membrane trafficking and morphogenesis (Rivero and Somesh, 2002). Rho GTPases are activated in primarily two ways: by soluble factors via cell receptors and by cell adhesion or integrin clustering (van Nieuw Amerongen and van Hinsbergh, 2001). Three major classes of proteins tightly control GTP and GDP cycling: GEFs which facilitates GTP displacement of GDP; GAPs which stimulate GTP hydrolysis, and guanine nucleotide exchange inhibitors which inhibit GDP dissociation, block interaction of the GTP-bound form with downstream effectors, and stimulate release of Rho GTPases from the plasma membrane (Rivero and Somesh, 2002; Raftopoulou and Hall, 2004). The three best-studied Rho GTPases involved in cell migration are Rho, Rac, and Cdc42. Activation of Rac and Cdc42 are identified with formation of lamellipodia and filopodia, respectively, as well as with focal complexes.

Activation of RhoA is involved in arrangement of stress fibers and associated focal adhesions (Kaverina *et al.*, 2002a; Kaverina *et al.*, 2002b). In support of their functional correlation, visualization of GTPase activity has localized Cdc42 to the tip of the leading edge, Rac1 just behind the edge, and RhoA mostly within the cytosol (Fukata *et al.*, 2003).

Cellular Targets of Rac and Cdc42

Rac and Cdc42 can stimulate new actin polymerization at the leading edge by activating Arp2/3 complex. Cdc42 does this by activating WASP and N-WASP directly (with the cooperation of PIP_2), while Rac activates Scar/WAVE proteins via its interaction with IRSp53. In addition to activating Arp2/3 complex, Rac promotes actin polymerization by inducing PIP_2, which in turn uncaps filaments and generates free polymerization-ready barbed ends (Ridley, 2001). Moreover, both Rac and Cdc42 activate the p21-activated kinase (PAK), which activates LIM kinase to phosphorylate (and thus inactivate) ADF/cofilin. Thus, Rac and Cdc42 stimulate formation as well as stabilization of new filaments (Pollard and Borisy, 2003). Activated PAK also acts by localizing to focal adhesions and promoting their disassembly (Ridley, 2001). Focal adhesion turnover is an essential characteristic of migrating cells and turnover allows the rear of the cell to retract during migration.

Cellular Targets of Rho

Rho activation is associated with formation of stress fibers and focal adhesions. Rho can affect stress fiber organization by two mechanisms: stimulation of actin-myosin interaction through phosphorylation of myosin (regulatory) light chain (MLC), and induction of actin polymerization through recruitment of mDia to the plasma membrane (Defilippi *et al.*, 1999). MLC phosphorylation, which leads to myosin activation and bundling of F-actin, is effected by MLCK and Rho-activation of Rho associated kinase (ROCK) (Ridley, 2001). ROCK stimulates MLC phosphorylation by directly phosphorylating MLC and by inactivating the MLC phosphatase myosin phosphatase type I (PP1M) (Ridley 2001). Additionally, ROCK activates LIM kinase, which phosphorylates cofilin and inhibits the actin-severing activity of ADF/cofilin thus stabilizing F-actin filaments (Maekawa *et al.*, 1999). MLCK is activated by Ca^{2+}-bound calmodulin and is inhibited through phosphorylation by PAK. This PAK inhibition provides another mechanism by which Rac and Cdc42 can regulate stress fiber assembly (van Hinsbergh and van Nieuw Amerongen, 2002). The Rho downstream effector mDia cooperates with ROCK to induce stress fiber formation by recruiting profilin to the site of Rho action. The tension generated by stress fibers leads to aggregation of integrins and results in maturation of small focal complexes into focal adhesions (Arthur *et al.*, 2002).

The Endothelial Microtubule Cytoskeleton

Structure and Assembly/Disassembly of Microtubules

Endothelial cells have microtubules that emanate from the microtubule organizing center, adjacent to the nucleus, and extend out to the periphery in the area of cortical actin. Microtubules are composed of $\alpha\beta$-tubulin heterodimers. Linear linkages, which connect tubulin dimers in a head-to-tail fashion, produce 'protofilaments'. Lateral linkages between protofilaments produce tubulin sheets, which curl into a cylinder to form the tubules. Since all heterodimers of a given protofilament are oriented in the same direction, microtubules are structurally polarized, with β subunits free at one end, and α subunits at the other. Assembly of microtubules occurs at nucleation sites by the centrosome called microtubule-organizing centers. As microtubules elongate, GTP-β subunits of free tubulin dimers are added onto GTP-α subunits. After binding, the β-bound-GTP is hydrolyzed (Carvalho et al., 2003). The α subunit binds GTP irreversibly and does not possess GTPase activity. In most microtubules, one end grows faster than the other and are designated the plus- and minus-ends, respectively. The orientation of microtubules with respect to the organizing center appears to be constant in most cell types, with the minus end being adjacent to the center. Microtubules undergo oscillating phases of growth and shrinkage, termed 'dynamic instability' (Kirschner and Mitchison, 1986; Mandelkow et al., 1991). 'Catastrophe' indicates the transition of microtubule to disassembly, while 'rescue' indicates the return to assembly (Walker et al., 1988). Like actin, microtubules exhibit 'treadmilling' in vitro. In vivo, dynamic instability is regulated by microtubule-associated proteins, plus-end binding proteins, and other soluble factors (Wittmann et al., 2003).

Dynamic microtubules may become stabilized and these microtubules are post-translationally modified, often by de-tyrosination. Stabilized microtubules have been shown to be oriented in the direction of cell migration (Fukata et al., 2003), and may be important in establishing cell polarity (Palazzo et al., 2001b; Small et al., 2002a). Unlike their dynamic counterparts, these microtubules do not undergo dynamic instability. Rho GTPase was found to stabilize a subset of microtubules at the leading wound edge in migrating fibroblasts (Cook et al., 1998). It was later reported that mDia, a member of the diaphanous-related formins class of Rho downstream effectors, promotes formation of stabilized, capped microtubules oriented towards the wound edge in repairing monolayers (Hollenbeck 2001; Palazzo et al., 2001a), and stabilization is regulated by integrin-mediated activation of focal adhesion kinase (FAK), which enables Rho to activate mDia (Palazzo et al., 2004). Furthermore, mDia regulates actin cytoskeleton by serving as a potent nucleator (Li and Higgs, 2003) and promoting polymerization of G-actin (Copeland and Treisman, 2002; Vicente-Manzanares et al., 2003) necessary for lamellipodia and filopodia formation. Thus, the ability of mDia to coordinately stimulate both actin and microtubules during cell migration establishes it as a likely mediator of crosstalk between the two cytoskeletal partners, as discussed below.

Microtubules and Cell Polarity

It has long been recognized that microtubules are essential for establishing cell polarity (Gotlieb et al., 1981), since disruption of microtubules in endothelial cells and fibroblasts led to loss of polarized cell shape and directional migration (Gotlieb et al., 1983; Rodriguez et al., 2003). How microtubules contribute to cell polarity is not entirely known; however, it is likely that complex spatiotemporal organization of cytoskeletal associated signaling pathways are involved. Cdc42, dynein, and dynactin control microtubule organizing center reorientation in cells at the wound edge independent of Rho-regulated microtubule stabilization (Palazzo et al., 2001b). Several hypotheses are currently being examined, as reviewed by Wittmann and Waterman-Storer (2001). Since microtubules extend out to the cell periphery and do explore the peripheral intracellular environment, they might serve as tracks for transporting substrates for formation of lamellipodia and filopodia at the leading edge of a cell (Nabi 1999). This was supported by inhibited cell migration, when blocking antibodies to the microtubule-motor, kinesin, were applied (Rodionov et al., 1993). Interestingly, recent studies support a model in which kinesins regulate microtubule growth by transporting regulatory molecules, such as microtubule plus-end tracking proteins, to the growing ends. Other proteins do bind microtubule plus ends without the motor dependent transport process.

Second, there is evidence suggesting that interaction of microtubules with cell-matrix adhesion plays an important role in cell migration (Small et al., 2002a). As a cell moves along a surface, it forms adhesions with the substratum near its front edge while retracting those at its rear. Balance between assembly and disassembly of adhesion sites is essential for migration; without dissolution of adhesions between the cell body and the underlying matrix, the cell cannot propel itself forward. Repetitive targeting of focal adhesions by microtubules reduces growth of the adhesions and promotes their dissolution at the retracting tail (Kaverina et al., 1999). Since microtubules may grow within nanometers to basal adhesion sites, it is suggested that microtubule tips induce signaling cascades leading to relaxation of adhesions (Kaverina et al., 1999; Krylyshkina et al., 2003; Small and Kaverina, 2003). A group of microtubule plus-end tracking proteins (+TIPs) regulates dynamic instability at the tip (Carvalho et al., 2003) and may play a role in disassembly. For instance, EB1 is a +TIP that mediates microtubule growth through recruitment of adenomatous polyposis coli protein (APC) to plus-ends, where it promotes polymerization of tubulin (Mimori-Kiyosue et al., 2000; Nakamura et al., 2001). Interestingly, APC also binds to Asef (a Rac-GEF) (Kawasaki et al., 2000), which increases activation of Rac. Elevation in Rac activity opposes Rho's effects at focal adhesions (Sander et al., 1999), resulting in relaxation of cell adhesion with the substratum (Small and Kaverina, 2003). Indeed, focal adhesions may represent sites of crosstalk between microtubules and actin cytoskeleton. Microtubules are guided towards focal adhesions by crosslinking to actin stress fibers (Krylyshkina et al., 2003; Rodriguez et al., 2003), and upon reaching the adhesions, microtubules become transiently stabilized potentially through downstream effectors of Rho (Palazzo et al., 2004).

Thirdly, microtubule growth may act as a signal to produce membrane protrusions (Wittmann and Waterman-Storer, 2001). Depolymerization of microtubules in fibroblasts led to a reduction in polarization of lamellipodia, which was corrected when repolymerization was allowed (Waterman-Storer et al., 1999). In addition, CLIP-170 (a +TIP) binds to plus-ends of growing microtubules where it sequesters IQGAP1, an important regulator of the actin cytoskeleton (Briggs and Sacks, 2003). It was shown that Rac1/Cdc42, IQGAP1 and CLIP-170 forms a complex at the plus-ends of microtubules (Erickson et al., 1997; Fukata et al., 2002). Rac1 and Cdc42 activation is involved in formation of protrusions in a migrating cell. Therefore, their sequestration at microtubule plus-ends, which faces the direction of cell migration, is significant. The binding of IQGAP1 to both F-actin (Mateer et al., 2002) and CLIP-170 bring microfilaments and microtubules into close proximity, thus facilitating cross-talk between the two (Briggs and Sacks, 2003). Furthermore, IQGAP1 inhibits the GTPase activity of Cdc42 (Brill et al., 1996; Hart et al., 1996; Ho et al., 1999), thereby preserving Cdc42 in its active form, which can go on to promote filopodia formation (Briggs and Sacks, 2003). Activity of IQGAP1 is in turn regulated in part by Cdc42. Activation of Cdc42 increases IQGAP1-induced crosslinking of F-actin *in vitro* (Fukata et al., 1997), and promotes binding to CLIP-170 (Fukata et al., 2002).

Crosstalk: Roles of Actin and Microtubules in Migrating Cells

For many years, it was thought that actin and microtubules played distinct, relatively independent roles. For instance, during cell migration, the actin cytoskeleton is responsible for protrusion and contraction, while microtubules direct cell polarization. Recently, however, their interaction and mutual regulation have been emphasized. The concept of endothelial cells migration involves overlapping roles for actin and microtubules, with extensive communication between the two (Lee and Gotlieb, 2002; Lee and Gotlieb, 2003b).

Although the actin cytoskeleton forms vital protrusive structures during cell migration, microtubules also contribute to the process by affecting signaling molecules involved in actin regulation. Maintenance of leading edge protrusions and cell polarity during migration requires an intact microtubule cytoskeleton (Vasiliev *et al.*, 1970). The polymerization state of microtubules influences Rho GTPase activity. Growth of microtubules activates Rac1, potentially because GTP-Rac1 is displaced from free tubulin dimers during polymerization (Waterman-Storer *et al.*, 1999). Release of GTP-Rac1 leads to actin polymerization at the leading edge and formation of lamellipodia. Conversely, depolymerization of microtubules promotes Rho activation (Ren *et al.*, 1999), potentially through release of Rho GEF-H1, which is inactive when bound to polymerized microtubules. Once free, GEF-H1 mediates activation of RhoA, promoting stress fiber assembly and activation of myosin II to enhance contractility (Krendel *et al.*, 2002). Another GEF, TrioD1, directly activates RhoG, which goes on to regulate the activity of Rac and Cdc42 (Blangy *et al.*, 2000). Constitutively active RhoG mutants led to increased formation of lamellipodia and filopodia (Wittmann and Waterman-Storer, 2001). Interestingly, the activity of TrioD1 requires binding with microtubules, and inhibition of microtubule polymerization resulted in a loss of RhoG localization to the cell periphery (Wittmann and Waterman-Storer, 2001). Therefore, although microtubules may not directly form lamellipodia and filopodia protrusions, they are nonetheless important by influencing the actin cytoskeleton via Rho GTPase signaling pathways.

Many variables in turn regulate the polymerization state of microtubules. At the leading edge, microtubules are coupled to retrograde flow of actin (Waterman-Storer and Salmon, 1997). In migrating newt lung epithelial cells, there are two subsets of microtubules associated with different growth dynamics. Centrosomal microtubules are organized at the microtubule organizing center and extend towards the leading edge of the cell. Non-centrosomal microtubules, however, are organized at the lamella rather than extending from the centrosome. Most microtubules grow perpendicularly to the leading edge, but some non-centrosomal microtubules that protrude into the lamellipodia may bend and become oriented in parallel to the leading edge. Perpendicular microtubules maintain a stable length through frequent and short cycles of dynamic instability, while parallel microtubules exhibit net increase in length due to fewer catastrophic events – a property special to plus-ends near leading edges. Furthermore, both groups of microtubules move towards the cell center continuously by crosslinking to actin retrograde flow, a process dependent on myosin. The rearward movement of microtubules parallel to the leading edge eventually leads to their breakage, generating non-centrosomal microtubules behind the lamella with free minus-ends undergoing rapid depolymerization (Gupton *et al.*, 2002). Therefore, microtubules nearest to the leading edge tend to experience net growth, while those produced by breakage behind the lamella undergoes shortening. Regional differences in microtubule polymerization states may in turn create gradients of Rho GTPase activity (Wittmann and Waterman-Storer, 2001), with Rac1 activation at, and RhoA activation behind the leading edge (Rodriguez *et al.*, 2003).

Table 1.
Linker proteins shown to interact with both microtubules and actin microfilaments.

Protein	System	Size	Function	References
Mip-90 (microtubule interacting protein)	Human fibroblasts	90 kDa	Associates with microtubules	(Gonzalez et al., 1998)
Crn1p (Coronin 1p)	Yeast	85 kDa	Actin-associated protein	(Goode et al., 1999)
D-CLIP-190 (homologue to cytoplasmic linker protein-170)	Embryos & Embryonic cultured cells	189 kDa	Localize to plus end of microtubules	(Lantz and Miller, 1998)
CLASPs (-1 and -2) (CLIP-associated proteins)	COS-1 cells & Swiss 3T3 fibroblasts	170 kDa (α) 140 kDa (β/γ)	Binds both CLIP and microtubules	(Akhmanova et al., 2001)
ACF7/kakapo (Plakin)	SW13 human adrenal carcinoma cells	600 kDa	Cytoskeletal linker protein Interact with intermediate filaments	(Karakesisoglou et al., 2000)
RHAMM/IHABP (Types A-D) (receptor for hyaluronic acid mediated motility/intracellular hyaluronic acid binding protein)	HeLa cells & Human cervical & mammary cancer cells	66, 75, 85-90kDa	Part of hyaluronic acid receptor complex (membrane-bound, multimeric protein complex)	(Assmann et al., 1999)
X-PAK5 (p21-activated kinase)	Xenopus cell lines	75 kDa	Rac and Cdc42 effector	(Cau et al., 2001)
MAP2c (microtubule-associated protein)	Melanoma cell lines	280 kDa	Microtubule-associated protein	(Cunningham et al., 1997)
mDia (mDiaphanous)	HeLa cells	140 kDa	Rho effector	(Ishizaki et al., 2001)
IQGAP1 and CLIP-170	Vero cells	185 kDa (IQGAP1) 170 kDa (CLIP-170)	IQGAP1: Binds actin, Rac and Cdc42 effector, and binds CLIP-170 CLIP-170: Binds to plus end of microtubules	(Fukata et al., 2002)
Arg (Abl-related gene)	Fibroblasts	135/140 kDa Doublet	Regulate cell migration (required for dynamic lamellipodial protrusion)	(Miller et al., 2004)

Interactions with the cytosolic proteins also regulate microtubule dynamic instability. Op18/stathmin (Op18) is a soluble microtubule-destabilizing protein. By binding to tubulin dimers, Op18 prevents polymerization and increases catastrophe frequency of microtubules (Horwitz et al., 1997; Gradin et al., 1998). Op18 is inactivated by phosphorylation due various protein kinases, including PAK1, which is a downstream effector of Rac1 GTPase (Wittmann et al., 2004). Thus, Rac1 activation leads to growth of microtubules through the Pak1/Op18 pathway. As a result, Rac1 controls not only actin, but also the microtubule cytoskeleton via downstream effectors. Indeed, Rac1 activation was shown in a PAK-dependent manner to promote retrograde flow, as well as growth of microtubules into lamellipodia by reducing catastrophe frequency and promoting polymerization (Wittmann et al., 2003). The extension of microtubules into the leading edge of the cell seems to be involved in cell polarization. In fact, many of the factors that regulate microtubule dynamics are implicated in establishing cell polarity during migration.

Evidence points to physical interactions between microtubules and actin microfilaments (Figure 1, inset). There are many examples where coordination between the actin and microtubule cytoskeleton systems seems to occur either through bifunctional linker proteins, or through a series of protein-protein interactions that form a physical bridge between microtubules and microfilaments (Table 1), some of which have been mentioned above. Proteins linking microfilaments and microtubules include the microtubule-interacting protein (Mip)-90 (Gonzalez et al., 1998), coronin1p (Crn1p) (Goode et al., 1999), D-CLIP-190 (Lantz and Miller, 1998), CLIP-associated proteins (CLASPs) (Akhmanova et al., 2001), plakins (Karakesisoglou et al., 2000), a receptor for hyaluronic acid mediated motility/ intracellular hyaluronic acid binding protein (GHAMM/IHABP) (Assmann et al., 1999), PAK (Cau et al., 2001), microtubule-associated protein-2c (MAP2c) (Cunningham et al., 1997), mDia (mDiaphanous) (Ishizaki et al., 2001; Palazzo et al., 2001a), IQGAP1 and CLIP-170 (Fukata et al., 2002), as well as Abl-related gene (Arg) (Miller et al., 2004).

It is now clear that the actin cytoskeleton and microtubules do not work independently in generating cell migration. Evidence points to Rho GTPases and downstream effectors as providing opportunities for cross talk between the two cytoskeletal elements. It seems that once the signal for migration is transmitted, multiple, communicating pathways are activated to direct coordinated movement of actin and microtubules. Although there is solid understanding of various processes involved in cell migration, a consensus on the initiation and overall integration of these processes remains to be reached. As research continues into this dynamic area, other molecular players may emerge as key regulators in cell migration.

Actin and Cell Adhesion

Cell-Cell adhesion

Cell-cell adhesion is an essential element of endothelial cells physiology; its functions include separation of intra- and extra- vascular compartments, mechanical connection of the cells, and maintenance of cell polarity (Schnittler, 1998). Endothelial cell-cell junctions become dissociated in several pathophysiological reactions, including inflammation and angiogenesis (Schnittler, 1998). As shown in Fig. 3 and discussed in detail in Chapter 9, several classes of specialized adhesive structures exist, all of which are connected to cytoskeletal components to mediate stable cell-cell adhesion (Braga, 2002). The two types of adhesive structures that associate with actin microfilaments are tight junctions and adherens junctions (Braga, 2002). These are both linked to the DPB, and formation of gaps between endothelial cells involves the dissociation of the DPB (Schnittler, 1998). Tight junctions seal the endothelium, preventing fluid from flowing freely across the cell layer, and contribute to maintenance of cell polarity. They are constructed from the transmembrane proteins occludin or claudin, which link to zonula occludens proteins (ZO-1 and ZO-2) on the cytosolic face of the membrane. ZO-1 and ZO-2 are, in turn, connected to other cytoskeletal proteins and actin filaments (Schnittler, 1998). Adherens junctions mechanically connect endothelial cells and are fundamental to maintenance of interendothelial mechanical stability. The primary structural component is cadherin, a transmembrane protein. Endothelial cells express N-, E- and vascular endothelial (VE)-cadherin, the latter being the predominant form (Breviario et al., 1995). Cadherin's extracellular region contains a Ca^{2+}-dependent adhesive domain that binds homotypically to cadherins on neighboring cells. The cytoplasmic domain links cadherin to actin filaments of the DPB system via its interaction with catenins (Gumbiner, 1996). β- and γ-catenin bind directly to cadherins, and they, in turn, are indirectly linked to the actin cytoskeleton via α-catenin (Lee and Gotlieb, 2002). α-catenin connects the cadherin-catenin complex to actin filaments by either directly interacting with actin or through actin-binding proteins such as α-actinin, an actin-bundling protein, ZO-1, ZO-2 and vinculin (Schnittler, 1998). This cadherin-cytoskeleton linkage is essential for cadherin adhesiveness, as a mechanism to strengthen the weak forces provided by the homophilic binding (Braga, 2000). There is prominent molecular reorganization of adherens junctions when endothelial cells undergo wound repair in order to allow for profound morphological changes to occur as the cells elongate and migrate forward (Lampugnani et al., 1995). Similarly, there is a partial disassembly of adherens junctions during shear induced shape change (Noria et al., 1999).

Cell-Matrix Adhesion

Linkage between the actin cytoskeleton and the extracellular matrix is mediated by integrins, a large family of αβ-heterodimeric transmembrane receptors, as discussed further in Chapter 8 (Petit and Thiery, 2000b). Several α- and β-subunit integrin isoforms exist, and the exact subunit combination dictates the binding specificity of the integrin to different endothelial cellsM components; in endothelial cells, the predominant combinations are $α_2β_1$ and $α_5β_1$ (Lee and Gotlieb, 2003b). The large extracellular domain of both the α and β subunits is involved in adhesion to the extracellular matrix components including collagens, laminin and fibronectin, whereas the cytoplasmic tail of the β subunit links integrins to the actin cytoskeleton (Petit and Thiery, 2000b).

When not bound to extracellular ligands, integrins are distributed diffusely over the surface and appear not to be linked to actin filaments. Extracellular ligand binding induces their association with the cytoskeleton and clustering into focal complexes and focal adhesions. Focal complexes are small adhesive structures still in the process of forming (e.g. ones formed at the tip of membrane protrusions) or the complexes present in quiescent endothelium; focal adhesions are the large, mature complexes found at the ends of stress fibers in activated or migrating endothelium. RhoA stimulates adhesion assembly by promoting actomyosin-interaction and/or contractility, which bundles actin filaments and produces tension, leading to integrin clustering (Schoenwaelder and Burridge, 1999). Assembly of focal adhesions involves linking of actin filaments to integrins as well as recruitment of cytoplasmic proteins to form a multimolecular adhesion complex (Brakebusch and Fassler, 2003). Some constituents participate in the structural link between integrins and the actin cytoskeleton, whereas others are signaling proteins (Petit and Thiery, 2000b). Recruitment of cytoplasmic proteins is important since it provides flexibility and changes in cell polarity and directional movement. An actin binding protein, as is the case of the SH3 domain protein Lasp-1, relocalizes to the leading edge of the migrating cell at FA and is required for cell migration (Lin et al., 2004).

Bridging of actin filaments to integrins is mediated by the focal adhesion proteins talin, α-actinin and paxillin; these link $β_1$-integrin and actin directly or indirectly through binding intermediate focal adhesion proteins, such as vinculin and tensin, that bind actin (Lee and Gotlieb, 2003b). Talin is a major structural component of focal adhesions, and its recruited early in focal adhesion formation (Calderwood et al., 2000). α-actinin connects actin filaments to integrins and crosslinks filaments to form bundles and networks (Brakebusch and Fassler, 2003). Paxillin, with many protein-protein binding sites (e.g. proline-rich regions, SH2 binding sites), is thought to act as an adapter protein to recruit other molecules into the complex (Petit and Thiery, 2000a). Vinculin does not bind integrins directly, but localizes to adhesions by interaction with talin, α-actinin, and paxillin (Calderwood et al., 2000). It couples newly engaged $β_1$-integrins to actin polymerization by recruiting Arp2/3 complex, which can initiate actin nucleation (Brakebusch and Fassler, 2003). Tensin crosslinks actin and may recruit tyrosine-phosphorylated structural and signaling proteins to focal adhesions through its SH2 domains (Lee and Gotlieb, 2003b). Mena/VASP proteins, also associated with focal adhesions, may target actin polymerization to nascent sites of integrin-extracellular matrix contact by interacting with profilin (Calderwood et al., 2000; Petit and Thiery, 2000b).

A mutual functional dependence between integrins and the actin cytoskeleton is established through bidirectional signaling events initiated at focal adhesions. Via 'inside-out' signaling, the cytoskeleton can affect the organization and function of integrins. Reciprocally, in 'outside-in' signaling, integrins initiate cytoskeletal reorganization after binding to the extracellular ligands, by activating signaling cascades that regulate formation, turnover and linkage of actin filaments (Brakebusch and Fassler, 2003). Integrin-mediated signaling involves several focal adhesion-associated kinases including FAK, Src-like kinases, PKC, PI3K, and integrin-linked kinase (ILK) (Petit and Thiery, 2000b). Integrin activation triggers formation of various phosphoproteins that can modify the cytoskeleton particularly by activating Rho GTPases (Brakebusch and Fassler, 2003).

Actin-Mediated Endothelial Integrity in Health & Disease

Permeability Barrier Dysfunction

Remodeling of the actin cytoskeleton and associated cell-cell and cell-matrix adhesion complexes has a very strong bearing on a major function of the endothelium – its role as a physical barrier between blood and tissues. Two major players in regulation of endothelial permeability are intercellular junctions and actomyosin-based cytoskeletal reorganization and contractility (Dudek and Garcia, 2001; Wojciak-Stothard and Ridley, 2002). Classically, the view has been that endothelial barrier function is governed by the equilibrium between adhesive forces generated by cell-cell and cell-matrix junctions that bind endothelial cells together, and centripetal tensile forces, initiated by actomyosin interactions, that pull endothelial cells apart (Birukova *et al.*, 2004a); however, other models based on actin reorganization and elongation are now emerging (Noria *et al.*, 2004), as outlined in Chapter 2. Inflammatory stimuli decrease intercellular adhesion and promote cell remodeling, leading to formation of interendothelial gaps and consequent vascular leakage (Murphy *et al.*, 2001). Endothelial permeability changes are accompanied by endothelial cells shape changes and actin redistribution – the actin cytoskeleton pattern characteristic of barrier dysfunction involves an increase in stress fiber density and a reduction or loss of the DPB (Ehringer *et al.*, 1999). Since localization of adherens junction to the cell border is highly dependent on the DPB, its disruption may be a mechanism that contributes to disassembly of junctions from the cell border (Patterson *et al.*, 2000; Lum and Roebuck, 2001). In general, actin cytoskeletal organization is a downstream effector of inflammatory stimuli, with the Rho family of GTPases acting as the intermediary mediators. Rho and Rac are key regulators, acting antagonistically to control endothelial barrier function – while Rac stabilizes endothelial cells junctions, Rho stimulates actomyosin contractility (Wojciak-Stothard and Ridley, 2002).

Rho plays a prominent role in the hyperpermeability-associated formation of stress fibers and generation of force in endothelial cells, by participating in MLC phosphorylation (van Nieuw Amerongen and van Hinsbergh, 2001). MLC phosphorylation is the best-characterized pathway leading to endothelial barrier dysfunction via stimulation of endothelial cells cytoskeletal remodeling and contractility (Wojciak-Stothard and Ridley, 2002). The phosphorylation state of MLC is affected by the Ca^{2+}/calmodulin-dependent action of MLCK, which directly phosphorylates MLC, and by Rho-activated ROCK, which promotes MLC phosphorylation mainly by inactivating the myosin phosphatase PP1M (Tinsley *et al.*, 2004). MLC phosphorylation leads to the interaction of myosin with actin, causing endothelial retraction and an increase in permeability (Ogunrinade *et al.*, 2002). The importance of MLC as a downstream effector causing hyperpermeability was highlighted in a recent study that examined MLC phosphorylation in rat lung endothelial cells following exposure to blood plasma from injured rats (which contains inflammatory mediators) (Tinsley *et al.*, 2004). The ensuing endothelial hyperpermeability was accompanied by MLC phosphorylation, and that it could be blocked by MLCK and Rho kinase inhibition (Tinsley *et al.*, 2004).

The role of MLC phosphorylation and Rho in endothelial permeability induced by specific vasoactive agents such as histamine and thrombin has also been studied. While histamine induces only a rapid and transient increase in permeability, thrombin causes prolonged hyperpermeability and plays a major role in the pathophysiology of acute lung injury (Patterson and Garcia, 1994; van Hinsbergh and van Nieuw Amerongen, 2002; Birukova *et al.*, 2004a). It has been found that histamine and thrombin induce an identical rise in cytoplasmic Ca^{2+} concentration of endothelial cells (Shasby *et al.*, 1997), and that the actin cytoskeleton rearrangements and endothelial cells contractility caused by both are dependent on MLCK activation (van Hinsbergh and van Nieuw Amerongen, 2002). However, prolonged elevated permeability triggered by thrombin additionally involves activation of RhoA and ROCK, leading to inhibition of MLC dephosphorylation (van Hinsbergh and van Nieuw Amerongen, 2002). Thus, in the thrombin-induced pathway, Ca^{2+}/calmodulin- and MLCK-dependent MLC phosphorylation can be prolonged or sensitized by activation of RhoA (van Nieuw Amerongen and van Hinsbergh, 2001). A recent *in vitro* study characterized the Rho-mediated pathway in thrombin-initiated pulmonary endothelial cells dysfunction (Birukova *et al.*, 2004a). For example, the guanosine nucleotide exchange factor p115-RhoGEF was implicated in the activation of Rho (Birukova *et al.*, 2004a). Moreover, inhibition of Rho abolished thrombin-induced stress fiber formation and MLC phosphorylation (Birukova *et al.*, 2004a). Another study, confirmed the importance of Rho in pulmonary endothelial barrier dysfunction and highlighted the relevance of crosstalk between microtubules and actin microfilaments in this process (Birukova *et al.*, 2004b). Use of the microtubule inhibitor, nocodazole, led to Rho activation, MLC phosphorylation and stress fiber formation, resulting in barrier disruption. In this study, Rho inhibition was able to attenuate barrier dysfunction. Thus, inhibition of MLC phosphorylation and Rho activation are both potential therapeutic strategies to prevent endothelial barrier dysfunction and pulmonary edema associated with acute lung injury (Tinsley *et al.*, 2004).

Inflammation

Under some pathobiological conditions, such as inflammation, enhanced endothelial permeability is, in fact, part of the normal response to injury. It is of interest that facilitating leukocyte extravasation is an endothelial function that also requires actin cytoskeleton remodeling. The transendothelial migration of leukocytes (e.g. neutrophils), a crucial part of an effective inflammatory and immune response, requires the active participation of endothelial cells, as will be further discussed in Chapter 11 (Vestweber 2002; Alevriadou, 2003). Signaling pathways must be initiated in endothelial cells that induce disruption of the endothelial monolayer, and stress fiber formation, and actomyosin-interactions dependent on MLC phosphorylation (Wojciak-Stothard and Ridley, 2002; Honing *et al.*, 2004). Leukocyte transmigration requires a transient increase in intracellular free Ca^{2+} within endothelial cells; this is required to activate MLCK, which in turn leads to endothelial cells retraction and formation of gaps between adjacent endothelial cells through which leukocytes can pass (Muller, 2003). Evidence also implicates Rho and ROCK in regulation of cytoskeletal reorganization associated with leukocyte extravasation (Saito *et al.*, 2002). The fact that engagement of the leukocyte-binding receptor ICAM-1 leads to Rho activation (Adamson *et al.*, 1999; Etienne-Manneville *et al.*, 2000) suggests a potential signal by which leukocyte adhesion to the endothelium initiates endothelial cells retraction to facilitate the subsequent transmigration of the leukocyte (Wojciak-Stothard and Ridley, 2002).

Atherosclerosis

The pathogenesis of atherosclerosis involves endothelial dysfunction and although atherosclerosis is not a prominent problem in the pulmonary circulation, a number of events are related to the remodeling associated with pulmonary hypertension and chronic hyperpermeability after severe injury, such as with ARDS. Studies have shown that cytoskeletal changes lead to endothelial dysfunction, characterized by gap formation between endothelial cells, poor endothelial repair, and even loss of endothelial cells from the surface of the vessel. In the latter case, this may lead to ulceration and fissure formation on the surface of the atherosclerotic plaques, which in turn may predispose to plaque rupture (Dickson and Gotlieb, 2003). Studies on the pathogenesis of atherosclerotic lesions have shown alterations in the pattern of microfilament distribution and organization in endothelial cells during development of atherosclerotic fatty streak-type lesions in hypercholesterolemic rabbits (Colangelo *et al.*, 1998). In the aorta, away from flow dividers, central stress fibers are not usually prominent in endothelial cells. Central microfilaments, however, became more prominent in the early stages of lesion development and then disappeared once the lesion was raised and contained several layers of macrophages. Thus, central microfilaments undergo dynamic changes during lesion formation. At flow divider sites, where prominent central microfilaments are normally present, microfilaments persisted after attachment of monocytes, but became thinner and eventually disappeared in many cells as the lesion progressed. Thus, the actin microfilament bundles undergo changes that initially may promote cell-substratum adhesions, but eventually further changes occur that are likely to result in dysfunction of cell-substratum adhesion due to a reduction in central actin microfilaments (Colangelo et al., 1998) and eventual endothelial cells detachment from the surface of the vessel.

It is interesting, however, that in initial stages of lesion development, there are prominent stress fibers present, and it is only once the lesion is elevated that central microfilaments are reduced. These findings suggest that mechanical factors impinging on the endothelial cells as the lesion expands may be an important factor in inducing cytoskeletal changes. It has also been shown that oxLDL, a risk factor for atherogenesis, delays endothelial cells migration during the repair of wounded endothelium (Murugesan et al., 1993). Since migration is dependent on actin microfilament organization, it is possible that oxidative stress may play a role in actin organization, possibly through an interaction at endothelial focal adhesion sites. The cytoskeletal changes induce an increase in endothelial permeability and thus promote further leakage of lipids and transmigration of monocyte/macrophages into the vascular wall. The eventual loss of the central microfilament bundles is associated with growth of atherosclerotic plaque and may be associated with endothelial loss, microthrombus formation, and further increases in permeability (Dickson and Gotlieb, 2003).

Another important consideration in the pathogenesis of atherosclerotic plaques is hemodynamic shear stress. Low shear, a well known risk factor, was shown to reduce central microfilaments in *in vivo* endothelial cells, resulting in loss of focal cells, and to upregulate VCAM-1 and macrophage adhesion to the endothelium (Walpola et al., 1995). This showed how important disruption of the actin cytoskeleton is in promoting conditions that initiate plaque formation. In addition, low shear resulted in profound dysfunction of the microfilaments and microtubules in the cells attempting to repair an *in vivo* arterial denuding injury (Vyalov et al., 1996), again showing that a dysfunctional endothelial cytoskeleton places the vascular wall at serious risk for vascular disease.

Perspective

Regulation and function of the cytoskeleton is clearly central to endothelial functional integrity and response to activation and injury. Much has been learned regarding the signaling and the cytoskeletal reorganization involved with cell motility. Since many of the studies have been conducted in epithelial cells and fibroblasts, which have some similar and some different properties compared to endothelium, there is a need to better understand these processes in endothelium, and particularly in pulmonary endothelium, which differes from systemic endothelium. Similarly, the interactions of the vast number of actin binding proteins, actin-junctional connections, and associated signaling molecules have been discovered, but many of these studies have been conducted in other tissues. Specific information is still needed to understand the intricate control of the actin cytoskeleton in endothelial activation and in its return to quiescence and barrier integrity. With information that has been accumulated on the endothelial actin and the microtubule cytoskeleton and the anticipated supplementation of this knowledge by ongoing investigations, it is expected that improved prevention strategies and treatment targets can be identified for combating acute and chronic pulmonary dysfunction that are based, at least in part, on endothelial dysfunction.

ACKNOWLEDGEMENTS

Supported in part by grant T#5475, Heart and Stoke Foundation of Ontario and Univerity of Toronto.

REFERENCES

Adamson, P., Etienne, S., Couraud, P., Calder, V., and Greenwood, J. (1999) Lymphocyte migration through brain endothelial cell monolayers involves signaling through endothelial ICAM-1 via a rho-dependent pathway. J. Immunol. 162:2964-2973.

Akhmanova, A., Hoogenraad, C., Drabek, K., Stepanova, T., Dortland, B., Verkerk, T., Vermeulen, W., Burgering, B., De Zeeuw, C., Grosveld, F., and Galjart, N. (2001) Clasps are CLIP-115 and -170 associating proteins involved in the regional regulation of microtubule dynamics in motile fibroblasts. Cell 104:923-935.

Alevriadou, B.R (2003) CAMs and Rho small GTPases: gatekeepers for leukocyte transendothelial migration. Amer. J. Physiol. 285:C250-C252.

Arthur, W.T, Noren, N., and Burridge, K. (2002) Regulation of Rho family GTPases by cell-cell and cell-matrix adhesion. Biol. Res. 35:239-246.

Ascough, K.R. (2004) Endocytosis: Actin in the driving seat. Curr. Biol. 14:R124-R126.

Assmann, V., Jenkinson, D., Marshall, J., and Hart, I.R. (1999) The intracellular hyaluronan receptor RHAMM/IHABP interacts with microtubules and actin filaments. J. Cell Sci. 112:3943-3954.

Azuma, T., Witke, W., Stossel, T., Hartwig, J., and Kwiatkowski, D.J. (1998) Gelsolin is a downstream effector of rac for fibroblast motility. EMBO J. 17:1362-1370.

Bear, J.E., Svitkina, T., Krause, M., Schafer, D., Loureiro, J., Strasser, G., Maly, I., Chaga, O., Cooper, J., Borisy, G., and Gertler, F.B. (2002) Antagonism between Ena/VASP proteins and actin filament capping regulates fibroblast motility. Cell 109:509-521.

Benesch, S., Lommel, S., Steffen, A., Stradal, T., Scaplehorn, N., Way, M., Wehland, J., and Rottner, K. (2002) PIP_2-induced vesicle movement depends on N-WASP and involves Nck, WIP, and Grb2. J. Biol. Chem. 277:37771-37776.

Bettinger, B.T., Gilbert, D., and Amberg, D.C. (2004) Actin up in the nucleus. Nat. Rev. Molec. Cell Biol. 5:410-415.

Birukova, A.A., Smurova, K., Birukov, K., Kaibuchi, K., Garcia, J., and Verin, A.D. (2004a) Role of Rho GTPases in thrombin-induced lung vascular endothelial barrier dysfunction. Microvasc. Res. 67:64-77.

Birukova, A.A., Smurova, K., Birukov, K., Usatyuk, P., Liu, F., Kaibuchi, K., Ricks-Cord, A., Natarajan, V., Alieva, I., Garcia, J., and Verin, A.D. (2004b) Microtubule disassembly induces cytoskeletal remodeling and lung vascular barrier dysfunction: Role of Rho-dependent mechanisms. J. Cell Physiol 201:55-70.

Blanchoin, L. and Pollard, T.D. (2002) Hydrolysis of ATP by polymerized actin depends on the bound divalent cation but not profilin. Biochem. 41:597-602.

Blangy, A., Vignal, E., Schmidt, S., Debant, A., Gauthier-Rouviere, C., and Fort, P. (2000) TrioGEF1 controls Rac- and Cdc42-dependent cell structures through activation of rhoG. J. Cell Sci. 113:729-739.

Borisy, G.G. and Svitkina, T.M. (2000) Actin machinery: pushing the envelope. Curr. Opin. Cell Biol. 12:104-112

Braga, V. M. (2000) Epithelial cell shape: cadherins and small GTPases. Exp. Cell Res. 261:83-90.

Braga, V.M. (2002) Cell-cell adhesion and signaling. Curr. Opin. Cell Biol. 14:546-556.

Brakebusch, C. and Fassler, R. (2003) The integrin-actin connection, an eternal love affair. EMBO J. 22:2324-2333.

Breviario, F., Caveda, L., Corada, M., Martin-Padura, I., Navarro, P., Golay, J., Introna, M., Gulino, D., Lampugnani, M., and Dejana, E. (1995) Functional properties of human vascular endothelial

cadherin (7B4/cadherin-5), an endothelium-specific cadherin. Arterioscl. Thromb. Vasc. Biol. 15:1229-1239.
Briggs, M.W. and Sacks, D.B. (2003) IQGAP proteins are integral components of cytoskeletal regulation. EMBO Rep. 4:571-574.
Brill, S., Li, S., Lyman, C., Church, D., Wasmuth, J., Weissbach, L., Bernards, A., and Snijders, A.J. (1996) The Ras GTPase-activating-protein-related human protein IQGAP2 harbors a potential actin binding domain and interacts with calmodulin and Rho family GTPases. Molec. Cell Biol. 16:4869-4878.
Calderwood, D.A., Shattil, S.J., and Ginsberg, M.H. (2000) Integrins and actin filaments: reciprocal regulation of cell adhesion and signaling. J. Biol. Chem. 275:22607-22610.
Caldwell, J.E., Heiss, S., Mermall, V., and Cooper, J.A. (1989) Effects of CapZ, an actin capping protein of muscle, on the polymerization of actin. Biochem. 28:8506-8514.
Carlier, M.F., Laurent, V., Santolini, J., Melki, R., Didry, D., Xia, G., Hong, Y., Chua, N., and Pantaloni, D. (1997) ADF/cofilin enhances filament turnover. J. Cell Biol. 136:1307-1322.
Carlier, M.F. and Pantaloni, D. (1986) Direct evidence for ADP-Pi-F-actin as the major intermediate in ATP-actin polymerization. Rate of dissociation of Pi from actin filaments. Biochem. 25:7789-7792.
Carvalho, P., Tirnauer, J., and Pellman, D. (2003) Surfing on microtubule ends. Trends Cell Biol. 13:229-237.
Cau, J., Faure, S., Comps, M., Delsert, C., and Morin, N. (2001) A novel p21-activated kinase binds the actin and microtubule networks and induces microtubule stabilization. J. Cell Biol. 155:1029-1042.
Clark, T.G. and Merriam, R.W. (1978) Actin in Xenopus oocytes. J. Cell Biol. 77:427-438.
Colangelo, S., Langille, B.L., and Gotlieb, A.I. (1994) Three patterns of distribution characterize the organization of endothelial microfilaments at aortic flow dividers. Cell Tissue Res. 278:235-242.
Colangelo, S., Langille, B., Steiner, G., and Gotlieb, A.I. (1998) Alterations in endothelial F-actin microfilaments in rabbit aorta in hypercholesterolemia. Arterioscl. Thromb. Vasc. Biol. 18:52-56.
Cook, T.A., Nagasaki, T., and Gundersen, G.G. (1998) Rho GTPase mediates the selective stabilization of microtubules induced by lysophosphatidic acid. J. Cell Biol. 141:175-185.
Cooke, R. (1975) The role of the bound nucleotide in the polymerization of actin. Biochem. 14:3250-3256.
Cooke, R. (1997) Actomyosin interaction in striated muscle. Physiol. Rev. 77:671-697.
Cooper, J.A. and Schafer, D.A. (2000) Control of actin assembly and disassembly at filament ends. Curr. Opin. Cell Biol. 12:97-103.
Copeland, J.W. and Treisman, R. (2002) The diaphanous-related formin mDia1 controls serum response factor activity through its effects on actin polymerization. Molec. Biol. Cell 13:4088-4099.
Cory, G.O., Garg, R., Cramer, R., and Ridley, A.J. (2002) Phosphorylation of tyrosine 291 enhances the ability of WASp to stimulate actin polymerization and filopodium formation. Wiskott-Aldrich Syndrome protein. J. Biol. Chem. 277:45115-45121.
Cunningham, C.C., Leclerc, N., Flanagan, L., Lu, M., Janmey, P., and Kosik, K.S. (1997) Microtubule-associated protein 2c reorganizes both microtubules and microfilaments into distinct cytological structures in an actin-binding protein-280-deficient melanoma cell line. J. Cell Biol. 136:845-857.
Davies, P.F., Robotewskyj, A., andGriem, M.L. (1994) Quantitative studies of endothelial cell adhesion. Directional remodeling of focal adhesion sites in response to flow forces. J. Clin. Invest. 93:2031-2038.
DeLa Cruz E.M., Mandinova, A., Steinmetz, M., Stoffler, D., Aebi, U., and Pollard, T.D. (2000) Polymerization and structure of nucleotide-free actin filaments. J. Molec. Biol. 295:517-526.
Defilippi, P., Olivo, C., Venturino, M., Dolce, L., Silengo, L., and Tarone, G. (1999) Actin cytoskeleton organization in response to integrin-adhesion. Microsc. Res. Tech. 47:67-78.
Dickson, B.C. and Gotlieb, A.I. (2003) Towards understanding acute destabilization of vulnerable atherosclerotic plaques. Cardiovasc. Pathol. 12:237-248.

Didry, D., Carlier, M.F., and Pantaloni, D. (1998) Synergy between actin depolymerizing factor/cofilin and profilin in increasing actin filament turnover. J. Biol. Chem. 273:25602-25611.
dos Remedios, C.G., Chhabra, D., Kekic, M., Dedova, I., Tsubakihara, M., Berry, D., and Nosworthy, N.J. (2003) Actin binding proteins: regulation of cytoskeletal microfilaments. Physiol. Rev. 83:433-473.
Dudek, S.M. and Garcia, J.G. (2001) Cytoskeletal regulation of pulmonary vascular permeability. J. Appl. Physiol 91:1487-1500.
Eden, S., Rohatgi, R., Podtelejnikov, A., Mann, M., and Kirschner, M.W. (2002) Mechanism of regulation of WAVE1-induced actin nucleation by Rac1 and Nck. Nature 418:790-793.
Ehringer, W.D., Yamany, S., Steier, K., Farag, A., Roisen, F., Dozier, A., and Miller, F.N. (1999) Quantitative image analysis of F-actin in endothelial cells. Microcirc. 6:291-303.
Erickson, J.W., Cerione, R.A., and Hart, M.J. (1997) Identification of an actin cytoskeletal complex that includes IQGAP and the Cdc42 GTPase. J. Biol. Chem. 272:24443-24447.
Etienne-Manneville, S., Manneville, J., Adamson, P., Wilbourn, B., Greenwood, J., and Couraud, P.O. (2000) ICAM-1-coupled cytoskeletal rearrangements and transendothelial lymphocyte migration involve intracellular Ca^{++} signaling in brain endothelial cell lines. J. Immunol. 165:3375-3383.
Ettenson, D.S. and Gotlieb, A.I. (1992) Centrosomes, microtubules, and microfilaments in the reendothelialization and remodeling of double-sided in vitro wounds. Lab. Invest. 66:722-733.
Ettenson, D.S. and Gotlieb, A.I. (1993) In vitro large-wound re-endothelialization. Inhibition of centrosome redistribution by transient inhibition of transcription after wounding prevents rapid repair. Arterioscl. Thromb. 13:1270-1281.
Fomproix, N. and Percipalle, P. (2004) An actin-myosin complex on actively transcribing genes. Exp. Cell Res 294:140-148.
Fukata, M., Kuroda, S., Fujii, K., Nakamura, T., Shoji, I., Matsuura, Y., Okawa, K., Iwamatsu, A., Kikuchi, A., and Kaibuchi, K. (1997) Regulation of cross-linking of actin filament by IQGAP1, a target for Cdc42. J. Biol. Chem. 272:29579-29583.
Fukata, M., Nakagawa, M., Kaibuchi, K. (2003) Roles of Rho-family GTPases in cell polarisation and directional migration. Curr. Opin. Cell Biol. 15:590-597.
Fukata, M., Watanabe, T., Noritake, J., Nakagawa, M., Yamaga, M., Kuroda, S., Matsuura, Y., Iwamatsu, A., Perez, F., and Kaibuchi, K. (2002) Rac1 and Cdc42 capture microtubules through IQGAP1 and CLIP-170. Cell 109:873-885.
Galarneau, L., Nourani, A., Boudreault, A., Zhang, Y., Heliot, L., Allard, S., Savard, J., Lane, W., Stillman, D., and Cote, J. (2000) Multiple links between the NuA4 histone acetyltransferase complex and epigenetic control of transcription. Molec. Cell 5:927-937.
Giannone, G., Dubin-Thaler, B., Dobereiner, H., Kieffer, N. Bresnick, A., Sheetz, M.P. (2004) Periodic lamellipodial contractions correlate with rearward actin waves. Cell 116:431-443.
Goldschmidt-Clermont, P.J., Furman, M., Wachsstock, D., Safer, D., Nachmias, V., Pollard, T.D. (1992) The control of actin nucleotide exchange by thymosin beta 4 and profilin. Molec. Biol. Cell 3:1015-1024.
Gonzalez, M., Cambiazo, V., and Maccioni, R.B. (1998) The interaction of Mip-90 with microtubules and actin filaments in human fibroblasts. Exp. Cell Res. 239:243-253.
Goode, B.L., Drubin, D.G., and Barnes, G. (2000) Functional cooperation between the microtubule and actin cytoskeletons. Curr. Opin. Cell Biol. 12:63-71.
Goode, B.L., Wong, J., Butty, A., Peter, M., McCormack, A., Yates, J., Drubin, D., and Barnes, G. (1999) Coronin promotes the rapid assembly and cross-linking of actin filaments and may link the actin and microtubule cytoskeletons in yeast. J. Cell Biol. 144:83-98.
Gotlieb, A.I., May, L., Subrahmanyan, L., and Kalnins, V.I. (1981) Distribution of microtubule organizing centers in migrating sheets of endothelial cells. J. Cell Biol. 91:589-594.
Gotlieb, A.I., Subrahmanyan, L., and Kalnins, V.I. (1983) Microtubule-organizing centers and cell migration: effect of inhibition of migration and microtubule disruption in endothelium. J. Cell Biol. 96:1266-1272.
Gradin, H.M., Larsson, N., Marklund, U., and Gullberg, M. (1998) Regulation of microtubule dynamics by extracellular signals: cAMP-dependent protein kinase switches off the activity of oncoprotein 18 in intact cells. J. Cell Biol. 140:131-141.

Grazi,E., Ferri, A., and Cino, S. (1983) The polymerization of actin. A study of the nucleation reaction. Biochem. J. 213:727-732.
Grazi, E., and Trombetta, G. (1985) Effects of temperature on actin polymerized by Ca2+. Direct evidence of fragmentation. Biochem J. 232:297-300.
Guan, J.Q., Almo, S., Reisler, E., and Chance, M.R. (2003) Structural reorganization of proteins revealed by radiolysis and mass spectrometry: G-actin solution structure. Biochem. 42:11992-12000.
Gumbiner, B.M. (1996) Cell adhesion: the molecular basis of tissue architecture and morphogenesis. Cell 84:345-357.
Gupton, S.L., Salmon, W.C., and Waterman-Storer, C.M. (2002) Converging populations of F-actin promote breakage of associated microtubules to spatially regulate microtubule turnover in migrating cells. Curr. Biol. 12:1891-9.
Harris, H.E. and Weeds, A.G. (1984) Gelsolin caps and severs actin filaments. FEBS Lett. 177:184-188.
Hart, M.J., Callow, M., Souza, B., and Polakis, P. (1996) IQGAP1, a calmodulin-binding protein with a rasGAP-related domain, is a potential effector for cdc42Hs. EMBO J. 15:2997-3005.
Helmke, B.P., Goldman, R.D., and Davies, P.F. (2000) Rapid displacement of vimentin intermediate filaments in living endothelial cells exposed to flow. Circ. Res. 86:745-752.
Ho, Y.D., Joyal, J., Li, Z., and Sacks, D.B. (1999) IQGAP1 integrates Ca2+/calmodulin and Cdc42 signaling. J. Biol. Chem. 274:464-470.
Hofmann, W., Reichart, B., Ewald, A., Muller, E., Schmitt, I., Stauber, R., Lottspeich, F., Jockusch, B., Scheer, U., Hauber, J., and Dabauvalle, M.C. (2001) Cofactor requirements for nuclear export of RRE- and CTE-containing retroviral RNAs. An unexpected role for actin. J. Cell Biol. 152:895-910.
Hollenbeck, P. (2001) Cytoskeleton: Microtubules get the signal. Curr. Biol. 11:R820-R823.
Honing, H., van den Berg, T., van der Pol, S., Dijkstra, C., van der Kammen, R., Collard, J. and de Vries, H.E. (2004) RhoA activation promotes transendothelial migration of monocytes via ROCK. J. Leuk. Biol. 75:523-528.
Horwitz, S.B., Shen, H., He, L., Dittmar, P., Neef, R., Chen, J., and Schubart, U.K. (1997) The microtubule-destabilizing activity of metablastin is controlled by phosphorylation. J. Biol. Chem. 272:8129-8132.
Huckaba, T.M. and Pon, L.A. (2002) Cytokinesis: rho and formins are ringleaders. Curr. Biol. 12:R813-R814.
Ishizaki, T., Morishima, Y., Okamoto, M., Furuyashiki, T., Kato, T., Narumiya, S. (2001) Coordination of microtubules and the actin cytoskeleton by the Rho effector mDia1. Nat. Cell Biol. 3:8-14.
Janmey, P.A., Iida, K., Yin, H., and Stossel, T.P. (1987) Polyphosphoinositide micelles and polyphosphoinositide-containing vesicles dissociate endogenous gelsolin-actin complexes and promote actin assembly from the fast-growing end of actin filaments blocked by gelsolin. J. Biol. Chem. 262:12228-12236.
Kang, F., Laine, R., Bubb, M., Southwick, F., and Purich, D.L. (1997) Profilin interacts with the Gly-Pro-Pro-Pro-Pro-Pro sequences of vasodilator-stimulated phosphoprotein (VASP): implications for actin-based Listeria motility. Biochem. 36:8384-8392.
Karakesisoglou, I., Yang, Y., and Fuchs, E. (2000) An epidermal plakin that integrates actin and microtubule networks at cellular junctions. J. Cell Biol. 149:195-208.
Kaverina, I., Krylyshkina, O., and Small, J.V. (1999) Microtubule targeting of substrate contacts promotes their relaxation and dissociation. J. Cell Biol. 146:1033-1044.
Kaverina, I.. Krylyshkina, O., Beningo, K., Anderson, K., Wang, Y., and Small, J.V. (2002a) Tensile stress stimulates microtubule outgrowth in living cells. J. Cell Sci. 115:2283-2291.
Kaverina, I., Krylyshkina, O., and Small, J.V. (2002b) Regulation of substrate adhesion dynamics during cell motility. Int. J. Biochem. Cell Biol. 34:746-761.
Kawasaki, Y., Senda, T., Ishidate, T., Koyama, R., Morishita, T., Iwayama, Y., Higuchi, O., and Akiyama, T. (2000) Asef, a link between the tumor suppressor APC and G-protein signaling. Sci. 289:1194-1197.

Kim, A.S., Kakalis, L., Abdul-Manan, N., Liu, G., and Rosen, M.K. (2000) Autoinhibition and activation mechanisms of the Wiskott-Aldrich syndrome protein. Nature 404:151-158.

Kim, D.W., Gotlieb, A.I., and Langille, B.L. (1989a) In vivo modulation of endothelial F-actin microfilaments by experimental alterations in shear stress. Arterioscler. 9:439-445.

Kim, D.W., Langille, B., Wong, M., and Gotlieb, A.I. (1989b) Patterns of endothelial microfilament distribution in the rabbit aorta in situ. Circ. Res. 64:21-31.

Kirschner, M. and Mitchison, T. (1986) Beyond self-assembly: from microtubules to morphogenesis. Cell 45:329-342

Kothakota, S., Azuma, T., Reinhard, C., Klippel, A., Tang, J., Chu. K., McGarry, T., Kirschner, M., Koths, K., Kwiatkowski, D., and Williams, L.T. (1997) Caspase-3-generated fragment of gelsolin: effector of morphological change in apoptosis. Sci. 278:294-298.

Krendel, M., Zenke, F.T., and Bokoch, G.M. (2002) Nucleotide exchange factor GEF-H1 mediates cross-talk between microtubules and the actin cytoskeleton. Nat. Cell Biol. 4:294-301.

Krylyshkina, O., Anderson, K., Kaverina, I., Upmann, I., Manstein, D., Small, J., and Toomre, D.K. (2003) Nanometer targeting of microtubules to focal adhesions. J. Cell Biol. 161:853-859.

Lampugnani, M.G., Corada, M., Caveda, L., Breviario, F., Ayalon, O., Geiger, B., and Dejana, E. (1995) The molecular organization of endothelial cell to cell junctions: differential association of plakoglobin, β-catenin, and α-catenin with VE-cadherin. J. Cell Biol. 129:203-217.

Langille, B.L., Graham, J., Kim, D., and Gotlieb, A.I. (1991) Dynamics of shear-induced redistribution of F-actin in endothelial cells in vivo. Arterioscl. Thromb. 11:1814-1820.

Lantz, V.A. and Miller, K.G. (1998) A class VI unconventional myosin is associated with a homologue of a microtubule-binding protein, cytoplasmic linker protein-170. J. Cell Biol. 140:897-910.

Le Clainche, C., Didry, D., Carlier, M., and Pantaloni, D. (2001) Activation of Arp2/3 complex by WASP is linked to enhanced binding of ATP to Arp2. J. Biol. Chem. 276:46689-46692.

Lee, J.S. and Gotlieb, A.I. (2002) Microtubule-actin interactions may regulate endothelial integrity and repair. Cardiovasc. Pathol. 11:135-140.

Lee, J.S. and Gotlieb, A.I. (2003a) Understanding the role of the cytoskeleton in the complex regulation of the endothelial repair. Histol. Histopath. 18:879-887.

Lee, T.Y. and Gotlieb, A.I. (2003b) Microfilaments and microtubules maintain endothelial integrity. Microsc. Res. Tech. 60:115-127.

Li, F. and Higgs, H.N. (2003) The mouse Formin mDia1 is a potent actin nucleation factor regulated by autoinhibition. Curr. Biol. 13:1335-1340.

Lin, K.M., Mejillano, M., and Yin, H.L. (2000) Ca2+ regulation of gelsolin by its C-terminal tail. J. Biol. Chem. 275:27746-27752.

Lin, Y.H., Park, Z., Lin, D., Brahmbhatt, A., Rio, M., Yates, J., and Klemke, R.L. (2004) Regulation of cell migration and survival by focal adhesion targeting of Lasp-1. J. Cell Biol. 165:421-432.

Loisel, T.P., Boujemaa, R., Pantaloni, D., and Carlier, M.F. (1999) Reconstitution of actin-based motility of Listeria and Shigella using pure proteins. Nature 401:613-616.

Louvet-Vallee, S. (2000) ERM proteins: from cellular architecture to cell signaling. Biol. Cell 92:305-316

Lum, H. and Roebuck, K.A. (2001) Oxidant stress and endothelial cell dysfunction. Amer. J. Physiol. Cell Physiol. 280:C719-C741.

Maekawa, M., Ishizaki, T., Boku, S., Watanabe, N., Fujita, A., Iwamatsu, A., Obinata, T., Ohashi, K., Mizuno, K., and Narumiya, S. (1999) Signaling from Rho to the actin cytoskeleton through protein kinases ROCK and LIM-kinase. Sci. 285:895-898.

Mandelkow, E.M., Mandelkow, E., and Milligan, R.A. (1991) Microtubule dynamics and microtubule caps: a time-resolved cryo-electron microscopy study. J. Cell Biol. 114:977-991.

Mateer, S.C., McDaniel, A., Nicolas, V., Habermacher, G., Lin, M., Cromer, D., King, M., and Bloom, G.S. (2002) The mechanism for regulation of F-actin binding of IQGAP1 by Ca^{++}/calmodulin. J. Biol. Chem. 277:12324-12333.

Miki, H., Yamaguchi, H., Suetsugu, S., and Takenawa, T. (2000) IRSp53 is an essential intermediate between Rac and WAVE in the regulation of membrane ruffling. Nature 408:732-735.

Millard, T.H., Sharp, S.J., and Machesky, L.M. (2004) Signaling to actin assembly via the WASP-family proteins and the Arp2/3 complex. Biochem. J. 380:1-17.
Miller, A.L., Wang, Y., Mooseker, M., and Koleske, A.J. (2004) The Abl-related gene (Arg) requires its F-actin-microtubule cross-linking activity to regulate lamellipodial dynamics during fibroblast adhesion. J. Cell Biol. 165:407-419.
Mimori-Kiyosue, Y., Shiina, N., and Tsukita, S. (2000) Adenomatous polyposis coli (APC) protein moves along microtubules and concentrates at their growing ends in epithelial cells. J. Cell Biol. 148:505-518.
Mudry, R.E., Perry, C., Richards, M., Fowler, V., and Gregorio, C.C. (2003) The interaction of tropomodulin with tropomyosin stabilizes thin filaments in cardiac myocytes. J. Cell Biol. 162:1057-1068.
Muller, W.A. (2003) Leukocyte-endothelial-cell interactions in leukocyte transmigration and inflammatory response. Trends Immunol. 24:327-334.
Mullins, R.D., Heuser, J.A., and Pollard, T.D. (1998) The interaction of Arp2/3 complex with actin: nucleation, high affinity pointed end capping, and formation of branching networks of filaments. PNAS USA 95:6181-6186.
Murphy, J.T., Duffy, S., Hybki, D., and Kamm, K. (2001) Thrombin-mediated permeability of human microvascular pulmonary endothelial cells is calcium dependent. J. Trauma 50:213-222.
Murugesan, G., Chisolm, G.M., and Fox, P.L. (1993) Oxidized low density lipoprotein inhibits the migration of aortic endothelial cells in vitro. J. Cell Biol. 120:1011-1019.
Nabi, I.R. (1999) The polarization of the motile cell. J. Cell Sci. 112:1803-1811.
Nakamura, M., Zhou, X.Z., and Lu, K.P. (2001) Critical role for the EB1 and APC interaction in the regulation of microtubule polymerization. Curr. Biol. 11:1062-1067.
Narumiya, S., Ishizaki, T., and Watanabe, N. (1997) Rho effectors and reorganization of actin cytoskeleton. FEBS Lett. 410:68-72.
Nishida, E. and Sakai, H. (1983) Kinetic analysis of actin polymerization. J. Biochem (Tokyo) 93:1011-1020.
Niwa, R., Nagata-Ohashi, K., Takeichi, M., Mizuno, K., and Uemura, T. (2002) Control of actin reorganization by Slingshot, phosphatases that dephosphorylate ADF/cofilin. Cell 108:233-246.
Noria, S., Cowan, D., Gotlieb, A.I., and Langille, B.L. (1999) Transient and steady-state effects of shear stress on endothelial cell adherens junctions. Circ. Res 85:504-514.
Noria, S., Xu, F., McCue, S., Jones, M., Gotlieb, A.I., and Langille, B.L. (2004) Assembly and reorientation of stress fibers drives morphological changes to endothelial cells exposed to shear stress. Amer. J. Pathol. 164:1211-1223.
Notarangelo, L.D. and Ochs, H.D. (2003) Wiskott-Aldrich Syndrome: a model for defective actin reorganization, cell trafficking and synapse formation. Curr. Opin. Immun. 15:585-591.
Ogunrinade, O., Kameya, G.T., and Truskey, G.A. (2002) Effect of fluid shear stress on the permeability of the arterial endothelium. Ann. Biomed. Eng. 30:430-446.
Ojala, P.J., Paavilainen, V., and Lappalainen, P. (2001) Identification of yeast cofilin residues specific for actin monomer and PIP2 binding. Biochem. 40:15562-15569.
Ono, S. (2003) Regulation of actin filament dynamics by actin depolymerizing factor/cofilin and actin-interacting protein 1: new blades for twisted filaments. Biochem. 42:13363-13370.
Ono, S. and Ono, K. (2002) Tropomyosin inhibits ADF/cofilin-dependent actin filament dynamics. J. Cell Biol. 156:1065-1076.
Palazzo, A.F., Cook, T., Alberts, A., and Gundersen, G.G. (2001a) mDia mediates Rho-regulated formation and orientation of stable microtubules. Nat. Cell Biol. 3:723-729.
Palazzo, A.F., Eng, C., Schlaepfer. D., Marcantonio. E., and Gundersen, G.G. (2004) Localized stabilization of microtubules by integrin- and FAK-facilitated Rho signaling. Sci. 303:836-839.
Palazzo, A.F., Joseph, H., Chen, Y., Dujardin, D., Alberts, A., Pfister, K., Vallee. R., and Gundersen, G.G. (2001b) Cdc42, dynein, and dynactin regulate microtubule organizing center reorientation independent of Rho-regulated microtubule stabilization. Curr. Biol. 11:1536-1541.
Pantaloni, D., Boujemaa, R., Didry, D., Gounon, P., and Carlier, M.F. (2000) The Arp2/3 complex branches filament barbed ends: functional antagonism with capping proteins. Nat. Cell Biol. 2:385-391.

Pantaloni,D., Hill, T., Carlier, M., and Korn, E.D. (1985) A model for actin polymerization and the kinetic effects of ATP hydrolysis. PNAS USA 82:7207-7211.
Papakonstanti, E.A., Vardaki, E.A., and Stournaras, C. (2000) Actin cytoskeleton: a signaling sensor in cell volume regulation. Cell Physiol. Biochem. 10:257-264.
Patterson, C.E. and Garcia, J.G.N. (1994) Regulation of thrombin-induced endothelial cell activation by bacterial toxins. Blood Coag. Fibrinol. 5:63-72.
Patterson, C.E., Lum, H., Verin, A., Schaphorst, K., and Garcia, J.G.N. (2000) Regulation of endothelial barrier function by the cAMP-dependent protein kinase. Endothelium 7:287-308.
Petit, V., Boyer, B., Lentz, D., Turner, C., Thiery, J., and Valles, A.M. (2000a) Phosphorylation of tyrosine residues 31 and 118 on paxillin regulates cell migration through an association with CRK in NBT-II cells. J. Cell Biol. 148:957-970.
Petit, V. and Thiery, J.P. (2000b) Focal adhesions: structure and dynamics. Biol. Cell 92:477-494.
Pollard, T.D. (1986) Rate constants for the reactions of ATP- and ADP-actin with the ends of actin filaments. J. Cell Biol. 103:2747-2754.
Pollard, T.D. and Borisy, G.G. (2003) Cellular motility driven by assembly and disassembly of actin filaments. Cell 112:453-465.
Pollard, T.D. and Mooseker, M.S. (1981) Direct measurement of actin polymerization rate constants by electron microscopy of actin filaments nucleated by isolated microvillus cores. J. Cell Biol. 88:654-659.
Pring, M., Evangelista, M., Boone, C., Yang, C., Zigmond, S.H. (2003) Mechanism of formin-induced nucleation of actin filaments. Biochem. 42:486-496.
Raftopoulou, M. and Hall, A. (2004) Cell migration:Rho GTPases lead the way. Dev. Biol. 265:23-32
Ren, X.D., Kiosses, W.B., and Schwartz, M.A. (1999) Regulation of the small GTP-binding protein Rho by cell adhesion and the cytoskeleton. EMBO J. 18:578-585.
Ridley, A.J. (2001) Rho GTPases and cell migration. J. Cell Sci. 114:2713-2722.
Rivero, F. and Somesh, B.P. (2002) Signal transduction pathways regulated by Rho GTPases in Dictyostelium. J. Muscle Res. Cell Motil. 23:737-749.
Robinson, R.C., Turbedsky, K., Kaiser, D., Marchand. J., Higgs, H., Choe, S., and Pollard, T.D. (2001) Crystal structure of Arp2/3 complex. Sci. 294:1679-1684.
Rodionov, V.I., Gyoeva, F., Tanaka, E., Bershadsky, A., Vasiliev, J., and Gelfand, V.I. (1993) Microtubule-dependent control of cell shape and pseudopodial activity is inhibited by the antibody to kinesin motor domain. J. Cell Biol. 123:1811-1820.
Rodriguez, O.C., Schaefer, A., Mandato, C., Forscher, P., Bement, W., and Waterman-Storer, C.M. (2003) Microtubule-actin interactions in cell movement and morphogenesis. Nat. Cell Biol. 5:599-609.
Rogers, S.L. and Gelfand, V.I. (2000) Membrane trafficking, organelle transport, and the cytoskeleton. Curr. Opin. Cell Biol. 12:57-62.
Safer, D., Sosnick, T.R., and Elzinga, M. (1997) Thymosin β4 binds actin in an extended conformation and contacts both the barbed and pointed ends. Biochem. 36:5806-5816.
Saito, H., Minamiya, Y., Saito, S., and Ogawa, J. (2002) Endothelial Rho and Rho kinase regulate neutrophil migration via endothelial myosin light chain phosphorylation. J. Leukoc. Biol. 72:829-836.
Sander, E.E., ten Klooster, J., van Delft, S., van der Kammen, R., and Collard, J.G. (1999) Rac downregulates Rho activity: reciprocal balance between both GTPases determines cellular morphology and migratory behavior. J. Cell Biol. 147:1009-1022.
Sarmiere, P.D. and Bamburg, J.R. (2004) Regulation of the neuronal actin cytoskeleton by ADF/cofilin. J. Neurobiol. 58:103-117.
Schnittler, H.J. (1998) Structural and functional aspects of intercellular junctions in vascular endothelium. Basic Res. Cardiol. 933:30-39.
Schoenwaelder, S.M. and Burridge, K. (1999) Bidirectional signaling between the cytoskeleton and integrins. Curr. Opin. Cell Biol. 11:274-286.
Selden, L.A., Kinosian, H., Estes, J., and Gershman, L.C. (1999) Impact of profilin on actin-bound nucleotide exchange and actin polymerization dynamics. Biochem. 38:2769-2778.

Sept, D. and McCammon, J.A. (2001) Thermodynamics and kinetics of actin filament nucleation. Biophys. J. 81:667-674.
Shasby, D.M., Stevens, T., Ries, D., Moy, A., Kamath, J., Kamath, A., and Shasby, S.S. (1997) Thrombin inhibits myosin light chain dephosphorylation in endothelial cells. Amer. J. Physiol 272:L311-L319.
Sheetz, M.P., Felsenfeld, D.P., and Galbraith, C.G. (1998) Cell migration: regulation of force on extracellular-matrix-integrin complexes. Trends Cell Biol. 8:51-54.
Shen, X., Mizuguchi, G., Hamiche, A., and Wu, C. (2000) A chromatin remodelling complex involved in transcription and DNA processing. Nature 406:541-544.
Small, J.V., Geiger, B., Kaverina, I., and Bershadsky, A. (2002a) How do microtubules guide migrating cells? Nat. Rev. Mol. Cell Biol. 3:957-964.
Small, J.V. and Kaverina, I. (2003) Microtubules meet substrate adhesions to arrange cell polarity. Curr. Opin. Cell Biol. 15:40-47.
Small, J.V., Stradal, T., Vignal, E., and Rottner, K. (2002b) The lamellipodium: where motility begins. Trends Cell Biol. 12:112-120.
Stebbings, H. (2001) Cytoskeleton-dependent transport and localization of mRNA. Int. Rev. Cytol. 211:1-31.
Sun, H.Q., Kwiatkowska, K., and Yin, H.L. (1995) Actin monomer binding proteins. Curr. Opin. Cell Biol. 7:102-110.
Sun, H.Q., Yamamoto, M., Mejillano, M., and Yin, H.L. (1999) Gelsolin, a multifunctional actin regulatory protein. J. Biol. Chem. 274:33179-33182.
Tinsley, J.H., Teasdale, N.R., and Yuan, S.Y. (2004) Myosin light chain phosphorylation and pulmonary endothelial cell hyperpermeability in burns. Amer. J. Physiol. 286:L841-L847.
Uruno, T., Liu, J., Zhang, P., Fan, Y., Egile, C., Li, R., Mueller, S., and Zhan, X. (2001) Activation of Arp2/3 complex-mediated actin polymerization by cortactin. Nat. Cell Biol. 3:259-266.
van Hinsbergh, V.W. and van Nieuw Amerongen, G.P. (2002) Intracellular signaling involved in modulating human endothelial barrier function. J. Anat. 200:549-560.
van Nieuw Amerongen, G.P. and van Hinsbergh, V.W. (2001) Cytoskeletal effects of rho-like small guanine nucleotide-binding proteins in the vascular system. Arterioscler. Thromb. Vasc. Biol. 21:300-311.
Vasiliev, J.M., Gelfand, I., Domnina, L., Ivanova, O., Komm, S., and Olshevskaja, L.V. (1970) Effect of colcemid on locomotory behaviour of fibroblasts. J. Embryol. Exp. Morph. 24:625-640.
Vestweber, D. (2002) Regulation of endothelial cell contacts during leukocyte extravasation. Curr. Opin. Cell Biol. 14:587-593.
Vicente-Manzanares, M., Rey, M., Perez-Martinez, M., Yanez-Mo, M., Sancho, D., Cabrero, J., Barreiro, O., de la Fuente, H., Itoh, K., and Sanchez-Madrid, F. (2003) The RhoA effector mDia is induced during T cell activation and regulates actin polymerization and cell migration. J. Immunol. 171:1023-1034.
Volkmann, N., Amann, K.J., Stoilova-McPhie, S., Egile, C., Winter, D., Hazelwood, L., Heuser, J., Li, R., Pollard, T., and Hanein, D. (2001) Structure of Arp2/3 complex in its activated state and in actin filament branch junctions. Sci. 293:2456-2459.
Vyalov, S., Langille, B.L., and Gotlieb, A.I. (1996) Decreased blood flow rate disrupts endothelial repair in vivo. Amer. J. Pathol. 149:2107-2118.
Walker, R.A., O'Brien, E., Pryer, N., Soboeiro, M., Voter, W., Erickson, H., Salmon, E.D. (1988) Dynamic instability of individual microtubules analyzed by video light microscopy: rate constants and transition frequencies. J. Cell Biol. 107:1437-1448.
Walpola, P.L., Gotlieb, A.I., Cybulsky, M., and Langille, B.L. (1995) Expression of ICAM-1 and VCAM-1 and monocyte adherence in arteries exposed to shear stress. Arterioscl. Thromb. Vasc. Biol. 15:2-10.
Wang, F., Sampogna, R.V., and Ware, B.R. (1989) pH dependence of actin assembly. Biophys J. 55:293-298.
Waterman-Storer, C.M. and Salmon, E.D. (1997) Actomyosin-based retrograde flow of microtubules in lamella of migrating epithelial cells influences microtubule dynamic instability and turnover and is associated with microtubule breakage and treadmilling. J. Cell Biol. 139:417-434.

Waterman-Storer, C.M., Worthylake, R., Liu, B., Burridge, K., and Salmon, E.D. (1999) Microtubule growth activates Rac1 to promote lamellipodial protrusion in fibroblasts. Nat. Cell Biol. 1:45-50.

Weaver, A.M., Karginov, A., Kinley, A., Weed, S., Li, Y., Parsons, J., and Cooper, J.A. (2001) Cortactin promotes and stabilizes Arp2/3-induced actin filament network formation. Curr. Biol. 11:370-374.

Weaver, A.M., Young, M., Lee, W., and Cooper, J.A. (2003) Integration of signals to the Arp2/3 complex. Curr. Opin. Cell Biol. 15:23-30.

Weber, A. (1999) Actin binding proteins that change extent and rate of actin monomer-polymer distribution by different mechanisms. Molec. Cell Biochem. 190:67-74.

Weber, A., Pennise, C., Babcock, G., and Fowler, V.M. (1994) Tropomodulin caps the pointed ends of actin filaments. J. Cell Biol. 127:1627-1635.

Weed, S.A. and Parsons, J.T. (2001) Cortactin: coupling membrane dynamics to cortical actin assembly. Oncogene 20:6418-6434.

Winder, S.J. (2003) Structural insights into actin-binding, branching, and bundling proteins. Curr. Opin. Cell Biol. 15:14-22.

Wittmann, T., Bokoch, G.M., and Waterman-Storer, C.M. (2003) Regulation of leading edge microtubule and actin dynamics downstream of Rac1. J. Cell Biol. 161:845-851.

Wittmann, T., Bokoch, G.M., and Waterman-Storer, C.M. (2004) Regulation of microtubule destabilizing activity of Op18/stathmin downstream of Rac1. J. Biol. Chem. 279:6196-6203.

Wittmann, T. and Waterman-Storer, C.M. (2001) Cell motility: can Rho GTPases and microtubules point the way? J. Cell Sci. 114:3795-3803.

Wojciak-Stothard, B. and Ridley, A.J. (2002) Rho GTPases and the regulation of endothelial permeability. Vascul. Pharm. 39:187-199.

Wong, A.J., Pollard, T., and Herman, I.M. (1983) Actin filament stress fibers in vascular endothelial cells in vivo. Sci. 219:867-869.

Wong, M.K. and Gotlieb, A.I. (1986) Endothelial cell monolayer integrity. I. Characterization of dense peripheral band of microfilaments. Arterioscl. 6:212-219.

Wong, M.K. and Gotlieb, A.I. (1990) Endothelial monolayer integrity. Perturbation of F-actin filaments and the dense peripheral band-vinculin network. Arterioscl. 10:76-84.

Wu, X., Jung, G., and Hammer, J.A., III (2000) Functions of unconventional myosins. Curr. Opin. Cell Biol. 12:42-51.

Yang, C., Huang, M., DeBiasio, J., Pring, M., Joyce, M., Miki, H., Takenawa, T., and Zigmond, S.H. (2000) Profilin enhances Cdc42-induced nucleation of actin polymerization. J. Cell Biol. 150:1001-1012.

Zalevsky, J., Lempert, L., Kranitz, H., Mullins, R.D. (2001) Different WASP family proteins stimulate different Arp2/3 complex-dependent actin-nucleation. Curr. Biol. 11:1903-1913.

Zimmerle, C.T. and Frieden, C. (1986) Effect of temperature on the mechanism of actin polymerization. Biochem. 25:6432-6438.

Chapter 8

Endothelial-Matrix Interactions in the Lung

Sunita Bhattacharya, M.S.,[1,2] Sadiqa Quadri, Ph.D.,[1,3] & Jahar Bhattacharya, M.D., Ph.D.[1,3]*

Lung Biology Laboratory,[1] Departments of Pediatrics[2] and Physiology & Cellular Biophysics,[3] College of Physicians & Surgeons, Columbia University, St. Luke's-Roosevelt Hospital Center, New York, NY 10019

CONTENTS:

Introduction

Integrins
Function and Composition
Unique Properties of $\alpha v \beta 3$ in Lung
Integrin Ligation and Calcium Signaling
Inside-Outside Signaling

Focal Adhesions-Structure
Formation
Focal Adhesion Kinase

Focal Adhesions in Endothelial Barrier Regulation
Agonists Effects
Hyperosmotic Effects

Focal Adhesions in Blood Vessels
Effects of Hyperventilation
Effects of Hyperosmolarity

Future Directions

Introduction

The lung microvascular bed supports not only the gas exchange function of the lung, but it also forms the major site both for lung liquid production and for rapid leukocyte recruitment. These non-gas exchange functions are critical. The liquid production maintains tissue hydration in the lung parenchyma and probably forms the source of airway liquid, while rapid leukocyte recruitment is essential for establishing the lung's innate immune defense. Exacerbation or dysregulation of these processes precipitates some of the most devastating forms of pulmonary disease including the acute lung injury syndrome and pulmonary edema. Endothelial cells are the primary cell type that determines these non-gas exchange functions in lung. Endothelial cells regulate the paracellular traffic of transvascular flux of liquid and inflammatory cells across intercellular junctions. The luminal endothelial membrane expresses leukocyte adhesion receptors that initiate the lung inflammatory processes. For these reasons, an understanding of endothelial mechanisms continues to be of interest and has been the focus of recent reviews on junctional mechanisms and leukocyte recruitment (Ulbrich *et al.*, 2003; Bazzoni and Dejana, 2004).

Here we consider these issues in the context of endothelial-matrix interactions that have received less attention. The matrix role is clearly important since the bulk of the lung microvascular bed comprises vessels that lack smooth muscle and in which endothelial cells lie immediately apposed to the surrounding interstitial matrix. It is long suspected that the endothelial-matrix association in the lung microvascular bed is very well developed. Classical data report that in capillaries lying immediately outside the alveolar septum, in so-called extra-alveolar vessels, the buttressing effect of the matrix prevents capillary collapse in the face of major decreases of vascular pressure (Sobin *et al.*, 1978; Sadurski *et al.*, 1994). Increasing implication that the matrix plays a direct role in endothelial regulation of barrier properties and proinflammatory responses now supports the older phenomenological evidence. In this chapter, we consider these matrix-related signaling mechanisms to the extent that they are known to apply to endothelial function in the adult lung.

Integrins

Function and Composition

Cell-matrix adhesion is mediated by integrins. These are a family of transmembrane α and β heterodimers with extracellular segments that contain matrix binding sites for proteins such as fibronectin, laminin, or collagen (Katsumi *et al.*, 2004). Intracellularly, they interact with a number of adaptor and signaling molecules and are linked to the actin cytoskeleton. Currently, about 18 different α subunits and 8 β subunits are identified. Some of the heterodimers in endothelial cells and fundamentals of structure were introduced in Chapter 2. The role of endothelial integrins has been discussed and recently reviewed largely in the context of cell adhesion, angiogenesis, wound healing, and mechanotransduction (Juliano, 2002; Ruegg and Mariotti, 2003; Katsumi *et al.*, 2004). However, their significance in the quiescent, non-proliferating lung vascular bed has received relatively less attention. Increasing evidence now implicates the integrin $\alpha v\beta 3$ in lung endothelial barrier regulation.

Unique Properties of α vβ 3 in Lung

Immunoelectronmicroscopy of lung capillaries indicates that the αvβ3 integrin is expressed both on the abluminal and luminal aspects of endothelial cells (Singh *et al.*, 2000). The evidence from immunohistochemistry study and the in situ polymerase chain reaction (in situ PCR) affirms that expressions for αvβ3 protein, and αv and β3 mRNAs are absent in systemic vessels (except liver), but present in lung in both microvascular and large vessel endothelium (Singh *et al.*, 2000). Southern blots on equal amounts of mRNA indicated that lung expression of αvβ3 is highest amongst major organs (Singh *et al.*, 2000). These findings indicate that the lung vascular bed is a preferred site of constitutive αvβ3 expression as opposed to systemic beds in which the integrin is probably expressed only during vessel proliferation, as in wound healing or tumor formation.

Since the luminal αvβ3 integrin of lung capillaries is exposed to blood-borne ligands, it is capable of ligating circulating products that contain αvβ3 ligands, such as vitronectin (Preissner, 1991). For example, the SC5b-9 complex that forms as an end-product of complement activation, and the thrombin-anti-thrombin-III complex that forms in clotting, both contain vitronectin (Preissner, 1991). Interactions of the αvβ3 integrin with these ligands could have pathological consequences, since exposure of lung capillaries to complement-activated serum, purified SC5b-9, or multimeric vitronectin, each increases capillary permeability, as quantified by the capillary hydraulic conductivity (Lp) (Ishikawa *et al.*, 1993; Tsukada *et al.*, 1995). Anti-αvβ3 antibodies block the Lp increases, thereby implicating the endothelial αvβ3 integrin as a barrier deteriorating receptor in lung capillaries.

Blood levels of SC5b-9 increase in complement-activated states such as sepsis (Langlois and Gawryl, 1988), raising the possibility that SC5b-9 ligation to αvβ3 may contribute to the pathological microvascular effects characteristic of sepsis. Vitronectin may also form the basis of αvβ3 ligation by Gram-positive and Gram-negative bacteria that bind vitronectin (Chhatwal *et al.*, 1987). Other inflammatory αvβ3 ligands include thrombospondin that is secreted by neutrophils and macrophages (Savill *et al.*, 1992), and osteopontin and von Willebrand factor that are excessively secreted by endothelial cells during lung injury (Kasper *et al.*, 1996; Berman *et al.*, 2004). The extent to which these αvβ3 ligations promote hyperpermeability in lung inflammatory diseases requires further understanding.

Although it is being understood that integrins aggregate following ligation (Miyamoto *et al.*, 1995; Miyamoto *et al.*, 1996), the physiological significance of integrin aggregation requires further clarification. Aggregation appears to be the critical step that initiates integrin-mediated cell signaling. Monomeric vitronectin, a normal plasma constituent that is evidently not pathogenic, ligates but does not aggregate the αvβ3 integrin. Accordingly, multimeric, but not monomeric vitronectin increases Lp (Tsukada *et al.*, 1995). Multivalent ligation of αvβ3 by multimeric vitronectin probably aggregates several αvβ3 dimers. The Lp increase attributable to SC5b-9, which contains multimeric vitronectin, is also the consequence of αvβ3 aggregation. Confocal images of lung endothelial cells reveal vitronectin- or SC5b-9-mediated αvβ3 aggregation as fluorescent clumps that localize mainly at the cell periphery at the apical, but not the basal surface (Bhattacharya *et al.*, 2001). The distribution of the fluorescent aggregates at the lung EC periphery indicates that αvβ3 aggregation localizes to inter-endothelial junctions where they may regulate endothelial barrier responses.

Integrin Ligation and Calcium Signaling

Endothelial hyperpermeability is attributed to Ca^{2+}-dependent endothelial contraction that increases the paracellular flux by widening inter-endothelial junctions. The present relevance is that the $\alpha v \beta 3$ ligands, multimeric vitronectin and SC5b-9, as well as crosslinking antibodies that aggregate the integrin, all increase cytosolic Ca^{2+} (Ca^{2+}cyt) in lung endothelial cells (Bhattacharya *et al.*, 2000). Interestingly, the Ca^{2+}cyt increase initiates at the cell periphery, the site of $\alpha v \beta 3$ aggregation, and then spreads centripetally. In contrast, histamine-induced Ca^{2+}cyt increases occur more globally and initiate at the cell center (Bhattacharya *et al.*, 2000). External Ca^{2+}-depletion blunts the $\alpha v \beta 3$-induced Ca^{2+}cyt increase, while thapsigargin inhibits the response completely, indicating that $\alpha v \beta 3$ ligation induces Ca^{2+} release from endosomal stores (Bhattacharya *et al.*, 2000). Although ligation of the $\alpha v \beta 3$ integrin does not directly activate a Ca^{2+} channel, the induced Ca^{2+}cyt increase activates a hyperpolarizing outward K^+ current (Kawasaki *et al.*, 2004). It is suggested that in lung endothelial cells, this $\alpha v \beta 3$-induced hyperpolarization may contribute to sustained Ca^{2+}cyt increases required for NO release (Kawasaki *et al.*, 2004).

The Ca^{2+}cyt response to $\alpha v \beta 3$ ligation is attributable to involvement of the phospholipase C-γ_1 (PLC-γ_1-inositol (3,4,5) triphosphate (InsP3) pathway. Induced aggregation of the $\alpha v \beta 3$ integrin enhances tyrosine phosphorylation of PLC-γ_1, which hydrolyses inositol bisphosphate to release InsP3. Ligation of InsP3 receptors releases Ca^{2+} from endosomal stores. The subsequent decrease in barrier properties may occur by mechanisms such as activation of Ca^{2+}-calmodulin dependent kinases that phosphorylate myosin light chain leading to actin-myosin dependent endothelial retraction. Activation of Ca^{2+}-dependent phospholipase A2 (cPLA2) induces the arachidonate pathway. Several products of this pathway may exert barrier-deteriorating effects. Genistein, a broad-spectrum tyrosine kinase inhibitor, abrogates tyrosine phosphorylation of PLC-γ_1, as well the $\alpha v \beta 3$-induced Ca^{2+}cyt increases in cultured endothelial cells and permeability increases in lung capillaries (Tsukada *et al.*, 1995; Bhattacharya *et al.*, 2000). These findings indicate that aggregation of the endothelial $\alpha v \beta 3$ integrin induces a rapid tyrosine phosphorylation-dependent increase of Ca^{2+}cyt that may underlie the integrins inflammatory role in lung blood vessels.

Inside-Out Signaling

According to the hypothesis of affinity-modulation of integrin-ligand binding (Shimizu *et al.*, 1990; Damsky and Werb, 1992; Crowe *et al.*, 1994), which is also known as the theory of "inside-out" signaling, cell activation induces binding of specific cytoplasmic proteins to conserved sequences in the cytoplasmic domains of the α and β subunits. This increases the extracellular binding affinity of integrins to specific ligands, presumably because of induced spatial or conformational changes in the extracellular integrin domains. Inside-out signaling has been documented for integrins of blood cells, but not for vascular integrins. The theory may be relevant to endothelial $\alpha v \beta 3$ integrin, because ligand binding to the platelet integrin $\alpha IIb \beta 3$, a homologue of $\alpha v \beta 3$, is affinity modulated (Chen *et al.*, 1994). Prolonged $\alpha v \beta 3$ ligation may activate inside-out signaling; thereby causing time dependent increases of binding affinity and possibly, secondary endothelial responses in barrier regulation or inflammation.

Focal Adhesions – Structure

Formation

On the cytosolic aspect, integrins form associations with multiple proteins to form the so-called focal complexes. A summary of the burgeoning literature on the dynamic properties and constitutive features of these structures is beyond the present scope, but is available in recent reviews (Bershadsky *et al.*, 2003; Schlaepfer and Mitra, 2004). Briefly, focal complexes are submicron sized structures that stabilize cells against the interstitial matrix. In cells exposed to mechanical stress, focal complexes mature into structures known as focal adhesions that may be several microns in size.

The growth of focal complexes is attributable to integrin ligation and aggregation that recruits focal adhesion kinase (FAK) to the cytoplasmic tail of the integrin β subunit. Subsequently, FAK phosphorylates paxillin and a cascade of tyrosine phosphorylations occur on vinculin, talin, ezrin, α-actinin and cortactin. In lung endothelial cells, αvβ3 ligation causes tyrosine phosphorylation on FAK, talin, ezrin, paxillin and cortactin, and the proteins co-localize with the clustered integrin (Bhattacharya *et al.*, 1995; Guan, 1997). Tyrosine phosphorylation of FAK and cytoskeletal proteins may be causally related, and the phoshorylations may promote actin stabilization and tension transmission from actin cytoskeleton to plasma membrane.

Focal Adhesion Kinase

Specific phosphorylated tyrosine residues in FAK are thought to be critical for signal relay by mediating complex formation between FAK and other signaling molecules. Y397 is the major FAK autophosphorylation site and phosphorylation at this residue creates a high affinity binding site for the SH2 domains of pp60c-src and pp59c-fyn (Schaller *et al.*, 1994), and phosphatidylinositol-3-kinase (Chen *et al.*, 1996). FAK is also phosphorylated *in vitro* by Src at tyrosine residues 407, 576, 577, 861, and 925 (Schlaepfer *et al.*, 1994; Calalb *et al.*, 1995; Calalb *et al.*, 1996). Phosphorylation at Y576 and Y577 in the activation loop enhances FAK kinase activity (Calalb *et al.*, 1995) and Y925 is a binding site for the SH2 domain protein GRB-2 (Schlaepfer *et al.*, 1994).

Phosphorylation of FAK at Y397 upon cell adhesion allows FAK to associate with Src, which triggers downstream signaling events such as phosphorylation of mitogen activated kinase (MAPK) (Schlaepfer *et al.*, 1998). FAK also binds the tyrosine kinase Shc, which is tyrosine phosphorylated in endothelial cells exposed to αvβ3 ligands (Schlaepfer *et al.*, 1999). Subsequent signaling events are significant for MAPK activation, since Shc recruits the adaptor protein Grb2, which binds nucleotide exchange factor, SOS. This allows conversion of Ras-GDP to Ras-GTP, hence activation of MAPK precursors and finally, MAPK. A potentially barrier-relevant consequence is that MAPK may phosphorylate and activate cPLA2. Barrier effects may then follow as discussed above for the PLCγ–IP3 mechanism.

The COOH-terminal domain of FAK is expressed in some tissues as an alternative transcript encoding a 41–43kDa protein called FRNK (for FAK-related non-kinase) (Schaller *et al.*, 1993), and this domain antagonizes FAK signaling by competing for binding to focal contacts (Richardson et al., 1996; Taylor et al., 2001). However, FRNK also binds the COOH-terminal domain of the FAK-like focal adhesion protein, PYK2 (Heidkamp et al., 2002), thereby introducing a certain amount of non-specificity in the interpretation of FRNK-induced inhibitory effects. FAK localization to focal adhesions is mediated primarily by the COOH-terminal focal adhesion targeting (FAT, residues 840–1052) domain (Hildebrand et al., 1993).

The non-catalytic NH2-terminal domain of FAK shares homology with the FERM domain of the ERM family of proteins including ezrin, radixin and moesin, which interact with the cortical cytoskeleton (Chishti et al., 1998; Girault et al., 1998; Girault et al., 1999). The proposed direct association of the FAK NH2-terminal domain with cytoplasmic domains of integrins (Schaller et al., 1995) remains unconfirmed in intact cells in which the FAK motif responsible for this interaction has not been identified. Increasing evidence suggests that the FAK NH2-terminal domain may be functionally distinct from the kinase and FAT domains (Sieg et al., 2000; Vial et al., 2000; Chen et al., 2001; Poullet et al., 2001), although the function remains unknown.

The intracellular targeting characteristics of the FAK NH2-terminal domain are not well understood. In HEK 293 and epithelial MDCK cells, the FAK NH2-terminal domain localizes to nuclei and cell–cell junctions with the tight junction integral membrane protein, occludin of (Stewart et al., 2002), suggesting a possible role for FAK in occludin-induced cell adhesion. FAK also associates with activated growth factor receptors through its NH2 -terminal domain and the FAK–Src complex is important in the regulation of growth factor–stimulated cell migration (Sieg et al., 2000). Activation of the FAK–Src complex facilitates the association with and/or phosphorylation of multiple signaling proteins (Schlaepfer et al., 1999).

Focal Adhesions in Endothelial Barrier Regulation

Agonists Effects

In lung endothelial cells, focal adhesions typically form near intercellular junctions, suggesting a role for these structures in endothelial barrier regulation (Ayalon and Geiger, 1997; Yuan, 2000; Lee et al., 2004). Studies in cultured endothelial cells indicate that permeability agonists such as thrombin, hydrogen peroxide, and vascular endothelial growth factor induce intercellular gaps that may account for the barrier deteriorative effects of these agents (Lum et al., 1993; Garcia et al., 1995; Abedi and Zachary, 1997; Schaphorst et al., 1997; Vepa et al., 1999; Carbajal et al., 2000; Alexander et al., 2001). Several of these agents have been shown to activate FAK and induce focal adhesion formation in an actin-dependent manner, since cytochalasin, an inhibitor of actin polymerization, prevents the FAK activation (Abedi and Zachary, 1997; Vepa et al., 1999), as also the induced increase in endothelial permeability (Phillips et al., 1989; Haselton et al., 1996; Waters, 1996). These findings give rise to the notion that loss of endothelial barrier function results from retraction of the endothelial plasma membrane at intercellular junctions. Hence, focal adhesions formed near the cell periphery may act as rivets that stabilize the cell membrane on the matrix, thereby counteracting the actin-myosin-induced membrane retraction that leads to increases of endothelial permeability.

Support for this notion comes from several recent studies on the role of FAK in barrier regulation in cultured lung endothelial cells. Taking advantage of the fact that thrombin not only causes endothelial barrier deterioration, but that it also forms focal adhesion complexes, the permeability responses was investigated in human pulmonary artery endothelial cells transfected with an antisense oligonucleotide for FAK (Mehta *et al.*, 2002). The resultant reduction in endothelial FAK expression increased both the extent as well as the duration of the thrombin-induced hyperpermeability. Thus, loss of FAK not only augmented barrier deterioration, but it also delayed barrier recovery in these endothelial monolayers. Several reports identify actin filament formation as the mechanism responsible for FAK translocation to focal adhesions (Abedi and Zachary, 1997; Vepa *et al.*, 1999; Bang *et al.*, 2000; Gerli *et al.*, 2000; Li *et al.*, 2002; Shikata *et al.*, 2003). Consistent with this view, barrier deteriorating effect of FAK depletion could be blocked by pre-treating endothelial cells with the inhibitor of actin polymerization, latrinculin A (Mehta *et al.*, 2002).

Using a different receptor-mediated approach, cultured pulmonary artery endothelial cells were exposed to the platelet secretion product, sphingosine-1-phosphate (S1P), which induces cell signaling by ligating Giα-linked Edg (endothelial differentiation gene) receptors (Garcia *et al.*, 2001). Interestingly, S1P enhanced endothelial barrier properties. Associated responses included a Rac1-GTPase- and cortactin-dependent recruitment of the actin regulatory protein, cofilin and assembly of cortical actin filaments (Garcia *et al.*, 2001; Dudek *et al.*, 2004). From the standpoint of focal adhesion involvement, an important finding was that S1P caused tyrosine phosphorylation of the focal adhesion protein, paxillin and that the platelet product caused a major redistribution of focal adhesion proteins to the cell periphery (Shikata *et al.*, 2003). Although a direct link between focal adhesion proteins and the S1P-selectin induced barrier enhancing response was not established in these studies, the remodeling of junctional sites lends notional support to the proposed role of focal adhesions in endothelial barrier stabilization.

Hyperosmotic Effects

Since translational displacement of the cell membrane against the matrix appears to be the initiating event for focal adhesion formation, the effect of passive membrane retraction was studied by exposing plated lung microvascular endothelial cells to sucrose-enriched hyperosmolar medium (Quadri *et al.*, 2003). The endothelial cells remodeled focal complexes into large focal adhesions at the cell periphery. Concomitantly, endothelial FAK activity increased and a dense thickening of cortical actin developed through a Rac1 GTPase-induced mechanism. Endothelial shrinkage should widen endothelial junctions, thereby increasing permeability. Contrary to this expectation, a bi-phasic permeability response was recorded, in which a transient barrier decrease was followed by a prolonged barrier enhancement. None of these effects occurred in the presence urea-induced hyperosmolarity that does not cause cell shrinkage, or in suspended cells, indicating that cell-matrix association is altogether important for establishing protective barrier responses.

These studies also revealed that increases in junctional E-cadherin content accompanied the hyperosmolarity-induced endothelial barrier enhancement (Quadri *et al.*, 2003). Suppression of endothelial focal adhesion formation and of FAK activity through expression of a kinase domain deleted FAK mutant completely blocked the E-cadherin response, while markedly blunting the barrier enhancement. These findings provide definitive evidence for a new role for focal adhesions, namely - as regulators of endothelial junctional proteins, hence of barrier properties. This relationship is interesting in light of the apparent reciprocal relationship of focal adhesions and cell-cell junctional integrity in agonists-induced barrier dysfunction in cultured pulmonary endothelium, as shown in Chapter 2, Figure 7 and reported previously (Alexander *et al.*, 2001). Further understanding is required to clarify the sequence of events occurring between the remodeling of focal adhesions and the increased insertion of E-cadherin in the endothelial membrane.

Focal Adhesions in Blood Vessels

Effects of Hyperventilation

Although reports are still relatively few, the understanding of endothelial focal adhesion function in intact blood vessels continues to accrue increasing interest. The technical difficulty of directly accessing endothelial cell signaling responses under intact conditions has been addressed by freshly isolating lung endothelial cells by immunosorting. In a new approach, collagenase treatment was applied at 4°C to block metabolic processes followed by immunosorting to obtain primary endothelial cell isolates from lungs previously exposed to experimental conditions (Bhattacharya *et al.*, 2003). These studies indicated that high volume mechanical ventilation, a procedure that imposes mechanical stretch on lung vessels, induces all features of focal adhesion activation, namely aggregation of $\alpha v \beta 3$ integrin, association of FAK with the $\alpha v \beta 3$ integrin, and tyrosine phosphorylation of paxillin. The ventilation stimulus increases expression of the leukocyte adhesion receptor, P-selectin on the endothelial surface and augments association of paxillin with P-selectin. As P-selectin is not externally expressed in quiescent endothelial cells, P-selectin expression signifies proinflammatory endothelial activation. Similarly, focal adhesion proteins also associate with expression of the leukocyte adhesion receptor, E-selectin (Yoshida *et al.*, 1996). The increased association of paxillin with P-selectin in ventilation challenged lungs indicates that focal adhesion proteins play a role in promoting the proinflammatory response in lung vessels.

Effects of Hyperosmolarity

The focal adhesion responses to hyperosmolar exposure have been replicated in lung capillaries to the extent that infusions of hyperosmolar sucrose enhance lung capillary barrier properties in association with increased E-cadherin deposition at endothelial junctions and enhanced formation of cortical actin filaments (Safdar *et al.*, 2003). Direct evidence for the presence of endothelial focal adhesions is now available through in situ immunofluorescence of vinculin (Safdar and Bhattacharya, unpublished findings). This immunofluorescence, though weakly evident under quiescent conditions, increases markedly after hyperosmolar sucrose infusion (Fig. 1), indicating that similar to cultured endothelial cell, hyperosmolarity increases focal adhesion formation in the intact lung capillary.

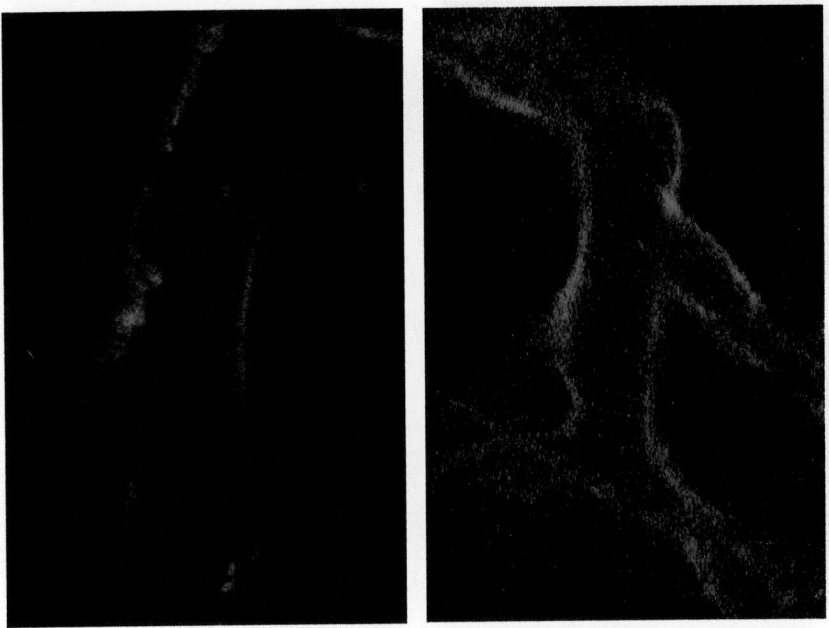

Figure 1. Lung venular capillary with evidence of vinculin staining.

Left: Lung venular capillaries viewed under baseline conditions show sparse vinculin staining.

Right: After 15 minutes of infusion of hyperosmolar sucrose vinculin staining is increased. Note that the vinculin fluorescence is patchy along the vascular wall.

Images of the hyperosmolarity-stimulated capillary reveal that vinculin distribution is typically non-uniform, being higher at capillary branch-points than at mid-segment. Confocal microscope optical sections reveal this unique distribution in that the fluorescence is clearly more pronounced at the point of entry of tributaries into trunk capillaries (Fig. 2). Although further functional evidence is required, the branch-point focal adhesion distribution suggests that the increased inflammatory potential of branch-point endothelial cells is determined by the higher density of focal adhesion proteins (Parthasarathi *et al.*, 2002; Ichimura *et al.*, 2003).

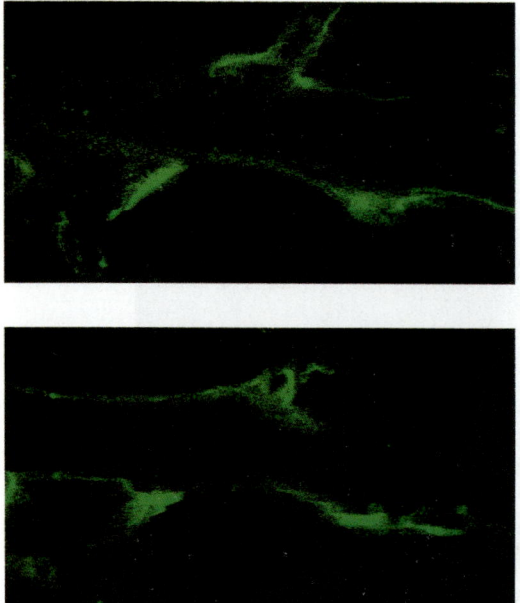

Figure 2. Optical sections through a lung venular capillary showing vinculin staining.
The same capillary was viewed by confocal microscopy at two levels 5 μm apart. A tributary (upper, arrow) entering the trunk capillary shows high vinculin fluorescence around the ostium in the deeper section (lower, arrow).

Future Directions

This review summarizes recent research that has led to recognition that endothelial focal adhesions are essential to barrier and inflammatory responses of lung blood vessels. Many conditions that promote lung endothelial barrier deterioration also stimulate endothelial focal adhesion formation. This concomitant juxtaposition of a negative barrier effect with positive enhancement of matrix adhesive properties signifies the barrier protective function of focal adhesion formation. Clearly, endothelial cells employ enlarged focal adhesions to counteract membrane retraction at the junction. Such a strategy counteracts the hyperpermeability stimulus and promotes restoration of the barrier. Evidently, more needs to be learned regarding the interplay of signaling mechanisms between focal adhesion proteins and constitutive junctional proteins that delineate overall permeability responses in lung blood vessels. The intriguing association of focal adhesion proteins in endothelial inflammatory responses, namely in the expressions of P- and E-selectin, also requires further clarification at the level of regulatory protein-protein interactions. Better understanding of these signaling mechanisms may lead to focal adhesion enhancing therapy to counteract pathology resulting from barrier deterioration and inflammatory activation of lung endothelium.

ACKNOWLEDGEMENTS

Supported by NIH grants HL36024, HL57556, HL69514, and HL54157.

REFERENCES

Abedi, H. and Zachary, I. (1997) Vascular endothelial growth factor stimulates tyrosine phosphorylation and recruitment to new focal adhesions of focal adhesion kinase and paxillin in endothelial cells. J. Biol. Chem. 272:15442-15451.

Alexander, J.S., Zhu, Y., Elrod, J., Alexander, B., Coe, L., Kalogeris, T., and Fuseler, J. (2001) Reciprocal regulation of endothelial substrate adhesion and barrier function. Microcirc. 8:389-401.

Ayalon, O. and Geiger, B. (1997) Cyclic changes in the organization of cell adhesions and the associated cytoskeleton, induced by stimulation of tyrosine phosphorylation in bovine aortic endothelial cells. J. Cell Sci. 110:547-556.

Bang, O.S., Kim, E., Chung, J., Lee, S., Park, T., and Kang, S.S. (2000) Association of focal adhesion kinase with fibronectin and paxillin is required for precartilage condensation of chick mesenchymal cells. Biochem. Biophys. Res. Comm. 278:522-529.

Bazzoni, G. and Dejana, E. (2004) Endothelial cell-to-cell junctions: molecular organization and role in vascular homeostasis. Physiol. Rev. 84:869-901.

Berman, J.S., Serlin, D., Li, X., Whitley, G., Hayes, J., Rishikof, D., Ricupero, D., Liaw, L., Goetschkes, M., and O'Regan, A.W. (2004) Altered bleomycin-induced lung fibrosis in osteopontin-deficient mice. Amer. J. Physiol. 286:L1311-L1318.

Bershadsky, A.D., Balaban, N.Q., and Geiger, B. (2003) Adhesion-dependent cell mechanosensitivity. Ann. Rev. Cell Dev. Biol. 19:677-695.

Bhattacharya, S., Fu, C., Bhattacharya, J., and Greenberg, S. (1995) Soluble ligands of the $\alpha v\beta 3$ integrin mediate enhanced tyrosine phosphorylation of multiple proteins in adherent bovine pulmonary artery endothelial cells. J. Biol. Chem. 270:16781-16787.

Bhattacharya, S., Patel, R., Sen, N., Quadri, S., Parthasarathi, K., and Bhattacharya, J. (2001) Dual signaling by the $\alpha v\beta 3$-integrin activates cytosolic PLA_2 in bovine pulmonary artery endothelial cells. Amer. J. Physiol. 280:L1049-L1056.

Bhattacharya, S., Sen, N., Yiming, M., Patel, R., Parthasarathi, K., Quadri, S., Issekutz, A., and Bhattacharya, J. (2003) High tidal volume ventilation induces proinflammatory signaling in rat lung endothelium. Amer. J. Respir. Cell Molec. Biol. 28:218-224.

Bhattacharya, S., Ying, X., Fu, C., Patel, R., Kuebler, W., Greenberg, S., and Bhattacharya, J. (2000) $\alpha v\beta 3$ integrin induces tyrosine phosphorylation-dependent Ca^{2+} influx in pulmonary endothelial cells. Circ. Res. 86:456-462.

Calalb, M.B., Polte, T.R. and Hanks, S.K. (1995) Tyrosine phosphorylation of FAK at sites in the catalytic domain regulates kinase activity: a role for Src family kinases. Molec. Cell Biol. 15:954-963.

Calalb, M.B., Zhang, X., Polte, T., and Hanks, S.K. (1996) FAK tyrosine-861 is a major site of phosphorylation by Src. Biochem. Biophys. Res. Comm. 228:662-668.

Carbajal, J.M., Gratrix, M., Yu, C., and Schaeffer, R.C., Jr. (2000) ROCK mediates thrombin's endothelial barrier dysfunction. Amer. J. Physiol. Cell Physiol. 279:C195-204.

Chen, H.C., Appeddu, P., Isoda, H., and Guan, J.L. (1996) Phosphorylation of tyrosine 397 in FAK is required for binding PI3K. J. Biol. Chem. 271:26329-26334.

Chen, R., Kim, O., Li, M., Xiong, X., Guan, J., Kung, H., Chen, H., Shimizu, Y., and Qiu, Y. (2001) Regulation of the PH-domain-containing tyrosine kinase Etk by FAK through the FERM domain. Nat. Cell Biol. 3:439-444.

Chen, Y.P., O'Toole, T., Ylanne, J., Rosa, J., and Ginsberg, M.H. (1994) A point mutation in the integrin $\beta 3$ cytoplasmic domain (S752-->P) impairs bidirectional signaling through $\alpha II_b \beta 3$

(platelet glycoprotein IIb-IIIa). Blood 84:1857-1865.

Chhatwal, G.S., Preissner, K., Muller-Berghaus, G., and Blobel, H. (1987) Specific binding of the human S protein (vitronectin) to streptococci, Staphylococcus aureus, and Escherichia coli. Infect. Immun. 55:1878-1883.

Chishti, A.H., Kim, A., Marfatia, S., Lutchman, M., Hanspal, M., Jindal, H., Liu, S., Low, P., Rouleau, G., Mohandas, N., Chasis, J., Conboy, J., Gascard, P., Takakuwa, Y., Huang, S., Benz, E., Bretscher, A., Fehon, R., Gusella, J., Ramesh, V., Solomon, F., Marchesi, V., Tsukita, S., Tsukita, S., Arpin, M., Louvard, D., Tonks, N., Anderson, J., Fanning, A., Bryant, P., Woods, D., and Hoover, K.B. (1998) The FERM domain: a unique module involved in the linkage of cytoplasmic proteins to the membrane. Trends Biochem. Sci. 23:281-282.

Crowe, D.T., Chiu, H., Fong, S., and Weissman, I.L. (1994) Regulation of the avidity of integrin $\alpha 4\beta 7$ by the $\beta 7$ cytoplasmic domain. J. Biol. Chem. 269:14411-14418.

Damsky, C.H. and Werb, Z. (1992) Signal transduction by integrin receptors for extracellular matrix: cooperative processing of extracellular information. Curr. Opin. Cell Biol. 4:772-781.

Dudek, S.M., Jacobson, J., Chiang, E., Birukov, K., Wang, P., Zhan, X. and Garcia, J.G. (2004) Pulmonary endothelial cell barrier enhancement by sphingosine 1-phosphate: roles for cortactin and myosin light chain kinase. J. Biol. Chem. 279:24692-24700.

Garcia, J.G., Liu, F., Verin, A., Birukova, A., Dechert, M., Gerthoffer, W., Bamberg, J., and English, D. (2001) Sphingosine 1-phosphate promotes endothelial cell barrier integrity by Edg-dependent cytoskeletal rearrangement. J. Clin. Invest. 108:689-701.

Garcia, J.G., Pavalko, F.M., and Patterson, C.E. (1995) Vascular endothelial cell activation and permeability responses to thrombin. Blood Coag. Fibrinol. 6:609-626.

Gerli, R., Solito, R., Weber, E., and Agliano, M. (2000) Specific adhesion molecules bind anchoring filaments and endothelial cells in human skin initial lymphatics. Lymph. 33:148-157.

Girault, J. A., Labesse, G., Mornon, J., and Callebaut, I. (1998) Janus kinases and FAK play in the 4.1 band: a superfamily important for cell structure and signal transduction. Molec. Med. 4:751-769.

Girault, J.A., Labesse, G., Mornon, J., and Callebaut, I. (1999) The N-termini of FAK and JAKs contain divergent band 4.1 domains. Trends Biochem. Sci. 24:54-57.

Guan, J. L. (1997) Focal adhesion kinase in integrin signaling. Matrix Biol. 16:195-200.

Haselton, F.R., Dworska, E., Evans, S., Hoffman, L., and Alexander, J.S. (1996) Modulation of retinal endothelial barrier in an in vitro model of the retinal microvasculature. Exp. Eye Res. 63:211-222.

Heidkamp, M.C., Bayer, A., Kalina, J., Eble, D., and Samarel, A.M. (2002) GFP-FRNK disrupts focal adhesions and induces anoikis in neonatal rat ventricular myocytes. Circ. Res. 90:1282-1289.

Hildebrand, J.D., Schaller, M.D., and Parsons, J.T. (1993) Identification of sequences required for the efficient localization of pp125FAK to cellular focal adhesions. J. Cell Biol. 123:993-1005.

Ichimura, H., Parthasarathi, K., Quadri, S., Issekutz, A., and Bhattacharya, J. (2003) Mechano-oxidative coupling by mitochondria induces proinflammatory responses in lung venular capillaries. J. Clin. Invest. 111:691-699.

Ishikawa, S., Tsukada, H., and Bhattacharya, J. (1993) Soluble complex of complement increases hydraulic conductivity in single microvessels of rat lung. J. Clin. Invest. 91:103-109.

Juliano, R.L. (2002) Signal transduction by cell adhesion receptors and the cytoskeleton: functions of integrins, cadherins, selectins, and immunoglobulin-superfamily members. Ann. Rev. Pharm. Tox. 42:283-323.

Kasper, M., Schobl, R., Haroske, G., Fischer, R., Neubert, F., Dimmer, V., and Muller, M. (1996) Distribution of vWf in capillary endothelial cells of rat lungs with pulmonary fibrosis. Exp. Toxicol. Pathol. 48:283-288.

Katsumi, A., Orr, A., Tzima, E., and Schwartz, M.A. (2004) Integrins in mechanotransduction. J. Biol. Chem. 279:12001-12004.

Kawasaki, J., Davis, G.E., and Davis, M.J. (2004) Regulation of Ca2+-dependent K+ current by $\alpha v \beta 3$ integrin engagement in vascular endothelium. J. Biol. Chem. 279:12959-12966.

Langlois, P.F. and Gawryl, M.S. (1988) Accentuated formation of the terminal C5b-9 complement

complex in patient plasma precedes ARDS. Amer. Rev. Respir. Dis. 138:368-375.
Lee, H.Z., Yeh, F.T., and Wu, C.H. (2004) The effect of elevated extracellular glucose on adherens junction proteins in cultured rat heart endothelial cells. Life Sci. 74:2085-2096.
Li, S., Butler, P., Wang, Y., Hu, Y., Han, D., Usami, S., Guan, J., and Chien, S. (2002) The role of the dynamics of focal adhesion kinase in the mechanotaxis of endothelial cells. PNAS USA 99:3546-3551.
Lum, H., Andersen, T., Siflinger-Birnboim, A., Tiruppathi, C., Goligorsky, M., Fenton, J., and Malik, A.B. (1993) Thrombin receptor peptide inhibits thrombin-induced increase in endothelial permeability by receptor desensitization. J. Cell Biol. 120:1491-1499.
Mehta, D., Tiruppathi, C., Sandoval, R., Minshall, R., Holinstat, M., and Malik, A.B. (2002) Modulatory role of FAK in regulating human pulmonary arterial endothelial barrier function. J. Physiol. 539:779-789.
Miyamoto, S., Akiyama, S.K., and Yamada, K.M. (1995) Synergistic roles for receptor occupancy and aggregation in integrin transmembrane function. Sci. 267:883-885.
Miyamoto, S., Teramoto, H., Gutkind, J., and Yamada, K.M. (1996) Integrins can collaborate with growth factors for phosphorylation of receptor tyrosine kinases and MAP kinase activation: roles of integrin aggregation and occupancy of receptors. J. Cell Biol. 135:1633-1642.
Parthasarathi, K., Ichimura, H., Quadri, S., Issekutz, A., and Bhattacharya, J. (2002) Mitochondrial ROS regulate spatial profile of proinflammatory responses in lung venular capillaries. J. Imm. 169:7078-7086.
Phillips, P.G., Lum, H., Malik, A.B., and Tsan, M.F. (1989) Phallacidin prevents thrombin-induced increases in endothelial permeability to albumin. Amer. J. Physiol. 257:C562-C567.
Poullet, P., Gautreau, A., Kadare, G., Girault, J., Louvard, D., and Arpin, M. (2001) Ezrin interacts with FAK and induces its activation independently of cell-matrix adhesion. J. Biol. Chem. 276:37686-37691.
Preissner, K.T. (1991) Structure and biological role of vitronectin. Ann. Rev. Cell Biol. 7:275-310.
Quadri, S.K., Bhattacharjee, M., Parthasarathi, K., Tanita, T., and Bhattacharya, J. (2003) Endothelial barrier strengthening by activation of focal adhesion kinase. J. Biol. Chem. 278:13342-13349.
Ruegg, C. and Mariotti, A. (2003) Vascular integrins: pleiotropic adhesion and signaling molecules in vascular homeostasis and angiogenesis. Cell. Molec. Life Sci. 60:1135-1157.
Sadurski, R., Tsukada, H., Ying, X., Bhattacharya, S., and Bhattacharya, J. (1994) Diameters of juxtacapillary venules determined by oil-drop method in rat lung. J. Appl. Physiol. 77:718-725.
Safdar, Z., Wang, P., Ichimura, H., Issekutz, A., Quadri, S., and Bhattacharya, J. (2003) Hyperosmolarity enhances the lung capillary barrier. J. Clin. Invest. 112:1541-1549.
Savill, J., Hogg, N., Ren, Y. and Haslett, C. (1992) Thrombospondin cooperates with CD36 and the vitronectin receptor in macrophage recognition of neutrophils undergoing apoptosis. J. Clin. Invest. 90:1513-1522.
Schaller, M.D., Borgman, C.A., and Parsons, J.T. (1993) Autonomous expression of a noncatalytic domain of the focal adhesion-associated protein tyrosine kinase pp125FAK. Molec. Cell Biol. 13:785-791.
Schaller, M.D., Hildebrand, J., Shannon, J., Fox, J., Vines, R., and Parsons, J.T. (1994) Autophosphorylation of pp125FAK directs SH2-dependent binding of pp60src. Molec. Cell Biol. 14:1680-1688.
Schaller, M.D., Otey, C., Hildebrand, J., and Parsons, J.T. (1995) FAK and paxillin bind to peptides mimicking beta integrin cytoplasmic domains. J. Cell Biol. 130:1181-1187.
Schaphorst, K.L., Pavalko, F., Patterson, C.E., and Garcia, J.G. (1997) Thrombin-mediated focal adhesion plaque reorganization in endothelium: role of protein phosphorylation. Amer. J. Respir. Cell Molec. Biol. 17:443-455.
Schlaepfer, D.D., Hanks, S., Hunter, T., and van der Geer, P. (1994) Integrin-mediated signal transduction linked to Ras pathway by GRB2 binding to FAK. Nature 372:786-791.
Schlaepfer, D.D., Hauck, C.R., and Sieg, D.J. (1999) Signaling through focal adhesion kinase. Prog. Biophys. Molec. Biol. 71:435-478.
Schlaepfer, D.D., Jones, K.C., and Hunter, T. (1998) Multiple Grb2-mediated integrin-stimulated signaling pathways to ERK2/mitogen-activated protein kinase: summation of both c-Src- and

focal adhesion kinase-initiated tyrosine phosphorylation events. Molec. Cell Biol. 18:2571-2585.
Schlaepfer, D.D. and Mitra, S.K. (2004) Multiple connections link FAK to cell motility and invasion. Curr. Opin. Genet. Dev. 14:92-101.
Shikata, Y., Birukov, K., Birukova, A., Verin, A., and Garcia, J.G. (2003) Involvement of site-specific FAK phosphorylation in S1P- and thrombin-induced focal adhesion remodeling: role of Src and GIT. FASEB J. 17:2240-2249.
Shikata, Y., Birukov, K.G., and Garcia, J.G. (2003) S1P induces FA remodeling in human pulmonary endothelial cells: role of Rac, GIT1, FAK, and paxillin. J. Appl. Physiol. 94:1193-1203.
Shimizu, Y., Van Seventer, G., Horgan, K., and Shaw, S. (1990) Regulated expression and binding of three VLA (beta 1) integrin receptors on T cells. Nature 345:250-253.
Sieg, D.J., Hauck, C., Ilic, D., Klingbeil, C., Schaefer, E., Damsky, C., and Schlaepfer, D.D. (2000) FAK integrates growth-factor and integrin signals to promote cell migration. Nat. Cell Biol. 2:249-256.
Singh, B., Fu, C., and Bhattacharya, J. (2000) Vascular expression of the $\alpha v \beta 3$-integrin in lung and other organs. Amer. J. Physiol. 278:L217-L226.
Sobin, S.S., Lindal, R., Fung, Y., and Tremer, H.M. (1978) Elasticity of the smallest noncapillary pulmonary blood vessels in the cat. Microvasc. Res. 15:57-68.
Stewart, A., Ham, C., and Zachary, I. (2002) The focal adhesion kinase amino-terminal domain localises to nuclei and intercellular junctions in HEK 293 and MDCK cells independently of tyrosine 397 and the carboxy-terminal domain. Biochem. Biophys. Res. Comm. 299:62-73.
Tsukada, H., Ying, X., Fu, C., Ishikawa, S., McKeown-Longo, P., Albelda, S., Bhattacharya, S., Bray, B., and Bhattacharya, J. (1995) Ligation of endothelial $\alpha v \beta 3$ integrin increases capillary hydraulic conductivity of rat lung. Circ. Res. 77:651-659.
Ulbrich, H., Eriksson, E.E., and Lindbom, L. (2003) Leukocyte and endothelial cell adhesion molecules as targets for therapeutic interventions in inflammatory disease. Trends Pharm. Sci. 24:640-647.
Vepa, S., Scribner, W., Parinandi, N., English, D., Garcia, J., and Natarajan, V. (1999) H_2O_2 stimulates tyrosine phosphorylation of FAK in vascular endothelial cells. Amer. J. Physiol. 277:L150-L158.
Vial, D., Okazaki, H., and Siraganian, R.P. (2000) The NH2-terminal region of focal adhesion kinase reconstitutes high affinity IgE receptor-induced secretion in mast cells. J. Biol. Chem. 275:28269-28275.
Waters, C.M. (1996) Flow-induced modulation of the permeability of endothelial cells cultured on microcarrier beads. J. Cell. Physiol. 168:403-411.
Yoshida, M., Westlin, W.F., Wang, N., Ingber, D.E., Rosenzweig, A., Resnick, N., and Gimbrone, M.A., (1996) Leukocyte adhesion to vascular endothelium induces E-selectin linkage to the actin cytoskeleton. J. Cell Biol. 133:445-455.
Yuan, S.Y. (2000) Signal transduction pathways in enhanced microvascular permeability. Microcirc. 7:395-403.

Chapter 9

Interendothelial Junctions and Barrier Integrity

Lopa Leach, Ph.D.,[1]* Carolyn E. Patterson, Ph.D.,[2] and Donna Carden, M.D.[3]

[1] Institute of Clinical Research, University of Nottingham, Nottingham, NG 7 2 UH, UK,
[2] Indiana University School of Medicine & The Roudebush VA Medical Center, Indianapolis, IN 46202 USA, and [3] Louisiana State University Health Sciences Center, Shreveport, LA 71130-3932 USA

CONTENTS:

Introduction

Barrier Integrity and Junctional Formations
Molecular Coupling of Endothelial Cells
Ultrastructural Variation
Assembly of the Functions

Adherens Junctions
Cadherin
The Catenins
Signaling in Adherens Junctions

Tight Junctions
Structure and Relationship to Monolayer Permeability
Claudin
Occludin
Junctional Adhesion Molecules
Anchoring Tight Junctions to the Cytoskeleton

Additional Intercellular Molecules
PECAM
Cell-Cell Integrins
Gap Junctions

Junctional Integrity, Dysfunction, and Disease
Inflammation and Junctional Compromise
Altered Junctional Protein Content in Inflammation and Disease

Coming Together

Introduction

The cornerstone of endothelial barrier integrity is formation and maintenance of the interendothelial junctions. Experimental findings *in vitro* suggest that molecular occupancy of such junctions regulate both barrier function and angiogenesis. Recent advances using transgenic mice support this hypothesis and allow dissection of signaling pathways behind the putative molecules. The challenge lies in discovering whether this is true in human vascular beds and how physiological or pathological changes affect phenotype and therefore vascular function in health and disease. Although transendothelial transport plays an important nutrient and regulatory role, untoward paracellular leakage of large solute molecules across the alveolar-capillary membrane is the single most critical factor in the development of pulmonary edema. Thus, defining the structure and regulation of endothelial junctional complexes is critical to understanding altered permeability of the lung vascular endothelium.

This chapter reviews work which demonstrates there are two distinct junctional phenotypes - "activated" and "stable" present in different vascular beds or even within the same the vascular tree, which allows different permeability status and angiogenic plasticity. The junctional phenotypes are reversible, as shown by experimental manipulation of vessels with vasoactive agents. Reductions in protein levels as well as post-translational events such as phosphorylation and loss of junctional localization of adhesion molecules result in increased permeability to macromolecules, whilst upregulation and re-targeting of these molecules inhibit cell proliferation, increases transendothelial resistance and confer quiescence. Ultimately, some of these molecules may be involved in cell survival. Furthermore, structural immaturity lends vulnerability to the junctional complexes, thus allowing activation and transduction of different signals. Interestingly, junctional immaturity appears to be at the expense of barrier function, thus angiogenic vessels are invariably leaky. These studies suggest junctional adhesion molecules define intrinsic phenotype of different vascular beds but more importantly can regulate barrier integrity, physiological angiogenesis and vascular remodeling in humans.

Barrier Integrity and Junctional Formations

Molecular Coupling of Endothelial Cells

Various mechanisms exist to provide a discriminatory barrier between tissue and blood that is responsive to location and tissue demand. Thus, endothelium varies from continuous monolayers in conduit vessels and microvessels of the blood-brain barrier, lung, muscle, skin, and placenta to fenestrated vessels in exocrine organs and to the discontinuous monolayers in hepatic sinusoids and lymphatic vessels. As introduced in Chapter 2 (see Figure 6, Chapt.2), there are several molecular structures which participate in this intercellular glue: adherens junctions, tight junctions, Ig-superfamily (PECAM), integrins, and gap junctions. Adherens junctions, composed of cadherins and catenins, are the primary adhesions between adjacent endothelial cells and as such, play an especially important roll in determination of barrier integrity (Lampugnani and Dejana, 1997; Nieset *et al.*, 1997). In addition, cadherin may also be linked to the actin cytoskeleton in endothelial cells by desmoplakin, γ-catenin, and vimentin complexes, referred to as a syndesmosome; but endothelial cells lack true epithelial desmosomes (Valiron *et al.*, 1996, Kowalczyk *et al.*, 1998). Tight junctions consist of the transmembrane proteins: occludin, members of the claudin family, and the more recently identified junctional adhesion molecules (JAMs), which are linked to the cellular cytoskeleton by ZO proteins (Fanning et al., 1998; Furuse et al., 1998; Martin-Padura et al., 1998; Matter and Balda, 1999). Although not nearly as definitive and orderly as in columnar epithelium, tight junctions are the most apical intercellular junctional structures in endothelium, followed by adherens junctions, and then by PECAM and integrin junctions in the more basal zone (Ayalon et al., 1994; Martin-Padura et al., 1998; Aurrand-Lions et al., 2002). Clear segregation of these structures is not always seen, especially in immature junctions, where adherens junction proteins may share location with tight junction proteins or with PECAM.

Ultrastructural Variation

Within continuous endothelium there is significant morphological variation in the specific organization of inter-endothelial or paracellular junctions. In the blood-brain and blood-retinal barriers, the paracellular clefts between adjoining endothelial cells contain complex tight junctions, which are fused for great extents of the intercellular cleft. These fusions involve the outer membrane leaflet of adjoining cell plasma membranes. In non-brain continuous capillaries, such as in lung, skeletal muscle, myocardium, mesentery, and placenta, tight junctions are not fused. There is a 4 nm gap between the two outer membrane leaflets of adjoining endothelial cells (Firth et al., 1983; Ward et al., 1988; Leach and Firth, 1992; Schulze & Firth, 1992; Adamson & Michel, 1993). Moreover, tight junctions here do not show as strong a positional preference (luminal or abluminal, Figure 1), as in epithelial cells. Rather, they are discontinuous along the cleft, typically 1-4 cross-linked tight junctional strands with intermittent discontinuities. At these areas of discontinuity the membrane separation widens, sometimes to the full 18-20 nm of the non-junctional parts of the paracellular cleft (Bundgaard, 1984; Huang et al., 1992; Adamson and Michel, 1993; Walker et al., 1994; Firth, 2002). In effect, in the non-brain continuous capillaries, the narrow clefts in tight junctions can be bypassed through the discontinuities. Therefore, tortuous paths can be traced through these wide discontinuities across the full tight junction thickness (Firth et al., 1983; Adamson and Michel, 1993). The most likely interpretation is that the discontinuities represent the open parts of the junction's "shutter" that limits the fraction of paracellular cleft available as a transport pathway for hydrophilic solutes (Firth, 2002). This allows adherens junctions, in discontinuities or wide zones, to play a greater role in determining paracellular permeability, integrity, and plasticity in non-CNS vessels. In fact, evidence indicates that disruption of cadherins within the adherens junction is critical in permeability lung edema evoked by inflammatory stimuli (Zhao et al., 2001; Lim et al., 2001).

Electron microscopic studies of term placenta have shown that the ultrastructural organization of tight junctions within paracellular clefts are remarkably similar throughout the placental vascular tree (Leach and Firth, 1992; Leach et al., 2000). Newly formed vessels in the first trimester, which arise by *de novo* vasculogenesis and angiogenesis, were shown to possess tight junctional appositions, reminiscent of that seen in term placenta (Leach et al., 2002). Paracellular structural organization alone (as visualized by transmission electron microscopy) does not completely reveal the rationale behind the functional differences of immature or exchange capillaries from mature or large conduit vessels. This ultrastructural junctional homology extends to all non-CNS vascular beds, suggesting that paracellular permeability is dictated at the molecular level, i.e. the presence or absence of specific adhesion molecules can dictate the strength, integrity and plasticity of the junctions.

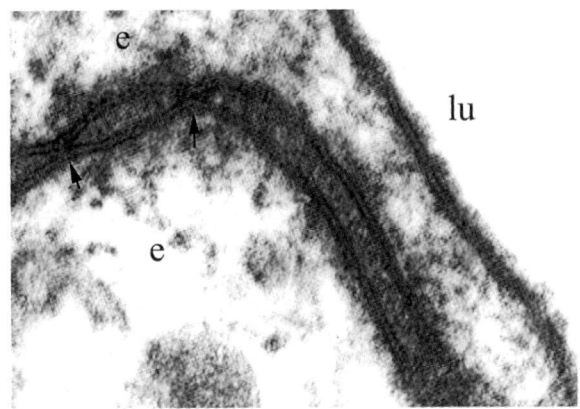

Figure 1. Organization of paracellular clefts. Electron micrograph showing organization of paracellular clefts in vessels of human feto-placental tissue. Tight junctions (arrows) are not fused (note separation between the adjoining membrane leaflets at the junctional appositions). They show no positional preference, being interspersed with wide zones along the cleft. e = endothelial cells, lu = lumen of blood vessels. Magnification x 210,000. Provided by L. Leach.

Assembly of the Junctions

Thus, barrier function is dictated by complexity of junctions, specifically those of the two main physical entities, tight and adherens junctions, and the extent of cleft occupied by these junctions, which act as "resistance in series". In the central nervous system of rats, the blood-brain barrier is not fully developed until postnatal day 24 (Schulze and Firth, 1992). Maturation of the clefts is accompanied with establishment of a characteristic ratio of "narrow zone" (complex tight junction) to wide zone (15-20 nm) where the adherens junctions reside. Freshly isolated endothelial cells and cultured cells have also been used to follow the temporal organization and formation of lateral adhesions. As the endothelial cells are first plated they form focal adhesion attachments to matrix proteins or to the plastic surface. Within 30 minutes they begin to extend lamellipodia and spread. Depending on plating density, they approach each other by lamellipodia extensions and form initial contacts by molecular interactions of the extracellular domains of membrane proteins (likely cadherins) and glycoproteins. Within 2 hrs of plating, adherens junctions form at intermittent areas of contact (Ayalon et al., 1994). The cadherin in these early junctions is highly tyrosine phosphorylated and primarily associated with β-catenin and p120 (Dejana, 2004). This formation of cell-to-cell junctions continues at the lateral edges until a nearly continuous ring of cadherin is observed by 24 hrs, during which time associated tyrosine phosphatases decrease phosphorylation, stabilize the cadherin, and increase γ-catenin association.

Once adherens junctions are formed, they initiate the formation of companion tight junctions (Gumbiner et al. 1988), the second physical entity which, in non-CNS vessels as stated previously, influences paracellular permeability by dictating the percentage of clefts open to solute transport. Over time, additional junctions are formed. PECAM-1 becomes recruited to the lateral edges and partially associates with the adherens junctions. In effect the cells "zip up" their lateral borders with progressive junctions until a barrier is established (Hinck et al., 1994). Other molecules are recruited to these preliminary junctions to stabilize the cell-to-cell adhesion and the endothelial cells are signalled in this process to convert from a migratory to a quiescent phenotype. A similar pattern of junctional adhesion was shown in freshly isolated rat brain microvessels using freeze-fracture techniques (Lane et al., 1995). Before cell isolation, junctional strands show segments of tightly packed intramembrane particles on the P face (the half of the bilipid layer that faces the protoplasm inside of the cell before fracture). In contrast, after isolation and 1 day in culture, tight junctions are sparse with incomplete tight junctional strands and some intramembrane particles fracturing on the E face (the portion of the lipid bilayer that faces extracellularly). After 4 to 10 days in culture, there was a progressive maturation of the junctional complexes with an increase in the number of tight junctions and microfilaments beneath the surface (maturation of the dense peripheral actin band). This shift of particles between the P face and E face has been shown to affect polarity (fence properties), but not permeability in the spontaneously hypertensive rat blood-brain barrier (Lippoldt et al. 2000a).

In vivo such large gaps between cells occur only in unusual circumstances, such as in wound healing. In the normal pattern of expansion or replacement, the cells remain in partial contact as they divide and spread. In formation of stable junctions, presence of the dense peripheral band and anchoring of the junctional adhesions to this band is critical. This inter-relationship is supported by observations that various means to disrupt the cortical actin result in dissolution of adherens junctions, whereas formation of a prominent actin band correlates with relocation of cadherin and catenins to the lateral borders (Quinlan and Hyatt, 1999). Concentration of cortactin, an actin binding protein involved in actin polymerization, to the cell periphery results in development of the dense peripheral actin band and subsequent increases in ZO1 and occludin within tight junctions in brain and pulmonary endothelium, suggesting the importance of the cortical band to formation and stabilization of cell-cell adhesion complexes (Romero et al., 2003; Jacobson et al., 2004). For instance, the increase in the dense peripheral band after treatment with sphingosine 1-phosphate parallels localization of VE-cadherin, α-, β- and γ-catenins to the junctions (Lee et al., 1999). However, the initial formation of junctions may also be important for the signaling and the spatial organization of the dense peripheral band.

Adherens Junctions

Cadherin

At endothelial adherens junctions, the key transmembrane protein is vascular endothelial (VE) cadherin (Lampugnani et al., 1992; Gumbiner, 1996). VE-cadherin is clustered at the junction and mediates cell-cell adhesion through homophilic binding to cadherin expressed on neighboring endothelial cells. Cadherins are a family of glycoproteins that play an important role in cell sorting and tissue pattern formation in embryologic development by formation of homologous bonds. By crystal structure analysis, the cell-cell recognition site common among the different cadherin subtypes was localized to a tripeptide HAV sequence in the first of five tandem repeats (~110 amino acids each) of the amino terminus (Lutz et al., 1995). However, endothelial VE-cadherin is unique in that it lacks this particular tripeptide sequence, but must express an alternative recognition sequence (Huber et al., 1996). Whereas VE-cadherin is endothelial specific, N-cadherin is non-specific and is expressed by endothelial and other types of cells. The N-cadherin is diffusely distributed rather than concentrated in junctional complexes (Salomon et al., 1992; Alexander et al., 1993; Leach et al., 1993).

These cadherin-to-cadherin bonds require Ca^{++}-dependent binding of the projecting extra-cellular domains. Internally the cadherins extend a cytoplasmic tail, which is connected to peri-junctional actin that is general contiguous with the dense peripheral band of actin. The bridges between the cytoplasmic tail and actin are formed by molecularly diverse catenins: β-catenin, γ-catenin (plakoglobin), α-catenin, and p120-catenin. Stable adherens junctions contain the full complement of these molecules. Moreover, formation of adherens junctional complexes results in contact inhibition and is thought to protect endothelial monolayers from proliferative signals (Lampugnani *et al.*, 2003). In confluent cells, inhibition of VEGF-induced proliferation requires VE-cadherin, β-catenin, and the phosphatase DEP-1/CD148, whereby VEGFR2 is thought to associate with the VE-cadherin- β-catenin complex upon VEGF stimulation (Lampugnani *et al.*, 2003). The receptor may then be inactivated by junctional phosphatases. This phenomenon is not observed in sparse cells or VE-cadherin-null cells which can be fully activated by VEGF. Given VEGF is a potent permeability enhancer, inactivation of the VEGF-R2 may protect against inflammatory insults.

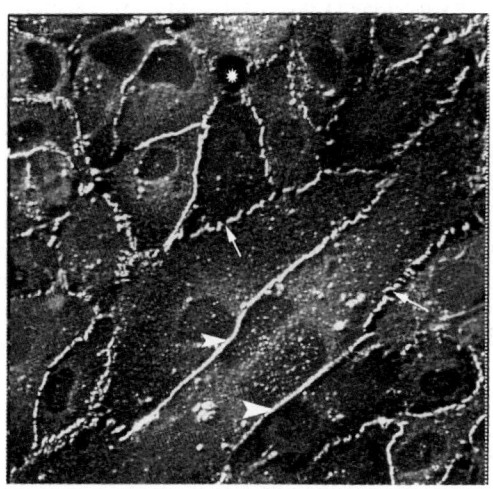

Figure 2. Cadherin gaps in histamine-challenged endothelial cell monolayers. The image is a confocal scanning micrograph showing effect of histamine (1 min; 10 μM) on VE-cadherin distribution in human dermal microvascular cell monolayers. VE-cadherin immunostaining ranges from continuous staining at cell-cell contacts (arrowhead) to discontinuous "stitch –like" pattern (arrows) where small gaps are beginning to appear to total loss of VE-cadherin from cell borders surrounding a frank pore (asterisk). Magnification x 240. Courtesy of Drs. Budworth, Clothier and Leach, University of Nottingham.

Early work revealed that truncated VE-cadherin (without the cytoplasmic tail for catenin-binding) can still promote homotypic recognition and weak cell to cell adhesion of VE-cadherin, but is unable to control paracellular permeability, indicating that barrier integrity depends upon the stable linkage of the cadherin to the actin cytoskeleton (Jou *et al.*, 1995; Navarro *et al.*, 1998). The importance of the cytoplasmic tail has been confirmed by *in vivo* studies. Transfection of a recombinant VE-cadherin (cytoplasmic domain only) into intact isolated coronary venules resulted in markedly elevated albumin permeability. This occurred due to competition of the recombinant VE-cadherin with endogenous VE-cadherin, which blocked the normal association with β-catenin; thus interfering with the maintenance of junctional integrity and microvascular integrity (Guo *et al.*, 2004). Indeed reassembly of VE-cadherin at junctions after calcium chelation has been shown to restore microvascular integrity in pulmonary microvessels (Gao *et al.*, 2000). Furthermore, the increase in solute permeability and resultant organ edema upon treatment of intact endothelium with anti-VE-cadherin monoclonal Ab (Corada *et al.*, 1999; Haselton and Heimark, 1997) or with anti-N-cadherin polyclonal Ab (Alexander *et al.*, 1993) underscores the importance of functional cadherin to barrier integrity. Likewise, dispersed endothelium treated with these antibodies exhibit impaired reformation of intercellular junctions. Paracellular permeability induced with inflammatory mediators, such as thrombin, histamine, TNFα, IFNγ, and VEGF, is accompanied by disruption of the VE-cadherin/catenin complex and loss of cadherin from the cell borders (Rabiet et al., 1996; Wong et al., 1999; Patterson et al., 2000; Leach et al. 1995). However, there is not usually an actual loss of the total cellular junctional protein content *per se* determined by whole cell western blot analysis. Figure 2 illustrates the three patterns of cadherin distribution. In this figure the cells have been treated for 1 minute with histamine and some borders retain the normal continuous band of cadherin, while most borders show the "stitch-like" or "picked fence" appearance of cadherin stands related to microgaps and typical of activated cells. The third pattern is the absence of cadherin and at this point there is a frank gap between the endothelial cells.

The Catenins

The cytoskeletal linkage of cadherin by the catenins stabilizes cadherin at the junctional site (Navarro et al., 1995; Takeichi, 1990). The role of catenins in regulating permeability became more evident with demonstration that the type of catenin present in the adherens junctions can dictate endothelial monolayer permeability. Actin binding is conferred by the α-catenin, which has homology to vinculin. The β-catenin and the γ-catenin, (plakoglobin) bind to α-catenin and to the cytoplasmic tail of cadherin. As mentioned above, at early stages of confluency, VE-cadherin is primarily linked to β-catenin, with barrier property at a minimum. Actually, catenin-catenin recognition and binding is not even dependent upon β-catenin, but the β-catenin aids in stabilization (Navarro et al., 1995). At full confluence, β-catenin partially detaches from the junctional complex and is replaced by plakoglobin with a concomitant decrease in monolayer permeability to macromolecules (Lampugnani et al., 1995). This differential expression of plakoglobin has also been seen in the developing human placenta, where plakoglobin is a feature of conduit vessels only in the last trimester of pregnancy (Leach, 2000). In primary cell cultures of human placental microvascular endothelial cells, plakoglobin can be targeted to the adherens junctions by growing the cells in the presence of cAMP enhancing agents (Dye et al., 2001). Conversely, permeability-inducing agents, such as thrombin, decrease the γ-catenin in the junction (Rabiet et al., 1996). Similarly, α-catenin is partially lost from cell borders of pulmonary endothelium with thrombin challenge, as seen in Figure 6.b. of Chapter 2. The appearance of catenin in this illustration bears striking similarity to the "stitched" or discontinuous appearance of cadherin in Figure 2 above. Moreover, in this same pulmonary endothelium, cAMP elevation prevented the thrombin-induced loss, whereas PKA inhibition caused loss of catenin from the junctions, indicating the important role of PKA in regulation of adherens junctions (Patterson et al., 2000). Cultured β-catenin (-/-) endothelial cells show a different organization of intercellular junctions with decreased α-catenin in favor of desmoplakin (Cattelino et al., 2003), resulting in decreased adhesion strength and increased paracellular permeability.

Catenin p120 also plays an important regulatory role in adherens junctions (Shibamoto et al., 1995). Recent evidence suggests that the core function of p120 is to regulate cadherin expression levels by controlling cadherin membrane trafficking (Review: Vincent et al., 2004). In pulmonary endothelial siRNA-induced loss of p120 resulted in a dramatic loss of endothelial barrier function (Iyer et al., 2004). This loss resulted in a corresponding loss of VE-cadherin, indicating a key role for p120 in modulating VE-cadherin accumulation. Furthermore, the level of p120 phosphorylation has been shown to play an important role in catenin function and cell-cell adhesion (Aono et al., 1999; Ohkubo and Ozawa, 2001).

Signaling in Adherens Junctions

It is to be expected that signaling influences the formation and the dissolution of adherens junctions, but adherens junctions also participate in cell signaling and even in transcription. First, we will discuss signaling to the junctions. All the resident molecules of adherens junctions are rich in tyrosine, serine and threonine residues and are therefore vulnerable to phosphorylation and protein kinases. Indeed numerous papers have shown, both in cell culture models and in perfused microvessels, that phosphorylation of adherens junctional adhesion molecules, in response to inflammatory mediators, such as histamine, oxidants, or VEGF results in junctional disassembly and decreased barrier function (Volberg *et al.*, 1992; Gloor *et al.*, 1997; Staddon *et al.*, 1997; Carbajal and Schaeffer, 1998; Esser *et al.*, 1998; Kevil *et al.*, 1998a; Knaus *et al.*, 1998; Yuan *et al.*, 1998; Cohen *et al.*, 1999; Cowell and Garrod, 1999; Ratcliffe *et al.*, 1999; Andropoliu *et al.*, 1999; Alexander *et al.*, 2000; Glover and Leach, 2003). For instance, tyrosine phosphorylation of adherens junction proteins results in loss of cell-cell contact and rounding in MDCK cells (Volberg *et al.*, 1992).

This type of inflammatory response is a feature of diseases, such as diabetes mellitus. Studies in diabetic retinopathy and proliferative retinopathy indicate that perturbation of junctional molecules is a complication of diabetes mellitus (Antonetti *et al.* 1998). The same appears to be true for feto-placental vessels in Type 1 diabetes (Leach *et al.*, 2004) where greater than 50% of placental microvessels show a complete loss of junctional VE-cadherin and β-catenin, which is not due to loss of total protein, but loss from designated microdomains (i.e. junctional regions). This loss is associated with increased phosphorylation of these molecules as well as increased percentage of vessels showing junctional phospho-tyrosine immunoreactivity. The immediate functional consequence of this junctional perturbation is extravasation of tracers (76 kDa dextrans), which in normal pregnancies are retained in the vascular compartment of the placental vessels (Figure 3). Rho GTPases are important in the regulation VE-cadherin and expression of mutant forms of Rho and Rac inhibit the ability of sphingosine1-phosphate to promote the formation of adherens junctions (Braga *et al.*, 1999; Lee *et al.*, 1999). Whereas an effector of small G-protein activation, IQGAP1, causes disassembly of α-catenin from cadherin and β-catenin with disruption of cell-cell adhesion, Cdc42 and Rac1 prevent this disruption and promote adhesion function of the adherens junction (Kaibuchi *et al.*, 1999). Comparable studies on junctional integrity have not been done in lung, but in a diabetic guinea pig model ultrastructural changes in pulmonary endothelium revealed numerous alterations, suggestive of an inflammatory process and indicating that the lung vasculature is subject to the effects of the high glucose and dyslipidemia associated with diabetes (Popov and Simionescu, 1997).

It is clear that the adhesion molecules that complex to form adherens junctions act beyond cell-cell adhesion (Dejana, 2004). That is, they are important signal transduction ligands, which influence diverse cellular processes including proliferation (Conacci-Sorrell *et al.*, 2002), paracellular permeability, endothelial cell survival, and apoptosis (Carmeliet *et al.*, 1999). There are numerous signaling molecules concentrated at the intracellular site of the adherens junction. These include kinases, such as src, lyn, and yes, components of the Ras pathways and small G-proteins as well, and regulatory tyrosine kinases and phosphatases (Dejana, 2004; Vincent *et al.*, 2004). Co-localization of adherens junctions and growth hormone receptors suggests an influence of adherens junctions on migratory-, proliferation-, and angiogenic-related transcription. Moreover, adherens junctions bind transcription factors, such as β-catenin, at the cell membrane preventing their nuclear translocation (Dejana, 2004).

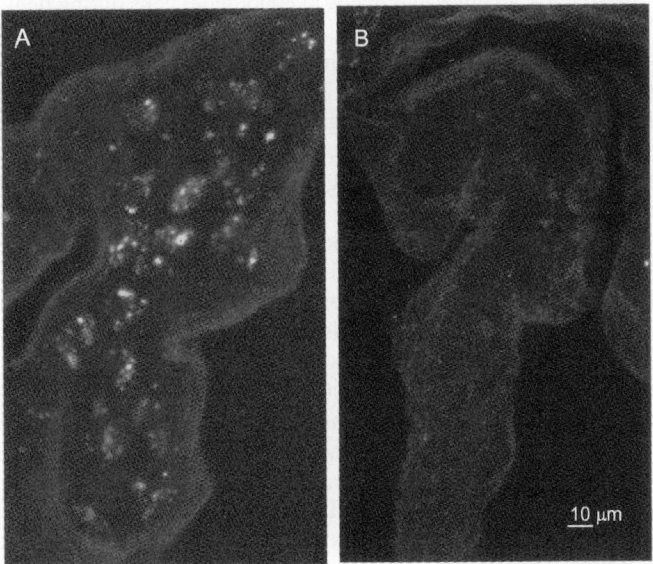

Figure 3. Endothelial barrier dysfunction in vessels in human placenta from diabetic patients. Confocal micrographs of villous biopsies taken from microvascular beds of placentae obtained from Type 1 diabetic and normal patients, which had been perfused with 76 kDa Dextran-TRITC for 10 min. The tracer can be seen trapped in perivascular regions in the diabetic (A) but not in the normal placenta (B). Bar = 10 µm. Kindly provided by L. Leach

The extracellular region of VE-cadherin, like all classical cadherins, is composed of 5 homologous domains, EC1-EC5. Monoclonal antibodies directed against EC1, EC3, EC3-4 increase paracellular permeability and block angiogenesis *in vitro* (Corada *et al.*, 2001). A novel antibody, E4G10, directed against a polypeptide sequence corresponding to the NH2-terminal region of VE-cadherin blocks angiogenesis, without affecting permeability (Liao *et al.*, 2002). The investigators hypothesized that this antibody recognizes an epitope on VE-cadherin that is accessible in neovasculature (actively proliferating endothelium) and becomes inaccessible once stable adherens junctions form in quiescent vasculature. This gives credence to the hypothesis of stable and dynamic junctions being present in the vasculature, thereby allowing different barrier restrictivity to exist.

In addition to its role in junctional integrity and regulation of paracellular permeability β-catenin can act as a transcription factor in the nucleus by serving as a co-activator of the lymphoid enhancing factor (LEF)/TCF family of DNA-binding proteins (Gottardi and Gumbiner, 2001). Transcription mediated by β-catenin is activated by the Wnt signaling pathway, which is crucial during embryonic development. Recruitment of β-catenin to adherens junctions antagonizes LEF/TCF transactivation, i.e. junctional sequestering inhibits proliferation (Conacci-Sorrell *et al.*, 2002). A Wnt-independent inhibitor of β-catenin binding to the TCF transcription factor, ICAT (Inhibitor of β-Catenin and TCF-4), was recently discovered (Gottardi and Gumbiner, 2004). Transfection of endothelial cells with Wnt-1 to activate β-catenin transcription caused increased endothelial proliferation (Wright *et al.*,

1999), while cytoplasmic β-catenin accumulates in endothelial cells during neovascularization after myocardial infarction (Blankesteijn *et al.*, 2000). A similar cytoplasmic accumulation (concomitant with loss of β-catenin from adherens junctions) occurs in human placental microvessels from pregnancies complicated by Type 1 diabetes where increased angiogenesis and increased vascular leakage are key pathological features (Leach *et al.*, 2004). Whilst fully differentiated and intact adherens junctions sequester its components within the complex, particularly β-catenin, and away from signaling pathways that lead to proliferation, ensuring quiescence, many activating mediators result in destabilization of the junctions. This may be partly related to signaling effects on the cytoskeleton, due to decreased cAMP levels, or by direct effects on the junctional components and assembly as noted above. When this release of β-catenin is sustained, it becomes available for transfer to the nucleus. Indeed, loss of junctional β-catenin and VE-cadherin have also been shown to be a very early event in VEGF-induced angiogenesis in human umbilical vein cells (Wright *et al.*, 2002).

Conversely, formation of the adherens junctions and the associated signaling is important in the contact inhibition that confines normal endothelium to a monolayer (Lampugnani *et al.*, 2003). In fact, VE-cadherin associated with β-catenin, but not p120, is capable of terminating the VEGF growth signal, as mentioned above, and β-catenin null endothelial cells are not contact inhibited (Lampugnani *et al.*, 2003). This occurs because VEGF engagement of its VEGFR2 results in association with the adhesion junctions is dependent upon β-catenin (Shay-Salit *et al.*, 2002). Despite activation of the tyrosine kinase activity of the receptor, the phosphatase (DEP-1) bound to the adherens junction reverses the tyrosine phosphorylation and terminates activation of the MAPK cascade. This results in contact inhibition and the poor responsiveness of confluent endothelium to VEGF-mediated proliferation. At the same time, the VEGFR2-adherens junction association selectively enhances the VEGF/VEGFR2-mediated PI3K-Akt activation and cell survival (Carmeliet *et al.*, 1999). Thus, β-catenin-mediated signaling appears to be critical for contact inhibition in endothelial monolayers. Importantly, physiologic shear stress also promotes association of the adherens junction and VEGFR2, resulting in Akt-survival signaling and nuclear translocation of the VEGFR2, apart from VEGF-stimulation of the receptor (Shay-Salit *et al.*, 2002). Additionally, the cadherin-p120 complex is capable of suppressing thrombin-induced MLC phosphorylation related to suppression of Rho kinase (Iyer *et al.*, 2004). Thus, overall the association of the adherens junction and VEGFR2 in fully confluent endothelium suppresses proliferation and enhances survival, whereas junctional disassembly permits unrestricted VEGF/VEGFR2- permeability and proliferation signaling to occur. The ultimate balance and regulation and the net effects of VEGF and β-catenin directed proliferation vs. adherens junction/VEGFR2 coupling with decreased proliferation and increased survival signaling, under differing conditions and states of endothelial confluence require further exploration.

Tight Junctions

Structure and Relationship to Monolayer Permeability

As stated above, the discontinuous network of tight junctional strands in systemic microvessels minimizes the contribution of tight junctions in regulating physiological permeability. The importance of tight junctions in the blood-brain and blood-retinal barrier is indisputable and beyond the focus of this chapter. We, therefore, recommend an excellent and comprehensive review by Harhaj and Antonetti (2004). Nevertheless, there is a role for the tight junctional complexes in permeability, signaling, and strength of inter-endothelial cell adhesion in non-CNS endothelium. Figure 1 above illustrates the presence of tight junctions between adjacent endothelium in such tissue. Tight junctions have been morphologically identified in lung capillary endothelium with increased number from the upper low-flow regions to the lower high-flow regions: although they are certainly less prominent than in lung epithelium (Simonescu, 1980; Yoneda, 1982; Walker et al., 1988; Walker et al., 1994; Vestweber, 2000) as seen in Figures 4 – 6 (from Walker et al., 1994). Tight junctions in pulmonary venous endothelium were less frequent than in capillaries, whereas they were more numerous in arterial segments (Schneeberger, 1982). The strand complexity of tight junctions is influenced by the molecules present therein, with varying levels of claudin, occludin, and junctional adhesion molecules (JAMs) (Furuse et al., 1998). Not surprisingly, there is an inverse relationship of angiogenesis and tight junction formation. SSeCKs, a PKC substrate, was found to regulate blood-brain barrier maturation by increasing tight junction formation and decreasing angiogenesis, mainly through decreased expression of VEGF and stimulated expression of Angiopoetin-1 (Lee et al., 2003). Although primary cultured lung endothelial cells did not have tight junctions (Craig et al., 1998), other studies found that cultured pulmonary artery endothelium developed tight junctions when cultured with a soluble factor from astrocyte conditioned medium or with dibutyryl cAMP, thus expression may be a function of maturation or complete quiescence (Shivers et al., 1988; Satoh et al., 1996). A correlation of tight junction expression level and barrier restriction in lung endothelium was shown, but the contribution of adherens junctions was not considered (Satoh et al., 1996).

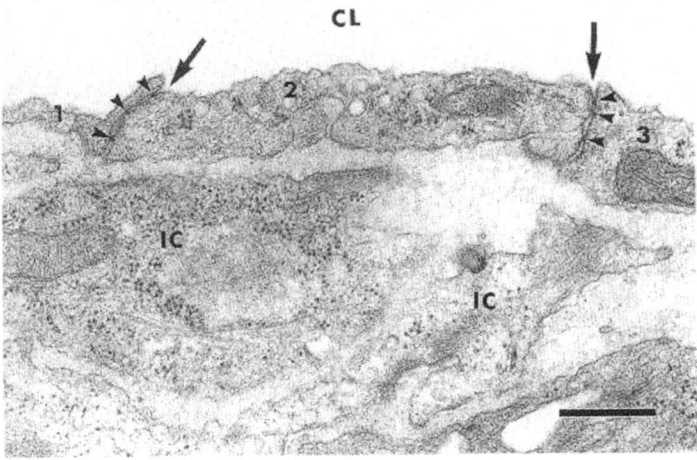

Figure 4. Pulmonary capillary endothelial tight junctions. Thin section through three (1,2,3) pulmonary capillary endothelial cells, near an alveolar corner, showing the increased depth and multiple points of close contact indicated by the arrowheads. Osmium tetroxide stained electronmicrograph. CL - capillary lumen. IC - interstitial cells. Scale bar = 0.5 µm. Figure from *Walker, D.C., MacKenzie, A., and Hosford, S. (1994) The structure of the tricellular region of endothelial tight junctions of pulmonary capillaries analyzed by freeze fracture. Microvascular Research 48:259-281.* by permission of Elsevier, Amsterdam.

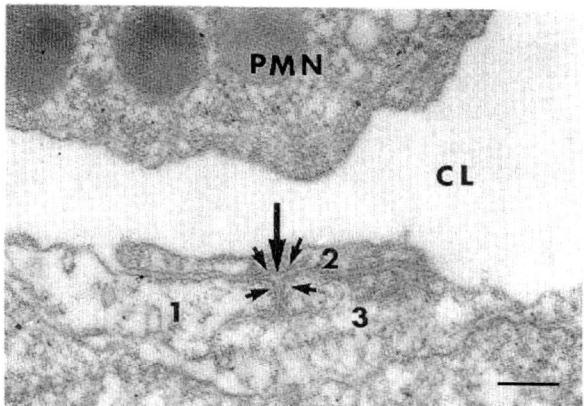

Figure 5. Pulmonary capillary endothelial tricellular flaps and tight junctions. Transverse section of a tricellular tight junction. Cells 1 and 3 are adjacent and cell 2 extends a flap over the junction. A neutrophil (PMN) is observed in the lumen. Scale bar = 0.25 µm. From *Walker, MacKenzie, & Hosford (1994) The structure of the tricellular region of endothelial tight junctions of pulmonary capillaries analyzed by freeze fracture. Microvasc. Res. 48:259-281.* by permission of Elsevier, Amsterdam.

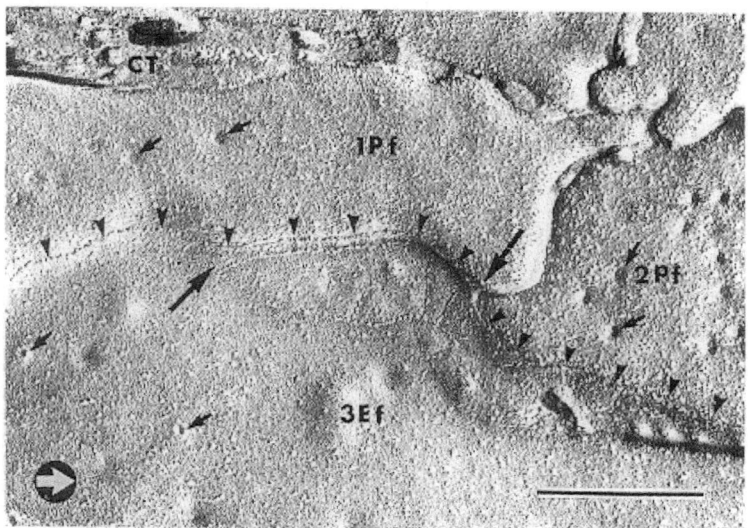

Figure 6. Scanning electronmicrograph of freeze fractured pulmonary capillary tight junctions. Cells 1 and 2 are intact as is the cleft between them, thus no details of the junctions in that cleft are visible. A freeze fracture plane has removed the adjacent cell 3 to leave the ectoplasmic half of its luminal membrane (3Ef). The tight junctions between cells 1 and 3 are visible as a double line of particles indicated by the arrowheads. The tight junctions between cells 2 and 3 are visible as the complex of furrows and particles, also indicated by arrowheads. The tight junctions are discontinuous between the two large arrows. The small arrows indicate the caveoli. Shadowing indicated by white arrow (scale bar = 0.5 µm). From *Walker, MacKenzie, & Hosford (1994) The structure of the tricellular region of endothelial tight junctions of pulmonary capillaries analyzed by freeze fracture. Microvascular Research 48:259-281.* by permission of Elsevier, Amsterdam.

Claudin

Claudin knockout mice form defective junctions (Mitic *et al.*, 2000). Epithelium and endothelium of different vascular beds show heterogeneity in expression of claudins (Lippoldt *et al.*, 2000b; Morita *et al.*, 1999). Claudin 1 is a component of continuous tight junctions, while claudin 2 is localized to discontinuous strands. In the lung and brain, Claudin-5 is expressed in endothelium of all vascular segments, but is absent from the epithelium (Morita *et al.*, 1999). In contrast in the kidney, Claudin-5 is located in arteries, but not capillaries or veins (Morita *et al.*, 1999). During vasculogenesis and angiogenesis of placental vessels, the early vessels form adherens junctions and tight junctions, but claudin is absent, indicating it is not essential for tight junction formation (Leach *et al.*, 2002). However, it is present in the placental vessels at later stage and in adult brain and skeletal muscle capillaries in mice, indicating that it is a later addition (Leach *et al.*, 2002; Nasdala *et al.*, 2002).

Occludin

Increased permeability correlates with decreased occludin expression in endothelium and epithelium (Balda *et al.*, 2000; Hirase *et al.*, 1997; Jiang *et al.*, 1999; Kevil *et al.*, 1998b). Histamine and cytokines disrupt occludin/ZO-1 junctions and decrease protein (Blum *et al.*, 1997; Gardner *et al.*, 1996). Decreased occludin and increased retinal vascular permeability was found in experimentally induced diabetic rats (Antonetti *et al.*, 1998). Moreover, VEGF reduces occludin content in retinal endothelial cells (Antonetti *et al.*, 1998). Overexpression of COOH-terminally truncated occludin in several cell lines results in discontinuous expression compared to the normally continuous border, increased paracellular permeability, and paradoxically, increased transepithelial resistance (Balda *et al.*, 1996, McCarthy *et al.*, 1996). In another study in a lung endothelial cell line, the presence or absence of occludin did not alter electrical resistance, but did direct development of the dense peripheral band, suggesting a role for occludin in control of cortical actin (Kuwabara *et al.*, 2001).

Despite this evidence, the role of occludin, one of the first transmembrane molecules located to tight junctions, as a structural molecule is under re-consideration. To this point, tight junctions have been found in occludin knockout cells (Saitou *et al.*, 1998), whereas cotransfection of fibroblasts with occludin results in incorporation of occludin with claudin-1 based strands (Furuse *et al.*, 1998). This has led to the hypothesis that occludin may be an accessory protein in some function of tight junction strands and it may be a feature of well-differentiated stable tight junctions. It is certainly present in epithelial tight junctions, blood-brain barrier and in tight junctions of mature conduit vessels. Although VE-cadherin, β-catenin and ZO-1 (the first classical marker of tight junctions) are present in newly formed vessels during vasculogenesis in the early human placenta, the tight junctions do not appear to display immunoreactivity to antibodies against occludin, claudin-1, or claudin-2 (Leach *et al.*, 2000). This profile may not be only characteristic of the temporal developmental stage; even in the full term human placenta there are two distinct vascular populations displaying differential expression of occludin. Whereas conduit vessels of stem and some intermediate villi contain stable occludin-rich junctional phenotype, the highly angiogenic terminal villous capillaries, which are the exchange vessels of the placenta, express the more labile junctional phenotype. These labile junctions are rich in β-catenin, but lack plakoglobin or occludin.

Using agents which enhance cAMP, occludin can be re-targeted back to cell-cell contacts in endothelial cells isolated from these microvessels, with accompanying increased transendothelial resistance, consistent with improved barrier function (Dye et al., 2001). Moreover, there is inhibition of proliferation (Dye et al., 2001). It is possible that cAMP increases expression of tight junction markers by stabilizing junctional complexes, by inducing rapid assembly of tight junction strands (Adamson et al., 1998). It is also possible that cAMP increases the synthesis of tight junction molecules, as described for the tight junction component 7H6 in lung endothelial cells (Satoh et al., 1996). The mechanisms behind these changes may also include inhibition of myosin light-chain kinase (Garcia and Schaphorst, 1995), but PKA activation may have many additional effects (Patterson and Lum, 2001). For instance, effects of cAMP may also result from inhibition of tyrosine kinase receptor signaling or reduced association of src family kinases with junctional complexes, rather than from cytoskeletal contractile effects. Additionally, cAMP/PKA may suppress cytoskeletal reorganization to the activated state since an increased presence of the dense peripheral actin bands are generally observed with elevation of cAMP, as described in Chapter 2. Physiological activation of adenylate cyclase by agents, which signal through G_s-coupled receptors (e.g. adenosine, serotonin, β-adrenergic agonists, and PGI_2) increase the number of tight junctions and decrease endothelial permeability (Langeler and van Hinsbergh, 1991; Baluk and McDonald, 1994; Adamson et al., 1998). Thus occludin absence from tight junctions may reflect a junctional phenotype of highly plastic, nascent or angiogenic vessels.

Junctional Adhesion Molecules

Junctional Adhesion Molecules (JAMs, 1, 2, & 3), members of the Ig superfamily, are transmembrane integral membrane proteins localized at the endothelial tight junctions, which have influence over paracellular permeability (Martin-Padura et al., 1998; Aurrand-Lions et al., 2001; Ebnet et al., 2003). JAM-2 appears to be expressed only by lymphatic endothelium (Aurrand-Lions et al., 2001). Serine phosphorylation of JAMs regulates their clustering within the cell-cell contacts and their subsequent binding of ZO-1 (Ebnet et al., 2003). Overexpression of JAM in CHO cells reduces paracellular permeability (Martin-Padura et al., 1998). JAM overexpression in endothelial cells increases lymphocytes transmigration (Aurrand-Lions et al., 2002), whilst treatment of human umbilical vein cells with a combination of TNF-α and IFN-γ results in redistribution of JAM from cell borders to other regions of cell surface and impairs leukocyte transmigration (Ozaki et al., 1999). Likewise, an antibody directed against JAM has been shown to inhibit both monocyte and lymphocyte transmigration *in vitro* and *in vivo*, suggesting that JAMs also participate in leukocyte diapedesis (Martin-Padura et al., 1998; Johnson-Leger et al., 2002).

Counter-receptors identified on platelets and leukocytes include other JAMs and the integrins that also bind ICAM and VCAM: LFA ($\alpha_L\beta_2$), Mac-1 ($\alpha_M\beta_2$), and VLA4 ($\alpha 4\beta 1$). The presence of these leukocyte counter receptors further indicates that JAMs assist in diapedesis as well as in controlling paracellular permeability (Cunningham et al., 2002; Santoso et al., 2002). Other studies have implicated PECAM rather than JAM in leukocyte transmigration, as leukocytes also express PECAM (Shaw et al., 2001). For instance, anti-JAM antibodies do not prevent leukocyte influx into the central nervous system after infection in mice (Lechner et al., 2000). This difference may be related to presence or absence of redundant systems in different types of endothelium. JAM-1 has been implicated in the ability of endothelial cells to respond to bFGF-induced ERK signaling with adherence and spreading on vitronectin (but not on fibronectin) substrate (Naik et al., 2003).

Anchoring Tight Junctions to the Cytoskeleton

Zonula occludens proteins (ZO1, ZO2, and ZO3) and also cingulin and 7H6 in endothelium form the intracellular link between the tight junction membrane proteins and the dense peripheral actin band in endothelial cells (Fanning *et al.*, 1998; Vestweber, 2000). ZO-1 is the main cytoplasmic linking molecule and is found in all endothelial junctions, regardless of position in vascular tree, developmental stage, or anatomical location. It can bind to the transmembrane tight junction molecules including occludin and JAM. As a member of the MAGUK family it contains PDZ domains, SH3 domains, guanylate kinase-like domains, and is a signaling molecule in its own right. (Review Vestweber, 2000; Harhaj and Antonetti, 2004). Phosphorylation (at tyrosine residues) of ZO-1 has been linked with decreased transcellular electrical resistance in endothelial cells (Staddon *et al.*, 1997; Antonetti *et al.*, 1998; Martin *et al.*, 2002). For instance, VEGF treatment of rat retinas mediates a four-fold increase in ZO-1 tyrosine phosphorylation and enhances vascular permeability (Antonetti *et al.*, 1999). Hepatocyte growth factor decreases the protein level of ZO-1, in addition to inducing tyrosine phosphorylation (Martin *et al.*, 2002). However, this molecule may not be exclusively involved in anchorage and signaling of tight junctions; but may also be involved in cadherin-based cell adhesion by functioning as a link between the cadherin/catenin complex and the actin-based cytoskeleton via its linkage to α-catenin (Itoh *et al.*, 1997).

Additional Intercellular Molecules

PECAM

PECAM, (CD31), a member of the Ig superfamily, is expressed by endothelial cells, platelets, and leukocytes. PECAM can form homotypic bonds with PECAM from the adjacent cell or can attach to glycosamineglycans (Dejana, 2004). As noted above, when endothelial cells become confluent, PECAM migrates from a random distribution over the entire cell surface to the lateral edges (Albelda *et al.*, 1991). In two dimensions, PECAM appears to associate with the adherens junctions, although not in discrete foci like cadherin. Whereas adherens junctions are somewhat apical within the lateral edge, PECAM is more basally located and is separated by a distance of at least 0.4 μm (Ayalon *et al.*, 1994; Leach *et al.* 1993). If the adherens junctions are then acutely disrupted by EGTA chelation of extracellular calcium, the PECAM attachments remain in place (Ayalon *et al.*, 1994). Yet, PECAM is much less tightly bound to the cytoskeleton than are the adherens proteins in complex. Treatment of endothelial cells with a combination of TNF and IFN causes a redistribution of PECAM away from the lateral edges and reduces the transmigration of leukocytes across the endothelium (Shaw *et al.*, 2001). Moreover, treating either leukocytes or endothelium with antibodies directed against PECAM reduces transmigration, supporting a role for homotypic binding of endothelial PECAM to leukocyte PECAM during diapedesis (Muller *et al.*, 1993; Shaw *et al.*, 2001). Notably, PECAM-1 has two tyrosine-based inhibitory motifs (ITIMs), which are able to bind src-homology 2 (SH2) when phosphorylated, and therefore able to contribute to a variety of cellular signaling events (Vestweber, 2000). Shear stress, aggregation of IgE receptor, ligation of PECAM, and VEGF, all result in tyrosine phosphorylation of PECAM-ITIM domains (Vestweber, 2000; Newman and Newman, 2003).

Cell-Cell Integrins

As described in Chapter 8, integrins are important in cell-to-matrix adhesion and, thus, in migration, spreading, and locational stability, but they also participate in lateral adhesion. Endothelial cells form intercellular junctions using specific integrins, α2β1, α3β1, and α5β1, that appear in the lateral borders (Lampugnani *et al.*, 1991; Yanez-Mo *et al.*, 1998). Interestingly, antibodies to these integrins disrupt the endothelial barrier (Lampugnani *et al.*, 1991). Possible ligands are the membrane spanning tetraspans (PETA-3, TAPA-1, and CD9), shown to locate at intercellular contact sites on endothelium and to bind each other and α3β1 integrins (Yanez-Mo *et al.*, 1998). The presence of PECAM at the lateral borders exhibits a positive cooperativity for location of the β-integrins to the lateral edges, whereas other Ig adhesion molecules lack this recruitment capacity (Fawcett *et al.*, 1995). Thus, these integrins would be late recruits to the interendothelial cleft. This is also apparent from the lack of integrins in the lateral borders of the placental vessels (Leach *et al.*, 1993). There is little known about their role in paracellular permeability or in signaling, but integrins elsewhere are associated with signaling molecules that provide responsiveness to the external environment. In quiescent cells there is a concentration of the focal adhesion components (i.e. paxillin and FAK) at the baso-lateral corners, whereas in activated cells these components are coalesced and localized to sites of strong focal adhesion (Schaphorst *et al.*, 1997; Patterson *et al.*, 2000). It is not known if these markers are specifically associated with integrins at the lateral positions, but it is likely given their normal binding to integrins. Further, it is not known if ligands for the integrins are matrix proteins or ligands expressed on the opposing cell, but their anchoring in the very corners suggests some structural adhesion unique to this position.

Gap Junctions

A limited number of gap junctions are found between endothelia (Curtis *et al.*, 1995; DePaola *et al.*, 1999). These tube-like junctions form from six connexin monomers (CX37, CX40, CX43), each cell contributing a hemi-channel. Pulmonary endothelia express all three connexins CX43 (Yeh *et al.*, 1998). Gap junctions are not thought to provide junctional strength or solute restriction, but serve to permit exchange of ions and small molecules between cells to promote coordinated function. The number and communication exchange of gap junctions is increased by cAMP and decreased by tyrosine phosphorylation (Loewenstein, 1985). Nitric oxide increases incorporation of CX 40, the dominant connexin, into the gap junctions in cultured endothelium and dye transfer between cells by a cAMP dependent mechanism (Hoffmann *et al.*, 2003). CX40 possesses phosphorylation sites for PKA, PKC, and PKG, whereas CX43 possesses MAPK and v-src sites, indicating differential regulation (Hoffmann *et al.*, 2003). Disturbed flow *in vitro* upregulates CX43 and disrupts the punctate distribution at lateral borders, interrupting cell-to-cell communication (DePaola *et al.*, 1999).

Junctional Integrity, Dysfunction, and Disease

Inflammation and Junctional Compromise

Maintenance of endothelial barrier integrity is critical to tissue health. Enhanced vascular permeability is a central feature of inflammation and is associated with fluid and solute flux across the vascular wall, development of tissue edema, and organ injury or failure. It is increasingly evident that a number of inflammatory mediators and disease processes evoke tissue edema and injury by inducing disruption, loss, or disorganization of the interendothelial junctional proteins that regulate the barrier between the vascular and interstitial space. For example, histamine released in response to acute or chronic inflammatory challenge evokes a reversible disruption of the vascular barrier by directly interfering with cadherin adhesion (Winter et al., 2004). In placental lobules, histamine changed junctional distribution of both PECAM and VE-cadherin and widened the tight junction zones (Leach et al., 1995). However, gaps were not noted, dimensions of wide zones in the clefts were unchanged, and there was no increase in albumin permeability; nevertheless, these subtle changes in protein distribution and tight junction formation could make the junction more vulnerable to subsequent inflammatory challenge. A number of inflammatory mediators evoke endocytosis of junctional cadherins and a consequent loss of cell-cell contact (Alexander et al., 2000) while simultaneously inducing enhanced endothelial adhesion to the extracellular matrix (Alexander et al., 2001). Thus, acute and chronic inflammatory states *in vivo* are likely to be associated with changes in barrier function through alterations in junctional distribution of proteins (Alexander et al., 1998; Kevil et al., 1998a; Alexander et al., 2000).

While abundant evidence exists documenting the role of VE-cadherin in the maintenance and regulation of vascular barrier integrity, it is also recognized that the tight junctional protein, occludin, plays a role in regulating the endothelial solute barrier (Kevil et al., 1998b). For example, in response to peroxide stimulation, occudin dissociates spatially from ZO-1, evoking loss of vascular barrier function (Kevil et al., 2000). These results suggest that inflammatory mediators are capable of disrupting junctional protein complex integrity and thus the restrictive properties of the vascular wall.

Altered Junctional Protein Content in Inflammation and Disease

While certain stimuli that cause disassembly of VE-cadherins are reversible (Gao et al., 2000) other inflammatory challenges have more disruptive effects on barrier integrity, possibly through proteolysis of junctional cadherins (Carden et al., 1998; Hermant et al., 2003) or catenins (Allport et al., 2000). Acute lung injury (ALI) and the acute respiratory distress syndrome (ARDS) are frequently fatal conditions associated with diverse inflammatory stimuli. The hallmark of ALI and ARDS is increased pulmonary vascular permeability, lung edema formation, and respiratory failure. The mechanisms responsible for the loss of vascular barrier integrity and edema formation observed in lungs subjected to an inflammatory challenge appear to be related to alterations in both the distribution and the content of junctional VE-cadherin (Zhao et al., 2001). Oxidant stress is generally associated with inflammatory stress, and induction of redox imbalance in perfused lung and lung microvascular cells results in both loss of cadherin from the junctions and in loss of total cellular cadherin protein, while the decrease in cadherin content and the permeability was prevented with sulfhydryl agents (Zhao et al., 2001). Leukocyte elastase degrades VE-cadherin following systemic inflammation, an event correlated with increased permeability lung edema (Carden et al., 1998). Moreover, plasma of ARDS patients contains soluble cadherin fragments. Thrombin or cytokine stimulation of lung endothelium not only disrupts VE-cadherin but also alters subcellular cytoskeletal proteins (Lim et al, 2001; Petrache et al., 2003). Other structural elements of the adherens junction also appear to be susceptible to proteolytic degradation in response to inflammatory insults that result in loss of barrier integrity (Bannerman et al., 1998; Allport et al., 2000).

Recent evidence suggests that disruption of junctional proteins underlie other clinical conditions associated with tissue edema and organ failure. Transplantation organ preservation is one condition in which organ failure may be limited by tissue edema. Organ preservation solutions were shown to elicit dramatic loss of endothelial occludin and cadherin content, changes in F-actin organization and loss of barrier function (Trocha et al., 1999). Similarly, exposure of lungs to heavy metals results in lung edema formation and concomitant loss of VE-cadherin in vascular endothelium (Pearson et al., 2003). Likewise, disruption of the blood brain barrier in patients with multiple sclerosis may also be related to downregulation of VE-cadherin and occludin (Minagar et al., 2003). Changes in cadherin content and distribution are associated with melanoma cell migration, survival, growth, and metatasis (McGary et al., 2002). It is now recognized that diabetes is a disease associated with chronic inflammation and endothelial dysfunction. Even relatives of diabetic, not yet insulin resistant, display early endothelial dysfunction (Cabellero et al., 1999; Cabellero, 2003). As shown above in Figure 3, placentae from diabetic patients exhibited increased vascular permeability, correlated with a marked loss of VE-cadherin and β-catenin in the junctions (Leach et al. 2004). Thus, elucidation of mechanisms by which loss or disorganization of junctional proteins leads to increased microvascular permeability may permit development of novel therapeutic strategies in conditions associated with inflammatory edema formation. Conversely, a thorough understanding of these same mechanisms may facilitate development of tools to transiently disrupt junctional proteins and barrier function in order to allow targeted tissue deliver of therapeutic agents (van Nieuw Amerongen and van Hinsbergh, 2002).

Coming Together

While much has been learned regarding the structure and function of tight junctions, adherens junctions, and other junctional components, there are many gaps in our knowledge of the assembly of these components in the pulmonary vasculature, in their dissassembly during transient challenge by bioactive agents and in acute lung injury, and in their relative contribution to the permeability dysfunction. For instance, the key participation of VE-cadherin in interendothelial cell adhesion is recognized, but the relative spatio-temporal importance of β–, γ-, and p120-catenin complex formation with VE-cadherin in assembly and function of the adherens junction is not resolved in pulmonary endothelium. Likewise, little is actually known regarding the role and composition of endothelial tight junctions in normal pulmonary barrier function *in vivo*. We are beginning to unravel the molecular regulatory interactions between junctional proteins and the cytoskeleton and the bi-directional signaling between junctional integrity and endothelial activation/quiescence. While there is a general correlation between phosphorylation of junctional proteins and loss of junctional integrity, phosphorylation of some members seems to help in initial assembly or in mechanotransduction signaling. The role of junctional proteins and assembly in control of endothelial survival, motility, and proliferation is coming to light, but again there is much to learn. It has become clear that destruction of junctional proteins by protease and oxidative assault from without and from within the endothelium is a key contributor to poor barrier selectivity and to poor reversibility of dysfunction. Resolution of these issues is critical to gaining control over the barrier dysfunction in acute lung injury.

ACKNOWLEDGEMENTS

This work was supported by a Merit Grant from the Veteran's Administration Medical Research Service (CEP). LL gratefully acknowledges support of the Wellcome Trust and the International Association of Cancer Research for funding the research reported here.

REFERENCES:

Adamson, R.H., Liu, B., Fry, N. Rubin, L.L., and Curry, F.E. (1998) Microvascular permeability and number of tight junctions are modulated by cAMP. Amer. J. Physiol. 274:H1885-H1894.
Adamson, R.H., and Michel, C.C. (1993) Pathways through the intercellular clefts of frog mesenteric capillaries. J. Phys 466: 303-327.
Albelda, S.M., Muller, W., Buck, C. Newman, P.J. (1991) Molecular and cellular properties of PECAM-1: a novel vascular cell adhesion molecule. J. Cell Biol. 114:1059-1068.
Alexander, J.S., Alexander, B., Eppihimer, L., Goodyear, N., Haque, R., Davis, C., Kalogeris, T., Carden, D.L., Zhu, Y., Kevil, C., Alexander, J.S. (2000) Inflammatory mediators induce sequestration of VE-cadherin in cultured human endothelial cells. Inflamm. 24:99-113.
Alexander, J.S., Blaschuk, O.W., and Haselton, F.R. (1993) An N-cadherin-like protein contributes to solute barrier maintenance in cultured endothelium. J. Cell.Physiol. 156:610-618.
Alexander, J.S., Jackson, S., Chaney, E., Kevil, C., Haselton, F.R. (1998) The role of cadherin endocytosis in endothelial barrier regulation: Involvement of PKC and actin-cadherin interactions. Inflamm. 22:419-433.
Alexander, J.S., Zhu, Y., Elrod, J., Alexander, B., Coe, L., Kalogeris, T., and Fuseler, J. (2001) Reciprocal regulation of endothelial substrate adhesion and barrier function. Microcirc. 8:389-401.
Allport, J.R., Muller, W.A., and Luscinskas, F.W. (2000) Monocytes induce reversible focal changes in vascular endothelial cadherin complex during transendothelial migration under flow. J. Cell Biol. 148:203-216.
Andriopoulou, P., Navarro, P., Zanetti, A., Lampugnani, M., and Dejana, E. (1999) Histamine induces tyrosine phosphorylation of endothelial cell-to-cell adherens junctions. Arterioscl. Thromb. Vasc. Biol. 19:2286-2297.
Antonetti, D.A., Barber, A. Hollinger, Z., Wolpert, E., and Gardner, T.W. (1999) Vascular endothelial growth factor induces rapid phosphorylation of tight junction proteins occludin and zonula occluden 1. A potential mechanism for vascular permeability in diabetic retinopathy and tumors. J. Biol. Chem. 274:23463-23467.
Antonetti, D.A., Barber, A., Khin, S., Lieth, E., Tarbell, J., and Gardner, T.W. (1998) Vascular permeability in experimental diabetes is associated with reduced endothelial occludin content: VEGF decreases occludin in retinal endothelial cells. Diabetes 47:1953-1959.
Aono, S., Nakagawa, S., Reynolds, A., and Takeichi, M. (1999) p120 acts as an inhibitory regulator of cadherin function in colon carcinoma cells. J. Cell. Biol. 145:551–562.
Aurrand-Lions, M., Johnson-Leger, C., Lamgna, C., Ozaki, H., Kita, T., and Imhof, B.A. (2002) Junctional adhesion molecules and interendothelial junctions. Cells Tissues Organs 172:152-160.
Aurrand-Lions, M., Johnson-Leger, C., Wong, C., Du Pasquier, L., and Imhof, B.A. (2001) Heterogeneity of endothelial junctions is reflected by differential expression and specific subcellular localization of the three JAM family members. Blood. 98:3699-3707.
Ayalon, O., Sabanai, H., Lampugnani, M., and Dejana, E. (1994) Spatial and temporal relationships between cadherins and PECAM-1 in cell-cell junctions of human enodthelial cells. J. Cell Biol. 126:247-258.
Balda, M.S., Maldonado, C., Cereijido, M., and Matter, K. (2000) Multiple domains of occludin are involved in regulation of paracellular permeability. J. Cell. Biochem. 78:85-96.
Balda, M.S., Whitney, J., Flores, C., Gonzalez, S., Cereijido, M., and Matter, K. (1996) Functional dissociation of paracellular permeability and transepithelial electrical resistance and disruption of

the apical-basolateral intramembrane diffusion barrier by expression of a mutant tight junction membrane protein. J. Cell Biol. 134:1031-1049.

Baluk, P. and McDonald, D.M. (1994) The β2-adrenergic receptor agonist formoterol reduces microvascular leakage by inhibiting endothelial gap formation. Amer. J. Physiol. 266:L461-L468.

Bannerman, D.D., Sathyamoorthy, M., and Goldblum, S.E. (1998) Bacterial lipopolysaccharide disrupts endothelial monolayer integrity and survival signaling events through caspase cleavage of adherens junction proteins. J. Biol. Chem. 273:35371-35380.

Blankesteijn, W.M., van Gijn, M., Essers-Janssen, Y., Daemen, M., and Smits, J.F. (2000) β-catenin, an inducer of uncontrolled cell proliferation and migration in malignancies, is localized in the cytoplasm of vascular endothelium during neovascularization after myocardial infarction. Amer. J. Pathol. 157:877-883.

Blum, M.S., Toninelli, E., Anderson, J., Balda, M., Zhou, J., ODonnell, L., Pardi, R., and Bender, J.R. (1997) Cytoskeletal rearrangement mediates human microvascular endothelial tight junction modulation by cytokines. Amer. J. Physiol. 42:H286-H294.

Braga, V.M., Del Maschio, A., Machesky, L., and Dejana, E. (1999) Regulation of cadherin function by Rho and Rac: modulation by junction maturation and cellular context. Molec. Biol. Cell 10:9-22.

Bundgaard, M. (1984) The three-dimensional organization of tight junctions in a capillary endothelium revealed by serial-section electron microscopy. J. Ultrastruct. Res. 88:1-17.

Caballero, A.E., Arora, S., Saouaf, R., Lim, S., Smakowski, P., Park, J., King, G., LoGerfo, F., Horton, E., and Veves, A. (1999) Microvascular and macrovascular reactivity in diabetes. Diab. 48:1856-1862.

Caballero, A.E. (2003) Endothelial dysfunction in obesity and insulin resistance: a road to diabetes and heart disease. Obes. Res. 11:1278-1289.

Carbajal, J.M. and Schaeffer, R.C. Jr. (1998) H_2O_2 and genistein differentially modulate protein tyrosine phosphorylation, endothelial morphology, and monolayer barrier function. Biochem. Biophys. Res. Comm. 249:461-466.

Carmeliet, P., Lampugnani, M., Moons, L., Breviario, F., Compernolle, V., Bono, F., Balconi, G., Spagnuolo, R., Oostuyse, B., Dewerchin, M., Zanetti, A., Angellilo, A., Mattot, V., Nuyens, D., Lutgens, E., Clotman, F., de Ruiter, M., Gittenberger-de Groot, A., Poelmann, R., Lupu, F., Herbert, J., Collen, D., and Dejana, E. (1999) Targeted deficiency or cytosolic truncation of the VE-cadherin gene in mice impairs VEGF-mediated endothelial survival and angiogenesis. Cell 98:147-157.

Carden, D., Xiao, F., Moak, C., Willis, B., Robinson-Jackson, S., and Alexander, S. (1998) Neutrophil elastase promotes lung microvascular injury and proteolysis of endothelial cadherins. Amer. J. Physiol. 275:H385-H392.

Cattelino, A., Liebner, S., Gallini, R., Zanetti, A., Balconi, G., Corsi, A., Bianchi, P., Wolburg, H., Moore, R., Oreda, B., Kemler, R., and Dejana, E. (2003) The conditional inactivation of the β-catenin gene in endothelial cells causes a defective vascular pattern and increases vascular fragility. J. Cell Biol. 15:1111-1122.

Cohen, A.W., Carbajal, J.M., and Schaeffer, R.C.J. (1999) VEGF stimulates tyrosine phosphorylation of beta-catenin and small-pore endothelial barrier dysfunction. Amer. J. Physiol. 277:H2038-H2049.

Conacci-Sorrell, M., Zhurinsky, J., and Ben-Ze'ev, A. (2002) The cadherin-catenin adhesion system in signaling and cancer. J. Clin. Invest. 109:987-991.

Corada, M., Marriotti, M., Thurston, G., Smith, K., Kunkel, R., Broackhaus, M., Lampugnini, M., Padura, I., Stoppaciaro, A., Ruco, L., McDonald, D., Ward, P., and Dejana, E. (1999) Vascular endothelial-cadherin is an important determinant of microvascular integrity in vivo. PNAS USA 96:9815-9820.

Cowell, H.E. and Garrod, D.R. (1999) Activation of PKC modulates cell-cell and cell-substratum adhesion of a human colorectal carcinoma cell line and restores 'normal' epithelial morphology. Internat'l J. Cancer 80:455-464.

Craig, L.E., Spelman, J., Strandberg, J., and Zink, M.C. (1998) Endothelial cells from diverse tissues exhibit differences in growth and morphology. Microvasc. Res. 55:65-76.
Cunningham, S.A., Rodriguez, J., Arrate, M., Tran, T., and Brock, T. (2002) JAM2 interacts with α4β1. Facilitation by JAM3. J. Biol. Chem. 277:27589-27592.
Curtis, T.M., McKeown, P., Vincent, P., Homan, S., Wheatley, E., and Saba, T.M. (1995) Fibronectin attenuates increased endothelial monolayer permeability after RGD peptide, anti-α5β1, or TNF-α exposure. Amer. J. Physiol. 269:L248-L260.
Dejana, E. (2004) Endothelial cell-cell junctions: happy together. Nat. Rev. Molec. Cell Biol. 5:261-270.
DePaola, N., Davies, P., Pritchard, W., Florez, L, Harbeck, N., and Polacek, D.C. (1999) Spacial and temporal regulation of gap junction connexin43 in vascular endothelial cells exposed to controlled disturbed flows. PNAS USA 96:3154-3159.
Dye, J.F., Leach, L., Clark, P., Firth, J.A. (2001) Cyclic AMP and acidic fibroblast growth factor have opposing effects on tight and adherens junctions in microvascular endothelial cells in vitro. Microvasc. Res. 62:94-113.
Ebnet, K., Aurrand-Lions, M., Kuhn, A., Kiefer, F., Butz, S., Zander, K., Meyer zu Brickwedde, M., Suzuki, A., Imhof, B., and Vestweber, D. (2003) The junctional adhesion molecule (JAM) family members JAM-2 and JAM-3 associate with the cell polarity protein PAR-3: a possible role for JAMs in endothelial cell polarity. J. Cell Sci. 116:3879-3891.
Esser, S., Lampugnani, M., Corada, M., Dejana, E., and Risau, W. (1998) VEGF induces VE-cadherin tyrosine phosphorylation in endothelial cells. J. Cell Sci. 111:1853-1865.
Fanning, A.S., Jameson, B., Jesaitis, L., and Anderson, J.M. (1998) The tight junction protein ZO-1 establishes a link between the transmembrane protein occludin and the actin cytoskeleton. J. Biol. Chem. 273:29745-29753.
Fawcett, J., Buckley, C., Holness, C., Bird, I., Spragg, J., Saunders, J., Harris, A., and Simmons, D.L. (1995) Mapping homotypic binding sites in CD31 and the role of CD31 adhesion in formation of interendothelial cell contacts. J. Cell Biol. 128:1229-1241.
Firth, J.A. (2002) Endothelial barriers: from hypothetical pores to membrane proteins. J. Anat. 200:541-548.
Firth, J.A., Bauman, K.F., and Sibley, C.P. (1983) The intercellular junctions of guinea-pig placental capillaries: a possible structural basis for endothelial solute permeability. J. Ultrastr. Res. 85:45-57.
Furuse, M., Fujita, K., Hiiragi, T., Fujimoto, K., and Tsukita, S. (1998) Claudin-1 and -2: novel integral membrane proteins localizing at tight junctions with no sequence similarity to occludin. J. Cell Biol. 141:1539-1550.
Gao, X., Kouklis, P., Xu, N., Minshall, R., Sandoval, R., Vogel, S., and Malik, A.B. (2000) Reversibility of increased microvessel permeability in response to VE-cadherin disassembly. Amer. J. Physiol. 279:L1218-L1225.
Gardner, T.W., Lesher, T., Khin, S., Vu, C., Barber, A., and Brennan, W.A. (1996) Histamine reduces ZO-1 tight-junction protein expression in cultured retinal microvascular endothelial cells. Biochem. J. 320:717-721.
Gloor, S.M., Weber, A., Adachi, N., and Frei, K. (1997) Interleukin-1 modulates protein tyrosine phosphatase activity and permeability of brain endothelial cells. Biochem. Biophys. Res. Comm. 239:804-809.
Glover, V. and Leach, L. (2003) Perfusion of human placental vessels with VEGF: Effect on leakage of 76 kDa dextran and junctional organisation. J. Vasc. Res. 40:306A.
Gottardi, C.J. and Gumbiner, B.M. (2001) Adhesion signaling: how beta-catenin interacts with its partners. Curr. Biol. 11:R792-R794.
Gottardi, C.J. and Gumbiner, B.M. (2004) Role for ICAT in beta-catenin-dependent nuclear signaling and cadherin functions. Amer. J. Physiol. 286:C747-C756.
Gumbiner, B.M. (1996) Cell adhesion: The molecular basis of tissue architecture and morphogenesis. Cell 84:345-357.

Gumbiner, B.M., Stevenson, B., and Grimaldo, A. (1988) The role of the cell adhesion molecule uvomorulin in the formation and maintenance of the epithelial junctional complex. J. Cell Biol. 107:1575-1587.

Guo, M., Wu, M., Grange, R., and Yuan, S.Y. (2003) Transference of recombinant VE-cadherin cytoplasmic domain alters endothelial junctional integrity and microvascular permeability. J. Physiol. 554:78-88.

Harhaj, N.S. and Antonetti, D.A. (2004) Regulation of tight junctions and loss of barrier function in pathophysiology. Int. J. Biochem. Cell Biol. 36:1206-1237.

Haselton, F.R. and Heimark, R.L. (1997) Role of cadherins 5 and 13 in the aortic endothelial barrier. J. Cell Physiol. 171:243-251.

Hermant, B., Bibert, S., Concord, E., Dublet, B., Weidenhaupt, M., Vernet, T., and Gulino-Debrac, D. (2003) Identification of proteases involved in the proteolysis of VE-cadherin during neutrophil transmigration. J. Biol. Chem. 278:14002-14012.

Hinck, L., Nathke, I., Papkoff, J., and Nelson, W.J. (1994) Dynamics of cadherin/catenin complex formation: novel protein interactions and pathways. J. Cell Biol. 125:1327-1340.

Hirase, T., Staddon, J., Saitou, M., Ando-Akatsuka, Y., Itoh, M., Furuse, M., Fujimoto, K., Tsukita, S., and Rubin, L.L. (1997) Occludin as a possible determinant of tight junction permeability in endothelial cells. J. Cell Sci. 110:1603-1613.

Huang, A.L., Jan, K.M., and Chien, S. (1992) Role of intercellular junctions in the passage of horseradish peroxidase across aortic endothelium. Lab. Invest. 67:201-209.

Huber, P., Dalmon, J., Engiles, J., Breviario, F., Gory, S., Siracusa, L.D., Buchberg, A., and Dejana, E. (1996) Genomic structure and chromosomal mapping of the mouse VE- cadherin gene (Cdh5). Genomics 32:21-28.

Itoh, M., Nagafuchi, A., Moroi, S., Tsukita, S. (1997) Involvement of ZO-1 in cadherin-based cell adhesion through binding to α–catenin and actin filaments. J. Cell Biol. 138:181-192.

Iyer, S., Ferreri, D., DeCocco, N., Minnear, F., and Vincent, P.A. (2004) VE-cadherin-p120 interaction is required for maintenance of endothelial barrier function. Amer. J. Physiol. 286:L1143-L1153.

Jacobson, J.R., Dudek, S., Birukov, K., Ye, S., Grigoryev, D., Girgis, R., and Garcia, J.G. (2004) Cytoskeletal activation and altered gene expression in endothelial barrier regulation by simvastatin. Amer. J. Respir. Cell Molec. Biol. 30:662-670.

Jiang, W.G., Martin, T., Matsumoto, K., Nakamura, T., and Mansel, R.E. (1999) HGF decreases expression of occludin and transendothelial resistance and increases paracellular permeability in human vascular endothelial cells. J. Cellular Physiol. 181:319-329.

Johnson-Leger, C.A., Aurrand-Lions, M., Beltraminelli, N., Fasel, N., and Imhof, B.A. (2002) Junctional adhesion molecule (JAM-2) promotes lymphocyte transendothelial migration. Blood 100:2479-2486.

Jou, T.S., Stewart, D., Stappert, J., Nelson, W., and Marrs, J.A. (1995) Genetic and biochemical dissection of protein linkages in the cadherin-catenin complex. PNAS USA 92:5067-5071.

Kaibuchi, K., Kuroda, S., Fukata, M., and Nakagawa, M. (1999) Regulation of cadherin-mediated cell-cell adhesion by the Rho family GTPases. Curr. Opin. Cell Biol. 11:591-596.

Kevil, C.G., Ohno, N., Gute, D., Okayama, N., Robinson, S., Chaney, E., and Alexander, J.S. (1998a) Role of cadherin internalization in hydrogen peroxide-mediated endothelial permeability. Free Rad. Biol. Med. 24:1015-1022.

Kevil, C.G., Okayama, N., Trocha, S., Kalogeris, T., Coe, L., Specian, R., Davis, C., and Alexander, J.S. (1998b) Expression of zonula occludens and adherens junctional proteins in human venous and arterial endothelial cells: role of occludin in endothelial solute barriers. Microcirc. 5:197-210.

Kevil, C.G., Oshima, T., Alexander, B., Coe, L., and Alexander, J.S. (2000) H_2O_2-mediated permeability: role of MAPK and occludin. Amer. J. Physiol 279:C21-C30.

Knaus, U.G., Wang, Y., Reilly, A., Warnock, D., and Jackson, J.H. (1998) Structural requirements for PAK activation by Rac GTPases. J. Biol. Chem. 273:21512-21518.

Kowalczyk, A.P., Navarro, P., Dejana, E., Bornslaeger, E., Green, K., Kopp, D., Borgwardt, J. (1998) VE-cadherin and desmoplakin are assembled into dermal microvascular endothelial intercellular junctions: a pivotal role for plakogloin. J. Cell Sci. 111:3045-3057.

Kuwabara, H., Kokai, Y., Kojima, T., Takkakuwa, R., Mori, M., and Sawada, N. (2001) Occludin regulates actin cytoskeleton in endothelial cells. Cell Struc. Func. 26:109-116.

Lampugnani, M.G., Corada, M., Caveda, L., Breviario, F., Ayalon, O., Geiger, B., and Dejana, E. (1995) The molecular organization of endothelial cell to cell junctions: differential association of plakoglobin, β-catenin, and α-catenin with vascular endothelial cadherin (VE-cadherin). J. Cell Biol. 129:203-217.

Lampugnani, M. G. and Dejana, E. (1997) Interendothelial junctions: structure, signaling and functional roles. Curr. Opin. Cell Biol. 9:674-682.

Lampugnani, M.G., Resnati, M., Dejana, E., and Marchisio, P.C. (1991) The role of integrins in maintenance of endothelial monolayer integrity. J. Cell Biol. 112:479-490.

Lampugnani, M.G., Resnati, M., Raiteri, M., Pigott, R., Pisacane, A., Houen, G., Ruco, L. , and Dejana, E. (1992) A novel endothelial-specific membrane protein is a marker of cell-cell contacts. J.Cell Biol. 118:1511-1522.

Lampugnani, M., Zanetti, A., Corada, M., Takahashi, T., Balconi, G., Breviario, F., Orsenigo, F., Cattelino, A., Kemler, R., Daniel, T., and Dejana, E. (2003) Contact inhibition of VEGF-induced proliferation requires vascular endothelial cadherin, β-catenin, and the phosphatase DEP-1/CD148. J. Cell Biol. 161:793-804.

Lane, N.J., Revest, P., Whytock, S., and Abbott, N.J. (1995) Fine structure investigation of rat brain microvascular endothelial cells:tight junctions and vesicular structures in freshly isolated and cultured preparations. J. Neurocytol. 24:347-360.

Langeler, E.G., Fiers, W., and van Hinsbergh, V.W. (1991) Effects of TNF on prostacyclin production and the barrier function of human endothelial cell monolayers. Arterioscler. Thromb. 11:872-881.

Leach, L., Babawale, M., Anderson, M., and Lammiman, M. (2002) Vasculogenesis, angiogenesis and molecular organisation of endothelial junctions in early human placenta. J. Vasc. Res. 39:246-259.

Leach, L., Clark, P., Lampugnani, M., Arroyo, A., Dejana, E., Firth, J.A. (1993) Immunoelectron characterisation of inter-endothelial junctions of human term placenta. J. Cell Sci. 104:1073-1081.

Leach, L. (2002) The phenotype of the human materno-fetal endothelial barrier: molecular occupancy of paracellular junctions dictate permeability and angiogenic plasticity. J. Anat. 200:599-606.

Leach, L., Eaton, B., Westcott, E., and Firth, J.A. (1995) Effect of histamine on endothelial permeability and structure and adhesion molecules of the paracellular junctions of perfused human term placental microvessels. Microvasc. Res. 50:323-337.

Leach, L. and Firth, J.A. (1992) Fine structure of the paracellular junctions of terminal villous capillaries in the perfused human placenta. Cell Tissue Res. 268:447-452.

Leach, L., Gray, C., Staton, S., Babawale, M., Gruchy, A., Foster, C., Mayhew, T. and James, D.K. (2004) Vascular endothelial cadherin and β-catenin in human fetoplacental vessels from pregnancies complicated by Type1 diabetes: associations with angiogenesis and perturbed barrier function . Diabetologia 47:695-709.

Leach, L., Lammiman, M., Babawale, M., Hobson, S., Bromilou, B., Lovat, S., and Simmonds, M.J.R. (2000) Molecular organization of tight and adherens junctions in the human placental vascular tree. Placenta 21:547-557.

Lechner, F., Sahrbacher, U., Suter, T., Frei, K., Brockhaus, M., Koedel, U., and Fontana, A., (2000) Antibodies to JAM cause disruption of endothelial cells and do not prevent leukocyte influx into the meninges after viral or bacterial infection. J. Infect. Dis. 182:978-982.

Lee, M.J., Thangada, S., Claffey, K., Ancellin, N., Liu, C., Kluk, M., Volpi, M., Sha'afi, R., and Hla, T. (1999) Vascular endothelial cell adherens junction assembly and morphogenesis induced by sphingosine-1-phosphate. Cell 99:301-312.

Lee, S.W., Kim, W., Choi, .K., Song, H., Son, M., Gelman, I., Kim, Y-J., and Kim, K-W. (2003) SSeCKS regulates angiogenesis and tight junction formation in blood-brain barrier. Nature Med. 9:900-906.

Liao, F., Doody, J., Overholser, J., Finnerty, B., Bassi, R., Wu, Y., Dejana, E., Kussie, P., Bohlen, P., and Hicklin, D.J. (2002) Selective targeting of angiogenic tumour vasculature by VE-cadherin antibody inhibits tumour growth without affecting vascular permeability. Cancer Res. 62:2567-2575.

Lim, M.J., Chiang, E., Hechtman, H., and Shepro, D. (2001) Inflammation-induced subcellular redistribution of VE-cadherin, actin, and γ–catenin in cultured human lung microvessel endothelial cells. Microvasc. Res. 62:366-382.

Lippoldt, A., Kniesel, U., Leibner, S., Kalbacher, H., Kirsch, T., Wolburg, H., Haller, H. (2000a) Structural alterations of tight junctions are associated with loss of polarity in stroke-prone spontaneously hypertensive rat blood-brain barrier endothelial cells. Brain Res. 885:251-261.

Lippoldt, A., Liebner, S., Andbjer, B., Kalbacher, H., Wolburg, H., Haller, H., and Fuxe, K. (2000b) Organization of choroid plexus epithelial and endothelial cell tight junctions and regulation by claudin-1, -2, and -5 expression by PKC. Molec. Neurosci. 11:1427-1431.

Loewenstein, W.R. (1985) Regulation of cell-to-cell communication by phosphorylation. Biochem. Soc. Symp. 50:43-58.

Lutz, K.L., Jois, S.D.S., and Siahaan, T.J. (1995) Secondary structure of the HAV peptide which regulates cadherin- cadherin interaction. J. Biomolec. Struc. Dyn. 13:447-455.

Martin, T.A., Mansel, R., and Jiang, W. (2002) Antagonistic effect of NK4 on HGF/SF induced changes in the transendothelial resistance and paracellular permeability of human vascular endothelial cells. J. Cell. Physiol. 192:268-275.

Martin-Padura, I., Lostaglio, S., Schneemann, M., Williams, L., Romano, M., Fruscella, P., Panzeri, C., Stoppacciaro, A., Ruco, L., Villa, A., Simmons, D., and Dejana, E. (1998) JAM, a novel member of the Ig superfamily that distributes at intercellular junctions and modulates monocyte transmigration. J. Cell Biol. 14:117-127.

Matter, K. and Balda, M.S. (1999) Occludin and functions of tight junctions. Int. Rev. Cyt. 186:117-146.

McCarthy, K.M., Skare, I., Stankewich, M., Furuse, M., Tsukita, S., Rogers, R., Lynch, R., and Schneeberger, E. (1996) Occludin is a functional component of the tight junction. J. Cell Sci. 109:2287-2298.

McGary. E.C., Lev, D.C., and Bar-Eli, M. (2002) Cellular adhesion pathways and metastatic potential of human melanoma. Canc. Biol. Therap. 1:459-465.

Minagar, A., Ostanin, D., Long, A., Jennings, M., Kelley, R., Sasaki, M., Alexander, J.S. (2003) Serum from patients with multiple sclerosis downregulates occludin and VE-cadherin expression in cultured endothelial cells. Mult. Scler. 9:235-238.

Mitic, L.M., Van Itallie, C.M., and Anderson, J.M. (2000) Tight junction structure and function: lessons from mutant animals and proteins. Amer. J. Physiol. 279:G250-G254.

Morita, K., Sasaki, H., Furuse, M., and Tsukita, S. (1999) Endothelial claudin: Claudin-5/TMVCF constitutes tight junction strands in endothelial cells. J. Cell Biol. 147:185-194.

Muller, W.A., Weigi, S., Deng, X., and Phillips, D.M. (1993) PECAM is required for transendothelial migration of leukocytes. J. Exp. Med. 178:449-460.

Naik, M.U., Vuppalanchi, D., and Naik, U.P. (2003) Essential role of JAM-1 in bFGF-induced endothelial cell migration. Arteriosc1. Thromb. Vasc. Biol. 23:2165-2171.

Nasdala, I., Wolburg-Buchholz, K., Wolburg, H., Kuhn, A., Ebnet, K., Brachtendorf, G., Samulowitz, U., Kuster, B., Engelhardt, B., Vestweber, D., and Butz, S. (2002) A transmembrane tight junction protein selectively expressed on endothelial cells and platelets. J. Biol. Chem. 277:16294-16303.

Navarro, P., Caveda, L., Breviario, F., Mandoteanu, I., Lampugnani, M., and Dejana, E. (1995) Catenin-dependent and -independent functions of VE-cadherin. J. Biol. Chem. 270:30965-30972.

Navarro, P., Ruco, L., and Dejana, E. (1998) Differential localization of VE- and N-cadherins in human endothelial cells. J. Cell Biol. 140:1475-1484.

Newman, P.J. and Newman, D.K. (2003) Signal transduction pathways mediated by PECAM-1. Arterioscler. Thromb. Vasc. Biol. 23:953-964.
Nieset, J.E., Redfield, A., Jin, F., Knudsen, K., Johnson, K., and Wheelock, M.J. (1997) Characterization of interactions of α–catenin with α- actinin and β-catenin/plakoglobin. J. Cell Sci. 110:1013-1022.
Ohkubo, T. and Ozawa, M. (2001) Tyrosine phosphorylation of p120 in v-Src transfected L cells depends on its association with E-cadherin and reduces adhesion activity. J. Cell Sci. 114:503-512.
Ozaki, H., Ishii, K., Horiuchi, H., Arai, H., Kawamoto, T., Okawa, K., Iwamatsu, A., Kita, T. (1999) Cutting edge: combined treatment of TNF-α and IFN-γ causes redistribution of junctional adhesion molecule in human endothelial cells. J. Immunol. 163:553-557.
Patterson, C.E. and Lum, H. (2001) Update on pulmonary edema: the role and regulation of endothelial barrier function. Endothelium 8:75-105.
Patterson, C.E., Lum, H., Verin, A., Schaphorst, K., and Garcia, J.G.N. (2000) Regulation of endothelial barrier function by the cAMP-dependent protein kinase. Endothel. 7:287-308.
Pearson, C.A., Lamar, P.C., and Prozialeck, W.C. (2003) Effects of cadmium on E-cadherin and VE-cadherin in mouse lung. Life Sci. 72:1303-1320.
Petrache, I., Birukova, A., Ramirez, S., Garcia, J., and Verin, A.D. (2003) The role of the microtubules in TNFα-induced endothelial cell permeability. Amer. J. Respir. Cell Molec. Biol. 28:574-581.
Popov, D. and Simionescu, M. (1997) Alterations of lung structure in experimental diabetes and diabetes associated with hyperlipidemia in hamsters. Eur. Resp. J. 10:1850-1858.
Quinlan, M.P. and Hyatt, J.L. (1999) Establishment of the circumferential actin filament network is a prerequisite for localization of the cadherin-catenin complex in epithelial cells. Cell Growth and Diff. 10:839-854.
Rabiet, M.J.P., Plantier, J., Rival, Y., Genoux, Y., Lampugnani, M., and Dejana, E. (1996) Thrombin-induced increase in endothelial permeability is associated with changes in cell-to-cell junction organization. Arterioscl. Thromb. Vasc. Biol. 16:488-496.
Ratcliffe, M.J., Smales, C., and Staddon, J.M. (1999) Dephosphorylation of the catenins p120 and p100 in endothelial cells in response to inflammatory stimuli. Biochem. J. 338:471-478.
Romero, I.A., Radewicz, K., Jubin, E., Michel, C., Greenwood, J., Couraud, P., and Adamson, P. (2003) Changes in cytoskeletal and tight junctional proteins correlate with decreased permeability induced by dexamethasone in cultured rat brain endothelial cells. Neurosci. Lett. 344:112-116.
Salomon, D., Ayalon, O., Patel-King, R., Hynes, R., and Geiger, B. (1992) Extrajunctional distribution of N-cadherin in cultures human endothelial cells. J. Cell Sci. 102:7-17.
Santoso, S., Sachs, U., Kroll, H., Linder, M., Ruf, A., Preissner, K., and Chavakis, T. (2002) The junctional adhesion molecule 3 (JAM-3) on human platelets is a counterreceptor for the leukocyte integrin Mac-1. J. Exp. Med. 196:679-691.
Satoh, H., Zhong, Y., Isomura, H., Saitoh, M., Enomoto, K., Sawada, N., and Mori, M. (1996) Localization of 7H6 tight junction-associated antigen along the cell border of vascular endothelial cells correlates with paracellular barrier function against ions, large molecules, and cancer cells. Exper. Cell Res. 222:269-274.
Schaphorst, K.L., Pavalko, F., Patterson, C.E., and Garcia, J.G.N. (1997) Thrombin-mediated focal adhesion plaque reorganization in endothelium: Role of protein phosphorylation. Amer. J. Respir. Cell Molec. Biol. 17:443-455.
Schneeberger, E.E. (1982) Structure of intercellular junctions in different segments of the intrapulmonary vasculature. Ann. NY Acad. Sci. 384:54-63.
Schulze, C. and Firth, J.A. (1992) Interendothelial junctions during blood-brain barrier development in the rat: morphological changes at individual tight junctional contacts. Dev. Brain Res. 69:85-95.
Shaw, S.K., Perkins, B., Lim, Y., Liu, Y., Nusrat, A., Schnell, F., Parkos, C., and Luscinskas, F.W. (2001) Reduced expression of JAM and PECAM-1 (CD31) at human vascular endothelial

junctions by cytokines TNF-α plus INF-γ does not reduce leukocyte transmigration under flow. Amer. J. Pathol. 159:2281-2291.

Shay-Salit, A., Shushy, M., Wolfovitz, E., Yahav, H., Brevario, F., Dejana, E., and Resnick, N. (2002) VEGFR2 and the adherens junctions as a mechanical transducer in vascular cells. PNAS USA 99:9462-9467.

Shibamoto, S., Hayakawa, M., Takeuchi, K., Hori, T., Miyazawa, K., Kitamura, N., Johnson, K., Wheelock, M., Matsuyoshi, N., Takeichi, M. (1995) Association of p120, a tyrosine kinase substrate, with E-cadherin/catenin complexes. J. Cell Biol. 128:949-957.

Shivers, R.R., Arthur, F.E., and Bowman, P.D. (1988) Induction of gap junctions and brain endothelium-like tight junctions in cultured bovine endothelial cells. J. Submicr. Cytol. Pathol. 20:1-14.

Simionescu, M. (1980) Ultrastructural organization of the alveolar-capillary unit. Ciba Found. Symp. 78:11-36.

Staddon, J., Ratcliffe, M., Morgan, L., Hirase, T., Smales, C., and Rubin, L. (1997) Protein phosphorylation and the regulation of cell-cell junctions in brain endothelial cells. Heart & Vessels 12:106-109.

Takeichi, M. (1990) Cadherins: a molecular family important in selective cell-cell adhesion. Ann. Rev. Biochem. 59:237-252.

Trocha, S.D., Kevil, C., Mancini, M., and Alexander, J.S. (1999) Organ preservation solutions increase endothelial permeability and promote loss of junctional proteins. Ann. Surg. 230:105-113.

Valiron, O., Chevrier, V., Usson, Y., Breviario, F., Job, D., and Dejana, E. (1996) Desmoplakin expression and organization at endothelial cell-to-cell junctions. J. Cell Sci. 109:2141-2149.

van Nieuw Amerongen, G.P. and van Hinsbergh, V.W. (2002) Targets for pharmacological intervention of endothelial permeability and barrier function. Vascul. Pharmacol. 39:257-272.

Vestweber, D. (2000) Molecular mechanisms control endothelial cell contact. J. Pathol. 190:281-291.

Vincent, P.A., Xiao, K., Buckley, K., and Kowalczyk, A.P. (2004) VE-cadherin: adhesion at arm's length. Amer. J. Physiol. 286:C987-C997.

Volberg, T., Zick, Y., Dror, R., Sabanay, L., Gilon, C., Levitzki, A., and Geiger, B. (1992) The effect of tyrosine-specific protein phosphorylation on adherens-type junctions. EMBO J. 11:1733-1742.

Walker, D.C., MacKenzie, A., and Hosford, S. (1994) The structure of the tricellular region of endothelial tight junctions of pulmonary capillaries analyzed by freeze fracture. Microvasc. Res. 48:259-281.

Walker, D.C., MacKenzie, A., Wiggs, B., Montaner, J., and Hogg, J.C. (1988) Assessment of tight junctions between pulmonary epithelial and endothelial cells. J. Appl. Physiol. 64:2348-2356.

Ward, B.J., Bauman, K.F., and Firth, J.A. (1988) Interendothelial junctions of cardiac capillaries in rats: structure and permeability properties. Cell Tissue Res. 252:57-66.

Winter, M.C., Shasby, S., Ries, D., Shasby, D.M. (2004) Histamine selectively interrupts VE-cadherin adhesion independent of capacitive calcium entry. Amer. J. Physiol. *In press*.

Wong, R.K., Baldwin, A.L., and Heimark, R.L. (1999) Cadherin-5 redistribution at sites of TNF-α and IFN-γ -induced permeability. Amer. J. Physiol. 276:H736-H748.

Wright, M., Aikawa, M., Szeto, W., and Papkoff, J. (1999) Identification of a Wnt-responsive signal transduction pathway in endothelium. Biochem. Biophys. Res. Comm. 263:384-388.

Wright, T.J., Leach, L., Shaw, P.E., Jones, P. (2002) Dynamics of VE-cadherin and β-catenin localisation by VEGF-angiogenesis in human umbilical vein cells. Exp. Cell Res. 280:159-168.

Yanez-Mo, M., Alfranca, A., Cabanas, C., Marazuela, M., Tejedor, R., Ursa, M., Ashman, L., de Landazuri, M., and Sanchez, F. (1998) Regulation of endothelial motility by complexes of tetraspan molecules with α3β1 integrin at endothelial lateral junctions. J. Cell Biol. 141:791-804.

Yeh, H-I, Rothery, S., Dupont, E., Coppen, S., and Severs, N.J. (1998) Individual gap junctions plaques contain multiple connexins in arterial endothelium. Circ. Res. 83:1248-1263.

Yoneda, K. (1982) Regional differences in the intercellular junctions of the alveolar-capillary membrane in the human lung. Amer. Rev. Respir. Dis. 126:893-897.

Zhao, X., Alexander, J., Zhang, S., Zhu, Y., Sieber, N., Aw, T., and Carden, D.L. (2001) Redox regulation of endothelial barrier integrity. Amer. J. Physiol. 281:L879-L886.

Chapter 10
Heterogeneity of Lung Endothelial Cells

Eric Thorin, Ph.D.,[1]* Troy Stevens, Ph.D.,[2] and Carolyn E. Patterson, Ph.D.[3]

[1]Univ. de Montréal, Institut de Cardiologie de Montréal, Montréal, Canada (PQ) H1T 1C8, [2]Univ. Southern Alabama College of Medicine, Mobile, AL 36688 USA, [3]Indiana University School of Medicine and Roudebush VA Medical Center, Indianapolis, IN 46202 USA

CONTENTS:

Introduction

The Origins of Endothelial Heterogeneity
The Intrinsic Hypothesis
The Extrinsic Hypothesis
Development and Aging

Endothelial Phenotypic Functional Differences: Arteries, Capillaries and Veins
Morphologic Heterogeneity
Functional Heterogeneity
Heterogeneity and Endothelial Cell Culture
Growth Heterogeneity of Endothelial Cells

Endothelial Cell Markers: Tissue and Physiologic State Heterogeneity
Markers are Functional Molecules
Commonly Expressed Markers
Markers of the Immune and Inflammatory Response
Markers of Adhesion, Growth, and Mobility

Endothelial Cell Markers: Phenotypic Differences in Arteries, Capillaries, & Veins
Segmental Differences in Coagulation Regulators and Ion Channels
Segmental Heterogeneity of Commonly Expressed Markers
Segmental Heterogeneity of Inducible Markers
Surface Glycoproteins

Flow Stress Effects on Endothelium
Architectural Responses
Nitric Oxide and Oxidant Production Responses

Pulmonary Vasculature *versus* Bronchial Circulation
Anatomical Differences
Functional Similarities and Differences

Endothelium – Variety is Thy Name

Introduction

The heterogeneity of endothelial cells is becoming increasingly more apparent and more appreciated. The phenotypic variation is characterized by differences in function, antigenic composition, metabolic properties, and in their response to growth factors (reviewed by Garlanda and Dejana, 1997; Thorin and Shreeve, 1998; Stevens et al., 2001; Aird, 2003; Gebb and Stevens, 2004). While species heterogeneity has been reported (Majno, 1965; Rhodin, 1968; Graier et al., 1996), of significantly more interest are differences seen within species where EC from assorted organs differ widely. Continuous endothelium is characteristic of brain, retina, and muscular capillaries. While endothelial cells in brain exhibit close junctional apposition to one another and a tight permeability barrier to fluid and solutes (i.e. the blood/brain barrier), fenestrated endothelium is found in endocrine glands and the kidney. The endothelium of the glomeruli is loosely connected and readily permits the trans-endothelial passage of relatively large solutes. Moreover, within an organ, the endothelial cells vary with the size, function, and location of the vessel and can even vary within discrete segments of a single microcirculatory loop (McClure, 1921; Simionescu et al., 1976; Keegan et al., 1982; Thilo-Korner et al., 1983; Fenyves et al., 1993).

As the lungs, receive both right heart – pulmonary circulation and left heart –bronchial circulation, the endothelium of these respective systems might be expected to exhibit diversity. Due to the large capillary surface area of pulmonary circulation and the extremely fine alveolar-capillary membrane, formed by juxtaposition of the thin endothelial cell and alveolar type I epithelial cells, there is a critical necessity for maintenance of barrier integrity in the pulmonary circulation. Therefore, the function of pulmonary endothelium has received considerable investigative attention (Patterson & Lum, 2001; Stevens et al., 2001). In this chapter the primary focus will be directed to heterogeneity of pulmonary endothelium. For many aspects, however, limited data are available and information from other vascular beds will be used for reference.

The Origins of Endothelial Heterogeneity

Normal endothelial cells maintain a delicate balance between growth promotion and inhibition (Nabel, 1991; Scott-Burden and Vanhoutte, 1994), vascular vasoconstriction and vasodilation (Furchgott and Zawadski, 1980; Dohi et al., 1992; Thorin and Atkinson, 1994; Chen et al., 1995; Hare and Colucci, 1995; Jino et al., 1996), blood cell adherence and non adherence (Vane et al., 1990; Kaul et al., 1994), anticoagulation and procoagulation (Gertler and Abbott, 1992; Wu and Thiagarajan, 1996), and exclusion and passage of blood borne solutes and cells. In this way, the endothelium modulates vasomotor tone, regulates vascular structure, controls blood flow, and mediates both inflammatory and immunologic responses. Thus, with different functional demands of site and condition, it is not surprising that there is variation among endothelial cells. Heterogeneity of the endothelium can develop from factors that are inherent in the cell or from clues in their environment (Boegehold, 1998; Stevens et al., 2001; Aird, 2003; Chi et al., 2003; Gebb and Stevens, 2004).

The Intrinsic Hypothesis

According to the intrinsic hypothesis organ-specific phenotypes are predetermined prior to migration from the mesoderm into their final vascular destinations. Studies in embryos demonstrated that precursor cells for endothelium could differentiate according to genetically determined programs. Vasculogenesis involves differentiation of particular splanchnopleuric mesenchymal cells, possibly hemangioblasts, and migration with organ development. Vasculogenesis is thought to account for the microvessels; whereas, developmental angiogenesis involves sprouting from blood islands or pre-existing vessels and tissue invasion. It is thought to give rise to the macrovessel endothelium (Korhonen *et al.*, 1994; Stevens *et al.*, 2001). In the developing lung, these independent macro- and micro-vessel systems converge to provide the intact pulmonary circulation system (deMello and Reid, 2000); although, other studies conclude that the vasculature in the developing lung arises either *via* vasculogenesis (Hall *et al.*, 2002) or *via* angiogenesis (Canis-Parera *et al.*, 2003). The divergent gene and antigen expression by capillary and large vessel endothelium would reflect a disparity in origin (Yamamoto, 1988; Chi *et al.*, 2003; Gebb and Stevens, 2004). Similarly, distinct embryonic origins for coronary endothelium and endocardial endothelium were shown in studies of chick embryo development (Mikawa and Fischman, 1992). Determination as to whether vascular tubes that form from endothelial islands develop into arteries or into veins is a highly important event in definition of the circulatory system.

It had generally held that the phenotypic difference between artery and vein endothelium was due to extrinsic factors, such as dissimilarities in subsequent pressures and flow, but now it is recognized that some markers of arterial/venous heterogeneity, including members of the VEGF receptor, ephrin, notch, gridlock and Bmx families, are clearly present prior to initiation of circulation and maturation of the vascular beds (Wang *et al.*, 1998; Adams *et al.*, 1999; Gerety *et al.*, 1999; Ekman *et al.*, 2000; Zhong *et al.*, 2000; Herzog *et al.*, 2001; Lawson *et al.*, 2001; Rajantie *et al.*, 2001; Chi *et al.*, 2003). Gene expression not only varies between large vessels and microvessels, but also between artery and vein endothelium (Chi *et al.*, 2003). For instance, the Hey2 transcription factor and CD44 are selectively expressed in artery-derived endothelium. Venous endothelial cells expressed genes typical of left/right developmental asymmetry, suggesting that vascular development is coordinated with body plan (Chi *et al.*, 2003). Cells from various organs and various sized vessels, cultured under the same conditions, maintained their unique characteristics, indicating intrinsic differences accounted for their heterogeneity (Craig *et al.*, 1998; Chi *et al.*, 2003 King *et al.*, 2004). Gene expression in the lung microvessel endothelium showed differences in global gene expression from microvessel endothelium from other organs, such as skin and intestine (Chi *et al.*, 2003). Thus, there is strong evidence emerging that intrinsic phenotypic distinction is maintained, even in culture. On the other hand, there are reports of endothelial cells being induced to differentiate into smooth muscle and cardiac muscle cells indicating a possible plasticity of phenotype expression under the right circumstances (Condorelli *et al.*, 2001; Frid *et al.*, 2002).

The Extrinsic Hypothesis

According to the extrinsic hypothesis, endothelial phenotypic heterogeneity is determined by signals arising from the interactions of the endothelial cells with local tissue derived-mediators and matrix proteins and from connections formed with its non-endothelial neighboring cells. Dynamic changes in endothelial cells that depend on the environmental factors have recently been reviewed (Bussolino *et al.*, 2001). Kidney endothelium and brain endothelium are derived respectively from nephrogenic mesenchyme and neuroectoderm and express very divergent phenotypes, suggesting an important influence of the local environment on development of the fenestrated renal endothelium and the tight junction, blood/brain barrier (Noden, 1990). Indeed, in some cases the local environment has been shown to directly affect endothelial phenotype. When endothelial cells derived from the aorta are cultured on a protein matrix derived from lung endothelium, they begin to express ICAM-1 like the pulmonary cells (Augustin-Voss *et al.*, 1991). When brain endothelial cells are placed in culture they begin to lose the tight junctions typical of *in vivo* brain endothelium; but if they are co-cultured with supportive astrocytes, they retain this unique property (Janzer and Raff, 1987; Laterra *et al.*, 1990). In an opposite functional differentiation, endothelial cells from the adrenal cortex begin to express glomerular-like fenestrae when cultured on a matrix derived from a kidney cell-line (Milici *et al.*, 1985). Cytokines and growth factors are well known to alter endothelial cell gene expression and phenotypes (Mantovani *et al.*, 1992).

Development and Aging

Probably, both hypotheses come into play and the combination of intrinsic and extrinsic factors ultimately govern phenotypic heterogeneity. Likely, the epigenetic cell memory influences the nature and extent of the responsiveness to the environment. The VEGF Receptor-1, Flt-1, is present in the associated mesenchymal clusters as early as the 11th day of gestation in the rat when the evagination of the foregut epithelium begins to form the lung (Gebb and Shannon, 2000). This early appearance of "endothelial" receptors suggests pre-programming. However, when these mesenchymal cells were isolated and cultured, they began to lose the Flt-1. If they were cultured in the presence of epithelial cells, they maintained this expression, indicating communication and extrinsic influence of the epithelium on development of the mesenchyme precursors into pulmonary endothelium. During angiogenesis and invasion of tissue by the new endothelium, the altered basement membrane and adhesion molecule profile may strongly influence endothelial cells sensitivity to factors. For example, the role of RhoA and Rac1 in regulating the formation and the dynamics of cell-to-matrix and cell-to-cell adhesive interactions is well established (Fukata and Kaibuchi, 2001; Sahai and Marshall, 2002; Sahai and Marshall, 2003). It has been shown that the different phases of the morphogenetic process according to which human capillary endothelial cells self-assembly into a vascular network require distinct spatial and temporal activation of Rac1 and RhoA (Cascone *et al.*, 2003). Moreover, during the different phases of this vascular morphogenetic process, there is a need for dynamic cell-to-cell adhesive contacts involving tight junctions (Bazzoni *et al.*, 1999; Bussolino *et al.*, 2001; Wang *et al.*, 2002).

As a final note, maturation and aging of endothelium affect its responsiveness and may alter marker distribution and other factors important in functional heterogeneity. Altered expression patterns of von Willebrand factor and CD34 were noted with aging in human tissues (Müller et al., 2002c). Functional responses to substance P are present only in adult pig resistance arteries, whereas bradykinin dilated all types of arteries at all ages (Boels et al., 1999; Boels et al., 2001). Wound healing repair (Gennaro et al., 2003) and angiogenesis (Rivard et al., 1999), both dependent on endothelial responses, are processes that are slowed in aging. Is there a heterogeneous change in function? This question remains unanswered.

Endothelial Phenotypic Functional Differences: Arteries, Capillaries and Veins

Morphologic Heterogeneity

There is a paucity of data focusing specifically on segmental heterogeneity of the endothelium in lung. Thus, common differences in circulatory level regardless of tissue origin will first be considered. *In situ*, anatomical and junctional heterogeneity between microvascular and large vessel endothelium, as well as among microvascular endothelia of diverse vascular beds, has been reported. Some of the earliest detailed work defining segmental divergence in endothelium was obtained with the use of freeze-cleavage and electron microscopy. Whereas capillaries were found to have an elaborate system of tight junctions, they had no communicating (gap) junctions (Simionescu et al., 1975; Simionescu et al., 1976). Venule endothelium had loosely organized, discontinuous occlusive junctions and no or sparse, small gap junctions, depending on organ source. Arterial and arteriole endothelium possessed both occluding and communicating gap junctional structures. In general, even in the lung, arterial endothelial cells were found to be thicker than those of capillaries and veins (Simionescu and Simionescu, 1977). Endothelial cells of systemic arteries in straight segments with laminar flow are typically orientated longitudinally in the direction of blood flow (Figure 1). The cells lining large vessels are polygonal; whereas, those lining the resistance microvessels are more flattened and elongated, although both cell types have been found to co-exist (Arey, 1963; Cornhill et al., 1980; Harris and Heath, 1986). In contrast to endothelium in the aorta, with its higher pressures and flow stresses, the endothelial cells in the pulmonary trunk are more rounded, although they are roughly the same size as arterial endothelium, as shown in Figure 1. When pulmonary hypertension was induced by 4 weeks of hypoxia, the pavement pattern resembled the aortic pattern: thus, demonstrating response to local environment (Harris and Heath, 1986). Also note that in the normal pulmonary and systemic circulation the endothelium of large veins cover more than twice the area as those in the artery (Silkworth et al., 1975; Harris and Heath, 1986). Generally capillary endothelium and post-capillary venules are rounder, flatter, and have more plasmalemmal vesicles (paralleled by more plasma membrane caveolae) than endothelium in large arteries (Simionescu and Simionescu, 1977). In pulmonary capillaries, the plasmalemma and endoplasmic reticulum are separated by ~ 87nm; whereas in pulmonary arteries, they are separated by only 8 nm (King et al., 2004). Consistent with these differences in architecture between arterial, microvessel, and venous endothelium, the external luminal surface glycocalyx varies, as reflected by differences in binding to specific lectins (Schnitzer et al., 1994; Brouland et al., 1999; King et al., 2004) (see section on markers below).

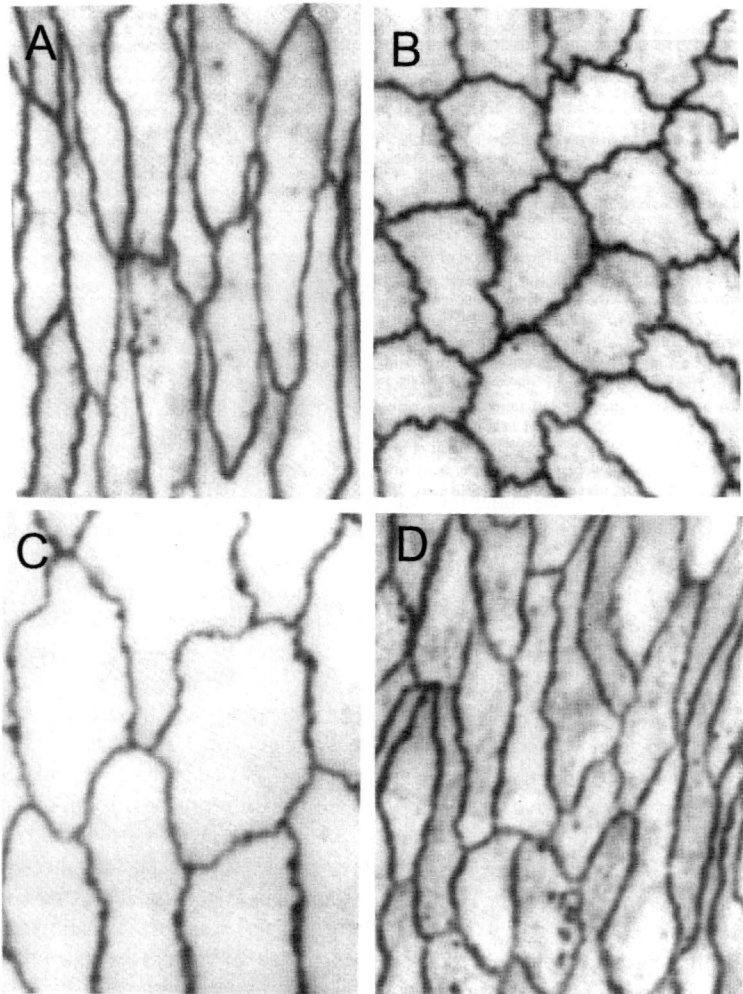

Figure 1. Pulmonary artery and venous endothelial pavement. Internal lining of large vessels in the rat were silver stained to define the shape patterns. **A.** Aorta – Endothelia are elongated in the direction of flow and smaller in surface area (476 µm^2), compared to rounded endothelium of the vena cave (679 µm^2, not shown). **B.** Pulmonary trunk - The pulmonary arterial endothelial cells are more polygonal and uniform than in systemic arteries, but similar in size (426 µm^2). **C.** Pulmonary vein - The endothelia are rounded or polygonal but are considerably larger (1022 µm^2) than the pulmonary artery endothelia. **D.** Pulmonary trunk with hypoxia-induced hypertension – After 4 weeks of maintenance at ½ atmosphere, resulting in global pulmonary hypoxic-vasoconstriction and pulmonary hypertension, the pulmonary artery and its endothelium had remodeled to resemble the systemic arterial pattern. From *The Human Pulmonary Circulation, 3rd. eds,. P. Harris and D. Heath, p60-61*, by permission of Churchill/Livingston/Elsevier, Amsterdam.

Functional Heterogeneity

Functional studies revealed heterogeneity in large lung artery and microvessel release of vasoactive mediators controlling smooth muscle and pericyte contraction (Boegehold, 1998). For instance, the bradykinin response of pulmonary conducting and resistance vessels is dissimilar; bradykinin-induced dilation is NO-dependent in conductance size arteries and veins whereas most of the dilation depends on EDHF in resistance size pulmonary vessels (Aschner et al., 2002). In addition, there are differences between lung arteries and veins. The efficacy of ACh-induced dilation is greater in veins than arteries in pig lungs (Arrigoni et al., 1999). Certainly, disease conditions, such as diabetes, alter endothelial mediated-smooth muscle relaxation/contraction factor balance (and other endothelial functions and characteristics), but segmental differences are not well studied.

Functional heterogeneity exists for barrier permeability as well. Venous endothelial cell heterogeneity in the lung has been rarely studied, but consistent with the observation of looser organization of junctions in venules in other organs above, many studies have shown that the post-capillary venules are the most permeable vessels in the lung. Lung microvascular endothelium has been consistently shown to be less permeable than arterial or venous endothelium (Schnitzer et al., 1994; Kelly et al., 1998; Chetham et al., 1999; Parker and Yoshikawa, 2002). Despite the tighter apposition of endothelium and lower hydraulic conductivity in the capillary bed, 40% of basal lung fluid flux occurs in the capillaries due to the large surface area (Parker and Yoshikawa, 2002). Endothelia in different organs and in different segments have distinctive responses to histamine, ranging from permeability to tightening of the barrier (Ikeda et al., 1999). This segmental heterogeneity is also true for Ca^{2+} mobilization in endothelial cells from pulmonary artery and pulmonary microvascular endothelium (Cioffi et al., 2002; Cioffi et al., 2003). Activation of store Ca^{2+} entry in response to thapsigargin, an inhibitor of the activity of the sarcoplasmic reticulum ATPase, induced perivascular edema in pre- and post-capillary vessels but not in the micro-vasculature (Chetham et al., 1999; Creighton et al., 2003). This was associated with the appearance of gaps and alteration of the barrier function in the macrovascular pulmonary circulation only. In addition, an equivalent rise in intracellular Ca^{2+} concentration induced by ionomycin led to increased permeability in pulmonary artery endothelial cells but not pulmonary microvascular endothelial cells (Kelly et al., 1998), suggesting a dissociation between intracellular Ca^{2+} and barrier function in pulmonary microvascular endothelium. Likewise, permeability responses to 200 nM thrombin were attenuated in microvessel pulmonary endothelium compared to pulmonary artery endothelium (unpublished data, C.E. Patterson) and the permeability response to higher 1 µM thrombin was more rapidly reversed in microvascular cells (Cioffi et al., 2003). These differences are related to distinctive cAMP regulation in pulmonary macro- and micro-vessels (Moore et al., 1998; Stevens et al., 1999), discussed in detail in Chapter 5.

Differences between cultured pulmonary artery, microvascular, and vein endothelium in oxidant susceptibility and damage repair were detected by decreased hybridization intensity of mitochondrial DNA, while nuclear DNA was not damaged (Grishko et al., 2001). The venous cells were the most susceptible to oxidant injury and the microvascular endothelial cells were the most resistant. When the oxidants were generated externally by exogenous xanthine oxidase, the arterial and microvascular cells were able to repair their DNA damage more quickly than the venous endothelial cells, as might be anticipated from their relative susceptibility. However, similar sensitivity to oxidant damage was found when the oxidants were generated internally in the mitochondria with menadione; yet, the venous and microvascular cells repair was much quicker than repair in the arterial endothelium.

Similarly, functional differences in endothelial interactions with leukocytes have been observed. Recently, Wang and co-workers (2002) reported that neutrophil adherence to tumor necrosis factor-α-treated human pulmonary microvascular endothelial cells induced the production of reactive oxygen species, which increased p38 MAP kinase activity. This led to profound cytoskeletal changes. This neutrophil response did not occur in pulmonary artery endothelia. Pulmonary microvascular endothelium is the site of neutrophil emigration and edema formation *in vivo*; thus expression patterns for genes involved with blood cell binding and trafficking were predictably higher in microvascular endothelium (Chi *et al.*, 2003). Likewise, microvascular endothelia expressed genes involved with lipid metabolism and transport, consistent with the function of capillaries in fuel provision to tissues.

Macrovessel and microvessel endothelium expressed different genes involved in matrix biosynthesis and remodeling (Chi *et al.*, 2003). Large vessel endothelium expressed genes involved with the dense connective tissues, characteristic of arteries, such as fibronectin, collagen 5α1, collagen 5α2, and osteonectin. Whereas microvascular endothelial cells expressed genes for basement membrane proteins, such as laminin, collagen 4α1, collagen 4α2, and collagen 4α-binding protein, and for proteins, such as CD36, α1 integrin, α4 integrin, α9 integrin, and β4 integrin, which interact with basement membrane proteins. Expression of these genes is unchanged by culture and passage (10-16 generations) on plastic with identical growth media, indicating dedication to phenotype.

Endothelial cells differ in their pro- and anti-fibrinolytic properties. Whereas human pulmonary artery and microvascular endothelium and umbilical vein endothelium expressed comparable levels of mRNA and protein for plasminogen activator inhibitor-1 (PAI-1), the pulmonary artery endothelium expressed much higher levels of PAI-2 and tissue plasminogen activator (tPA) (Muth *et al.*, 2004). In contrast, the lung microvascular endothelium expressed the highest levels of urokinase-type plasminogen activator (u-PA). Challenge of both macro- and microvascular pulmonary endothelium with TNFα or lipopolysaccharide resulted in upregulation of PAI-1 and PAI-2 and down-regulation of t-PA in, which might favor local fibrin deposition (Muth *et al.*, 2004).

Heterogeneity and Endothelial Cell Culture

Aside from *in vivo* studies such as those reported above, it is difficult to obtain reliable data from cultured endothelium since standard isolation techniques have not usually permitted the clear distinction between capillary, post-capillary venular, and venous endothelial cells. Therefore, data obtained for cultured lung microvascular endothelium generally represent a mix population. Cultured endothelial cells isolated from large conductance vessels from diverse tissues look similar under a constant culture environment (Craig *et al.*, 1998). This is not true, however, for microvascular endothelial cells and at least five phenotypes have been observed: cobblestone, polymorphic, spindle shape, round, and phase dense appearance in phase contrast microscopy (Spanel-Borowski and Fenyves, 1994). While these have been considered culture-dependent morphological artifacts, some authors propose that such heteromorphology may represent innate differences maintained in culture (Rone and

Goodman, 1987). Most cells differentiate to some degree throughout culture and the culture conditions, even at the first passage (Kawanami, 1997). The potential for differentiation is even greater when cells are transformed. Thus, studies with cultured cells must always be interpreted with this culture-induced heterogeneity in mind, regardless of fidelity of origin.

Growth Heterogeneity of Endothelial Cells

Variations in cell size have been observed among cultured large vessel endothelia within the same species. Ohbayashi and co-workers (1994) reported that endothelium in pig aorta were 30% smaller than coronary artery cells. Indirect evidence also suggests that endothelial cells isolated from human cerebral arteries are larger than those isolated from peripheral vessels of the same size (Thorin *et al.*, 1997a). Rat lung pulmonary microvascular and arterial endothelium were similar in size when grown under the same conditions (King *et al.*, 2004)

Endothelial growth is another parameter that may vary in culture. Endocardial endothelial cells were reported to grow at 25% higher density than vascular endothelial cells in association with an increase in cellular overlaps and interdigitations (Mebazaa *et al.*, 1995). Furthermore, cerebral endothelium has a longer doubling time than cells isolated from peripheral vessels (Thorin *et al.*, 1997a). Data are not available to directly compare growth characteristics of endothelium from lung large arteries and veins, but microvascular endothelium from rat lung grew faster than pulmonary artery endothelium under the same conditions (King *et al.*, 2004). Moreover, beyond the typical 2 day lag period after seeding in culture, the microvascular cells continued to grow when serum was reduced to 0.1%, whereas pulmonary artery endothelium went into growth arrest, indicating intrinsic differences in growth factor dependency and proliferative ability (King *et al.*, 2004).

Although *in situ* endothelium is typically in a quiescent state in a normal and healthy adult, there is evidence that in some vessels endothelial cells can become plastic with respect to proliferative capacity and differentiation. For example, proliferation of venule and capillary endothelium is essential to vascularization of transplanted limbs and tumors (Risau, 1995). However, the endothelia of larger vessels do not participate in neovascularization. This may be due, at least in part, to variabilities in the expression of vascular endothelial cell growth factor (VEGF) receptors, which are rapidly induced in venule and capillary endothelium but not arterial endothelial cells (for review, see Ferrara *et al.*, 1992).

Endothelial Cell Markers: Tissue and Physiologic State Heterogeneity

Markers are Functional Molecules

Antigens expressed by different endothelia provide compelling evidence for endothelial cell heterogeneity (Garlanda and Dejana, 1997; Zetter, 1988). There is heterogeneity of marker expression, depending on: 1) the marker, 2) the site - as discussed later in the section on phenotypic variations in arteries, capillaries, and veins, and 3) the cytokine/growth factor activation. Nevertheless, taking the segmental differences into consideration, some markers are commonly expressed by normal, confluent endothelium. Importantly, these markers are functional molecules and their differential expression reflects and determines variations in endothelial cell function. Some of these are: 1) the endothelial specific Weibel Palade bodies (containing von Willebrand Factor and variable levels of vWF-associated proteins, such as coagulation factor VIII, p-selectin, tissue plasminogen activator, and interleukin-8), 2) thrombomodulin (CD141), 3) VE-cadherin, 4) angiotensin converting enzyme (ACE, CD143, also expressed by some leukocytes and epithelial cells), 5) Type I scavenger for acetylated-LDL uptake (CD 36, shared with some other cells), 6) platelet endothelial cell adhesion molecule (CD31, PECAM; also expressed by platelets and some other blood cells), 7) CD34 (specific to endothelial cells and precursor cells), and 8) Tie-1 & Tie-2 receptors associated with angiopoietin. Although, commonly found in certain quiescent endothelium, the amounts and differential expression of these markers can be altered by activation and inducing stimuli. Other markers are less common is quiescent endothelium and their expression is considered "inducible", such as: 1) ICAM (CD54), 2) VCAM (CD 106), 3) ELAM-1 (E-selectin), and 4) VEGF Receptor 1 (Flt-1) (Garlanda and Dejana, 1997; Mutin et al., 1997). Markers and vascular distribution (discussed in the next section) are summarized in Table 1 and Table 2. Finally, other markers are thought to be more or less specific for particular endothelial cell types, such as VEGF Receptor-4 (Glt-4, for lymphatic endothelium), LuECAM (for pleural, perivenous, and peribronchial capillaries and venular endothelium), or P-glycoprotein (mdr 1a, for brain endothelium) (Zhu and Pauli, 1993; Garlanda and Dejana, 1997).

Commonly Expressed Markers

Distinction between basal and inducible expression by certain normal confluent endothelium is not clear-cut and there is lack of agreement on particular classifications. Markers commonly expressed may undergo upregulation under particular circumstances; while in some tissues in some species other inducible markers may be present without apparent activating factors. Some markers may be normal in endothelia at certain sites, but not others. Release of some markers from endothelia into plasma (vWF, thrombomodulin, ICAM-1) is taken to indicate endothelial activation, inflammation, and damage. Patterns also change during embyogenesis and fetal development. **VEGF receptor-1** is highly expressed and essential during embyogenesis and fetal development but switched off at vasculature maturity (Risau, 1994). Then it is induced during repair of endothelial wounds or tumor growth (Dvorak et al., 1995). **Tie** - In contrast to the decrease in VEGFR-1 expression after vessel maturation, endothelial-specific angiopoeitin-linked receptors are further increased in some organs after gestation (especially lung) and associated with tightening of the endothelial barrier. Interestingly, Tie-1 is expressed in adult lung at levels 10-100 times that found in other tissues (Taichman et al., 2003). **vWF** –von Willebrand factor is a complex glycoprotein expressed under basal conditions that binds and stabilized factor VIII of the thrombogenic cascade and ushers other proteins, such as tissue plasminogen activator and p-selectin, into storage in the Weibel Palade bodies for potential secretion from endothelium. Additionally and to a lesser extent, smaller dimeric forms are localized to endoplasmic reticulum and extracellularly to endothelial basal lamina matrix (Jones and van Rij, 2000). *In vivo,* shear stress, inflammation, irradiation, fibrosis, pulmonary hypertension, and angiogenesis increase vWF expression (Meyrick and Reid, 1978; Kasper et al., 1996; Sun et al., 2000; Jin et al., 2001; Müller et al., 2002d). TNF decreases expression of vWF and the Weibel Palade bodies in cultured endothelium (Sun et al., 2000).

Table 1: Endothelial functional and adhesion marker expression heterogeneity

Antigen	Segment			Expression conditions	Expression by other cells
	Art	MV	Vein		
ACE, CD143	+	+	+	Systemic capillary expression is poor while pulmonary capillary expression is normally strong	Mesothelium, leukocytes, & epithelial cells
CD34, L-selectin ligand	+	+	+	Decreases in MV with age, while increasing in large vessels; trans locates from luminal to sproutin areas of the abluminal surface in tumor EC	Specific to EC and precursor cells, and hematopoietic cells
ELAM-1, E-selectin	+	-	++	Inducible; not in capillaries, but small arterioles and venules have strongest expression	
ICAM-1, CD54	+	+	+	Cytokine inducible; controversial distribution	Epithelium, intimal smooth muscle, fibroblast, leukocytes
Lu-ECAM-1	-	+	+	Mainly in pleural, perivenous & peribronchial venules	EC specific & related to melanoma metastases to lung
MHC-I	-	+	-	Inducible	
MHC-II	-	+	-	Inducible; expressed in coronary arteries	
CD36, LDL scavenger	-	+	-	Inducible	
PECAM, CD 31	+	+	+	Basally expressed, moves from surface to lateral edge with confluency	Platelets, leukocytes, mast cells, & megakaryocytes
Muc18/S-Endo-1, CD146	+	+	+	Basally expressed in junctions	
P-selectin, CD62	+	-	++	Inducible; Absent in capillaries, but present in arterioles and venules	Platelets
VCAM, CD106	+	-	+	Cytokine inducible; absent in capillaries, but present in arterioles and venules	Mesothelium, macrophages
VE Cadherin	+	++	+	Expressed in confluent monolayers	Some macrophages
Von Willebrand Factor, vWF	+	-	++	Some in arterioles and venules, but poor in capillaries; increases with activation, increases in all segments with age	Megakaryocytes and platelets, although most platelet carried vWF is EC-derived

1) Markers are listed with common, equivalent names. 2) MV may represent a mixture of capillary, pre- and post-capillary microvascular endothelia. 3) + indicates usual detection and – indicates detection is not usual, irregular, or sparse, but may be induced. 4) for a more extensive listing see Garlanda and Dejana, 1997 & Mutin, 1997.

ACE - Angiotensin converting enzyme is normally expressed and located on the luminal surface of endothelium in caveolae. It can also be found on leukocytes, epithelium, and smooth muscle, particularly when upregulated. ACE has received considerable research attention due to the pathogenic role of angiotensin II in hypertension, atherosclerosis, and diabetes and the effectiveness of ACE inhibitors in ameliorating endothelial dysfunction and attenuating progression and instability of atherosclerotic lesions. **CD34** – Although expressed in all pulmonary vascular endothelia, expression was increased in lungs specimens from patients with pulmonary hypertension (Müller et al., 2002b). **PECAM** - PECAM is involved with platelets and leukocyte adhesion to endothelium, cadherin regulation and localization, inter-endothelial junctions, and cell migration (Newman, 1997). It is expressed by platelets, leukocytes, and endothelium, where it is consistently found from large vessels to capillaries in both systemic and pulmonary vessels (Danilov et al., 2001). While present on the advancing surface of migrating endothelium, it is upregulated and localized to lateral borders in confluent, quiescent endothelium (Raychaudry et al., 2001). There was homogeneous PECAM expression in all pulmonary vessels in normal human lung and in lungs of patients who died from ARDS (respiratory failure with loss of alveolar/capillary barrier function) (Müller et al., 2002a). Similarly, PECAM was not altered by pulmonary hypertension (Müller et al, 2002 b). **JAM** – Junctional adhesion molecules (JAM-1, JAM-2, and JAM-3) are associated with endothelial and epithelial tight junctions and involved in leukocyte transmigration with unique tissue distribution and functional properties (Aurrand-Lions et al., 2003). Whereas JAM-1 is abundant in lung, heart, and liver and involved in re-sealing of junctions, its expression in lymph nodes and spleen is low (Aurrand-Lions et al., 2001). In contrast, mouse-JAM-2 (equivalent to human JAM-3) expression is low in heart, lower still in lung and liver, but high in the lymph nodes. Moreover, JAM-2 expression in MDCK cells, but not JAM-1 expression, increased their permeability to macromolecules five fold (Aurrand-Lions et al., 2001).

Markers of the Immune and Inflammatory Response

Endothelial cells actively contribute to the development of local immune and inflammatory responses including the initiation of coagulation by von Willebrand factor, participation as antigen-presenting cells, and modulation of leukocyte-vessel wall adhesion. They express divergent patterns of surface immunity/inflammation antigens in different anatomic compartments of the liver (Fukuda et al., 1986; Nagura et al., 1986), lung (Yamamoto et al., 1988) and kidney (Tsutomu et al., 1989; Fleming and Jones, 1989). Thus, they share various phenotypic and functional properties with antigen-presenting cells of the macrophage/monocyte lineage, including expression and induction of human class II antigens i.e. HLA-DR (Hirschberg et al., 1979, Hirschberg et al., 1980; Pober and Gimbrone, 1982; Yamamoto et al., 1988; Rose et al., 1990), induction of Fc and complement receptors (Ryan et al., 1981), expression of monocyte cell antigens (Knowles et al., 1984; Cerilli et al., 1985), antigen-induced T-cell proliferation (Hirschberg et al., 1980; Wagner et al., 1985), and presentation of peptides to primed T cells (Nunez et al., 1983). Many of these antigen-presenting cell-like properties (Major Histocompatibility Complex-MHC I, MHC II, and CD36) are mostly confined *in vivo* to capillaries and are not present in large vessels, indicating that the capillary endothelium is more efficient at antigen presentation on one hand, but are more susceptible to immune attack on the other. In contrast to the usual pattern, normal coronary arteries are unique among the large vessels in that they expresses both MHC II and VCAM (Page et al., 1992). This antigen-presenting role is particularly important for the alveolar capillary endothelium, as the lung is a major site of exposure to xenobiotics and infectious agents. In fact, alveolar capillaries from human lung express HLA-DR, CD36, and CD116, in common with monocytes/macrophages, and frequently were positive for IL-1 but only weakly stained for vWF (Yamamoto et al., 1988). Endothelial cells can be stimulated to secrete immunoregulatory factors such as interleukin-1 (Miossec et al., 1986) and interleukin-6 (Sironi et al., 1989) that can further alter their expression of antigen-presenting and leukocyte binding molecules in an autocrine fashion.

Modulation of leukocyte-vessel wall adhesion is thought to be mediated by basal or cytokine-upregulated expression of some or all of the following vascular endothelial molecules: ICAM-1 (Van-Epps et al., 1989), VCAM-1 (Osborn et al., 1989), endothelial leukocyte adhesion molecule-1 (ELAM-1; E-selectin) (Bevilaqua et al., 1987), and PECAM-1 (CD31) (Parums et al., 1990). A heterogeneous distribution of vascular adhesion molecule was observed in vascular endothelium from diverse regions of the human cardiovascular system such as the aorta, pulmonary and coronary arteries and umbilical cord (Page et al., 1992; see discussion in next section). Staining for markers, PECAM-1 and vWF, was not noticeably different in human lung vessels from with and without septic shock and ARDS. However, the leukocyte adhesion molecules, ICAM (as shown in Figure 2), ELAM (shown in Figure 3), and VCAM, were dramatically upregulated, (Müller et al., 2002a). Similarly, ICAM upregulation was reported in rat lung after bleomycin injury (Sato et al., 2000). Isolated pulmonary microvascular and arterial endothelial cells, both responded to TNF with upregulation of ICAM (Wang et al., 2002). See discussion of segmental heterogeneity in next section and Chapter 11 for more detailed discussion of blood cell adhesion.

Markers of Adhesion, Growth, and Mobility

Hyaluronic acid (HA) keeps tissues hydrated, maintains osmotic balance, and regulates cellular processes such as adhesion, migration and proliferation (Laurent and Fraser, 1992). Endothelial surface HA-receptors or HA-binding proteins mediate the biological function of HA. HA binds to human pulmonary artery endothelium and pulmonary microvascular endothelium with similar affinity and Bmax. Thus, HA induces adhesion and motility, although with better efficacy in pulmonary microvascular endothelial cells (Lokeshwar and Selzed, 2000). In contrast, human umbilical vein endothelial cells neither bond nor responded to HA. Endothelial cell adhesion molecules and cell-to-cell junctions (Dejana et al., 1995) are important regulators of cell permeability (Cattelino et al., 2003; Lampugnani et al., 2003). As discussed in Chapters 2 and 9, VE cadherin is an endothelial specific component of the adherens junctions between adjacent cells and is normally expressed in confluent cells. However, it can also be expressed by precursor monocytes, which are induced to differentiate toward an endothelial phenotype (Allport et al., 2000). When the endothelial cells are activated or migrating, the VE cadherin is lost from the endothelial plasma membrane.

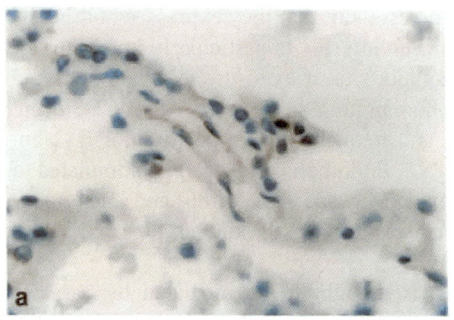

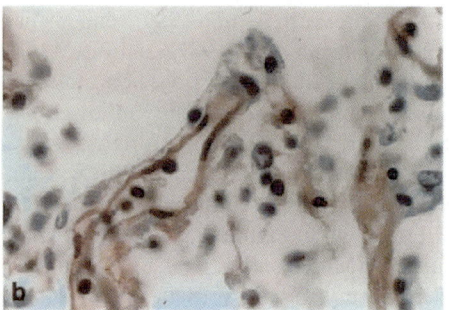

Figure 2. ICAM expression in human lung microvessels. a. Homogeneous mild positive immunostaining (brown, avidin-biotin-peroxide system) was found for the leukocyte adhesion molecule, ICAM, on endothelium in all the microvessels of normal human lung (H&E counter stained). **b.** Dramatic upregulation of ICAM expression was found by strong staining in the lung microvessels from patients who had died from septic shock, with clinically defined ARDS. From *Müller, A.M., Cronen, C., Müller, K.M., and Kirkpatrick, C.J. (2002) Heterogeneous expression of cell adhesion molecules by endothelial cells in ARDS. J. Pathol. 198:270-275* by permission of Dr. Annette Muller. Copyright Pathological Society of Great Britain and Ireland. Reproduced with permission granted by John Wiley & Sons Ltd. on behalf of Path Soc.

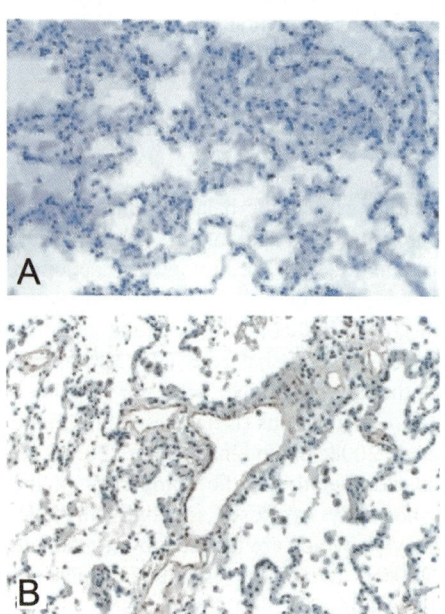

Figure 3. ELAM upregulation in ARDS in human lung. As in Figure 2, lung tissue from 5 controls and 5 patients who died from sepsis and ARDS were obtained and immunostained with H&E counterstain. **A.** No staining for ELAM (E-selectin) was observed in any of the lung microvessels in normal human lung. **B.** Moderate to strong staining for ELAM was found in arteries, veins, and pre- and post-capillary microvessels in the lungs from the septic shock patients. However, only random, sparse staining was observed in alveolar capillaries. From *Müller, A.M., Cronen, C., Müller, K.M., and Kirkpatrick, C.J. (2002) Heterogeneous expression of cell adhesion molecules by endothelial cells in ARDS. J. Pathol. 198:270-275* by kind permission of Dr. Muller Copyright Pathological Society of Great Britain and Ireland. Reproduced with permission granted by John Wiley & Sons Ltd. on behalf of the Pathology Society.

Table 2. Endothelial receptor marker expression

Antigen	Segment			Expression conditions	Expression by other cells
	Art	MV	Vein		
Ac-LDL R-1	+	+	+	Expressed by normal endothelium	Mesothelium, macrophages, fibroblast, & smooth muscle
Tie-1	+	+		Late embryonic and fetal development; inducible in wound healing angiogenesis; highest in adult lung	Hematopoietic cells; platelets
Tie-2		+		Embryonic and fetal development and angiogenesis; normally expressed	
Thrombomodulin, CD141	-	+	-	Consistent expresssion in normal microvessels	
VEGF R-1, Flt-1	+	+	+	Embryonic and fetal development; inducible	Epithelium, macrophages
VEGF R-2, KDR, Flk-1	+	+	+	Embryonic and fetal development; Inducible	Clara cells, mast cells
VEGF R-4				Embryonic and fetal development	Lymph endothelium only

Notes: 1) Markers are listed with their several common, equivalent names 2) MV reports a mixture of data from capillary and microvascular endothelium; 3) + indicates usual reported detection and – indicates detection is not usual or is irregular and sparse, however in special condition the endothelium may be induced to express the marker; blanks indicate specific information is not generally available or is not resolved 4) Embryonic and fetal development expression at those times is a normal finding, but these markers may be increased in special conditions of angiogenesis, such as wound healing, and in tumor angiogenesis.

Recently, gene expression by normal mouse lung endothelium was studied by rapid isolation of the cells (Favre *et al.*, 2003). This was accomplished by fluorescence activated cell sorting (FACS) of endothelium, which had taken up rhodamine-labeled liposomes. They were separated from contaminating leukocytes (labeled with leukocyte-specific fluorescein CD18). As might be expected, they expressed RNA for VE-cadherin, claudin 5, connexin 37, CD31, and CD34. Interestingly, these normal endothelium also expressed message for the VEGF, VEGFR-1, VEGFR-2, angiopoietin-2, Tie-1, Tie-2, and Edg1 receptor genes, that are usually associated with embryonic, growing blood vessels, and repair of endothelial wounds. Of course, gene expression and protein expression do not always go hand in hand; still these cells appear ready to translate if activated or presented with the ligand. Hypoxia and VEGF treatment induced expression of Tie-1 protein in bovine aortic endothelium (McCarthy *et al.*, 1998). Tie-2 was expressed at high levels and, during wound repair in mouse skin, was increased only slightly; while Tie-1 expression increased dramatically (Kampfer *et al.*, 2001).

The increase in Tie-1 with injury was blunted and sustained in diabetic mice, while Tie-2 expression was abrogated. These data are just one example of the complex and multifactor alterations in endothelium with presentation of cytokines, growth factors, mediators, reactive oxygen species, and other stressors or pathologic changes. These effects lead to a dynamic heterogeneity in the endothelium that may be important in proper adaptation of function, but may also contribute to ongoing dysfunction.

Extensive investigation of angiogenesis in tumors and the evolving development of anti-cancer therapies based on targeting circulation supporting tumor growth has been reviewed (Carmeliet and Jain, 2000) and here the primary focus on barrier function will be maintained. However, we will briefly note that in primary lung adenocarcinoma, the capillary endothelium lose their normal expression of thrombomodulin and gained expression of vWF (Jin et al., 2001). They also gained expression of VEGF165 and VEGFR-2 (KDR), consistent with new vessel growth paralleling and supporting the tumor growth. The neoplastic cells meanwhile increased VEGF expression. The new capillaries were fenestrated (and leaky), rather than continuous like endothelium of the surrounding lung tissue. Also, human umbilical vein endothelial cells express antigens or CD34 antigen relocation to the abluminal surface typical of tumor-induced blood vessels (Cui et al., 1983; Schlingemann et al., 1990).

Endothelial Cell Markers: Phenotypic Differences in Arteries, Capillaries, and Veins

Due to differences in the organism studied, the differences in the endothelium of systemic and pulmonary circulation, the physiologic state of the vasculature, and the fact that the microvascular endothelium is not always resolved into pre-capillary arterioles, post-capillary venules, and capillaries in these investigations, interpretation of the literature available on the segmental distribution of markers is not simple and not always straightforward. There are conflicting data or interpretations even in the same tissue, such as normal lung. Moreover, most marker studies are conducted by immunohistochemistry or immunoblotting and the vicissitudes of antibody specificities and the variable *in situ* availability of antigens for binding contribute additional uncertainty. Nevertheless, we have attempted a summary of the segmental vascular distribution of functional and adhesion markers in Table 1, receptors in Table 2, and surface glycoprotein markers in Table 3. Specific references and information are given below and more detailed discussion of adhesion molecules is found in Chapter 11.

Segmental Differences in Coagulation Regulators and Ion Channels

Segmental differences occur in the relative number of Weibel-Palade bodies, a unique endothelial specific inclusion providing storage for granular vWF and associated factorVIII. Weibel-Palade bodies are lowest in microvessels (Hibbs et al., 1958; Weibel and Palade, 1964) and increase in number as blood vessels converge toward to the heart. In studies on the antigen presenting properties of human lung endothelium, vWF stained medium and small vessels but capillary staining was weak (Yamamoto et al., 1988). As shown in Figure 4,

Kawanami and co-workers expanded this observation on heterogeneity of the lung endothelium, to show that the alveolar capillaries, while negative for vWF, were positively stained for thrombomodulin (Kawanami et al., 2000; Jin et al., 2001). Large vessels and vessels > 10 µm in the bronchial connective tissue and in the vaso vasorum, on the other hand, were vWF positive and thrombomodulin negative. Their elegant and detailed studies also described that certain juxta-alveolar microvessels and some peribronchial capillaries displayed an intermediate, mosaic pattern of vWF and thrombomodulin (Fig. 4). This opposite expression of thrombomodulin and vWF in diverse segments of lung circulation is interesting as these markers are associated with the anti-thrombotic and pro-thrombotic influences of endothelium, respectively. Müller and co-workers (2002d) also reported on expression of the vWF in human lung. CD31-specific antibody was used as a control stain for endothelial cells and produced equally strong staining reactions in all pulmonary endothelial cells. In contrast to the consistent CD31 presence, vWf-specific antibody yielded negative or weakly positive staining reactions in capillary endothelium and staining intensities that increased with the increase vessel caliber in agreement with the observations on the Weibel-Palade bodies in systemic vessels and the studies on lung vWF segmental distribution above. This distribution is also consistent with poor capillary staining of vWF in normal human lung capillaries when contrasted to consistent staining with lectin staining with *Ulex europaeus* -1 agglutin (Feuerhake et al., 1998). Thus, there is solid agreement on segmental heterogeneity of vWF expression in lung endothelium. Interestingly, this pattern was unchanged in septic lungs (Müller et al., 2002a). However, in alveolar fibrosis and in neovascularization in lung adenocarcinoma, the capillary pattern becomes reversed with loss of the anti-coagulant thrombomodulin and gain of the pro-coagulant vWF (Kawanami et al., 2000; Jin et al., 2001). Segmental differences were also noted for the receptor for activated protein C, an important anticoagulant enzyme (Laszik et al., 1997). While expression was strong in arteries and veins, most arterioles, and in some post-capillary venules, it was absent in capillary endothelium.

This variation in expression of anti-thrombotic and pro-thrombotic markers suggests segmental difference in endothelium in the tendency toward clot formation. Along this line, there is an interesting difference between macrovascular endothelium and microvascular endothelium related to pulmonary coagulation or occlusion in sickle cell disease. Rather than the simple classic view of physical plugging due to the deformed erythrocytes, which would be expected predominantly in pre-capillary sites, specific binding of the sickled (and even normal) cells predominates in the post-capillary microvascular venule, which initiates retrograde cell accumulation and development of the clot (Hebble, 2000). The red cells express factors that bind CD36, integrin $\alpha_v\beta_5$, and vWF-glycoproteins complexes on the endothelium. Expression of these binding molecules by the endothelium and release of vWF has been related to increases in cytosolic Ca^{++} linked to activation of G_q-linked receptors for a variety of agonists. A recent study has explained the segmental nature of the microvascular vaso-occlusion in sickle cell disease. Apparently, microvascular pulmonary endothelial cells, but not macrovascular endothelium, express a T- type voltage-gated Ca^{++} channel, which is opened when G_q-receptor activation initiates membrane depolarization (Wu et al., 2003). Pharmacologic inhibition of this channel was further shown to abrogate the retention of sickle cells in isolated perfused lungs in response to stimulation with activated neutrophils (Wu et al., 2003). In contrast, store operated Ca^{++} entry channels are functional in the pulmonary macrovascular, but not microvascular endothelium, as mentioned above (Cioffi et al., 2003).

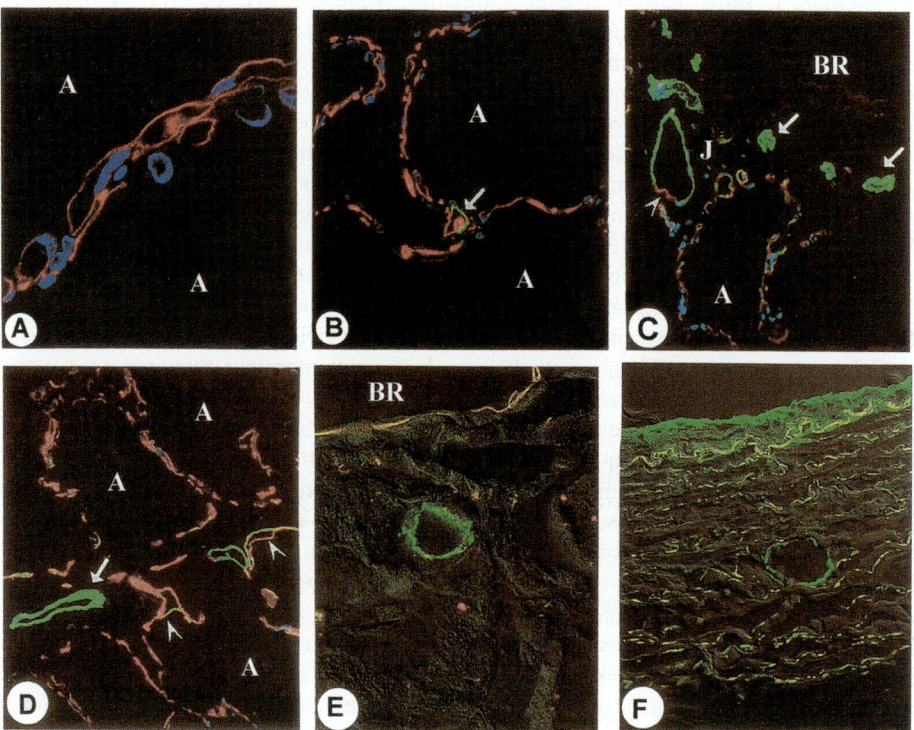

Figure 4. Heterogeneous distribution of von Willebrand Factor and thrombomodulin in human lung vasculature. Normal lung tissue from patients was immunostained for vWf (green fluorescence), thrombomodulin (red), and nuclei in some sections by DAPI (blue). **A.** Capillary loops in the inter-alveolar wall (A=alveoli) are thrombomodulin positive and vWf negative. **B.** Alveolar capillaries are exclusively stained for thrombomodulin in this low power image, while the pulmonary venule (arrow) at the junction of emerging capillaries shows some green staining for vWF, as well as for thrombomodulin. **C.** At lower power, the two larger microvessel (to the lower right of "J") in the juxtaalveolar zone (J) bordering the bronchiole (BR) show a mosaic pattern of thrombomodulin and vWf staining, while interalveolar capillaries only stain for thrombomodulin. The large vessel (left of "J") is mainly positive for vWF, except in the region most adjacent to alveolar capillaries (arrow head). The larger vessel closer to the bronchiole (upper left of "J") is only vWf positive. The microvessels in the connective tissue zone below the bronchiole are exclusively vWf stained. **D.** An interlobule venule (arrow) is stained for vWf, while the capillaries are thrombomodulin positive. Microvessels at the junctions of connective tissue and alveolar areas (arrow heads) show mosaic staining. **E.** A microvessel in the submucosal zone beneath the bronchial epithelium is exclusively vWF positive. **F.** The surface of the large pulmonary artery is exclusively vWF positive, as is the vasovasorem in the arterial wall. From *Kawanami et al., 2000* by kind permission of Dr. Kawanami, the Journal of the Nippon Medical School, and Kyorinsha Publishers, Japan.

Segmental Heterogeneity of Commonly Expressed Markers

Angiotensin converting enzyme - ACE is a consistent marker for endothelium, but in systemic rat vascular endothelium, it is most strongly expressed in arteries and arterioles, moderately expressed in veins, and poorly expressed in capillaries (Danilov et al., 2001). In contrast, pulmonary capillary endothelium exhibited high levels ACE staining and activity and high selective uptake of infused anti-ACE antibodies (Danilov et al., 2001). ACE activity during perfusion reflects both capillary surface area and functional health of the pulmonary capillary endothelium. In prior studies in human tissue, very large arteries and veins displayed poor ACE staining compared to microvascular endothelium. **Antigen presentation markers** - As noted, lung endothelia share some markers with peripheral monocytes/macrophages, which are involved in presentation of antigens to trigger lymphocyte reaction. These markers, including HLA-DR+, CD116, and CD36, were strongly expressed in alveolar capillaries, but were weakly expressed or absent in endothelium of small and medium vessels, further indicating heterogeneity of function reflected by the segmental distribution of the markers (Yamamoto et al. 1988). **PECAM** - As mentioned above, PECAM is found consistently throughout the vascular endothelium in both systemic and pulmonary vessels. Even in failing lung, PECAM immunostaining was similar to normal (Müller et al., 2002a). **CD34** – As noted above CD34, an L-selectin ligand, is expressed in all human pulmonary endothelium, but staining was also found to be strongest in the capillaries (Müller et al., 2002b). They also reported that no unique gender-related characteristics were detected, but staining in the large vessels increased with age, while CD34 decreased in the microvasculature. Thus, even in markers normally expressed by confluent quiescent endothelium, there is heterogeneity is distribution along the vascular beds and alterations in expression levels with age and with activation or disease. In contrast to the uniform expression in the segments in pulmonary circulation, CD34 was constitutively expressed in coronary microvascular, but not macrovascular endothelium, even after TNF-stimulation (Zakrzewicz et al., 1997).

Segmental Heterogeneity of Inducible Markers

ICAM - In various tissues of the cardiovascular system, ICAM and CD36 were reported to be strongly expressed in the capillaries and weakly expressed in larger vessels (Page et al., 1992). In patients with nephropathy, ICAM-1 was found in the glomerular capillaries as well as throughout the kidney vasculature (Ogawa et al., 1997). ICAM staining was fairly consistent in all segments of systemic vessels in various organs of the rat (Danilov et al., 2001). They also reported ICAM positive capillaries in the rat lung. Sato et al. (2000) demonstrated ICAM in veins and capillaries, but little in arteries in normal rat lung by intravital confocal luminescence. Two days after bleomycin injury, the poor arterial ICAM staining was unchanged but there was a two-fold increase in signal in capillaries and venules (Sato et al., 2000). Then there was a secondary rise in ICAM in the venules at 21 days post-injury when fibrosis occurred. In contrast to the above studies, Feuerhake et al. (1998) detected ICAM-1 in the venules/arterioles > veins > arteries, but not in the capillaries in normal human lung. So, segmental distribution of ICAM in normal and activated systemic and lung tissue is not resolved. **VCAM** -Similar to ICAM-1, VCAM was expressed most strongly in venules and arterioles, moderately expressed in veins and arteries, and not expressed in capillaries in normal human lung (Feuerhake et al., 1998). Müller et al. (2002a) only found sporadic staining for VCAM in the largest vessels in normal lung, but in septic lungs of the ARDS patients, there was upregulation in all segments but the capillaries. **ELAM** – Similar to VCAM distribution, the larger vessels expressed ELAM-1 in various tissues, but it was not present in capillaries (Page et al., 1992; Ogawa et al., 1997). In human lung tissue, ELAM was detected most strongly in venules and veins, consistently found in arteries and arterioles, but none was detected in capillaries (Feuerhake et al., 1998). Müller et al. (2002a) did not detect ELAM in any vessels in normal human lung specimens, but showed moderate to strong upregulation in arteries, veins, arterioles, and venules, but not capillaries (other than a few sporadic cells) in the septic lung of ARDS patients. **P-selectin** - Like ELAM, p-selectin (GMP-140) staining was strong in venules and veins, moderate in arteries and arterioles, and absent in capillaries in human lungs (Feuerhake et al., 1998, Müller et al., 2002a). This is also consistent with its association with vWF.

Table 3. Heterogeneity of endothelial surface glycoproteins

Antigen	Segment			Expression by other cells
	Art	MV	Vein	
Lectin binding – *Arachis hypogaea*	-	+	+	
Lectin binding - *Bandiera (Griffonia) simplicifolia*	?	?	?	Macrophages & Pericytes
Lectin binding – *Caragana arborescens*	+	-	+	
Lectin binding – *Concanavalia ensiformis*	++	++	++	
Lectin binding – *Erythrina cristagalli*	+	+	++	
Lectin binding – *Glycine max*	-	+	-	
Lectin binding – *Helix pomatia*	+	-	+	
Letin binding – *Lens culinaris*	+	+	+	
Lectin binding – *Lycopersicon esculentum*	-	+	-	EC specific
Lectin binding – *Limax Flavus*	+	-	++	
Lectin binding – *Ricinus communis*	+	+	+	
Lectin binding – *Ulex europaeus* I agglutinin	-	+	-	
P-glycoprotein, mdr-1a				Brain EC only
Thy-1 glycoprotein	-	++	-	Numerous other cells

Note: 1) MV are a mix of capillary and microvascular endothelium and controversy may exist depending on organ source and specific cells in MV mix; 2) + indicates usual detection and – indicates detection is not usual, irregular, or sparse; however, the endothelium may be induced to express the marker; blanks indicate specific information is not generally available or is not resolved.

Surface Glycoproteins

In addition to the various functional and adhesion molecules that serve as markers of endothelial heterogeneity, the surface glycoproteins bind various lectins in fairly specific and diverse ways (see Table 3). For instance, *Glycine max* bound alveolar capillaries and post-capillary venules but not arterioles or the large lung vessels (King et al., 2004). Whereas, *Caragana arborescens* and *Helix pomatia* bound arterial and venous endothelium more strongly than the microvascular endothelium in lung (Schnitzer et al., 1994; King et al., 2004). As shown in Figure 5, *Lycopersicon esculentum* binds the lung microvessels, but neither artery nor vein, while *Arachis hypogaea* binds microvessels and veins but not lung arteries

(Schnitzer et al., 1994). On the other hand, the segmental-specific binding of *Griffonia (Bandeiraea) simplicifonia* has proved controversial. It has been reported to bind endothelium from rat and mouse aorta (Nicosia et al., 1994; Bastaki et al., 1997) as well as heart microvasculature (Porter et al., 1990; Bastaki et al., 1997). Whereas it bound mouse brain microvascular endothelium (Bastaki et al., 1997); it was absent from rat brain microvasculatur (Porter et al., 1990). In the lung, it was reported to bind macrovascular but not microvascular endothelium from bovine lung (Schnitzer et al., 1994), but it was also reported to bind microvascular, but not macrovascular endothelium from rat lung (Bankston et al., 1991; King et al., 2004). Thus, the glycoprotein surface differs between large vessels and the microvasculature, but the differences between organ and segmental expression are not fully resolved and may require staining of intact beds rather than cultured cells, which may be altered by conditions of isolation and growth.

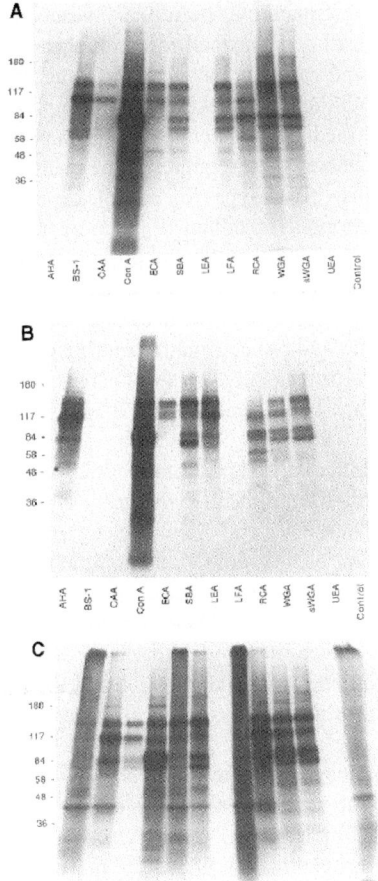

Figure 5.
Heterogeneity of lectin-binding to endothelium.
Bovine endothelial cells were isolated, passaged only once, and the surface glycoproteins were radiolabeled with iodine by lactoperoxidase and hydrogen peroxide. The detergent-soluble phase of the cell lysates was precipitated with the indicated biotinylated lectin and processed by SDS-PAGE. AHA= *Arachis hypogaea*, BS-1= *Bandeiraea simlicifolia*, CAA= *Caragana arborescens*, Con A= *Concanavalia ensiformis*, ECA= *Erythrina cristagalli*, SBA= *Glycine max*, LEA= *Lycopersicon esculentum*, LFA= *Limax flavus*, RCA= *Ricinus communis*, WGA= *Triticum vulgare*, sWGA succinylated-*Triticum vulgare*, UEA= *Ulex europaeus*). **A.** Bovine pulmonary artery endothelium **B.** Bovine lung microvessel endothelium **C.** Bovine pulmonary vein endothelium. Although grown under the same conditions, the endothelial cells from the different segments expressed different patterns of surface glycoproteins. Reprinted from Schnitzer, J.E., Siflinger-Birnboim, A., DelVecchio, P., and Malik, A.B. (1994) *Segmental differentiation of permeability, protein glycosylation, and morphology of cultured bovine lung vascular endothelium. Biochem. Biophys. Res. Comm. 199:11-19* by kind permission from Dr. Jan Schnitzer and Elsevier, Amsterdam.

Flow Stress Effects on Endothelium

Architectural Responses

Simple anatomical differences in endothelium in vascular beds with high shear stress vs. those with low shear stress indicate stress adaptation (Fig. 1). Continuous laminar flow is recognized as protective for endothelium, but disorganized and high flow is, well, stressful and will be the focus of this discussion. Moreover, shear stress changes are translated into both acute release of endothelial factors such as vWF as well as chronic increases in the vWF storage Weibel bodies (Sun *et al.*, 2000). As discussed in Chapter 2, pulmonary vascular barrier functional integrity is dependent on the adhesive cell-cell and cell-matrix tethering forces, upon the structural framework and basal tensegrity, and upon the balance of contractile forces. These forces are intimately linked through the endothelial cytoskeleton, a complex network of actin microfilaments, microtubules, and intermediate filaments, which impart shape to the cell and anchors junctional complexes. Altogether, they regulate shape change and transduce signals within and between endothelial cells. Two types of complexes are responsible for intercellular contacts along the endothelial monolayer, adherens junctions and tight junctions (Dejana *et al.*, 1995). They link to the actin cytoskeleton to provide both mechanical stability and transduction of extracellular signals into the cell. Focal adhesions provide additional adhesive forces in barrier regulation by forming a critical bridge for bi-directional signal transduction between the actin cytoskeleton and the cell-matrix interface, as explored in Chapter 8. Increasingly, effects of mechanical forces such as shear stress and ventilator-induced stretch on endothelial barrier function are being recognized (Berk *et al.*, 2001; Davies *et al.*, 2003; Dudek and Garcia, 2001).

Viscous drag generated at the luminal surface of endothelial cells by blood flow (wall shear stress) and cyclic strain of the vascular wall that results from pulsatile changes in blood pressure are particularly important in this context (for review: Ali and Schumacker, 2002; Busse and Fleming, 2003). Active cytoskeletal rearrangement begins rapidly and continues to occur over several hours as endothelial cells orient themselves to reduce both peak shear stresses and shear stress gradients (Barbee *et al.*, 1995; Galbraith *et al.*, 1998; Birukov *et al.*, 2003). Both apical actin stress fibers linked to cell-cell contact sites and integrin-mediated signal transduction are involved in sensing flow and transducing its signal (Chen *et al.*, 1999; Kano *et al.*, 2000). When cultured endothelium is exposed to shear stress, multiple signaling pathways implicated in cytoskeletal rearrangement are stimulated, including Ca^{2+} mobilization, G-protein activation, increased tyrosine phosphorylation, and MLCK and MAPK activation (Busse and Fleming, 2003). Shear stress helps maintain endothelial monolayers through the increase of focal adhesions (Urbich *et al.*, 2000). However, endothelial cells exposed to high shear gradients, or turbulent flow, develop increased permeability relative to areas of either constant laminar flow or no flow (Phelps and DePaola, 2000). One mechanism by which shear stress may alter barrier function is by inhibition of endothelial cell apoptosis (Dimmeler *et al.*, 1996); however, Petrache and co-workers (2001) recently reported that in a TNF model under static conditions, distinct signaling and cytoskeletal are involved in cytokine-induced apoptosis and permeability. It should be noted that the majority of these studies were performed with systemic endothelium, and pulmonary endothelial cell-specific responses to flow are not well understood. Moreover, effects of increased shear stress in pulmonary hypertension on endothelial function are not well defined.

Acute shear stress was reported to induce cortical cytoskeletal rearrangement in human pulmonary artery endothelial cells that was characterized by myosin light chain kinase (MLCK)- and Rho-associated kinase (RhoK)-dependent accumulation of diphosphorylated regulatory myosin light chains (MLC) in the cortical actin ring and junctional protein tyrosine phosphorylation (Birukov et al., 2002). Additionally, there was transient peripheral translocation of cortactin, an actin-binding protein involved in the regulation of actin polymerization. Shear stress-induced cortactin translocation was independent of Erk-1/2 MAP kinase, p60 Src, MLCK, or RhoK activities. However, shear stress-dependent Rac GTPase-induced p21-activated kinase (PAK)-1 activation was required for transient cortactin translocation and cytoskeletal reorientation in response to sustained shear stress by pulmonary artery endothelial cells.

Nitric Oxide and Oxidant Production Responses

Sensitivity of the endothelium to flow is also responsible for changes in protein expression and thus subsequent responses to variation in shear stress or autocoid environment. It is well known that eNOS expression is maintained by laminar flow in endothelial cells (Le Cras et al., 1998). Flow-adapted (24 h) rat pulmonary microvascular endothelial cells depolarize upon stop of flow, a response that is inhibited by a K_{ATP} channel opener and prevented by addition of cycloheximide to the medium during the flow adaptation period (Chatterjee et al., 2003). Induction of the K_{ATP} channel with flow adaptation has also been observed in bovine pulmonary artery endothelium (Chatterjee et al., 2003). The increase in fluid shear stress at birth, resulting from increased pulmonary blood flow, is an important mediator of post-natal gene expression. Endothelial nitric oxide synthase (eNOS) mRNA and protein levels increase during late gestation and then decrease in the post-natal life in lung parenchyma (Black et al., 1997). Recently, Wedgwood and co-workers (2003) reported that the transcriptional activity from a 1,600-bp eNOS promoter fragment increased in both fetal and adult pulmonary artery endothelial cells exposed to shear stress. Conversely, activity driven from an 840-bp promoter fragment containing a putative activator protein (AP)-1 binding site was increased only in fetal pulmonary artery endothelium. The AP-1 protein c-Jun was localized to the cytosol in static adult pulmonary artery endothelial cells and to the nucleus in static fetal pulmonary artery endothelial cells. After shear, c-Jun was localized in the nucleus of both cell types. However, transcriptionally active phosphorylated c-Jun was only elevated in the nuclei of sheared fetal pulmonary artery endothelium and resting levels of eNOS and NO were 2- and 20-fold higher, respectively, in fetal cells, whereas shear increased eNOS and NO approximately 2.5-fold more in fetal pulmonary artery endothelial cells. Phosphorylation of Akt and eNOS was evident in sheared fetal but not adult pulmonary artery endothelial cells. This therefore provides a mechanism by which shear stress at birth increases eNOS regulation and enzyme activity in the fetal pulmonary arterial circulation. But a large number of genes are regulated by shear stress, as recently reported by Dekker and co-workers

(2002). This group reported, however, that the lung Kruppel-like factor is an endothelial transcription factor (in human umbilical vein endothelial cells) that is uniquely induced by flow and might therefore be at the molecular basis of the physiological healthy, flow-exposed state of the endothelium.

A role for mechanotransduction of shear stress change to cellular biochemical adaptation was also shown with the model of non-hypoxic ischemia in pulmonary arterial endothelial cells (Wei *et al.*, 1999). The cells were first flow adapted and then flow was abruptly stopped, while oxygen tension was maintained by continued abluminal perfusion. In the first hour of no flow, generation of reactive oxygen species increased almost 2 fold. This was accompanied by an activation of transcription factors, NF-κB and AP-1 and increased DNA synthesis. If the cells were simultaneously treated with N-acetylcysteine to scavenge the free radicals produced or were treated with diphenyleneiodium to inhibit superoxide production by NADPH/NADH oxidase (and possibly from the mitochondrial respiratory chain and NOS), the transcription factors were not activated, implying the involvement of the reactive oxygen species in the shear stress response signal transduction apart from cellular hypoxia.

More studies are required to assess the heterogeneity of the response to shear stress in pulmonary artery endothelial cells and pulmonary microvascular endothelial cells. Based on our knowledge arising from the consequences of pulmonary hypertension on endothelial permeability, it is almost predictable to detect high sensitivity of the pulmonary microvascular endothelial cells to variation in shear stress and pressure. Whether these changes are qualitatively similar to those observed in large vessel endothelium is undefined.

Pulmonary Vasculature *versus* Bronchial Circulation

Anatomical Differences

Although this review has primarily focused on pulmonary endothelial heterogeneity, it is interesting to highlight the differences in endothelial phenotype and structure between pulmonary and bronchial vessels. The divergence in gross anatomy is obvious: bronchial vessels are subjected to normal systemic working pressures, with increased shear stress. As such, they are not different from endothelium in other systemic organs. Yet, they receive the highest pO_2 of the systemic vessels, which gives uniqueness to the bronchial circulation. Naturally, the morphology of bronchial and pulmonary endothelium would be expected to differ. Whereas the pulmonary capillary endothelium is smooth, thin, and continuous, the bronchial capillary endothelium has been described as "fenestrated along their thin cytoplasmic segments," providing clues as to their different origin (Hirabayashi and Yamamoto, 1984; Kawanami, 1997). On the other hand sheep lung microvessel capillary endothelial cells (Craig *et al.*, 1998) and the microvessel bronchial endothelial cell in culture displayed the characteristic cobblestone morphology of endothelial monolayers and thus, resemble endothelium from other microvessels (Arey, 1963; Cornhill *et al.*, 1980). There is also evidence for anastomoses between pulmonary and bronchial circulation at the capillary level, but how this affects pressures and phenotypes at the micro level is not known.

Functional Similarities and Differences

The bronchial endothelium shares another distinguishing feature common to the endothelial phenotype - the ability to metabolize Ac-LDL (Grooby *et al.*, 1997; Craig *et al.*, 1998). Histological assessment of vWF expression demonstrated granular cytoplasmic staining in both macro- and micro- vessels of bronchial endothelial cells that faded with passage, but neither contained Weibel-Palade bodies, the usual storage site for vWF (Craig *et al.*, 1998; Moldobaeva and Wagner, 2002). Similarly, bronchial endothelium in normal human lung was positive for vWF (Kawanami *et al.*, 2000). Capillary staining of vWF was weak or absent in pulmonary capillaries as discussed above (Yamamoto *et al.*, 1988; Feuerhake *et al.*, 1998; Kawanami *et al.*, 2000; Müller *et al.*, 2002d). The bronchial endothelium is unusual among systemic vessels because of a consistent expression of tissue plasminogen activator (Levin *et al.*, 1998). Even though it is absent in the aorta, it is present immediately in the main bronchus and appears consistently in all bronchial arteries, regardless of size. The pulmonary vessels have a segmental determined expression of this tissue plasminogen activator. While absent in the large vessels and in the capillary endothelium, it is found in vessels between 7 and 30 um in diameter. In addition to these differences, bronchial endothelial cells, but not pulmonary endothelial cells, are associated with the basement membrane containing collagen VII (Kawanami, 1997).

Endothelial segmental heterogeneity in bronchial circulation was studied by successful isolation and culture of bronchial artery (BAEC) and bronchial microvascular endothelial cells (BMVEC) from the sheep (Moldobaeva and Wagner, 2002). The cultured BAEC and BMVEC exhibited distinct characteristics in growth rate and response to inflammatory factors, such as bradykinin and thrombin. The doubling time for BAEC (39 h) was within the range reported for large vessel endothelial culture (17–42 h) (Sadovnikova *et al.*, 1990) and similar to peripheral arteries, such as the superficial temporal (47 h) and omental artery (43 h) (Thorin *et al.*, 1997a). Although growth of BMVEC lagged that of BAEC, doubling time of BMVEC (65 h) was similar to growth characteristics of other microvascular endothelium (67 h) (Folkman *et al.*, 1979). Yet, segmental differences in growth behavior were opposite that of pulmonary endothelium (Stevens, 2002), where microvascular cells grow faster than macrovessel endothelium.

Permeability of BMVEC monolayers increased after thrombin treatment; however, BAEC were resistant at the applied thrombin concentration (Moldobaeva and Wagner, 2002). This is opposite to observed permeability sensitivity to thrombin in pulmonary macro-and micro-vessel endothelium (unpublished data, C.E. Patterson). Additionally, basal permeability characteristics differ, with bronchial microvessel permeability greater than macrovessel permeability (Moldobaeva and Wagner, 2002); whereas basal pulmonary microvessel permeability is less than macrovessel permeability (Schnitzer *et al.*, 1994; Stevens, 2002). In contrast to the selective response to thrombin, bradykinin caused a prompt and substantial increase in permeability that was similar in both BAEC and BMVEC.

There is a paucity of data concerning bronchial endothelium. This should be highlighted and further studies are required to appreciate the unique role of these cells and circulatory loop in the regulation of lung permeability. This is of importance in connection with the changes that occur during pulmonary hypertension and during impaired integrity of the permeability associated with inflammation and septic shock, for example.

Endothelium – Variety is Thy Name

Within the same species, endothelium can differ in phenotype, growth, antigenic composition, basal release of endothelial derived factors and responsiveness to agonists. What is less clear is why endothelial cells are heterogeneous. To a large degree, the gene expression pattern is dedicated in endothelium from diverse organs and from different vascular segments. This variability is associated with the diverse functions and roles the endothelium has in the various regions of the vasculature. How endothelial cells also adapt to meet unique functional requirements due to changes in external environment within these site-specific gene patterns is not well understood. Moreover, several disease states selectively target and alter the function of certain regions of the endothelium. A clear understanding of these differences and recognition of the anatomic variations could not only lead to more insight into cardiovascular physiology, but also open new therapeutic approaches that could selectively target drugs to particular endothelia or particular organs as will be explored in Chapters 15 and 16.

ACKNOWLEDGEMENTS

E. Thorin is grateful for the support of the Foundation of the Montreal Heart Institute, the Heart and Stroke Foundation of Quebec and the Canadian Institute for Health Research. E. Thorin is a scholar of the Heart and Stroke Foundation of Canada. This work was also supported by a Merit Award from the Veteran's Administration Medical Research Service (CEP), HL66299 (TS), and HL60024 (TS).

REFERENCES

Adams, R.H., Wilkinson, G.A., Weiss, C., Diella, F., Gale, N.W., Deutsch, U., Risau, W. and Klein, R. (1999) Roles of ephrinB ligands and EphB receptors in cardiovascular development: demarcation of arterial/venous domains, vascular morphogenesis, and sprouting angiogenesis. Genes. Dev. 13:295-306.
Aird, W.C. (2003) Endothelial cell heterogeneity. Crit. Care Med. 31:S221-S230.
Ali, M.H and Schumacker, P.T. (2002) Endothelial responses to mechanical stress: Where is the mechanosensor? Crit. Care Med. 30:S198 –S206.
Allport, J.R., Muller, W., and Luscinskas, F.W. (2000) Monocytes induce reversible focal changes in vascular endothelial cadherin complex during transendothelial migration under flow. J. Cell Biol. 148:203-216.
Arrigoni, F.I., Hislop, A., Haworth, S., and Mitchell, J.A. (1999) Newborn intrapulmonary veins are more reactive than arteries in normal and hypertensive piglets. Amer. J. Physiol. 277:L887-L892.
Arey, L.B. (1963) The development of peripheral blood vessels. In: Orbison, J.L. and Smith, D.E. (Eds) The peripheral blood vessels. Williams and Williams, Baltimore, pp 1-16.

Aschner, J.L., Smith, T.K., Kovacs, N., Pinheiro, J.M. and Fuloria, M. (2002) Mechanisms of bradykinin-mediated dilation in newborn piglet pulmonary conducting and resistance vessels. Amer. J. Physiol. 283:L373-L382.
Augustin-Voss, H.G., Johnson, R., and Pauli, B.U. (1991) Modulation of endothelial cell surface glycoconjugate expression by organ-derived biomatrices. Exp. Cell Res. 192:346-351.
Aurrand-Lions, M., Johnson-Leger, C., Imhof, B.A. (2003) Role of interendothelial adhesion molecules in the control of vascular functions. Vasc. Pharm. 39:239-246.
Aurrand-Lions, M., Johnson-Leger, C., Wong, C., Du Pasquier, L., and Imhof, B.A. (2001) Heterogeneity of endothelial junctions is reflected by differential expression and specific subcellular localization of the three JAM family members. Blood 98:3699-3707.
Bankston, P.W., Porter, G.A., Milici, A.J., and Palade, G.E. (1991) Differential and specific labeling of epithelial and vascular endothelial cells of the rat lung by lectins. Eur. J. Cell Biol. 54:187-195.
Barbee, K.A., Mundel, T., Lal, R. and Davies, P.F. (1995) Subcellular distribution of shear stress at the surface of flow-aligned and nonaligned endothelial monolayers. Amer. J. Physiol. 268:H1765–H1772.
Bastaki, M., Nelli, E., Dell'Era, P., Rusnati, M., Molinari-Tosatti, M., Parolini, S., Auerbach, R., Ruco, L., Possati, L., and Presta, M. (1997) Basic fibroblast growth factor-induced angiogenic phenotype in mouse endothelium. A study of aortic and microvascular endothelial cell lines. Arterioscl. Thromb. Vasc. Biol. 17:454-464.
Bazzoni, G., Martinez Estrada, O., and Dejana, E. (1999) Molecular structure and functional role of vascular tight junctions. Trends Cardiovasc Med. 9:147-152.
Berk, B.C., Abe, J., Min, W., Surapisitchat, J., and Yan, C. (2001) Endothelial atheroprotective and anti-inflammatory mechanisms. Ann. NY Acad. Sci. 947:93-109.
Bevilaqua, M.P., Pober, J.S., Mendrix, D.L., Cotran, R.S., and Gibrone, M.A. (1987) Identification of an inducible endothelial leucocyte adhesion molecule. PNAS USA 84:9238-9242.
Birukov, K.G., Birukova, A., Dudek, S., Verin, A., Crow, M., Zhan, X., DePaola, N. and Garcia, J.G. (2002) Shear stress-mediated cytoskeletal remodeling and cortactin translocation in pulmonary endothelial cells. Amer. J. Respir. Cell Mol. Biol. 26:453-464.
Birukov, K.G., Jacobson, J., Flores, A., Ye, S., Birukova, A., Verin, A., and Garcia, J.G. (2003) Magnitude-dependent regulation of pulmonary endothelial cell barrier function by cyclic stretch. Amer. J. Physiol. 285:L785-L797.
Black, S.M., Johengen, M., Ma, Z., Bristow, J., and Soifer, S.J. (1997) Ventilation and oxygenation induce endothelial nitric oxide synthase gene expression in the lungs of fetal lambs. J. Clin. Invest. 100:1448-1458.
Boegehold, M.A. (1998) Heterogeneity of endothelial function within the circulation. Curr. Opinion Neph. Hyperten. 7:71-78.
Boels, P.J., Deutsch, J., Gao, B., and Haworth, S.G. (1999) Maturation of the response to bradykinin in resistance and conduit pulmonary arteries. Cardiovasc. Res. 44:416-428.
Boels, P.J., Deutsch, J., Gao, B., and Haworth, S.G. (2001) Perinatal development influences mechanisms of bradykinin-induced relaxations in pulmonary resistance and conduit arteries differently. Cardiovasc Res. 51:140-150.
Brouland, J.P., Gilbert, M., Bonneau, M., Pignaud, G., Bal di Solier, C., and Drouet, L. (1999) Macro and micrheterogeneity in normal endothelial cells. Endothelium 6:251-262.
Busse, R. and Fleming, I. (2003) Regulation of endothelium-derived vasoactive autacoid production by hemodynamic forces. Trends Pharmacol. Sci. 24:24-29.
Bussolino, F., Serini, G., Mitola, S., Bazzoni, G., and Dejana, E. (2001) Dynamic modules and heterogeneity of function: a lesson from tyrosine kinase receptors in endothelial cells. EMBO Rep. 2:763-767.
Canis-Parera, M., Grosveld, F., Tibboel, D., Rottier, R., and Erasmus, M.C. (2003) Lung vascular morphogenesis at early pseudoglandular stage: important role for vascular remodeling. Amer. J. Respir. Cell Molec. Biol. 167:A380.
Carmeliet, P. and Jain, R.K. (2000) Angiogenesis in cancer and other diseases. Nature 407:249-257.

Cascone, I., Giraudo, E., Caccavari, F., Napione, L., Bertotti, E., Collard, J.G., Serini, G., and Bussolino, F. (2003) Temporal and spatial modulation of Rho GTPases during in vitro formation of capillary vascular network. Adherens junctions and myosin light chain as targets of Rac1 and RhoA. J. Biol. Chem. 10:1074-1081.

Cattelino, A., Liebner, S., Gallini, R., Zanetti, A., Balconi, G., Corsi, A., Bianco, P., Wolburg, H., Moore, R., Oreda, B., Kemler, R., and Dejana, E. (2003) The conditional inactivation of the {beta}-catenin gene in endothelial cells causes a defective vascular pattern and increased vascular fragility. J. Cell Biol. 162:1111-1122.

Cerilli, J., Brasile, L., Galouzis, T., Lempert, N., and Clarke, J. (1985) The vascular endothelial cell antigen system. Transplantation 39:286-289.

Chatterjee, S., Al-Mehdi, A-B, Levitan, I., Stevens, T., and Fisher, A.B. (2003) Shear stress increases expression of a K_{ATP} channel in rat and bovine pulmonary vascular endothelial cells. Amer. J. Physiol. 285:C959–C967.

Chen, K.D., Li, Y., Kim, M., Li, S., Yuan, S., Chien, S., and Shyy, J.Y.J. (1999) Mechanotransduction in response to shear stress. J. Biol. Chem. 274:18393–18400.

Chen, L., McNeill, J., Wilson, T., and Gopalakrishnan, V. (1995) Heterogeneity in vascular smooth muscle responsiveness to angiotensin II. Role of endothelin. Hypertension 26:83-88.

Chetham, P.M., Baba, P., Bridges, J., Moore, T., and Stevens T. (1999) Segmental regulation of pulmonary vascular permeability by SOC-Ca^{2+} entry. Amer. J. Physiol. 276:L41–L50.

Chi, J.T., Chang, H., Haraldsen, G., Jahnsen, F., Troyanskaya, O., Chang, D., Wang, Z., Rockson, S., van de Rijn, M., Botstein, D., and Brown, P.O. (2003) Endothelial cell diversity revealed by global expression profiling. PNAS USA 100:10623-10628.

Cioffi, D.L., Moore, T., Schaack, J., Creighton, J., Cooper, D., and Stevens, T. (2002) Dominant regulation of interendothelial cell gap formation by calcium-inhibited type 6 adenylyl cyclase. J. Cell Biol. 157:1267-1278.

Cioffi. D.L., Wu, S., and Stevens, T. (2003) On the endothelial cell I_{SOC}. Cell Cal. 33:323-336.

Condorelli, G., Borello, U., DeAngelis, L., Latronico, M., Sirabella, D., Coletta, M., Galli, R., Balconi, G., Frati, G., Cusella, M., Gioglio, L., Amuchastegui, S., Adorini, L., Naldini, L., Vescovi, A., Dejana, E., and Cossu, G. (2001) Cardiomyocytes induce endothelial cells to trans-differentiate into cardiac muscle:implications for myocardial regeneration. PNAS USA 98:10733-10738.

Cornhill, J.F., Levesque, M.J., Herderick, E.E., Nerem, R.M., Kilman, J.W., and Vasko, J.S. (1980) Quantitative study of the rabbit aortic endothelium using vascular casts. Atherosclerosis 35:321-337.

Craig, L.E., Spelman, J., Strandberg, J.D., and Zink, M.C. (1998) Endothelial cells from diverse tissues exhibit differences in growth and morphology. Microvasc. Res. 55:65-76.

Creighton, J.R., Masada, N., Cooper, D., and Stevens, T. (2003) Coordinate regulation of membrane cAMP by Ca2+-inhibited adenylyl cyclase and phosphodiesterase activities. Amer. J. Physiol. 284:L100-L107.

Cui, Y., Tai, P.C., Gatter, K.C., Mason, D.Y., and Spry, C.J. (1983) A vascular endothelial cell antigen with restricted distribution in human foetal, adult and malignant tissues. Immunology 49:183-189.

Danilov, S.M., Gavrilyuk, V., Franke, F., Pauls, K., Harshaw, D., McDonald, T., Miletich, D., and Muzykantov, V.R. (2001) Lung uptake of antibodies to endothelial antigens: key determinants of vascular immunotargeting. Amer. J. Physiol. 280:L1335-L1347.

Davies, P.F., Zilberberg, J., and Helmke, B.P. (2003) Spatial microstimuli in endothelial mechanosignaling. Circ. Res. 92:359-370.

Dejana, E., Corada, M., and Lampugnani, M.G. (1995) Endothelial cell-to-cell junctions. FASEB J. 9:910-918.

Dekker, R.J., van Soest, S., Fontijn, R., Salamanca, S., de Groot, P., VanBavel, E., Pannekoek, H., and Horrevoets, A.J. (2002) Prolonged fluid shear stress induces a distinct set of endothelial cell genes, most specifically lung Kruppel-like factor. Blood. 100:1689-1698.

DeMello, D.E. and Reid, L.M. (2000) Embyonic and early fetal development of human lung vasculature and its functional implications. Pediat. Dev. Pathol. 3:439-449.

Dimmeler, S., Haendeler, J., Rippmann, V., Nehls M., and Zeiher, A.M. (1996) Shear stress inhibits apoptosis of human endothelial cells. FEBS Lett. 399:71–74.
Dohi, Y., Hahn, A., Boulanger, C., Bülher, F., and Lüscher, T.F. (1992) Endothelin stimulated by angiotensin II augments contractility of spontaneously hypertensive rat resistance arteries. Hypertension 19:131-137.
Dudek, S.M. and Garcia, J.G. (2001) Cytoskeletal regulation of pulmonary vascular permeability. J. Appl. Physiol. 91:1487-1500.
Dvorak, H.F., Detmar, M., Claffey, K., Nagy, J., van der Water, L., and Senger, D.R. (1995) Vascular permeability factor/vascular endothelial growth factor: an important mediator of angiogenesis in malignancy and inflammation. Int. Arch. Allergy Immunol. 107:233-235.
Ekman, N., Arighi, E., Rajantie, I., Saharinen, P., Ristimaki, A., Silvennoinen, O., and Alitalo, K. (2000) The Bmx tyrosine kinase is activated by IL-3 and G-CSF in a PI-3K dependent manner. Oncogene 19:4151-4158.
Favre, C.J., Mancuso, M., Maas, K., McLean, J., Baluk, P., and McDonald, D.M. (2003) Expression of genes involved in vascular development and angiogensis in endothelial cells in freshly isolated from adult lungs. Amer. J. Physiol. 285:H1917-H1938.
Fenyves, A.M., Behrens, J., and Spanel-Browski, K. (1993) Cultured microvascular endothelial cells differ in cytoskeleton, expression of cadherins and fibronectin matrix. J. Cell Sci. 106:879-890.
Ferrara, N., Houck, K., Jakeman, L., and Leung, D.W. (1992) Molecular and biological properties of the VEGF family of proteins. Endocrine Rev. 13:18-32.
Feuerhake, F., Fushsl, G., Bals, R., and Welsh, U. (1998) Expression of inducible cell adhesion molecules in normal human lung. Histochem. Cell Biol. 110:387-394.
Fleming, S. and Jones, D.B. (1989) Antigenic heterogeneity of renal endothelium. J. Pathol. 158:319-323.
Folkman, J., Haudenschild, C.C., and Zetter, B.R. (1979) Long-term culture of capillary endothelial cells. PNAS USA 76:5217–5221.
Frid, M.G., Kale, V.A., and Stenmark, K.R. (2002) Mature vascular endothelium can give rise to SMC via endothelial-mesenchymal transdifferentiation. Circ. Res. 90:1189-1196.
Fukata, M. and Kaibuchi, K. (2001) Rho-family GTPases in cadherin-mediated cell-cell adhesion. Nat. Rev. Mol. Cell Biol. 2:887-897.
Fukuda, Y., Nagura, H., Imoto, M., and Koyama, Y. (1986) Immunohistochemical studies on structural changes of the hepatic lobules in liver diseases. Amer. J. Gast. 81:1149-1155.
Furchgott, R.F. and Zawadzki, J.V. (1980) The obligatory role of endothelial cells in the relaxation of arterial smooth muscle by acetylcholine. Nature 288:373-376.
Galbraith, C.G., Skalak, R., and Chien, S. (1998) Shear stress induces spatial reorganization of the endothelial cytoskeleton. Cell Motil. Cytosk. 40:317–330.
Garlanda, C. and Dejana, E. (1997) Heterogeneity of endothelial cells. Specific markers. Arterioscler. Thromb. Vasc. Biol. 17:1193-1202.
Gebb, S. and Shannon, J.M. (2000) Tissue interactions mediate early events in pulmonary vasculogenesis. Devel. Dynam. 217:159-169.
Gebb, S. and Stevens, T. (2004) On lung endothelial cell heterogeneity. Microv. Res. 68:1-12.
Gennaro, G., Ménard, C., Michaud, S.E., and Rivard, A. (2003) Age-dependent impairement of reendothelialization after arterial injury. Circulation 107:230-233.
Gerety, S.S., Wang, H., Chen, Z., and Anderson, D.J. (1999) Symmetrical mutant phenotypes of the receptor EphB4 and its specific transmembrane ligand ephrin-B2 in cardiovascular development. Mol. Cell. 4:403-414.
Gertler, J.P. and Abbott, W.M. (1992) Prothrombotic and fibrinolytic function of normal and perturbed endothelium. J. Surg. Res. 52:52-89.
Graier, W.F., Holzmann, S., Hoebel, B., Kukovetz, W., and Kostner, G.M. (1996) Mechanisms of L-NG nitroarginine/indomethacin-resistant relaxation in bovine and porcine coronary arteries. Br. J. Pharmacol. 119;1177-1186.
Grazia Lampugnani, M., Zanetti, A., Corada, M., Takahashi, T., Balconi, G., Breviario, F., Orsenigo, F., Cattelino, A., Kemler, R., Daniel, T., and Dejana, E. (2003) Contact inhibition of VEGF-

induced proliferation requires vascular endothelial cadherin, beta-catenin, and the phosphatase DEP-1/CD148. J. Cell. Biol. 161:793-804.

Grishko, V., Solomon, M., Wilson, G., LeDoux, S., and Gillespie, M.N. (2001) Oxygen radical-induced mitochondrial DNA damage and repair in pulmonary vascular endothelial cell phenotypes. Amer. J. Physiol. 280:L1300-L1308.

Grooby, W.L., Krishnan, R., and Russ, G.R. (1997) Characterization of ovine umbilical vein endothelial cells and their expression of cell adhesion molecules: comparative study with human endothelial cells. Immunol. Cell Biol. 75:21–28.

Hall, S.M., Hislop, A., and Haworth, S.G. (2002) Origin, differentiation, and maturation of human pulmonary veins. Amer. J. Respir. Cell Molec. Biol. 26:333-340.

Hare, J.M. and Colucci, W.S. (1995) Role of nitric oxide in the regulation of myocardial function. Prog. Cardiovasc. Dis. 38:155-166.

Harris, P. and Heath, D. (1986) The Human Pulmonary Circulation. 3rd edition, Churchill Livingston/Elsevier, Amsterdam, pp. 60-62.

Hebbel, R.P. (2000) Blockade of adhesion of sicle cells to endothelium by monoclonal antibodies. New Eng. J. Med. 342:1910-1912.

Herzog, Y., Kalcheim, C., Kahane, N., Reshef, R., and Neufeld, G. (2001) Differential expression of neuropilin-1 and –2 in arteries and veins. Mech. Dev. 109:115-119.

Hibbs, R.G., Burch, G., and Phillips, J.H. (1958) The small structure of the blood vessels of normal human dermis subcutis. Amer. Heart J. 56:662-670.

Hirabayashi, M. and Yamamoto, T. (1984) An electron microscopic study of the endothelium in mammalian bronchial microvasculature. Cell & Tiss. Res. 326:19-25.

Hirschberg, H., Bergh, O., and Thorsby, E. (1980) Antigen presenting properties of human vascular endothelial cells. J. Exp. Med. 152:249s-255s.

Hirschberg, H., Moen, T., and Thorsby, E. (1979) Specific destruction of human endothelial cells by antisera. Transplantation 28:116-120.

Hojo, Y., Saito, Y., Tanimoto, T., Hoefen, R.J., Baines, C.P., Yamamoto, K., Haendeler, J., Asmis, R., and Berk, B.C. (2002) Fluid shear stress attenuates H_2O_2-induced c-Jun NH2-terminal kinase activation via a GSH reductase- mechanism. Circ. Res. 91:712-718.

Ikeda, K., Utoguchi, N., Makimoto, H., Mizuguchi, H., Nakagawa, S., and Mayumi, T. (1999) Different reactions of aortic and venular endothelial cell monolayers to histamine on macromolecular permeability: role of cAMP, cytosolic Ca^{2+} and F-actin. Inflam. 23:87-97.

Janzer, R.C. and Raff, M.C. (1987) Astrocytes induce blood-brain barrier properties in endothelial cells. Nature 325:253-257.

Jin, E., Ghazizadeh, J.E., Fujiwara, M., Nagashima, M., Simizu, H., Ohaki, Y., Arai, S., Gomibuchi, M., Takemura, T., and Kawanami, O. (2001) Angiogenesis and phenotypic alteration of alveolar capillary endothelium in areas of neoplastic cell spread in primary lung adenocarcinoma. Path. Internt'l. 51:691-700.

Jino, H., Kurahashi, K., Usui, H., Nakata, Y. and Shimizu, Y. (1996) Possible involvement of endothelial leukotrienes in acetylcholine-induced contraction in rabbit coronary artery. Life Sci. 59:961-967.

Kano, Y., Katoh, K., and Fujiwara, K. (2000) Lateral zone of cell-cell adhesion as the major fluid shear stress-related signal transduction site. Circ. Res. 86:425–433.

Kaul, S., Padgett, R.C., and Heistad, D.D. (1994) Role of platelets and leukocytes in modulation of vascular tone. Ann. N. Y. Acad. Sci. 714:122-135.

Kawanami, O. (1997) The endothelium of the pulmonary microvessels. J. Nippon Med. School 64:495-511.

Kawanami, O, Jin, E., Ghazizadeh, M., Fujiwara, M., Jiang, L., Nagashima, M., Shimizu, H., Takemura, t., Ohaki, T., Arai, S., Gomibuchi, M., Takeda, K., Yu, Z., and Ferrens, V. (2000) Heterogeneous distribution of thrombomodulin and vaon Willebrand factor in endothelial cells in human pulmonary microvessels. J. Nippon Med. School 67, 118-125.

Kelly, J.J., Moore, T., Babal, P., Diwan, A., Stevens, T., and Thompson, W.J. (1998) Pulmonary microvascular and macrovascular endothelial cells: differential regulation of Ca^{2+} and permeability. Amer. J. Physiol. 274:L810-L819.

King, J., Hamil, T., Creighton, J., Wu, S., Bhat, P., McDonald, F., and Stevens, T. (2004) Structural and functional characteristics of lung macro- and microvascular endothelial cell phenotypes. Microv. Res. 67:139-151.

Knowles, D.M., Toudjian, B., Marboe, C., D'Agati, V., Grimes, M., and Chess, L. (1984) Monoclonal antihuman monocyte antibodies OKM1 and OKM5 possess distinctive tissue distributions: differential reactivity with vascular endothelium. J. Immunol. 132:2170-2173.

Korhonen J., Polvi, A., Partanen, J., and Alitalo, K. (1994) The mouse tie receptor tyrosine kinase gene: expression during embryonic angiogenesis. Oncogene 9:395-403.

Laszik, Z., Mitro, A., Taylor, F., Ferrell, G., and Esmon, C.T. (1997) Human protein C receptor is present primarily on endothelium of large blood vessels: implications for the control of the protein C pathway. Circ. 96:3633-3640.

Laterra, J., Guerin, C., and Goldstein G.W. (1990) Astrocytes induce neural microvascular endothelial cells to form capillary-like structures in vitro. J. Cell Physiol. 144:204-215.

Laurent, T.C. and Fraser, J.R. (1992) Hyaluronan. FASEB J. 6:2397-2404.

Lawson, N.D., Scheer, N., Pham, V., Kim, C., Chitnis, A., Campos-Ortega, J., and Weinstein, B.M. (2001) Notch signaling is required for arterial-venous differentiation during embryonic vascular development. Development. 128:3675-3683.

Le Cras, T.D., Tyler, R., Horan, M., Morris, K., Tuder, R., McMurtry, I., Johns, R., and Abman, S.H. (1998) Effects of chronic hypoxia and altered hemodynamics on endothelial nitric oxide synthase expression in the adult rat lung. J. Clin Invest. 101:795–801.

Levin, E.G., Osborn, K., and Schleuning, W. (1998) Tissue plasminogen activator expression is limited to bronchial arteries and pulmonary vessels of discrete size. Chest 114:68S.

Lokeshwar, V.B. and Selzer, M.G. (2000) Differences in hyaluronic acid-mediated functions and signaling in arterial, microvessel, and vein-derived human endothelial cells. J Biol. Chem. 275:27641-27649.

Majno, G. (1965) Ultrastructure of the vascular membrane. In: Hamilton, W.F. and Dow, P. (Eds) Handbook of physiology, vol. 2, Williams and Wilkins, Baltimore, pp. 2293-2375.

Mantovani, A., Bussolino, F., and Dejana, E. (1992) Cytokine regulation of endothelial cell function. FASEB J. 6:2591-2599.

Mebazaa, A., Wetzel, R., Cherian, M., and Abraham, M. (1995) Comparison between endocardial and great vessel endothelial cells: morphology, growth, and prostaglandin release. Amer. J .Physiol. 268:H250-H259.

Meyrick, B. and Reid, L. (1978) The effect of continued hypoxia an rat pulmonary arterial circulation. Lab. Invest. 38:188-200.

Mikawa, T. and Fischman, D.A. (1992) Retroviral analysis of cardiac morphogenesis: discontinuous formation of coronary vessels. PNAS USA 89:9504-9508.

Milici, A.J., Furie, M., and Carley, W.W. (1985) The formation of fenestrations and channels by capillary endothelium in vitro. PNAS USA 82:6181-6185.

Miossec, P., Cavender, D., and Ziff, M. (1986) Production of interleukin-1 by human endothelial cells. J. Immunol. 136:2486-2491.

Moldobaeva A. and Wagner, E.M. (2002) Heterogeneity of bronchial endothelial cell permeability. Amer. J. Physiol. 283:L520–L527.

Moore, T.M., Chetham, P., Kelly, J., and Stevens, T. (1998) Signal transduction and regulation of lung endothelial cell permeability. Amer. J. Physiol. 275:L203–L222.

Müller, A.M., Cronen, C., Müller, K.M., and Kirkpatrick, C.J. (2002a) Heterogeneous expression of cell adhesion molecules by endothelial cells in ARDS. J. Pathol. 198:270-275.

Müller, A.M., Nesslinger, M., Skipka, G., and Müller, K.M. (2002b) Expression of CD34 in pulmonary endothelial cells. Pathobiology 70:11-17.

Müller, A.M., Skrzynski, C., Nesslinger, M., Skipka, G., and Müller, K.M. (2002c) Correlation of age with in vivo expression of endothelial markers. Exper. Gerontol. 37:713-719.

Müller, A.M., Skrzynski, C., Skipka, G., and Müller, K-M. (2002d) Expression of von Willebrand factor by human pulmonary endothelial cells in vivo. Respiration 69:526–533.

Muth, H., Maus, U., Wygrecka, M., Lohmeyer, J., Grimminger, F., Seeger, W., and Gunther, A. (2004) Pro- and antifibrinolytic properties of human pulmonary microvascular versus artery endothelial cells: impact of endotoxin and TNFα. Crit. Care Med. 32:217-226.

Mutin, M., Dignat-George, F., and Sampol, J. (1997) Immunologic phenotype of cultured endothelial cells. Tissue Antigens. 50:449-458.

Nabel, E.G. (1991) Biology of the impaired endothelium. Amer. J. Cardiol. 68:6C-8C.

Nagura, H., Koshikawa, T., Fukuda, Y., and Isai, J. (1986) Hepatic vascular endothelial cells heterogeneously express surface antigens associated with monocytes, macrophages and T lymphocytes. Virchows Arch. 409:407-416.

Newman, P.J. (1997) The biology of PECAM-1. J. Clin. Invest. 99:3-8.

Nicosia, R.F., Villaschi, S., and Smith, M. (1994) Isolation and characterization of vasoformative endothelial cells from the rat aorta. In Vitro Cell. Devel. Biol. 30:394-399.

Noden, D.M. (1989) Embryonic origins and assembly of blood vessels. Amer. Rev. Respir. Dis. 140:1097-1103.

Ogawa, T., Yorioka, N., Ito, T., Ogata, S., Kumagai, J., Kawanishi, H., and Yamakido, M. (1997) Precise ultrastructural localization of ELAM-1 VCAM-1, and ICAM-1 in patients with IgA nephropathy. Nephron 75:54-64.

Ohbayashi, A., Hiraga, T., Okubo, M., Murase, T., Matsushita, H., and Hara, M. (1994) Characteristics of porcine coronary artery endothelial cells in culture: comparison with aortic endothelium. Biochem. Biophys. Res. Commun. 202:504-511.

Osborne, J.A., Siegman, M.J., Sedar, A.E., Moore, S.U. and Lefer, A.M. (1989) Lack of endothelium-dependent relaxation in coronary resistance arteries of cholesterol-fed rabbit. Amer. J. Physiol. 256:C591-C597.

Page, C., Rose, M., Yacoub, M., and Pigott, R. (1992) Antigenic heterogeneity of vascular endothelium. Amer. J. Pathol. 141:673-683.

Parker, J.C. and Yoshikawa, S. (2002) Vascular segmental permeabilites at high peak inflation pressue in isolated lungs. Amer. J. Physiol. 283:L1203-L1209.

Parums, D.V., Cordell, J., Micklem, K., Heryet, A., Gatter, K., and Mason, D.Y. (1990) A new monoclonal antibody that detects vascular endothelium associated antigen on routinely processed tissue sections. J. Clin. Pathol. 43:752-757.

Patterson C.E. and Lum H. (2001) Update on pulmonary edema: the role and regulation of endothelial barrier function. Endothelium 8:75-105.

Petrache, I., Verin, A.D., Crow, M.T., Birukova, A., Liu, F., and Garcia, J.G.N. (2001) Differential effect of MLC kinase in TNF-ainduced endothelial cell apoptosis and barrier dysfunction. Amer. J. Physiol. 280:L1168–L1178.

Phelps, J.E. and DePaola, N. (2000) Spatial variations in endothelial barrier function in disturbed flows in vitro. Amer. J. Physiol. 278:H469–H476.

Pober, J.S. and Gimbrone, M.A. (1982) Expression of Ia-like antigens by human vascular endothelial cells is inducible in vitro. Demonstration by monoclonal antibody binding and immunoprecipitation. PNAS USA 79:6641-6645.

Porter, G.A., Palade, G., and Milici, A.J. (1990) Differential binding of the lectins Griffonia simplicifolia I and Lycopersicon esculentum to microvascular endothelium: organ-specific localization and partial glycoprotein characterization. Euro. J. Cell Biol. 51:85-95.

Rajantie, I., Ekman, N., Iljin, K., Arighi, E., Gunji, Y., Kaukonen, J., Palotie, A., Dewerchin, M., Carmeliet, P., and Alitalo, K. (2001) Bmx tyrosine kinase has a redundant function downstream of angiopoietin and vascular endothelial growth factor receptors in arterial endothelium. Mol Cell Biol. 21:4647-4655.

RayChaudhury, A., Elkins, M., Kozien, D., and Nakada, M.T. (2001) Regulation of PECAM-1 in endothelial cells during cell growth and migration. Exper. Biol. & Med. 226:686-689.

Rhodin, J.A.G. (1968) Ultrastructure of mammalian venous capillaries, venules and small collecting veins. J. Ultrastruct. Res. 25:452-500.

Risau, W. (1994) Angiogenesis and endothelial cell function. Arzneim. Forsch. 44:416-417.

Risau, W. (1995) Differentiation of the endothelium. FASEB J. 9:926-933.

Rivard, A., Fabre, J., Silver, M., Chen, D., Murohara, T., Kearney, M., Magner, M., Asahara, T., and Isner, J.M. (1999) Age-dependent impairment of angiogenesis. Circ. 99:111-120.
Rone, J.D. and Goodman, A.L. (1987) Heterogeneity of rabbit aortic endothelial cells in primary culture. Proc. Soc. Exp. Biol. Med. 184:495-503.
Rose, M.L., Page, C., Hengstenberg, C., and Yacoub, M.H. (1990) Identification of antigen presenting cells in normal and transplanted human heart. Human Immunol. 28:179-185.
Sadovnikova, E., Martynov, A.V., and Danilov, S.M. (1990) Long-term serial cultivation of human vascular endothelial cells. Biomed Sci Instrum 1:199–205.
Sahai, E. and Marshall, C. J. (2002) ROCK and Dia have opposing effects on adherens junctions downstream of Rho. Nature Cell. Biol. 4:408-415.
Sahai, E. and Marshall, C. J. (2003) Differing modes of tumour cell invasion have distinct requirements for Rho/ROCK signaling and proteolysis. Nat. Cell Biol. 5:711-719.
Sato, N., Suzuki, Y., Nishio, K., Suzuki, K., Naoki, K., Takeshita, K., Kudo, H., Miyao, N., Tsumura, H., Serizawa, H., Suematsu, M., and Yamaguchi, K. (2000) Roles of ICAM-1 for abnormal leukocyte recruitment in the microcirculation of bleomycin-induced fibrotic lung injury. Amer. J. Respir. Crit. Care Med. 161:1681-1688.
Schlingemann, R.O., Rietveld, F., de Waal, R., Bradley, N., Skene, A., Davies, A., Greaves, M., Denekamp, J., and Ruiter, D.J. (1990) Leukocyte antigen CD34 is expressed by a subset of cultured endothelial cells and on endothelial abluminal microprocesses in the tumor stroma. Lab. Invest. 62:690-696.
Schnitzer, J.E., Siflinger-Birnboim, A., DelVecchio, P., and Malik, A.B. (1994) Segmental differentiation of permeability, protein glycosylation, and morphology of cultured bovine lung vascular endothelium. Biochem. Biophys. Res. Comm. 199:11-19.
Scott-Burden, T. and Vanhoutte, P.M. (1994) Regulation of smooth muscle cell growth by endothelium-derived factors. Texas Heart Inst. J. 21:91-97.
Silkworth, J.B. and Stehbens, W.E. (1975) The shape of endothelial cells in en face preparations of rabbit blood vessels. Angiology 26:474-487.
Simionescu, N. and Simionescu, M. (1977) The cardiovascular system. Histology. In: Weiss, L. and Greep, R.O. (Eds) McGraw Hill, New York, pp 373-431.
Simionescu, N., Simionescu, M., and Palade G.E. (1975) Segmental differentiations of cells junctions in the vascular endothelium: The microvasculature. J. Cell. Biol. 67:863-885.
Simionescu, N., Simionescu, M., and Palade, G.E. (1976) Segmental differentiations of cells junctions in vascular endothelium: Arteries and veins. J. Cell. Biol. 68:705-723.
Sironi, M., Breviario, F., Proserpio, P., Biondi, A., Vecchi, A., Van Damme, J., Dejana, E., and Mantovani, A. (1989) IL-1 stimulates IL-6 production in endothelial cells. J. Immunol. 142:549-553.
Spanel-Borowski, K. and Fenyves, A. (1994) The heteromorphology of cultured microvascular endothelial cells. Drug Res. 44:385-391.
Stevens, T. (2002) Bronchial endothelial cell phenotypes and the form:function relationship. Amer. J. Physiol. 283:L518-L519.
Stevens, T., Creighton, J., and Thompson, W. (1999) Control of cAMP in lung endothelial cell phenotypes. Implications for control of barrier function. Amer. J. Physiol. 277:L119-L126.
Stevens, T., Rosenberg, R., Aird, W., Quertermous, T., Johnson, F.L., Garcia, J.G., Hebbel, R.P., Tuder, R.M., and Garfinkel, S. (2001) Endothelial cell phenotypes in heart, lung, and blood diseases. Amer. J. Physiol. 281:C1422-C1433.
Sun, R.J., Muller, S., Wang, X., Zhuang, F., and Stoltz, J.F. (2000) Regulation of vWF of human endothelial cells exposed to laminar flows. Clin. Hemo. Microcirc. 23:1-11.
Taichman, D.B., Schachtner, S., Yixun, L., Puri, M., Bernstein, A., and Baldwin, H.S. (2003) A unique pattern of Tie1 expression in developing murine lung. Exp. Lung Res. 29:113-122.
Thilo-Korner, D.G.S., Heinrich, D., and Temme, H. (1983) Endothelial cells in culture. In: The endothelial cell - a pluripotent control cell of the vessel wall. Thilo-Korner, D.G.S. and Freshney, R.I. (eds) Karger, Basel, pp 202-258.
Thorin, E. and Atkinson, J. (1994) Modulation by the endothelium of sympathetic vasoconstriction in an in vitro preparation of rat tail artery. Br. J. Pharmacol. 111:351-357.

Thorin, E., Shatos, M.A., Shreeve, S.M., Walters, C.L., and Bevan, J.A. (1997) Human vascular endothelium heterogeneity: a comparative study of cerebral and peripheral cultured vascular endothelial cells. Stroke 28:375-381.
Thorin, E. and Shreeve, S.M. (1998) Heterogeneity of vascular endothelial cells in normal and disease states. Pharm. Ther. 78 :155-166.
Tsutomu, K., Takashi, M., Miyake, K., and Nagura, H. (1989) Phenotypic heterogeneity of vascular endothelial cells in the human kidney. Cell Tissue Res. 256:27-34.
Urbich, C., Walter, D., Zeiher, A., and Dimmeler, S. (2000) Laminar shear stress upregulates integrin expression. Role in endothelial cell adhesion and apoptosis. Circ Res. 87:683–689.
Vane, J.R., Anggard, E., and Botting, R.M. (1990) Regulatory functions of the vascular endothelium. N. Engl. J. Med. 323:27-36.
Van-Epps, D.E., Potter, J., Vachula, M., Smith, C., and Anderson, D.C. (1989) Suppression of lymphocyte chemotaxis and transendothelial migration by anti-LFA-1 antibody. J. Immunol. 143:3207-3210.
Wagner, C.R., Vetto, M., and Burger, D.R. (1985) Subcultured human endothelial cells can function as fully competent antigen presenting cells. Hum. Immunol. 13:33-47.
Wang, H.U., Chen, Z., and Anderson, D.J. (1998) Molecular distinction and angiogenic interaction between embryonic arteries and veins revealed by ephrin-B2 and its receptor Eph-B4. Cell 93:741-753.
Wang, Q., Pfeiffer, G., Stevens, T. and Doerschuk, C.M. (2002) Lung microvascular and arterial endothelial cells differ in their responses to intercellular adhesion molecule-1 ligation. Amer. J. Respir. Crit. Care Med. 166:872-877.
Wedgwood, S., Mitchell, C., Fineman, J., and Black, S.M. (2003) Developmental differences in the shear stress-induced expression of endothelial NO synthase: changing role of AP-1. Amer. J. Physiol. 284:L650-L662.
Wei, Z., Costa, K., Al-Mehdi, A., Dodia, C., Muzykantov, V., and Fisher, A. B. (1999) Simulated ischemia in flow-adapted endothelial cells leads to generation of reactive oxygen species and cell signaling. Circ. Res. 85:682-689.
Weibel, E.R. and Palade, G.E. (1964) New cytoplasmic components in arterial endothelia. J. Cell. Biol. 23:101-112.
Wu, K.K. and Thiagarajan, P. (1996) Role of endothelium in thrombosis and hemostasis. Ann. Rev. Med. 47:315-331.
Wu, S., Haynes, J., Taylor, J., Obiako, B., Stubbs, J., Li, M., and Stevens, T. (2003) $Ca_v3.1$ (α_{1G}) T-type Ca^{2+} channels mediate vaso-occlusion of sickled erythrocytes in lung microcirculation. Circ. Res. 93:346-353.
Yamamoto, M., Shimokata, K., and Nagura, H. (1988) Immunohistochemical study on phenotypic heterogeneity of human pulmonary vascular endothelial cells. Virchows Arch. 412:479-486.
Zakrzewicz, A., Grafe, M., Terbeek, D., Bongrazio, M., Auch-Schwelk, W., Walzog, B., Graf, K., Fleck, E., Ley, K., and Gaehtgens, P. (1997) L-selectin-dependent leukocyte adhesion to microvascular but not to macrovascular endothelial cells of the human coronary system. Blood 89:3228-3235.
Zetter, B.R. (1988) Endothelial heterogeneity: influence of vessel size, organ localization, and species specificity on the properties of cultured endothelial cells. In: Endothelial cell II. Ryan, U.S. (Ed) Boca Raton, FL, CRC Press, pp 63-79.
Zhong, T.P., Rosenberg, M., Mohideen, M., Weinstein, B., and Fishman, M.C. (2000) Gridlock, an HLH gene required for assembly of the aorta in zebrafish. Science 287:1820-1824.
Zhu, D. and Pauli, B.U. (1993) Correlation between the lung distribution pattersn of Lu-ECAM-1 and melanoma metastases. Intern. J. Cancer 53:628-633.

Chapter 11.
Interaction of Pulmonary Endothelial Cells with Blood Elements

Qin Wang, Ph.D.,* Inkyung Kang, M.S., & Claire M. Doerschuk, M.D.

Division of Integrative Biology, Department of Pediatrics, Case Western Reserve University
and Rainbow Babies and Children's Hospital, Cleveland, OH 44106

CONTENTS:

Introduction

Adhesion Molecules: Key to Interactions of Endothelium and Leukocytes
Neutrophil Margination and Emigration
Adhesion Molecules and Neutrophil Emigration
Monocyte Kinetics

Regulation of Adhesion Molecules during Acute Lung Inflammation
Expression of Neutrophil CD11/CD18
Expression of Endothelial ICAM-1

Endothelial Responses to Leukocyte Adhesion and Transmigration
Early Signaling Events
Neutrophils Alter Endothelial Oxidant Production and Modulate Coagulation
Other Leukocytes Affect the Endothelium

Adhesion Molecules as Signaling Molecules
ICAM-1
Other Endothelial Adhesion Molecules

ICAM-1 Signaling and Neutrophil Migration
Endothelial Activation Affects Neutrophil Migration
Possible Mechanisms

Interactions of Endothelium with other Blood-borne Elements
Platelets
Albumin

Conclusions

Introduction

The lung is anatomically equipped for efficient gas exchange. The human pulmonary vasculature contains about 3×10^{11} capillary segments that spread over 10^8 alveoli (Hogg and Doerschuk, 1995). Lining the inner surface of these capillaries is a single layer of endothelial cells. Approximately 8600 liters of blood pass through the pulmonary circulation of an adult human in 24 hr. Thus, the pulmonary capillary endothelium is well situated to interact with the blood constituents including leukocytes and plasma proteins, primarily through the interactions of molecules that are either constitutively expressed or induced under pathological conditions. The interaction of endothelial cells with leukocytes and plasma proteins plays important roles in regulating leukocyte emigration, vascular homeostasis, as well as endothelial permeability.

This chapter focuses on the interaction of leukocytes and plasma proteins with endothelial cells in the lung. During pulmonary inflammation, endothelial cells as well as blood cells respond to the released mediators, resulting in enhanced interaction between endothelial cells and blood cells, including leukocytes. Recent studies suggest that this interaction results in changes in both endothelial cells and leukocytes, in part due to signaling mechanisms induced by the ligation of adhesion molecules. These events that occur in both endothelial cells and leukocytes during adhesion likely influence the migration of leukocytes and other inflammatory processes. This chapter will discuss the interaction of endothelial cells with blood elements, with an emphasis on the interaction of endothelial cells with neutrophils, the most numerous type of polymorphonuclear leukocytes, during acute pulmonary inflammation.

Adhesion Molecules: Key to interactions of Endothelium and Leukocytes

Neutrophil Margination and Emigration

In healthy lungs, the capillary blood contains a higher concentration of neutrophils than the blood in large vessels. This marginated pool in the pulmonary capillaries exists even in the absence of an inflammatory stimulus and is due primarily to the anatomical characteristics of pulmonary capillaries and the biomechanical properties of neutrophils. In a single pass through the pulmonary capillary bed, neutrophils on average need to transit through 50 to 100 capillary segments (Hogg *et al.*, 1994; Hogg and Walker, 1995). The narrow diameter of these capillaries ($7.5 \pm 2.3 \mu m$) relative to the average diameter of spherical neutrophils ($6.8 \pm 0.8 \mu m$) requires that neutrophils deform to pass through at least 40-60% of the capillary segments (Doerschuk *et al.*, 1993). This anatomical restriction, along with the decreased deformability compared to red blood cells, contributes to the delayed neutrophil transit time through the pulmonary capillary bed, and thus, neutrophil margination in the pulmonary capillaries. This neutrophil margination in a noninflamed pulmonary capillary bed does not require adhesion molecules. Blockade of L-, E- or P-selectin or of CD11/CD18 does not affect this marginated pool of neutrophils (Yoder *et al.*, 1990; Gebb *et al.*, 1995; Doerschuk *et al.*, 1996; Qin *et al.*, 1996; Doyle *et al.*, 1997).

Neutrophil emigration in response to an acute inflammatory stimulus occurs primarily in the pulmonary capillaries. This process includes the sequestration of neutrophils in the pulmonary capillaries, adhesion of neutrophils to the pulmonary capillaries, transmigration of neutrophils across endothelium, and migration of neutrophils through extracellular matrix and across alveolar epithelial cells into the alveolar space (reviewed by Doerschuk, 2001; Burns et al., 2003). Thus, compared to what happens in the systemic circulation, neutrophil emigration occurs at different sites in the pulmonary circulation (capillaries vs. post-capillary venules), and neutrophil rolling does not occur (Gebb et al., 1995). For this reason, the molecular mechanisms pertinent to neutrophil rolling will be not discussed in detail in this chapter.

The initial sequestration of neutrophils in the pulmonary capillary bed is thought to occur as a result of decreased deformability of neutrophils in response to inflammatory mediators (Doerschuk, 2001). None of the known adhesion molecules is required for this initial sequestration of neutrophils. However, L-selectin and CD11/CD18 are required to keep neutrophils in the pulmonary capillaries.

Adhesion Molecules and Neutrophil Emigration

Both neutrophils and endothelial cells respond to inflammatory mediators to upregulate adhesion molecules on their surface. The binding of these adhesion molecules to their ligands allows neutrophils to adhere to capillary endothelial cells (Table 1). The ligation of these adhesion molecules also induces signaling events into both neutrophils and endothelial cells, as will be discussed in detail below. Events that occur during neutrophil adhesion to endothelial cells likely play important roles in modulating subsequent neutrophil migration on the endothelial surface to reach the borders and/or transmigration across the endothelial layer.

TABLE 1. Leukocyte/Endothelial Cell Adhesion Molecules

Leukocyte	Endothelial Cell
L-selectin	PSGL-1, PNAd, MAdCAM-1
PSGL-1	P-selectin, E-selectin
CD11a/CD18 ($\alpha_L\beta_2$)	ICAM-1, ICAM-2, JAM-A
CD11b/CD18 ($\alpha_M\beta_2$)	ICAM-1, JAM-C
CD49d/CD29 (VLA 4, $\alpha_4\beta_1$)	VCAM-1, JAM-B
$\alpha_4\beta_7$ (LPAM)	MAdCAM-1, VCAM-1
JAM-C	JAM-B
CD226 (DNAM-1)	CD155 (PVR)
CD99	CD99
PECAM-1	PECAM-1
?	VAP-1

Table 1: Adhesion molecules that mediate the interaction between leukocytes and endothelial cells. The adhesion molecules expressed by leukocytes and their corresponding ligands expressed by endothelial cells are included in the respective column. PSGL-1: P-selectin glycoprotein ligand-1; PNAd: peripheral lymph node addressin; MAdCAM-1: mucosal addressin cell adhesion molecule-1; JAM: junctional adhesion molecule; LPAM: lymphocyte Peyer patch adhesion molecule; DNAM-1: DNAX accessory molecule 1; PVR: poliovirus receptor; VAP-1: vascular adhesion protein-1.

The lung is unique compared to other organs in that neutrophil adhesion and migration can occur through CD18-dependent or -independent mechanism, depending on the stimulus (Doerschuk, 2001). The mechanisms regulating CD18-independent neutrophil migration are not understood. The process regulating CD18-dependent neutrophil emigration has been examined in many studies at molecular and cellular levels. Inflammatory cytokines including TNF-α and IL-1β, by binding to their receptors, induce signaling events in cells including capillary endothelial cells, resulting in upregulation of adhesion molecules such as ICAM-1 on the endothelial surface. In addition, inflammatory mediators generated during inflammatory responses stimulate neutrophils to upregulate their expression of CD11/CD18. The roles of these early inflammatory cytokines and mediators in modulating neutrophil emigration have been reviewed and will not be repeated here (Ward, 1996; Zhang et al., 2000; Doerschuk, 2001; Mizgerd, 2002; Strieter et al., 2002; Guo and Ward, 2002).

The role for CD11/CD18 and ICAM-1 in mediating neutrophil emigration in the lung has been established using functional blocking antibodies or techniques to inhibit the expression of these adhesion molecules. Functional blocking antibodies to CD18 or CD18-deficiency of neutrophils inhibit neutrophil emigration in response to some stimuli including *E. coli*, *E. coli* LPS or *P. aeruginosa*, but not to others, including *S. pneumoniae* or *S. aureus* (Doerschuk et al., 1990b; Qin et al., 1996; Ramamoorthy et al., 1997; Mizgerd et al., 1999). Thus, CD18-dependent and -independent neutrophil emigration can occur in the lung, depending on the stimulus. In response to a stimulus that elicits CD18-dependent neutrophil emigration, ICAM-1 is also required for neutrophil emigration. Blocking antibodies to ICAM-1 or antisense oligonucleotide to ICAM-1 prevent neutrophil emigration in response to these stimuli (Kumasaka et al., 1996; Qin et al., 1996). These data suggest that ICAM-1 binding to CD18 is required for neutrophil emigration when CD18 is utilized.

The role of other adhesion molecules in mediating neutrophil emigration into the lungs during the acute inflammatory process has been examined. For example, very late antigen (VLA)-5 and VLA-6 may mediate migration of neutrophils from the interstitium into alveoli in response to *E. coli* LPS (Ridger et al., 2001). Blocking VLA-4 has a small inhibitory effect on neutrophil emigration induced by *S. pneumoniae*, but the majority of neutrophil emigration was not affected (Tasaka et al., 2002). Mice deficient in platelet endothelial cell adhesion molecule-1 (PECAM-1, CD31) or rats treated with a blocking anti-PECAM-1 antibody show no defect in neutrophil emigration in lungs in response to *E. coli*, *S. pneumoniae* or acid aspiration, while blocking PECAM-1 partially inhibits neutrophil emigration into peritoneum induced by glycogen (Tasaka et al., 2003; Albelda et al., 2004).

Monocyte Kinetics

Similar to neutrophils, there is a marginated pool of monocytes in pulmonary capillaries (Ohgami et al., 1991). During pulmonary inflammation, neutrophil emigration is often followed by subsequent accumulation of monocytes in the lung (Ohgami et al., 1992; Li et al., 1998; Maus et al., 2002). Monocyte accumulation in the lung also requires coordinated interaction between monocyte adhesion molecules including CD11/CD18 and VLA-4 and their ligands including those expressed on endothelium such as ICAM-1 and VCAM-1. In response to intratracheal instillation of *E. coli* LPS, monocyte accumulation in the pulmonary capillaries is partially inhibited by anti-CD18 antibody, while monocyte accumulation in the alveolar space is inhibited by combined CD18 and VLA-4 antibodies (Li et al., 1998). In response to intratracheal instillation of low dose *E. coli* LPS plus monocyte chemoattractant

JE/monocyte chemotactic protein (MCP)-1, monocyte emigration into the alveolar space is inhibited by blocking antibodies against CD11/CD18, ICAM-1, VLA-4, and a ligand for VLA-4, VCAM-1 (Maus *et al.*, 2002).

Regulation of Adhesion Molecules during Acute Lung Inflammation

Expression of Neutrophil CD11/CD18

CD11/CD18 on circulating neutrophils exists in a quiescent state of low binding affinity to their ligands. Many inflammatory mediators including chemokines increase the expression of CD11/CD18 on the surface of neutrophils, largely due to the mobilization of secretary granules to the plasma membrane (Graves *et al.*, 1992; Hughes *et al.*, 1992; Sengelov *et al.*, 1995). The adhesiveness of CD11/CD18 expressed on neutrophils is regulated by so-called "inside-out" signaling events induced by inflammatory mediators as well as by the ligation of other adhesion molecules on neutrophils (for example, Simon *et al.*, 2000; Seo *et al.*, 2001). These inside-out signaling events induce two forms of changes in CD11/CD18 that result in enhanced adhesiveness: 1) increases in the binding affinity of CD11/CD18 to their ligands due to conformational changes; 2) increases in the binding avidity of CD11/CD18 due to lateral motility of these molecules leading to the formation of high avidity clusters (Harris *et al.*, 2000; Harris *et al.*, 2001; Hogg *et al.*, 2002; Laudanna *et al.*, 2002). The regulation of CD11/CD18 by inside-out signaling mechanisms is physiologically important. Patients whose leukocytes including neutrophils have defective integrin activation induced by inflammatory agonists suffer from severe bacterial as well as viral infections, despite normal expression levels of integrins on leukocytes (Harris *et al.*, 2001; McDowall *et al.*, 2003).

Expression of Endothelial ICAM-1

ICAM-1 is normally expressed at very low levels on pulmonary capillary endothelial cells (Burns *et al.*, 1994). During acute inflammatory responses, ICAM-1 can be upregulated through at least three mechanisms; increased adhesiveness through phosphorylation of the cytoplasmic tail, increased expression through enhanced transcription, and increased adhesiveness through dimerization and/or association with the cytoskeleton. First, short term exposure to cytokines such as TNF-α in cultured pulmonary artery endothelial cells results in phosphorylation of the cytoplasmic domain of ICAM-1 through PKCζ. This modulation enhances the adhesiveness of ICAM-1 prior to de novo synthesis (Javaid *et al.*, 2003). Second, the expression of ICAM-1 is transcriptionally upregulated in response to a variety of stimuli. Stresses such as hypoxia as well as inflammatory mediators such as TNF-α and IL-1β induce transcriptional upregulation, resulting in increased ICAM-1 expression on the endothelial surface, as demonstrated by numerous studies using cultured endothelial cells. The expression of ICAM-1 increases on pulmonary capillary endothelia during *E. coli* pneumonia *in vivo* (Burns *et al.*, 1994). The promoter of ICAM-1 has binding sites for transcription factors including NF-κB, and activation of NF-κB signaling cascade is pivotal for ICAM-1 transcription (Roebuck and Finnegan, 1999). Intratracheal instillation of *E. coli* LPS results in nuclear translocation of NF-κB subunits, p65 and p50, in the lung (Blackwell *et al.*, 1999). Mice deficient in both TNF-α and p65 have reduced ICAM-1 mRNA expression induced by *E. coli* LPS, when compared with wild type animals or mice singly deficient in

TNF-α (Alcamo et al., 2001). NF-κB-independent transcription of ICAM-1 has also been described (Takizawa et al., 1999; Gorgoulis et al., 2003). For example, two functional p53-responsive elements are present within the introns of the *ICAM-1* gene, and the binding of p53 to these two elements induces expression of ICAM-1 during irradiation (Gorgoulis et al., 2003). These studies indicate that transcriptional regulation of ICAM-1 is complex and that NF-κB-independent upregulation can occur. Third, at least a fraction of ICAM-1 expressed on endothelium exists as dimers, and dimeric ICAM-1 is more adhesive than monomeric ICAM-1 (Miller et al., 1995; Reilly et al., 1995; Jun et al., 2001). Moreover, the cytoplasmic domain of ICAM-1 binds several actin-binding proteins including ezrin and α-actinin, and linkage of ICAM-1 to the cytoskeleton through these actin-binding proteins likely modulates adhesive functions of ICAM-1 (Carpen et al., 1992; Heiska et al., 1998; Barreiro et al., 2002).

Endothelial Responses to leukocyte Adhesion and Transmigration

Early Signaling Events

Firm adhesion, mediated by the interaction of neutrophil CD11/CD18 and endothelial ICAM-1, is often required for subsequent transmigration across endothelium. Neutrophil transmigration across endothelial cells often occurs at endothelial junctions as observed *in vivo* and *in vitro*, although neutrophil migration through endothelial cells in response to intradermal injection of fMLP has also been demonstrated (Walker et al., 1995; Burns et al., 1997; Feng et al., 1998; Shaw et al., 2001). Leukocytes including neutrophils and monocytes migrate on the surface of cultured endothelial cells before transmigration occurs (Luu et al., 2000; Rochon et al., 2000; Schenkel et al., 2004). Recent studies suggest that during leukocyte adhesion and transmigration, outside-in signaling occurs in both neutrophils and endothelial cells within minutes, and that these early signaling events may very likely influence leukocyte crawling on the surface of endothelial cells to the junctions and transmigration across endothelial cells. These early signaling events occurring in endothelial cells will be discussed in detail below. They include intracellular changes in Ca^{++}, oxidant production, changes in the cytoskeleton and the cell's biomechanical properties, and changes in transcription factors. Most of these studies are conducted using cultured endothelial cells derived from different vascular beds *in vitro*, and tissue-specific differences may exist.

Huang and colleagues demonstrated that fMLP-activated neutrophils induce an increase in intracellular Ca^{++} in cultured human umbilical vein endothelial cells (Huang et al., 1993). A recent study by Su and colleagues extended this observation by measuring calcium responses at the single-cell level in human umbilical vein endothelial cells during neutrophil transmigration induced by fMLP (Su et al., 2000). Interestingly, increases in calcium in human umbilical vein endothelial cells are associated with neutrophil transmigration, but not adhesion (Su et al., 2000). Chelation of intracellular Ca^{++} inhibits neutrophil transmigration across endothelial cells without inhibiting neutrophil adhesion, suggesting that Ca^{++}-dependent events in endothelial cells are required for transmigration (Huang et al., 1993). One of the Ca^{++}-dependent events is the phosphorylation of myosin light chain. Indeed, fMLP-stimulated neutrophils induce myosin light chain phosphorylation and isometric tension generation in cultured human umbilical vein endothelial cells, pulmonary arterial endothelial cells, and pulmonary arterial rings, and inhibition of MLCK in endothelial cells prevents neutrophil transmigration (Patterson et al., 1989; Patterson et al., 1992; Hixenbaugh *et al.*, 1997; Garcia *et al.*, 1998; Saito *et al.*, 1998; Yuan *et al.*, 2002; Breslin and Yuan, 2004). These studies suggest that activated neutrophils induce Ca^{++} increases in large vessel endothelia, leading to MLCK activation and increased actin and myosin interactions, and these events may be essential for neutrophil transmigration across endothelial cells.

Neutrophil adhesion and/or transmigration also induce changes in cytokine-pretreated endothelial cells. In cytokine-treated human umbilical vein endothelial cells, increases in Ca^{++} occur during neutrophil adhesion and transmigration (Huang *et al.*, 1993; Ziegelstein *et al.*, 1994; Lorenzon *et al.*, 1998). In TNF-α-pretreated endothelial cells, neutrophils induce changes in the biomechanical properties and F-actin cytoskeleton of endothelial cells within 2 minutes of adhesion (Wang and Doerschuk, 2000; Wang *et al.*, 2001). These endothelial cytoskeletal changes consist of increased thickness of microfilaments and focal F-actin aggregates. These changes require actin rearrangement, since they are prevented by agents that either disrupt or stabilize F-actin (Wang *et al.*, 2001). These cytoskeletal changes, however, do not appear to require Ca^{++} or myosin light chain phosphorylation, but occur through a phosphatidylinositol-dependent mechanism (Wang *et al.*, 2001). This cytoskeletal remodeling in endothelial cells is accompanied by an increase in neutrophil migration toward endothelial borders (Wang *et al.*, 2001). These studies demonstrating that neutrophil adhesion induces cytoskeletal changes in pulmonary microvascular endothelial cells that do not require Ca^{++} or MLCK are in contrast to previously described studies in large vessel endothelial cells.

These apparent discrepancies likely reflect unappreciated and exciting differences in mechanisms underlying the migratory process in different cell types. First, the changes in endothelial cells induced by neutrophil adhesion may be differentially regulated compared to those occurring during transmigration. Since neutrophil adhesion and migration likely engage different adhesion molecules, it is conceivable that different responses will be elicited in endothelial cells. For example, a recent study showed that neutrophil transmigration across IL-1β-pretreated human umbilical vein endothelial cells in the presence of an exogenous chemoattractant, PAF, induces a decrease in the nuclear NF-κB in these cells (Cepinskas *et al.*, 2003). Neutrophil adhesion alone is not sufficient to induce these changes in endothelial cells. Rather, this change appears to require transmigration of neutrophils across endothelial cells, and is prevented by an anti-PECAM-1 antibody, but not by an anti-ICAM-1 antibody (Cepinskas *et al.*, 2003). Thus, neutrophil adhesion and transmigration induce different signaling events into endothelial cells, likely due to the fact that different adhesion molecules are engaged during these two processes. Second, the responses in endothelial cells induced by neutrophil adhesion may depend on the type of endothelial cells. There may be different mechanisms regulating neutrophil-induced cytoskeletal changes in endothelial cells derived from different vascular beds or from large vessels compared to microvessels of the same tissue. Indeed, when endothelial cells isolated from rat pulmonary arteries and microvessels are compared, neutrophil-induced cytoskeletal changes occur only in pulmonary microvascular endothelial cells, and not in the arterial cells (Wang *et al.*, 2002).

Neutrophils Alter Endothelial Oxidant Production and Modulate Coagulation

Neutrophil adhesion to endothelial cells also induces intracellular oxidant production in endothelial cells. Adhesion of activated neutrophils to arterial endothelial cells induces conversion of xanthine dehydrogenase to its active form, xanthine oxidase (Phan *et al.*, 1989; Phan *et al.*, 1992; Wakabayashi *et al.*, 1995). The conversion of xanthine dehydrogenase to xanthine oxidase is inhibited by an anti-CD18 antibody, suggesting that neutrophil adhesion is required (Phan *et al.*, 1989; Wakabayashi *et al.*, 1995). In addition, neutrophil adhesion to TNF-α-pretreated rat pulmonary microvascular endothelial cells induces an increase in the activity of xanthine oxidase in these endothelial cells (Wang *et al.*, 2002). Since activation of xanthine oxidase represents an important mechanism for superoxide production in endothelial cells (Granger, 1988; Grisham and Granger, 1989), this conversion to xanthine oxidase likely results in increases in the production of superoxide and other reactive oxygen species (ROS) in endothelial cells. Indeed, neutrophil adhesion to TNF-α-pretreated pulmonary microvascular endothelial cells induces oxidant production in endothelial cells within minutes, but not in neutrophils (Wang and Doerschuk, 2000). This increase in oxidant production in endothelial cells is partially inhibited by allopurinol, a xanthine oxidase inhibitor, suggesting that xanthine oxidase contributes to oxidant production in endothelial cells induced by neutrophil adherence. Because the inhibition was partial, xanthine oxidase may not be solely responsible for oxidant production, and other pathways contribute as well. One possible candidate is NADPH oxidase. This enzyme complex is expressed in lung endothelial cells, and its activation in endothelial cells is responsible for oxidant generation during conditions including lung ischemia (Fisher *et al.*, 1999; Wei *et al.*, 1999; Fisher *et al.*, 2003; Parinandi *et al.*, 2003; Gertzberg *et al.*, 2004). This increase in oxidant production in pulmonary microvascular endothelial cells is required for the endothelial stiffening response induced by neutrophil adherence, suggesting that oxidant production soon after neutrophil adhesion may serve as signaling molecules in endothelial cells and result in subsequent cytoskeletal changes (Wang and Doerschuk, 2000).

Neutrophil adhesion to bovine aorta endothelial cells induces an increase in the tissue factor-dependent coagulation activity in endothelial cells (Watanabe *et al.*, 2001; Watanabe *et al.*, 2002). This increase is blocked with an anti-ICAM-1 antibody (Watanabe *et al.*, 2001). Although the mechanisms underlying this increase in tissue factor activity and the physiological functions of this response are unclear, these studies suggest that neutrophil adhesion induces changes in endothelial cells that may regulate coagulation.

These changes in endothelial cells induced by neutrophil adhesion may occur through at least two mechanisms, neutrophil-derived mediators and ligation-induced signaling. First, neutrophils are capable of releasing ROS and elastase in response to adhesion and ligation of CD11/CD18 (Nathan, 1987; Nathan *et al.*, 1989; Shappell *et al.*, 1990; Berton *et al.*, 1992; Zhou *et al.*, 1993). These mediators may act on endothelial cells and induce subsequent responses. For example, in pulmonary arterial endothelial cells, a role for neutrophil elastase was identified in mediating conversion of xanthine dehydrogenase to xanthine oxidase in response to adhesion by PMA-activated neutrophils (Phan *et al.*, 1992). This requirement for elastase seems to also depend on the endothelial type, since a similar role for elastase was not found in carotid arterial endothelial cells (Wakabayashi *et al.*, 1995). Nevertheless, neutrophil-derived mediators are very likely to play important roles in mediating endothelial responses during neutrophil adhesion, and high concentrations of these mediators may accumulate at the site of adhesion. In this context, endothelial adhesion molecules may play a role in neutrophil activation as ligands for neutrophil CD11/CD18. Second, ligation of endothelial adhesion molecules during neutrophil adhesion may directly initiate signaling events in the endothelial cells. Studies supporting this mechanism will be discussed below.

Other Leukocytes Affect the Endothelium

Endothelial adhesion by other leukocytes including monocytes and lymphocytes also induces changes in endothelial cells. For example, monocyte adhesion to TNF-α-activated human umbilical vein endothelial cells induces activation of Rho and F-actin rearrangement in endothelial cells (Wojciak-Stothard *et al.*, 1999; Honing *et al.*, 2004). Adhesion of monocyte to cytokine-stimulated human umbilical vein endothelial cells induces increases in intracellular Ca^{2+} in endothelial cells (Lorenzon *et al.*, 1998; Kielbassa-Schnepp *et al.*, 2001). In addition, lymphocyte adhesion to cytokine-pretreated endothelial cells induces increases in cytosolic Ca^{2+}, NADPH-dependent oxidant production, actin rearrangements, as well as transcription of cytokines, chemokines, and tissue factor in endothelial cells (Pfau *et al.*, 1995; Matheny *et al.*, 2000; Monaco *et al.*, 2002).

Adhesion Molecules as Signaling Molecules

ICAM-1

ICAM-1-initiated signaling events can be induced through ligation of ICAM-1 by neutrophils, monocytes, lymphocytes, antibodies or fibrinogen in different cell types, resulting in Ca^{2+} increases, cytoskeletal changes and gene transcription. In pulmonary microvascular endothelial cells, neutrophil adherence-induced production of reactive oxygen species and cytoskeletal changes are completely inhibited by an anti-ICAM-1 antibody and can be mimicked by crosslinking ICAM-1 with antibodies, suggesting that ICAM-1 is required and sufficient to initiate signaling events into endothelial cells (Wang and Doerschuk, 2000).

The signaling pathways induced by ligation of ICAM-1 have been examined in various cell types. In cultured pulmonary microvascular endothelial cells, ligation of ICAM-1 leads to sequential activation of xanthine oxidase, Src tyrosine kinases and p38 MAP kinases (MAPK), as shown in Figure 1. Activation of Src tyrosine kinases induced by ICAM-1 crosslinking is inhibited by allopurinol, a xanthine oxidase inhibitor, and various antioxidants (Wang *et al.*, 2003). These data suggest that activation of Src tyrosine kinases occurs downstream of activation of xanthine oxidase and production of reactive oxygen species. Activation of Src tyrosine kinases also requires a tyrosine phosphatase, SH-2-domain-containing tyrosine phosphatase-2 (SHP-2), possibly through dephosphorylation of Tyr519 near the C-terminus of Src tyrosine kinases (Wang *et al.*, 2003). Inhibition of Src tyrosine kinases by PP2 prevents activation of p38 MAPK, suggesting p38 MAPK activation occurs downstream of Src tyrosine kinases. In addition, activation of p38 MAPK is inhibited by allopurinol, an xanthine oxidase inhibitor, while inhibition of p38 MAPK has no effect on ROS production, further indicating that ICAM-1-induced p38 MAPK activation occurs downstream of xanthine oxidase and Src tyrosine kinases (Wang *et al.*, 2001). Activation of p38 MAPK in turn induces phosphorylation of heat shock protein 27 (HSP27), an actin binding protein that may induce actin polymerization when phosphorylated and is required for the cytoskeletal rearrangement induced by neutrophil adherence or ICAM-1 crosslinking (Wang and Doerschuk, 2001). These studies demonstrate that ligation of ICAM-1 in pulmonary microvascular endothelial cells induces a sequence of signaling events including production of ROS, activation of Src tyrosine kinases and p38 MAPK, and HSP27 phosphorylation, and that these signaling events play important roles in mediating the cytoskeletal changes in endothelial cells induced by neutrophil adherence.

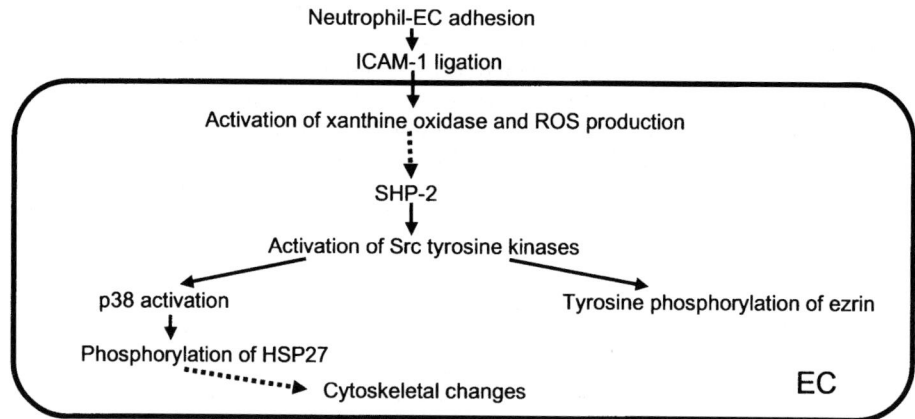

Figure 1: Neutrophil adhesion-induced signaling in pulmonary microvascular endothelium. ICAM-1-initiated signaling events in TNF-α-pretreated pulmonary microvascular endothelial cells during neutrophil adhesion. The dotted lines indicate postulated pathways.

Signaling through ICAM-1 initiated by crosslinking antibodies has also been reported in other endothelial cells. Crosslinking ICAM-1 with antibodies in brain endothelial lines or venular endothelial cells induces increases in intracellular Ca^{2+} and activation of Src tyrosine kinases, RhoA and PKC (Durieu-Trautmann et al., 1994; Clayton et al., 1998; Wojciak-Stothard et al., 1999; Etienne-Manneville et al., 2000; Thompson et al., 2002). These signaling pathways act upon several actin-associated proteins including cortactin, FAK, paxillin and p^{130} Cas, which in turn may induce changes in the actin cytoskeleton of these endothelial cells (Durieu-Trautmann et al., 1994; Etienne et al., 1998; Adamson et al., 1999). In addition, crosslinking ICAM-1 induces transcription of VCAM-1 and ICAM-1 through activation of ERK-1 and AP-1 (Clayton et al., 1998; Lawson et al., 1999). Moreover, activation of RhoA stimulates transcription of *c-fos* and *RhoA* (Thompson et al., 2002). These studies demonstrate that ICAM-1-induced signaling events result in changes in endothelial cells including cytoskeletal rearrangement and gene transcription that may likely modulate leukocyte migration and other inflammatory responses.

CAM-1 signaling also occurs in cell types other than endothelial cells. In astrocytes, ligation of ICAM-1 induces expression of proinflammatory cytokines that requires activation of Erk and p38 MAP kinase (Lee et al., 2000). In human renal fibroblasts, ICAM-1 ligation induces expression of ICAM-1 and RANTES (Blaber et al., 2003). Signaling through ICAM-1 in B and T lymphocytes and fibroblasts has also been reported, and is described in review (Hubbard and Rothlein, 2000). In addition, ICAM-1 signaling can also be initiated through ligation by fibrinogen. Fibrinogen-induced activation of ERK-1, Src tyrosine kinases, and cell proliferation is mediated through ICAM-1 (Gardiner et al., 1997; Gardiner et al., 1999; Pluskota et al., 2000). Whether different ligands for ICAM-1 initiate a different set of signaling pathways remains to be defined, as well as how ICAM-1 signaling differs depending on the cell type.

How ICAM-1 ligation initiates oxidant production and downstream signaling events in endothelial cells remains an important question. ICAM-1 is a glycosylated protein that belongs to the superfamily of immunoglobulin (Ig)-like proteins (Staunton et al., 1990). The common form of ICAM-1 has five extracellular Ig domains, a transmembrane domain and a short cytoplasmic domain. The signaling events induced by antibodies often require crosslinking by a secondary antibody, suggesting that ICAM-1 clustering may be required for ICAM-1 signaling. Indeed, ICAM-1 crosslinking induces formation of ICAM-1 clusters and aggregates, which is regulated by Rho family GTPases, and ICAM-1 clustering is observed at the site of monocyte adhesion (Wojciak-Stothard et al., 1999). The cytoplasmic domain of ICAM-1 is composed of 28 amino acids: RQRKIKKYRLQQAQKGTPMKPNTQATPP (478-505) (Staunton et al., 1988; Carpen et al., 1992). This cytoplasmic domain is required for the activation of RhoA induced by ICAM-1 ligation (Sans et al., 2001, Greenwood et al., 2003; Lyck et al., 2003). In addition, this domain is essential for leukocyte transmigration across endothelial cells. Deletion of the ICAM-1 cytoplasmic domain completely inhibits neutrophil transmigration, but not adhesion, in CHO cells, suggesting a role for ICAM-1 signaling in neutrophil migration (Sans et al., 2001). Moreover, expression of a truncated form ICAM-1 missing its cytoplasmic domain in brain endothelial cells is sufficient to support T-lymphocyte adhesion, but not transmigration (Greenwood et al., 2003; Lyck et al., 2003).

The cytoplasmic domain of ICAM-1 does not have intrinsic kinase activity or Src homology domains (SH) that can recruit tyrosine phosphorylated proteins (Staunton et al., 1988). However, this domain does interact with other molecules including actin-binding proteins, suggesting that ICAM-1-induced signaling may be initiated at the membrane-cytoskeletal interface. The intracellular domain of ICAM-1 is linked to an actin binding protein, α-actinin (Vogetseder and Dierich, 1991; Carpen et al., 1992). In addition, this domain can bind phosphatidylinositol 4,5-biphosphate (PtdIns(4,5)P$_2$), a molecule implicated in various signaling cascades (Heiska et al., 1998). This domain can also bind ezrin/radixin/moesin (ERM) proteins, and this interaction is facilitated by the presence of PtdIns(4,5)P$_2$ (Heiska et al., 1998). ERM proteins function as plasma membrane-actin cytoskeleton linkers and may also regulate signal propagation. ERM proteins are tyrosine phosphorylated in response to stimulation by growth factors, and tyrosine phosphorylated ERM proteins recruit and activate Src homology 2 (SH2)-containing kinases such as phosphatidylinositol 3-kinase (PI 3-kinase) by binding to their SH2 domain (Gautreau et al., 1999). In pulmonary microvascular endothelial cells, tyrosine phosphorylation of ezrin occurs in response to ICAM-1 crosslinking as a consequence of activation of Src tyrosine kinases (see Figure 1) (Wang et al., 2003). Thus, the interaction of ICAM-1 with PtdIns(4,5)P$_2$ may result in binding to the ERM proteins, which in turn recruit additional signaling molecules. Upon ICAM-1 crosslinking, ICAM-1 clusters indeed colocalize with the ERM proteins (Wojciak-Stothard et al., 1999).

In addition, in response to fibrinogen ligation, ICAM-1 is tyrosine phosphorylated (most likely at Y^{485} in the cytoplasmic domain), possibly via Src activation, and tyrosine phosphorylated ICAM-1 binds to the SH2-containing tyrosine phosphatase-2 (SHP-2) (Pluskota et al., 2000). A dominant negative form of SHP-2 inhibits activation of Ras and MAPK induced by growth factors (Bennett et al., 1994; Deb et al., 1998). Thus, tyrosine phosphorylation of ICAM-1 and recruitment of SHP-2 may be yet another mechanism through which ICAM-1-induced signaling events are initiated.

Other Endothelial Adhesion Molecules

Other endothelial adhesion molecules also signal, although signaling events and physiological importance in lung remain to be elucidated. E-selectin and P-selectin play important roles in mediating neutrophil rolling on post-capillary venules in systemic circulation. Neutrophil adherence to endothelial cells induces association of E-selectin with the cytoskeleton (Yoshida *et al.*, 1996) and monocyte adhesion to endothelial cells induces E-selectin clustering (Wojciak-Stothard *et al.*, 1999), suggesting that E-selectin may signal into endothelial cells. Indeed, both E-selectin and P-selectin function as signal transducers (Aplin *et al.*, 1998). Ligation of E-selectin and P-selectin with monoclonal antibodies induces signaling events in human umbilical vein endothelial cells that include: 1) increases in intracellular Ca^{2+} concentration; 2) stress fiber formation and shape changes; 3) activation of phospholipase Cγ; 4) changes in the phosphorylation state of the cytoplasmic tail of E-selectin; 5) activation of ERK and subsequent transcription of *c-fos* (Kaplanski *et al.*, 1994; Lorenzon *et al.*, 1998; Yoshida *et al.*, 1998; Hu *et al.*, 2000; Hu *et al.*, 2001; Kiely *et al.*, 2003).

VCAM-1 is a ligand for VLA-4. Monocyte adhesion on endothelial cells induces VCAM-1 clustering (Wojciak-Stothard *et al.*, 1999). Ligation of VCAM-1 by lymphocytes or antibodies induces intracellular Ca^{2+} increases, activation of NADPH oxidase, production of ROS activation of Rac1 and p38 MAPK in endothelial cells (Lorenzon *et al.*, 1998; Matheny *et al.*, 2000; van Wetering *et al.*, 2003; Alevriadou *et al.*, 2003). These signaling events are required for actin cytoskeletal changes and paracellular gap formation in endothelial cells, as well as for lymphocyte transmigration across endothelial cells (Lorenzon *et al.*, 1998; Matheny *et al.*, 2000; van Wetering *et al.*, 2003).

ICAM-1 Signaling and Neutrophil Migration

Endothelial Activation Affects Neutrophil Migration

Ligation of ICAM-1 during leukocyte adhesion clearly induces signaling events in endothelial cells, resulting in an array of downstream responses that include cytoskeletal changes, oxidant production, and transcription of genes. These downstream responses likely modulate inflammatory responses including leukocyte migration on endothelial cells and/or transmigration across endothelial cells. Neutrophil migration on the endothelial surface in particular may be regulated by ICAM-1 signaling. ICAM-1-dependent changes in endothelial cells induced by neutrophil adhesion are accompanied by the crawling of neutrophils to endothelial borders, as illustrated in Figure 2 (Wang and Doerschuk, 2001). This migration is reduced when endothelial cells are pretreated with SB203580, a p38 inhibitor that also inhibits the cytoskeletal changes and the stiffening of endothelial cells (Wang and Doerschuk, 2001). These studies suggest that ICAM-1-dependent activation of p38 MAPK and its downstream events may regulate neutrophil migration on endothelial surface toward the junctions, where transmigration occurs.

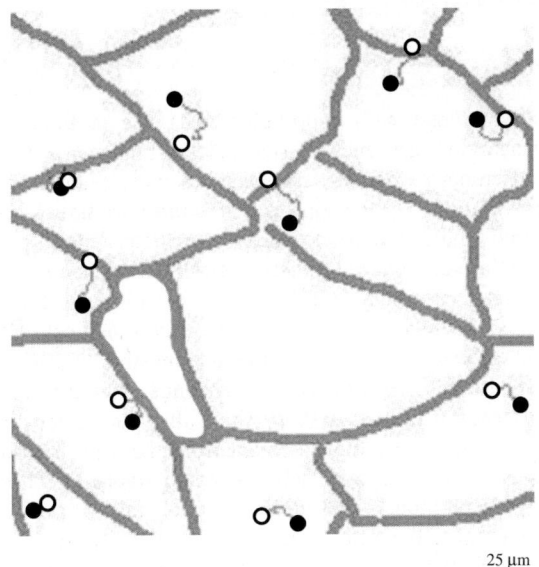

Figure 2: Neutrophil migration along EC surface. TNF-α-pretreated pulmonary microvascular endothelial cells were washed, and purified neutrophils from venous blood were allowed to adhere for 1 minute before non-adherent cells were washed. The migration of neutrophils along EC surface was imaged every 30 seconds for 10 minutes. The path of neutrophil migration was overlaid with traces of EC borders. Filled circles represent the position of neutrophils at the beginning of measurement, and open circles represent the position of neutrophils at the end of the 10 minute measurement.

Possible mechanisms

How leukocyte-induced signaling events in endothelial cells including p38 activation may influence migration of the cells on the endothelial surface and/or transmigration across endothelial cells is unknown. These signaling events may: 1) induce redistribution of ICAM-1 on the endothelial surface and association of ICAM-1 with cytoskeletal proteins such as the ERM proteins, which in turn affect neutrophil adhesion and/or migration. Indeed, ICAM-1 ligation induces clustering of ICAM-1 and formation of microvilli-like membrane projections on endothelial cells, redistribution of ICAM-1 into low-density, detergent-insoluble "raft" domains, and association with actin binding proteins such as moesin, ezrin and α-actinin in cultured endothelial cells (Amos *et al.*, 2001; Barreiro *et al.*, 2002; Tilghman *et al.*, 2002; Carman *et al.*, 2003); 2) influence the topography of the endothelial surface on which neutrophils crawl; 3) induce increases in endothelial stiffness, which may enhance neutrophil migration to endothelial borders; 4) alter the junctional functions in endothelial cells that may regulate neutrophil emigration during inflammation. Whether these signaling events influence junctional proteins implicated in neutrophil transmigration such as PECAM-1, JAMs and CD99 remains to be determined (Palmeri *et al.*, 2000; Ostermann *et al.*, 2002; Schenkel *et al.*, 2002).

Whether these ICAM-1-initiated signaling events in endothelial cells during neutrophil adherence are required for neutrophil emigration into the alveolar space during inflammatory processes *in vivo* also remains to be determined. Generation of genetically altered animal models such as animals expressing a truncated form of ICAM-1 without its cytoplasmic domain required for signaling will help elucidate the physiological significance of ICAM-1-initiated signaling events in mediating neutrophil emigration in lung inflammation *in vivo*.

Interactions of Endothelium with Other Blood-borne Elements

Platelets

Circulating platelets do not normally interact with endothelial cells due to the anti-thrombotic nature of the endothelial surface (Ware and Heistad, 1993; Lo et al., 1998; Ruggeri, 2003; Marcus et al., 2003). Different approaches have been taken to examine the transit of platelets through the pulmonary circulation. One approach is to infuse radioisotope-labeled platelets and red blood cells into animals and measure the platelets' first pass extraction through pulmonary circulation relative to red blood cells, as well as the percentage of infused platelets remaining in the lung after 10 minutes (Doerschuk et al., 1990a). During the first pass through the lungs, platelets show a delayed transit relative to red blood cells. In another study, fluorescently-labeled platelets were infused into animals, and the velocity of platelets relative to red blood cells was measured over a period of 5 seconds using intravital microscopy (Eichhorn et al., 2002). The results from this study show that the velocity of platelets transiting through the pulmonary circulation is not different from that of red blood cells, and that the majority of platelets pass the capillaries without being trapped. These discrepancies may reflect differences in the pulmonary vascular physiological state of the animals and may suggest that platelet kinetics is influenced by vascular flow and pressure.

Sequestration of platelets in pulmonary microcirculation occurs within hours during pathological conditions such as endotoxemia or infusion of endothelin-1 (Helset et al., 1996; Kiefmann et al., 2004). The exact mechanisms regulating the interactions between platelets and endothelial cells during acute pulmonary inflammatory responses are yet to be fully understood, although selectins and enhanced secretion of von Willebrand factor by endothelial cells, as well as binding through fibrinogen via $\alpha_{IIb}\beta_3$ on platelets to ICAM-1 on endothelium, are all potential mechanisms (Cheresh et al., 1989; Frenette et al., 1995; Andre et al., 2000; Frenette et al., 2000; Kiefmann et al., 2004).

Sequestered and activated platelets can attract more platelets and neutrophils, through the release of agonists such as thromboxane A_2 and the interaction of selectins and their ligands (Valles et al., 1993; Marcus et al., 2003). Activated platelets also release metabolites including sphingosine 1-phosphate and lysophosphatidic acid that can modulate endothelial barrier integrity (Haselton and Alexander, 1992; Luscher, 1993; Alexander et al., 1998; Yatomi et al., 2000; Gainor et al., 2001; Sachais, 2001; Schaphorst et al., 2003). In addition, platelets release products that can be metabolized into other products by cells in the proximity (reviewed by Marcus et al., 2003). One such example is the eicosanoid pathway (Marcus et al., 2003). Activated platelets can process arachidonic acid, resulting in release of thromboxane A_2, or PGG_2 and PGH_2, which can be further processed by endothelial cells into PGI_2. Arachidonic acid released from platelets can be processed by activated neutrophils to generate LTB_4. In addition, platelets release ADP, which is metabolized into AMP by CD39 on endothelial surface (Marcus et al., 2003). Through another enzyme localized on endothelial surface, 5'-nucleotidase, AMP is metabolized into adenosine. Both PGI_2 and adenosine inhibit platelet activation. Thus, activated platelets, through adhesion molecules and the release of these metabolites, play important roles in inflammatory responses and homeostasis. Platelet-adhered endothelial cells act to modulate platelet activation and recruitment.

Albumin

The physiological significance of albumin in plasma includes the following roles. First, albumin plays important roles in regulating vascular homeostasis through maintaining the osmotic pressure (Victorino et al., 2003). Secondly, albumin is the major acid/base buffer in plasma and pH has profound effects on endothelial function. Thirdly, albumin is a major carrier protein in plasma that binds bioactive mediators including fatty acids. Fatty acids binding to albumin can be transported across the lung capillary endothelium through transcytosis (Galis et al., 1988). Fourth, albumin and other proteins in serum bind to the endothelial luminal surface, reducing its permeability to other macromolecules possibly by blocking binding sites (Turner et al., 1983; Schneeberger et al., 1984; Lum et al., 1991; He and Curry, 1993). Fifth, albumin modulates the inflammatory responses by acting as an anti-oxidant and by reducing cytokine-induced adhesion molecule expression on endothelial cells (Zhang and Frei, 2002; Lang et al., 2004).

Albumin also interacts with the endothelial surface through specific binding proteins. The pulmonary capillary endothelium indeed contains specific albumin-binding sites that are restricted mainly to vesicles (Ghitescu et al., 1986). Albumin-binding proteins including gp60 (albodin), gp30 and gp18 are expressed on the surface of endothelial cells, and binding of these proteins to albumin can activate transcytosis of albumin across endothelium through intracellular signaling mechanisms that include activation of heterotrimeric G proteins and Src tyrosine kinases (Shasby and Shasby, 1985; Ghinea et al., 1988; Schnitzer et al., 1988; Ghinea et al., 1989; Schnitzer, 1992; Schnitzer and Oh, 1994; Tiruppathi et al., 1996; Tiruppathi et al., 1997; Minshall et al., 2000; Vogel et al., 2001; John et al., 2003). Antibody to albodin reduces tissue uptake of ^{125}I-BSA in the isolated perfused rat lung, although uptake of modified albumin is not affected (Schnitzer and Oh, 1994). Thus, these specific albumin-binding proteins on the endothelial surface may modulate albumin transport across endothelium. Despite cellular uptake and transcytosis, albumin reduces the total albumin flux and permeability of the endothelium to other macromolecules (Lum et al., 1991; He and Curry, 1993; Pizurki et al., 2003). Moreover, albumin decreases hydraulic conductivity of the endothelium (Dull et al., 1991). This albumin effect on improving the barrier integrity is likely related to reduction in intracellular Ca^{++} and strengthening of the intercellular junctions, which is of increasing importance in the paracellular permeability encountered in pathologic conditions (He and Curry, 1993).

Conclusions

Pulmonary capillary endothelial cells actively engage in interactions with blood elements, including leukocytes and plasma proteins. During acute inflammation in the lung, capillary endothelial cells and neutrophils usually respond to inflammatory mediators to upregulate adhesion molecules on their surface. These adhesion molecules such as CD11/CD18 on neutrophils and ICAM-1 on endothelial cells allow neutrophils to adhere to the endothelial cells. Ligation of ICAM-1 on endothelial cells also induces signaling events into endothelial cells, resulting in downstream responses that include oxidant production, cytoskeletal rearrangements, and transcriptional regulation. We are beginning to understand the signaling events induced during adhesion and the physiological significance of these events. Understanding how this signal influences neutrophil crawling on the endothelial surface and transmigration across endothelial cells will further our understanding of the mechanisms regulating neutrophil emigration during inflammatory responses in the lung.

ACKNOWLEDGEMENTS

The authors would like to acknowledge the following grant support: a Parker B. Francis Fellowship from the Francis Families Foundation (QW), NIH HL070009 (QW), HL48160 (CMD), HL52466 (CMD) and a Clinical Scientist Award in Translational Research from the Burroughs Wellcome Fund (CMD).

REFERENCES

Adamson, P., Etienne, S., Couraud, P., Calder, V., and Greenwood, J. (1999) Lymphocyte migration through brain endothelial monolayers involves signaling through endothelial ICAM-1 via a rho-dependent pathway. J. Immunol. 162: 2964-2973.

Albelda, S.M., Lau, K., Chien, P., Huang, Z., Arguiris, E., Bohen, A., Sun, J., Billet, J. Christofidou-Solomidou, M., Indik, Z., and Schreiber, A.D. (2004) Role for PECAM in macrophage Fcγ receptor function. Amer. J. Respir. Cell Molec. Biol. 31:246-255.

Alcamo, E., Mizgerd, J., Horwitz, B., Bronson, R., Beg, A., Scott, M., Doerschuk, C.M., Hynes, R.O., and Baltimore, D. (2001) Targeted mutation of TNF receptor I rescues the RelA-deficient mouse and reveals a critical role for NF-kappa B in leukocyte recruitment. J. Immunol. 167:1592-1600.

Alevriadou, B.R. (2003) CAMs and Rho small GTPases: gatekeepers for leukocyte transendothelial migration. Focus on "VCAM-1-mediated Rac signaling controls endothelial cell-cell contacts and leukocyte transmigration". Amer. J. Physiol. 285:C250-C252.

Alexander, J.S., Patton, W., Christman, B., Cuiper, L., and Haselton, F.R. (1998) Platelet-derived lysophosphatidic acid decreases endothelial permeability *in vitro*. Amer. J. Physiol. 274:H115-H122.

Amos, C., Romero, I., Schultze, C., Rousell, J., Pearson, J., Greenwood, J., and Adamson, P. (2001) Cross-linking of brain endothelial ICAM-1 induces association of ICAM-1 with detergent-insoluble cytoskeletal fraction. Arterioscl. Thromb. Vasc. Biol. 21:810-816.

Andre, P., Denis, C.V., Ware, J., Saffaripour, S., Hynes, R., Ruggeri, Z., and Wagner, D.D. (2000) Platelets adhere to and translocate on vWf presented by endothelium in stimulated veins. Blood 96:3322-3328.

Aplin, A.E., Howe, A., Alahari, S., and Juliano, R.L. (1998) Signal transduction and signal modulation by cell adhesion receptors: the role of integrins, cadherins, immunoglubulin-cell adhesion molecules, and selectins. Pharm. Rev. 50:197-263.

Barreiro, O., Yanez-Mo, M., Serrador, J., Montoya, M., Vicente-Manzanares, M., Tejedor, R., Furthmayr, H. and Sanchez-Madrid, F. (2002) Dynamic interaction of VCAM-1 and ICAM-1 with moesin and ezrin in a novel endothelial docking structure for adherent leukocytes. J. Cell Biol. 157:1233-1245.

Bennett, A.M., Tang, T., Sugimoto, S., Walsh, C., and Neel, B.G. (1994) Protein-tyrosine-phosphatase SHPTP2 couples platelet-derived growth factor β to Ras. PNAS USA 91:7335-7339.

Berton, G., Laudanna, C., Sorio, C., and Rossi, F. (1992) Generation of signals activating neutrophil functions by leukocyte integrins: LFA-1 and gp 150/95, but not CR3, are able to stimulate the respiratory burst of human neutrophils. J. Cell Biol. 116:1007-1017.

Blaber, R., Stylianou, E., Clayton, A., and Steadman, R. (2003) Selective regulation of ICAM-1 and RANTES gene expression after ICAM-1 ligation on human renal fibroblasts. J. Amer. Soc. Nephrol. 14:116-127.

Blackwell, T.S., Lancaster, L., Blackwell, T., Venkatakrishnan, A., and Christman, J.W. (1999) Differential NF-κB activation after intratracheal endotoxin. Amer. J. Physiol. 277:L823-L830.
Breslin, J.W. and Yuan, S.Y. (2004) Involvement of RhoA and Rho kinase in neutrophil-stimulated endothelial hyperpermeability. Amer. J. Physiol. 286:H1057-H1062.
Burns, A.R., Smith, C., and Walker, D.C. (2003) Unique structural features that influence neutrophil emigration into the lung. Physiol. Rev. 83:309-336.
Burns, A.R., Takei, F., and Doerschuk, C.M. (1994) Quantitation of ICAM-1 expression in mouse lung during pneumonia. J. Immunol. 153:3189-3198.
Burns, A.R., Walker, D., Brown, E., Thurmon, L., Bowden, R., Keese, C., Simon, S., Entman, M., and Smith, C.W. (1997) Neutrophil transendothelial migration is independent of tight junctions and occurs preferentially at tricellular corners. J. Immuno. 159:2893-2903.
Carman, C.V., Jun, C., Salas, A., and Springer, T.A. (2003) Endothelial cells proactively form microvilli-like membrane projections upon intercellular adhesion molecule 1 engagement of leukocyte LFA-1. J. Immunol. 171:6135-6144.
Carpen, O., Pallai, P., Staunton, D., and Springer, T.A. (1992) Association of ICAM-1 with actin-containing cytoskeleton and alpha-actinin. J. Cell Biol. 118:1223-1234.
Cepinskas, G., Savickiene, J., Ionescu, C., and Kvietys, P.R. (2003) PMN transendothelial migration decreases nuclear NFκB in IL-1β-activated endothelial cells: role of PECAM-1. J Cell Biol. 161:641-651.
Cheresh, D.A., Berliner, S., Vicente, V., and Ruggeri, Z.M. (1989) Recognition of distinct adhesive sites on fibrinogen by related integrins on platelets and endothelial cells. Cell 58:945-953.
Clayton, A., Evans, R., Pettit, E., Hallett, M., Williams, J., and Steadmean, R. (1998) Cellular activation through the ligation of ICAM-1. J. Cell Sci. 111:443-453.
Deb, T.B., Wong, L., Salomon, D., Zhou, G., Dixon, J., Gutkind, J., Thompson, S., and Johnson, G.R. (1998) A common requirement for the catalytic activity and both SH2 domains of SHP-2 in MAPK activation by the ErbB family of receptors. J. Biol. Chem. 273:16643-16646.
Doerschuk, C.M. (2001) Mechanisms of leukocyte sequestration in inflamed lungs. Microcirc. 8:71-88.
Doerschuk, C.M., Beyers, N., Coxson, H., Wiggs, B., and Hogg, J.C. (1993) Comparison of neutrophil and capillary diameters and their relation to neutrophil sequestration in lung. J. Appl. Physiol. 74:3040-3045.
Doerschuk, C.M., Downey, G., Doherty, D., English, D., Gie, R., Ohgami, M., Worthen, G., Henson, P., and Hogg, J.C. (1990) Leukocyte and platelet margination within microvasculature of rabbit lungs. J. Appl. Physiol. 68:1956-1961.
Doerschuk, C.M., Quinlan, W., Doyle, N., Bullard, D., Vestweber, D., Jones, M., Takei, F., Ward, P., and Beaudet, A.L. (1996) The role of P-selectin and ICAM-1 in acute lung injury as determined using blocking antibodies and mutant mice. J. Immunol. 157:4609-4614.
Doerschuk, C.M., Winn, R., Coxson, H., and Harlan, J.M. (1990) CD18-dependent and -independent mechanisms of neutrophil emigration in the pulmonary and systemic microcirculation of rabbits. J. Immunol. 144:2327-2333.
Doyle, N.A., Bhagwan, S., Meek, B., Kutkoski, G., Steeber, D., Tedder, T., and Doerschuk, C.M. (1997) Neutrophil margination, sequestration, and emigration in the lungs of L-selectin-deficient mice. J. Clin. Invest. 99:526-533.
Dull, R.O., Jo, H., Sill, H., Hollis, T., and Tarbell, J.M. (1991) The effect of vaying albumin concentration and hydrostatic pressure on hydraulic conductivity and albumin permeability of cultuted endothelial monolayers. Microv. Res. 41:390-407.
Durieu-Trautmann, O., Chaverot, N., Cazaubon, S., Strosberg, A., and Couraud, P. (1994) ICAM-1 activation induces tyrosine phosphorylation of the cytoskeleton-associated protein cortactin in brain microvessel endothelial cells. J. Biol. Chem. 269:12536-12540.
Eichhorn, M.E., Ney, L., Massberg, S., and Goetz, A. (2002) Platelet kinetics in the pulmonary microcirculation *in vivo* assessed by intravital microscopy. J. Vasc. Res. 39:330-339.
Etienne, S., Adamson, P., Greenwood, J., Strosberg, A.D., Cazaubon, S., and Couraud, P. (1998) ICAM-1 signaling pathways associated with Rho activation in microvascular brain endothelial cells. J. Immunol. 161:5755-5761.

Etienne-Manneville, S., Manneville, J., Adamson, P., Wilbourn, B., Greenwood, J., and Couraud, P.O. (2000) ICAM-1-coupled cytoskeletal rearrangements and transendothelial lymphocyte migration involve intracellular calcium signaling in brain endothelial lines. J. Immunol. 165:3375-3383.

Feng, D., Nagy, J., Pyne, K., Dvorak, H., and Dvorak, A.M. (1998) Neutrophils emigrate from venules by a transendothelial pathway in response to FMLP. J. Exp. Med. 187:903-915.

Fisher, A.B., Al-Mehdi, A.B., and Muzykantov, V. (1999) Activation of endothelial NADPH oxidase as the source of a reactive oxygen species in lung ischemia. Chest 116:25S-26S.

Fisher, A.B., Al-Mehdi, A.B., Wei, Z., Song, C., Manevich, Y. (2003) Lung ischemia: endothelial cell signaling by reactive oxygen species. Adv. Exp. Med. Biol. 510:343-347.

Frenette, P.S., Johnson, R., Hynes, R., and Wagner, D.D. (1995) Platelets roll on stimulated endothelium *in vivo*: an interaction mediated by endothelial P-selectin. Proc. Nat'l. Acad. Sci. USA 92:7450-7454.

Frenette, P.S., Denis, C., Weiss, L., Jurk, K., Subbarao, S., Kehrel, B., Hartwig, J., Vestweber, D., and Wagner, D.D. (2000) P-Selectin glycoprotein ligand 1 (PSGL-1) is expressed on platelets and can mediate platelet-endothelial interactions *in vivo*. J. Exp. Med. 191:1413-1422.

Gainor, J.P., Morton, C., Roberts, J., Vincent, P., and Minnear, F.L. (2001) Platelet-conditioned medium increases endothelial electrical resistance independently of cAMP/PKA and cGMP/PKG. Amer. J. Physiol. 281:H1992-H2001.

Galis, Z., Ghitescu, L., and Simionescu, M. (1988) Fatty acids binding to albumin increases its uptake and transcytosis by the lung capillary endothelium. Eur. J. Cell Biol. 47:358-365.

Garcia, J.G.N., Verin, A., Herenyiova, M., and English, D. (1998) Adherent neutrophils activate endothelial myosin light chain kinase: role in transendothelial migration. J. Appl. Physiol. 84:1817-1821.

Gardiner, E.E. and D'Souza, S.E. (1997) A mitogenic action for fibrinogen mediated through intercellular adhesion molecule-1. J. Biol. Chem. 272:15474-15480.

Gardiner, E.E. and D'Souza, S.E. (1999) Sequences within fibrinogen and ICAM-1 modulate signals required for mitogenesis. J. Biol. Chem. 274:11930-11936.

Gautreau, A., Poullet, P., Louvard, D., and Arpin, M. (1999) Ezrin, a plasma membrane-microfilament linker, signals cell survival through the PI3K/Akt pathway. PNAS USA 96:7300-7305.

Gebb, S.A., Graham, J., Hanger, C., Godbey, P., Capen, R., Doerschuk, C.M., and Wagner, W.W. (1995) Sites of leukocyte sequestration in the pulmonary microcirculation. J. Appl. Physiol. 79:493-497.

Gertzberg, N., Neumann, P., Rizzo, V., and Johnson, A. (2004) NAD(P)H oxidase mediates the endothelial barrier dysfunction induced by TNF-α. Amer. J. Physiol. 286:L37-L48.

Ghinea, N., Eskenasy, M., Simionescu, M., and Simionescu, N. (1989) Endothelial cell albumin binding proteins are membrane-associated components exposed on the cell surface. J. Biol. Chem. 264:4755-4758.

Ghinea, N., Fixmanm, A., Alexandru, D., Popov, D., Hasu, M., Ghitescu, L., Eskenasy, M., Simionescu, M., and Simionescu, N. (1988) Identification of albumin binding proteins in capillary endothelial cells. J. Cell Biol. 107:231-239.

Ghitescu, L., Fixman, A., Simionescu, M., and Simionescu, N. (1986) Specific binding sites for albumin restricted to plasmalemmal vesicles of continuous capillary endothelium: receptor-mediated transcytosis. J. Cell Biol. 102:1304-1311.

Gorgoulis, V.G., Zacharatos, P., Kotsinas, A., Kletsas, D., Mariatos, G., Zoumpourlis, V., Ryan, K., Kittas, C., and Papavassiliou, A.G. (2003) p53 activates ICAM-1 (CD54) expression in an NFkB-independent manner. EMBO J. 22:1567-1578.

Granger, D.N. (1988) Role of xanthine oxidase and granulocytes in ischemia-reperfusion injury. Amer. J. Physiol. 255:H1269-H1275.

Graves, V., Gabig, T., McCarthy, L., Strour, E., Leemhuis, T., and English, D. (1992) Simultaneous mobilization of Mac-1 (CD11b/CD18) and formyl peptide chemoattractant receptors in human neutrophils. Blood 80:776-787.

Greenwood, J., Amos, C., Walters, C., Couraud, P., Lyck, R., Engelhardt, B., and Adamson, P. (2003) Intracellular domain of brain endothelial ICAM-1 is essential for T lymphocyte-mediated signaling and migration. J. Immunol. 171:2099-2108.

Grisham, M.B. and Granger, D.N. (1989) Metabolic sources of reactive oxygen metabolites during oxidant stress and ischemia with reperfusion. Clin. Chest Med. 10:71-81, 1989.
Guo, R.F. and Ward, PA. (2002) Mediators and regulation of neutrophil accumulation in inflammatory responses in lung: insights from the IgG immune complex model. Free Rad. Biol. Med. 33:303-310.
Harris, E.S., McIntyre, T., Prescott, S., and Zimmerman, G.A. (2000) The leukocyte integrins. J. Biol. Chem. 275:23409-23412.
Harris, E.S., Shigeoka, A., Li, W., Adams, R., Prescott, S., McIntyre, T., Zimmerman, G., and Lorant, D.E. (2001) A novel syndrome of variant leukocyte adhesion deficiency involving defects in adhesion mediated by β1 and β2 integrins. Blood 97:767-776.
Haselton, F.R. and Alexander, J.S. (1992) Platelets and a platelet-released factor enhance endothelial barrier. Amer. J. Physiol. 263:L670-L678.
He, P. and Curry, F.E. (1993) Albumin modulation of capillary permeability: role of endothelial $[Ca^{2+}]_i$. Amer. J. Physiol. 265: H74-H82.
Heiska, L., Alfthan, K., Gronholm, M., Vilja, P., Vaheri, A., and Carpen, O. (1998) Association of ezrin with ICAM-1 and ICAM-2. Regulation by phosphatidylinositol 4, 5-bisphosphate. J. Biol. Chem. 273:21893-21900.
Helset, E., Lindal, S., Olsen, R., Myklebust, R., and Jorgensen, L. (1996) Endothelin-1 causes sequential trapping of platelets and neutrophils in pulmonary microcirculation in rats. Amer. J. Physiol. 271:L538-L546.
Hixenbaugh, E.A., Goeckeler, Z., Papaiya, N., Wysolmerski, R., Silverstein, S., and Huang, A.J. (1997) Stimulated neutrophils induce myosin light chain phosphorylation and isometric tension in endothelial cells. Amer. J. Physiol. 273:H981-H988.
Hogg, J.C., Coxson, H., Brumwell, M., Beyers, N., Doerschuk, C.M., MacNee, W., and Wiggs, B.R. (1994) Erythrocyte and polymorphonuclear cell transit time and concentration in human pulmonary capillaries. J. Appl. Physiol. 77:1795-1800.
Hogg, J.C. and Doerschuk, C.M. (1995) Leukocyte traffic in the lung. Ann. Rev. Physiol. 57:97-114.
Hogg, N., Henderson, R., Leitinger, B., McDowall, A., Porter, J., and Stanley, P. (2002) Mechanisms contributing to the activity of integrins on leukocytes. Immunol. Rev. 186:164-171.
Hogg, J.C. and Walker, B.A. (1995) Polymorphonuclear leucocyte traffic in lung inflammation. Thorax. 50:819-820.
Honing, H., van den Berg, T., van der Pol, S., Dijkstra, C., van der Kammen, R., Collard, J., and de Vries, H.E. (2004) RhoA activation promotes transendothelial migration of monocytes via ROCK. J. Leuk. Biol. 75:523-528.
Hu, Y., Kiely, J., Szente, B., Rosenzweig, A., and Gimbrone, M.A. (2000) E-selectin-dependent signaling via the MAPK pathway in vascular endothelial cells. J. Immunol. 165:2142-2148.
Hu, Y., Szente, B., Kiely, J., and Gimbrone, M.A. (2001) Molecular events in transmembrane signaling via E-selectin: SHP2 association, adaptor protein complex formation and ERK1/2 activation. J. Biol. Chem. 276:48549-48553.
Huang, A., Manning, J., Bandak, T., Ratau, M., Hanser, K., and Silverstein, S.C. (1993) Endothelial cytosolic free Ca^{++} regulated neutrophil migration across endothelial monolayers. J. Cell Biol. 120:1371-1380.
Hubbard, A.K. and Rothlein, R. (2000) ICAM-1 expression and cell signaling cascades. Free Rad. Biol. Med. 28:1379-1386.
Hughes, B.J., Hollers, J., Crockett-Torabi, E., and Smith, C.W. (1992) Recruitment of CD11b/CD18 to neutrophil surface and adherence-dependent cell locomotion. J. Clin. Invest. 90:1687-1696.
Javaid, K., Rahman, A., Anwar, K., Frey, R., Minshall, R., and Malik, A.B. (2003) TNFα induces early-onset endothelial adhesivity by PKCζ-dependent activation of ICAM-1. Circ. Res. 92:1089-1097.
John, T.A., Vogel, S., Tiruppathi, C., Malik, A.B., and Minshall, R.D. (2003) Quantitative analysis of albumin uptake and transport in the rat microvessel endothelial monolayer. Amer. J. Physiol. 284:L187-L196.
Jun, C.D., Shimaoka, M., Carman, C., Takagi, J., and Springer, T.A. (2001) Dimerization and the effectiveness of ICAM-1 in mediating LFA-1-dependent adhesion. PNAS USA 98:6830-6835.

Kaplanski, G., Farnarier, C., Benoliel, A., Foa, C., Kaplanski, S., and Bongrand, P. (1994) A novel role for E- and P-selectins: shape control of endothelial monolayers. J. Cell Sci. 107: 2449-2457.
Kiefmann, R., Heckel, K., Schenkat, S., Dorger, M., Wesierska-Gadek, J., and Goetz, A.E. (2004) Platelet-endothelial interaction in pulmonary micro-circulation: the role of PARS. Thromb Haemost. 91:761-770.
Kielbassa-Schnepp, K., Strey, A., Janning, A., Missiaen, L., Nilius, B., and Gerke, V. (2001) Endothelial intracellular Ca2+ release following monocyte adhesion is required for the transendothelial migration of monocytes. Cell Calcium 30:29-40.
Kiely, J.M., Hu, Y., Garcia-Cardena, G., and Gimbrone, M.A. (2003) Lipid raft localization of cell surface E-selectin required for ligation-induced activation of PLCγ. J. Immunol. 171:3216-3224.
Kumasaka, T., Quinlan, W., Doyle, N., Condon, T., Sligh, J., Takei, F., Beaudet, A., Bennett, C., and Doerschuk, C.M. (1996) Role of the ICAM-1 in endotoxin-induced pneumonia evaluated using ICAM-1 antisense oligonucleotides, anti-ICAM-1 monoclonal antibodies, and ICAM-1 mutant mice. J. Clin. Invest. 97:2362-2369.
Lang, J.D., Figueroa, M., Chumley, P., Aslan, M., Hurt, J., Tarpey, M., Alvarez, B., Radi, R., and Freeman, B.A. (2004) Albumin and hydroxyethyl starch modulate oxidative inflammatory injury to vascular endothelium. Anesthesiol. 100:51-58.
Laudanna, C., Kim, J., Constantin, G., and Butcher, E. (2002) Rapid leukocyte integrin activation by chemokines. Immunol. Rev. 186:37-46.
Lawson, C., Ainsworth, M., Yacoub, M., and Rose, M. (1999) Ligation of ICAM-1 on endothelial cells leads to expression of VCAM-1 via a NFκB-independent mechanism. J. Immunol. 162:2990-2996.
Lee, S.J., Drabik, K., Van Wagoner N., Lee, S., Choi, C., Dong, Y., and Benveniste, E.N. (2000) ICAM-1-induced expression of proinflammatory cytokines in astrocytes: involvement of extracellular signal-regulated kinase and p38 MAPK pathways. J. Immunol. 165:4658-4666.
Li, X.C., Miyasaka, M., and Issekutz, T.B. (1998) Blood monocyte migration to acute lung inflammation involves both CD11/CD18 and very late activation antigen-4-dependent and independent pathways. J. Immunol. 161(11):6258-6264.
Lo, S.K., Burhop, K., Kaplan, J., and Malik, A.B. (1988) Role of platelets in maintenance of pulmonary vascular permeability to protein. Amer. J. Physiol. 254:H763-H771.
Lorenzon, P., Vecile, E., Nardon, E., Ferrero, E., Harlan, J.M., Tedesco, F., and Dobrina, A. (1998) Enodthelial E- and P-selectin and vascular cell adhesion molecule-1 function as signaling receptors. J. Cell Biol. 142:1381-1391.
Lum, H., Siflinger-Birnboim, A., Blumenstock, F., and Malik, A.B. (1991) Serum albumin decreases transendothelial permeability to macromolecules. Microv. Res. 42:91-102.
Luscher, T.F. (1993) Platelet-vessel wall interaction: role of nitric oxide, prostaglandins and endothelins. Baillieres Clin. Haematol. 6:609-627.
Luu, N.T., Rainger, G.E., and Nash, G.B. (2000) Differential ability of exogenous chemotactic agents to disrupt transendothelial migration of flowing neutrophils. J. Immunol. 164:5961-5969.
Lyck, R., Reiss, Y., Gerwin, N., Greenwood, J., Adamson, P., and Engelhardt, B. (2003) T-cell interaction with ICAM-1/ICAM-2 double-deficient brain endothelium *in vitro*: the cytoplasmic tail of endothelial ICAM-1 is necessary for transendothelial migration of T cells. Blood. 102:3675-3683.
Marcus, A.J., Broekman, M., Drosopoulos, J., Islam, N., Pinsky, D., Sesti, C., and Levi, R. (2003) Heterologous cell-cell interactions: thromboregulation, cerebroprotection and cardioprotection by CD39 (NTPDase-1). J. Thromb. Haemost. 1:2497-2509.
Matheny, H.E., Deem, T.L., and Cook-Mills, J.M. (2000) Lymphocyte migration through monolayers of endothelial lines involves VCAM-1 signaling via endothelial NADPH oxidase. J. Immunol. 164:6550-6559.
Maus U, Huwe J, Ermert L, Ermert M, Seeger W, Lohmeyer J. (2002) Molecular pathways of monocyte emigration into the alveolar air space of intact mice. Amer. J. Respir. Crit. Care Med. 165:95-100.
McDowall, A., Inwald, D., Leitinger, B., Jones, A., Liesner, R., Klein, N., and Hogg, N. (2003) A novel form of integrin dysfunction involving β1, β2, and β3 integrins. J. Clin. Invest. 111:51-60.

Miller, J., Knorr, R., Ferrone, M., Houdei, R., Carron, C., and Dustin, M.L. (1995) Intercellular adhesion molecule-1 dimerization and its consequences for adhesion mediated by lymphocyte function associated-1. J. Exp. Med. 182:1231-1241.

Minshall, R.D., Tiruppathi, C., Vogel, S., Niles, W., Gilchrist, A., Hamm, H., and Malik, A.B. (2000) endothelial-surface gp60 activates vesicle formation and trafficking via G_i-coupled *Src* kinase signaling pathway. J. Cell Biol. 150:1059-1069.

Mizgerd, J.P. (2002) Molecular mechanisms of neutrophil recruitment elicited by bacteria in the lungs. Semin. Immunol. 14:123-132.

Mizgerd, J.P., Horwitz, B., Quillen, H., Scott, M., and Doerschuk, C.M. (1999) Effects of CD18 deficiency on the emigration of murine neutrophils during pneumonia. J. Immunol. 163:995-999.

Monaco, C., Andreakos, E., Young, S., Feldmann, M., and Paleolog, E. (2002) T cell-signaling to vascular endothelium: induction of cytokines, chemokines, and tissue factor. J. Leukoc. Biol. 71:659-668.

Nathan, C.F. (1987) Neutrophil activation on biological surfaces. J. Clin. Invest. 80:1550-1560.

Nathan, C., Srimal, S., Farber, C., Sanchez, E., Kabbash, L., Asch, A., Gailit, J., and Wright, S.D. (1989) Cytokine-induced respiratory burst of human neutrophils: dependence on extracellular matrix proteins and CD11/CD18 integrins. J. Cell Biol. 109:1341-1349.

Ohgami, M., Doerschuk, C.M., Gie, R., English, D., and Hogg, J.C. (1991) Monocyte kinetics in rabbits. J. Appl. Physiol. 70:152-157.

Ohgami, M., Doerschuk, C.M., Gie, R.P., English, D., and Hogg, J.C. (1992) Late effects of endotoxin on the accumulation and function of monocytes in rabbit lungs. Amer. Rev. Respir. Dis. 146:190-195.

Ostermann, G., Weber, K., Zernecke, A., Schroder, A., and Weber, C. (2002) JAM-1 is a ligand of the β2 integrin LFA-1 involved in transendothelial migration of leukocytes. Nat. Immunol. 3:151-158.

Palmeri, D., van Zante, A., Huang, C., Hemmerich, S., and Rosen, S.D. (2000) Vascular endothelial junction-associated molecule, a novel member of the immunoglobulin superfamily, is localized to intercellular boundaries of endothelial cells. J. Biol. Chem. 275:19139-19145.

Parinandi, N.L., Kleinberg, M., Usatyuk, P., Cummings, R., Pennathur, A., Cardounel, A., Zweier, J., Garcia, J., and Natarajan, V. (2003) Hyperoxia-induced NAD(P)H oxidase activation and regulation by MAP kinases in human lung endothelial cells. Amer. J. Physiol. 284:L26-L38.

Patterson, C.E., Barnard, J.W., LaFuze, J.E., Hull, M.T., Baldwin, S.J., and Rhoades, R.A. (1989) The role of activation of neutrophils and microvascular pressure is acute pulmonary edema. Amer. Rev. Resp. Dis. 140:1052-1062.

Patterson, C.E., Jin, N., Packer, C.S., and Rhoades, R.A. (1992) Activated neutrophils alter contractile properties of the pulmonary artery. Amer. J. Respir. Cell Mol. Biol. 6:260-269.

Pfau, S., Leitenberg, D., Rinder, H., Smith, B., Pardi, R., and Bender, J.R. (1995) Lymphocyte adhesion-dependent calcium signaling in human endothelial cells. J. Cell Biol. 128:969-978.

Phan, S.H., Gannon, D., Varani, J., Ryan, U., and Ward, P.A. (1989) Xanthine oxidase activity in rat pulmonary artery endothelial cells and its alteration by activated neutrophils. Amer. J. Pathol. 134:1201-1211.

Phan, S.H., Gannon, D., Ward, P., and Karmiol, S. (1992) Mechanism of neutrophil-induced xanthine dehydrogenase to xanthine oxidase conversion in endothelium: evidence of a role for elastase. Amer. J. Respir. Cell Mol. Biol. 6:270-278.

Pizurki, L., Zhou, Z., Glynos, K., Roussos, C., and Papapetropoulos, A. (2003) Angiopoietin-1 inhibits endothelial permeability, neutrophil adherence and IL-8 production. Brit. J. Pharmacol. 139:329-336.

Pluskota, E., Chen, Y., and D'Souza, S.E. (2000) Src homology domain 2-containing tyrosine phosphatase 2 associates with ICAM-1 to regulate cell survival. J. Biol. Chem. 275:30029-30036.

Qin, L., Quinlan, W., Doyle, N., Graham, L., Sligh, J., Takei, F., Beaudet, A., and Doerschuk, C.M. (1996) The roles of CD11/CD18 and ICAM-1 in acute Pseudomonas aeruginosa-induced pneumonia in mice. J. Immunol. 157:5016-5021.

Ramamoorthy, C., Sasaki, S., Su, D., Sharar, S., Harlan, J., and Winn, R.K. (1997) CD18 adhesion blockade decreases bacterial clearance and neutrophil recruitment after intrapulmonary E. coli, but not after S. aureus. J. Leukoc. Biol. 61:167-172.

Reilly, P.L., Woska, J., Jeanfavre, D., McNally, E., Rothlein, R., and Bormann, B.J. (1995) The native structure of ICAM-1 is a dimer. Correlation with binding to LFA-1. J. Immunol. 155:529-532.

Ridger, V.C., Wagner, B., Wallace, W., and Hellewell, P.G. (2001) Differential effects of CD18, CD29, and CD49 integrin subunit inhibition on neutrophil migration in pulmonary inflammation. J. Immunol. 166:3484-3490.

Rochon, Y.P., Kavanagh, T.J., and Harlan, J.M. (2000) Analysis of integrin (CD11b/CD18) movement during neutrophil adhesion and migration on endothelia. J. Microsc. 197:15-24.

Roebuck, K.A. and Finnegan, A. (1999) Regulation of ICAM-1 (CD54) gene expression. J. Leukoc. Biol. 66:876-888.

Ruggeri, Z.M. (2003) Von Willebrand factor, platelets and endothelial interactions. J. Thromb. Haemost. 1:1335-1342.

Sachais, B.S. (2001) Platelet-endothelial interactions in atherosclerosis. Curr. Atheroscl. Reports 3:412-416.

Saito, H., Minamiya, Y., Kitamura, M., Saito, S., Enomoto, K., Terada, K., and Ogawa, J. (1998) Endothelial MLCK regulates neutrophil migration across human umbilical vein endothelial monolayer. J. Immunol. 161:1533-1540.

Sans, E., Delachanal, E., and Duperray, A. (2001) Analysis of the roles of ICAM-1 in neutrophil transmigration using a reconstituted mammalian cell expression model: implication of ICAM-1 cytoplasmic domain and Rho-dependent signaling pathway. J. Immunol. 166:544-551.

Schaphorst, K.L., Chiang, E., Jacobs, K., Zaiman, A., Natarajan, V., Wigley, F., Garcia, J.G. (2003) Role of sphingosine-1 phosphate in the enhancement of endothelial barrier integrity by platelet-released products. Amer. J. of Physiol. 285:L258-L267.

Schenkel, A.R., Mamdouh, Z., Chen, X., Liebman, R., and Muller, W.A. (2002) CD99 plays a major role in the migration of monocytes through endothelial junctions. Nat. Immunol. 3:143-150.

Schenkel, A.R., Mamdouh, Z., and Muller, W.A. (2004) Locomotion of monocytes on endothelium is a critical step during extravasation. Nat. Immunol. 5:393-400.

Schneeberger, E.E. and Hamelin, M. (1984) Interaction of serum proteins with lung endothelial glycocalyx: its effect on endothelial permeability. Amer. J. Physiol. 247:H206-H217.

Schnitzer, J.E. (1992) Gp60 is an albumin-binding glycoprotein expressed by continuous endothelium involved in albumin transcytosis. Amer. J. Physiol. 262: H246-H254.

Schnitzer, J.E., Carley, W.W., and Palade, G.E. (1988) Albumin interacts specifically with a 60-kDa microvascular endothelial glycoprotein. PNAS USA 85:6773-6777.

Schnitzer, J.E. and Oh, P. (1994) Albondin-mediated capillary permeability to albumin. Differential role of receptors in endothelial transcytosis and endocytosis of native and modified albumins. J. Biol. Chem. 269:6072-6082.

Sengelov, H., Follin, P., Kjeldsen, L., Lollike, K., Dahlgren, C., and Borregaard, N. (1995) Mobilization of granules and secretory vesicles during *in vivo* exudation of neutrophils. J. Immunol. 154:4157-4165.

Seo, S.M., McIntire, L.V., and Smith, C.W. (2001) Effects of IL-8, Gro-α, and LTB$_4$ on the adhesive kinetics of LFA-1 and Mac-1 on human neutrophils. Amer. J. Physiol. 281:C1568-C1578.

Shappell, S.B., Toman, C., Anderson, D., Taylor, A., Entman, M., and Smith, C.W. (1990) Mac-1 (CD11b/CD18) mediates adherence-dependent hydrogen peroxide production by human and canine neutrophils. J. Immunol. 144:2702-2711.

Shasby, D.M. and Shasby, S.S. (1985) Active transendothelial transport of albumin. Circ. Res. 57:903-908.

Shaw, S.K., Bamba, P., Perkins, B., and Luscinskas, F.W. (2001) Real-time imaging of VE-cadherin during leukocyte transmigration across endothelium. J. Immunol. 167:2323-2330.

Simon, S.I., Hu, Y., Vestweber, D., and Smith, C.W. (2000) Neutrophil tethering on E-selectin activates β2 integrin binding to ICAM-1 through a MAPK signal transduction pathway. J. Immunol. 164:4348-4358.

Staunton, D.E., Dustin, M., Erickson, H., and Springer, T.A. (1990) The arrangement of the immunoglobulin-like domains of ICAM-1 and the binding sites for LFA-1 and rhinovirus. Cell 61:243-254.
Staunton, D.E., Marlin, S., Stratowa, C., Dustin, M., and Springer, T.A. (1988) Primary structure of ICAM-1 demonstrates interaction between immunoglobulin and integrin supergene families. Cell 52:925-933.
Strieter, R.M., Belperio, J.A., and Keane, M.P. (2002) Cytokines in innate host defense in the lung. J. Clin. Invest. 109:699-705.
Su, W.H., Chen, H., Huang, J., and Jen, C.J. (2000) Endothelial $[Ca^{2+}]_i$ signaling during transmigration of polymorphonuclear leukocytes. Blood. 96:3816-3822.
Takizawa, K., Kamijo, R., Ito, D., Hatori, M., Sumitani, K., and Nagumo, M. (1999) Synergistic induction of ICAM-1 expression by cisplatin and 5-fluorouracil in a cancer cell line via a NF-κB independent pathway. Br. J. Cancer. 80:954-963.
Tasaka, S., Qin, L., Saijo, A., Albelda, S., DeLisser, H., and Doerschuk, C.M. (2003) PECAM-1 in neutrophil emigration during acute pneumonia in mice and rats. Amer. J. Respir. Crit. Care Med. 167:164-170.
Tasaka, S., Richer, S., Mizgerd, J., and Doerschuk, C.M. (2002) Very late antigen-4 in CD18-independent neutrophil emigration during acute pneumonia in mice. Amer. J. Respir. Crit. Care Med. 166:53-60.
Thompson, P.W., Randi, A.M., and Ridley, A.J. (2002) ICAM-1, but not ICAM-2, activates RhoA and stimulates c-fos and rhoA transcription in endothelial cells. J. Immunol. 169:1007-1013.
Tilghman, R.W. and Hoover, R.L. (2002) E-selectin and ICAM-1 are incorporated into detergent-insoluble membrane domains following clustering in endothelial cells. FEBS Lett. 525:83-87.
Tiruppathi, C., Finnegan, A., and Malik, A.B. (1996) Isolation and characterization of a cell surface albumin binding protein from vascular endothelial cells. PNAS USA 93:250-254.
Tiruppathi, C., Song, W., Bergenfeldt, M., Sass, P., and Malik, A.B. (1997) Gp60 activation mediates albumin transcytosis in endothelial cells by a tyrosine kinase-dependent pathway. J. Biol. Chem. 272:25968-25975.
Turner, M.R., Clough, G., and Michel, C.C. (1983) The effects of cationised ferritin and native ferritin upon the filtration coefficient of single frog capillaries. Evidence that proteins in the endothelial coat influence permeability. Microvasc. Res. 25:205-222.
Valles, J., Santos, M., Marcus, A., Safier, L., Broekman, M., Islam, N., Ullman, H., and Aznar, J. (1993) Downregulation of human platelet reactivity by neutrophils. Participation of lipoxygenase derivatives and adhesive proteins. J. Clin. Invest. 92:1357-1365.
van Wetering, S., van den Berk, N., van Buul, J., Mul, F., Lommerse, I., Mous, R., ten Klooster, J., Zwaginga, J., and Hordijk, P.L. (2003) VCAM-1-mediated Rac signaling controls endothelial cell-cell contacts and leukocyte transmigration. Amer. J. Physiol. 285:C343-C352.
Victorino, G.P., Newton, C.R., and Curran, B. (2003) The impact of albumin on hydraulic permeability: comparison of isotonic and hypertonic solutions. Shock. 20:171-175.
Vogel, S.M., Minshall, R., Pilipovic, M., Tiruppathi, C., and Malik, A.B. (2001) Activation of 60 kDa albumin-binding protein (gp60) stimulates albumin transcytosis in the intact pulmonary microvessel. Amer. J. Physiol. 281:L1512-L1522.
Vogetseder, W. and Dierich, M.P. (1991) ICAM-1 (CD54) is associated with actin filaments. Immunobiol. 182:143-151.
Wakabayashi, Y., Fujita, H., Morita, I., Kawaguchi, H., and Murota, S. (1995) Conversion of xanthine dehydrogenase to an oxidase in bovine carotid artery endothelium induced by activated neutrophils: involvement of adhesion molecules. Biochem. Biophy. Acta 1265:103-109.
Walker, D.C., Behzad, A.R., and Chu, F. (1995) Neutrophil migration through preexisting holes in the basal laminae of alveolar capillaries and epithelium during pneumonia. Microvasc. Res. 50:397-416.
Wang, Q., Chiang, E., Lim, M., Lai, J., Rogers, R., Janmey, P., Shepro, D., and Doerschuk, C.M. (2001) Changes in biomechanical properties of neutrophils and endothelia during adhesion. Blood 97:660-668.

Wang, Q., and Doerschuk, C.M. (2000) Neutrophil-induced changes in the biomechanical properties of endothelial cells: the roles of ICAM-1 and oxidants. J. Immunol. 164:6487-6494.

Wang, Q. and Doerschuk, C.M. (2001) The role of p38 MAPK in mediating changes in the biomechanical properties of pulmonary microvascular endothelial cells upon ICAM-1 ligation. J. Imm. 166:6877-6884.

Wang, Q., Pfeiffer, G.R., and Gaarde, W.A. (2003) Activation of SRC tyrosine kinases in response to ICAM-1 ligation in pulmonary microvascular endothelial cells. J. Biol. Chem. 278:47731-47743.

Wang, Q., Pfeiffer, G., Stevens, T., and Doerschuk, C.M. (2002) Lung microvascular and arterial endothelial cells differ in their responses to ICAM-1 ligation. Amer. J. Respir. Crit. Care Med. 166:872-877.

Ward, P.A. (1996) Role of complement in lung inflammatory injury. Amer. J. Pathol. 149:1081-1086.

Ware, J.A. and Heistad, D.D. (1993) Platelet-endothelium interactions. N. Eng. J. Med. 328:628-635.

Watanabe, T., Tokuyama, S., Yasuda, M., Sasaki, T., and Yamamoto, T. (2001) Changes of tissue factor-dependent coagulant activity mediated by adhesion between polymorphonuclear leukocytes and endothelial cells. Jpn. J. Pharmacol. 86:399-404.

Watanabe, T., Tokuyama, S., Yasuda, M., Sasaki, T., and Yamamoto, T. (2002) Involvement of adenosine A2 receptors in the changes of tissue factor-dependent coagulant activity induced by polymorphonuclear leukocytes in endothelial cells. Jpn. J. Pharmacol. 88:407-413.

Wei, Z., Costa, K., Al-Mehdi, A., Dodia, C., Muzykantov, V., and Fisher, A.B. (1999) Simulated ischemia in flow-adapted endothelial cells leads to generation of ROS and cell signaling. Circ. Res. 85:682-689.

Wojciak-Stothard, B., Williams, L., and Ridley, A.J. (1999) Monocyte adhesion and spreading on human endothelial cells is dependent on Rho-regulated receptor clustering. J. Cell Biol. 145:1293-1307.

Yatomi, Y., Ohmori, T., Rile, G., Kazama, F., Okamoto, H., Sano, T., Satoh, K., Kume, S., Tigyi, G., Igarashi, Y., and Ozaki, Y. (2000) S1P as a major bioactive lysophospholipid that is released from platelets and interacts with endothelial cells. Blood 96:3431-3438.

Yoder, M.C., Checkley, L., Giger, U., Hanson, W., Kirk, K., Capen, R., and Wagner, W.W. (1990) Pulmonary microcirculatory kinetics of neutrophils deficient in leukocyte adhesion-promoting glycoproteins. J. Appl. Physiol. 69:207-213.

Yoshida, M., Szente, B., Kiely, J., Rosenzweig, A., and Gimbrone, M.A. (1998) Phosphorylation of the cytoplasmic domain of E-selectin regulated during leukocyte-endothelial adhesion. J. Imm. 161:933-941.

Yoshida, M., Westlin, W., Wang, N., Ingber, D., Rosenzweig, A., Resnick, N., and Gimbrone, M.A. (1996) Leukocyte adhesion to vascular endothelium induces E-selectin linkage to the actin cytoskeleton. J. Cell Biol. 133:445-455.

Yuan, S.Y., Wu, M., Ustinova, E., Guo, M., Tinsley, J., De Lanerolle, P., and Xu, W. (2002) Myosin light chain phosphorylation in neutrophil-stimulated coronary microvascular leakage. Circ. Res. 90:1214-1221.

Zhang, P., Summer, W., Bagby, G., and Nelson, S. (2000) Innate immunity and pulmonary host defense. Immunol. Rev. 173:39-51.

Zhang, W.J. and Frei, B. (2002) Albumin selectively inhibits TNF α-induced expression of VCAM-1 in human aortic endothelial cells. Cardiovasc. Res. 55:820-829.

Zhou, M. and Brown, E.J. (1993) Leukocyte response integrin and -associated protein act as a signal transduction unit in generation of a phagocyte respiratory burst. J. Exp. Med. 178:1165-1174.

Ziegelstein, R.C., Corda, S., Pili, R., Passaniti, A., Lefer, D., Zweier, J., Fraticelli, A., and Capogrossi, M.C. (1994) Initial contact and subsequent adhesion of human neutrophils or monocytes to human aortic endothelial cells releases endothelial intracellular Ca^{++} store. Circ. 90:1899-1907.

Chapter 12

Endothelial Cell Injury and Defense

Hedwig S. Murphy, M.D., Ph.D.*, James Varani, Ph.D., and Peter A. Ward, M.D.

Department of Pathology, Univiversity of Michigan, Ann Arbor, MI, 48109 USA

CONTENTS:

Introduction

Source of Injurious Agents in Inflammation
Leukocytes – The Primary Purveyor
Reactive Oxygen Species
Neutrophil Enzymes

Nonlethal injury
Oxidized Lipids
Effects of Neutrophil Enzymes
Cationic Proteins, Cytokines, and Arachidonic Acid Metabolites

Leukocyte Products and Apoptosis
Oxidants and Apoptosis
Proteinases and Apoptosis

Cytotoxic Endothelial Injury
Physiologic vs. Pathologic Interactions
Biochemical Events in Neutrophil-mediated Injury
Interaction between Oxidants and Other Mediators in Acute Injury

Endothelial Cell Protective Mechanisms
Antioxidants
Antiproteinases
Coagulation Proteins

Acute Endothelial Cell Injury Promotes Chronic Lung Disease
Fibrosis in ARDS
Acute Lung Injury Preceding Chronic Pulmonary Disease

Conclusions

Introduction

The vascular endothelium, as gatekeeper to the tissue compartment, both selectively facilitates an enhanced vascular and tissue response to injury and inflammation, and presents a barrier to excess injurious cells and agents, thereby protecting the vasculature and tissue. In inflammation, the detrimental cooperation of neutrophils and endothelial cells results in vascular permeability (Wedmore and Williams, 1981). Initial exposure to low concentrations of leukocyte products, systemic mediators, and local tissue factors results in a dynamic interaction of endothelial cells, tissue cells, and inflammatory cells. The activated endothelium responds with pro-inflammatory as well as self- and tissue-protective mechanisms: expression of cell surface molecules, release of soluble inflammatory and vasoactive mediators, and generation of detoxifying agents. However, continued assault on the vascular endothelium by increasing concentrations of inflammatory mediators and cells overwhelms the endothelial defenses and non-cytotoxic injury progresses to endothelial cell death, promoting a continued acute tissue response. In some settings, this continued cell damage can proceed to an unmitigated tissue response and chronic pulmonary disease with pulmonary fibrosis.

The type and extent of endothelial response is determined by the local milieu of cells and inflammatory products. Leukocytes, primarily neutrophils but also eosinophils, macrophages and mast cells, contribute activating and toxic products including inflammatory cytokines and chemokines, proteases, lipases, and oxidants that directly and indirectly alter endothelial cell function. Vascular injury and barrier disruption occur when the production of these injurious products exceeds the capacity of the endothelium to detoxify injurious agents with endogenous mechanisms. Loss of endothelial cells can result from both cytotoxic injury and altered balance of cell survival and cell apoptotic events. The net pulmonary tissue response is directly linked to the extent of impairment of this vascular endothelial cell sentinel function.

Sources of Injurious Agents in Inflammation

Leukocytes – The Primary Purveyor

The events that initiate inflammatory tissue damage in the lung may occur locally as in the case of microbial infection or systemically as in sepsis, ischemia-reperfusion, burn, and trauma. Regardless of the initiating event, there is a rapid accumulation of neutrophils in the lung capillaries. Early inflammatory responses in lung cells, mainly in the resident macrophages, involve production of chemokines and cytokines (Guo and Ward, 2002; Vozzelli *et al.*, 2004). As depicted in Figure 1 chemokines, such as macrophage inflammatory protein (MIP)-2 and cytokine-inducible neutrophil chemoattractant (CINC), signal neutrophil migration to the lung. Cytokines, such as IL-1, IL-8, and TNFα, and inflammatory molecules, such as PAF, induce upregulation of leukocyte adhesion molecules in endothelium (Ward, 1996; Drost and MacNee, 2002). Expression of these adhesion molecules on the luminal surface of the endothelium then results in firm binding and transendothelial migration of neutrophils as described in Chapter 11. Complement activation, particularly C5a production, is a major upstream event, which leads to activation of the macrophages and the neutrophil influx, activation, and binding in the pulmonary microvessels, as well as direct activation of the endothelium and upregulation of adhesion molecules (Guo and Ward, 2002). Thus, neutrophil sequestration and activation in the pulmonary microvasculature precedes emigration into the tissue space.

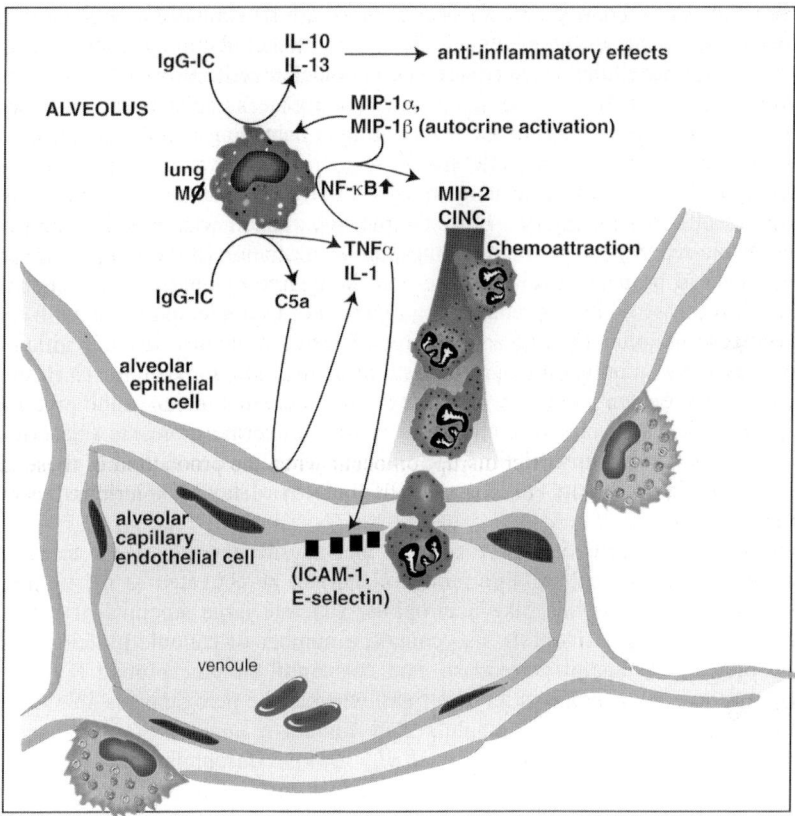

Figure 1. Schematic depiction of the pathogenesis of acute lung injury in response to IgG-IC deposition. Cytokines, such as TNFα and IL-1, induce upregulation of leukocyte adhesion factors, such as ICAM-1 and E-selectin, on endothelium allowing neutrophil adhesion and transmigration. The CXC chemokines, such as macrophage inflammatory protein (MIP-2) and cytokine-inducible neutrophil chemoattractant (CINC) create a gradient to guide the transmigration. Complement activation product, C5a, is an additional chemoattractant, but it also functions to activate endothelial cells, increase permeability, and upregulate leukocyte adhesion molecules. Reproduced from *Guo, R. and Ward, P.A. (2002) Mediators and regulation of neutrophil accumulation in inflammatory responses in lung: Insights from the IgG immune complex model. Free Rad. Biol. Med. 33:303-310,* with permission from Elsevier, Amsterdam.

Importantly, these neutrophils responding to an inflammatory stimulus are, in large measure, responsible for the damage to host tissue that occurs during acute inflammation. It is within this vascular space prior to emigration, as well as in the tissue space after transmigration, that neutrophils release their products in proximity to the vascular endothelium. Leukocyte-derived reactive oxygen species, cationic peptides, eicosanoids, and proteolytic enzymes are primarily directed towards antimicrobial suppression but also damage vascular and resident tissue cells. When the neutrophil establishes primary contact with activated endothelium, secretory vesicles and granules are sequentially mobilized to the cell surface, thereby allowing the contents of the these intracellular organelles to reach and accumulate in the extracellular space adjacent to endothelial cells (Borregaard and Cowland, 1997; Murphy et al., 2003). At the same time, membrane associated enzyme complexes generate reactive oxygen species capable of activating or damaging endothelial cells.

In animal models *in vivo*, multiple studies exploring three avenues provide evidence linking neutrophil products to acute lung injury. First, in animal injury models, systemic inflammation results in lung injury. For example, reperfusion with blood or plasma after intestinal ischemia-reperfusion results in neutrophil sequestration in the lung which parallels increased elastase in bronchoalveolar lavage and the degree of lung injury (Carden et al., 1998). Second, protease inhibitors reduce lung injury. For example, inhibition with secretory leukocyte protease inhibitor (SLPI), an inhibitor of serine proteases, reduces inflammatory lung injury caused by an intrapulmonary deposition of IgG-immune complexes (Mulligan et al., 1993). The third evidential pathway utilizes animals deficient in neutrophil products. For example, mice deficient in one or both of the neutrophil serine proteases (elastase and/or cathepsin G) have extended survival time and enhanced resistance to diffuse alveolar damage in lipopolysaccharide-endotoxin (LPS) model of endotoxic shock (Warner et al., 2001a; Warner et al., 2001b).

Other inflammatory cells also have effector mechanisms that contribute to cell injury. Lung macrophages are present in large numbers and can be activated as the inflammatory process is initiated. Eosinophils, like neutrophils, generate large amounts of extracellular oxygen radicals. Human eosinophils also contain a number of cationic proteins, including major basic protein, eosinophil peroxidase, and eosinophil cationic protein (Gleich et al., 1979). Each of these is cytotoxic in its own right, whereas the peroxidase is able to generate hypohalous acids from H_2O_2. Eosinophils also contain a number of potent proteolytic enzymes in stored granules (Charlesworth et al., 1989). Basophils are a rich source of histamine and other mediators that affect vascular tone and permeability. Thus, although these cells may not *per se* be as highly cytotoxic to endothelial cells, a role in allergic-type inflammatory responses may be critical.

Reactive Oxygen Species

Central to the development of the tissue response to neutrophils is the generation and release of reactive oxygen species, resulting in activation of resident tissue cells and promotion of intracellular signaling events as well as cell injury and death with continued generation of reactive metabolites. Activation of neutrophils with a variety of agents leads to the surface organization of the components of NADPH oxidase into an active enzyme complex which converts molecular oxygen into the superoxide anion radical ($O_2^{\cdot}$) (Babior et al., 1977). This radical is then broken down spontaneously or extremely rapidly in the presence of the enzyme superoxide dismutase (SOD), to hydrogen peroxide, which in turn is a central intermediate in a number of different metabolic pathways. In the presence of catalase (a mitochondrial heme enzyme) or glutathione peroxidase (a cytoplasmic selenium-containing enzyme), two enzymes that contribute to the endothelial cell antioxidant barrier, hydrogen peroxide can be reduced to molecular oxygen and water. Myeloperoxidase, a neutrophil product with a very strong cationic charge, is secreted from granules during exocytosis. In the presence of chloride ion, it catalyzes the conversion of hydrogen peroxide to hypochlorous acid. Although hypochlorous acid is a strong oxidant, it is probably not primarily responsible for cell injury. Rather, studies suggest that iron-catalyzed hydrogen peroxide reduction, leading to formation of the hydroxyl radical (·OH) is the most injurious pathway (Varani et al., 1992a). Moreover, in inflammation upregulation of inducible nitric oxidase synthase (iNOS) in leukocytes produces high amounts of the NO free radical (Blackford et al., 1994; Wang et al., 2001; Kobayashi et al., 2002).

The neutrophil is not the only source of reactive oxygen species; the vascular endothelial cell contributes its own repertoire of intra- and extra-cellular oxygen radicals, which interact with those generated by neutrophils. Although superoxide is generated in 10-100-fold lower concentrations in endothelial cells compared to neutrophils, this radical has important pathological functions as a modulator of endothelial cell signal transduction elements and by interacting with iron and with intracellular and extracellular nitric oxide to generate toxic radicals (Siebenlist et al., 1994; Karin, Liu et al., 1997). Superoxide is generated by several enzyme systems present in mammalian cells including mitochondrial electron transport enzymes, cytochrome P450 enzymes present in endoplasmic reticulum as well as in biologic membranes, amino acid oxidases (peroxisomes), and lipid-metabolizing oxidases (e.g., cyclooxygenase and lipoxygenase) in biologic membranes. Two enzymes systems make particularly significant contributions to endothelial cell superoxide: NADPH oxidase and xanthine oxidase. The multi-component neutrophil-like NADPH oxidase, responsible for generation of extracellular superoxide (and probably intracellular superoxide as well) is involved in pulmonary vasoconstriction and nitric oxide scavenging (Jones et al., 1996; Murphy et al., 1999; Murphy et al., 2000). The cytosolic purine-metabolizing enzyme xanthine oxidase catalyzes oxidation of hypoxanthine to xanthine and is a source of intracellular superoxide, which under physiologic conditions may be neutralized by intracellular antioxidant systems. Under appropriate conditions, however, the superoxide generated by these enzymes participates in cytotoxic reactions. In perfused lung, where endothelium was the primary cellular source of oxidants, and in cultured endothelial cells, xanthine oxidase is a major ROS contributor in anoxia/hypoxia when ATP levels fall. But NADPH oxidase is the major contributor in ischemia/reperfusion when O_2 and ATP levels are maintained (Al Medhi et al., 1998; Wei et al., 1999).

Another critical oxidant is nitric oxide (·NO), synthesized in endothelial cells either by a constitutively expressed Ca^{++}-calmodulin dependent isoform of nitric oxide synthase, eNOS (endothelial NOS), or iNOS. NO functions as a vasoregulatory molecule but inflammatory cytokines increase expression of NO as well (Murphy et al., 1999; Speyer et al., 2003). In lung sepsis, the inducible NOS gene has an early response to induction by LPS, increasing in vascular endothelial cells, airway epithelial cells and inflammatory cells (Liu et al., 1997). As indicated above NO produced by endothelial cells can react with superoxide to produce peroxynitrite anion, which is toxic. On the other hand, endogenous NO interferes with neutrophil adhesion to endothelial cells; the loss of this function enhances neutrophil adhesiveness and potentiate the inflammatory response (Roman et al., 2004).

Neutrophil Enzymes

In addition to oxidants, leukocytes generate a variety of enzymes having a broad range of cellular and noncellular substrates. These proteinases, including serine proteases and matrix metalloproteinases (MMPs), are now known to facilitate activation of inflammatory mediators as well as acting on extracellular matrix proteins (Kang et al., 2001). When endothelial cells are subjected to sublethal, reversible injury by exposure to purified serine proteinases (elastase and cathepsin G) from human neutrophils, cells are not killed but are detached from their underlying extracellular matrix, decreasing the permeability barrier. *In vivo*, the close contact between activated neutrophils and the endothelium provides a "protective environment" that can exclude serum proteinase inhibitors. In this environment, neutrophil proteinases degrade inter-endothelial cell junctional proteins, endothelial glycocalyx, basement membrane components, and connective tissue of the subendothelial extracellular matrix. The resulting endothelial cell retraction increases permeability, and detachment exposes the extracellular matrix to further damage with the net effect of disrupting microvascular barrier integrity. Accordingly, the protease inhibitory activity of leukocyte mutants correlated with *in vivo* protective effects, evidenced by reduced lung albumin leak, decreased neutrophil accumulation, and suppression of intrapulmonary activation of NF-κB (Mulligan et al., 2000). These secretory leukocyte protease inhibitor mutants expressed the critical binding site for of chymotrypsin, elastase, and trypsin.

Nonlethal Injury

For a variety of reasons it is believed that damage to the endothelium, even without cell death, results in a temporary increase in permeability and initiates inflammatory tissue injury. Non-cytotoxic injury to the vascular endothelial cells induces cell retraction and exposure of the underlying basement membrane to neutrophil enzymes including serine and matrix metalloproteinases. Damage to the basement membrane and underlying extracellular connective tissue matrix compromises barrier integrity.

Oxidized Lipids

Reactive oxygen species generated in neutrophils and/or endothelial cells have the potential to interact with proteins, lipids, and DNA. Peroxynitrate, the product of superoxide and nitric oxide interaction, mediates lipid peroxidation, resulting in generation of reactive aldehydes and nitrogen oxides, oxidized low-density lipoproteins (oxLDL) and nitrated proteins (Nedeljkovic et al., 2003). Peroxidation of linoleic or arachidonic acid yields smaller aldehydes such as 4-hydrosynonenal with prominent cytotoxic effects on endothelial cells (Herbst et al., 1999). Oxidized phospholipids in serum or cell membranes promote leukocyte binding to endothelial cells (Hessler et al., 1983; Berliner and Heinecke, 1996; Leitinger et al., 1999). Moreover, oxidized low-density lipoprotein stimulates expression of ICAM-1 (Navab et al., 1991), directly inactivates nitric oxide (Chin et al., 1992), stimulates vascular smooth muscle cell proliferation, and upregulates tissue factor and plasminogen activator inhibitor-1 expression. In addition to lipoprotein oxidation, reaction of oxygen species with cell membrane bound fatty acids can promote a vicious cycle of continued oxidative damage, resulting in alterations in cell membrane permeability and functional impairment in cellular transport and signaling (Wolin, 2000).

Effects of Neutrophil Enzymes

Neutrophil enzymes independently, as well as in synergy with oxidants, affect a wide range of intracellular as well as extracellular proteins, with a spectrum of effects from cell activation to cell injury and death. The serprocidin family of serine protease homologues stored in the primary granules of neutrophils have broad-spectrum antimicrobial activity and include proteinase 3, neutrophil elastase (human leukocyte elastase), cathepsin G, and the non-proteolytic cationic protein (azurocidin – see next section) (Borregaard and Cowland, 1997). Proteinase 3 is the main autoantigen to antineutrophil cytoplasmic antibodies (ANCA) in patients with Wegener's granulomatosis, a disease characterized by granulomatous inflammation of the respiratory tract (Witko-Sarsat *et al.*, 1999). Proteinase 3 induces endothelial cell detachment and cytolysis, apoptosis, enhanced IL-8 and MCP-1 production, and increases expression of ICAM-1, promoting further neutrophil recruitment and adhesion thereby augmenting the inflammatory insult (Ballieux *et al.*, 1994; Berger *et al.*, 1996; Yang *et al.*, 1996; Taekema-Roelvink *et al.*, 1998; Taekema-Roelvink *et al.*, 2001). Neutrophil elastase and cathepsin G have overlapping microbicidal activity and substrate specificities and thus have many similar effects on the pulmonary vasculature (Rest, 1988; Shapiro 2002). Activated neutrophils as well as purified cathepsin G, elastase and proteinase 3 induce early endothelial cell membrane changes and release of the transmembrane glycoprotein thrombomodulin (MacGregor *et al.*, 1997; Boehme *et al.*, 2002). Prolonged exposure to these agents directly, as well as the endothelial cell detachment they induce, results in subsequent endothelial cell apoptosis (Re *et al.*, 1994; Yang *et al.*, 1996; Boehme *et al.*, 2002).

Neutrophil elastase is increased in lavage fluid from acute respiratory distress syndrome (ARDS) patients (Lee *et al.*, 1981; Cochrane *et al.*, 1983; Idyll *et al.*, 1985), where it can cleave surfactant proteins and impair the ability of surfactant to reduce tension (Ryan *et al.*, 1991; Liau *et al.*, 1996). Early elevated peripheral blood elastase correlates with subsequent requirement for mechanical ventilation as an indicator of lung injury, thereby implicating this enzyme as a key mediator of disruption in vascular integrity during the development of acute lung injury (Donnelly *et al.*, 1995). Regulation of cell migration and neutrophil recruitment, generation of chemoattractants from precursors, organization and modulation of adhesion molecules and endothelial cell junctions, and degradation of basement membrane and other connective tissues are facilitated by elastase. Its potential substrates include almost all components of the extracellular matrix, as well as proteins as diverse as matrix metalloproteinases, coagulation factors, immunoglobulins, complement, cytokines and many protease inhibitors (Lee and Downey, 2001; Moraes *et al.*, 2003).

One target of neutrophil proteases is cadherin an integral membrane protein that contributes to the solute barrier by forming homotypic bonds between adjacent endothelial cells (Alexander *et al.*, 1993). Activated human neutrophils, as well as purified elastase proteolyze recombinantly expressed VE-cadherins on human umbilical vein endothelial cells. Cleavage of VE-cadherin induces formation of gaps through which neutrophils transmigrate and compromises the integrity of the vascular barrier, promoting endothelial and epithelial permeability and increasing pulmonary interstitial edema (Carden *et al.*, 1998; Hermant *et al.*, 2003). Plasma of ARDS patients contains soluble cadherin fragments (Carden *et al.*, 1998), suggesting proteolysis of these molecules is related to vascular injury.

Once endothelial cell retraction has occurred, the basement membrane and extracellular connective tissue stroma are exposed to leukocyte products. The endothelial basement membrane is composed mainly of type IV collagen, elastin, proteoglycans, and laminin and its integrity is maintained by a balance between synthesis and degradation. Elastase, gelatinase, and matrix metalloproteinases degrade these components of the basement membrane as well as the underlying extracellular matrix. Emerging evidence suggests that matrix metalloproteinase production is modulated by other inflammatory mediators and that the upregulation and activation of matrix metalloproteinases in the lung is ultimately responsible for much of the resulting tissue injury. By influencing recruitment of inflammatory cells into the lung, matrix metalloproteinases have been implicated in the pathogenesis of emphysema (Ohnishi *et al.*, 1998), ARDS (Pugin *et al.*, 1999), and asthma (Mautino *et al.*, 1997). Neutrophils and macrophages are the primary sources, but endothelial cells also generate matrix metalloproteinases, particularly MMP-2 and MMP-9 as well as MT-MMP, all of which are among the most potent inducers of vascular permeability (Partridge *et al.*, 1993; Delclaux *et al.*, 1997; Chun *et al.*, 2004). Matrix metalloproteinases not only degrade extracellular matrix components but are also involved in the recruitment of inflammatory cells into the lung, most likely by modulation of leukocyte chemotactic factors. Reactive oxygen species such as hydrogen peroxide upregulate matrix metalloproteinase production directly and through activation of AP-1 (Pinkus *et al.*, 1996; Rajagopalan *et al.*, 1996). In patients with asthma or ARDS there is an overproduction of matrix metalloproteinases in the lung as compared to levels of the TIMP matrix metalloproteinase inhibitors, suggesting that accelerated matrix metalloproteinase production overwhelms the endogenous tissue defense (Mautino *et al.*, 1997; Ohnishi *et al.*, 1998; Pugin *et al.*, 1999).

Proteolytic activities are not just directed toward the resident cells and extracellular matrix. Soluble inflammatory mediators are activated by proteolytic digestion of precursor molecules, rendering them then capable of activating or injuring endothelial and epithelial cells. On one hand, cathepsin G, elastase and Proteinase 3 release angiotensin I from angiotensinogen and angiotensin II from angiotensin I and angiotensinogen (Ramaha and Patston, 2002). But on the other hand, the normal conversion of circulating angiotensin I to angiotensin II by pulmonary capillary endothelium-bound angiotensin converting enzyme (ACE) is suppressed (Catravas *et al.*, 1990; Cziraki *et al.*, 2000). In fact, alteration of ACE is an early, sensitive, and quantifiable lung injury index in animal models (Orfanos *et al.*, 2001).

Complement activation products are among the several mediators of leukocyte activation and recruitment into extravascular tissue during inflammation. Neutrophil proteases, in particular elastase, cleave C5 to generate functionally active, chemotactic C5a-like fragment and a C5b6-like complex, which proceeds to form the lytic C5b-9, membrane attack complex (Vogt, 2000). Complement-derived products evoke activation of endothelial cells with release of soluble mediators and increased adhesion of leukocytes, correlating with upregulation of endothelial cell as well as leukocyte surface adhesion molecules. Elastase also interacts with plasma proteins to generate proinflammatory peptides such as IL-2, IL-6 and IL-8 and activating enzymes such as matrix metalloproteinases, all of which may indirectly affect vascular endothelial cells and other resident tissue cells (Moraes *et al.*, 2003).

Cationic Proteins, Cytokines, and Arachidonic Acid Metabolites

Other products may not be directly toxic to endothelial cells, but may activate endothelial cells and activate pathways leading to endothelial injury. One such important product is the 37 kDa, HBP/CAP37 (heparin binding protein HBP/CAP37), also known as azurocidin. It is released from neutrophils upon binding of the PMN CD11b/CD18 to endothelial ICAM-1. It is a homologue of the serine proteases, elastase and cathepsin G, but lacks proteolytic activity, due to substitutions in the serine protease active site, a histidine to serine and serine to glycine. Nevertheless, it is a multifunctional protein, mediates neutrophil and monocyte adherence to endothelial monolayers, upregulates adhesion molecules, protects endothelial cells from apoptosis, mediates endothelial contraction, and tyrosine phosphorylation of junctional proteins (Østergaard and Flodgaard, 1992; Heinzelmann *et al.*, 1999; Olofsson *et al.*, 1999; Gautam *et al.*, 2001; Lee *et al.*, 2003). *In vitro*, transmigration of neutrophils and transendothelial permeability increase upon addition of either elastase, cathepsin G, or HBP/CAP37 to endothelial cell monolayers (Edens and Parkos, 2003). HBP/CAP37 is a strong cationic protein with eight cysteine residues and intrinsic heparin-binding characteristics, which can potentially bind surface glycosaminoglycans (GAGs), including heparin/heparan sulfate-like glycosaminoglycans (HSGAGs), components of the glycocalyx. The glycocalyx, an integral component of the endothelial cell barrier, is a thick surface layer of proteoglycans with glycosaminoglycans and heparin sulfate-glycosaminoglycans whose anionic side chains form an entangled meshwork on the cell surface (Pries *et al.*, 2000). One group of the endothelial cell heparin sulfate-glycosaminoglycans in particular, the syndecans, are known to influence cytoskeletal organization, cell-cell adhesion, and motility, processes associated with changes in vascular permeability (Zimmermann and David, 1999; Dudek and Garcia, 2001). Glycosaminoglycans directly mediate cationic peptide-induced signaling by neutrophil products such as HBP/CAP37 that lead to cytoskeletal reorganization and subsequent increases in endothelial permeability (Dull *et al.*, 2003). HBP/CAP37 also directly binds and neutralizes lipopolysaccharides and it acts as an opsonin to increase phagocyte removal of bacteria and superoxide production.

As mentioned above, a number of cytokines, such as IL-1, IL-8, and TNFα, are released from activated leukocytes, which have profound influences on propagation of the inflammatory pathway, but these also have direct effects on endothelial cells and play an important role in lung injury (Krishnadasan *et al.*, 2003). For instance, macrophage-TNFα release orchestrates early events in ischemia-reperfusion lung injury (Naidu *et al.*, 2004). Leukocytes also produce a variety of lipid mediators, such as eicosanoids and PAF (Camussi *et al.*, 1981), which have direct effects on endothelial cells (Farmer *et al.*, 2002; Victorino *et al.*, 2004). Table 1 summarizes injurious products and defenses.

In summary, the interaction of leukocyte products, endothelial cell metabolites, and tissue and plasma factors mediate changes in the vascular endothelium which, while not lethal to the endothelial cell nevertheless result in alteration of the sentinel function of these cells, opening a door into the extravascular tissue compartment and leading the way to tissue injury.

Table 1. Agents of Endothelial Cell Injury

INJURIOUS AGENTS		PROTECTIVE RESPONSE
Agent	*Source*	
OXIDANTS		
$O_2^\bullet$	PMN, EC	SOD, $^\bullet$NO
H_2O_2, HOCl, $^\bullet$OH	PMN	Catalase, glutathione peroxidase
PROTEASES		
Neutrophil elastase	PMN	α1-antiprotease, Secretory leukoprotease inhibitor (SLPI), $α_2$-macroglobulin
HBP/CAP37	PMN	
Proteinase 3	PMN	EPCR
MMPs	PMN, Macs, fibroblasts	TIMP-1, TIMP-2
SURFACE MOLECULES		
Adhesion molecules	PMN, EC	NO, Oxidized phospholipids, EPCR
Tissue Factor	EC	TFPI
Oxidized lipid	EC, resident cells	Aldehyde products of lipid oxidation, EPCR
SOLUBLE FACTORS		
Cytokines/chemokines	EC, leukocytes, resident cells	Adehyde products of lipid oxidation, APC
Complement	Plasma	
Coagulation proteins	Plasma	TFPI, PGI_2, thrombomodulin, ATII, heparin sulfate
thrombin	Plasma	
ATII	EC	heparin-sulfate-antithrombin, APC,

KEY:
EC: endothelial cell   SOD: superoxide dismutase   EPCR: protein C receptor
PMN: neutrophil   GSH: glutathione   APC: Activated Protein C
Macs: macrophage PGI_2: Prostaglandin   ATII: Angiotensin II
TFPI: Tissue Factor Pathway Inhibitor

Leukocytes Products and Endothelial Apoptosis

Oxidants and Apoptosis

Reactive oxygen species initiate pulmonary endothelial cell damage leading to an increase in endothelial permeability and pulmonary edema. Hydrogen peroxide-induced apoptosis of endothelial cells is associated with uptake of cellular iron and activation of an intracellular cascade including phosphorylation of ASK-1 (Apoptosis signal-regulating kinase), p38 MAPK and JNK, and activation of caspase-3 (Machino *et al.*, 2003). This oxidant induced apoptosis is inhibited by nitric oxide (Kotamraju *et al.*, 2003). In atherosclerotic lesions as well as in endothelial cell cultures, oxidized low-density lipoproteins induce apoptosis of vascular cells and the ubiquitin-proteasome pathway appears to play a role in cellular defenses against oxidized low-density lipoprotein-induced toxicity. Oxidized low-density lipoprotein-induced apoptosis involves modification of cell proteins by oxidized lipids and inhibition of this proteasome pathway (Chen et al., 2003; Vieira et al., 2000). Neutrophil products alter the normal cell turnover in such a way as to tip the balance in favor of a higher rate of apoptosis.

Proteinases and Apoptosis

Detachment of endothelial cells from their basement membranes not only exposes this structure to proteolytic enzymes but also activates a sequence of endothelial intracellular events culminating in apoptosis. Leukocyte proteases, however, may also directly influence endothelial cell apoptosis. *In vitro*, proteinase 3 and elastase directly cause apoptosis of endothelial cells (Yang et al., 1996; Taekema-Roelvink et al., 1998). In contrast to proteinase 3, HBP/CAP37 is internalized by endothelial cells, where it targets to perinuclear compartments. This markedly reduces caspase-3 activation and protects endothelial cells from apoptosis. A balance of proteases, then, appears to be required for sustained viability of endothelial cells in the context of locally activated neutrophils (Olofsson et al., 1999).

Cytotoxic Endothelial Injury

Physiologic vs. Pathologic Interactions

The acute inflammatory response is the first line of defense against most pathogenic microorganisms. As noted above, the neutrophil is the major effector cell in the acute phase of inflammation. A variety of mediators can induce neutrophil adhesion and motility without precipitating vascular injury. Other inflammatory mediators, including precipitating immune complexes and opsinized particles, promote massive changes in the neutrophil surface. The result is that in addition to increased adhesiveness, the stimulated cells also produce large amounts of superoxide anion through the NADPH oxidase and release of secretory granule contents (including various proteases, phospholipases, cationic proteins, etc.) into the microenvironment.

The physiological counterpart to this is the phagocytosis of the invading microorganism that presumably initiated the acute inflammatory response. In the case of phagocytosis, the eliciting particle is completely engulfed by the neutrophil and incorporated into a phagosome. Combination of the phagosome with neutrophil granules produces a phagolysosome with oxidant generation and granule content release within the enclosed environment. In contrast, when the oxidant-generating enzymes are activated at the cell surface and granule contents are released to the external environment, severe tissue damage is the result. Figure 2 provides ultrastructural views of neturophil – endothelial cell interactions under conditions in which trafficking occurs (panel a) and under conditions in which target cell injury results (panel b). Evident in panel b, but not a, is the interdigitation of neutrophil and endothelial cell membranes. Close apposition of this type between effector and target cell insures that inflammatory mediators produced by the neutrophils will interact with the endothelial cell as the target. Although neutrophils have a number of cytotoxic moieties, past studies clearly identified oxidants, and in particular, hydrogen peroxide, as an essential component in the endothelial cell injury process (Sacks *et al.*, 1978; Weiss *et al.*, 1981; Martin, 1984).

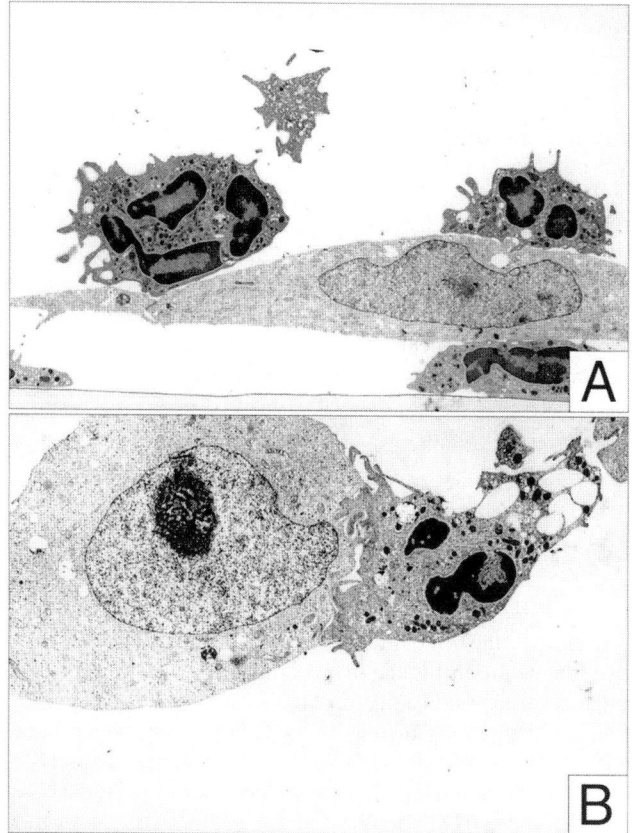

Figure 2. Neutrophil interactions with endothelial cells.
A. Ultrastructural image of neutrophil trafficking.
B. Ultrastructural image of neutrophil-mediated endothelial cell injury.

Biochemical Events in Neutrophil-mediated Injury

Because of the central role of endothelial cells in acute inflammatory injury, much effort has gone into understanding how these cells are injured by activated neutrophils. Based on years of study in several laboratories, we now have a reasonably clear picture of some of the critical events that occur. An overall view is depicted in Figure 3. Essential events include: i) tight adhesive interaction between the effector cell and target, providing a sequestered environment, ii) generation of superoxide anion by the effector cell (neutrophil) and conversion to hydrogen peroxide, and iii) reduction of the peroxide by transition-state (Fe^{2+}) iron to the highly cytolytic hydroxyl radical. A key concept in the injury process is that the target cell actively participates in its own destruction. It provides a source of reduced iron and a source of intracellular reducing equivalents in the form of superoxide anion. In the presence of reduced iron, the neutrophil-derived hydrogen peroxide is converted to the hydroxyl radical. Superoxide anion then "regenerates" transition iron from the oxidized form. The two-step process is referred to as the Fenton reaction.

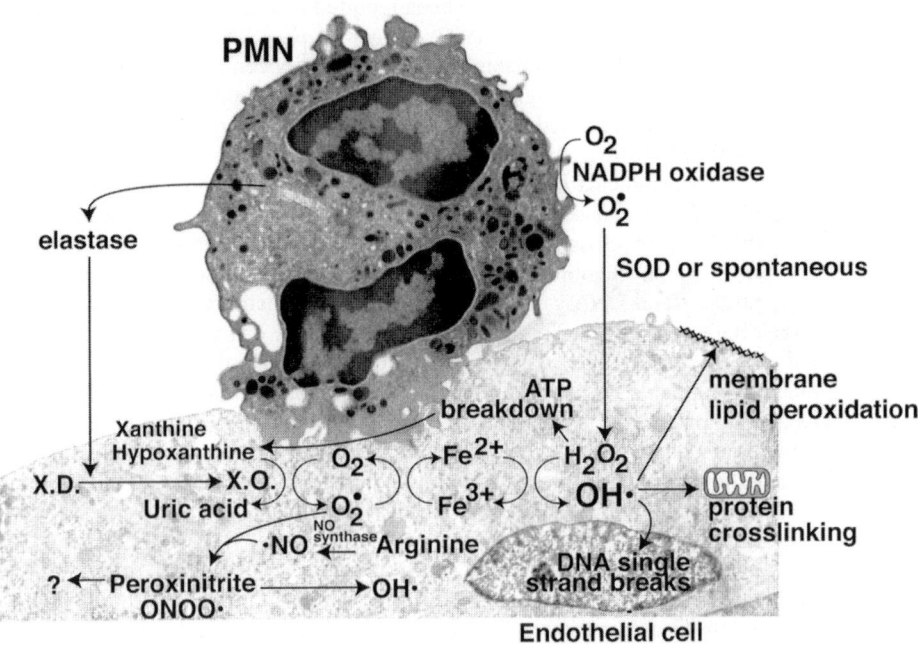

Figure 3. Biochemical events during neutrophil-mediated endothelial cell injury. Ultrastructural depiction of a neutrophil and portions of an endothelial cell, demonstrating release of elastase and generation of superoxide ($O_2^{\cdot}$), and subsequent intra-endothelial events leading to injury.

Although there are a number of possible enzymatic sources of superoxide anion generation within the target cell, the enzyme xanthine oxidase is likely to be one of the major sources. This enzyme normally converts xanthine to uric acid without superoxide generation, but when converted, this process generates superoxide anion from molecular oxygen. While NADPH is the preferred electron acceptor for the dehydrogenase form of the enzyme, the oxidase form uses molecular oxygen as its preferred electron acceptor. A number of events take place in the target cell under attack to facilitate the generation of superoxide anion by this enzyme. First, there is conversion of the enzyme from the dehydrogenase form to the oxidase form due to oxidation of the tetrahydobiopterin cofactor, which uncouples the enzyme (Verhaar *et al.*, 2004), involving a conformational shape change and a subsequent proteolytic cleavage. Concomitant with activation of the oxidant-generating enzyme, exposure of the target cell to H_2O_2 causes a rapid and dramatic consumption of ATP. This results in the build-up of ATP breakdown products including xanthine and hypoxanthine - substrates for xanthine oxidase. Thus, conditions are favorable inside the target cell for the conversion of neutrophil-derived hydrogen peroxide to a more toxic radical (hydroxyl radical) and subsequent injury. Thus, the target cell is not an innocent bystander but actively participates in its own destruction. These observations have been reported in a series of reports (Varani et al., 1985; Markey et al., 1990; Phan et al. 1992; Varani et al., 1992a) and summarized in reviews (Varani and Ward, 1994). Mechanisms of injury have been studied with other cell types, among them hepatocytes, mesangial cells and renal tubule epithelium, and common mechanistic features to those identified in endothelial cells have been reported (Starke and Farber, 1985; Andreoli and McAtter, 1990; Varani, et al., 1992b).

Exactly how hydroxyl radicals lethally injure endothelial cells is not fully understood. This potent oxygen radical can cause destruction of even the most stable covalent bonds. In cells, this frequently leads to cross-linking of proteins, single-strand breaks in DNA and peroxidation of membrane phospholipids (Halliwell and Gutteridge, 1986; Redl et al., 1993). While damage to proteins and DNA can, themselves, be lethal, membrane phospholipids are an especially important target. The loss of membrane function can quickly leads to ionic imbalances that are lethal to the cell. Furthermore, unlike protein and DNA damage, where one target molecule is damaged by one free radical molecule, the process of lipid peroxidation initiates a chain reaction. That is, the phospholipid degradation product is, itself, a free radical. It can initiate damage to an adjacent phospholipid, which can, in turn, damage another molecule of phospholipid. In this way, one hydroxyl radical can be responsible for destruction of several phospholipid molecules. Membrane damage can quickly escalate to a "clinically" relevant state.

While the scenario outlined above describes a sequence of events thought to be responsible for much of the injury that occurs in endothelial cells during acute inflammation, there are several variations on the theme. For example, while the hydroxyl radical appears to be the major cytotoxic oxygen free radical, the metabolism of hydrogen peroxide via myeloperoxidase to hypochlorous acid also produces a cytotoxic oxygen species (Babior et al., 1977). Additionally, xanthine oxidase is not the only enzyme capable of generating intracellular superoxide anion. Other enzymes including the arachidonic acid-metabolizing cyclooxygenase and lipoxygenase enzymes contribute to the intracellular oxidant burden. With activation by a variety of inflammatory mediators, NADPH oxidase is also assembled and activated in endothelium and in lung this has been shown to be a major oxidant source, as discussed above. There is also evidence for participation of mitochondrial oxidant-generating enzymes. Of particular interest is the role of nitric oxide in the endothelial cell injury process.

Nitric oxide is generated in endothelial cells and can be metabolized through interaction with superoxide anion to peroxynitrite and then to the hydroxyl radical. Generation of hydroxyl radical through this pathway could be expected to contribute to the overall oxidant burden. Alternatively, consumption of superoxide anion through interaction with nitric oxide reduces its availability to participate in the Fenton reaction (Rubanyi et al., 1991; Gutierrez and Nieves, 1996). This could lead to a lessening of the overall oxidant burden. Experimentally, it has been demonstrated that exposure of "resting" endothelial cells to a nitric oxide-generating system results in lethal cell injury. However, when the same cells are exposed to a hydrogen peroxide-generating system and a nitric oxide-generating system concomitantly, the overall level of injury in the presence of both enzyme systems is less than the level of injury seen in the presence of the hydrogen peroxide-generating system alone (Murphy *et al.*, 1999). These data suggest that under some conditions at least, nitric oxide can be protective against oxidant injury.

Respiratory epithelial cells and interstitial fibroblasts, like endothelial cells, are damaged in acute lung inflammation. In chronic inflammatory conditions such as asthma, the epithelial cells rather than the endothelial cells may be the primary target. While both epithelial cells and fibroblasts are susceptible to oxidant injury (Simon *et al.*, 1981), these cells are quantitatively more resistant to injury than endothelial cells. Anti-oxidant levels (superoxide anion dismutase, catalase and glutathione peroxidase) are higher in respiratory epithelial cells and fibroblasts than in endothelial cells. It may be that these cells (respiratory epithelial cells, at least) by virtue of their constant association with a highly-oxygenated environment have acquired defenses against oxidant injury that are not as well developed on the endothelial side of the alveolar wall.

Interaction Between Oxidants and Other Mediators in Acute Injury

Although cytotoxic endothelial cell injury is primarily the result of oxidant generation, a role for other leukocyte effector mechanisms has also been shown. A number of studies have shown that cytotoxic injury can be produced *in vitro* by exposure of endothelial cells to sources of proteolytic enzymes (Pontremoli *et al.*, 1986; Smedly *et al.*, 1986). Although proteolytic enzyme exposure can produce cytotoxic endothelial cell injury, endothelial cells that have been exposed to sub-lytic concentrations of oxidants demonstrate increased sensitivity to lethal cell injury by proteolytic enzymes (Rodell *et al.*, 1987; Varani *et al.*, 1989). The mechanism by which a combination of oxidants and proteolytic enzymes brings about injury under conditions in which neither oxidants nor proteolytic enzymes are effective alone is not fully understood. One prominent possibility is that serine proteinases, especially leukocyte elastase and cathepsin G, are very effective in collaborating with oxidants to injure endothelial cells. In contrast, matrix metalloproteinases are ineffective. As noted above, the serine proteinases from neutrophils are very effective in remodeling (damaging) the endothelial cell surface. While healthy cells have the capacity to "repair" membrane damage, concomitant oxidant injury makes cellular repair processes less efficient. This may be due to the dramatic and sustained reduction in ATP and total cellular energy that follows exposure to hydrogen peroxide (Hyslop *et al.*, 1988). Similar results are seen in cells exposed to damaging enzymes after treatment with metabolic poisons (Kristensen, 1989). The collaboration between oxidants and proteolytic enzymes in producing endothelial cell injury is not unique to this combination of effector molecules. Neutrophil cationic proteins and neutrophil phospholipases both produce injury to oxidant-exposed endothelial cells under conditions in which healthy cells are not affected (Ginsburg *et al.*, 1992).

In a similar manner, it has been demonstrated that various microbial metabolites also have a detrimental effect on endothelial cells. In each case, oxidant-injured cells are significantly more susceptible to injury than are cells not exposed to oxidants. Studies on the pathophysiology of Group A streptococcal infection demonstrated a strong synergy between streptococcal phospholipases (particularly Streptolysin S) and neutrophil-derived hydrogen peroxide in bringing about endothelial cell injury (Ginsburg and Varani, 1993). Secreted metabolites were not the only microbial product with detrimental effects. While the bacterial cells themselves did not induce cell injury, viable bacteria binding to the endothelial cell layer was found to precipitate injury as neutrophils responded to the surface bound bacteria and tried to phagocytose the bacteria (Ofek *et al.*, 1990; Ginsburg and Varani, 1993). Protection against injury was seen with peroxide inhibitors, confirming the role of oxidants in the injury process. Thus, it appears that metabolites from the microbial invaders and the responding neutrophils combined to produce damage to the endothelial cell layer.

To carry this paradigm of oxidant plus additional mediator combinations as underlying cytotoxic cell injury further, it has been demonstrated that products from the dead and dying cells, themselves, interact with oxidants in the microenvironment to exacerbate injury to adjacent cells. Enzymes (proteases and phospholipases) are among the products released from the injured tissue cells. Additionally, DNA fragmentation occurs in the dead cells and releases large amounts of (cationic) histones. The importance of cationic products to the injury process is suggested by the finding that anionic substances are as effective as anti-oxidants or anti-proteases in exerting a protective effect (Ginsburg *et al.*, 1989). Finally, even cationic antibiotics used to control the microbial infection can bind to the surface of endothelial cells and foster cellular injury (Ginsburg *et al.*, 1989). As with other cytotoxic moieties, lower antibiotic concentrations are needed in the presence of oxidants.

Up until now, the discussion has been on how interactions between oxidants and additional mediators of injury increase endothelial cell sensitivity to cytotoxic effects. However, many of the same degradative enzymes, cationic proteins, and bacterial cell wall components that bind to the endothelial cell surface to increase sensitivity to injury also bind the neutrophil surface resulting in an enhanced oxidative burst. Thus, many products in the microenvironment enhance endothelial cell sensitivity, while at the same time increasing the elaboration of the oxidants that initiate cell injury. Damage to the vasculature may enhance cytotoxicity in another manner as well. Large numbers of red blood cells accumulate at sites of vascular damage and release transition-state iron from hemoglobin. As indicated above, iron in the reduced form is necessary for conversion of hydrogen peroxide to the hydroxyl radical. The presence of dead and dying red blood cells potentiates this rate-limiting event in cellular injury (Bauer *et al.*, 1995).

In summary, it is clear that at least *in vitro*, endothelial cells are susceptible to cytotoxic injury by a number of mechanisms. Exposure of cells to hydrogen peroxide or to hydrogen peroxide-generating enzyme systems can produce substantial injury in the absence of other mediators. A variety of non-oxidant surface-active agents including proteolytic enzymes, phospholipases, cationic proteins, bacterial cell wall extracts and debris from dead and dying host tissue cells can also produce lethal injury to endothelial cells *in vitro*. For the most part, such mediators are effective in the absence of oxidants only at high concentrations. When endothelial cells are exposed to a combination of oxidant and non-oxidant mediator, lethal injury occurs at much lower concentrations.

Is it possible to extrapolate from *in vitro* injury models to what occurs in a tissue during an acute inflammatory response? Does neutrophil-mediated endothelial cell killing *in vitro* provide insight into tissue injury during acute inflammation? The answer to this question, we believe, is yes. Both *in vivo* and *in vitro*, injury only occurs if neturophils become fully activated while in contact with target endothelial cells in the vascular wall. In both cases it appears that oxidants are a major contributor to lethal cell injury but that multiple mediators function in concert to exert maximal damage. Using soluble mediators to mimic the environment, we can identify the combinations of mediators that bring about cell death and elucidate the biochemical pathways involved. Although mediators of injury are used in solution to study the pathophysiology of injury, it is important to keep in mind that injury to endothelial cells in a vascular bed does not occur as a result of interaction between the target cell and soluble mediators in the plasma. Mediators released into the plasma would: i) be rapidly diluted, ii) be inactivated by inhibitors and iii) find alternative targets. Rather, endothelial cell injury occurs as a consequence of direct interaction between the target cell and the effector cell in a sequestered environment.

Endothelial Cell Protective Mechanisms

Inflammation is a regulated process with mechanisms in place to reduce, as well as to amplify, the tissue response. The balance between oxidants and antioxidants, proinflammatory and anti-inflammatory cytokines, together with proteases and antiproteases in the inflamed lung largely determines the outcome of an inflammatory insult. A number of endogenous anti-inflammatory mediators are upregulated during lung inflammation, including cytokines such as IL-10 and IL-13, anti-oxidants and protease inhibitors. The vasoprotective function of endothelial cells is associated with biosynthesis and release of nitric oxide, prostacyclin, prostaglandin E2, carbon monoxide, and plasminogen activator. These mediators reduce platelet and leukocyte activation, prevent the occurrence of thrombotic events, promote thrombolysis, maintain tissue perfusion, reduce pulmonary vascular permeability, and attenuate endothelial cell damage and, ultimately, tissue injury during acute inflammation. Elevation of leukocyte cAMP attenuates activation and, thus, normal release of mediators from quiescent endothelium, like PGI_2, has a calming paracrine effect on leukocytes as it does on platelets (Sato, 2004). Potentially, therapeutic elevation of leukocyte cAMP might have similar benefits

Antioxidants

Vascular endothelial cells have a central position in maintaining the delicate intra- and extracellular oxidant balance important for the maintenance of tissue homeostasis. During inflammation, this oxidant balance is tipped either towards cell and tissue injury or towards protection against oxidant mediated injury and maintenance of tissue integrity. Endothelial cells generate endogenous antioxidants including superoxide dismutase, reduced glutathione (GSH) and other thiol containing compounds. These protect cells from oxidative stress due to reactive oxygen metabolites such as hydrogen peroxide and hydroxyl radical, limiting redox-regulated signal transduction, cell proliferation, apoptosis, and inflammation, thereby presenting a defense against oxidant-mediated injury (Abello *et al.*, 1994; Droge and Breitkreutz, 2000). Superoxide dismutase catalyzes the metabolism of superoxide to hydrogen peroxide, which in turn either combines with transition metals to generate hydroxyl radical or to be detoxified to H_2O by glutathione peroxidase or catalase. Reduced glutathione appears to be a specific target of oxidative damage in ARDS patients who demonstrate reduced alveolar fluid reduced glutathione and increased oxidized glutathione (GSSG) (Pacht *et al.*, 1991). In cultured endothelial cells as well as in lungs of animals subjected to oxidant-mediated injury, exogenous glutathione reverses redox-induced microvascular dysfunction (Zhao *et al.*, 2001).

In addition to intra- and extra-cellular enzyme systems, anti-oxidant mechanisms include scavenging systems by other reactive molecules. As mentioned above, nitric oxide is one molecule which plays very diverse roles in the physiology and pathophysiology of the vascular system, earning the conflicting reputation of being both protective and deleterious to vascular function (Niu et al., 1996). The roles of nitric oxide as a vasodilator and as a co-factor in inflammation are well established and nitric oxide appears to be an important mediator during sepsis (Belvisi et al., 1995; Murray et al., 2000). While nitric oxide generated by endothelial nitric oxide synthase (eNOS) functions as a vasoregulatory molecule in large vessels, in the microvasculature where it is largely generated by inducible iNOS, nitric oxide likely serves a very different purpose. Impairment of nitric oxide bioactivity is related to pathological conditions in the macrovasculature, but in the microvasculature it may play a role in regulating oxidant balance to prevent oxidative cell damage. Inflammatory cytokines increase expression of iNOS and in lung sepsis the iNOS gene has an early induction response to LPS, in vascular endothelial cells, airway epithelial cells, and inflammatory cells (Xue and Johns, 1995). Adverse affects of excess nitric oxide include its contribution to septic vascular hypocontractility, organ hypoperfusion, and interaction with superoxide to yield the highly toxic peroxynitrite as well as, ultimately, the hydroxyl radical. At the same time, nitric oxide reacts with a number of intra- and extra-cellular pathways to limit the extent of oxidant-mediated cell injury. Nitric oxide scavenges other oxygen radicals, ultimately reducing superoxide available for cell damage, but also modulating oxidant-mediated signal transduction events such as activation of NF-κB leading to adhesion molecule expression (Rubanyi et al., 1991). Nitric oxide prevents platelet adherence and aggregation at sites of vascular injury, compensating for disruption of microvascular integrity (Radomski et al., 1987). Inhibition of leukocyte and monocyte adhesion to endothelial cells by exogenously added nitric oxide suggests that nitric oxide limits the potential for injury by producing quiescence in leukocytes and platelets thus preventing interaction of inflammatory cells with the vascular endothelial cells (Kubes et al., 1991; MacMicking et al., 1995; Hollenberg et al., 2000). Nitric oxide decreases the sensitivity of endothelium to LPS-induced apoptosis (Tang et al., 2002). Finally, retardation of smooth muscle proliferation and migration, and stimulation of endothelial cell proliferation as well as nitric oxide inhibition of stress induced endothelial cell apoptosis serve to preserve endothelial as well as tissue function.

Nitric oxide has also been linked to increased expression of another family of protective molecules, the heat shock proteins. Nitric oxide upregulates heat shock protein expression and protects against apoptosis. Protection mediated by heat shock proteins is generally attributed to the molecular chaperone function resulting in, among other effects, increased protein expression due to enhanced folding of nascent proteins, suppression of NADPH oxidase and decrease in pro-inflammatory cytokines (Benjamin and McMill, 1998). Vascular endothelial cells express the ubiquitous HSP70 and HSP90 as well as HSP25. HSP70 in particular is increased in lungs of patients with ARDS and asthma and appears to protect during sepsis and metabolic stress.

While the products of oxidative stress are critical factors in the induction of injury to the vascular endothelium, some evidence suggests that the formation of lipid peroxidation products such as aldehydes may also reduce the endothelial cell response to inflammatory stimuli. These small molecules, which react with biological proteins, have been shown to attenuate endothelial cell adhesion molecule expression including VCAM-1, ICAM-1 and E-selectin, to suppress the generation of MCP-1 accompanied by reduction of NF-kB activation, and to reduce levels of intracellular GSH (Minekura et al., 2001). Oxidized phospholipid also interferes with the ability of LPS to bind to LPS-binding protein (LBP) and to CD14. The unopposed LPS–CD14 interaction results in widespread gene activation *via* engagement of Toll-like receptor-4 (TLR-4), p38 mitogen-activated protein kinase (MAPK) and NF-κB activation, and in endothelial cells, expression of adhesion molecules ICAM-1, E-selectin and VCAM-1. *In vivo,* systemic administration of oxidized phospholipid attenuated adhesion-molecule expression by endothelial cells by inhibiting the LPS-induced gene activation. As a result, organ injury was prevented, and survival rates were greatly improved (Bochkov et al., 2002). It seems then that oxidized phospholipids have been endowed with both pro-inflammatory and anti-inflammatory activities.

Antiproteinases

Antiproteinases present a protective shield at the site of inflammation. Host tissues are protected from unregulated proteolysis by elastase and other proteases by multiple proteinase inhibitors including α1-antiprotease, secretory leukoprotease inhibitor (SLPI), α_2-macroglobulin, and eglin (Weiss, 1989; Rice and Weiss, 1990). Antiprotease activity directed, perhaps, at the intracellular proteases may be responsible for the suppressive effects of SLPI on the intrapulmonary activation of NF-κB and neutrophil recruitment into the lung (Mulligan et al., 2000). However, in a manner similar to oxidants, a delicate balance between proteases and antiproteases is maintained. The antiproteinase defense may be overwhelmed by high concentration of proteases, and even low concentrations of oxygen radicals can not only neutralize the protective antiproteinases at the site of inflammation but may also increase the activity of neutrophil derived proteases. Neutrophil-mediated endothelial cell injury is inhibited by both oxidant inhibitors (catalase) and protease inhibitors (Soybean trypsin inhibitor), and these appear to function synergistically (Varani et al., 1989).

Coagulation Proteins

Abnormalities of fibrin turnover occur in sepsis, resulting in increased procoagulant response. Activated Protein C (APC) at the vascular endothelial cell surface is uniquely positioned to control the acute inflammatory response. Composed of thrombin and protein C bound to endothelial cell thrombomodulin (TM) and protein C receptor (EPCR) (Laszik et al., 1997), activated protein C possesses anticoagulant, vasoregulatory and anti-inflammatory properties. It also inhibits inflammatory cytokine release to prevent LPS-induced pulmonary vascular injury (Murakami et al., 1997). APC has profibrinolytic properties, and is an inhibitor of factors Va and VIIIa (Abraham, 2000). Leukocyte adhesion to activated endothelium is reduced by complex of soluble protein C receptor bound to proteinase 3, which in turn binds to CD11b/CD18 on activated neutrophils (Hirose et al., 2000; Kurosawa et al., 2000). Thrombin-antithrombin complexes are elevated during acute lung injury (Wenzel et al., 2002). At the same time, the heparin-sulfate-antithrombin anticoagulant systems stimulate increased endothelial PGI_2, a potent inhibitor of leukocyte activation, and inhibit inflammatory cytokine and tissue factor production by endothelial cells (Wiedermann and Romisch, 2002). Tissue Factor Pathway Inhibitor (TFPI), also synthesized by endothelial cells under basal conditions and constitutively bound to the endothelial cell, inhibits tissue factor thereby preventing association with Factor VII and further activation of the coagulation system. Endogenous mechanisms, therefore, are in place to limit endothelial cell injury and prevent progression to acute tissue injury.

Acute Endothelial Cell Injury Promotes Chronic Lung Disease

Fibrosis in ARDS

In the lung, acute as well as chronic injuries lead to a final common pathway resulting in pulmonary fibrosis. ARDS is traditionally divided into three phases: exudative, proliferative, and fibrotic. Damage to the endothelial and epithelial surfaces leads to exudation and inflammation; fibroproliferation and the abnormal and excessive deposition of extracellular matrix proteins, in particular collagen, then ensues. Unabated, this process leads to established fibrosis and the obliteration of alveolar spaces with a dense irregular matrix. Excessive interstitial and intra-alveolar fibrosis is characteristic then of the more advanced stages of ARDS (Zapol *et al.*, 1979; Marshall *et al.*, 1998).

A number of mechanisms exist which could lead to an early activation of the fibroproliferative response in ARDS including release of fibrogenic cytokines such as TNF-α, IL-1β, IL-2, IL-4, IL-6 and IL-8 from leukocytes, and products of the coagulation cascade such as tissue factor/factor VII, thrombin, and fibrin (Gunther *et al.*, 2000; Idell, 2002). The coagulation cascade is initiated by endothelial cell injury and subsequent release of tissue factor by endothelial cells as well as parenchymal and inflammatory cells, creating a hypercoagulable environment (Idell, 2002). Tissue factor, a membrane bound glycoprotein, associates with factor VIIa in a dynamic complex regulated by tissue factor pathway inhibitor (TFPI). This complex activates a sequence culminating in generation of thrombin and fibrin, which in turn regulate endothelial cell functions including expression of adhesion molecules and release of soluble inflammatory mediators. In patients with lung injury, endothelial cell disturbance is related to activation of the coagulation system and accumulation of extravascular thrombin and fibrin. Fibrin is a major component of the hyaline membrane and both thrombin and fibrin are mitogenic for fibroblasts and stimulate collagen synthesis in these cells (Dawes *et al.*, 1993). Fibrin provides a critical provisional matrix on which cells can proliferate, organize, and carry out specialized functions. A variety of cell types including endothelial cells, specifically bind to and migrate on fibrin matrices, utilizing integrin and non-integrin receptors (Degen, 1999). Disordered pathways of fibrin turnover have been implicated in fibrotic repair in the injured lung, lung remodeling, and pulmonary fibrosis (Idell, 1995; Marshall *et al.*, 1998; Idell, 2002; Wenzel *et al.*, 2002). MMP-2 and MMP-9, generated by endothelial cells as well as epithelial and inflammatory cells, may play an anti-fibroproliferative role in the pathogenesis of ARDS and airway remodeling by degrading the extracellular matrix components synthesized by fibroblasts (Lechapt-Zalcman and Escudier, 2000; Lanchou *et al.*, 2003). Long-term outcome may then be influenced by the balance or imbalance between matrix metalloproteinaseses and TIMP, and by the extent of activation of the coagulation system during acute lung injury.

Acute Lung Injury Preceding Chronic Pulmonary Disease

While the etiology of chronic lung diseases is varied, several diseases demonstrate an acute, endothelial cell damaging component. In a number of animal models of chronic pulmonary disease, a preceding inflammatory phase with capillary endothelial cell injury as well as type I alveolar cell injury is followed by type II cell proliferation and subsequent fibrosis (Dodge *et al.*, 1991). The endothelial cell injury may be humorally mediated and associated with deposition of complement factors and immunoglobulin or arise in the setting of viral pneumonia, or introduction of endogenous (hemosiderosis) or exogenous (silica, bleomycin) factors. In chronic interstitial lung diseases as well as in experimental chronic injury models, acute lung injury associated with endothelial cell activation/injury precedes the chronic fibrosis. For example, in bleomycin injured lung, sustained entrapment of leukocytes in capillaries is attributed to augmented ICAM-1 expression by activated microvascular endothelium (Sato *et al.*, 2000). Similarly, in lungs of rats given a single intratracheal injection of silica particles, early lesions are characterized by accumulations of macrophages and neutrophils in alveolar lumen and interstitium and by damage to alveolar capillaries and epithelial cells (Kawanami *et al.*, 1995). Cellular injury in idiopathic pulmonary fibrosis may be mediated by oxygen radicals produced by infiltrating inflammatory cells (Berk, 1999; Magro *et al.*, 2003). Rat lungs subjected to radiation showed a statistically significant elevation of von Willebrand Factor expression, a marker of endothelial injury, and the sequelae of radiation-induced fibrosis (Kasper *et al.*, 1996). In systemic sclerosis characterized by excess collagen formation, pulmonary endothelial dysfunction occurs early in the disease as evidenced by depression of pulmonary capillary angiotensin converting enzyme activity (Orfanos *et al.*, 2001). In these and other chronic lung diseases, endothelial cells injured by leukocytes in the microvasculature may contribute to the unbridled tissue response, thereby promoting chronic disease. Persistent recruitment of leukocytes by activated endothelium may then amplify tissue damage by releasing a variety of injurious substances including oxygen free radicals, and proteolytic enzymes as well as fibrogenic agents. Some studies have suggested that pulmonary fibrosis may, in fact, be the sequelae of acute lung injury occurring sequentially throughout the lung

Fibrosis, the hallmark of chronic interstitial lung disease, is characterized by deposition of excessive collagen within the terminal alveolar unit. The fibrotic process in the lung appears to result from a complex interaction between fibroblasts, other lung parenchymal cells, and macrophages. Injury to the epithelium and basement membranes appears to be necessary for the fibrotic process to occur. Fibroblasts migrate into areas of acute lung injury and are stimulated to secrete collagen and other matrix proteins. These cells also release various proteases that have the capacity to degrade and remodel these matrix proteins. The stimuli that activate fibroblasts to remodel the lung are not well defined but include components of the coagulation cascade. Just as in ARDS, discussed above, in some forms of chronic lung disease (i.e., bleomycin-induced) alveolar fibrin deposition is persistent and precedes parenchymal lung scarring. Matrix degradation products and mediators (like transforming growth factor beta, platelet-derived growth factor, plasminogen activator factor, angiotensin II) released from macrophages, endothelial cells and lung parenchymal cells promote fibroblast activation. Increased numbers of leukocytes in BAL are associated with progression to pulmonary fibrosis, suggesting that ongoing endothelial activation, with continued leukocyte recruitment and subsequent promotion of endothelial and epithelial injury is central to the continued inflammatory response (Ward and Hunninghake, 1998).

Conclusions

During acute lung injury, influx of neutrophils is usually a key event in the development of endothelial cell and tissue injury. Release of proteolytic enzymes, generation of reactive oxidants, and metabolism of soluble proteins, by neutrophils in particular, serve to activate endothelial cells of the pulmonary microvasculature and to amplify the inflammatory process. Endothelial cells exposed to these products thus have a central role in the tissue response. Activated endothelial cells are capable of presenting a defense against harmful agents, and when successful, this reduces tissue damage and prevents progression to life-threatening outcome or chronic disease. When these defenses are overwhelmed, however, endothelial cell injury and death ensue with the net result of disruption of the vascular barrier and subsequent exposure of parenchymal tissue to inflammatory products. Continuous cell injury then leads not only to tissue injury and system failure, but repeated insults may lead to chronic disease.

REFERENCES

Abello, P.A., Fidler, S., Bulkley, G., and Buchman, T.G. (1994) Antioxidants modulate induction of programmed endothelial cell death (apoptosis) by endotoxin. Arch. Surg. 129:134-140.

Abraham, E. (2000) Tissue factor inhibition and clinical trial results of tissue factor pathway inhibitor in sepsis. Crit. Care Med. 28:S31-S33.

Alexander, J.S., Blaschuk, O.W., and Haselton, F.R. (1993) An N-cadherin-like protein contributes to the maintenance of solute barrier in cultured bovine aortic endothelial cells. J. Cell. Physiol 156: 610-618.

Al-Mehdi, A., Zhao, G., Dodia, C., Tozawa, K., Costa, K., Muzykantov, V., Ross, C., Blecha, F., Dinauer, M., and Fisher, A. (1998) Endothelial cell NADPH oxidase as source of oxidants in lungs exposed to ischemia. Circ. Res. 83:730-737.

Andreoli, S.P. and J.A. McAtter (1990) Reactive oxygen molecule-mediated injury in endothelial and renal tubular epithelial cells in vitro. Kidney Int. 38:785-794.

Babior, B.M., Kipnes, R.S., and Curnutte, J.T. (1977) Biological defense mechanism: the production by leukocytes of superoxide, a potent bacteriocidal agent. J. Clin. Invest. 52:741-746.

Ballieux, B.E., Hiemstra, P., Klar-Mohamad, N., Hagen, E., van Es, L., van der Woude, F., and Daha, M.R. (1994) Detachment and cytolysis of human endothelial cells by proteinase 3. Euro. J. Immunol. 24:3211-3215.

Bauer, M., Feucht, K., Ziegenfuss, T., and Marzi, I. (1995) Attenuation of shock-induced hepatic microcirculatory disturbances by the use of a starch-deferoxamine conjugate for resuscitation. Crit. Care Med. 23:316-322.

Belvisi, M., Barnes, P., Larkin, S., Yacoub, M., Tadjkarimi, S., Williams, T., and Mitchell, J.A. (1995) Nitric oxide synthase activity is elevated in inflammatory lung disease in humans. Eur. J. Pharmacol. 283:255-258.

Benjamin, I.L. and McMill, D.R. (1998) Stress (Heat Shock) proteins: Molecular chaperones in cardiovascular biology and disease. Circ. Res. 83:117-132.

Berger, S.P., Seelen, M., Hiemstra, P., Gerritsma, J., Heemskerk, E., van der Woude, F., and Daha, M.R. (1996) Proteinase 3, the major autoantigen of Wegener's granulomatosis, enhances IL-8 production by endothelial cells in vitro. J. Amer. Soc. Neph. 7:694 -701.

Berk, B.C. (1999) Redox signals that regulate the vascular response to injury. Thromb. Haemost. 82:810-817.

Berliner, J.A. and Heinecke, J.W. (1996) The role of oxidized lipoproteins in atherogenesis. Free Radic Biol Med 20:707-27.

Blackford, J.A. Jr., Antonini, J., Castranova, V., and Dey, R.D. (1994) Intratracheal instillation of silica up-regulates iNOS gene expression and increases NO production in alveolar macrophages and neutrophils. Amer. J. Respir. Cell Molec. Biol. 11:426-431.

Bochkov, V.N., Kadl, A., Huber, J., Gruber, F., Binder, B., and Leitinger, N. (2002) Protective role of phospholipid oxidation products in endotoxin-induced tissue damage. Nat. Med. 419:77-81.

Boehme, M.W., Galle, P., Stremmel, W. (2002) Kinetics of thrombomodulin release and endothelial cell injury by neutrophil-derived proteases and oxygen radicals. Immunol. 107:340-9.

Borregaard, N. and Cowland, J.B. (1997) Granules of the human neutrophilic polymorphonuclear leukocyte. Blood 89:3503-3521.

Camussi, G., Tetta, C., Bussolino, F., Caligaris, F., Coda, R., Masera, C., and Segoloni, G. (1981) Mediators of immune-complex-induced aggregation of polymorphonuclear neutrophils. II. Platelet-activating factor as the effector substance of immune-induced aggregation. Internat'l. Arch. Allergy Appl. Immunol. 64:25-41.

Carden, D., Xiao, F., Moak, C., Willis, B.H., Robinson-Jackson, S., and Alexander, S. (1998) Neutrophil elastase promotes lung microvascular injury and proteolysis of endothelial cadherins. Amer. J. Physiol. 275:H385-H392.

Catravas, J.D., Ryan, J., Chung, A., Quinn, N., and Anthony, B.L. (1990) Inhibition of endothelial-bound angiotensin converting enzyme in vivo. Br. J. Pharmacol. 101:121–127.

Charlesworth, E.N., Hood, A., Soter, N., Kagey-Sobotka, A., Norman, P., and Lichtenstein, L.M. (1989) Cutaneous late-phase response to allergen: mediator release and inflammatory cell infiltration. J. Clin. Invest. 83:1519-1526.

Chen, C.H., Jiang, T., Yang, J., Jiang, W., Lu, J., Marathe, G., Pownall, H., Ballantyne, C., McIntyre, T., Henry, P., and Yang, C.Y. (2003) Low-density lipoprotein in hypercholesterolemic human plasma induces vascular endothelial cell apoptosis by inhibiting fibroblast growth factor 2 transcription. Circ. 107:2102-2108.

Chin, J.H., Azhar, S., and Hoffman, B.B. (1992) Inactivation of endothelium-derived relaxing factor by oxidized lipoproteins. J. Clin. Invest. 89:10–18.

Chun, T.H., Sabeh, F., Weiss, S.J. (2004) MTI-MMP-dependent Neovessel formation within the confines of the 3-dimensional extracellular matrix. Genes & Devel. *In press*.

Cochrane, C.G., Spragg, R., Revak, S., Cohen, A., and McGuire, W.W. (1983) The presence of neutrophil elastase and evidence of oxidation activity in bronchoalveolar lavage fluid of patients with adult respiratory distress syndrome. Amer. Rev. Respir. Dis. 127:S25–S27.

Cziraki, A., Horvath, I.G., and Papp, L. (2000) Quantification of pulmonary capillary endothelium-bound angiotensin converting enzyme inhibition in man. Gen. Pharm. 35:213-218.

Dawes, K.E., Gray, A.J, and Laurent, G.J. (1993) Thrombin stimulates fibroblast chemotaxis and replication. Eur. J. Cell Biol. 61:126-130.

Degen, J.L. (1999) Hemostatic factors and inflammatory disease. Thromb. Haemost. 82:858-864.

Delclaux, C., d'Ortho, M., Delacourt, C., Lebargy, F., Brun-Buisson, C., Brochard, L., Lemaire, F., Lafuma, C., and Harf, A (1997) Gelatinases in epithelial lining fluid of patients with adult respiratory distress syndrome. Amer. J. Physiol. 272:L442 -L 451.

Dodge, A.B., Patton, W., Yoon, M., Hechtman, H., and Shepro, D. (1991) Organ and species specific differences in cytoskeletal protein profiles of cultured microvascular endothelial cells. Comp. Biochem. Physiol. B. 98:461-470.

Donnelly, S.C., MacGregor, I, Zamani, A., Gordon, M., Robertson, C., Steedman, D., Little, K., and Haslett, C. (1995) Plasma elastase levels and the development of the adult respiratory distress syndrome. Amer. J. Respir. Crit. Care Med. 151:1428-1433.

Droge, W. and Breitkreutz, R. (2000) Glutathione and immune function. Proc. Nutr. Soc. 59:595-600.
Drost, E.M. and MacNee, W. (2002) Potential role of IL-8, PAF, and TNFα in the sequestreation of neutrophils in the lung: effects on neutrophil deformability, adhesion receptor expression, and chemotaxis. Eur. J. Immunol. 32:393-403.
Dudek, S.M. and Garcia, J.G.N. (2001) Cytoskeletal regulation of pulmonary vascular permeability. J. Appl. Physiol. 91:1487-1500.
Dull, R.O., Dinavahi, R., Schwartz, L., Humphries, D., Berry, D., Sasisekharan, R., and Garcia, J.G. (2003) Lung endothelial heparan sulfates mediate cationic peptide-induced barrier dysfunction: a new role for the glycocalyx. Amer. J. Physiol. 285:L986-L995.
Edens, H.A. and Parkos, C.A. (2003) Neutrophil transendothelial migration and alteration in vascular permeability: neutrophil-derived azurocidin. Curr. Opin. Hematol. 10:25-30.
Gautam, N., Olofsson, A., Herwald, H., Iversen, L., Lundgren, E., Hedqvist, P., Arfors, K., Flodgaard, H., and Lindbom, L. (2001) Heparin-binding protein (HBP/CAP37): a missing link in neutrophil-evoked alteration of vascular permeability. Nature Med. 7:1123-1127.
Ginsburg, I., Gibbs, D., Schuger, L., Johnson, K., Ryan, U., Ward, P., and Varani, J. (1989) Vascular endothelial cell killing by combinations of membrane-active agents and hydrogen peroxide. Free Rad. Biol. Med. 7:369-376.
Ginsburg, I. and Varani, J. (1993) Interaction of viable group A Streptococci and hydrogen peroxide in killing of vascular endothelial cells. Free Rad. Biol. Med. 14:495-500.
Ginsburg, I., Misgav, R., Pinson, A., Varani, J., Ward, P.A., and Kohen, R. (1992) Synergism among oxidants, proteinases, phospholipases, microbial hemolysins, cationic proteins and cytokines: A possible major cause of cell and tissue injury in inflammation. Inflam. 16:519-538.
Gleich, G.J., Frigas, E., Loegering, D., Wassom, D., and Steinmuller, D. (1979) Cytotoxic properties of the eosinophil major basic protein. J. Immunol. 123:2925-2927.
Gunther, A., Mosavi, P., Heinemann, S., Ruppert, C., Muth, H., Markart, P., Grimminger, F., Walmrath, D., Temmesfeld-Wollbruck, B., and Seeger, W. (2000) Alveolar fibrin formation caused by enhanced procoagulant and depressed fibrinolytic capacities in severe pneumonia: Comparison with the acute respiratory distress syndrome. Amer. J Respir. Crit. Care Med. 161:454-462.
Guo, R. and Ward, P.A. (2002) Mediators and regulation of neutrophil accumulation in inflammatory responses in lung: Insights from the IgG immune complex model. Free Rad. Biol. Med. 33:303-310.
Gutierrez, H.H. and Nieves, B. (1996) Nitric oxide regulation of superoxide-dependent lung injury: oxidant-protective actions of endogenously-produced and exogenously-added nitric oxide. Free Rad. Biol. Med. 21:43-52.
Halliwell, B. and Gutteridge, J.M.C. (1986) Oxygen free radicals and iron in relation to biology and medicine: Some problems and concepts. Arch. Biochem. Biophys. 246:501-514.
Hamacher, J., Lucas, R., Lijnen, H., Buschke, S., Dunant, Y., Wendel, A., Grau, G., Suter, P., and Ricou, B. (2002) Tumor necrosis factor-alpha and angiostatin are mediators of endothelial cytotoxicity in bronchoalveolar lavages of patients with acute respiratory distress syndrome. Amer. J. Respir. Crit. Care Med. 166:651-656.
Heinzelmann, M., Platz, A., Flodgaard, H., Polk, H., and Miller, F.N. (1999) Endocytosis of heparin-binding protein (CAP37) is essential for the enhancement of lipopolysaccharide-induced TNF-production in human monocytes. J. Immunol. 162:4240–4245.
Herbst, U., Toborek, M., Kaiser, S., Mattson, M., and Hennig, B. (1999) 4-Hydroxynonenal induces dysfunction and apoptosis of cultured endothelial cells. J. Cell. Physiol. 181:295-303.
Hermant, B., Bibert, S., Concord, E., Dublet, B., Weidenhaupt, M., Vernet, T., and Gulino-Debrac, D. (2003) Identification of proteases involved in the proteolysis of vascular endothelium cadherin during neutrophil transmigration. J. Biol. Chem. 278:14002-14012.
Hessler, J.R., Morel, D., Lewis, L., and Chisolm, G.M. (1983) Lipoprotein oxidation and lipoprotein-induced cytotoxicity. Arteriosclerosis 3:215-222.

Hirose, K., Okajima, K., Taoka, Y., Uchiba, M., Tagami, H., Nakano, K., Utoh, J., Okabe, H., and Kitamura, N. (2000) Activated Protein C reduces the ischemia/reperfusion -induced spinal cord injury in rats by inhibiting neutrophil activation. Ann. Surg. 232:272-280.

Hollenberg, S.M., Broussand, M., Osman, J., and Parrillo, J.E. (2000) Increased microvascular reactivity and improved mortality in septic mice lacking inducible nitric oxide synthase. Circ.Res. 86:774-779.

Hyslop, P.A., Hinshaw, D., Halsey, W., Schraufstatter, I., Sauerheber, R., Spragg, R., Jackson, J., and Cochrane, C.G. (1988) Mechanisms of oxidant-mediated cell killing: The glycolytic and mitochondrial pathways of ADP phosphorylation are major targets of H2O2-mediated injury. J. Biol. Chem. 263:1665-1675.

Idell, S. (1995) Coagulation, fibrinolysis and fibrin deposition in lung injury and repair. In Pulmonary Fibrosis. Eds. S.H. Phan and R.S. Thrall. New York, Marcel Dekker. pp. 743-776.

Idell, S. (2002) Endothelium and disordered fibrin turnover in the injured lung: Newly recognized pathways. Crit. Care Med. 30:S274-S280.

Idell, S., Kucich, U., Fein, A., Kueppers, F., James, H., Walsh, P., Weinbaum, G., Colman, R., and Cohen, A.B. (1985) Neutrophil elastase-releasing factors in BAL from patients with ARDS. Amer. Rev. Respir. Dis. 132:1098-1105.

Jones, S.A., O'Donnel, V., Wood, J., Broughton, J., Hughes, E., and Jones, O.T. (1996) Expression of phagocyte NADPH oxidase components in human endothelial cells. Amer. J. Physiol. 271:H1626-H1634.

Kang, T., Yi, J., Guo, A., Wang, X., Overall, C., Jiang, W., Elde, R., Borregaard, N., and Pei, D. (2001) Subcellular distribution and cytokine- and chemokine-regulated secretion of leukolysin/MT6-MMP/MMP-25 in neutrophils. J. Biol. Chem. 276:21960-21968.

Karin, M., Liu, Z., and Zandi, E. (1997) AP-1 function and regulation. Curr. Opin. Cell Biol. 9:240-246.

Kasper, M., Schobl, R., Haroske, G., Fischer, R., Neubert, F., Dimmer, V., and Muller, M. (1996) Distribution of von Willebrand factor in capillary endothelial cells of rat lungs with pulmonary fibrosis. Exper. Toxicol. Path. 48:283-288.

Kawanami, O., Jiang, H., Mochimaru, H., Yoneyama, H., Kudoh, S., Ohkuni, H., Ooami, H., and Ferrans, V.J. (1995) Alveolar fibrosis and capillary alteration in experimental pulmonary silicosis in rats. Amer. J. Respir. Crit. Care Med. 151:1946-1955.

Kobayashi, H., Cui, T., Ando, M., Hataishi, R., Imasaki, T., Mitsufuji, H., Hayashi, I., and Tomita, T. (2002) Nitric oxide released from iNOS in PMNs makes them deformable in an autocrine manner. Nitric Oxide 7:221-227.

Kotamraju, S., Tampo, Y., Keszler, A., Chitambar, C., Joseph, J., Haas, A., and Kalyanaraman, B. (2003) Nitric oxide inhibits H_2O_2-induced transferrin receptor-dependent apoptosis in endothelial cells: Role of ubiquitin-proteasome pathway. PNAS USA 100:10653-10658.

Krishnadasan, B., Naidu, B., Byrne, K., Fraga, C., Verrier, E., and Mulligan, M.S. (2003) The role of proinflammatory cytokines in lung ischemia-reperfusion injury. J. Thorac. Cardiovasc. Surg. 125:261-272.

Kristensen, S.R. (1989) A critical appraisal of the association between energy charge and cell damage. Biochim. Biophys. Acta 1012:2725-2731.

Kubes, P., Suzuki, M., Granger, D.N. (1991) Nitric oxide: An endogenous modulator of leukocyte adhesion. PNAS USA 88:4651-4655.

Kurosawa, S., Esmon, C., Stearns-Kurosawa, D.J. (2000) The soluble endothelial protein C receptor binds to activated neutrophils: Involvement of proteinase-3 and CD11b/CD18. J. Immunol. 165:4697-4703.

Lanchou, J., Corbel, M., Tanguy, M., Germain, N., Boichot, E., Theret, N., Clement, B., Lagente, V., and Malledant, Y. (2003) Imbalance between matrix metalloproteinases (MMP-9 and MMP-2) and tissue inhibitors of metalloproteinases (TIMP-1 and TIMP-2) in acute respiratory distress syndrome patients. Crit. Care Med. 31:536-542.

Laszik, Z., Mitro, A., Taylor, F., Ferrell, G., and Esmon, C.T. (1997) Human protein C receptor is present primarily on endothelium of large blood vessels: implications for the control of the protein C pathway. Circ. 96:3633-3649.

Lechapt-Zalcman, E. and Escudier, E. (2000) Implication of extracellular MMP in the course of chronic inflammatory airway diseases. Morphol. 84:45-49.

Lee, T., Gonzalez, M., Kumar, P., Grammas, P., and Pereira, H.A. (2003) CAP37, a neutrophil-derived inflammatory mediator, augments leukocyte adhesion to endothelial monolayers. Microvasc. Res. 66:38-48.

Lee, W.L. and Downey, G. (2001) Leukocyte Elastase: Physiological functions and role in acute lung injury. Amer. J. Respir. Crit. Care Med. 164:896–904.

Leitinger, N., Tyner, T., Oslund, L., Rizza, C., Subbanagounder, G., Lee, H., Shih, P., Mackman, N., Tigyi, G., Territo, M., Berliner, J., and Vora, D.K. (1999) Structurally similar oxidized phospholipids differentially regulate endothelial binding of monocytes and neutrophils. PNAS USA 96:12010-12015.

Liau, D.F., Yin, N., Huang, J., and Ryan, S.F. (1996) Effect of human PMN elastase on surfactant proteins in vitro. Biochim. Biophys. Acta 1302:117-128.

Liu, S.F., Barnes, P., and Evans, T.W. (1997) Time course and cellular localization of lipopolysaccharide-induced inducible nitric oxide synthase messenger RNA expression in the rat in vivo. Crit. Care Med. 25: 512-518.

MacGregor, I.R., Perrie, A., Donnelly, S., and Haslett, C. (1997) Modulation of human endothelial thrombomodulin by neutrophils and their release products. Amer. J. Respir. Crit. Care Med. 155:47-52.

Machino, T., Hashimoto, S., Maruoka, S., Gon, Y., Hayashi, S., Mizumura, K., Nishitoh, H., Ichijo, H., and Horie, T. (2003) Apoptosis signal-regulating kinase 1-mediated signaling pathway regulates hydrogen peroxide-induced apoptosis in human pulmonary vascular endothelial cells. Crit. Care Med. 31:2776-2781.

MacMicking, J.D., Nathan, C., Hom, G., Chartrain, N., Fletcher, D., Trumbauer, M., Stevens, K., Xie, Q., Sokol, K., and Hutchinson, N. (1995) Altered responses to bacterial infection and endotoxic shock in mice lacking inducible nitric oxide synthase. Cell 81:641-650.

Magro, C.M., Allen, J., Pope-Harman, A., Waldman, W., Moh, P., Rothrauff, S., and Ross, P. Jr. (2003) The role of microvascular injury in the evolution of idiopathic pulmonary fibrosis. Amer. J. Clin. Pathol. 119:556-567.

Markey, B.A., Phan, S., Varani, J., Ryan, U., and Ward, P.A. (1990) Inhibition of H2O2 and neutrophil-induced endothelial cell cytotoxicity by intracellular SOD supplementation. Free Rad. Biol. Med 9:307-314.

Marshall, R., Bellingan, G., Laurent, G. (1998) The acute respiratory distress syndrome: Fibrosis in the fast lane. Thorax 53:815-817.

Martin, W.J. (1984) Neutrophils kill pulmonary endothelial cells by a hydrogen peroxide-dependent pathway. An in vitro model of neutrophil-mediated lung injury. Amer. Rev. Respir. Dis. 130:209-215.

Mautino, G., Oliver, N., Chanez, P., Bousquet, J., and Capony, F. (1997) Increased release of MMP9 in bronchoalveolar lavage fluid and by alveolar macrophages of asthmatics. Amer. J. Respir. Cell Mol. Biol. 17:583-591.

Minekura, H., Kumagai, T., Kawamoto, Y., Nara, F., and Uchida, K. (2001) 4-Hydroxy-2-nonenal is a powerful endogenous inhibitor of endothelial response. Biochem. Biophys. Res. Comm. 282:557-561.

Moraes, T.J., Chow, C., Downey, G.P. (2003) Proteases and lung injury. Crit. Care Med. 31:S189-S194.

Mulligan, M.S., Desrochers, P., Chinnaiyan, A., Gibbs, D., Varani, J., Johnson, K., and Weiss, S.J. (1993) In vivo suppression of immune complex-induced alveolitis by secretory leukoproteinase inhibitor and tissue inhibitor of MMP2. PNAS USA 90:11523-11527.

Mulligan, M.S., Lentsch, A., Huber-Lang, M., Guo, R., Sarma, V., Wright, C., Ulich, T., and Ward, P.A. (2000) Anti-inflammatory effects of mutant forms of secretory leukocyte protease inhibitor. Amer. J. Pathol. 156:1033-1039.

Murakami, K., Okajima, K., Uchiba, M., Johno, M., Nagaki, T., Okabe, H., and Takatsuki, K. (1997) Activated protein C prevents LPS-induced pulmonary vascular injury by inhibiting cytokine production. Amer. J. Physiol. 272:L197-L202.

Murphy, H.S., Varani, J., Ward, P.A. (2003) Biology of Endothelial Cells: Role of the Endothelium in Lung Inflammation. in Middleton's Allergy: Principles and Practice. Ed. N.F. Adkinson. St Louis, Mosby.

Murphy, H.S., Warner, R., Bakopoulos, N., Dame, M., Varani, J., Ward, P.A. (1999) Endothelial cell determinants of susceptibility to neutrophil-mediated killing. Shock 12:111-117.

Murphy, H. S., Yu, C., and Quddus, J. (2000) Functional expression of NAD(P)H oxidase p47 in lung microvascular endothelial cells. Biochem. Biophys. Res. Comm. 278:584-589.

Murray, P.T., Wylam, M., Umans, J.G. (2000) Nitric oxide and septic vascular dysfunction. Anesth. Analges. 90:89-101.

Naidu, B.V., Woolley, S., Farivar, A., Thomas, R., Fraga, C., Goss, C., and Mulligan, M.S. (2004) Early tumor necrosis factor-alpha release from the pulmonary macrophage in lung ischemia-reperfusion injury. J. Thor. Cardiovasc. Surg. 127:1502-1508.

Navab, M., Imes, S., Hama, S., Hough, G., Ross, L., Bork, R., Valente, A., Berliner, J., Drinkwater, D., and Laks, H. (1991) Monocyte transmigration induced by modification of low density lipoprotein in cocultures of human aortic wall cells is due to induction of monocyte chemotactic protein 1 synthesis and is abolished by high density lipoprotein. J. Clin. Invest. 88:2039–2046.

Nedeljkovic, Z.S., Gokce, N., Loscalzo, J. (2003) Mechanisms of oxidative stress and vascular dysfunction. Postgrad. Med. J. 79:195-199.

Niu, X.F., Ibbotson, G., and Kubes, P. (1996) A balance between NO and oxidants regulates mast cell- dependent neutrophil-endothelial cell interactions. Circ. Res. 79:992-999.

Ofek, I., Zafriri, D., Goldhar, J., and Eisenstein, B.I. (1990) Inability of toxin inhibitors to neutralize enhanced toxicity caused by bacterial adherence to tissue culture cells. Infect. Immun. 58:3737-3742.

Ohnishi, K., Takagi, M., Kurokawa, Y., Satomi, S., and Konttinen, Y.T. (1998) Matrix metalloproteinase-mediated extracellular matrix protein degradation in human pulmonary emphysema. Lab. Invest. 78:1077-1087.

Olofsson, A.M., Vestberg, M., Herwald, H., Rygaard, J., David, G., Arfors, K., Linde, V., Flodgaard, H., Dedio, J., Muller-Esterl, W., and Lundgren-Akerlund, E. (1999) Heparin-binding protein targeted to mitochondrial compartments protects endothelial cells from apoptosis. J. Clin. Invest. 104:885-894.

Orfanos, S.E., Psevdi, E., Stratigis, N., Langleben, D., Catravas, J., Kyriakidis, M., Moutsopoulos, H., Roussos, C., and Vlachoyiannopoulos, P.G. (2001) Pulmonary capillary endothelial dysfunction in early systemic sclerosis. Arth. Rheum. 44:902-911.

Østergaard, E. and Flodgaard, H. (1992) A neutrophil-derived proteolytic inactive elastase homologue (hHBP) mediates reversible contraction of fibroblasts and endothelial cell monolayers and stimulates monocyte survival and thrombospondin secretion. J. Leuk. Biol. 51:316-323.

Pacht, E.R., Timerman, A., Lykens, M., and Merola, A.J. (1991) Deficiency of alveolar fluid glutathione in patients with sepsis and the adult respiratory distress syndrome. Chest 100:1397-1403.

Partridge, C.A., Jeffrey, J., Malik, A.B. (1993) A 96-kDa gelatinase induced by TNF- contributes to microvascular endothelial permeability. Amer. J. Physiol 265:438-447.

Phan, S.H., Gannon, D., Ward, P.A., Karmiol, S. (1992) Mechanism of neutrophil-induced xanthine dehydrogenase to xanthine oxidase conversion in endothelial cells: Evidence of a role for elastase. Amer. J. Respir. Cell Mol. Biol. 6:270–278.

Pinkus, R., Weiner, L., and Daniel, V. (1996) Role of oxidants and antioxidants in the induction of AP-1 and NF-kB and glutathione S-transferase gene expression. J. Biol. Chem. 271:13422-13429.

Pontremoli, S., Melloni, E., Michetti, M., Sacco, O., Sparatore, B., Salamino, F., Damiani, G., Horecker, B.L. (1986) Cytolytic effects of neutrophils: Role for a membrane-bound neutral protease. PNAS USA 83:1685-1689.

Pries, A.R., Secomb, T., Gaehtgens, P. (2000) The endothelial surface layer. Pflügers Arch 440:653-666.

Pugin, J., Verghese, G., Widmer, M., and Matthay, M.A (1999) The alveolar space is the site of intense inflammatory and profibrotic reactions in the early phase of acute respiratory distress syndrome. Crit. Care Med. 27:304-312.

Radomski, M.W., Palmer, R., and Moncada, S. (1987) Endogenous nitric oxide inhibits human platelet adhesion to vascular endothelium. Lancet 2:1057-1058.

Rajagopalan, S., Meng, X., Ramasamy, S., Harrison, D., and Galis, Z.S. (1996) Reactive oxygen species produced by macrophage-derived foam cells regulate the activity of vascular matrix metalloproteinases in vitro. J. Clin. Invest. 98:2572-2579.

Ramaha, A. and Patston, P.A. (2002) Release and degradation of angiotensin I and II from angiotensinogen by neutrophil serine proteinases. Arch. Biochem. Biophys. 397:77-83.

Re, F., Zanetti, A., Sironi, M., Polentarutti, N., Lanfrancone, L., Dejana, E., and Colotta, F. (1994) Inhibition of anchorage-dependent cell spreading triggers apoptosis in cultured human endothelial cells. J. Cell Biol. 127:537-546.

Redl, H., Gasser, H., Sallstrom, S. (1993) Radical related cell injury. in Pathophysiology of Shock, Sepsis and Organ Failure. Eds. G. Schlag and H. Redl. Heidelberg, Springer-Verlag. pp. 92-110.

Rest, R.F. (1988) Human neutrophil and mast cell proteases implicated in inflammation. Meth.Enzymol. 163:309-327.

Rice, W.G. and Weiss, S.J. (1990) Regulation of proteolysis at the neutrophil-substrate interface by secretory leukoprotease inhibitor. Science 249:178-181.

Rodell, T.C., Cheronis, J., Ohnemus, C., Piermattei, D., and Repine, J.E. (1987) Xanthine oxidase mediates elastase induced injury to isolated lungs and endothelium. J. Appl. Physiol. 63:2159-2164.

Roman, A., Legallo, R., and McGahren, E.D. (2004) Blocking of endogenous nitric oxide increases white blood cell accumulation in rat lung. J. Ped. Surg. 39:48-52.

Rubanyi, G.M., Ho, E., Cantor, E., Lumma, W., and Botelho, L.H. (1991) Cytoprotective function of nitric oxide: inactivation of superoxide radicals produced by human leukocytes. Biochem. Biophys. Res. Comm. 181:1392-1397.

Ryan, S.F., Ghassibi, Y., and Liau, D.F. (1991) Effect of activated PMNs upon pulmonary surfactant in vitro. Amer. J. Respir. Cell Molec. Biol. 4:33-41.

Sacks, T., Moldow, C., Craddock, P., Bowers, T., and Jacob, H.S. (1978) Oxygen radicals mediate endothelial cell damage by complement-stimulated granulocytes: An in vitro model of immune complex vasculitis. J. Clin. Invest. 61:1161-1165.

Sato, Y. (2004) Modulation of PMN-endothelial cells interactions by cyclic nucleotides. Curr. Pharm. Design 10:163-170.

Sato, N., Suzuki, Y., Nishio, K., Suzuki, K., Naoki, K., Takeshita, K., Kudo, H., Miyao, N., Tsumura, H., Serizawa, H., Suematsu, M., and Yamaguchi, K. (2000) Roles of ICAM-1 for abnormal leukocyte recruitment in the microcirculation of bleomycin-induced fibrotic lung injury. Amer. J. Respir. Crit. Care Med. 161:1681-1688.

Shapiro, S.D. (2002) Neutrophil elastase ; path clearer, pathogen killer, or just pathologic? Amer. J. Respir. Cell Mol. Biol. 26:266-268.

Siebenlist, U., Franzoso, G., Brown, K. (1994) Structure, regulation and function of NF-κB. Ann. Rev. Cell Biol. 10:405-455.

Simon, R.H., Scoggin, C., and Patterson, D. (1981) Hydrogen peroxide causes the fatal injury to human fibroblasts exposed to oxygen radicals. J. Biol. Chem. 256:7181-7186.

Smedly, L.A., Tonneson, M., Sandhaus, R., Haslett, C., Guthrie, L., Johnston, R., Henson, P., and Worthen, G.S. (1986) Neutrophil-mediated injury to endothelial cells: Enhancement by endotoxin and essential role of neutrophil elastase. J. Clin. Invest. 77:1233-1243.

Speyer, C.L., Neff, T., Warner, R., Guo, R., Sarma, J., Riedemann, N., Murphy, M., Murphy, H., and Ward, P.A. (2003) Regulation of chemokine expression by inducible nitric oxide synthase. Amer. J. Pathol. 63:2319-2328.

Starke, P.E. and Farber, J.L. (1985) Ferric iron and superoxide ions are required for the killing of cultured hepatocytes by H_2O_2: Evidence for the participation of hydroxyl radicals formed by an iron-catalyzed Haber-Weiss reaction. J. Biol. Chem. 260:10099-10106.

Taekema-Roelvink, M.E., Kooten, C., Kooij, S., Heemskerk, E., and Daha, M.R. (2001) Proteinase 3 enhances endothelial monocyte chemoattractant protein-1 production and induces increased adhesion of neutrophils to endothelial cells by upregulating ICAM-1. J. Amer. Soc. Neph. 12:932-940.

Taekema-Roelvink, M.E., Kooten, C., Janssens, M., Heemskerk, E., and Daha, M.R. (1998) Effect of anti-neutrophil cytoplasmic antibodies on proteinase 3-induced apoptosis of human endothelial cells. Scand. J. Immunol. 48:37-43.

Tang, Z.L., Wasserloos, K., Liu, X., Stitt, M., Reynolds, I., Pitt, B., and St. Croix, C.M. (2002) Nitric oxide decreases the sensitivity of pulmonary endothelial cells to LPS-induced apoptosis in a zinc-dependent fashion. Molec. Cell. Biochem. 234:211-217.

Varani, J., Dame, M., Gibbs, D., Taylor, C., Weinberg, J., Shayevitz, J., and Ward PA (1992a) Human umbilical vein endothelial cell killing by activated neutrophils. Loss of sensitivity to injury is accompanied by decreased iron content during in vitro culture and is restored with exogenous iron. Lab. Invest. 66:708-712.

Varani, J., Fligiel, S., Till, G., Kunkel, R., Ryan, U., and Ward, P.A. (1985) Pulmonary endothelial cell killing by human neutrophils: Possible involvement of hydroxyl radical. Lab. Invest. 53:656-663.

Varani, J., Ginsburg, I., Schuger, L., Gibbs, D., Bromberg, J., Johnson, K., Ryan, U., and Ward, P.A. (1989) Endothelial cell killing by neutrophils: synergistic interaction of oxygen products and proteases. Amer. J. Pathol. 135:435-438.

Varani, J., Taylor, C., Riser, B., Shumaker, D., Yeh, K., Dame, M., Gibbs, D., Todd, R., Dumler, F., Bromberg, J., and Ward, P.A. (1992b) Mesangial cell killing by leukocytes: Role for leukocyte oxidants and proteolytic enzymes. Kidney Int. 42:1169-1177.

Varani, J. and Ward, P.A. (1994) Mechanisms of leukocyte dependent and leukocyte-independent endothelial cell injury. Biol. Signals 3:1-14.

Verhaar, M.C., Westerweel, P., van Zonneveld, A., and Rabelink, T.J. (2004) Free radical production by dysfunctional eNOS. Heart (Brit. Card. Soc.) 90:494-495.

Vieira, O., Escargueil-Blanc, I., Jurgens, G., Borner, C., Almeida, L., Salvayre, R., and Negre-Salvayre, A. (2000) Oxidized LDLs alter the activity of the ubiquitin-proteasome pathway: potential role in oxidized LDL-induced apoptosis. FASEB J. 14:532-542.

Vogt, W. (2000) Cleavage of the fifth component of complement and generation of a functionally active C5b6-like complex by human leukocyte elastase. Immunobiol. 201:470-477.

Vozzelli, M.A., Mason, S., Whorton, M., and Auten, R. (2004) Antimacrophage chemokine treatment prevents neutrophil and macrophage influx in hyperoxia-exposed newborn rat lung. Amer. J. Physiol. 286:L488-L493.

Wang, C.H., Lin, H., Liu, C., Huang, K., Huang, T., Yu, C., and Kuo, H.P. (2001) Upregulation of inducible nitric oxide synthase and cytokine secretion in peripheral blood monocytes from pulmonary tuberculosis patients. Internat'l. J. Tuberc. Lung Dis. 5:283-291.

Ward, P.A. (1996) Role of complement, chemokines, and regulatory cytokines in acute lung injury. Ann. N.Y. Acad. Sci. 796:104-112.

Ward, P.A. and Hunninghake, G.W. (1998) Lung inflammation and fibrosis. Amer. J. Respir. Crit. Care Med. 157:S123-S129.

Warner, R.L., Beltran, L., Younkin, E., Lewis, C., Weiss, S., Varani, J., and Johnson, K.J. (2001a) Role of stromelysin 1 and gelatinase B in experimental acute lung injury. Amer. J. Respir. Cell Molec. Biol. 24:537-544.

Warner, R.L., Lewis, C., Beltran, L., Younkin, E., Varani, J., and Johnson, K.J. (2001b) The role of metalloelastase in immune complex-induced acute lung injury. Amer. J. Pathol. 158:2139-2144.

Wedmore, C.V. and Williams, T.J. (1981) Control of vascular permeability by polymorphonuclear leukocytes in inflammation. Nature Med. 289:646-650.

Wei, Z., Costa, K., Al-Mehdi, A., Dodia, C., Muzykantov, V., and Fisher, A. (1999) Simulated ischemia in flow-adapted EC leads to ROS and cell signaling. Circ. Res. 85:682-689.

Weiss, S.J. (1989) Tissue destruction by neutrophils. New Eng. J. Med. 320:365-375.

Weiss, S.J., Young, J., LoBuglio, A., Slivka, A., and Nimeh, N.F. (1981) Role of hydrogen peroxide in neutrophil-mediated destruction of cultured endothelial cells. J. Clin. Invest. 68:714-719.

Wenzel, C., Kofler, J., Licker, G., Laczika, K., Quehenberger, P., Frass, M., and Knobl, P. (2002) Endothelial cell activation and blood coagulation in critically ill patients with lung injury. Wiener Klin. Wochen. 114:853-858.

Wiedermann, C. and Romisch, J. (2002) The anti-inflammatory action of antithrombin - a review. Acta Med. Autst. 29:89-92.

Witko-Sarsat, V., Cramer, E., Hieblot, C., Guichard, J., Nusbaum, P., Lopez, S., Lesavre, P., and Halbwachs-Mecarelli, L. (1999) Presence of proteinase 3 in secretory vesicles: evidence of a novel, highly mobilizable intracellular pool distinct from azurophil granules. Blood 94:2487-2496.

Wolin, M. (2000) Interactions of oxidants with vascular signaling systems. Arterioscler. Thromb. Vasc. Biol. 20:1430-1442.

Xue, C. and Johns, R. (1995) Endothelial nitric oxide synthase in the lungs of patients with pulmonary hypertension. New Eng. J. Med. 333:1642-1645.

Yang, J. J., Kettritz, R., Falk, R., Jennette, J., and Gaido, M.L. (1996) Apoptosis of endothelial cells induced by the neutrophil serine proteases proteinase 3 and elastase. Amer. J. Pathol. 149:1617-1626.

Zapol, W.M., Trelstad, R., Coffey, J., Tsai, I., and Salvador, R.A. (1979) Pulmonary fibrosis in severe acute respiratory failure. Amer. Rev. Respir. Dis. 119:547-554.

Zhao, X., Alexander, J., Zhang, S., Zhu, Y., Sieber, N., Aw, T., and Carden. D/L. (2001) Redox regulation of endothelial barrier integrity. Amer. J. Physiol. 281:L879-L886.

Zimmermann, P. and David,G. (1999) The syndecans, tuners of transmembrane signaling. FASEB J. 13: S91-S100.

Chapter 13

Endothelial Injury Due to Infectious Agents

Stefan Hippenstiel, M.D.* and Norbert Suttorp, M.D.

Charité University Medicine Berlin, Department of Internal Medicine/Infectious Diseases,
13353 Berlin, Germany

CONTENTS:

Introduction

Pathogen Attack of the Endothelium: Access, Adhesion, and Activation
Gaining Access to the Endothelium
The First Contact: Pathogen – Endothelial Cell Membrane Interaction
Pathogen Adhesion Alters the Endothelium
Endotoxin and Toll-like Receptor Recognition of Bacterial Products

Pathogen Invasion of the Endothelium
Mechanisms of Cell Entry and Monolayer Penetration
Bacterial Entry into Endothelium
Intracellular Interactions and Replication
Viral Invasion

Pathophysiological Consequences of Endothelial Infection
Endothelial Activation and Signaling
Permeability
Leukocyte Recruitment and Adhesion
Vasoactive Mediator Generation
Pro-thrombosis
Necrosis and Apoptosis
Vascular Proliferation

Clinical Consequences of Endothelial Infection
Acute Infections
Chronic Infections

Perspective

Introduction

Infection, whether present as pneumonia or in a distant site, has dire consequences for the pulmonary endothelium and is the leading cause of ARDS (Hudson et al., 1995; Ware and Matthay, 2000; Brun-Buisson, et al., 2004). During an inflammatory reaction, the large endothelial surface is exposed to both pathogens and their products, as well as to agents of the activated host defense. Endothelium is not only specifically targeted by important infective agents such as *Rickettsiae* (Faure et al., 2000; Valbuena et al., 2002) or *Bartonella* (Dehio, 2001; Dehio, 2003), but it is also involved in most, if not all, acute inflammatory responses. Pathogens attack the endothelium by a broad variety of strategies, as different as activation of preformed receptor-mediated pathways, release of pore-forming exotoxins, intracellular replication, and chronic parasitism. These pathophysiological forces affect the endothelial phenotype, resulting in barrier dysfunction, increased leukocyte-endothelial interaction, mediator release, and procoagulant activity. Furthermore, depending on the pathogen and the inflammatory response, the endothelial response may range from apoptosis or necrosis to proliferation. As the endothelium acts on the invading pathogen, as well as on the host defence system, a complex and dynamic interaction occurs. Reconstitution of physiological circumstances includes normalization of endothelial functions and may involve support by progenitor cells. Overall, endothelial activation makes considerable contribution to inflammation and resulting clinical manifestations. In this context the endothelium is not just a passive victim, rather it aggravates the ongoing hassle with the pathogen.

Pathogen Attack of the Endothelium: Access, Adhesion, and Activation

Gaining Access to the Endothelium

A number of different, highly sophisticated strategies have evolved by which pathogens enter and colonize the host. Most epithelia, which build an important cellular barrier against bacteria and viruses, are normally exposed only to a defined subset of pathogens, e.g. the urogenital epithelium may be exposed to *Escherichia coli* but never to *Streptococcus pneumoniae*. In contrast, after entering the bloodstream and distribution in all capillary beds, virtually all pathogens and their products come in close contact with the endothelium (Hippenstiel and Suttorp, 2003). Hence, it follows that the endothelium is not only attacked by endothelium-specialized pathogens, but also by a vast variety of other pathogens and their specific ways of attacking host cells as shown in Figure 1. Moreover, in the lung the alveolar epithelial surface is extensive and if pathogens gain access, the capillary endothelium is merely one thin epithelial cell away from the infection. Pathogens and pathogen-derived products may use pre-formed endothelial signaling pathways by binding to cell-surface receptors or alter endothelial function by attacking membrane integrity. Both endothelial-matrix and intercellular junctions may be altered by pathogens. Due to the vast endothelial surface in the lung and the fragility of the alveolar-capillary membrane, such pathogen-endothelial interactions can cause life threatening lung failure.

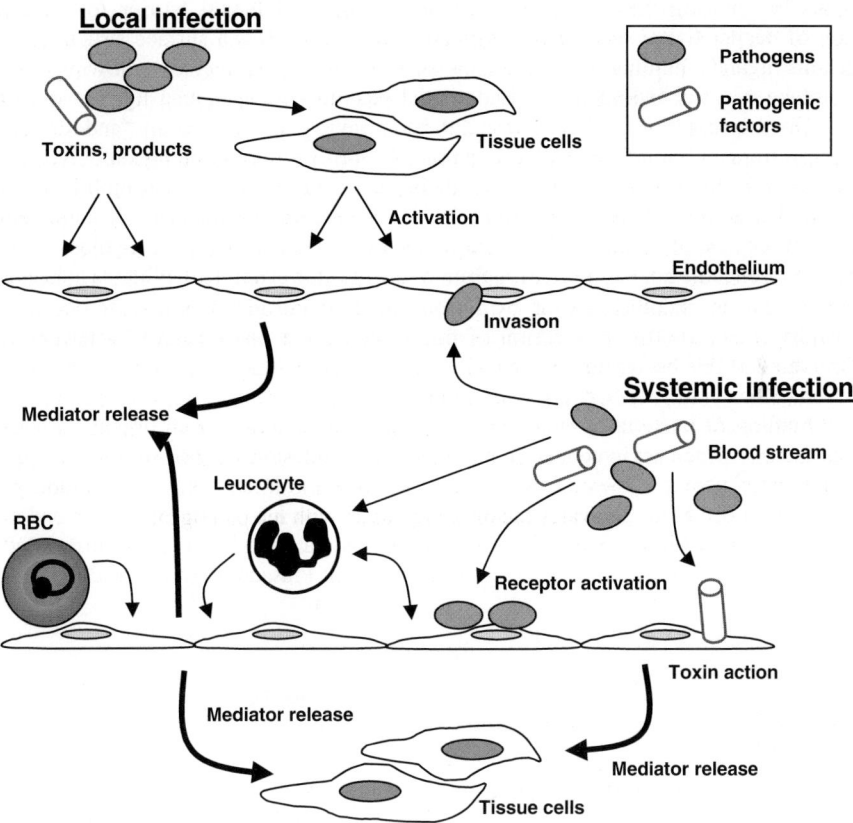

Figure 1: Pathogen attack of the endothelium. After entering the bloodstream or the capillary area, virtually all pathogens and their released products are in close contact with the endothelium. In local infection, pathogens or products like exotoxins can activate the endothelium from the basolateral site. Pro-inflammatory products released by stimulated tissue cells also affect endothelial function. Systemic infections give direct access of pathogens and their virulence factors to the endothelium. Stimulated or infected blood cells perpetuate endothelial infection. Mediators released by endothelia may lead to auto- and paracrine stimulation, activation of tissue cells, and systemic effects. Overall, a complex and highly dynamic interaction occurs at infection sites, with endothelium as a key player.

The First Contact: Pathogen – Endothelial Cell Membrane Interaction

A complex and highly diverse set of pathogen and host membrane structures mediates the adhesion process. Pathogen adhesion to endothelial cells normally is a prerequisite for endothelial activation and transformation to a pro-inflammatory phenotype. Such adhesion is important for entering the cell or for overcoming the endothelial barrier for subsequent infection of deeper tissue structures. Typically, specific pathogen surface structures mimic endogenous ligands binding to endothelial surface structures, thereby allowing adhesion and/or uptake into the endothelium. Endothelial receptors for pathogen ligands seem to be differentially expressed in different vascular beds paving the way to an "endothelial cell-based organ tropism" in infection. For example, *E. coli* S fimbria-binding sialoglycoproteins are expressed on brain microvascular endothelial cells but not on human umbilical vein or aortic endothelial cells (Prasadarao *et al.*, 1997). However, the majority of pathogens use ubiquitously expressed endothelial surface molecules for adhesion, such as those shown in Figure 2. Fibronectin-binding protein A from *S. aureus*, for example, facilitates adhesion and temporal uptake in endothelial cells by fibronectin bridging to integrin $\alpha_5\beta_1$ (Sinha *et al.*, 1999; Sinha *et al.*, 2000). Expression of this protein by a non-invasive bacterium causes cellular uptake of this bacterium by the endothelium, highlighting the power of this molecular interaction (Sinha *et al.*, 1999; Sinha *et al.*, 2000).

The binding of bacteria to endothelial cells may depend on a well-organized subset of bacterial factors, which act in concert to allow bacterial adhesion. *E. coli* strain K1, one of the main causes for neonatal meningitis, express fimbriae FimH and outer membrane protein OmpA (Kim, 2002; Kim, 2003). Initial binding of *E. coli* K1 to microvascular endothelial cells occurs through FimH. Subsequently, OmpA binds a gp96 homolog on the endothelial cell, stabilizing the pathogen-host interaction (Kim, 2003; Prasadarao *et al.*, 2003).

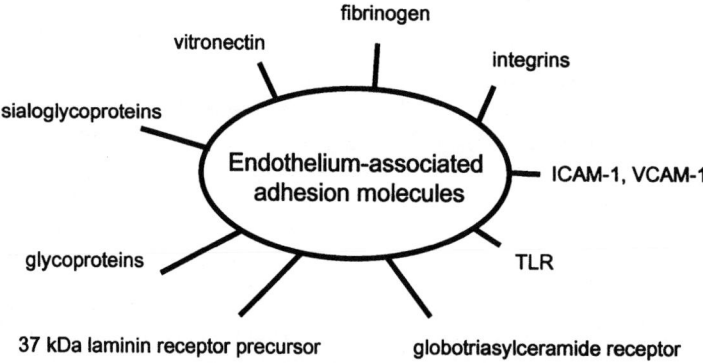

Figure 2: Pathogen interaction with the endothelial cell membrane. Constitutively expressed endothelial cell surface molecules like glycoproteins or TLR, as well as cell surface exposure of molecules expressed during the inflammatory reaction (e.g. ICAM-1, VCAM-1) may serve as anchors for attacking pathogens. In addition, endothelial matrix proteins like vitronectin or deposited fibronectin act as adhesion molecules for pathogens. However, it should be noted that the existence of more, yet not identified, structures mediating pathogen-endothelial adhesion is highly supposable.

After impairment of the endothelial barrier, pathogens may attach directly to exposed extracellular matrix components, contributing to the disease progress. For example, *S. aureus* binds fibrinogen, vitronectin, and elastin (Flock, 1999; Sinha *et al.*, 1999; Sinha *et al.*, 2000; Massey *et al.*, 2001). It is important to note that bacterial adhesion proteins also may function as inhibitors of leukocyte recruitment to the site of inflammation (Chavakis *et al.*, 2002; Harraghy *et al.*, 2003). As shown recently, *S. aureus* secreted extracellular adherence protein efficiently blocks β_2-integrin-dependent neutrophil recruitment (Chavakis *et al.*, 2002). Despite effective attachment of pathogens to the endothelium, the magnitude of bacteraemia may also be critical for disease development, as shown for meningitis (Pfister *et al.*, 1994).

Pathogen Adhesion Alters the Endothelium

Adhesion of pathogens to endothelial cells is sufficient to activate signaling pathways, underscoring the important role of the initial contact for the following inflammatory reaction. Simple attachment of *Chlamydia pneumoniae* (Krull *et al.*, 1999) or *Bartonella henselae* (Fuhrmann *et al.*, 2001) to endothelial cells induces lipopolysaccharide (LPS)-independent activation of protein kinases within minutes. Remarkably, inactivated *C. pneumoniae* induced NF-κB-dependent gene expression without active infection of the endothelium (Baer *et al.*, 2003). Current evidence suggests that bacterial outer membrane proteins contribute significantly to this rapid activation. Isolated outer membrane protein from *B. henselae* (Fuhrmann *et al.*, 2001) as well as polymorphic PMP20 and PMP21 from *C. pneumoniae* (Niessner *et al.*, 2003) activated expression of pro-inflammatory genes in endothelial cells.

Sequestration of infected erythrocytes in cerebral microvascular beds is a central pathogenetic issue for development of malaria caused by *Plasmodium falciparum* infection (Miller *et al.*, 2002). The *P. falciparum* erythrocyte membrane protein 1 (PfEMP1) is directly involved in the adhesive interaction with host cells. Rolling and tethering of infected erythrocytes is mediated *via* endothelial adhesion receptors like ICAM-1 and VCAM-1, finally resulting in CD36 and chondroitin sulphate A mediated stable adhesion (Miller *et al.*, 2002). A recombinant protein (PpMC-179) encoding a sequence of PfEMP1, critically involved in PfEMP1-CD36 interaction, inhibited and reversed adhesion of infected erythrocytes to endothelia *in vitro* and *in vivo* (Yipp *et al.*, 2003a). Further experiments suggested a prominent role of endothelial signaling pathways in the firm adhesion process. Initial attachment of the infected erythrocytes to CD36 activated Src kinase-signaling, which promoted IE adhesion to CD36 by increasing ectoalkaline phosphatase activity. The resultant dephosphorylation increased CD36 binding affinity to the erythrocytes (Yipp *et al.*, 2003b). Thus, the endothelium itself actively contributes to the adhesion process of erythrocytes by increasing receptor affinity of CD36. It would be interesting to search for comparable receptor affinity modifying signaling events in e.g. bacteria receptor mediated adhesion.

Due to the large number of immunocompromised patients, *Candida albicans* infections have increased significantly (Clark and Hajjeh, 2002). Infection occurs through bloodstream dissemination, adherence, and invasion of endothelial cells and subendothelial matrix. This is a prerequisite for fungal invasion of organs, including lungs and brain, frequently observed in the disease (Klotz and Maca, 1988; Mayer *et al.*, 1992; Orozco *et al.*, 2000). A plethora of *Candida* virulence factors, including dimorphism, adhesion molecules, and proteinases facilitate fungus binding and invasion. For example, *C. albicans* germ tubes adhere to $\alpha v\beta 3$ integrins of the endothelial cells (Santoni *et al.*, 2001). However, for the most pathogens interacting with the endothelium, identification of the adhesion molecules is still missing.

Highly aggressive bacteria perturb integrity of the endothelial plasma membrane by exotoxins, secreted in the host environment, as seen in Figure 3 (Bhakdi *et al.*, 1994; Karch, 2001; Zysk *et al.*, 2001; Grandel and Grimminger, 2003). The ability of bacteria to produce and release toxin contributes significantly to its ability to cause disease. *S. aureus* producing α-toxin are more pathogenic than staphylococci not producing toxin (Bhakdi *et al.*, 1994; Grandel and Grimminger, 2003). Typical toxins of this class are *S. aureus* α-toxin and *E. coli* hemolysin (HlyA). Pore forming exotoxins, amphipathic proteins, insert and oligomerize into the membrane by interacting with lipid chains or nonpolar segments of integral membrane proteins, thereby creating an inner hydrophilic cavity. These toxins punch exactly defined, small holes in the plasma membrane of eukaryotic cells, allowing flux of ions but not macromolecules (Bhakdi *et al.*, 1994; Grandel and Grimminger, 2003). The important role of *S. aureus* α-toxin and HlyA for disturbance of endothelial cell function is well documented. Injection of a single bolus of HlyA is sufficient to induce massive endothelial permeability *in vitro* (Suttorp *et al.*, 1988; Suttorp *et al.*, 1993b; Suttorp *et al.*, 1996; Hippenstiel *et al.*, 2002) as well as edema formation and microvascular perfusion disturbance in isolated lung (Seeger *et al.*, 1989) or ileum preparations (Mayer *et al.*, 1992). Moreover, these toxins activated surface expression of endothelial adhesion molecules and subsequently enhanced endothelial-leukocyte interaction promoting the inflammatory response (Krull *et al.*, 1996).

Besides pore forming exotoxins, which integrate into the host cell membrane, some bacterial exotoxins bind endothelial cell receptors, resulting in toxin translocation into the cytosol. This process is a major virulence factor of enterohemorrhagic *E. coli* causing postdiarrheal haemolytic-uremic syndrome (Karch, 2001; LeBlanc, 2003). Kidneys of these patients typically show renal glomerular capillary thrombotic microangiopathy. Such *E. coli* strains produce Shiga toxin 1 and 2 (Nakao and Takeda, 2000). After intestinal infection with the enterohemorrhagic *E. coli*, Shiga toxins overcome the intestinal barrier and bind to the endothelial membrane globotriasylceramide receptor, which facilitates endocytosis of the toxin (Karch, 2001; Hughes *et al.*, 2002; Stricklett *et al.*, 2002). Inside the cell, Shiga toxin subunit A hydrolyses an adenine residue of the 60S ribosomal subunit to shut down protein translation (Nakao and Takeda, 2000). The cytokine milieu in which endothelial cells are exposed to Shiga toxin modulates the resulting endothelial response and the toxin-induced expression of pro-inflammatory molecules, e.g. IL-8, IL-16, MCP-1, ICAM-1 (Matussek *et al.*, 2003). Glomerular capillary as well as microvascular cortical renal endothelial cells have been shown to be highly susceptible for Shiga toxin induced cell damage.

Other bacterial toxins affect endothelial function during different stages of the infection process. *L. monocytogenes* hemolysin (listeriolysin O, LLO), a cholesterol-dependent, pore forming toxin (Drevets, 1998; Vazquez-Boland *et al.*, 2001), can affect endothelial function when it is released extracellularly, as well after pathogen uptake. Extracellular LLO attack of endothelial cell outer membrane induced a phosphatidylinositol response and lipid mediator generation (Rose *et al.*, 2001), as well as adhesion molecule expression (Krull *et al.*, 1997). As discussed below, after invasion of endothelia by Listeria, LLO liberation is critical for Listeria release from phagosomes (Vazquez-Boland *et al.*, 2001).

Besides these well-characterized examples of endothelial cell activation by pathogen-derived toxins, less is known about the function of other important toxins. For example, the central role of pneumolysin from *S. pneumoniae* and endothelial damage is noted, but the cellular biology of pneumolysin-endothelial interaction has to be elucidated. In addition, the role of the injection of bacterial proteins into endothelial cells by the type III secretion system (TTS) of Gram-negative bacteria has to be analysed. Either extracellular bacteria or bacteria localized into phagosomes inject powerful bacterial effectors in the host cell to take control of the cellular machinery of the target cell (Cornelis, 2002). *Pseudomonas aerogenosa* is an important pathogen causing septicaemia in immunocompromised or critically ill individuals. In contrast to bacteria with an intact TTS which kills endothelial cells *in vitro* within hours, bacteria with impaired TTS did not (Saliba *et al.*, 2002). Gene disruption of *exsA* (controls the expression of the most TTS related genes) or *pscC* (encodes a protein from the Pseudomonas secretion machinery) reduced the cell killing ability of the bacteria significantly (Saliba *et al.*, 2002). In addition, the YopE protein of *Yersinia*, injected into endothelial cells by TTS, interferes with regulation of the endothelial cytoskeleton (Andor *et al.*, 2001). Since bacteria which chronically infect endothelial cells like *C. pneumoniae* possibly have the power of a TTS machinery, a new focus of interest may arise (Molestina *et al.*, 2002).

Released viral proteins are also capable of activating endothelial cells. For example, in contrast to HIV itself, HIV-1 Tat protein secreted from HIV-1 infected T cells efficiently activates a pro-inflammatory and angiogenic program after membrane binding (Bussolino *et al.*, 2001; Rusnati and Presta, 2002). Tat protein is a 86-101 amino acid polypeptide that acts as the main transactivator of HIV (Gatignol and Jeang, 2000). Binding of Tat to endothelial $\alpha v \beta 3$ integrins as well as VEGF-2/KDR was followed by, angiogenic, chemotactic, and mitotic activity of endothelial cells (Rusnati and Presta, 2002). HIV Tat also caused endothelial dysfunction of porcine pulmonary endothelium, evidenced by lack of endothelial-derived relaxation in pulmonary arterial rings (Paladugu *et al.*, 2003).

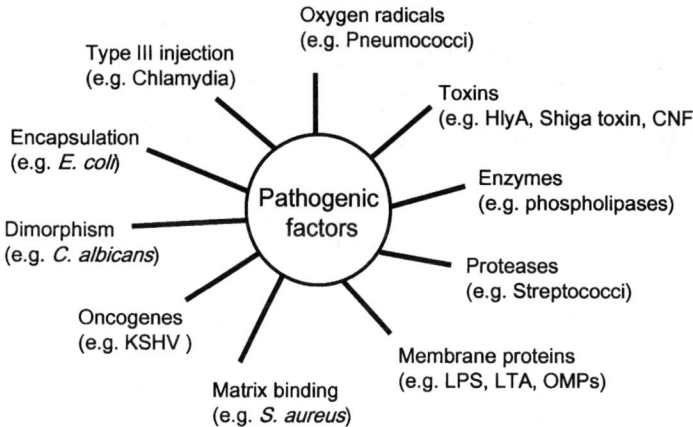

Figure 3: Pathogen products, that affect endothelial integrity. A highly diverse array of pathogenic factors directly attacks the endothelium. For example, released factors like exotoxins, or proteases as well as membrane-bound pathogenic constituents like LPS can activate the endothelium. Moreover, pathogens may inject enzymes or toxins into endothelia by type III injection system. In addition, transfer of oncogenes may lead to chronic impairment of endothelial function.

Endotoxin and Toll-like Receptor Recognition of Bacterial Products

Gram-negative bacteria shed cell wall lipopolysaccharides into circulation, especially upon death. The lipid A portion of LPS is particularly potent in causing many of the clinical manifestations of sepsis, i.e. endothelial activation and injury, vascular leak, intravascular coagulation, and shock (Bannerman and Goldblum, 1999). LPS (endotoxin) effects are mediated by activation of the host-response system and leukocytes and also by direct effects on the endothelium. In isolated systems with absence of host response or leukocytes, direct LPS effects on macrovessel- and microvessel-endothelium include: release of IL-1, Il-6, and IL-8; release of tissue factor; increased expression of ICAM-1; barrier dysfunction; and cell death (Bannerman and Goldblum, 1999). Gap formation and increased monolayer protein permeability is associated with cytoskeletal reorganization and disassembly of adherens junctions. In numerous *in vivo* studies, LPS injection caused lung permeability edema, with pulmonary hypertension and lung sequestration of leukocytes, that exacerbate direct effect on the endothelium; yet the importance of direct activation is underscored by failure of leukocyte or platelet depletion to block LPS-induced lung edema (Pingelton *et al.*, 1975; Snapper *et al.*, 1984; Winn *et al.*, 1987). Intravenous administration or intratracheal instillation of LPS is a common means to model the lung injury and barrier dysfunction of ARDS (Dhingra *et al.*, 2001; Powers *et al.*, 2002; Mikawa *et al.*, 2003; Nagase, 2003; Powers *et al.*, 2003; Kuklin *et al.*, 2004; Laffey *et al.*, 2004). LPS also causes endothelial apoptosis as discussed later (Bannerman and Goldblum, 2003). The *toll*-like receptors (TLR), consisting of TLR1-11, plays an essential role in recognition of extracellular microbial components, including LPS (Underhill and Ozinsky, 2002; Beutler *et al.*, 2003; Janssens and Beyaert, 2003).

Many host cells including endothelia recognize "pathogen-associated molecular patterns" by "pattern recognition receptors", which clearly distinguish between self and conserved microbial structures. TLR1-10 have been detected in endothelial cells and upregulation of TLR2 and TLR4 was demonstrated after stimulation with pro-inflammatory agents and in atherosclerotic lesions (Edfeldt *et al.*, 2002). There is a growing repertoire of TLR ligands. TLR4 is not only activated by LPS, but also by heat shock protein 60 from *Chlamydia*, and a fusion protein of respiratory syncytial virus. Bacterial lipoproteins and peptidoglycan activate TLR2 (+TLR6 or TLRx), and flagellin TLR5, while unmethylated CpG motifs found in microbial DNA bind to TLR9 (Underhill and Ozinsky, 2002; Beutler *et al.*, 2003; Janssens and Beyaert, 2003). However, the situation is further complicated by the fact that cells may use multiple TLRs to detect several features of a microbe simultaneously.

Moreover, endothelial TLR expression may be regulated in inflammation. LPS, IFNγ, or TNFα increased TLR2 and 4 expression in endothelial cells (Faure *et al.*, 2001). In contrast, hypoxia may diminish the expression of endothelial TRL4 due to oxidants generated by mitochondria (Ishida *et al.*, 2002). A recently recognized mechanism connects recruitment of neutrophils to the endothelium with regulation of TLR expression on endothelial cells (Fan *et al.*, 2003). LPS activated neutrophils can induce endothelial TLR2 expression, thereby sensitizing these cells to TLR2 ligands. The critical signal leading to enhanced TLR2 expression was release of free oxygen radicals, resulting from a CD18-dependent cell-cell interaction. The role of endothelial TLRs in LPS related inflammation and lung injury is highlighted by reduced lung leucocyte sequestration in the LPS-treated TLR4 (-/-) mice with endothelial, but not leukocyte knock-out; whereas, knock out of leukocyte, but not endothelial, TLR4 had no effect on leukocyte sequestration (Andonegui et al., 2003).

In endothelial cells, which do not express CD14 on their cell surface, a complex of serum-derived soluble CD14, LPS-binding protein, and LPS bind to an arrangement of TLR4 and the protein MD-2 allowing the transduction of the signal into the cell (Triantafilou and Triantafilou, 2002). Activation of TLRs induce a complex signaling cascade leading to the transformation of endothelial cells into a pro-inflammatory phenotype (Zhao et al., 2001; Ahn et al., 2003). Keeping in mind that the endothelium is exposed to nearly all conceivable ligands of TLRs, a detailed analysis of TLR function warrants further investigation.

Pathogen Invasion of the Endothelium

Mechanisms of Cell Entry and Monolayer Penetration

Bacteria, fungi and viruses have developed a great variety of sophisticated mechanisms to enter host cells. Two main methods for bacterial internalization are a trigger mechanism secondary to the formation of membrane ruffles and a zipper mechanism involving a direct contact between a bacterial ligand and a cellular receptor (Steele-Mortimer et al., 2000; Gruenheid and Finlay, 2003). Although only few bacteria, including, *Rickettsiae* (Valbuena et al., 2002), *Bartonella henselae* (Dehio, 2001; Dehio, 2003), *Chlamydia pneumoniae* (Summersgill et al., 2000) and *Listeria* (Vazquez-Boland et al., 2001) persist and replicate in the endothelium, many pathogens, including *S. pneumoniae* (Ring et al., 1998), and *Neisseria meningitides* (Hoffmann et al., 2001), enter endothelial cells as a transient process. Whether pathogens parasite in endothelial cells for longer periods, such as *Rickettsiae*, or for shorter periods of time, such as *Staphylococcus aureus*, endothelial activation results. Rather than enter the endothelium *per se*, fungi like *C. albicans* penetrate the endothelial layer to gain access to tissue parenchyma (Zink et al., 1996). Different viruses (e.g. Filoviridae (Schnittler and Feldmann, 1999) and Kaposi's sarcoma associated herpesvirus (Poole et al., 2002)) infect and replicate efficiently within the endothelium.

Bacterial Entry into Endothelium

A central strategy for entering host cells involves direct alteration of the cytoskeleton by proteins derived from the pathogen. To do so, pathogens release effector molecules into the extracellular space or Gram-negative bacteria inject proteins by the TTS, as mentioned above. Through actions of these pathogen-derived proteins, many bacteria affect the microfilament system by modulation of small GTP-binding Rho proteins which act as molecular switches in the regulation of the cytoskeleton (Steele-Mortimer et al., 2000; Takai et al., 2001; Gruenheid and Finlay, 2003). Alternatively, *N. meningitides* enters endothelial cells by interaction of bacterial pili with a cellular receptor leading to Rho protein dependent microvilli formation (Eugene et al., 2002). Recruitment of a tyrosine receptor and subsequent phosphorylation of the cytoskeletal protein cortactin promote bacterial invasion into endothelial cells. These toxins and bacterial components have, in fact, been extremely useful for basic investigation of actin polymerization and regulation of cytoskeletal assembly. By assuming control of the endothelial cytoskeleton, the pathogens, such as *Salmonella*, can induce ruffles at the cell membrane, with the formation of actin tails and pedestals in non-phagocytotic cells to induce their uptake by engulfment (Steele-Mortimer *et al.*, 2000; Gruenheid and Finlay, 2003).

E. coli invade microvascular endothelium using a zipper-like mechanism and transmigrate through the cell enclosed in a vacuole without replication. Cytotoxic necrotizing factor 1 (CNF1) contributed significantly to the invasion of *E. coli* K1 into the endothelium of blood brain barrier. CNF1 host cell uptake is mediated by the 37 kDa laminin receptor precursor (Chung *et al.*, 2003). Endothelial entrance by *E. coli* K1 is also mediated by Ibe proteins of the bacteria (Wang and Kim, 2002; Kim, 2003). Isogenic mutants of the bacteria lacking these genes were less invasive *in vitro* and *in vivo* (Wang and Kim, 2002). CNF1 deamidates glutamine 63 of RhoA and glutamine 61 of Rac and Cdc42, respectively, resulting in permanent GTPase activation (Boquet, 2001). Subsequent actin reorganization is accompanied by phosphatidylinositol 3-kinase as well as focal adhesion kinase activation and phosphorylation of cytoskeletal proteins. Interestingly, CNF1 modification of Rho GTPases targets these molecules for ubiquitin-proteasome dependent degradation thereby restricting GTPase activation and allowing host cell migration and effective pathogen invasion (Doye *et al.*, 2002), However, it is not known if this later mechanism actually does occur in endothelial cells.

An additional uptake mechanism, called the invasome, may be used by Bartonella in endothelial cells (Dehio *et al.*, 1997; Dehio, 2001; Dehio, 2003). The human-specific species *Bartonella bacilliformis*, *B. quintana*, and *B. henselae* invade and colonize endothelial cells. This is considered as a prerequisite for building of *Bartonella*-triggered angioproliferative lesions leading to verruga peruana (Kaposi`s sarcoma-like skin lesions; *B. bacilliformis*) or bacillary angiomatosis (*B. quintana*, *B. henselae*), as discussed below (Dehio, 2001; Dehio, 2003). As shown for *B. henselae*, depending on the microfilament system, a bacterial aggregate on the cell membrane is formed, engulfed, and internalized. However, uptake of *B. bacilliformis* into endothelial cells seems to depend on Rho proteins Rac and Cdc42, but it is unclear how this "classical" uptake pathway is shared by the other angioproliferative Bartonella (Verma *et al.*, 2000; Verma and Ihler, 2002). Besides directly hijacking the microfilament system, some pathogens may utilize the endothelial caveolae to mediate uptake (Shin and Abraham, 2001a; Shin and Abraham, 2001b; Minshall *et al.*, 2003). For example, group A streptococci (Rohde *et al.*, 2003) as well as Japanese encephalitis virus (Liou and Hsu, 1998), were shown to enter endothelia by using caveolae.

Intracellular Interactions and Replication

Once inside the host cell, control of the cytoskeleton is used by the invading pathogen to affect its own intracellular movement and survival. Internalized bacterial pathogens may either remain in a vacuole, such as *Chlamydia* (Summersgill *et al.*, 2000) or escape into the cytoplasm, like *Listeria* (Vazquez-Boland *et al.*, 2001). One strategy to ensure intracellular survival is to avoid fusion of endosomes with lysosomes. For example, the K1 capsule of *E. coli* modulates the maturation process of bacteria containing vacuoles. Whereas isogenic K1 capsule-deletion mutants are degraded after fusion of late endosomes with lysosomes, K1 capsule positive bacteria prevents fusion in endothelial cells (Kim *et al.*, 2003a).

In contrast to *E. coli*, which seem not to replicate inside endothelial cells, *C. pneumoniae* do so efficiently (Summersgill et al., 2000; Wyrick, 2000). The chlamydial growth cycle is initiated when an infectious elementary body (EB) entered the target cell by mechanisms, possibly involving endo- or pinocytosis (Wyrick, 2000). As evidenced by studies with epithelial cells, EB-vesicles fuse with each other and escape from the endocytotic to the exocytic pathway by translocation to the peri-Golgi region (Al Younes et al., 1999; Wyrick, 2000). After 6-40 hours, infectious EB transit to metabolically active, non-infectious reticular bodies. Protected by the surrounding inclusion membrane, the bacteria grow. *Chlamydia* utilize sphingomyelin, cholesterol and phospholipids from the Golgi-apparatus for the inclusion membrane (Wolf and Hackstadt, 2001). The molecular basis of early chlamydial gene expression for vacuole modification leading to vesicle location in the exocytic pathway is unknown (Hackstadt et al., 1997). Moreover, the role of needle-like projections of Chlamydia EB (Wyrick, 2000), considered to be part of a type III-like secretion apparatus (Molestina et al., 2002), have to be investigated. Finally, it is unknown which signals and which molecular mechanisms regulate the maturation of reticular bodies back to the EB. Nor is it known if and how EB are released from an infected endothelial cell to perpetuate the infectious cycle.

Although there is evidence that Listeria also prevent phagosome maturation to the phagolysosomal stage (Alvarez-Dominguez et al., 1997; Vazquez-Boland et al., 2001), Listeria move out of the vacuoles into the endothelial cytosol. About 30 minutes after uptake, bacteria begin to disrupt the phagosome membrane. Listerial LLO and phopholipases act in concert to lyse the membrane (Vazquez-Boland et al., 2001). Phosphatidylinositol-specific PLC and phosphatidylcholine-specific PLC are expressed by *L. monocytogenes* (Vazquez-Boland et al., 2001). As shown for LLO, these bacterial PLCs not only contribute to vacuole disruption, they activated pro-inflammatory signaling pathways in endothelial cells (Schwarzer et al., 1998). Once free in endothelial cytosol, *L. monocytogenes* can recruit endothelial actin cytoskeleton machinery elements to allow active movement in the cell as well as cell to cell spread (Vazquez-Boland et al., 2001). Similar to *Listeria*, *Rickettsiae* spp. (*R. rickettsii*, *R. conorii*, *R. prowazekii*) invade endothelial cells, lyse the phagosomal membrane, and spread in a actin-dependent manner from endothelial to endothelial cell (Valbuena et al., 2002; Walker et al., 2003). Interestingly, although Rickettsia are obligatory intracellular pathogens and one of the most pathogenic bacteria replicating efficiently in endothelial cells, there is limited knowledge of its intracellular life cycle.

Dependent on the particular invading pathogen species, different intracellular endothelial phospholipases and kinases are needed for sufficient uptake. Endothelial invasion by *L. monozytogenes* required activation of Src kinases but not FAK or cPLA$_2$ (Das et al., 2001). Uptake of group B streptococci was independent of PI-3 Kinase and cPLA$_2$ (Nizet et al., 1997). Therefore, additional studies addressing the role of specific signaling pathways activated by particular pathogen entering the endothelium are needed.

Viral Invasion

Besides bacteria and fungi a great variety of viruses infect endothelial cells. Viruses causing viral hemorrhagic fevers like Filoviridae (Marburg, Ebola), Arenaviridae (e.g. Lassa fever), Bunyaviridae (e.g. Rift Valley fever) or Flavivirus (Yellow fever), efficiently enter and replicate in endothelial cells (Schnittler and Feldmann, 1999). Some of these viruses induce massive direct endothelial damage (e.g. Rift Valley fever virus) but others, like Hanta- or Arenavirus do not (Schnittler and Feldmann, 1999). In all hemorrhagic fevers, the massive host response, with high levels of pro-inflammatory cytokines, contributes significantly to disease pathogenesis and endothelial malfunction (Schnittler and Feldmann, 1999; Peters and Zaki, 2002). Dengue-3 virus causing dengue or dengue hemorrhagic fever infects endothelial cells *in vitro* although their role as primary targets *in vivo* is controversial (Halstead, 2002). Interestingly, large scale gene expression analysis was used to demonstrate that *in vitro* dengue virus infection resulted in a complex endothelial gene expression pattern, including stress, defense, immune, cell adhesion, wounding, inflammatory, as well as antiviral pathways (Warke *et al.*, 2003). Therefore, it has to be considered that dengue virus mediates endothelial activation by both direct infection and cytokine induction.

The ubiquitous herpesvirus, human cytomegalovirus (HCMV), considered to cause infection only in neonates, has gained importance as a pathogen in immunocompromised and immunocompetent humans. The lung is an important target for primary infection or reactivation in immunocompromised hosts, associated with serious morbidity or mortality (Sinzger *et al.*, 1995). Importantly, besides in macrophages, HCMV replicate and persist in human endothelial cells (Jarvis and Nelson, 2002a; Jarvis and Nelson, 2002b). Accumulation in brain endothelium resulted in cell lysis, whereas it failed to accumulate in and lyse aortic endothelial cell (Fish *et al.*, 1998). Thus, endothelial cells of different origin may differ as a HCMV reservoir. Since several viral strains displayed variable endothelial-cell tropism, both viral and endothelial factors likely determine the infection process (Sinzger *et al.*, 1999; Sinzger *et al.*, 2000). In acute infection or reactivation, virus-related CXC chemokine liberation by endothelium may lead to neutrophil recruitment and neutrophil-endothelial contact, paving the way to new virus dissemination (Grundy *et al.*, 1998). In contrast, CMV transmission to differentiating and invasive placental cytotrophoblasts occurred directly from infected uterine microvascular endothelial cells, causing the fetal infection (Maidji *et al.*, 2002). Moreover, vascular injury related to the development of thrombotic microangiopathy in patients after bone marrow transplantation may be caused by CMV-virus infection (Takatsuka *et al.*, 2003).

Although the endothelium shows a low and discrete susceptibility to HIV-1 infection, HIV-related damage of the endothelium contributes significantly to AIDS-related vasculitis in the central and peripheral nervous system, and manifestation of bacillary angiomatosis (Bussolino *et al.*, 2001). AIDS patients are at particular risk for Kaposi's sarcoma. Kaposi´s sarcoma-associated herpesvirus (KSHV), also known as human herpesvirus 8, is suggested to be the etiologic agent of Kaposi´s sarcoma. Infection of human endothelial cells with KSHV significantly alters cellular pattern of genes related to tumorgenesis, angiogenesis, host defense, cell growth, differentiation, transcription and metabolism (Moses *et al.*, 2002a; Moses *et al.*, 2002b; Poole *et al.*, 2002). KSHV-infected endothelial cells downregulated major histocompatibility complex class I, CD31, and ICAM-1 expression on the cell surface suggesting a mode of immune evasion employed by KSVH (Tomescu *et al.*, 2003). Several virus-encoded proteins may act together to transform endothelial cells. Up-regulation of c-kit proto-oncogene, encoding a receptor tyrosine kinase with stem cell factor as a ligand, was sufficient to induce spindle cell formation *in vitro* (Moses *et al.*, 2002b). In addition, KSHV G protein-coupled receptor oncogene immortalizes endothelial cells by activation of the VEGF-2/KDR signaling pathway (Bais *et al.*, 2003).

Pathophysiological Consequences of Endothelial Infection

Endothelial Activation and Signaling

As seen above, a great variety of diverse pathogens, using numerous pathologic mechanisms, act on the endothelium. Besides enzymes of pathogens, which may be incorporated in the endothelial signaling machinery, activation of central endothelial signaling pathways contribute to the response, as summarized in Figure 4. While victims of the attacking pathogen, the endothelium may also contribute to the ongoing pathology due to the complexity of the inflammatory response (Hack and Zeerleder, 2001). As illustrated in Figure 5, these endothelial activation and pro-inflammatory responses are diverse and include: mediator release (Hack and Zeerleder, 2001), leukocyte recruitment (Granger and Kubes, 1994), pro-coagulant activity (Keller *et al.*, 2003; Peters *et al.*, 2003), and breakdown of endothelial barrier function (Stevens *et al.*, 2000). A number of markers of endothelial damage, including ICAM-1, VCAM, E-selectin, IL-6, procalcitonin, thrombomodulin, and von Willebrand factor, correlate with severity of the pathophysiology in sepsis (Reinhart *et al.*, 2002). These responses are general in nature and a strict pathogen-specific response of the endothelium has not been elucidated. In addition, the response of the endothelium to inflammatory stimuli may differ grossly among different microvasculature sites of the body. Importantly, infection may disturb endothelial function, leading to either cell death, often realized by apoptosis, or proliferation.

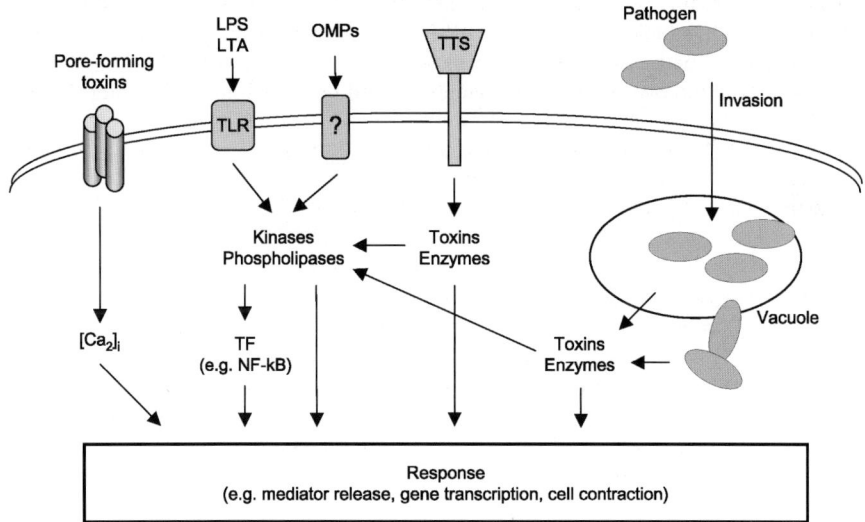

Figure 4: Pathogens induce complex signaling in endothelial cells. Pore forming bacterial toxins induced ion influx leading to cell activation. Receptor-mediated activation (e.g. TLR) leads to stimulation of endothelial signaling pathways. In addition, injected enzymes of pathogens or factors released in endothelial vacuoles or cytosol highjack and manipulate endothelial signaling machinery. Sometimes pathogens activate multiple pathways leading to a complex endothelial response.

Among the plethora of signaling pathways activated in infected endothelial cells, induction of NF-κB-dependent gene transcription is considered to be one of the most important (De Martin *et al.*, 2000; Bierhaus and Nawroth, 2003). There is little knowledge about the molecular connection between pathogens and NF-κB activation, although the signaling pathways leading to NF-κB activation are well known for receptor-mediated stimuli like LPS or TNFα (Dixit and Mak, 2002; Li and Verma, 2002). In the canonical pathway as described in Chapter 6, stimulated receptors, like TNF receptor 1, recruit adapter molecules with subsequent activation of the IKK-complex. IKKα and IKKβ phosphorylate IκB molecules, which normally sequester NF-κB in the cytosol. The IκB serine phosphorylation is followed by its ubiquitination and degradation, allowing translocation of NF-κB into the nucleus. It is likely that bacterial molecules like outer membrane proteins (Fuhrmann *et al.*, 2001; Niessner *et al.*, 2003) or bacteria without infection, such as inactivated *Chlamydia*, (Baer *et al.*, 2003)) activate NF-κB by using molecules of the classic pathway. However, pathogens may have additional measures to assure NF-κB activation. For example, *L. monocytogenes* phosphatidylinositol-PLC and phosphatidylcholine-PLC lead to ceramide generation, thereby activating this transcription factor (Schwarzer *et al.*, 1998). A yet unknown *R. rickettsii* related molecule seems to interact directly with inactive cytosolic IκB-NF-κB complex leading to transcription factor activation in an unique way (Sahni *et al.*, 2003). Furthermore, pathogens and their products may activate basic elements of the host cell transcription machinery. For example, a wide array of specific covalent modification of accessible N-terminal histone tails are decisive for transcription repression or gene activity in host cells (Grewal and Moazed, 2003). LPS-induction of IL-8 transcription required p38 MAP kinase-dependent modification of histones H3 and H4 (Saccani *et al.*, 2002). However, there are no data available regarding epigenomic effects of pathogens in endothelial cells. Therefore, signaling pathways activated by pathogens should be analyzed carefully in order to disclose the special interactions between pathogens and host cells, including endothelial cells.

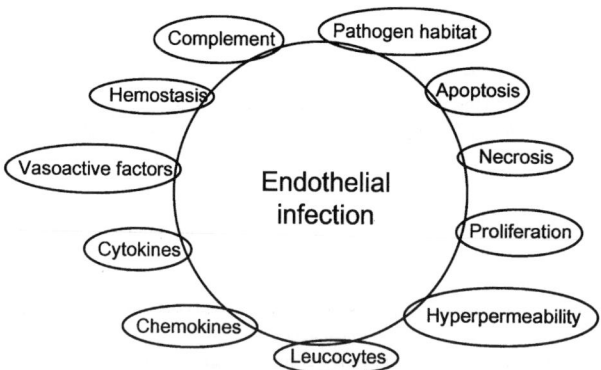

Figure 5: Pathogens affect endothelial cell functions. Endothelial infection may have very diverse local and systemic consequences. In acute infections, rapid endothelial stimulation with temporary stimulation of cytokine and vasoactive factor release as well as perturbation of hemostasis may be the dominant effect. Leukocyte recruitment, hyperpermeability, and apoptosis lead to edema formation. In addition, pathogens persistent in endothelia may have the power to induce endothelial proliferation and inhibition of apoptosis. These variable responses may reflect the highly diverse nature of endothelial behaviour in infection and inflammation.

Permeability

Endothelial hyperpermeability with subsequent edema formation is a hallmark of acute inflammation. It is well established that opening of intercellular gaps between endothelial cells allowing paracellular fluid flux is the dominating route for the extravasation of fluids and macromolecules in acute inflammatory reactions (Stevens *et al.*, 2000; Patterson and Lum, 2001). A great variety of very diverse pathogen-related factors like pore-forming exotoxins (Suttorp *et al.*, 1988), endotoxin (Bannerman and Goldblum, 1999), or direct alterations of the endothelial cytoskeleton during pathogen-endothelium interaction result in loss of barrier function. Moreover, viral proteins, such as HIV Tat (Bussolino *et al.*, 2001), as well as endogenous mediators of the inflammatory response to infection, such as TNF-α or leukocyte-derived agents (oxygen radicals, proteases) attack the endothelium finally resulting in barrier dysfunction (Stevens *et al.*, 2000). Although the activation modes differ greatly from each other, pathogen interactions with the endothelial cytoskeleton, as discussed above in entry mechanisms, have important consequences for endothelial barrier function. Many of the activated signaling pathways converge at myosin light chain phosphorylation, leading to increased actin-myosin interaction, but other mechanisms have not been well explored. Some pathogens directly modify junctional components, while others cause disassembly of tight junctions *via* their effects on the cytoskeleton (Sears, 2000; Gruenheid and Findlay, 2003). For example, Clostridium difficile toxins A and B inactivated RhoA, which induced permeability in pulmonary artery endothelial cells (Hippenstiel *et al.*, 1997), also cause remodeling of the cytoskeletal architecture and disassociation of the cortical actin band from the components of tight junctions (Sears, 2000). Bacterial or host proteases can promote endothelial hyperpermeability by directly degrading cell-cell contacts or cell-matrix-adhesions. As a consequence, exposed sub-endothelial matrix is accessible to bacterial adhesins and becomes a vulnerable target of bacterial proteases, thus promoting bacterial colonialization of host tissue (Ljungh and Wadstrom, 1996; Lahteenmaki *et al.*, 2001).

The important role of bacterial pathogenic factors in induction of hyperpermeability is exemplified by *E. coli*. While HlyA producing *E coli* (as well as purified hemolysin) induced endothelial barrier dysfunction *in vivo* and *in vitro*, non-producing bacteria did not (Suttorp *et al.*, 1988; Seeger *et al.*, 1989). Mechanisms utilized to disrupt barrier function include: myosin light chain phosphorylation, ezrin phosphorylation, occludin dephosphorylation, and occludin and ZO-1 dissasociation from the tight junctions (Gruenheid and Findlay, 2003). The pathway of bacterial attack and entry may be important for their effects on barrier function, since *E coli* pneumonia induced by intratracheally challenge did not necessarily induce hyperpermeability (Ong *et al.*, 2003). Permeability of cultured lung microvascular endothelium was shown for Group B streptococci, a leading cause of pneumonia and alveolar edema in newborns, which produced beta-hemolysin, but not for the strains with no hemolytic activity (Gibson *et al.*, 1999). *Staphylococcus aureus* alpha toxin at low levels induced progressive leak in isolated perfused lungs at the blood-gas exchange area (Seeger *et al.*, 1990). This permeability occurred despite inhibition of thromboxane-induced hypertension and was later followed by observed detachment of endothelial cells from the basal lamina, indicating a primary effect on endothelium. *S. pneumoniae* and purified pneumococcal cell wall components induced VEGF liberation from human neutrophils (van Der *et al.*, 2000), but overall, the role of VEGF for permeability regulation in acute or chronic infections is not resolved (Hippenstiel *et al.*, 1998; Clauss *et al.*, 2001; Issbrucker *et al.*, 2003).

Leukocyte Recruitment and Adhesion

Upon stimulation by a pathogen or pathogen components, such as endotoxin, as well as by pro-inflammatory factors produced by the host defense, an array of inflammatory mediators is produced by endothelial cells as outlined in Chapter 3. This pro-inflammatory response is nonspecific, rather than pathogen-specific. This response includes production of cytokines like chemotactic CXCL8, lipid mediators like prostaglandins or platelet-activated factor, and complement factors (Keller *et al.*, 2003; Peters *et al.*, 2003). CXCL8 expression was induced in endothelial cells by diverse stimuli such as LPS (Hippenstiel *et al.*, 2000), shiga toxin-2 (Zoja *et al.*, 2001), *C. pneumoniae* (Molestina *et al.*, 1998), *S. aureus* (Yao *et al.*, 1996), dengue virus (Bosch *et al.*, 2002), and HIV-tat protein (Hofman *et al.*, 1999). Anti-inflammatory mediators like TGF-β1 may also be released, although most research has focused on release of pro-inflammatory mediators. In addition, some pathogens may directly counteract the endothelial pro-inflammatory response. The periodontal disease causing bacteria *Prophyromonas gingivalis*, which infects and invades endothelial cells, stimulates CXCL8 expression, but expression of bacterial lysine-specific proteinase gingipain K degrades the cytokine (Nassar *et al.*, 2002; Kobayashi-Sakamoto *et al.*, 2003).

Besides liberation of cytokines, expression of adhesion molecules by the endothelium contributes significantly to the recruitment of leukocytes to the arena. Marginalization, rolling, adhesion, and transmigration through the endothelial monolayer into underlying tissue is based on the interaction of specific pairs of ligands on both cell types, as described in detail in Chapter 11. *L. moncytogenes* (Krull *et al.*, 1997; Schwarzer *et al.*, 1998), *B. henselae* (Fuhrmann *et al.*, 2001), *C. pneumonia* (Krull *et al.*, 1999), bacterial exotoxins (*E. coli* hemolysin and *S. aureus* α-toxin (Krull *et al.*, 1996)) or viruses (e.g. coxsackie B virus (Zanone *et al.*, 2003)) increase adhesion molecule expression on endothelial cells. Different virulence factors of a pathogen may address different adhesion molecules in the inflammatory process, illustrating the complex nature of this process. While listerial phospholipases induced expression of E-selectin but not of P-selectin, LLO up-regulated P-selectin but not ICAM-1 or E-selectin (Krull *et al.*, 1997; Schwarzer *et al.*, 1998). Organ-specific mechanisms of leukocyte recruitment by infective agents also have to be considered (Liu and Kubes, 2003). Mice deficient in endothelial P-selectin did not exhibit vascular plugging, hemmorrhage, or edema in cerebral vessels when infected with *Plasmaodium berghei* ANKA, which causes cerebral malaria; whereas P-selectin deficiency in platelets had no protective effect, indicating the importance of the endothelial adherence response (Combes *et al.*, 2004). Overall, recruitment of immune cells to the site of infection contributes to the localization of infection and clearing of pathogens. However, in cases of severe infection with a systemic inflammatory response or long-lasting close contact of endothelial cells with leukocytes, negative effects may outweigh possible benefits for the host. Release of leukocyte-derived mediators, including superoxide, NO, hydroxyl radicals, and proteases, cause collateral injury to endothelial cells localized at the site of action (Hardy *et al.*, 1994; Nakatani *et al.*, 2001). Mice genetically lacking endothelial selectins were shown to be resistant to the lethality in septic shock (Matsukawa *et al.*, 2002). Although pathogens seem to promote leukocyte-endothelial interaction, it is important to note that opposing effects have also been shown. An extracellular adherence protein secreted by *S. aureus* efficiently blocked β_2-integrin-dependent neutrophil recruitment (Chavakis *et al.*, 2002) and KSHV downregulated adhesion molecule expression on endothelial cells (Tomescu *et al.*, 2003). Thus, a situation-specific estimation of the role of endothelial adhesion molecules is necessary.

Vasoactive Mediator Generation

Infection and inflammation effects on endothelium alter endothelial production of vasoactive compounds, that influence blood pressure and local perfusion (Vanhoutte and Mombouli, 1996; Keller *et al.*, 2003; Peters *et al.*, 2003). Pore forming exotoxins like *E. coli* HlyA or staphylococcal α-toxin, stimulate NOS1-related NO-release in a Ca^{++}-dependent manner (Suttorp *et al.*, 1993a). In addition, *L. moncytogenes* activate endothelial NO-release in a LLO-depending way, while other listerial pathogenic factors like PLCs display no effect (Rose *et al.*, 2001). Pro-inflammatory stimulation by cytokines or LPS leads to iNOS expression, resulting in Ca^{++}-independent liberation of large quantities of NO into the vascular and perivascular space. Pre-exposure of vascular beds to low doses of LPS may sensitize the endothelium for a second perturbation by pore forming exotoxins resulting in massive release of vasoactive and pro-inflammatory mediators as shown for a rabbit lung preparation (Walmrath *et al.*, 1994). Besides regulating vascular tone, high NO may contribute to pathogen killing (Fang, 1997). In general, NO seems to stabilize endothelial function and acts as a survival factor (Dimmeler and Zeiher, 1999). In terms of survival of the host, however, NO and related radicals may act as a double-edged sword, as indicated by studies in iNOS-deficient mice (Nathan, 1997) and in experiments using pharmacological NOS inhibitors (Benjamim *et al.*, 2002). Although there is no doubt about the central role of iNOS in pathogen-host (and particularly endothelial) interaction, the signaling molecules involved in pathogen-related iNOS-expression are largely unknown.

Similar to NO, prostacyclin is a potent vasodilator and inhibitor of platelet aggregation. Stimulation of endothelial cells with pore-forming exotoxins, such as *S. aureus* α-toxin (Suttorp *et al.*, 1985), induced rapid and massive liberation of prostacyclin in a COX-1 dependent manner, but the net effect in isolated blood free perfused lungs was hypertension due to thromboxane release (Seeger *et al.*, 1990). Endothelial cell stimulation with bacterial components like LPS (Suttorp *et al.*, 1987; Schmeck *et al.*, 2003), *B. henselae* outer membrane proteins, intact bacteria like *L. monocytogenes* (our unpublished results), as well as with endogenous pro-inflammatory mediators like TNFα (Schmeck *et al.*, 2003) activate COX2-expression with release of large amounts of prostacyclin. Expression of COX2 seems to be dependent on small GTP binding Rho protein mediated activation of NF-κB in these cells (Schmeck *et al.*, 2003). In contrast to the above-mentioned induction of prostacyclin, *P. aerogenosa* inhibited prostacyclin release from endothelium, suggesting a pathogen-specific response (Kamath *et al.*, 1995). This may be clinically important since adherence of *Candida albicans* to endothelial cells was inhibited by prostacyclin (Klotz, 1994). In addition to their important role in the regulation of vascular tone, it must be taken in mind that NO and prostacyclin influence additional functions in the vascular system, including: leukocyte and platelet adhesion, coagulation, apoptosis, and pathogen adherence.

Furthermore, recent studies suggest a pivotal role for the multifunctional peptide adrenomedullin in the inflammatory response. Adrenomedullin is a 52 amino acid peptide of the calcitonin gene peptide family involved in the regulation of blood pressure (Wang, 1998). Several studies demonstrated elevated plasma levels in septic humans and animal models of sepsis (Wang, 1998; Nikitenko *et al.*, 2002). Interestingly, the endothelium is suggested to be a major source of adrenomedullin liberated in inflammatory reactions. Adrenomedullin stabilizes endothelial barrier function *in vitro* (Hippenstiel *et al.*, 2002), in isolated rabbit lungs (Hippenstiel et al., 2002), as well as in an isolated rat ileum preparation (unpublished results). Moreover, it protects mice from LPS-induced septic shock (Shindo et al., 2000), and inhibited endothelial apoptosis (Sata et al., 2000). Thus, endothelial derived adrenomedullin may act as a counter-regulatory peptide in severe infections like sepsis. Further research is needed to clarify the role of this peptide in inflammatory reactions.

Although production of vasodilators by infected or pathogen factor-challenged endothelial cells seems to be the dominant reaction, decreased release of dilators and liberation of potent vasoconstrictors was also observed (Bhagat et al., 1996; Petkova et al., 2001). S. aureus α-toxin exposure of pulmonary artery endothelial cells resulted in strong synthesis of platelet-activated factor (PAF) (Suttorp et al., 1992). Tat protein, associated with HIV-infection, decreased endothelium-dependent vasorelaxation and eNOS mRNA and protein expression in endothelial cells of porcine coronary arteries (Palduga et al., 2003). And *Trypanosoma cruzi* infection of endothelium causes vasospasm, reduced blood flow, focal ischemia, platelet thrombi, and generalized vasculitis due to elevated plasma levels of thromboxane A_2 and endothelin-1 (Camargos et al., 2002).

Pro-thrombosis

The endothelium is intimately involved in the complex initiation and regulation of both coagulation and fibrinolysis. Under normal conditions, the endothelium combats thrombosis by expressing anticoagulant agents, such as thrombomodulin, plasminogen activator, or heparin sulphate (Levi et al., 2002; Levi et al., 2003; Vallet, 2003). Loss of anticoagulant properties (reduction in antithrombin or activated protein C) in combination with enhanced endothelial expression of procoagulant factors, especially tissue factor and von Willebrand factor, leads to fibrin formation. All three major pathways of the coagulation system, tissue factor mediated thrombin generation, loss of anticoagulant pathways, and blocked fibrinolysis, contribute to endothelial-related coagulation dysfunction, which is highly important in the pathologic manifestations in sepsis. This multistep activation is triggered by agents such as complete bacteria like *Chlamydia*, viruses, pathogen components like LPS, or mediators of the host response (Levi et al., 2002; Levi et al., 2003; Vallet, 2003).

Anchoring leukocytes and juxtacrine signaling by the interaction of adhesion receptors may aggravate expression of prothrombotic agonists (McEver, 2001). Increased tissue factor expression by the pathogen-activated endothelium, tethering of monocytes bearing tissue factor and leukocyte-derived microparticles together may also activate the coagulation cascade (McEver, 2001). Thrombin activation subsequently elicits endothelial responses like hyperpermeability, adhesion molecule expression, and cytokine production (Stevens et al., 2000; Coughlin, 2001). Endothelial products of the coagulation process perpetuate the inflammatory process by aggravating the host response. For example, thrombin or factor Xa exposed endothelial cells expressed pro-inflammatory cytokines like monocyte chemotactic protein 1 and IL-8 (Senden et al., 1998; Szaba and Smiley, 2002). Moreover, platelet activation by pathogens may contribute to formation of platelet-bacteria thrombi on the endothelial surface. *S. aureus*, for example, interacts with platelets by several surface proteins (Patti et al., 1994), thereby promoting platelet aggregation (O'Brien et al., 2002). The importance of this interaction is highlighted by the formation of platelet-bacteria thrombi on the valve surface in infective endocarditis (Sullam et al., 1996). Although, endothelial NO and prostacyclin, liberated in inflammatory conditions, inhibit adhesion of activated platelets to endothelial cells, other liberated pro-adhesive factors give the stronger signals.

In general, activation of the coagulation cascade throws up a molecular fishing net for immobilizing invading pathogens. Some microorganisms try to counteract the host response by activation of the plasminogen system. Streptokinase (secreted by group A, C, and C streptococci) and staphylokinase activate plasminogen by formation of 1:1 complexes with it (Lahteenmaki *et al.*, 2001). Plasminogen receptors of *E. coli*, *S. aureus*, *Borrelia burgdorferi*, or *C. albicans* immobilize plasminogen on the bacterial surface; thus enhancing plasminogen activation and turning bacteria into a "proteolytic organism" (Lahteenmaki *et al.*, 2001). This proteolytic activity not only counteracts the host's fishing strategy, it promotes invasion of pathogens by degradation of extracellular matrices (Coleman *et al.*, 1995; Jong *et al.*, 2003).

In severe infection with gram-negative bacteria with massive release of LPS in the bloodstream, endothelial injury paves the way to disseminated intravascular coagulation. Activation of endothelium and platelets, as well as fibrin formation, leads to consumption of platelets, coagulation factors, and activation of the fibrinolytic system, resulting in a life-threading syndrome (Levi *et al.*, 2002; Levi *et al.*, 2003; Vallet, 2003). Endotoxin induced endothelial cells to release PAF, thromboxane, and prostacyclin, von Willebrand factor, tissue factor expression, and plasminogen activator inhibitor- I, while tissue factor inhibitor expression was decreased (Bannerman and Goldblum, 1999). Each of the changes is pro-coagulant alone and the combination is devastating. Lung microvascular endothelium normally has high expression of u-PA, while the pulmonary artery endothelium has high t-PA expression. TNFα and LPS challenge caused down-regulation of t-PA in both macro- and microvascular pulmonary endothelium and upregulated plasminogen activator inhibitor- 1 and -2, favoring local fibrin deposition (Muth *et al.*, 2004). Thus, it is easy to understand the association of coagulation pathologies and sepsis. The importance of endothelial dysfunction and coagulation in sepsis and acute lung injury is underscored by the successful use of anticoagulant therapy of septic patients to reduce mortality, where numerous other strategies have failed as discussed in Chapter 15 (Patel *et al.*, 2003; Schultz *et al.*, 2003).

Necrosis and Apoptosis

Endothelium displays low cell turnover under physiological conditions but under pathophysiologic conditions both increased cell death and proliferation are noted (Dimmeler and Zeiher, 2000; Kockx and Herman, 2000). Cumulating evidence suggests that modulation of endothelial survival is a central issue in infection processes. Many toxins, which produce barrier dysfunction early in the process or at low levels, lead to cell death at higher levels and with longer exposure. For instance, β-hemolysin from Group B streptocci, which causes direct pulmonary endothelial permeability and exudative edema, will result in trypan blue nuclear staining and lactate dehydrogenase release, indicating cellular injury (Gibson *et al.*, 1999). Although both endothelial necrosis and apoptosis are initiated by pathogens, recent research has focused on the regulation of apoptosis. Apoptosis, or programmed cell death, is morphologically characterized by cell shrinkage, membrane blebbing, and chromatin condensation (Hengartner, 2000). An array of anti-apoptotic proteins controls the pre-formed, ready to go, death machinery, including members of the Bcl-2 family and the FLICE-like inhibitor protein (FLIP) (Desagher and Martinou, 2000; Hengartner, 2000). Pre-formed zymogens, cysteinyl aspartate-specific proteases (caspases), are activated in an orchestrated manner to execute the cells suicide (Earnshaw *et al.*, 1999). Receptor-mediated activation (e.g. FAS/CD95, TLR4, TNFR) or the release of mitochondrial factors (e.g. cytochrome C) activate the caspases (Earnshaw *et al.*, 1999; Desagher and Martinou, 2000; Hengartner, 2000). Finally, the cells body fragments into small membrane-bound bodies, which are cleared by professional and non-professional phagocytes without initiation of inflammation. In contrast to the ATP-dependent apoptotic process, cell necrosis is energy-independent and characterized by cell swelling and lysis. Release of intracellular components of the dying cells induces inflammation in surrounding viable tissue.

The diversity of stimuli for endothelial apoptosis reflects the complexity of the various pathways. Several pathogens or pathogen-related products, including *S. aureus* (Menzies and Kourteva, 2000), *T. cruzi* (Zhang *et al.*, 1999), bacterial exotoxins, LPS (Bannerman and Goldblum, 2003), viral proteins (Park *et al.*, 2001; Kim *et al.*, 2003b), antibodies against dengue virus (Lin *et al.*, 2003) or *P. falciparum* infected erythrocytes (Pino *et al.*, 2003), induce endothelial apoptosis. Shiga toxin induced apoptosis of endothelial cells of different origin and reduced expression of anti-apoptotic Blc-2 protein Mcl-1; blocking of proteasome-dependent Mcl-1 degradation reduced apoptosis (Zhang *et al.*, 1999; Brigotti *et al.*, 2002; Erwert *et al.*, 2002; Ergonul *et al.*, 2003;Erwert *et al.*, 2003). Besides using traditional pathways, shiga toxin enzymatic activity also damaged nuclear DNA directly by adenine removal (Brigotti *et al.*, 2002). In addition to Mcl-1-reduction, shiga toxin reduced the expression of FLIP, an anti-apoptotic protein with structural antagonism of caspase-8 (Erwert *et al.*, 2002; Erwert *et al.*, 2003). Constitutive FLIP expression likely contributes to the relative resistance of human endothelial cells against apoptosis (Bannerman *et al.*, 2001) and shiga toxin-mediated FLIP degradation and reduced synthesis results in sensitization of human endothelium to apoptosis. Thus, shiga toxin producing *E coli* have all the tools to induce massive endothelial apoptosis.

Although evidence suggests significant LPS related endothelial apoptosis during severe gram-negative infections (Bannerman and Goldblum, 2003), mechanisms of LPS-induced endothelial apoptosis are still unclear, especially the observation that LPS exposed bovine and ovine endothelial cells undergo massive apoptosis, while human endothelial cells are relatively resistant as mentioned above (Bannerman and Goldblum, 2003). Moreover, additional anti-apoptotic proteins like the Bcl-2 homolog A1, the zink-finger protein A20, and cellular inhibitors of apoptosis proteins were expressed in LPS treated endothelium in a NF-κB dependent manner (Bannerman and Goldblum, 2003). Despite these powerful anti-apoptotic defence mechanisms, LPS-mediated decreases in expression of anti-apoptotic proteins of the Bcl-2 family and increased protein levels of pro-apoptotic members resulted in caspase activation (Bannerman and Goldblum, 2003). Overall, it is easy to imagine that in the aggressive milieu of LPS-related inflammation *in vivo,* the simultaneous exposure of endothelial cells to additional pro-apoptotic factors (e.g. hydroxyl radicals from granulocytes, cytokines, complement) may lead to substantial programmed endothelial cell death.

Viruses or virus-related proteins also induce endothelial apoptosis. HIV gp120 protein induced apoptosis in a caspase-3 dependent manner (Park *et al.*, 2001). In addition, the immune response against these pathogens causes endothelial damage. Antibodies against the dengue virus non-structural protein 1, for example, induced endothelial programmed cell death by interfering with Bcl-2 protein expression and caspase activation (Lin *et al.*, 2002).

Sometimes pathogens strengthen endothelial vitality. In *R. rickettsii* infection, NF-κB dependent inhibition of apoptosis allowed survival of infected cells, while NF-κB inhibition resulted in caspase-8, -9-, and –3 dependent programmed death of Rickettsia infected cells

(Joshi et al., 2003). Moreover, these bacteria induced strong expression of heme oxygenase 1, possibly serving a protective function against oxidative injury in the inflammation process (Rydkina et al., 2002). B. henselae and B. quintana seemed to inhibit caspase activation and DNA-fragmentation (Kirby and Nekorchuk, 2002). It is probable that pathogens persisting in endothelial cells, like Chlamydia, interfere with apoptosis regulation; however, data is limited. In viral infection, it is known that KSHV infection of endothelial cells activates several strong survival signals. Infection by Kasposi's sarcoma-associated herpesvirus was accompanied by G protein coupled receptor-mediated activation of Akt/protein kinase B pathway leading to cell survival (Montaner et al., 2001). In addition, this receptor immortalizes human endothelial cells by activation of VEGFR2/KDR (Bais et al., 2003). Another KSHV gene product, latency-associated nuclear antigen, also prolongs endothelial life span, suggesting multiple survival factors in KSHV infection (Watanabe et al., 2003). Nevertheless, inference of molecules like these with classical pro-apoptotic pathways is unknown. The question about life or death in pathogen-endothelial interaction is further complicated by the fact that *in vivo* the pathogen never is alone in contact with the endothelium, there will be other cellular players (e.g. blood cells, pericytes, smooth muscle cells), humoral factors, and cytokines as well. In addition, endothelial cells from different vascular beds may show differential responses with respect to apoptosis (Hooper 03).

Vascular Proliferation

After colonialization of endothelium, bacteria may stimulate endothelial proliferation and migration necessary for effective angiogenesis (Dehio, 2001; Dehio, 2003). Bartonella, including *B. bacilliformis*, *B. quintana*, and *B. henselae*, inhibited apoptosis and induced angioproliferation (Dehio, 2003). In *B. bacilliformis* infection, secreted GroEL seemed to be involved in mitotic stimulation of endothelia (Minnick et al., 2003). In contrast, GroEL was not involved in *B. henselae* related mitosis. Thus, additional undetermined soluble, secreted factors appear to be involved. Bartonella infection results in only low VEGF production by endothelial cells. Recruitment of leucocytes to infected endothelium allows establishment of a paracrine loop of endothelial proliferation by stimulation of VEGF-release by Bartonella infected immune cells (Dehio, 2003). In specimens of Verruga peruana (hemangioma-like lesions induced by *B. bacilliformis*), Bartonella induced high expression of VEGF receptor and angiopoetin-2 (Cerimele et al., 2003). Not only was endothelial proliferation started, but rapid bacterial rRNA synthesis and replication was noted inside the endothelium (Kempf et al., 2000). Site-specific and transposon mutagenesis should be helpful to identify Bartonella pathogenic factors involved (Schulein and Dehio, 2002). A fascinating hypothesis has been proposed that the plant pathogen *Agrobacterium tumefaciens* and Bartonella share the same strategy to survive by triggering the promotion of their own habitat (Kempf et al., 2002).

In addition to Bartonella, KSHV is known to induce endothelial proliferation. This virus oncogene encodes a G protein coupled receptor with constitutive VEGFR2/KDR activation, VEGF secretion, and stimulation of angiogenesis (Bais et al., 2003). Subsequently, the PI3–kinase/AKT kinase pathway was activated resulting in an oncogenic immortalizing effect (Bais et al., 2003). Furthermore, expression of virus encoded kinase c-Kit promoted proliferation and transformation of infected endothelia (Moses et al., 2002b). Finally, infected endothelial cells *in vitro* recapitulate aspects of the Kasposi's angiogenic phenotype in human endothelial cells, underlining the role of KSHV in this disease.

Clinical Consequences of Endothelial Infection

Acute Infections

The central role of the endothelium for the inflammatory process in acute infections is highlighted by the endothelial contribution to the classical signs of infection: rubor, dolor, calor, and tumour. Whether acute infection occurs primarily in the lung or in other tissues, endothelial malfunction has fundamental impact on the disease process and contributes in a multitude of ways to the pathologic process as discussed above. One very important manifestation is the loss of barrier function. Overcoming the endothelial barrier may be a key step in the distribution of pathogens into specially protected compartments, like the brain with its tight endothelial blood-brain barrier. Barrier dysfunction also has direct physiologic consequences of clinical significance and is a prerequisite for the perilous lung edema formation in ARDS, frequently observed in sepsis (Ware and Matthay, 2000; Groeneveld, 2003; MacArthur *et al.*, 2004). Moreover, the most frequent precipitating event in ARDS is sepsis; the second is pneumonia. Barrier dysfunction followed by endothelial injury is fundamental to the pulmonary edema and the alveolar hemorrhage of Group B streptococcal infection, the most common cause of pneumonia and sepsis in newborns (Gibson *et al.*, 1999). Moreover, in severe systemic infection, endothelial dysfunction contributes to the development of multiple organ failure (Levi *et al.*, 2002; Keller *et al.*, 2003; Peters *et al.*, 2003; Vallet, 2003). LPS endotoxin-induced endothelial injury and inflammation is a common cause of life-threatening ARDS and diseminated intravascular coagulation (Bannerman and Goldblum, 1999). Parameters indicating endothelial activation in systemic infection include soluble adhesion molecules, coagulation factors, IL-6, and procalcitonin (Rubin *et al.*, 1990; Ware *et al.*, 2001; Atabai and Matthay, 2002; Reinhart *et al.*, 2002).

Circulating endothelial cells were detected in septic humans, indicating damage and sloughing, but whether this correlated with necrotic or apoptotic death of endothelium or with barrier dysfunction was not determined (Mutunga *et al.*, 2001). Endothelial damage and activation of prothrombotic signaling induced by enterohemorrhagic *E. coli* -produced Shiga toxin is recognized as the key trigger for development of microangiopathic processes leading to haemolytic-uremic syndrome (Karch, 2001). *Rickettsiae* infection of endothelium causes acute, potentially lethal diseases like Rocky Mountain spotted fever (*R. rickettsii*), louse-borne epidemic typhus (*R. prowazekii*), or Mediterranean spotted fever (*R. conorii*), with multi-organ involvement (Valbuena *et al.*, 2002). Multiple factors of the immune response acting on the endothelium, as well as direct virus-related activation, participate in the often fatal course of viral hemorrhagic fevers (Schnittler and Feldmann, 1999). Unfortunately, many studies of sepsis lack specificity for infection-specific endothelial damage.

Chronic Infections

There is increasing attention focused on the role of chronic endothelial infection in disease development. In numerous infections, local and systemic activation and disturbance of endothelial function contributes significantly to organ dysfunction and disease complications. Chronic viral infection by human herpesvirus 8 likely contributes to the disease process in primary pulmonary hypertension (Cool et al., 2003). Whether there is a notable role of chronic viral or bacterial infections for the onset of atherosclerosis is still unclear, but there is support for a contributory role of infection and a primary role for endothelial dysfunction (Noll, 1998; Keichl et al., 2001; Mahoney and Coombes, 2001; Coombes et al., 2002; Belland et al., 2004; Campbell and Kuo, 2004). Endothelial dysfunction was deemed "the most plausible link between infection, inflammation, and atherosclerosis" in HIV infection, but it is unclear how both infection and the metabolic derangements due to highly active antiretroviral therapy independently contribute to endothelial activation (de Gaetano et al., 2004). Chronic bacterial infection with *Bartonella* species is considered as causative in cat-scratch disease, urban trench fever, and bacillary angiomatosis-peliosis (Dehio, 2001; Dehio, 2003). Infection of endothelial cells with Kaposi's sarcoma associated herpesvirus is believed to be etiological for the formation of Kaposi's sarcoma (Poole et al., 2002). Adult human heart microvessel endothelial cells were permissive to non-lytic CMV infection and may contribute to formation of cardiac disorders by inducing chronic release of pro-inflammatory factors (Ricotta et al., 2001). Especially in the immunocompromised host the endothelium may serve as a reservoir for HCMV resulting in recurrent, possible life-threading infections (Bruggeman, 1999). The role of parvovirus B19 as well as enterovirus-induced endothelial cells activation for disease development is widely unknown (Bultmann et al., 2003; Saijets et al., 2003). Overall, ranging from simple, local infection to severe, systemic infection and inflammation of the host, there is no doubt that the endothelium is critically involved in the development of the resulting disease.

Perspective

It is well known that the endothelium, located as an interactive barrier between the blood and tissue, has a major place in pathogen-host interactions and is a primary target as well as a key player in the inflammatory response. The diverse pathogens, pathogen-derived products, and immune responses, which act on the endothelium, are poorly understood. Several pathogen-activated signaling pathways in the endothelium have been identified, but the picture is incomplete. Current research often focuses on initiation of inflammation, but does not address termination of the response. How endothelial expression of anti-inflammatory cytokines contributes to barrier dysfunction is unresolved. Which cells and signals contribute to the repair of the injured endothelial monolayer after infection is unclear. Recent evidence suggesting an important role for the mobilization of progenitor cells for endothelial repair (Hibbert et al., 2003) is of considerable interest in light of the observation that endothelial NOS is essential for mobilization of stem and progenitor cells (Aicher et al., 2003).
   Investigation of the endothelium using culture methods, which mimic characteristics of the local environment of a vascular bed in an organ, may help identify factors leading to pathogen tropism and the differentiated endothelium response. Genotyping of pathogens may aid in understanding vascular tropism for bacteria (Gieffers et al., 2003) and viruses (Bolovan-Fritts and Wiedeman, 2002). Moreover, most *in vitro* models analyze bacterial adhesion and bacterial-endothelial interactions under static conditions, ignoring effects of shear stress and hydrostatic pressure, yet lectin-mediated adhesion of bacterial pili may increase under flow conditions (Isberg and Barnes, 2002). In contrast, shear stress prevented

fibronectin binding protein-mediated *S. aureus* adhesion to resting endothelium (Reddy and Ross, 2001). Furthermore, shear stress alone induced significant alterations in endothelial gene transcription, which must be considered in experimental interpretation (Ohura *et al.*, 2003). Experiments using isolated endothelial cells and pathogens add small pieces of knowledge of high analytical impact, but results must be verified in more complex models.

Although progress has been made in the molecular analysis of the host response on viral or bacterial invasion, little work has been done with respect to effects of host entry on the pathogen. Using large-scale analysis of the transcriptome and proteome of pathogens and endothelial cells under different inflammatory conditions may help to discriminate pathogen-specific responses from undifferentiated stress response (Huang *et al.*, 2002). Identification of specific target genes and signaling pathways affected in a specific infection could lead to the development of new, pathogen-specific therapeutics. Overall, pathogen-endothelial interaction is a fascinating field of research with impact for understanding inflammatory disease and of great importance in the permeability and inflammatory consequences of pulmonary edema as seen in Adult Respiratory Distress Syndrome.

ACKNOWLEDGEMENTS

This work was supported by the German Federal Research Ministry (BMBF-NBL3 to S.H, BMBF-CAPNETZ/C6 to N.S.) and Deutsche Forschungsgemeinschaft (DFG HI 789/5-1 to S.H and N.S., and SSP 1130 to N.S.).

REFERENCES

Ahn, S.K., Choe, T., and Kwon, T.J. (2003) The gene expression profile of human umbilical vein endothelial cells stimulated with lipopolysaccharide using cDNA microarray analysis. Int. J. Mol. Med. 12:231-236.

Aicher, A., Heeschen, C., Mildner-Rihm, C., Urbich, C., Ihling, C., Technau-Ihling, K., Zeiher, A., and Dimmeler, S. (2003) Essential role of endothelial nitric oxide synthase for mobilization of stem and progenitor cells. Nat. Med. 9:1370-1376.

Al Younes, H.M., Rudel, T., and Meyer, T.F. (1999) Characterization and intracellular trafficking pattern of vacuoles containing *Chlamydia pneumoniae* in human epithelial cells. Cell Microbiol. 1:237-247.

Alvarez-Dominguez, C., Roberts, R., and Stahl, P.D. (1997) Internalized *Listeria monocytogenes* modulates intracellular trafficking and delays maturation of the phagosome. J. Cell Sci. 110:731-743.

Andonegui, G., Bonder, C., Green, F., Mullaly, S., Zbytnuik, L., Raharjo, E., and Kubes, P. (2003) Endothelium-derived Toll-like receptor-4 is the key molecule in LPS-induced neutrophil sequestration into lungs. J. Clin. Invest. 111:1011-1020.

Andor, A., Trulzsch, K., Essler, M., Roggenkamp, A., Wiedemann, A., Heesemann, J., and Aepfelbacher, M. (2001) YopE of Yersinia, a GAP for Rho GTPases, selectively modulates Rac-dependent actin structures in endothelial cells. Cell. Microbiol. 3:301-310.

Atabai, K. and Matthay M.A. (2002) The pulmonary physician in critical care: Acute lung injury and the acute respiratory distress syndrome: definitions and epidemiology. Thorax 57:452-458.

Shiga toxin-induced response patterns in human vascular endothelial cells. Blood 102:1323-1332.
Mayer, C.L., Filler, S.G., and Edwards, J.E. (1992) Candida albicans adherence to endothelial cells. Microvasc. Res. 43:218-226.
McEver, R.P. (2001) Adhesive interactions of leukocytes, platelets, and the vessel wall during hemostasis and inflammation. Thromb. Haemost. 86:746-756.
Menzies, B.E. and Kourteva, I. (2000) *Staphylococcus aureus* alpha-toxin induces apoptosis in endothelial cells. FEMS Immunol. Med. Microbiol. 29:39-45.
Mikawa, K., Nishina, K., Takao, Y., and Obara, H. (2003) ONO-1714, a NOS inhibitor, attenuates endotoxin-induced acute lung injury in rabbits. Anesthesia & Analgesia. 97:1751-1755.
Miller, L.H., Baruch, D., Marsh, K., and Doumbo, O.K. (2002) The pathogenic basis of malaria. Nature 415:673-679.
Minnick, M.F., Smitherman, L.S., and Samuels, D.S. (2003) Mitogenic effect of *Bartonella bacilliformis* on human vascular endothelial cells and involvement of GroEL. Infect. Immun. 71:6933-6942.
Minshall, R.D., Sessa, W., Stan, R., Anderson, R., and Malik, A.B. (2003) Caveolin regulation of endothelial function. Amer. J. Physiol. 285:L1179-L1183.
Molestina, R.E., Dean, D., Miller, R., Ramirez, J., and Summersgill, J.T. (1998) Characterization of a strain of *Chlamydia pneumoniae* isolated from a coronary atheroma by analysis of the omp1 gene and biological activity in human endothelial cells. Infect. Immun. 66:1370-1376.
Molestina, R.E., Klein, J., Miller, R., Pierce, W., Ramirez, J., and Summersgill, J.T. (2002) Proteomic analysis of differentially expressed *Chlamydia pneumoniae* genes during persistent infection of HEp-2 cells. Infect. Immun. 70:2976-2981.
Montaner, S., Sodhi, A., Pece, S., Mesri, E., and Gutkind, J.S. (2001) The Kaposi's sarcoma-associated herpesvirus G protein-coupled receptor promotes endothelial cell survival through the activation of Akt/protein kinase B. Cancer Res. 61:2641-2648.
Moses, A.V., Jarvis, M., Raggo, C., Bell, Y., Ruhl, R., Luukkonen, B., Griffith, D., Wait, C., Druker, B., Heinrich, M., Nelson, J., and Fruh, K. (2002a) A functional genomics approach to Kaposi's sarcoma. Ann. NY Acad. Sci. 975:180-191.
Moses, A.V., Jarvis, M., Raggo, C., Bell, Y., Ruhl, R., Luukkonen, B., Griffith, D., Wait, C., Druker, B., Heinrich, M., Nelson, J., and Fruh, K. (2002b) Kaposi's sarcoma-associated herpesvirus-induced upregulation of the c-kit proto-oncogene, as identified by gene expression profiling, is essential for the transformation of endothelial cells. J. Virol. 76:8383-8399.
Muth, H., Maus, U., Wygrecka, M., Lohmeyer, J., Grimminger, F., Seeger, W., and Gunther, A. (2004) Pro- and antifibrinolytic properties of human pulmonary microvascular versus artery endothelial cells: impact of endotoxin and tumor necrosis factor-alpha. Crit. Care Med. 32:217-226.
Mutunga, M., Fulton, B., Bullock, R., Batchelor, A., Gascoigne, A., Gillespie, J., and Baudouin, S.V. (2001) Circulating endothelial cells in patients with septic shock. Amer. J. Respir. Crit. Care Med. 163:195-200.
Nagase, T., Uozumi, N., Aoki-Nagase, T., Terawaki, K., Ishii, S., Tomita, T., Yamamoto, H., Hashizume, K., Ouchi, Y., and Shimizu, T. (2003) A potent inhibitor of cytosolic phospholipase A2, arachidonyl trifluoromethyl ketone, attenuates LPS-induced lung injury in mice. Amer. J. Physiol. 284:L720-L726.
Nakao, H. and Takeda, T. (2000) *Escherichia coli* Shiga toxin. J. Nat. Toxins 9:299-313.
Nakatani, K., Takeshita, S., Tsujimoto, H., Kawamura, Y., and Sekine, I. (2001) Inhibitory effect of serine protease inhibitors on neutrophil-mediated endothelial cell injury. J. Leuk. Biol. 69:241-247.
Nassar, H., Chou, H., Khlgatian, M., Gibson, F., Van Dyke, T., and Genco, C.A. (2002) Role for fimbriae and lysine-specific cysteine proteinase gingipain K in expression of interleukin-8 and monocyte chemoattractant protein in *Porphyromonas gingivalis*-infected endothelial cells. Infect. Immun. 70:268-276.
Nathan, C. (1997) Inducible nitric oxide synthase: what difference does it make? J. Clin. Invest. 100:2417-2423.

Niessner, A., Kaun, C., Zorn, G., Speidl, W., Turel, Z., Christiansen, G., Pedersen, A., Birkelund, S., Simon, S., Georgopoulos, A., Graninger, W., De Martin, R., Lipp, J., Binder, B., Maurer, G., Huber, K., and Wojta, J. (2003) Polymorphic membrane protein 20 and PMP 21 of *Chlamydia pneumoniae* induce proinflammatory mediators in human endothelial cells in vitro by activation of the NFκB pathway. J. Infect. Dis. 188:108-113.

Nikitenko, L.L., Smith, D., Hague, S., Wilson, C., Bicknell, R., and Rees, M.C. (2002) Adrenomedullin and the microvasculature. Trends Pharm. Sci 23:101-103.

Nizet, V., Kim, K., Stins, M., Jonas, M., Chi, E., Nguyen, D., and Rubens, C.E. (1997) Invasion of brain microvascular endothelial cells by group B streptococci. Infect. Immun. 65:5074-5081.

Noll, G. (1998) Pathogenesis of atherosclerosis: a possible relation to infection. Atherosclerosis 140:S3-S9.

O'Brien, L., Kerrigan, S., Kaw, G., Hogan, M., Penades, J., Litt, D., Fitzgerald, D., Foster, T., and Cox, D. (2002) Multiple mechanisms for the activation of human platelet aggregation by *Staphylococcus aureus*: roles for the clumping factors ClfA and ClfB, the serine-aspartate repeat protein SdrE and protein A. Mol. Microbiol. 44:1033-1044.

Ohura, N., Yamamoto, K., Ichioka, S., Sokabe, T., Nakatsuka, H., Baba, a., Shibata, M., Nakatsuka, T., Harii, K., Wada, Y., Kohro, T., Kodama, T., and Ando, J. (2003) Global analysis of shear stress-responsive genes in vascular endothelial cells. J Atheroscler. Thromb. 10:304-313.

Ong, E.S., Gao, X., Xu, N., Predescu, D., Rahman, A., Broman, M., Jho, D., and Malik, A.B. (2003) *E. coli* pneumonia induces CD18-independent airway neutrophil migration in the absence of increased lung vascular permeability. Amer. J. Physiol. 285:L879-L888.

Orozco, A.S., Zhou, X., and Filler, S.G. (2000) Mechanisms of the proinflammatory response of endothelial cells to *Candida albicans* infection. Infect. Immun. 68:1134-1141.

Paladugu, R., Fu, W., Conklin, B., Lin, P., Lumsden, A., Yao, Q., and Chen, C. (2003) Hiv Tat protein causes endothelial dysfunction in porcine coronary arteries. J. Vasc. Surg. 38:549-555.

Park, I.W., Ullrich, C., Schoenberger, E., Ganju, R., and Groopman, J. (2001) HIV-1 Tat induces microvascular endothelial apoptosis through caspase activation. J. Immun. 167:2766-2771.

Patel, G.P., Gurka, D.P., and Balk, R.A. (2003) New treatment strategies for severe sepsis and septic shock. Curr. Opinion in Crit. Care 9:390-396.

Patterson, C.E. and Lum, H. (2001) Update on pulmonary edema: the role and regulation of endothelial barrier function. Endothelium 8:75-105.

Patti, J.M., Allen, B., McGavin, M., and Hook, M. (1994) MSCRAMM-mediated adherence of microorganisms to host tissues. Ann. Rev. Microbiol. 48:585-617.

Petkova, S.B., Huang, H., Factor, S., Pestell, R., Bouzahzah, B., Jelicks, L., Weiss, L., Douglas, S., Wittner, M., and Tanowitz, H.B. (2001) The role of endothelin in the pathogenesis of Chagas' disease. Intern'l J. Parasitol. 31:499-511.

Peters, C.J. and Zaki, S.R. (2002) Role of the endothelium in viral hemorrhagic fevers. Crit. Care Med. 30:S268-S273.

Peters, K., Unger, R., Brunner, J., and Kirkpatrick, C. J. (2003) Molecular basis of endothelial dysfunction in sepsis. Cardiovasc. Res. 60:49-57.

Pfister, H.W., Fontana, A., Tauber, M., Tomasz, A., and Scheld, W.M. (1994) Mechanisms of brain injury in bacterial meningitis: workshop summary. Clin. Inf. Dis. 19:463-479.

Pingelton, W.W., Coalson, J.J., and Guenter, C.A. (1975) Significance of leukocyte s in endotoxic shock. Exp. Mol. Pathol. 22:183-194.

Pino, P., Vouldoukis, I., Kolb, J., Mahmoudi, N., Desportes-Livage, I., Bricaire, F., Danis, M., Dugas, B., and Mazier, D. (2003) *Plasmodium falciparum*-infected erythrocyte adhesion induces caspase activation and apoptosis in human endothelial cells. J. Infect. Dis. 187:1283-1290.

Poole, L.J., Yu, Y., Kim, P., Zheng, Q., Pevsner, J., and Hayward, G.S. (2002) Altered patterns of cellular gene expression in dermal microvascular endothelial cells infected with Kaposi's sarcoma-associated herpesvirus. J. Virol. 76:3395-3420.

Powers, K.A., Kapus, A., Khadaroo, R., He, R., Marshall, J., Lindsay, T., and Rotstein, O.D. (2003) Twenty-five percent albumin prevents lung injury following shock/resuscitation. Crit. Care Med. 31:2355-2363.

Powers, K.A., Kapus, A., Khadaroo, R., Papia, G., and Rotstein, O.D. (2002) 25% Albumin modulates adhesive interactions between neutrophils and the endothelium following shock/resuscitation. Surgery. 132:391-398.
Prasadarao, N.V., Srivastava, P., Rudrabhatla, R., Kim, K., Huang, S., and Sukumaran, S.K. (2003) Cloning and expression of the *Escherichia coli* K1 outer membrane protein A receptor, a gp96 homologue. Inf. Immun. 71:1680-1688.
Prasadarao, N.V., Wass, C.A., and Kim, K.S. (1997) Identification and characterization of S fimbria-binding sialoglycoproteins on brain microvascular endothelial cells. Inf. Immun. 65:2852-2860.
Reddy, K. and Ross, J.M. (2001) Shear stress prevents fibronectin binding protein-mediated *Staphylococcus aureus* adhesion to resting endothelial cells. Inf. Immun. 69:3472-3475.
Reinhart, K., Bayer, O., Brunkhorst, F., and Meisner, M. (2002) Markers of endothelial damage in organ dysfunction and sepsis. Crit. Care Med. 30:S302-S312.
Ricotta, D., Alessandri, G., Pollara, C., Fiorentini, S., Favilli, F., Tosetti, M., Mantovani, A., Grassi, M., Garrafa, E., Dei, C., Muneretto, C., and Caruso, A. (2001) Adult human heart microvascular endothelial cells are permissive for non-lytic infection by human cytomegalovirus. Cardiovasc. Res. 49:440-448.
Ring, A., Weiser, J., and Tuomanen, E.I. (1998) Pneumococcal trafficking across the blood-brain barrier. Molecular analysis of a novel bidirectional pathway. J. Clin. Invest. 102:347-360.
Rohde, M., Muller, E., Chhatwal, G.S., and Talay, S.R. (2003) Host cell caveolae act as an entry-port for group A streptococci. Cell Microbiol. 5:323-342.
Rose, F., Zeller, S., Chakraborty, T., Domann, E., Machleidt, T., Kronke, M., Seeger, W., Grimminger, F., and Sibelius, U. (2001) Human endothelial cell activation and mediator release in response to *Listeria monocytogenes* virulence factors. Inf. Immun. 69:897-905.
Rubin, D.B., Wiener-Kronish, J., Murray, J., Green, D., Turner, J., Luce, J., Montgomery, A., Marks, J., Matthay, M.A. (1990) Elevated von Willebrand factor antigen is an early plasma predictor of acute lung injury in nonpulmonary sepsis syndrome. J. Clin. Invest. 86:474-80.
Rusnati, M. and Presta, M. (2002) HIV-1 Tat protein and endothelium: from protein/cell interaction to AIDS-associated pathologies. Angiogen. 5:141-151.
Rydkina, E., Sahni, A., Silverman, D., and Sahni, S.K. (2002) *Rickettsia rickettsii* infection of cultured human endothelial cells induces heme oxygenase 1 expression. Inf. Immun. 70:4045-4052.
Saccani, S., Pantano, S., and Natoli, G. (2002) p38-Dependent marking of inflammatory genes for increased NF-kappa B recruitment. Nat. Immunol 3:69-75.
Sahni, S.K., Rydkina, E., Joshi, S., Sporn, L., and Silverman, D.J. (2003) Interactions of *Rickettsia rickettsii* with endothelial nuclear factor-kappaB in a "cell-free" system. Ann. NY Acad. Sci. 990:635-641.
Saijets, S., Ylipaasto, P., Vaarala, O., Hovi, T., and Roivainen, M. (2003). Enterovirus infection and activation of human umbilical vein endothelial cells. J. Med. Virol. 70:430-439.
Saliba, A.M., Filloux, A., Ball, G., Silva, A., Assis, M., and Plotkowski, M.C. (2002) Type III secretion-mediated killing of endothelial cells by *Pseudomonas aeruginosa*. Microb. Pathog. 33:153-166.
Santoni, G., Spreghini, E., Lucciarini, R., Amantini, C., and Piccoli, M. (2001) Involvement of alpha(v)beta3 integrin-like receptor and glycosaminoglycans in *Candida albicans* germ tube adhesion to vitronectin and to a human endothelial cell line. Microb. Pathog. 31:159-172.
Sata, M., Kakoki, M., Nagata, D., Nishimatsu, H., Suzuki, E., Aoyagi, T., Sugiura, S., Kojima, H., Nagano, T., Kangawa, K., Matsuo, H., Omata, M., Nagai, R., and Hirata, Y. (2000) Adrenomedullin and nitric oxide inhibit human endothelial cell apoptosis via a cyclic GMP-independent mechanism. Hyperten. 36:83-88.
Schmeck, B., Brunsch, M., Seybold, J., Krull, M., Eichel-Streiber, C., Suttorp, N., and Hippenstiel, S. (2003) Rho protein inhibition blocks cyclooxygenase-2 expression by proinflammatory mediators in endothelial cells. Inflamm. 27:89-95.
Schnittler, H.J. and Feldmann, H. (1999) Molecular pathogenesis of filoviral infections: role of macrophages and endothelial cells. Curr. Top. Microbiol. Imm 235:175-204.:175-204.

Schulein, R. and Dehio, C. (2002) The VirB/VirD4 type IV secretion system of Bartonella is essential for establishing intraerythrocytic infection. Mol. Microbiol. 46:1053-1067.

Schultz, M.J., Levi, M., and van der Poll, T. (2003) Anticoagulant therapy for acute lung injury or pneumonia. Curr. Drug Targets 4:315-321.

Schwarzer, N., Nost, R., Seybold, J., Parida, S., Fuhrmann, O., Krull, M., Schmidt, R., Newton, R., Hippenstiel, S., Domann, E., Chakraborty, T., and Suttorp, N. (1998) Two distinct phospholipases C of *Listeria monocytogenes* induce ceramide generation, nuclear factor-kappa B activation, and E-selectin expression in human endothelial cells. J. Immunol. 161:3010-3018.

Sears, C.L. (2000) Molecular physiology and pathophysiology of tight junctions V. assault of the tight junction by enteric pathogens. Amer. J. Physiol. 279:G1129-G1134.

Seeger, W., Birkemeyer, R., Ermert, L., Suttorp, N., Bhakdi, S., and Duncker, H.R. (1990) *Satphylococcus aureus* alpha toxin-induced vascular leakage in isolated perfused rabbit lungs. Lab. Invest. 63:341-349.

Seeger, W., Walter, H., Neuhof, H., Suttorp, N., and Bhakdi, S. (1989) *Escherichia coli* hemolysin causes thromboxane-mediated hypertension and vascular leakage in rabbit lungs. Prog. Clin. Biol. Res. 308:67-72.

Senden, N.H., Jeunhomme, T. Heemskerk, J. Wagenvoord, R., van't Veer, C., Hemker, H., and Buurman, W.A. (19989 Factor Xa induces cytokine production and expression of adhesion molecules by human umbilical vein endothelial cells. J. Imm. 161:4318-4324.

Shin, J.S. and Abraham, S.N. (2001a) Caveolae as portals of entry for microbes. Microbes. Infect. 3:755-761.

Shin, J.S. and Abraham, S.N. (2001b) Cell biology. Caveolae--not just craters in the cellular landscape. Science 293:1447-1448.

Shindo, T., Kurihara, H., Maemura, K., Kurihara, Y., Kuwaki, T., Izumida, T., Minamino, N., Ju, K., Morita, H., Oh-hashi, Y., Kumada, M., Kangawa, K., Nagai, R., and Yazaki, Y. (2000) Hypotension and resistance to lipopolysaccharide-induced shock in transgenic mice overexpressing adrenomedullin in their vasculature. Circ. 101:2309-2316.

Sinha, B., Francois, P., Que, Y., Hussain, M., Heilmann, C., Moreillon, P., Lew, D., Krause, K., Peters, G., and Herrmann, M. (2000) Heterologously expressed *Staphylococcus aureus* fibronectin-binding proteins are sufficient for invasion of host cells. Inf. Immun. 68:6871-6878.

Sinha, B., Francois, P., Nusse, O., Foti, M., Hartford, O., Vaudaux, P., Foster, T., Lew, D., Herrmann, M., and Krause, K.H. (1999) Fibronectin-binding protein acts as *Staphylococcus aureus* invasin via fibronectin bridging to integrin alpha5beta1. Cell Microbiol. 1:101-117.

Sinzger, C., Grefte, A., Plachter, B., Gouw, A., The, T., and Jahn, G. (1995) Fibroblasts, epithelial cells, endothelial cells and smooth muscle cells are major targets of human cytomegalovirus infection in lung and gastrointestinal tissues. J. Gen. Virol. 76:741-750.

Sinzger, C., Kahl, M., Laib, K., Klingel, K., Rieger, P., Plachter, B., and Jahn, G. (2000) Tropism of human cytomegalovirus for endothelial cells is determined by a post-entry step dependent on efficient translocation to the nucleus. J. Gen. Virol. 81:3021-3035.

Sinzger, C., Schmidt, K., Knapp, J., Kahl, M., Beck, R., Waldman, J., Hebart, H., Einsele, H., and Jahn, G. (1999) Modification of human cytomegalovirus tropism through propagation in vitro is associated with changes in the viral genome. J. Gen. Virol. 80:2867-2877.

Snapper, J.R., Hinson, J., Hutchison, A., Lefferts, P., Ogeltree, M., and Brigham, K.L. (1984) Effects of platelet depletion on the unanesthetized sheep's pulmonary response to endotoxemia. J. Clin. Invest. 74:1782-1791.

Steele-Mortimer, O., Knodler, L.A., and Finlay. B.B. (2000) Poisons, ruffles and rockets: bacterial pathogens and the host cell cytoskeleton. Traffic 1:107-118.

Stevens, T., Garcia, J.G., Shasby, D., Bhattacharya, J., and Malik, A.B. (2000) Mechanisms regulating endothelial cell barrier function. Amer. J. Physiol. 279:L419-L422.

Stricklett, P.K., Hughes, A., Ergonul, Z., and Kohan, D.E. (2002) Molecular basis for up-regulation by inflammatory cytokines of Shiga toxin 1 cytotoxicity and globotriaosylceramide expression. J. Inf. Dis. 186:976-982.

Sullam, P.M., Bayer, A., Foss, W., and Cheung, A.L. (1996) Diminished platelet binding in vitro by *Staphylococcus aureus* is associated with reduced virulence in a rabbit model of infective endocarditis. Inf. Immun. 64:4915-4921.
Summersgill, J.T., Molestina, R., Miller, R., and Ramirez, J.A. (2000) Interactions of *Chlamydia pneumoniae* with human endothelial cells. J. Inf. Dis. 181:S479-S482.
Suttorp, N., Buerke, M., and Tannert-Otto, S. (1992) Stimulation of PAF-synthesis in pulmonary artery endothelial cells by *Staphylococcus aureus* alpha-toxin. Thromb. Res. 67:243-252.
Suttorp, N., Fuhrmann, M., Tannert-Otto, S., Grimminger, F., and Bhadki, S. (1993a) Pore-forming bacterial toxins potently induce release of nitric oxide in porcine endothelial cells. J. Exp. Med. 178:337-341.
Suttorp, N., Galanos, C., and Neuhof, H. (1987) Endotoxin alters arachidonate metabolism in pulmonary endothelial cells. Amer. J. Physiol. 253:C384-C390.
Suttorp, N., Hessz, T., Seeger, W., Wilke, R., Koob, F., Lutz, F., and Drenckhahn, D. (1988) Bacterial exotoxins and endothelial permeability for water and albumin in vitro. Amer. J. Physiol. 255:C368-C376.
Suttorp, N., Hippenstiel, S., Fuhrmann, M.,Krull, M., and Podzuweit, T. (1996) Role of nitric oxide and phosphodiesterase isoenzyme II for reduction of endothelial hyperpermeability. Amer. J. Physiol. 270:C778-C785.
Suttorp, N., Seeger, W., Dewein, E., Bhakdi, S., and Roka, L. (1985) Staphylococcal alpha-toxin-induced PGI2 production in endothelial cells: role of calcium. Amer. J. Physiol. 248:C127-C134.
Suttorp, N., Weber, U., Welsch, t., and Schudt, C. (1993b) Role of phosphodiesterases in the regulation of endothelial permeability in vitro. J. Clin. Invest. 91:1421-1428.
Szaba, F.M.. and Smiley, S.T. (2002) Roles for thrombin and fibrin(ogen) in cytokine/chemokine production and macrophage adhesion in vivo. Blood 99:1053-1059.
Takai, Y., Sasaki, T., and Matozaki, T. (2001) Small GTP-binding proteins. Physiol. Rev. 81:153-208.
Takatsuka, H., Wakae, T., Mori, A., Okada, M., Fujimori, Y., Takemoto, Y., Okamoto, T., Kanamaru, A., and Kakishita, E. (2003) Endothelial damage caused by cytomegalovirus and human herpesvirus-6. Bone Marrow Transp. 31:475-479.
Tomescu, C., Law, W.K., and Kedes, D.H. (2003) Surface downregulation of major histocompatibility complex class I, PE-CAM, and ICAM-1 following infection of endothelial cells with Kaposi's sarcoma-associated herpesvirus. J. Virol. 77:9669-9684.
Triantafilou, M. and Triantafilou, K. (2002) Lipopolysaccharide recognition: CD14, TLRs and the LPS-activation cluster. Trends Immun. 23:301-304.
Underhill, D.M. and Ozinsky, A. (2002) Toll-like receptors: key mediators of microbe detection. Curr. Opin. Immun. 14:103-110.
Valbuena, G., Feng, H.M., and Walker, D.H. (2002) Mechanisms of immunity against rickettsiae. New perspectives and opportunities offered by unusual intracellular parasites. Microbes. Infect. 4:625-633.
Vallet, B. (2003) Bench-to-bedside review: endothelial cell dysfunction in severe sepsis: a role in organ dysfunction? Crit. Care 7:130-138.
van Der, F.M., Coenjaerts, F., Kimpen, J., Hoepelman, A., and Geelen, S.P. (2000) *Streptococcus pneumoniae* induces secretion of vascular endothelial growth factor by human neutrophils. Inf. Immun. 68:4792-4794.
Vanhoutte, P.M. and.Mombouli, J.V. (1996) Vascular endothelium: vasoactive mediators. Prog. Cardiovasc. Dis. 39:229-238.
Vazquez-Boland, J.A., Kuhn, M., Berche, P., Chakraborty, T., Dominguez-Bernal, G., Goebel, W., Gonzalez-Zorn, B., Wehland, J., and Kreft, J. (2001) Listeria pathogenesis and molecular virulence determinants. Clin. Microbiol. Rev. 14:584-640.
Verma, A., Davis, G.E., and Ihler, G.M. (2000) Infection of human endothelial cells with *Bartonella bacilliformis* is dependent on Rho and results in activation of Rho. Inf. Immun. 68:5960-5969.
Verma, A., and Ihler, G.M. (2002) Activation of Rac, Cdc42, and downstream signaling molecules by *Bartonella bacilliformis* during entry into human endothelial cells. Cell Microbiol. 4:557-569.

Walker, D.H., Valbuena, G.A., and Olano, J.P. (2003) Pathogenic mechanisms of diseases caused by Rickettsia. Ann. NY Acad. Sci. 990:1-11.
Walmrath, D., Ghofrani, H., Rosseau, S., Schutte, H., Cramer, A., Kaddus, W., Grimminger, F., Bhakdi, S., and Seeger, W. (1994) Endotoxin "priming" potentiates lung vascular abnormalities in response to *Escherichia coli* hemolysin: synergism between endo- and exotoxin. J. Exp. Med. 180:1437-1443.
Wang, P. (1998) Adrenomedullin in sepsis and septic shock. Shock 10:383-384.
Wang, Y. and Kim, K.S. (2002) Role of OmpA and IbeB in *Escherichia coli* K1 invasion of brain microvascular endothelial cells in vitro and in vivo. Ped. Res. 51:559-563.
Ware, L.B., Conner, E., and Matthay, M.A. (2001) von Willebrand factor antigen is an independent marker of poor outcome in patients with early acute lung injury. Crit. Care Med. 29:2325-2331.
Ware, L.B. and Matthay, M.A. (2000) The acute respiratory distress syndrome. New Eng. J. Med. 342:1334-1349.
Warke, R.V., Xhaja, K., Martin, K., Fournier, M., Shaw, S., Brizuela, N., De Bosch, N., Lapointe, D., Ennis, F., Rothman, A., and Bosch, I. (2003) Dengue virus induces novel changes in gene expression of human umbilical vein endothelial cells. J. Virol. 77:11822-11832.
Watanabe, T., Sugaya, M., Atkins, A., Aquilino, E., Yang, A., Borris, D., Brady, J., and Blauvelt, A. (2003) Kaposi's sarcoma-associated herpesvirus latency-associated nuclear antigen prolongs the life span of primary human umbilical vein endothelial cells. J. Virol. 77:6188-6196.
Winn, R., Maunder, R., Chi, E., and Harlan, J. (1987) Neutrophil depeletion does not prevent lung edema after endotoxin infusion in goats. J. Appl. Physiol. 62:116-121.
Wolf, K. and T. Hackstadt. (2001) Sphingomyelin trafficking in *Chlamydia pneumoniae*-infected cells. Cell Microbiol. 3:145-152.
Wyrick, P.B. (2000) Intracellular survival by Chlamydia. Cell Microbiol. 2:275-282.
Yao, L., Lowy, F.D., and Berman, J.W. (1996) Interleukin-8 gene expression in *Staphylococcus aureus*-infected endothelial cells. Inf. Immun. 64:3407-3409.
Yipp, B.G., Baruch, D., Brady, C., Murray, A., Looareesuwan, S., Kubes, P., and Ho, M. (2003a) Recombinant PfEMP1 peptide inhibits and reverses cytoadherence of clinical *Plasmodium falciparum* isolates in vivo. Blood 101:331-337.
Yipp, B.G., Robbins, S., Resek, M., Baruch, D., Looareesuwan, S., and Ho, M. (2003b) Src-family kinase signaling modulates the adhesion of *Plasmodium falciparum* on human microvascular endothelium under flow. Blood 101:2850-2857.
Zanone, M.M., Favaro, E., Conaldi, P., Greening, J., Bottelli, A., Perin, P., Klein, N., Peakman, M., and Camussi, G. (2003) Persistent infection of human microvascular endothelial cells by coxsackie B viruses induces expression of adhesion molecules. J. Immun. 171:438-446.
Zhang, J., Andrade, Z., Yu, Z., Andrade, S., Takeda, K., Sadirgursky, M., and Ferrans, V.J. (1999) Apoptosis in a canine model of acute Chagasic myocarditis. J. Mol. Cell Cardiol. 31:581-596.
Zhao, B., Bowden, R., Stavchansky, S., and Bowman, P.D. (2001) Human endothelial cell response to gram-negative lipopolysaccharide assessed with cDNA microarrays. Amer. J. Physiol. 281:C1587-C1595.
Zink, S., Nass, T., Rosen, P., and Ernst, J.F. (1996) Migration of the fungal pathogen *Candida albicans* across endothelial monolayers. Inf. Immun. 64:5085-5091.
Zoja, C., Morigi, M., and Remuzzi, G. (2001) The role of the endothelium in hemolytic uremic syndrome. J. Nephrol. 14:S58-S62.
Zysk, G., Schneider-Wald, B., Hwang, J., Bejo, L., Kim, K., Mitchell, T., Hakenbeck, R., and Heinz, H.P. (2001) Pneumolysin is the main inducer of cytotoxicity to brain microvascular endothelial cells caused by *Streptococcus pneumoniae*. Inf. Immun. 69:845-852.

Chapter 14

Chronic Lung Vascular Hyperpermeability

Geerten P. van Nieuw Amerongen,[1*] Victor W.M. van Hinsbergh,[1] and Bradford C. Berk[2]

[1]Laboratory for Physiology, Institute for Cardiovascular Research, VU Medical Center, Amsterdam, The Netherlands and [2]Center for Cardiovascular Research and Department of Medicine, University of Rochester Medical Center, Rochester, NY USA

CONTENTS:

Introduction

Stages of Endothelial Hyperpermeability
Acute Permeability
Prolonged Permeability
Chronic Responses

Mechanisms and Mediators of Endothelial Hyperpermeability
Chronic Endothelial Hyperpermeability
Thrombin
VEGF
Tumor Necrosis Factor-α
Oxidants
Hypoxia
Mechanical Forces

Chronic Hyperpermeability, Vascular Remodeling, and Transcription
Chronic Endothelial Hyperpermeability and Remodeling
β-Catenin
HIF1
NF-κB

Chronic Vascular Changes in Pulmonary Disease
ARDS
Pulmonary Hypertension
Airway Disease and Chronic Obstructive Pulmonary Disease
Vascular Leakage, VEGF, and Hypoxia
Therapeutic Possibilities

Conclusions

Introduction

During the last two decades, molecular mechanisms of immediate vascular hyperpermeability responses have largely been elucidated and signaling cascades have been resolved in great detail as described earlier in Chapters 4 and 6. Therapies aimed at intervention with these signaling pathways, however, were in general not very successful. Systematic detailed studies of the molecular mechanisms of the more clinically relevant chronic forms of vascular leakage are in the pioneering phase. These studies are making a valuable contribution to our understanding of the subsequent stages in the development of vascular leakage. We now understand, that signals of many vascular hyperpermeability inducers are transduced at the level of gene transcription, resulting in chronic vascular adaptation. Here, chronic vascular leakage is accompanied by the formation of new blood vessels ('angiogenesis') and/or the remodeling of existing ones, as exemplified in pulmonary diseases like asthma. Such observations have led to the paradigms that vascular leakage stimulates both angiogenesis, by laying down a provisional matrix, and vascular remodeling, by exposure of the underlying layers to blood-borne growth factors. In addition, an often-observed characteristic of newly-formed vessels is their leakiness. However, in several chronic pulmonary pathologies, chronic leakage and vascular remodeling do not accompany each other. In ARDS and sepsis, massive vascular leakage is observed without overt formation of new blood vessels. Conversely, dramatic remodeling of the vasculature and disordered angiogenesis are characteristics of pulmonary hypertension in the absence of clinical symptoms of vascular leakage. Understanding the signaling of these divergent phenomena offers promising targets for novel therapy. In this chapter we will discuss the chronic signaling responses involved in regulation of barrier function and dysfunction, including altered protein synthesis and gene induction, and how they relate to responses induced by mediators like thrombin, cytokines like TNFα, growth factors like VEGF, and condition like mechanical stress or hypoxia.

Stages of Endothelial Hyperpermeability

Our understanding of the processes that impair the vascular barrier function has considerably improved by the recognition that various forms of vascular leakage occur. This is most evident in inflammatory diseases like asthma, but also encountered in many other pulmonary diseases. Depending on the severity and duration of an inflammatory reaction, three main steps contribute to vascular leakage: initial acute permeability, prolongation of inflammation, and remodeling. These subsequent stages (transient -> prolonged -> chronic) of barrier dysfunction form a continuum in the development of chronic vascular leakage. Distinction of different types of vascular leakage is of importance for clinical practice as they probably require different pharmacological approaches, as will be described later.

Acute Permeability

At an early stage of inflammation, various inflammatory mediators (including vasoactive agents) induce a transient vascular leakage. In general two routes for transendothelial passage of proteins and hormones can be discriminated. In healthy endothelium macromolecules will predominantly pass *via* vesicular exchange. As discussed extensively in the preceding chapters, inflammatory mediators activate the endothelium and induce an increase in barrier permeability due to intercellular gaps that appear at cell-cell junctions, creating exchange of macromolecules *via* the minute pores. This is particularly evident in postcapillary venules, where application of one single inflammatory mediator induces a transient hyperpermeability response (minutes) (Baluk *et al.*, 1997; van Hinsbergh and Van Nieuw Amerongen, 2002). An identical response can be observed by exposing endothelial monolayers *in vitro* to vasoactive agents like histamine, bradykinin, and ATP (Van Nieuw Amerongen *et al.*, 1998). According to the classical ideas, the formation of small gaps between neighboring cells occurs *via* a Ca^{2+}-dependent MLCK-driven contractile process at the margins of the endothelial cells (Moy *et al.*, 2001). Intracellular actin–myosin interaction and loss of junctional integrity are pivotal processes in enhancing endothelial permeability. Many pharmacological agents were developed to target the histamine-type response, and were, as we now understand, far less active than was expected, since most clinical problems involve prolonged types of leakage.

Prolonged Permeability

Subsequently, activated leukocytes become involved, as described in Chapters 11 and 12. Leukocytes extravasate *via* the cell junctions (Van Nieuw Amerongen and van Hinsbergh, 2002); however, it should be noted that leukocyte extravasation does not necessarily induce vascular leakage *per se*. But the excessive and sustained presence of activated leukocytes contributes to prolongation of vascular leakage, primarily by the release of a plethora of inflammatory mediators that act on the endothelium (van Buul and Hordijk, 2004). Importantly, this inflammatory reaction can set the stage for a chronic response.

Chronic Responses

Chronic forms of pulmonary vascular leakage often are accompanied by the formation of new blood vessels from existing ones (i.e. angiogenesis) and extensive remodeling of the vasculature. In addition to the angiogenesis-related remodeling of the microvasculature (including capillaries), remodeling of larger vessels is often observed in pulmonary diseases, especially those in which the vascular resistance of the pulmonary vascular bed has increased. Chronic vascular leakage and vascular remodeling cannot be viewed as single entities, because chronic leakage only occurs under circumstances in which vascular adaptation occurs. *Vice versa*, one of the hallmarks of the process of angiogenesis *in vivo* is that, already in its initial phase, it is accompanied by an increase in endothelial permeability. This increase in permeability often results in formation of a fibrinous exudate and deposition of a provisional matrix, providing a situation favoring the ingrowth of endothelial cells. It has been thought for a long time that increased vascular permeability is 'a cardinal feature of pathological angiogenesis (Bates and Harper, 2002). Recent data support the idea that changes at the molecular level in microvascular endothelium, which contribute to increased permeability, also facilitate a pro-angiogenic state of the endothelium. However, the mutual relationship between vascular leakage and vascular remodeling is not straightforward. Chronic hyperpermeability is not *per se* accompanied by angiogenesis, and extensive remodeling of the pulmonary vasculature can occur in the absence of clinically manifest edema. Yet, both processes share similarities in underlying signaling mechanisms. It is therefore of importance to understand the differences in signaling mechanisms underlying the various pulmonary pathologies in which vascular leakage and/or vascular remodeling occur.

The dramatic remodeling in this third phase is initiated by angiogenic factors (Joris *et al.*, 1990). These angiogenic growth factors affect not only the integrity of the cell junctions by induction of endothelial migration, but some of them, particularly vascular endothelial growth factor (VEGF), induce a hyperpermeable status of the vasculature by themselves. VEGF has been shown to induce transendothelial pathways by formation of so-called vesiculo-vacuolar organelles, or VVOs, which are interconnected chains of vesicles forming a kind of a pore through endothelial cells (Dvorak *et al.*, 1996).

Mechanisms and Mediators of Endothelial Hyperpermeability

Chronic Endothelial Hyperpermeability

To understand how permeability relates to angiogenesis at the molecular level, it is necessary to delineate signaling pathways involved in development of endothelial hyperpermeability. As noted above, transient permeability primarily involves opening of paracellular gaps, but in angiogenic vessels leakage occurs both *via* open junctions and *via* pores that originate from the fusion of intracellular vesicles. A more prolonged response, more closely resembling long-lasting pathological hyperpermeability, can be induced by mediators like thrombin *in vitro* (Grand *et al.*, 1996). Four signaling pathways in the endothelium have been identified that are involved in this more prolonged type of endothelial barrier dysfunction (Alexander, 2000). 1) A transient Ca^{2+}-dependent endothelial response *via* phosphorylation of MLC by MLCK results in increased actin/myosin interaction contributing to formation of small gaps, similar to histamine-type response (Goeckeler and Wysolmerski, 1995). 2) Prolongation of this process occurs *via* RhoA and one of its distal effectors Rho kinase and its inhibition of myosin phosphatase (Essler *et al.*, 1998a; Van Nieuw Amerongen *et al.*, 1998; Knapp *et al.*, 1999; Wojciak-Stothard *et al.*, 2001). 3) Recently, a third pathway involving PKCζ in thrombin-induced permeability was reported, which is independent of Ca^{2+}- and RhoA-signaling and, interestingly, is prevented by angiopoietin-1 (Li et al., 2004). 4) Activation of protein tyrosine kinases, which probably act via the disruption of intercellular junctions and formation of focal adhesions (Andriopoulou et al., 1999; Bogatcheva et al., 2002; Nawroth et al., 2002; Van Nieuw Amerongen et al., 2004).

RhoA/Rho kinase-mediated hyperpermeability is not limited to thrombin, but also applies to permeability induced by factors like H_2O_2, VEGF, lysophosphatidic acid, and many bacterial toxins, all highly relevant to permeability in the lung (Essler et al., 1998b; Van

Nieuw Amerongen et al., 2000b; Chiba et al., 2001; Van Nieuw Amerongen et al., 2003). Rho/Rho kinase has been implicated in many vascular disorders associated with remodeling, altered cell contractility, and cell migration, including hypertension, vasospasm and vascular leakage (Van Nieuw Amerongen and Van Hinsbergh, 2001; Etienne-Manneville and Hall, 2002). Inhibition of myosin phosphatase can also sensitize endothelium to stimuli that activate MLCK and prolong their action. The finding that RhoA and Rho kinase play an important role in endothelial permeability was based on use of specific inhibitors and dominant negative and constitutive active mutants. These findings enabled us to demonstrate a potential role for the widely-used cholesterol-lowering statin drugs as modulators of vascular leakage via interference with RhoA function (Van Nieuw Amerongen et al., 2000a). Statins have been shown to reduced vascular leakage in a rat lung model of ischemia-reperfusion (Naidu et al., 2003) and in a preliminary clinical trial (Dell'Omo et al., 2000).

Recent work in epithelial cells has shown that cell-cell adhesion mediated by cadherin engagement inhibits RhoA activity *via* p120catenin, possibly contributing to stabilization of the epithelial barrier (Fukata *et al.*, 1999). Conversely, RhoA activation disruption of junctional complexes reduces barrier function. Given the importance of Rho-family GTPases in regulation of the F-actin cytoskeleton, it is probable that Rho proteins regulate junctions by inducing changes in the actin cytoskeleton. However, RhoA might also affect adherens junction proteins directly, as has been shown for the tight junction protein occludin (Hirase *et al.*, 2001). In contrast to RhoA, the related Rho-like small GTPases Rac and Cdc42 have been mainly implicated in enhancement of barrier function and recovery of disturbed barrier integrity (Wojciak-Stothard *et al.*, 2001; Lampugnani *et al.*, 2002; Waschke *et al.*, 2004). But, some studies point to a role of Rac in mediating a hyperpermeability response as well (van Wetering *et al.*, 2002; Eriksson *et al.*, 2003). How these rapid signaling events link to *chronic* hyperpermeability largely remains to be investigated, but initial results for several permeability inducers will be described below, with emphasis on changes in gene expression.

Thrombin

Thrombin activation of endothelial cells results in multiple phenotypic changes, including alterations in cell contractility, cell permeability, leukocyte trafficking, migration, proliferation, and angiogenesis. Thrombin induces a myriad of signaling pathways, as described above. In addition, the thrombin signal is also transduced at the level of gene transcription, resulting in chronic endothelial adaptation. Recently, it was shown that 12 out of 34 genes upregulated at 1 hour after thrombin challenge were transcription factors, uncovering important transcriptional networks/gene programs activated by thrombin (Minami *et al.*, 2004). Interestingly, histamine was also shown to induce upregulation of leukocyte-binding cellular adhesion molecules (Shimamura *et al.*, 2004) and alterations in gene expression in specific conditions (Diks *et al.*, 2003). However, these thrombin and histamine data have not been linked to chronic changes in vascular permeability. Furthermore, the *in vivo* observations on thrombin-induced vascular leakage may be more complex than the *in vitro* thrombin effects (Vogel *et al.*, 2000). Massive thrombin activation has been reported in bacterial infection-induced acute lung injury, a condition in which both endothelial and epithelial permeability increases dramatically (Kipnis *et al.*, 2004). High dose of antithrombin III reduced endotoxin-induced lung hyperpermeability and vascular leakage caused by ischemia–reperfusion injury in animals (Ostrovsky *et al.*, 1997; Dickneite and Kroez, 2001).

Those findings suggest an important role for thrombin at least in specific forms of chronically altered pulmonary vascular barrier function. There is a strong need to learn more about *in vivo* conditions in which thrombin is involved. Remarkably, thrombin activation of endothelial cells has also been shown to promote angiogenesis (Haralabopoulos *et al.*, 1997). Recently, these effects of thrombin were shown to be mediated in least in part by HIF1-mediated VEGF upregulation in endothelial cells (Dupuy *et al.*, 2003).

VEGF

The number of growth factors specifically acting on vasculature is quite small, the most important of them being the VEGF and angiopoietin families. They are not redundant, but instead have distinct and complementary roles, with VEGF acting early to initiate vessel growth, and angiopoietins acting subsequently to promote maturation and maintenance of vessels. VEGF-A (further indicated as VEGF) was initially identified as VPF (vascular permeability factor) and was shown in some circumstances to be one of the most potent endogenous hyperpermeability inducers, being 50,000 times more potent than histamine (Feng *et al.*, 1996). VEGF activities include endothelial proliferation, enhanced survival, migration, and tube formation. In addition, it acts as a proinflammatory cytokine inducing expression of cellular adhesion molecules (Kim *et al.*, 2001). Vessels that result from overexpression of VEGF are leaky under baseline conditions despite their relatively normal appearance, as might be expected based on the leak-producing action of VEGF. The leaky vessels are abnormally sensitive to inflammatory stimuli. Angiopoietin has the opposite effect on permeability, being barrier protective (Thurston *et al.*, 2000). These hyperpermeability and angiogenic effects of VEGF are associated with many aspects of pulmonary pathobiology, as will be discussed in the section on pulmonary disease.

Many of the molecular mechanisms underlying the chronic increase in permeability remain to be defined, but we have begun to unravel some of the signaling involved (Zachary and Gliki, 2001). PLCγ activation leads to PKC activation and thence to NF-κB, resulting in increased expression of ICAM-1, VCAM, and E-selectin (Kim *et al.*, 2001). VEGF activates several other transcription factors, including Ets1, Egr1, NFAT, Stat3 and 5 (Zachary, 2001).

Tumor Necrosis Factor-α

TNF-α induces an increase in pulmonary permeability in cultured pulmonary arterial monolayers, in the isolated lung, and *in vivo* (Lo *et al.*, 1992; Gertzberg *et al.*, 2004). TNF-α exerts some of its effects by upregulated expression of cellular adhesion molecules (CAMs) (Mulligan *et al.*, 1993). CAM expression requires an intact F-actin cytoskeleton (VandenBerg *et al.*, 2004). Furthermore, once CAMs are expressed and leukocytes bind, clustering of CAMs requires Rho kinase-dependent but not MLCK-dependent MLC phosphorylation. This provides an interesting link between the acute and chronic effects of Rho kinase activation on endothelial hyperpermeability. In addition, some of the direct effects of TNF-α are mediated by NAD(P)H oxidase-produced oxidants (Gertzberg *et al.*, 2004). Furthermore, TNF-α increases the sensitivity of pulmonary endothelial cells to oxidants *in vitro* by a reduction of intracellular glutathione content (Ishii *et al.*, 1992).

Oxidants

It is generally agreed that oxidants act as regulators of gene expression through redox-sensitive pathways. However, as outlined previously in Chapters 4 and 6, the signaling mechanisms for oxidant-induced barrier dysfunction remain controversial (Patterson and Lum, 2001; Zhao *et al.*, 2001). In addition, upregulation/surface mobilization of leukocyte-adhesion molecules (P-selectin, ICAM-1, and VCAM-1) contributes to oxidant-induced barrier dysfunction (Cooper *et al.*, 2002). Interestingly, elevation of intracellular cyclic GMP levels protects the barrier against oxidant injury in some endothelium (Lofton *et al.*, 1991).

Hypoxia

During periods of hypoxia, endothelial cells acquire a characteristic pattern of responses that can be either adaptive or pathologic (Ten and Pinsky, 2002). In general, hypoxia shifts the endothelial phenotype towards one in which leukocyte adhesion and permeability are increased and anticoagulant properties are diminished. Such mechanisms probably contribute to endothelial dysfunction in high altitude pulmonary edema (Rodway *et al.*, 2003; Bartsch *et al.*, 2004), similar to findings demonstrated for the related high-altitude cerebral edema (Xu and Severinghaus, 1998). However, exceptions exist. For example, in primary pulmonary hypertension, good evidence for hypoxia-induced endothelial barrier dysfunction is not available, despite the presence of (severe) hypoxia. A major system to sense hypoxia is the activity status of the transcription factor HIF-1, discussed below.

Mechanical Forces

The two most important mechanical forces perceived by pulmonary endothelial cells are shear stress exerted by the flowing blood and cyclic stretch caused by breathing and phasic blood flow (systole vs. diastole). Mechanical stresses are transmitted across the plasma membrane and converted into chemical signals and cytoskeletal changes *via* mechanotransduction. Well-studied candidate mechanosensors in endothelium are caveolae, G proteins, focal adhesions, ion channels, integrins, VEGFR2 and PECAM-1 (Osawa *et al.*, 2002; Katsumi *et al.*, 2004). Remarkably, both moderate shear stress and cyclic stretch have beneficial effects on endothelial cells in general, and are protective regarding the endothelial barrier function in particular; whereas, an increased extent of either one results in impairment of in endothelial function, altered gene expression and chronic barrier dysfunction (Davies *et al.*, 2003; Fujiwara, 2003).

When endothelial cells in static culture are exposed to shear stress, multiple signaling pathways implicated in cytoskeletal rearrangement are stimulated, including Ca^{2+}-mobilization, G-protein activation, increased tyrosine phosphorylation, and MLC kinase and MAP kinase activation. These pathways interact downstream to produce the complex cellular effects of flow. For example, during shear stress, the GTPases RhoA and Cdc42 combine to activate MAP kinases; however, individually, Rho is necessary for flow-induced stress fiber formation and cell alignment and Cdc42 activates transcription factors and regulates cell polarity (Katsumi et al., 2004). The integrated effects of these shear-induced signals on barrier function are variable depending on the magnitude, duration, and gradient of flow. Shear stress maximally increases protein expression of integrins after 12 h of exposure and significantly enhances cell-matrix attachment, suggesting that flow helps to maintain the EC monolayer through augmentation of focal adhesions. However, endothelial cells exposed to high shear gradients, or turbulent flow, develop increased permeability relative to areas of either constant laminar flow or no flow. The majority of these studies have been performed using cells obtained from the systemic circulation, and pulmonary endothelial cell-specific responses to flow are not well understood. One other mechanism by which shear stress may alter barrier function is by inhibition of endothelial apoptosis (Haga et al., 2003).

Breathing causes lung tissue, and thus the endothelium of lung tissue, to stretch cyclically (Fujiwara, 2003). Ventilator-induced lung injury is a highly morbid clinical entity that is believed to be caused by excessive mechanical stretch of pulmonary airways and vasculature, producing fluid flux across capillaries primarily through an active endothelial response. Pathologically relevant levels of cyclic stretch enhanced, whereas physiological levels of cyclic stretch protected against a thrombin assault both in magnitude of barrier disruption and in barrier recovery time (Birukov et al., 2003). Interestingly, chronic cyclic stretch induces several genes, including the gene coding for the small GTPase RhoA, known to be involved in regulation of barrier function.

Chronic Hyperpermeability, Vascular Remodeling, and Transcription

Chronic Endothelial Hyperpermeability and Remodeling

As outlined above, in inflammation-related vascular leakage, an early phase can be distinguished from a more chronic phase. In general one can say, that in the early phase blood vessels dilate and become hyperpermeable and diapedesis of leukocytes occurs. In the late phase more structural changes in the microvessels occur. There is remodeling of existing vessels and formation of new vessels from existing ones (angiogenesis). Also in the case of endothelial activation by VEGF, permeability and edema formation is an initial event, followed by the formation of new blood vessels. These findings suggest a relationship between permeability and vascular remodeling. Indeed, microvascular hyperpermeability plays a mechanistic role in angiogenesis. In fact, angiogenesis was believed for a long time to be accompanied in almost all states by increases in vascular permeability (Bates and Harper, 2002). The original observation of this link was the leak of India ink from capillaries growing into a recent wound (Abell, 1946). The most extensive evidence for an association between microvascular hyperpermeability and angiogenesis comes from studies of solid tumors. Tumor vessels are hyperpermeable and the plasma proteins, which extravasate from these leaky vessels, form a new provisional extravascular fibrin matrix that permits and indeed favors the inward migration of endothelial cells.

This mechanism is probably not specific to formation of tumor blood vessels, but also applies to pulmonary vascular remodeling, as many similarities exist. Many pulmonary pathologies are characterized by remodeling of the existing vasculature, as will be discussed in the section on pulmonary diseases. Not only does a leaky endothelial monolayer favor angiogenesis and vascular remodeling, but the remodeled and/or newly formed vessels themselves are also leaky, aggravating plasma extravasation. Several factors are likely to participate in the leakiness of remodeled blood vessels (McDonald, 2001). 1) There is an increase in endothelial permeability from focal separations ~400 nm in diameter between the endothelial cells. 2) There is an increase in luminal surface area created by angiogenesis and microvascular enlargement. 3) The enlargement of arterioles may lower upstream resistance and increase the transmural driving force for leakage. Remodeled pulmonary vessels in rats after *M. pulmonalis* infection were abnormally sensitive to stimuli that evoke plasma leakage (McDonald, 2001). Depending on the strain, proliferating endothelial cells in capillaries convert to venules or form new capillaries upon *M. pulmonalis* infection. The remodeled capillaries and venules were the sites of focal plasma leakage and leukocyte adherence.

Thus, increased vascular permeability is now recognized as a cardinal feature of pathological angiogenesis (Bates & Harper 2002). As outlined, endothelial hyperpermeability favors the formation of new blood vessels; yet unexpectedly, chronic pulmonary vascular leakage does not always result in angiogenesis. And a clinically manifest edema is not always observed in all cases of lung angiogenesis. The absence of (clinically manifest) edema as evidence for an intact vascular barrier function, however, has to be interpreted with caution, as it does not guarantee barrier integrity *per se*. First of all, as long as clearance *via* lymphatics suffices, edema will not necessarily form, despite a disturbed endothelial barrier. Second, methods used to measure leakage might not be sensitive enough to detect focal sites of leakage (McDonald, 2001). In addition, leakage occurring at the micro scale may have profound effects in the absence of edema. Despite these restrictions, good evidence is available, that not all forms of chronic leakage are accompanied by angiogenesis and *vice versa*. Two such examples in which this coexistence is absent: ARDS and pulmonary hypertension will be described under diseases. The angiogenesis, which occurs in many chronic pulmonary diseases, involves chronic changes in endothelial cells and altered gene expression. A remarkable common feature of chronic vascular leakage and angiogenesis is activation of similar gene programs by shared transcription factors and some of the most relevant transcription factors will be discussed.

β-Catenin

Control of junctional integrity and gap formation is governed by the tethering forces of junctional proteins and locally acting contracting forces in the cell. In contrast to the belt of tight junctions in epithelial cells and brain endothelial cells, which largely determine barrier properties of these cells, the tight junctions of most endothelial cells are incomplete mosaic-like structures with a more limited barrier function, as discussed in Chapter 9. Whereas in endothelium, adherens junctions are thought to play a more important role in the regulation of permeability and disruption of this complex results in loss of barrier function (Lampugnani *et al.*, 1992; Carmeliet *et al.*, 1999; Firth, 2002). The intracellular domain of VE-cadherin is linked to the cytoskeleton *via* catenins, of which β-catenin has particular importance in chronic signaling. β-catenin knock-out mice suffer from paracellular endothelial permeability edema, resulting in fluid accumulation in the pericardial cavity (Cattelino *et al.*, 2003). In addition, disruption of junctional complexes favors the translocation of β-catenin to the nucleus, where β-catenin (together with other transcription factors) regulates expression of genes associated with migration and proliferation.

Recent studies suggest, for example, that VEGF can destabilize endothelial β-catenin association with catenin, increasing its cytoplasmic stabilization and nuclear signaling (Behzadian et al., 2003). This is of interest because β-catenin is also generated by the activation of Wnt-frizzled signal transduction (Bienz and Clevers, 2003). In the canonical Wnt signaling pathway, β-catenin accumulates in the cytoplasm and translocates into the nucleus. During human embryonic development, when vessels proliferate, nuclear and/or cytoplasmic β-catenin can be seen in several parts of the vascular bed (Goodwin and D'Amore, 2002). Cells of normal the adult vasculature rarely if ever accumulate β-catenin in the cytoplasm or nucleus (Goodwin and D'Amore, 2002). In several non-endothelial cell lines β-catenin nuclear signaling was initiated by the inflammatory molecule, histamine (Diks et al., 2003). An increasing body of evidence indicates that β-catenin is frequently observed in the cytoplasm and nucleus of vascular cells during angiogenesis and vascular remodeling in disease states. Among the proteins known to be up-regulated by β-catenin, matrix metalloproteinase-7, cyclooxygenase-2, cyclin D1, c-myc and fibronectin are of particular interest for vascular remodeling (Goodwin and D'Amore, 2002). Thus, β-catenin might form an important link between (chronic) hyperpermeability and vascular remodeling/angiogenesis (Dixelius et al., 2003).

HIF-1α

An important regulatory system that responds to hypoxia is the transcription factor HIF-1α-mediated signaling cascade (Schofield and Ratcliffe, 2004). Under normoxic conditions, HIF-1α is hydroxylated on a conserved prolyl residue in a reaction with molecular oxygen. The hydroxylated prolyl group allows the Von Hippel-Lindau protein to bind and polyubiquinate the molecule, thereby earmarking HIF-1α for destruction by the proteasome. However, in hypoxia, this hydroxyprolination is reduced, allowing HIF-1α to accumulate and translocate to the nucleus to activate transcription of genes that assist in adapting the organism to an environment in which oxygen is limiting, such as VEGF (explaining many key aspects of hypoxia-related lung diseases), several angiopoietins, EPO, and inducible NOS (Yamakawa et al., 2003). More than 40 HIF-target genes have been characterized and large scale gene expression arrays indicate that several hundreds of genes are positively or negatively regulated by hypoxia (whether direct or indirect *via* HIF1 is unclear). The main HIF targets are angiogenic signaling, energy metabolism, cell migration, growth and apoptosis (Semenza, 2003). It is of interest to note that RhoA is upregulated by hypoxia probably *via* HIF-1 (McMurtry et al., 2003). Inhibition of Rho kinase attenuates the effects of acute and chronic hypoxia on the pulmonary circulation of mice (Fagan et al., 2004), possibly by diminishing vasoconstriction and remodeling. Mice with chronic and acute hypoxia have increased HIF mRNA (Wiener et al., 1996). Other transcription factors known to be induced by hypoxia include AP-1, nuclear factor IL-6 (NF-IL6), and early growth response gene 1 (Egr1) that control transcriptional activation of genes encoding growth factors and other mediators of mitogenesis (Faller, 1999; Semenza, 2000; Jeffery and Wanstall, 2001).

NF-κB

Most of the genes activated *in vitro* during the endothelial stress response are controlled by at least two transcription factor families: AP-1 and NF-κB. NF-κB has gained wide interest as a target in inflammatory diseases as it seems to be invariably upregulated. A variety of agents including cytokines and reactive oxygen stress initiate signal transduction events that lead to phosphorylation of IκB, which results in ubiquination of IκB, its dissociation from the complex, and subsequent degradation by the 26S proteasome. IκB – dissociation allows NF-κB to translocate to the nucleus and initiate gene transcription (Lentsch and Ward, 2001). NF-κB activation in the endothelium results in the expression of adhesion molecules and the release of chemotactic factors for inflammatory cells. As a consequence, leukocytes bind and accumulate in the affected vessel and contribute to local activation and leakage. *In situ* nuclear translocation of NF-κB has been found in the vessel wall near regions of disturbed blood flow, like bifurcations, curvatures, and branching points. In the lung, NF-κB has been shown to be important in many inflammatory diseases including ARDS (Lentsch and Ward, 2001). In conclusion, transcription factors share the common theme of cytoplasmic stabilization upon activation, followed by targeting to the nucleus where they activate gene transcription programs.

Chronic Vascular Changes in Pulmonary Disease

Thus, hyperpermeability can be both the forerunner and the consequence of vascular remodeling. Although, as noted above, in some conditions the connection between permeability and angiogenesis is not readily evident. The bearing of these observations on several important pulmonary diseases will be explored.

ARDS

The pathologic features of the lung in ARDS derive from severe injury of the alveolar-capillary unit (Tomashefski, 2000; Ware and Matthay, 2000). Extravasation of intravascular fluid dominates the onset of the disease and the term ARDS is often simplistically equated with permeability pulmonary edema. As the process unfolds, however, edema is overshadowed by cellular necrosis, inflammation, epithelial hyperplasia, and fibrosis. The damage can be divided in three interrelated and overlapping phases of the disease: 1) the exudative phase of edema and hemorrhage; 2) the proliferative phase of organization and repair; and 3) end-stage fibrosis (Tomashefski, 2000). In the exudative phase there is evidence of endothelial injury such as endothelial swelling and widening of inter-endothelial junctions. Frequently the extent of interstitial edema, however, is disproportionate to the mild ultrastructural changes in a generally intact endothelial layer. The subtle morphological changes in the endothelial cells may however underlie a significant loss of barrier integrity.

In the proliferative phase, permanent pulmonary structural changes and extensive remodeling of the vascular bed are observed. There is severe destruction of the capillary network, with fibrous obliteration of microcirculation. There is arterial muscularization, with fibrocellular intimal proliferation and arterial tortuosity. Increased contraction of vessels in patients with late stage ARDS probably does not result from arteriogenesis, but represents the combined effect of abnormally dilated vessels and increased profiles of tortuous vessels. Although endothelial cells may proliferate, evidence for angiogenesis is scarce or absent.

Thus, we have seen that in ARDS massive vascular leakage occurs in the absence of angiogenesis. Evidence is available that angiogenesis is even suppressed in ARDS. Increased levels of angiostatin, a powerful inhibitor of angiogenesis *in vivo*, were measured in bronchoalveolar lavage fluids from ARDS patients (Luca *et al.*, 2002). This finding suggests that angiostatin may contribute to endothelial damage probably *via* an increased permeability of the alveolar capillary barrier. The contribution of VEGF and the VEGFR2 to acute lung injury is well established (Lassus *et al.*, 2001). Thus, the expression of VEGF might be involved in the pathogenesis of pulmonary vascular remodeling at early stages of this disease. Furthermore, plasma levels of VEGF are elevated in experimental animals as well as in the clinical setting, but local lung VEGF levels are reduced in ARDS (Bhatia and Moochhala, 2004). Also, treatment of mice with VEGF actually prevented fatal respiratory distress in premature mice (Compernolle *et al.*, 2002). More work is clearly needed.

Pulmonary Hypertension

Pulmonary remodeling is an important feature of pulmonary hypertension, leading to increased pulmonary resistance and reduced compliance. Stimuli responsible for remodeling include transmural pressure, stretch, shear stress, hypoxia, and stimulation by vasoactive agents like growth factors and inflammatory mediators, many of them being hyperpermeability inducers as well. The pathogenesis of pulmonary hypertension involves a complex and multifactorial process and pulmonary hypertension is a poorly understood disease (Fedullo *et al.*, 2001; Newman *et al.*, 2004). Pulmonary hypertension is a progressive, life threatening disorder, characterized by raised pulmonary arterial pressure due to pathological changes in structure and function of pre-capillary arterioles, and the formation of so-called plexiform lesions ultimately leading to cardiac right ventricular hypertrophy. Vascular remodeling involves all three layers of the blood vessel wall, the adventitia, the media and the intima due to hypertrophy and hyperplasia (Jeffery and Wanstall, 2001). Endothelial dysfunction seems to play an integral role in mediating the structural changes in the pulmonary vasculature, both in initiation and in progression of severe pulmonary hypertension. Dysfunction of endothelial cells is reflected in what is called 'disordered proliferation and angiogenesis', the former contributing to arteriolar remodeling, the latter resulting in formation of plexiform lesions. In addition to extensive vascular remodeling in pulmonary hypertension, recent evidence suggests occurrence of angiogenesis as well. Chronic hypoxia was shown to lead to increased blood vessel volume and length in the pulmonary circulation, increased capillary surface area, and an increased number of vascular endothelial cells when compared with control animals maintained in normal oxygen conditions. These results indicate that new vessel formation had occurred in response to hypoxia and is the first reported demonstration of hypoxia-induced angiogenesis in the adult pulmonary circulation (Howell *et al.*, 2003). Notably, this extensive remodeling and disordered angiogenesis occurs without overt endothelial hyperpermeability.

The endothelium plays is critical role in pulmonary vascular remodeling. First, it acts as a physical barrier, protecting the underlying smooth muscle from blood-borne mediators (Dawes *et al.*, 1994). Second, it detects both alterations in mechanical forces that are generated by the flowing blood and blood pressure (Davies *et al.*, 2003) and reductions in the ambient oxygen tension. Third, many of the mediators of vascular remodeling are produced and released from the endothelium, such as the antimitogenic substances - NO and prostacyclin or pro-mitotic substances such as PDGF. All of them potentially can be involved in the development of pulmonary hypertension. Evidence for a clinically detected disturbed endothelial barrier in pulmonary hypertension is limited; however, Rabinovitch proposed a model in which 'serum leak' contributes to development of pulmonary hypertension (Rabinovitch, 2001). In this model the physiological function of endothelial cells to shield the underlying layers from growth factors in the blood is compromised by the endothelial injury. This results in leakage of proliferative mediators, such as fibroblast growth factor, into the sub-endothelial tissue (Rabinovitch, 2001). The exuberant production of VEGF by alveolar epithelium in response to hypoxia or from inflammatory mediators, cytokines and oxidants, initiates and/or exacerbates the leakage and the response. Endogenous vascular elastase may permeate the vascular wall and initiate, *via* degradation of matrix elements, growth signals to medial smooth muscle cells. In additionally, as noted above, activated endothelial cells produce a number of factors that stimulate smooth muscle proliferation. Together with shear stress-induced endothelial matrix protein synthesis and adaptive hypertrophy of smooth muscle cells and adventitia, these phenomena constitute 'remodeling' (Budhiraja *et al.*, 2004).

HIF mediates the angiogenic response to hypoxia by upregulating the expression of multiple angiogenic factors. Activation of the angiopoietin/Tie-2 system may play a role in the ability of HIF to induce hypervascularity without causing excessive permeability. Transgenic expression of a constitutively stable HIF-1 mutant increases the number of vessels without vascular leakage, tissue edema or inflammation, by upregulating multiple angiogenic factors, including Ang4. This might explain the differences observed in several lung pathologies. Many of them are accompanied by increased VEGF expression, but not all of them display increased permeability. However, the initial observation that HIF induces hypervascularity without excessive permeability indicated that this leakage resistance was not caused by up-regulation of angiopoietin-1 or angiopoietin-2 (Elson *et al.*, 2001). Other factors upregulated by HIF that might contribute to the leakage resistance include VEGFR1, NOS-2, and adrenomedullin (Bruick and McKnight, 2001).

Overexpression of angiostatin, an angiogenesis inhibitor, in the lung by intratracheal adenovirus administration, aggravates pulmonary hypertension in chronically hypoxic mice (Pascaud *et al.*, 2003). This suggests that lung angiogenesis counteracts development of hypoxic pulmonary hypertension. Expressions of VEGF (by alveolar epithelium, lung endothelium and platelets), HIF1α, and VEGFR-2 are enhanced in pulmonary arteries of patients with severe pulmonary hypertension especially in the endothelial plexiform lesions. It is therefore proposed that these lesions may form by a process of disordered angiogenesis (Tuder *et al.*, 2001). Gene transfer of VEGF was also shown to reduce pulmonary hypertension in rats (Campbell *et al.*, 2001). This demonstrates the complex and dual nature of VEGF. VEGF blockade results in severe pulmonary hypertension and overexpression of VEGF is protective against the disease; whereas, VEGF contributing to the etiology of the disease by stimulating plexiform lesion formation (Tuder *et al.*, 2001).

Airway Disease and Chronic Obstructive Pulmonary Disease

Plasma leakage is a key feature of chronic inflammatory airway diseases and is associated with edema in the airway wall and narrowing of the airway lumen. For instance, excessive plasma extravasation may aggravate life-threatening obstruction of respiratory airways during asthma and related pulmonary disorders (Groeneveld, 2002). Plasma extravasation and airway edema are thought to be a part of the inflammatory response in asthma. Airway vascularity is significantly increased in patients with asthma and correlates with mast cell number. Airway hypervascularity was significantly reduced by treatment with corticosteroids (Chetta *et al.*, 2003). Some studies showed a relationship between vascular remodeling and asthma severity although, but not all studies agree at this point (Salvato, 2001). These findings support the view that angiogenesis and vascular remodeling of the airway wall can be one of the main aspects characterizing the chronic inflammatory changes in bronchial asthma (Chetta *et al.*, 2003). In asthmatic airways, newly generated blood vessels are leaky, immature and unstable. Those alterations contribute to the subsequent thickening of the airway wall mucosa. The ultimate result is a narrowing of the airway lumen (Wilson and Kotsimbos, 2003).

VEGF expression increases in the airways of subjects with asthma and correlates with mucosal vascularity (Hoshino *et al.*, 2001). Eosinophilic bronchitis shares a similar eosinophilic inflammation with asthma, but in the former there is no airflow obstruction, whereas VEGF release was only elevated in the latter (Kanazawa *et al.*, 2004). VEGF is thought to cause leakage of the mucosal and submucosal capillary beds and to induce airway wall thickness. Likewise, VEGF expression in pulmonary arteries is increased in early phases of COPD and smokers and correlates with vessel wall thickness (Santos *et al.*, 2003). Furthermore, a low degree of chronic leakage can also occur very locally, and as such can cause a burden for the distal tissue, including the arterial wall. This is observed after oxidant or metabolic stress, such as in cigarette smokers and diabetic patients (Hansen *et al.*, 1992; Parving *et al.*, 1996).

Vascular Leakage, VEGF, and Hypoxia

Finally, VEGF expression is upregulated in donor lung grafts. Transplantation-induced hypoxia increased vascular permeability in such donor lung grafts (Abraham *et al.*, 2004). Donor graft hypoxia is a major cause of injury after lung transplantation and pulmonary edema is the main clinical manifestation (Arcasoy and Kotloff, 1999). It has been shown that increased vascular permeability under those conditions is associated with upregulated VEGF-A in human donor grafts. In addition, upregulated HIF-1 protein expression and HIF-1 binding to HIF-binding site of the VEGF promotor suggest an important role for hypoxia and HIF-1 in mediating VEGF upregulation in donor lung tissue (Abraham *et al.*, 2004).

In conclusion, we are at the beginning of understanding vascular leakage in molecular terms, but still we are far from giving definitive answers or predicting when vascular leakage results will be accompanied by angiogenesis and *vice versa*. Yet, good evidence exists that the balance between pro-leakage and pro-angiogenic effects of mediators like VEGF on the one hand and the anti-leakage effects of angiopoietins and anti-angiogenic effects of mediators like angiostatin on the other hand determines final outcome.

Therapeutic Possibilities

Advances in the treatment of pulmonary vascular leakage have been reviewed recently and will be discussed in Chapter 15 (Groeneveld, 2002; Matthay *et al.*, 2003). In general, the clinical results of selective receptor antagonists are poor, undoubtedly because there are many different inflammatory mediators that act in concert to sustain inflammation and endothelial hyperpermeability. Thus, it is unlikely that a selective receptor agonist will be the practical solution to completely prevent leakage in chronic disease states. Antihistamines have an advantage in this regard over other selective receptor antagonists, as they act as a modulator of several aspects of the inflammatory process (Van Nieuw Amerongen and van Hinsbergh, 2002; Togias, 2003). In addition to preventing the histamine effects on the endothelium, they interfere with airway smooth muscle contraction, vascular tone, production of cytokines and antibodies, and antigen receptor-mediated proliferative responses. Glucocorticoids are another exception, as they act as broad spectrum inhibitors under inflammatory conditions (Masferrer and Seibert, 1994). Glucocorticoids also reversed the leakiness of airway vasculature produced by *M. pulmonalis* infection, indicating that even severe structural remodeling and leakiness in inflammation of the airway microvasculature are reversible (McDonald, 2001). This is of major importance, as we have seen that hyperpermeability inducers like VEGF play an important role in inflammatory remodeling (Hoshino *et al.*, 2001; Salven *et al.*, 2002). However, the toxicity and side effects of glucocorticoids are often very problematic and they certainly are not successful in all cases. Although corticosteroids worked in some animal models of ARDS (Metz and Sibbald, 1991), human studies on early administration of steroids in ARDS have largely failed to show efficacy (Groeneveld, 2002). More specific measures interfering with proinflammatory responses include administration of anti-TNFα agents, PAF blockers, or other immunomodulating compounds, which have been successful in amelioration of sepsis-induced lung injury in animals (Chang *et al.*, 1987), yet these compounds were also ineffective in clinical trials for sepsis (Reinhart *et al.*, 2001). β2-adrenergic receptor agonists are used in the treatment of chronic airway disease and are known to inhibit plasma leakage evoked by a variety of stimuli including antigens, substance P, bradykinin, and PAF (Rippe and Grega, 1978; McDonald, 2001). The second messenger cyclic AMP mediates the barrier protection and is one of few endogenous molecules known to enhance barrier integrity (Langeler and van Hinsbergh, 1991; Patterson and Lum, 2001). However, effects on other cell types cannot be excluded, for example, β2-adrenergic agonists can reduce the release of leak-inducing inflammatory mediators from sensory nerves and mast cells (McDonald, 2001). Moreover, dobutamine and dopexamine, ameliorated murine lung injury by intratracheal endotoxin installation by decreasing proinflammatory cytokine release and neutrophils entrapped in the injured lungs (Dhingra *et al.*, 2001).

The recent insight that different types of vascular leakage exist, even within the development of a single disease will be of tremendous help in designing novel therapies. Now we understand why Rho kinase inhibitors are less effective in experimental histamine-induced vascular, are partially effective in leukotriene-induced leakage, and are highly effective in H_2O_2- and bacterial toxin-induced lung edema (Chiba *et al.*, 2001; Tokuyama *et al.*, 2002). Availability of effective tools for treatment of pulmonary vascular leakage is still limited and a strong need exists to learn more about the mechanisms underlying chronic forms of vascular leakage. Promising novel targets for therapy have been identified. Enhancement of the endogenous barrier stabilizers angiopoietin-1, adrenomedullin, and sphingosine 1-

phosphate and inhibition of the protein kinases Src, Rho kinase, and PKCβ may aid in treatment of chronic hyperpermeability (Van Nieuw Amerongen and van Hinsbergh, 2002). These and other potential advances for treatment and prevention of acute and chronic endothelial barrier dysfunction will be discussed in the following chapter.

Conclusions

In conclusion, it has become clear now, that signals of many vascular hyperpermeability inducers are transduced at the level of gene transcription, resulting in chronic vascular adaptation. Translating these fascinating findings into novel pharmacological agents for the treatment of chronic forms of vascular leakage remains a major challenge for the near future. Recent studies have provided valuable advances in our understanding of the subsequent stages in the development of vascular leakage, and of the different types of vascular leakage that can be discriminated. The statement of McDonald, in discussing vascular leakage in airways, holds true for many pulmonary diseases with endothelial involvement: "mechanisms and therapeutic implications of alterations in airway blood vessels are just beginning to be elucidated and changes in the microvasculature still represent an important gap in the understanding of the pathophysiology of asthma and other chronic inflammatory airway diseases" (McDonald, 2001). We are at the beginning of understanding vascular leakage in molecular terms, but still we are far from giving definitive answers or predicting when vascular leakage will result and when it will be accompanied by angiogenesis and *vice versa*. Yet, good evidence exists that the balance between pro-leakage and pro-angiogenic effects of mediators like VEGF on the one hand and the anti-leakage effects of angiopoietins and anti-angiogenic effects of mediators like angiostatin on the other hand determines final outcome.

ACKNOWLEDGEMENTS:

GPvNA was supported by a grant from the NHS (T2003-0032). Critical reading of the manuscript by A. Vonk Noordegraaf M.D. Ph.D. (Dept Pulmonology, VU medical center, Amsterdam) was greatly appreciated.

REFERENCES

Abell, R.G. (1946) The permeability of blood capillary sprouts and newly formed blood capillaries as compared to older capillaries. Amer. J. Physiol. 147:231-241.
Abraham, D., Krenn, K., Seebacher, G., Paulus, P., Klepetko, W., and Aharinejad, S. (2004) Upregulated HIF-1 DNA binding activity to the VEGF-A promoter mediates increased vascular permeability in donor lung grafts. Ann. Thorac. Surg. 77:1751-1755.
Alexander, J.S. (2000) Rho, tyrosine kinase, Ca(2+), and junctions in endothelial hyperpermeability. Circ. Res. 87:268-271.

Andriopoulou, P., Navarro, P., Zanetti, A., Lampugnani, M.G., and Dejana, E. (1999) Histamine induces tyrosine phosphorylation of endothelial cell-to-cell adherens junctions. Arterioscler. Thromb. Vasc. Biol. 19:2286-2297.

Arcasoy, S.M. and Kotloff, R.M. (1999) Lung transplantation. N. Eng. J. Med. 340:1081-1091.

Baluk, P., Hirata, A., Thurston, G., Fujiwara, T., Neal, C.R., Michel, C.C., and McDonald, D.M. (1997) Endothelial gaps: time course of formation and closure in inflamed venules of rats. Amer. J. Physiol. 272:L155-L170.

Bartsch, P., Swenson, E., and Maggiorini, M. (2004) Update: High altitude pulmonary edema. Adv Exp Med Biol XX:89-106.

Bates, D.O. and Harper, S.J. (2002) Regulation of vascular permeability by vascular endothelial growth factors. Vascul. Pharmacol. 39:225-237.

Behzadian, M.A., Windsor, L., Ghaly, N., Liou, G., Tsai, N., and Caldwell, R.B. (2003) VEGF-induced paracellular permeability in cultured endothelial cells involves urokinase and its receptor. FASEB J. 17:752-754.

Bhatia, M. and Moochhala, S. (2004) Role of inflammatory mediators in the pathophysiology of acute respiratory distress syndrome. J. Pathol. 202:145-156.

Bienz, M. and Clevers, H. (2003) Armadillo/beta-catenin signals in the nucleus--proof beyond a reasonable doubt? Nat. Cell Biol. 5:179-182.

Birukov, K., Jacobson, J., Flores, A., Ye, S., Birukova, A., Verin, A., and Garcia, J.G. (2003) Magnitude-dependent regulation of pulmonary endothelial cell barrier function by cyclic stretch. Amer. J. Physiol. Lung Cell. Mol. Physiol. 285:L785-L797.

Bogatcheva, N.V., Garcia, J.G., and Verin, A.D. (2002) Role of tyrosine kinase signaling in endothelial cell barrier regulation. Vasc. Pharmacol. 39:201-212.

Bruick, R.K. and McKnight, S.L. (2001) Building better vasculature. Genes Dev. 15:2497-2502.

Budhiraja, R., Tuder, R.M., and Hassoun, P.M. (2004) Endothelial dysfunction in pulmonary hypertension. Circ. 109:159-165.

Campbell, A.I., Zhao, Y., Sandhu, R., and Stewart, D.J. (2001) Cell-based gene transfer of VEGF attenuates monocrotaline-induced pulmonary hypertension. Circ. 104:2242-2248.

Carmeliet, P., Lampugnani, M., Moons, L., Breviario, F., Compernolle, V., Bono, F., Balconi, G., Spagnuolo, R., Oostuyse, B., Dewerchin, M., Zanetti, A., Angellilo, A., Mattot, V., Nuyens, D., Lutgens, E., Clotman, F., de Ruiter, M., Gittenberger-de, G., Poelmann, R., Lupu, F., Herbert, J., Collen, D., and Dejana, E. (1999) Targeted deficiency or cytosolic truncation of the VE-cadherin gene in mice impairs VEGF-mediated endothelial survival and angiogenesis. Cell 98:147-157.

Cattelino, A., Liebner, S., Gallini, R., Zanetti, A., Balconi, G., Corsi, A., Bianco, P., Wolburg, H., Moore, R., Oreda, B., Kemler, R., and Dejana, E. (2003) The conditional inactivation of the beta-catenin gene in endothelial cells causes a defective vascular pattern and increased vascular fragility. J. Cell Biol. 162:1111-1122.

Chang, S.W., Feddersen, C., Henson, P., and Voelkel, N.F. (1987) Platelet-activating factor mediates hemodynamic changes and lung injury in endotoxin-treated rats. J. Clin. Invest. 79:1498-1509.

Chetta, A., Zanini, A., Foresi, A., Del Donno, M., Castagnaro, A., D'Ippolito, R., Baraldo, S., Testi, R., Saetta, M., and Olivieri, D. (2003) Vascular component of airway remodeling in asthma is reduced by high dose of fluticasone. Amer. J. Respir. Crit. Care Med. 167:751-757.

Chiba, Y., Ishii, Y., Kitamura, S., and Sugiyama, Y. (2001) Activation of rho in the mechanism of H_2O_2-induced lung edema in isolated perfused rabbit lung. Microvasc. Res. 62:164-171.

Compernolle, V., Brusselmans, K., Acker, T., Hoet, P., Tjwa, M., Beck, H., Plaisance, S., Dor, Y., Keshet, E., Lupu, F., Nemery, B., Dewerchin, M., Van Veldhoven, P., Plate, K., Moons, L., Collen, D., and Carmeliet, P. (2002) Loss of HIF-2α and inhibition of VEGF impair fetal lung maturation, whereas VEGF prevents fatal respiratory distress in premature mice. Nat. Med. 8:702-710.

Cooper, D., Stokes, K., Tailor, A., and Granger, D.N. (2002) Oxidative stress promotes blood cell-endothelial cell interactions in the microcirculation. Cardiovasc.Toxicol. 2:165-180.

Davies, P.F., Zilberberg, J., and Helmke, B.P. (2003) Spatial microstimuli in endothelial mechanosignaling. Circ. Res. 92:359-370.

Dawes, K.E., Peacock, A., Gray, A., Bishop, J., and Laurent, G.J. (1994) Characterization of fibroblast mitogens and chemoattractants produced by endothelial cells exposed to hypoxia. Amer. J. Respir. Cell Mol. Biol. 10:552-559.

Dell'Omo, G., Bandinelli, S., Penno, G., Pedrinelli, R., and Mariani, M. (2000) Simvastatin, capillary permeability, and acetylcholine-mediated vasomotion in atherosclerotic, hypercholesterolemic men. Clin. Pharm. Ther. 68:427-434.

Dhingra, V.K., Uusaro, A., Holmes, C., and Walley, K.R. (2001) Attenuation of lung inflammation by adrenergic agonists in murine acute lung injury. Anesthesiol 95:947-953.

Dickneite, G. and Kroez, M. (2001) Treatment of porcine sepsis with high-dose antithrombin III reduces tissue edema and effusion. Blood Coagul. Fibrinol. 12:459-467.

Diks, S.H., Hardwick, J., Diab, R., van Santen, M., Versteeg, H., van Deventer, S., Richel, D., and Peppelenbosch, M.P. (2003) Activation of the canonical beta-catenin pathway by histamine. J. Biol. Chem. 278:52491-52496.

Dixelius, J., Cross, M., Matsumoto, T., and Claesson-Welsh, L. (2003) Endostatin action and intracellular signaling: beta-catenin as a potential target? Cancer Lett. 196:1-12.

Dupuy, E., Habib, A., Lebret, M., Yang, R., Levy-Toledano, S., and Tobelem, G. (2003) Thrombin induces angiogenesis and VEGF expression in human endothelial cells: possible relevance to HIF-1alpha. J. Thromb. Haemost. 1:1096-1102.

Dvorak, A.M., Kohn, S., Morgan, E., Fox, P., Nagy, J., and Dvorak, H.F. (1996) The vesiculo-vacuolar organelle (VVO): a distinct endothelial cell structure that provides a transcellular pathway for macromolecular extravasation. J. Leukoc. Biol. 59:100-115.

Elson, D.A., Thurston, G., Huang, L., Ginzinger, D., McDonald, D., Johnson,R., and Arbeit, J.M. (2001) Induction of hypervascularity without leakage or inflammation in transgenic mice overexpressing hypoxia-inducible factor-1α. Genes Dev. 15:2520-2532.

Eriksson, A., Cao, R., Roy, J., Tritsaris, K., Wahlestedt, C., Dissing, S., Thyberg, J., and Cao, Y. (2003) Small GTP-binding protein Rac is an essential mediator of vascular endothelial growth factor-induced endothelial fenestrations and vascular permeability. Circ. 107:1532-1538.

Essler, M., Amano, M., Kruse, H., Kaibuchi, K., Weber, P., and Aepfelbacher, M. (1998a) Thrombin inactivates myosin light chain phosphatase via Rho and its target Rho kinase in human endothelial cells. J. Biol. Chem. 273:21867-21874.

Essler, M., Hermann, K., Amano, M., Kaibuchi, K., Heesemann, J., Weber, P., and Aepfelbacher, M. (1998b) Pasteurella multocida toxin increases endothelial permeability via rho kinase and MLC phosphatase. J. Immunol. 161:5640-5646.

Etienne-Manneville, S. and Hall, A. (2002) Rho GTPases in cell biology. Nature 420:629-635.

Fagan, K.A., Oka, M., Bauer, N., Gebb, S., Ivy, D., Morris, K., and McMurtry, I.F. (2004) Attenuation of acute hypoxic pulmonary vasoconstriction and hypoxic pulmonary hypertension in mice by inhibition of Rho-kinase. Amer. J. Physiol. Lung Cell. Mol. Physiol. *In press.*

Faller, D.V. (1999) Endothelial cell responses to hypoxic stress. Clin. Exp. Pharm. Physiol. 26:74-84.

Fedullo, P.F., Auger, W., Kerr, K., and Rubin, L.J. (2001) Chronic thromboembolic pulmonary hypertension. N. Eng. J. Med. 345:1465-1472.

Feng, D., Nagy, J., Hipp, J., Dvorak, H., and Dvorak, A.M. (1996) Vesiculo-vacuolar organelles and the regulation of venule permeability to macromolecules by vascular permeability factor, histamine, and serotonin. J. Exp. Med. 183:1981-1986.

Firth, J.A. (2002) Endothelial barriers: from hypothetical pores to membrane proteins. J. Anat. 200:541-548.

Fujiwara, K. (2003) Mechanical stresses keep endothelial cells healthy: beneficial effects of physiologic cyclic stretch on endothelial barrier function. Amer. J. Physiol. 285:L782-L784.

Fukata, M., Nakagawa, M., Kuroda, S., and Kaibuchi, K. (1999) Cell adhesion and Rho small GTPases. J. Cell Sci. 112:4491-4500.

Gertzberg, N., Neumann, P., Rizzo, V., and Johnson, A. (2004) NAD(P)H oxidase mediates the endothelial barrier dysfunction induced by TNF-α. Amer. J. Physiol. 286:L37-L48.

Goeckeler, Z.M. and Wysolmerski, R.B. (1995) MLCK-regulated endothelial cell contraction: the relationship between isometric tension, actin polymerization, and myosin phosphorylation. J. Cell Biol. 130:613-627.
Goodwin, A.M. and D'Amore, P.A. (2002) Wnt signaling in the vasculature. Angiogen. 5:1-9.
Grand, R.J., Turnell, A.S., and Grabham, P.W. (1996) Cellular consequences of thrombin-receptor activation. Biochem. J. 313:353-368.
Groeneveld, A.B. (2002) Vascular pharmacology of acute lung injury and acute respiratory distress syndrome. Vasc. Pharm. 39:247-256.
Haga, M., Chen, A., Gortler, D., Dardik, A., and Sumpio, B.E. (2003) Shear stress and cyclic strain may suppress apoptosis in endothelial cells by different pathways. Endothelium 10:149-157.
Hansen, C., Sorensen, L., Asmussen, I., and Autrup, H. (1992) Transplacental exposure to tobacco smoke in human-adduct formation in placenta and umbilical cord blood vessels. Teratog. Carcinog. Mutagen. 12:51-60.
Haralabopoulos, G.C., Grant, D., Kleinman, H., and Maragoudakis, M.E. (1997) Thrombin promotes endothelial cell alignment in Matrigel in vitro and angiogenesis in vivo. Amer. J. Physiol. 273:C239-C245.
Hirase, T., Kawashima, S., Wong, E., Ueyama, T., Rikitake, Y., Tsukita, S., Yokoyama, M., and Staddon, J.M. (2001) Regulation of tight junction permeability and occludin phosphorylation by RhoA-p160ROCK-dependent and -independent mechanisms. J. Biol. Chem. 276:10423-10431.
Hoshino, M., Nakamura, Y., and Hamid, Q.A. (2001) Gene expression of VEGF and its receptors and angiogenesis in bronchial asthma. J. Allergy Clin. Immunol. 107:1034-1038.
Howell, K., Preston, R.J., and McLoughlin, P. (2003) Chronic hypoxia causes angiogenesis in addition to remodelling in the adult rat pulmonary circulation. J. Physiol 547:133-145.
Ishii, Y., Partridge, C., Del Vecchio, P., and Malik, A.B. (1992) TNFα-mediated decrease in GSH increases sensitivity of pulmonary vascular endothelial cells to H_2O_2. J. Clin. Invest. 89:794-802.
Jeffery, T.K. and Wanstall, J.C. (2001) Pulmonary vascular remodeling: a target for therapeutic intervention in pulmonary hypertension. Pharm. Ther. 92:1-20.
Joris, I., Cuenoud, H., Doern, G., Underwood, J., and Majno, G. (1990) Capillary leakage in inflammation. A study by vascular labeling. Amer. J. Pathol. 137:1353-1363.
Kanazawa, H., Nomura, S., and Yoshikawa, J. (2004) Role of microvascular permeability on physiologic differences in asthma and eosinophilic bronchitis. Amer. J. Respir. Crit. Care Med. 169:1125-1130.
Katsumi, A., Orr, A., Tzima, E., and Schwartz, M.A. (2004) Integrins in mechanotransduction. J. Biol. Chem. 279:12001-12004.
Kim, I., Moon, S., Kim, S., Kim, H., Koh, Y., and Koh, G.Y. (2001) VEGF expression of ICAM-1, VCAM-1, and E-selectin through NFκB in endothelial cells. J. Biol. Chem. 276:7614-7620.
Kipnis, E., Guery, B., Tournoys, A., Leroy, X., Robriquet, L., Fialdes, P., Neviere, R., and Fourrier, F. (2004) Massive alveolar thrombin activation in pseudomonas aeruginosa-induced acute lung injury. Shock 21:444-451.
Knapp, J., Boknik, P., Luss, I., Huke, S., Linck, B., Luss, H., Muller, F., Muller, T., Nacke, P., Noll, T., Piper, H., Schmitz, W., Vahlensieck, U., and Neumann, J. (1999) The protein phosphatase inhibitor cantharidin alters vascular endothelial cell permeability. J. Pharmacol. Exp. Ther. 289:1480-1486.
Lampugnani, M.G., Resnati, M., Raiteri, M., Pigott, R., Pisacane, A., Houen, G., Ruco, L., and Dejana, E. (1992) A novel endothelial-specific membrane protein is a marker of cell-cell contacts. J. Cell Biol. 118;1511-1522.
Lampugnani, M.G., Zanetti, A., Breviario, R., Balconi, G., Orsenigo, F., Corada, M., Spagnuolo, R., Betson, M., Braga, V., and Dejana, E. (2002) VE-cadherin regulates endothelial actin activating Rac and increasing membrane association of Tiam. Molec. Biol. Cell 13:1175-1189.
Langeler, E.G. and van Hinsbergh, V.W. (1991) Norepinephrine and iloprost improve barrier function of human endothelial cells: role of cAMP. Amer. J. Physiol. 260:C1052-C1059.
Lassus, P., Turanlahti, M., Heikkila, P., Andersson, L., Nupponen, I., Sarnesto, A., and Andersson, S. (2001) Pulmonary VEGF and Flt-1 in fetuses, in acute and chronic lung disease, and in persistent pulmonary hypertension of the newborn. Amer. J. Respir. Crit. Care Med. 164:1981-1987.

Lentsch, A.B. and Ward, P.A. (2001) Regulation of experimental lung inflammation. Respir. Physiol. 128:17-22.
Li, X., Hahn, C., Parsons, M., Drew, J., Vadas, M., and Gamble, J.R. (2004) Role of protein kinase C {zeta}in thrombin-induced endothelial permeability changes: Inhibition by Angiopoietin-1. Blood *Published online ahead of print.*
Lo, S.K., Everitt, J., Gu, J., and Malik, A.B. (1992) TNF mediates experimental pulmonary edema by ICAM-1 and CD18-dependent mechanisms. J. Clin. Invest. 89:981-988.
Lofton, C.E., Baron, D., Heffner, J., Currie, M., and Newman, W.H. (1991) Atrial natriuretic peptide inhibits oxidant-induced increases in endothelial permeability. J. Mol. Cell Cardiol. 23:919-927.
Luca, R., Lijnen, H., Suffredini, A., Pepper, M., Steinberg, K., Martin, T., and Pugin, J. (2002) Increased angiostatin levels in bronchoalveolar lavage fluids from ARDS patients and from human volunteers after lung instillation of endotoxin. Thromb. Haemost. 87:966-971.
Masferrer, J.L. and Seibert, K. (1994) Regulation of prostaglandin synthesis by glucocorticoids. Receptor 4:25-30.
Matthay, M.A., Zimmerman, G., Esmon, C., Bhattacharya, J., Coller, B., Doerschuk, C., Floros, J., Gimbrone, M., Hoffman, E., Hubmayr, R., Leppert, M., Matalon, S., Munford, R., Parsons, P., Slutsky, A., Tracey, K., Ward, P., Gail, D., and Harabin, A.L. (2003) Future research directions in acute lung injury. Amer. J. Respir. Crit Care Med. 167:1027-1035.
McDonald, D.M. (2001) Angiogenesis and remodeling of airway vasculature in chronic inflammation. Amer. J. Respir. Crit. Care Med. 164:S39-S45.
McMurtry, I.F., Bauer, N., Fagan, K., Nagaoka, T., Gebb, S., and Oka, M. (2003) Hypoxia and Rho/Rho-kinase signaling. Lung development versus hypoxic pulmonary hypertension. Adv. Exp. Med. Biol. 543:127-137.
Metz, C. and Sibbald, W.J. (1991) Anti-inflammatory therapy for ALI. Chest 100:1110-1119.
Minami, T., Sugiyama, A., Wu, S., Abid, R., Kodama, T., and Aird, W.C. (2004) Thrombin and phenotypic modulation of the endothelium. Arterioscler. Thromb. Vasc. Biol. 24:41-53.
Moy, A.B., Blackwell, K., and Kamath, A. (2001) Differential effects of histamine and thrombin on endothelial barrier function through actin-myosin tension. Amer. J. Physiol. 282:H21-H29.
Mulligan, M.S., Vaporciyan, A., Miyasaka, M., Tamatani, T., and Ward, P.A. (1993) TNFα regulates in vivo intrapulmonary expression of ICAM-1. Amer. J. Pathol. 142:1739-1749.
Naidu, B.V., Woolley, S., Farivar, A., Thomas, R., Fraga, C., and Mulligan, M.S. (2003) Simvastatin ameliorates injury in an experimental model of lung ischemia-reperfusion. J. Thorac. Cardiovasc. Surg. 126:482-489.
Nawroth, R., Poell, G., Ranft, A., Kloep, S., Samulowitz, U., Fachinger, G., Golding, M., Shima, D., Deutsch, U., and Vestweber, D. (2002) VE-PTP and VE-cadherin ectodomains interact to facilitate regulation of phosphorylation and cell contacts. EMBO J. 21:4885-4895.
Newman, J.H., Fanburg, B., Archer, S., Badesch, D., Barst, R., Garcia, J., Kao, P., Knowles, J., Loyd, J., McGoon, M., Morse, J., Nichols, W., Rabinovitch, M., Rodman, D., Stevens, T., Tuder, R., Voelkel, N., and Gail, D.B. (2004) Pulmonary arterial hypertension: future directions. Circ. 109:2947-2952.
Osawa, M., Masuda, M., Kusano, K., and Fujiwara, K. (2002) Evidence for PECAM-1 in endothelial cell mechanosignal transduction: a mechanoresponsive molecule? J. Cell Biol. 158:773-785.
Ostrovsky, L., Woodman, R.C., Payne, D., Teoh, D., and Kubes, P. (1997) Antithrombin III prevents and rapidly reverses leukocyte recruitment in ischemia/reperfusion. Circ. 96:2302-2310.
Parving, H.H., Nielsen, F., Bang, L., Smidt, U., Svendsen, T., Chen, J., Gall, M., and Rossing, P. (1996) Macro-microangiopathy and endothelial dysfunction in NIDDM patients with and without diabetic nephropathy. Diabetologia 39:1590-1597.
Pascaud, M.A., Griscelli, F., Raoul, W., Marcos, E., Opolon, P., Raffestin, B., Perricaudet, M., Adnot, S., and Eddahibi, S. (2003) Lung overexpression of angiostatin aggravates pulmonary hypertension in chronically hypoxic mice. Amer. J. Respir. Cell Molec. Biol. 29:449-457.
Patterson, C.E. and Lum, H. (2001) Update on pulmonary edema: the role and regulation of endothelial barrier function. Endothelium 8:75-105.
Rabinovitch, M. (2001) Pathobiology of pulmonary hypertension. Extracellular matrix. Clin.Chest Med. 22:433-449.

Reinhart, K., Menges, T., Gardlund, B., Harm, Z., Smithes, M., Vincent, J., Tellado, J., Salgado-Remigio, A., Zimlichman, R., Withington, S., Tschaikowsky, K., Brase, R., Damas, P., Kupper, H., Kempeni, J., Eiselstein, J., and Kaul, M. (2001) Randomized, placebo-controlled trial of the anti-TNF antibody fragment afelimomab in hyperinflammatory response during severe sepsis: The RAMSES Study. Crit. Care Med. 29:765-769.

Rippe, B. and Grega, G.J. (1978) Effects of isoprenaline and cooling on histamine induced changes of capillary permeability in the rat hindquarter vascular bed. Acta Physiol. Scand. 103:252-262.

Rodway, G.W., Hoffman, L.A., and Sanders, M.H. (2003) High-altitude-related disorders--Part I: Pathophysiology, differential diagnosis, and treatment. Heart Lung 32:353-359.

Salvato, G. (2001) Quantitative and morphological analysis of the vascular bed in bronchial biopsy specimens from asthmatic and non-asthmatic subjects. Thorax 56:902-906.

Salven, P., Hattori, K., Heissig, B., and Rafii, S. (2002) IL-1α promotes angiogenesis in vivo via VEGFR-2 pathway by inducing inflammatory cell VEGF synthesis and secretion. FASEB J. 16:1471-1473.

Santos, S., Peinado, V., Ramirez, J., Morales-Blanhir, J., Bastos, R., Roca, J., Rodriguez-Roisin, R., and Barbera, J.A. (2003) Enhanced expression of VEGF in pulmonary arteries of smokers and patients with moderate COPD. Amer. J. Respir. Crit. Care Med. 167:1250-1256.

Schofield, C.J. and Ratcliffe, P.J. (2004) Oxygen sensing by HIF hydroxylases. Nat. Rev. Molec. Cell Biol. 5:343-354.

Semenza, G.L. (2000) Oxygen-regulated transcription factors and their role in pulmonary disease. Respir. Res. 1:159-162.

Semenza, G.L. (2003) Targeting HIF-1 for cancer therapy. Nat. Rev. Cancer 3:721-732.

Shimamura, K., Takashiro, Y., Akiyama, N., Hirabayashi, T., and Murayama, T. (2004) Expression of adhesion molecules by S1P and histamine in endothelium. Eur. J. Pharmacol. 486:141-150.

Ten, V.S. and Pinsky, D.J. (2002) Endothelial response to hypoxia: physiologic adaptation and pathologic dysfunction. Curr. Opin. Crit. Care 8:242-250.

Thurston, G., Rudge, J., Ioffe, E., Zhou, H., Ross, L., Croll, S., Glazer, N., Holash, J., McDonald, D., and Yancopoulos, G.D. (2000) Angiopoietin-1 protects the adult vasculature against plasma leakage. Nat. Med. 6:460-463.

Togias, A. (2003) H1-receptors: localization and role in airway physiology and in immune functions. J. Allergy Clin. Immunol. 112:S60-S68.

Tokuyama, K., Nishimura, H., Iizuka, K., Kato, M., Arakawa, H., Saga, R., Mochizuki, H., and Morikawa, A. (2002) Effects of Y-27632, a Rho/Rho kinase inhibitor, on leukotriene D(4)- and histamine-induced airflow obstruction and airway microvascular leakage in guinea pigs in vivo. Pharmacol. 64:189-195.

Tomashefski, J.F., Jr. (2000) Pulmonary pathology of ARDS. Clin. Chest Med. 21:435-466.

Tuder, R.M., Chacon, M., Alger, L., Wang, J., Taraseviciene-Stewart, L., Kasahara, Y., Cool, C., Bishop, A., Geraci, M., Semenza, G., Yacoub, M., Polak, J., and Voelkel, N.F. (2001) Expression of angiogenesis-related molecules in plexiform lesions in severe pulmonary hypertension: evidence for a process of disordered angiogenesis. J. Pathol. 195:367-374.

van Buul, J.D. and Hordijk, P.L. (2004) Signaling in leukocyte transendothelial migration. Arterioscler. Thromb. Vasc. Biol. 24:824-833.

van Hinsbergh, V.W. and Van Nieuw Amerongen, G.P. (2002) Intracellular signalling involved in modulating human endothelial barrier function. J. Anat. 200:549-560.

van Nieuw Amerongen, G.P., Draijer, R., Vermeer, M., and Van Hinsbergh, V.W.M. (1998) Transient and prolonged increase in endothelial permeability induced by histamine and thrombin. Role of protein kinases, Calcium, and RhoA. Circ. Res. 83:1115-1123.

van Nieuw Amerongen, G.P., Koolwijk, P., Versteilen, A., and van Hinsbergh, V.W. (2003) Involvement of RhoA/Rho kinase signaling in VEGF-induced endothelial cell migration and angiogenesis in vitro. Arterioscler. Thromb. Vasc. Biol. 23:211-217.

van Nieuw Amerongen, G.P., Natarajan, K., Yin, G., Hoefen, R.J., Fujiwara, K., Haendeler, J., Ridley, A., Van Hinsbergh, V., and Berk, B.C. (2004) GIT1 mediates thrombin signaling in endothelial cells: role in turnover of RhoA-type focal adhesions. Circ. Res. 94:1041-1049.

van Nieuw Amerongen, G.P. and van Hinsbergh, V.W.M. (2001) Cytoskeletal effects of rho-like small guanine nucleotide-binding proteins in the vascular system. Arterioscler. Thromb. Vasc. Biol. 21:300-311.

van Nieuw Amerongen, G.P. and van Hinsbergh, V.W.M (2002) Targets for pharmacological intervention of endothelial hyperpermeability and barrier function. Vascul. Pharm. 39:257-272.

van Nieuw Amerongen, G.P., Vermeer, M., Negre-Aminou, P., Lankelma, J., Emeis, J., and Van Hinsbergh, V.W.M. (2000a) Simvastatin improves disturbed endothelial barrier function. Circ. 102:2803-2809.

van Nieuw Amerongen, G.P., Vermeer, M.A., and van Hinsberg, V.W. (2000b) Role of RhoA and Rho kinase in LPA-induced endothelial barrier dysfunction. Arterioscler. Thromb. Vasc. Biol. 20:E127-E133.

van Wetering, S., van Buul, J., Quik, S., Mul, F., Anthony, E., Ten Klooster, J., Collard, J., and Hordijk, P.L. (2002) Reactive oxygen species mediate Rac-induced loss of cell-cell adhesion in primary human endothelial cells. J. Cell. Sci. 115:1837-1846.

VandenBerg, E., Reid, M., Edwards, J., and Davis, H.W. (2004) The role of the cytoskeleton in CAM expression in TNF-stimulated endothelial cells. J. Cell. Biochem. 91:926-937.

Vogel, S.M., Gao, X., Mehta, D., Ye, R., John, T., Andrade-Gordon, P., Tiruppathi, C., and Malik, A.B. (2000) Abrogation of thrombin-induced increase in pulmonary microvascular permeability in PAR-1 knockout mice. Physiol. Genomics 4:137-145.

Ware, L.B. and Matthay, M.A. (2000) The ARDS. N. Eng. J. Med. 342:1334-1349.

Waschke, J., Baumgartner, W., Adamson, R., Zeng, M., Aktories, K., Barth, H., Wilde, C., Curry, F., and Drenckhahn, D. (2004) Requirement of Rac activity for maintenance of capillary endothelial barrier properties. Amer. J. Physiol. 286:H394-H401.

Wiener, C.M., Booth, G., and Semenza, G.L. (1996) In vivo expression of mRNAs encoding hypoxia-inducible factor 1. Biochem. Biophys. Res. Commun. 225:485-488.

Wilson, J.W. and Kotsimbos, T. (2003) Airway vascular remodeling in asthma. Curr. Allergy Asthma Rep. 3:153-158.

Wojciak-Stothard, B., Potempa, S., Eichholtz, T., and Ridley, A.J. (2001) Rho and Rac but not Cdc42 regulate endothelial cell permeability. J. Cell Sci. 114:1343-1355.

Xu, F. and Severinghaus, J.W. (1998) Rat brain VEGF expression in alveolar hypoxia: possible role in high-altitude cerebral edema. J. Appl. Physiol. 85:53-57.

Yamakawa, M., Liu, L., Date, T., Belanger, A., Vincent, K., Akita, G., Kuriyama, T., Cheng, S., Gregory, R., and Jiang, C. (2003) Hypoxia-inducible factor-1 mediates activation of cultured vascular endothelial cells by inducing multiple angiogenic factors. Circ. Res. 93:664-673.

Zachary, I. (2001) Signaling mechanisms mediating vascular protective actions of VEGF. Amer. J. Physiol. 280:C1375-C1386.

Zachary, I. and Gliki, G. (2001) Signaling transduction mechanisms mediating biological actions of the VEGF family. Cardiovasc. Res. 49:568-581.

Zhao, X., Alexander, J., Zhang, S., Zhu, Y., Sieber, N., Aw, T., and Carden, D.L. (2001) Redox regulation of endothelial barrier integrity. Amer. J. Physiol. 281:L879-L886.

Chapter 15
Advances in Protection of Endothelial Barrier Function

Carolyn E. Patterson, Ph.D.,[1]* Hazel Lum, Ph.D.,[2] and A.B. Johan Groeneveld, M.D., Ph.D., FCCP, FCCM[3]

[1]Depts of Medicine & Physiology, Indiana Univ. School of Medicine, Indianapolis IN., 46202, USA;
[2]Dept. of Pharmacology, Rush Univ. Presbyterian St. Luke's Med. Center, Chicago, IL, 60612, USA &
[3] Dept. Intensive Care & Inst. Cardiovasc. Res., Vrije Univ., 1081 HV, Amsterdam, The Netherlands

CONTENTS:

Introduction

Normal Endothelial Barrier Formation
Study of Endothelial Cultures
Growth Factors
Interactions between the Endothelium and the Matrix
Cellular Signaling of State Transition
Autocrine and Blood-borne Signals

Blocking Endothelial Activating Agonists and Receptors
Endotoxin
Leukocyte Interactions and Inflammatory Mediators
Complement and Coagulation Factors
Bioactive Mediators

Blocking Activation Signaling
Calcium Signaling, Glucocorticoids, and Eicosanoids
PKA Inactivation
Kinases and Small GTPases
Nuclear Signals

Maximizing Intrinsic Protective Pathways
Feedback Inhibition and Autocrine Signaling
PKA Activation
PKG Activation and Nitric Oxide
Angiopoietin-1
PPARγ

Antioxidants
Decreasing Oxidant Effects and Oxidant Generation
Albumin

Prevention and Reversal of Injury
Cell Death and Replacement
Matrix Destruction, Prevention, and Repair
Surfactant and Acidosis

Protecting the Dam on Multiple Fronts

Introduction

Preceding chapters described the vital role of endothelial integrity in maintaining lung function, the etiology and mechanisms leading to endothelial dysfunction and injury, and the clinical consequences of pulmonary edema. Whether pulmonary dysfunction is a result of remote pathology or direct insult to the lung, management of edema and arterial hypoxia is vital for successful outcome. Reduced ventilation volumes and positive end expiratory pressure ventilation to avoid exacerbating mechanical damage have been forward steps in management (Artigas *et al*., 1998; Eisner *et al*., 2001; Malarkkan *et al*., 2003). Recommendations to address sepsis, inflammation, and cardiovascular abnormalities, which underlie the complex pathophysiological processes resulting in lung edema, have been reviewed (Wiedemann *et al*., 2003). Improvements in early identification and treatment of sepsis, a major factor in development of ARDS, have reduced development and progression of ARDS (Ware and Matthay, 2000; Marshall *et al*., 2003; Toh *et al*., 2003; MacArthur *et al*., 2004). One of the more recent and effective interventions is the use of activated protein C to attenuate inflammation (Bernard *et al*., 2001; Opal, 2003; Vincent *et al*., 2003). Other methods to block leukocyte activation have shown promise. Thus, strategies to intervene in originating pathologies, prevent leukocyte sequestration and activation, and manage the pulmonary and hemodynamic dysfunction have wrought improvements in treatment of acute lung injury and ARDS.

Although, endothelial dysfunction is recognized as important in pathophysiologic processes, most studies have not addressed the direct contribution of endothelial barrier dysfunction to the disease state, nor have the studies determined explicit effects of individual treatments on endothelial function in the clinical setting. In this chapter, we will focus on the cellular and biochemical mechanisms that normally usher in and maintain the intact, quiescent, barrier-competent state as well as those factors that directly interfere with or reverse endothelial activation and injury (Figure 1). Beneficial effects of the endothelial protective agents on lung injury and ARDS will be explored in particular. Notably, many of the barrier promoting agents to be discussed, also result in smooth muscle relaxation that would improve perfusion, lower pulmonary hypertension, and reduce driving pressure for filtration. In fact, this has been a main focus of many clinical studies of pulmonary hypertension with or without lung edema. Additionally, these factors may assuage primary or exacerbating effects of leukocyte activation and sequestration, a central concern of the investigations even when there is little pulmonary involvement. Yet, promoting endothelial quiescence and barrier integrity directly can significantly contribute to these goals, normalize endothelial control of coagulation, and get at the fundamental abnormality that results in edema due to the central role of endothelium in regulation of all these processes. Thus, many of these biochemical targets represent new opportunities for potential prevention and treatment of pulmonary edema at the cellular level.

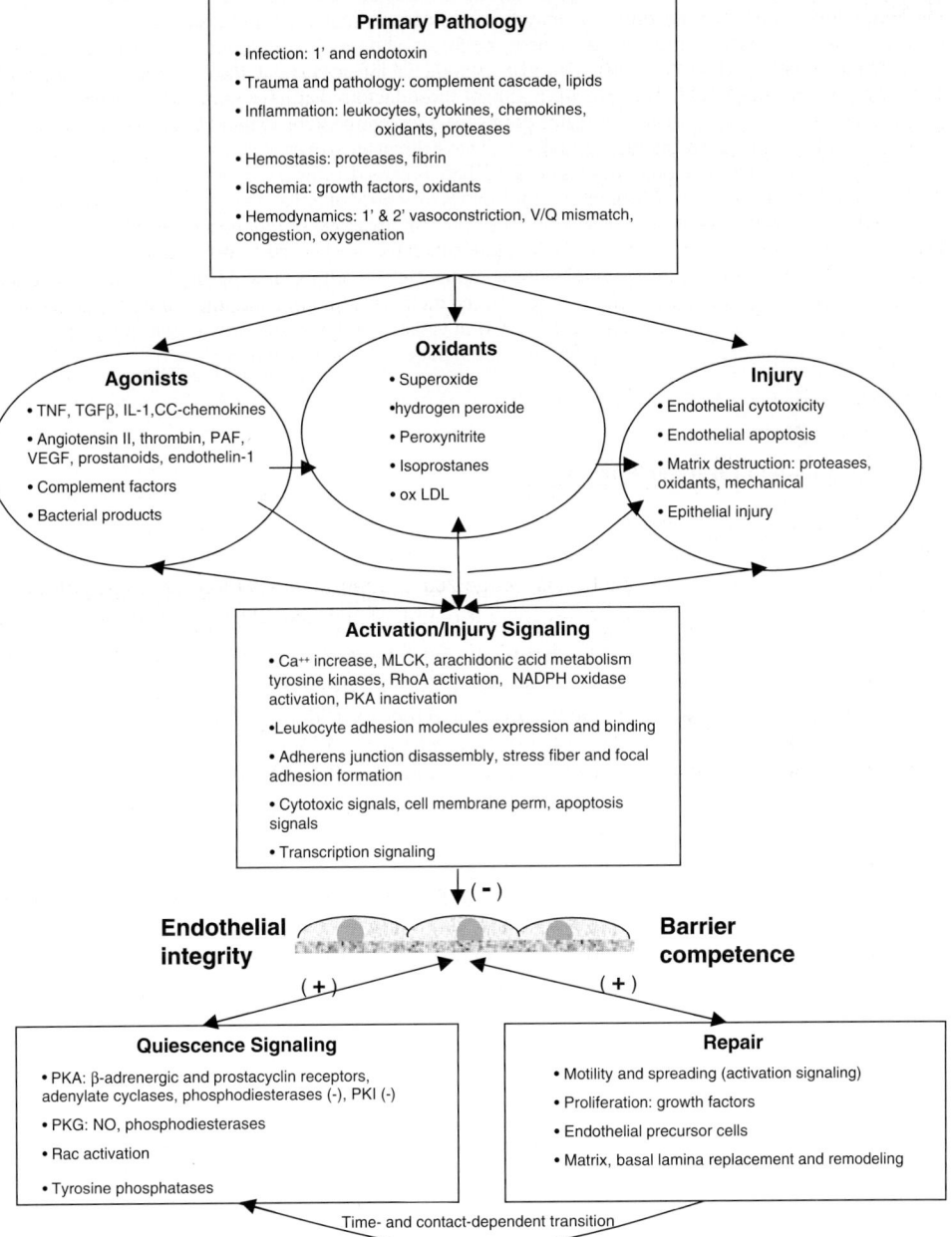

Figure 1: Paradigm for preventing activation and injury and promoting endothelial barrier function. Primary pathologies and trauma leading to loss of endothelial integrity in the lungs and to pulmonary dysfunction may be either remote or pulmonary in origin and may arise from a wide variety of etiologies that trigger inflammation, coagulation, and hemodynamic changes. These untoward effects on the lung endothelium may be direct or may be secondary to leukocyte sequestration and activation, airway/alveolar or vascular mechanical trauma, perfusion abnormality, or injury to other lung tissues. Treating the initiating events is critical to ultimate endothelial integrity and to the outcome in patients with acute lung injury and ARDS, but here we highlight those factors that have more direct bearing on the endothelial cell, which are potential or tested points for intervention to be explored in this chapter. These protective and restorative treatments, naturally, will have direct crossover benefits on leukocytes, vascular smooth muscle, platelets, and other tissues because there is considerable commonality in mechanisms and mediators and also because the endothelium contributes to control of leukocytes, vascular smooth muscle, and platelets. Nevertheless, in some cases non-specificity of treatment or inopportune timing can produce undesirable effects and in many cases single treatments have not proved to be effective in preventing incidence of ARDS or improving survival in the complex clinical setting, as suggested by the multifaceted and interactive pathophysiology in the schema. Based on this paradigm, experimental studies of specific interventions of the listed factors will be described in isolated endothelium, in vascular preparations, in isolated lung, in animal models of lung failure, and in patients with lung injury.

Normal Endothelial Barrier Formation

Study of Endothelial Cultures

Much of our understanding of the regulation of endothelial cell barrier function has come from studying the transition of growing, motile endothelial cells in culture to a competent monolayer and the reverse transition to barrier dysfunction with activation (Lum and Malik, 1994; Baldwin and Thurston, 2001; Patterson and Lum, 2001). When detached cells are placed on a compatible surface, the cells first attach, spread, and then extend long projections. As described in Chapter 9, when point contact is made between adjacent cells, connections are formed and the cells proceed to form a mutual wall of junctional connections and transform from the motile phenotype to the quiescent state. Alternatively and concomitantly, endothelial cells are dividing and junctional connections are maintained between parent and daughter cells, resulting in proliferating islands of cells. Outward growth from these isles most resembles the growth when a strip of cultured confluent cells is scraped away to form a wound and, presumably, when cells are stripped away with wounding *in vivo* (Ettenson and Gotlieb, 1992). Two external factors stand out in their ability to promote growth and formation of an endothelial monolayer: growth factors and the character of the matrix material to which the endothelial cells are attached.

Growth Factors

Most cultured endothelial cells require serum, with its incompletely defined and variable growth factors and hormones. Serum-free media, with defined partial replications of these factors, have limited, variable success in promoting growth for some cell lines, but more generally result in growth arrest. Even serum-supplemented cells usually require additional factors, such as Vascular Endothelial Growth Factor (VEGF), which is absolutely essential for fetal vessel development, Angiopoietin-1, acidic Fibroblast Growth Factor (FGF), and Hepatocyte Growth Factor (HGF) (Millauer et al., 1993; Terramani et al., 2000; Dye et al., 2001; Liu et al., 2002). Heparin added to culture sometimes reduces the amount of growth factor needed for optimal proliferation. After cells appear to be a confluent monolayer, there is a further transition to a tight, cobblestoned barrier. Not all types of endothelial cells obtain this morphology upon culture and in 3D culture some endothelia form tubules. The growth factors, not only promote proliferation, but also the ultimate formation of an effective barrier to proteins, depending on which factors and conditions are studied. For instance, when human umbilical vein endothelial cells are grown and passaged in media with 20% serum and 20 µg/ml Endothelial Growth Supplement (mixed growth factors), they permit very limited albumin clearance across the confluent monolayer. However, if they are passaged just once with serum, but without the growth factors, they appear equally confluent and have equal cell density and size, but are three times leakier to albumin (Patterson, unpublished data). This positive effect on barrier function may be explained by angiopoietin-1 via its Tie2 receptor, VEGF via the VEGF-1 receptor, by unknown factors, or by delay of growth and transition from confluent cells to barrier intact monolayers (Jones, 2003; Pizurki et al., 2003). Most likely, angiopoietin-1 is the responsible factor as it has been shown to reduce basal permeability of cultured human umbilical vein endothelium and promote formation of inter-endothelial cell adhesions (Gamble et al., 2000). Although the Tie2 receptor signals cell survival and is essential for embryologic and fetal development, expression of the Tie1 receptor increases dramatically in lung in late gestation and on into adulthood, when it is ten-fold higher in lung than in other organs, and knock out of the Tie1 receptor results in death from pulmonary edema and hemorrhage the first day after birth (Taichman et al., 2003). Thus, angiopoietin-1 and -2 appear to be important for this support and maturation of barrier function during both development and in the adult.

On the other hand, a number of studies have demonstrated increased permeability when some confluent endothelial cells were treated with VEGF, also termed VPF or vascular permeability factor (Connolly, 1991; Hippenstiel et al., 1998; Cohen et al., 1999; Wu et al., 1999; Chang et al., 2000; Bates et al., 2002). Not all endothelium respond to VEGF, even at high levels, with increased permeability (Connolly, et al., 1991; Hippenstiel et al, 1998; Murohara et al., 1998; Fischer et al, 1999; Chang et al., 2000; Patterson, unpublished data). This response divergence may reflect differential expression of the different VEGF receptors, but also the resistance of highly confluent endothelium to terminate the VEGF signaling by cadherin-based phosphatase activation (Esser et al., 1998; Lampugnani et al., 2003). VEGF is implicated in induction of the leaky vessels associated with neoplasms (Dvorak et al., 1995; Carmeliet and Jain, 2000). VEGF-induced permeability occurs with a slow onset and requires membrane-TNFα for its barrier disrupting effect, but not for its proliferation effect (Clauss et al., 2001). Possibly a secondary release of permeability-inducing agonists, such as NO or prostaglandins is needed (Murohara et al., 1998). Furthermore, while permeability seems intuitively contradictory to growth and barrier promotion, it is actually not surprising that a growth factor may cause transient permeability. In order to divide, grow, spread, and move to re-establish the monolayer, some disassembly of the cell-cell junctions, the cell/matrix junctions, and the cytoskeleton must occur in order to re-assemble cellular components and connections in new ways. Current clinical drug studies related to endothelial growth are directed at preventing angiogenesis in cancer and restenosis in vascular grafts and, thus, focus on suppression of uncontrolled proliferation. Conversely, experimental gene transfer studies with VEGF have shown benefit in peripheral diabetic neuropathy by supporting focal angiogenesis, but similar studies in myocardium resulted in tumor development (Lee et al., 2000; Schratzberger et al., 2001). Notably, VEGF overexpression alone in skin resulted in more numerous, but leaky vessels, whereas combinations of angiopoietin and VEGF expression resulted in reduced permeability (Thurston, 2002). In permeability edema, the concern is not for angiogenesis but for providing optimal support for maintaining and re-establishing endothelial health and is an area worth consideration.

Interactions between the Endothelium and the Matrix

The two-way relationship of endothelial cells and the sub-endothelial matrix has been elucidated by use of cell culture and studies of angiogenesis. As described in Chapters 2 and 8, it is clear that the nature of the basement substrate influences endothelial growth and function and that the endothelium influences and modifies the matrix composition (Madri and Williams, 1983; Pratt et al., 1985; Iivanainen et al., 2003). Molecular connections between matrix proteins and particular integrins spanning the basement membrane provide anchoring for both the quiescent and the motile cell and serve to signal to the cell with regard to the external environment. Surface heparan sulfate proteoglycans also form interconnections with matrix proteins. Spreading and cell motility involves coordinated formation and release of the cytoskeletal protein/integrin/matrix protein complexes (focal adhesions) and cytoskeletal extensions of lamellipodia (Chapter 7). Some matrix proteins, such as fibronectin, promote attachment and spreading, while others, such as laminin, promote migration. Normally, the basal lamina, directly under the endothelium, consists of type IV collagen, laminins, and proteoglycans, whereas the deeper interstitium consists of fibrillar collagens and fibronectin. Fibronectin also exists in a soluble form in the plasma and its incorporation into the matrix is proposed to play a protective and reparative role in inflammation. When the endothelial cells become attached to each other, focal adhesion kinase, paxillin, and other molecules in the focal adhesion complexes are then re-distributed from clustered, basement membrane-localized focal adhesions to more diffuse complexes at the basolateral corners. These complexes are associated with signaling molecules that vary in activation, depending partly on their organization and location, which communicate information re. the external environment to the cell and are involved in transition from motility to formation of an intact monolayer (Vestweber, 2000; Juliano, 2002).

Cellular Signaling of State Transition

When endothelial cells come into close apposition, there is a slow transition from the motile phenotype to the barrier phenotype with characteristic growth inhibition, cell-cell junctions, and release or redistribution of focal adhesions (Chapters 2 and 8). Initial cell-cell contacts are primarily due to Ca^{++}-dependent cadherins, which must be present at the surface. Progressive formation of adherens junctions involves de-phosphorylation of the cadherin, internal binding with α- and β-catenins, and a close association of the adherens junction to the actin band. Kinases, tyrosine phosphatases, Ras GTPases, and VEGFR2 are localized at these junctions which regulate formation of the cell-cell junctions, their connections to the cytoskeleton, and organization of the cytoskeleton, but formation of the junctions also signal the cell to initiate and maintain the quiescent, non-proliferative state (Braga et al., 1999; Vestweber, 2000; Baldwin and Thurston, 2001; Shay-Salit et al., 2002; Lampugnani et al., 2003). Binding of β-catenin to the complex may be one important signal, as the unbound β-catenin has nuclear signaling function. Finally, association of γ-catenin (plakoglobin) with the adherens junction correlates with improved strength and barrier function. Several hours subsequent to first formation of the patchy cadherin connections, PECAM-1 (Ig family) becomes progressively associated with more diffuse lateral contacts below the adherens junctions (Ayalon et al., 1994), similar and perhaps mixed with the proteins of the former focal adhesions that also relocate to these basolateral corners. Over 24hrs the junctions progress from random patches to a continuous lateral belt, imparting mechanical strength to the cell-cell contact. On one hand the dense peripheral actin band is a prerequisite for localization of the cadherin-catenin complex (Quinlan and Hyatt, 1999), but on the other hand appropriate clustering of the junctions was shown to be required to provide the focus for the reorganization of actin (Braga et al., 1999). Lastly, formation of lateral tight junctions toward the apical surface and gap junctions fully establish the tight, restrictive, barrier-intact endothelial border and endothelial quiescence. Moreover, the confluent endothelium exhibits contact-dependent growth inhibition ascribed to VE-cadherin attenuation of the VEGF receptor 2 signaling (Lampugnani et al., 2003).

The Rho family of small GTPases is important in the coordinated signaling of cytoskeletal rearrangements, that function in forward and backward transitions between activation, motility, and quiescence. The current 22 identified family members have different activators and targets (Asperstrom et al., 2004). For instance, in endothelium, RhoA is implicated in activation of stress fiber formation; whereas, Rac mediates stress fiber disassembly and is required for formation and maintenance of the dense peripheral band, which anchors the cell-cell junctional complexes necessary for barrier integrity (Waschke et al., 2004). Formation of the adherens junction complexes augments Rac activity, but diminishes Rho (Grazia-Lampugnani et al., 2002; Juliano, 2002; Grazia-Lampugnani et al., 2003). This altered signaling perpetuates solidification of the junctional complexes. The forward, interactive signaling finally results in a new biochemical, morphologic, and functional steady state – the quiescent endothelial monolayer with barrier integrity. Compared to signaling with activation described in Chapter 4, the new steady state corresponds to high PKA activity, low $[Ca^{++}]_i$ levels, low tyrosine kinase activity, low MLCK activity, low RhoA-kinase activity, constitutive NO production, low NAD(P)H oxidase activity, and low expression of leukocyte adhesion molecules. The levels and balance of arachidonic acid products changed. Released mediators result in further autocrine signaling of quiescence and *in vivo* they also transmit paracrine signals for smooth muscle relaxation and platelet quiescence.

Many studies have shown that elevation of cAMP and, thus, activation of PKA correlates with improved basal barrier function, whereas decreased cAMP relates to loss of integrity (Rippe and Grega, 1978; Stelzner et al., 1989; Langeler et al., 1991a; Langeler et al., 1991b; Patton et al., 1991; Patterson and Garcia, 1994; Adamson et al., 1998; Patterson et al., 2000; Waypa et al., 2000). On the other hand, PKA activation inhibits proliferation, thus timing is critical (Moodie and Martin, 1991; Dye et al., 2001). Intracellular cAMP levels are primarily determined by the balance of its synthesis *via* adenylate cyclase and its breakdown *via* phosphodiesterases, as explained in Chapter 5. Much has been learned regarding the regulation of cAMP and its interrelated regulation with Ca^{++} in various endothelial cells and in various states of endothelial activation, covered in detail in Chapter 5. Here we will focus on the downstream regulation and effects of PKA. One important factor in PKA activation and specificity of its phosphorylation effects is its anchoring and compartmentation within the cell (Coghlan et al., 1993). This compartmentation has not been well studied in endothelium, especially in regard to barrier function. The interaction of PKA with its endogenous inhibitor, PKI, counters cAMP activation of PKA (Collins and Uhler, 1997; Lum et al., 2002). Overexpression of PKI in pulmonary endothelial cells to reduce PKA activity results in mild basal barrier dysfunction and exacerbates agonist-induced permeability (Patterson et al., 2000). Thus, effective PKA activity is controlled by both the cAMP level and PKI interaction with PKA. While much is known regarding cAMP regulation, relatively little is know regarding regulation of PKI.

One PKA target is MLCK. PKA-mediated phosphorylation of MLCK attenuates MLCK-phosphorylation of MLC (Moy et al., 1993; Patterson et al., 1994a; Garcia et al., 1995a). Lower MLC phosphorylation correlates with lower levels of contractile interactions of cytoskeletal proteins, but it is apparent that this does not account for all of the pro-barrier effects of PKA (Patterson et al., 1995). Some other possible PKA phosphorylation targets with relevance to endothelial barrier function include IP_3 receptors (Moore et al., 1998; Wojcikiewicz and Luo, 1998), filamin (Hastie et al., 1997), serine/threonine protein phosphatases (Usui et al., 1998), Ras and RhoA (Lang et al., 1996; Arimura et al., 1997; Manganello et al., 2003; Qiao et al., 2003), and the vasodilator-stimulated phosphoprotein (VASP), (Comerford et al., 2002; Sudo et al., 2003). VASP is a regulator of actin dynamics, which localizes strongly to intercellular tight junctions and diffusely to the actin cytoskeleton in endothelial cells upon PKA activation or to focal adhesions and the cytoskeleton when the cells resemble the activated state (Comerford et al., 2002). In smooth muscle cAMP-relaxation is primarily attributed to its ability to reduce $[Ca^{++}]_i$, but in endothelium PKA activation does not alter $[Ca^{++}]_i$ (Buchan and Martin, 1991; Patterson et al., 1994a; Stevens et al., 1997; Patterson et al., 2000).

Autocrine and Blood-borne Signals

When endothelial cells are confluent and quiescent, they do not produce as many total eicosanoid products, but importantly, the ratio of PGE_1 to $PGF_{2\alpha}$ or of prostacyclin (PGI_2) to thromboxane is higher than in the activated state. Released PGE_1 and PGI_2 act as autocrine regulators (in addition to paracrine regulation of smooth muscle and platelets) and signal PKA activation, promoting establishment and maintenance of barrier integrity. Additionally, the quiescent endothelium produce low levels of NO, but the ratio of NO to endothelin-1 is high compared to the activated state. NO signals *via* activation of guanylate cyclase, increased $[cGMP]_i$, and activation of cGMP kinase (PKG). This endothelial-derived relaxing factor is a main mechanism controlling vascular smooth muscle relaxation, but NO also has a possible autocrine effect on the endothelium as well (Boulanger et al., 1990). Additionally, NO serves as a scavenger of superoxide. Constitutive release of NO from mesenteric microvascular endothelium was found to be responsible for improved barrier function (Victorino et al., 2001), but whether this effect was *via* cGMP elevation or *via* oxidant scavenging or some other mechanisms was not determined. Elevation of cGMP decreased basal permeability in aortic endothelium (Holschermann et al., 1997). On the other hand, other studies failed to show an effect of cGMP on basal barrier function (Patterson et al., 1994a; Draijer et al., 1995). As indicated above these mediators from barrier-competent endothelium also signal platelet quiescence, a non-thrombotic state, and smooth muscle relaxation.

In vivo, the endothelial apical surface is bathed with blood, which has beneficial effects, as discussed in Chapter 11. In addition to its importance as the major plasma pH buffer, a carrier for fatty acids and other substances, and the principal provider of plasma oncotic pressure, albumin has direct interactions with the endothelium, which promote cell health and barrier function (Dull et al., 1991; Lum et al. 1991; He and Curry, 1993a; Zoellner et al., 1999; Victorino et al., 2003). Negatively charged albumin does not appreciably bind to the negative glycocalyx, but its positive arginine moieties form some glycocalyx connections and glycocalyx expansion in the presence of albumin indicates its incorporation (Baldwin and Thurston, 2001). More importantly, albumin binds specific receptors in the caveoli (Schnitzer et al., 1992). Albumin promotes endothelial barrier stability and lowers the basal $[Ca^{++}]_i$, by effects unrelated to its stabilization of pH (He and Curry, 1993a; Patterson, microscopy observation; Pizurki et al., 2003).

Sphingosine-1 phosphate and possibly lysophosphatidic acid, produced by platelets, promote endothelial barrier function (Haselton and Alexander, 1992; Alexander et al., 1998; Yatomi et al., 2000; Schaphorst et al., 2003). These lipids are bound to albumin and exert their effect independently of cAMP/PKA and cGMP/PKG (Gainor et al., 2001; Minnear et al., 2001). Moreover, upon stimulation with thrombin, platelets release angiopoietin-1, which has barrier-promoting effects, discussed below in protective mediators (Jones, 2003). Exogenous ATP is another factor that promotes improved baseline barrier function. Apparently, this effect is not related to the adenosine receptors, changes in $[Ca^{++}]_i$, altered cyclic nucleotide levels, or PKC (Noll et al., 1999). Adenosine is released by activated neutrophils or formed from 5'-AMP released from the neutrophils in a paracrine feedback mechanism to close the inter-endothelial gaps during diapedesis. In cultured endothelium adenosine decreased endothelial permeability (Haselton et al., 1993) and adenosine induced pulmonary vasodilation in norepinephrine contracted lung vasculature (Roepke et al., 1991). Adenosine-promotion of barrier function was found to be due to activation of its receptor-2, PKA activation, phosphorylation of VASP, and sealing of tight junctions (Comerford et al., 2002), as well as by PKA-dependent up-regulation of ecto-5'-nucleotidase, the surface enzyme that metabolizes 5'-AMP to adenosine (Narravula et al., 2000). Yet, the protective effect of adenosine against oxidant injury in endothelium was thought to be independent of cAMP (Richard et al., 1998). High levels of adenosine induced dysfunction of the endothelial glycocalyx that contributed to increased permeability via receptor-3 activation (Platts and Duling, 2004).

Blocking Endothelial Activating Agonists and Receptors

Endotoxin

Endotoxin, lipopolysaccharide from gram-negative bacteria, partly exerts it untoward effects by activation of leukocytes and endothelium (Chapter 13). Macrophage activation results in cytokine production and occurs via binding of the endotoxin to a binding protein, interaction of this complex with the CD14 receptor, and activation of TLR-4 (toll-like receptors). Endothelial activation also occurs via binding of the endotoxin to MD-2 accessory protein and activation of the TLR-4 receptor (Bochud and Calandra, 2003; Peters et al., 2003). Other toll-like receptors bind and signal other bacterial components (Peters, et al., 2003). Endothelial activation and permeability related to caspase cleavage of adherens junction proteins (Bannerman and Goldblum, 1999) and apoptosis (Peters et al., 2003). Leukocyte-independent effects of endotoxin on vascular permeability in a hamster cheek pouch model were blocked with lipoxygenase or PAF inhibitors, suggesting that these agents mediated endotoxin-induced endothelial permeability

(Klabunde and Cavello, 1995). Notwithstanding the direct effects of endotoxin on endothelial cells as well as the compelling correlation of endotoxin levels and pulmonary dysfunction and the promise of benefit by early trials on survival from sepsis (Greenman *et al.*, 1991), clinical trials to date have not demonstrated protection of anti-endotoxins on pulmonary consequences or survival in severe sepsis (Marshall, 2003; Opal and Gluck, 2003).

Leukocyte Interactions and Inflammatory Mediators

Leukocytes are important in protection from bacterial and fungal insults. Thus, impaired leukocyte function can result in life threatening diseases, such as chronic granulomatous disease and AIDS. Long lasting, broad intervention in leukocyte function would make the lung more susceptible to infection. However, leukocyte sequestration, activation, and inappropriate function can harm the endothelium and lung tissue, as noted in Chapters 1, 3, 11, and 12. Thus, the trick is to block pathologic changes in specific locales, while leaving necessary physiologic functions intact. Neutrophil sequestration and radiation-induced lung injury were attenuated in ICAM-1 knock out mice (Hallahan and Virudachalam, 1997). Likewise, isograft rejection was prevented by antisense ICAM-1 (Toda *et al.*, 2000) and reperfusion lung injury was prevented with blocking antibodies to selectins (Demertzis *et al.*, 1999). Leukocyte binding to endothelium increases presentation of leukocyte-derived mediators, oxidants, and proteases, so prevention of binding would have direct implication for endothelial activation. Prevention of leukocyte-endothelial binding also has implication for leukocyte diapedesis and influx into the sub-endothelial lung tissue, where more damage might occur to the interstitium, epithelium, or other cells such as fibroblast, when leukocytes are not actually needed to ward off infection.

Evidence that cytokines play an important role in endothelial injury and pulmonary dysfunction is overwhelming (Chapters 3, 11, 12, & 13). Lung lavage fluid from ARDS patients contains elevated cytokines, which negatively correlate with survival (Meduri *et al.*, 1995; Dhainaut *et al.*, 2003; Fahy *et al.*, 2003). As endothelial activation generally occurs *via* receptors by agonists released in pathological states, it is obvious that neutralizing the agonists or blocking their receptors would be protective if properly timed and targeted (Chapters 1, 3, and 4). For instance, IL-8 directly causes permeability in endothelium, which is blocked by antibodies to CXCR1 and CXCR2 receptors (Biffl *et al.*, 1995; Schraufstatter *et al.*, 2001; Wang *et al.*, 2004). Antibodies against IL-8 *per se* effectively prevented lung endothelial and epithelial permeability in rabbit models of acid and smoke lung injury (Folkesson *et al.*, 1995; Laffon *et al.*, 1999; Modelska *et al.*, 1999). Similarly, IL-6 antibodies greatly attenuated the upregulation of C5-a receptors and improved survival in septic mice (Riedermann *et al.*, 2003) and treatment with TNF antibodies improved oxygenation and survival in rats challenged by tracheal administration of *E. coli* and *S. aureus* (Figure 2) (Karzai *et al.*, 2003). Yet clinical trials with TNF neutralizing aptamers, anti-TNFα antibody fragments, or soluble receptor fragments to bind TNF were ineffective in ameliorating lung injury in sepsis (Reinhart *et al.*, 2001; Butty *et al.*, 2003; Reimold, 2003). Although, IL-1 antagonism reduced lung injury in a rat oxidant model of ARDS (Leff *et al.*, 1994), blocking IL-1 was ineffective (Opal *et al.*, 1997; Remick, 2003). This is probably due to limitations in blocking single mediators in a complex set of processes, the importance of timing and targeting, and the protective role of the immune response in infectious conditions. TGFβ is considered an inhibitory cytokine with regard to stem sell differentiation, but can serve pro-inflammatory pathologic roles, including induction of endothelial monolayer permeability

(Goldberg *et al.*, 2002) and pulmonary fibrosis (Dhainaut *et al.*, 2003; Fahy *et al.*, 2003). Antibodies to TGFβ inhibited protein exudation and improved oxygenation in rat lung injury induced by intratracheal instillation of IL-1, but did not alter neutrophil sequestration or transmigration (Hybertson *et al.*, 2003). TGFβ blockade also prevented lung edema and pathology after hemorrhagic shock in mice (Shenkar *et al.*, 1994). TGFβ inhibition or knock-out of the integrin that interacts with TGFβ prevented lung permeability in bleomycin-injured mice (Pittet *et al.*, 2001). CC-chemokines receptor antagonism attenuated lung injury due to acute pancreatitis in mice (Bhatia *et al.*, 2003) and a new broad-spectrum chemokine inhibitor was recently shown to attenuate ischemia reperfusion lung injury and leukocyte increase (Naidu *et al.*, 2004).

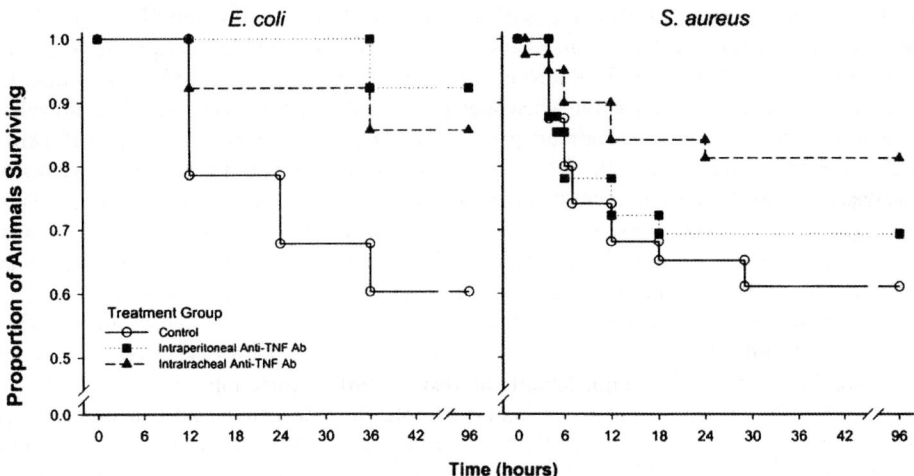

Figure 2: Anti-TNF antibody improves survival in sepsis. Rats were challenged by intratracheal instillation of *Escherichia coli* or *Staphylococcus aureus* followed by: intratracheal treatment with antibody to tumor necrosis factor (*closed triangles*), intraperitoneal treatment with antibody to tumor necrosis factor (*closed squares*), or control serum by both routes (*open circles*). From *Karzai, W., Cui, X., Mehlhorn, B., Straube, E., Hartung, T., Gerstenberger, E., Banks, S., Natanson, C., Reinhart, K., and Eichacker, P. (2003) Protection with antibody to TNF differs with similarly lethal Escherichia coli versus Staphylococcus aureus pneumonia in rats. Anesthesiology 99:81-89.* With kind permission of Dr. Karzai and Lippincott, Williams, and Wilkins, publishers of Anesthesiology.

Complement and Coagulation Factors

The complement cascade plays an important role in disease fighting, but also can have profound self-directed injurious effects *via* impaired leukocyte function, leukocyte-mediated oxidant production, and direct effects on endothelium (Ward, 1996; Tedesco *et al.*, 1999; Acosta *et al.*, 2004). Treatment of lung-injury rodent models with anti-complement receptor improved barrier function and survival (Ward *et al.*, 2003; von Dobschuetz *et al.*, 2004). Clinical trials in ARDS patients have shown safety, but no reports on efficacy have appeared (Zimmerman *et al.*, 2000). Thrombin and possibly other activators of Protease Activated Receptors (PARs), such as trypsin, tryptase, and plasmin, induce barrier dysfunction in cultured cells (Garcia *et al.*, 1986; Minnear *et al.*, 1989; Lum *et al.*, 1992; Nadel, 1992; Rabbani *et al.*, 1993; Patterson *et al.*, 1994a; Garcia, 1995; Shasby *et al.*, 1997; Van Nieuw Amerongen *et al.*, 2000a). Normally, these proteases are present in an inactive form or are sequestered in mast cells. Moreover, there are serum and cellular mechanisms to prevent the interactions of the proteases with endothelium, but when thrombi are trapped or formed in the lung, active thrombin in the clot can be presented directly to the endothelium and in allergic and inflammatory reactions, lung mast cells can release large amount of tryptase (Johnson *et al.*, 1982; Schaeffer *et al.*, 1987; Lo *et al.*, 1990; Nadel, 1992).

Yet, thrombin-induced edema in isolated perfused lung was ascribed to hemodynamic changes, rather than to permeability changes (Waypa *et al.*, 1996). Secondarily, thrombin-activation of leukocytes sustains and amplifies the initial direct effect on endothelium (Lo *et al.*, 1990; Roemisch *et al.*, 2002). Despite the direct and indirect thrombin-mediated increase in endothelial permeability, results *in vivo* to block thrombin are less clear. Treatment with antithrombin III, normally produced by liver but active when bound to the endothelial surface, reduced endothelial-leukocyte binding and improved capillary perfusion in endotoxin-treated hamsters (Hoffmann *et al.*, 2002; Roemish *et al.*, 2002). These effects were ascribed to antithrombin effects on prostacyclin release by the endothelium but permeability was not studied. Antithrombin III also improved survival in septic baboons (Minnema *et al.*, 2000); however, the thrombin-antagonist, hirudin, failed to alter either microcirculation perfusion or endothelial-leukocyte binding in the endotoxin-challenged hamster model (Hoffmann *et al.*, 2000).

The importance of uncontrolled coagulation in sepsis and other causes of ARDS is recognized, but clinical trials with agents to neutralize thrombin did not measure specific effects on the endothelium or on the lung and mortality was not clearly altered. For instance, trials with anti-thrombin III and trials with tissue factor pathway inhibitor in sepsis and disseminated intravascular coagulation were disappointing (Osterman, 2002; Abraham, *et al.*, 2003; Eichacker and Natanson, 2003; Opal, 2003). This is probably due to the complex feedback effects within the clotting cascade. In isolated endothelium, anti-thrombins or anti-coagulation agents reduced Protein C activation, which normally occurs when thrombin binds thrombomodulin on the endothelium (Linder *et al.*, 2003). If Protein C activation is reduced, the normal check on coagulation is removed, and thus, blocking thrombin may paradoxically actually increase widespread thrombosis. Reduced Protein C correlated with the severity of inflammation and organ damage (Reinhart *et al.*, 2002) and as mentioned earlier, clinical trials with activated Protein C have shown benefit in sepsis (Bernard *et al.*, 2001; Opal, 2003; Vincent *et al.*, 2003). It might be assumed that its inhibition of thrombin generation and coagulation is the major mechanism; yet, the other agents, tissue factor inhibitor and antithrombin III, were not effective although they have equal ability to prevent coagulation. Thus, effects of activated protein C, such as inhibition of inflammatory signaling in leukocytes, may be even more important than its anti-coagulation properties. Of particular interest is the demonstration that activated protein C inhibited NFκB activation in endothelial cells, thereby suppressing expression of ICAM-1, VCAM-1, and E-selectin, and leukocyte binding (Joyce and Grinnell, 2002). Moreover, activated protein C suppressed endothelial apoptosis, a known problem in sepsis. In a study of lung injury in combined smoke inhalation and *Pseudomonas aeruginosa* sepsis, high dose heparin did not prevent ensuing dysfunction, but nebulized heparin attenuated cellular infiltrates, lung edema, poor gas exchange, and congestion (Murakami *et al.*, 2002; Murakami *et al.*, 2003), suggesting delivery mode is critical to maximize benefit and minimize side effects of anti-coagulant therapy.

Bioactive Mediators

A number of bioactive mediators with multiple inflammatory, vasoactive, and coagulation effects are released by activated endothelium (and other cells), which perpetuate endothelial activation and barrier dysfunction (Chapter 3). Whereas, PGI_2 and PGE_1 are barrier promoting, PAF, endothelin-1, TxA_2, $PGF_{2\alpha}$, and isoprostanes are barrier disruptive (Victorino *et al.*, 2004). PAF effects on adherens junction and tight junction disruption and barrier dysfunction in endothelium were attenuated experimentally by PGE_1 (Farmer *et al.*, 2002) and by removal of oxidants with exogenous superoxide dismutase (SOD) (Zhang *et al.*, 2003). Whereas the PAF antagonist, lexipafant, did not prevent lung dysfunction after intestinal ischemia/reperfusion injury (Borjesson *et al.*, 2002) or after cardiopulmonary bypass (Taggart, 2001), it did improved lung function and wet/dry lung weight in a swine model of lung transplantation (Kelly, 2000). Importantly, PAF antagonism ameliorated lung injury in patients with acute pancreatitis (Galloway and Kingsworth, 1996) and treatment with PAF acetylhydrolase to catabolize PAF, improved survival in patients with severe sepsis and ARDS (Schuster *et al.*, 2003). An antagonist to endothelin-1, tezosentan, reduced lung hypertension, pulmonary edema, and improved oxygenation in an endotoxin-infused sheep model of ARDS (Kuklin *et al.*, 2004). Thromboxane inhibition initially yielded positive outcomes in incidence of acute lung injury and in mortality in ARDS (Yu and Tomasa, 1993, Schilling *et al.*, 2001), but a larger study did not demonstrate improvements (ARDS Network, 2000).

Endothelial cells are primarily responsible for conversion of angiotensin I to active angiotensin II due of their expression of angiotensin converting enzyme (ACE) and their large surface area. ACE-inhibitors, widely used to attenuate vasoconstrictive/hypertensive effects of angiotensin II, are now recognized to have salutary effects on the endothelium beyond lowering of blood pressure. They prevent loss of endothelial-mediated vasorelaxation and block activation of NADPH oxidase-superoxide production in endothelium (Zhang *et al.*, 1999; Zhang *et al.*, 2003). Similarly, blockers of angiotensin type I receptor, AT1, such as losartan, stabilized endothelium (Zhang *et al.*, 1999). While there are a number of studies describing reduction in total ACE activity with lung injury (Wiedemann and Gillis, 1986; Cziraki *et al.*, 2002), studies of ACE on lung injury have been rather limited despite direct activation of endothelium by angiotensin. In an oleic acid-injury model, ACE inhibition reduced pulmonary pressures, improved arterial oxygenation, reduced the number of injured circulating endothelial cells, and restored lung wet/dry weight to normal levels (Liu and Zhou, 2002). And in a radiation injury model, endothelial activation and later fibrosis was inhibited by ACE inhibition (Ward *et al.*, 1989).

Because of its permeability inducing effects VEGF blockade may at times be desirable. However, VEGF in combination with angiopoietin (to counter permeability) also produces cell survival benefits and at times endothelial proliferation may be desirable. Nevertheless, there are several compounds available with varying specificity for the VEGFRA and VEGFR2 receptors (Whittles *et al.*, 2002; Bates and Harper, 2003; van Nieuw Amerongen and van Hinsbergh, 2003). Clinical trial with another inhibitor, not specific for the three VEGF receptors, as anti-tumor agents revealed that the tumor vessel leaks were greatly diminished and the drug had minimal side effects (Morgan *et al.*, 2003). An antibody to VEGF A has also been tested in clinical trials, but side effects were noted (Dvorak, 2002). Another naturally occurring VEGF blocker, Neovastat, is also in current trial (van Nieuw Amerongen and van Hinsberg, 2003).

Blocking Activation Signaling

Calcium Signaling, Glucocorticoids, and Eicosanoids

As activation signals are common to many cell types and intracellular control heavily depends on temporal and spatial organization of signaling, targeting of drugs or interventions to modify specific intracellular signaling is extremely problematic. Endothelial cell activation and increased permeability consists of Ca^{++}-dependent and Ca^{++}-independent mechanisms as highlighted in Chapters 2-4. *In vitro* use of Ca^{++} release blockers and chelators have attenuated agonist-induced permeability (Siflinger-Birnboim *et al.*, 1996; Chetham *et al.*, 1997; Kelly *et al.*, 1998). Inhibitors of myosin light chain kinase and protein kinase C, activated with the rise in Ca^{++}, have been used to prevent barrier dysfunction in cultured cells, but it is not likely that either Ca^{++} or kinase inhibition would be clinically useful without very specific targeting. Membrane receptor signaling and the rise in Ca^{++} also trigger activation of phospholipase A_2, leading to release and metabolism of arachidonic acid. Inhibition of phospholipase A_2 ameliorated endotoxin-induced lung injury in mice (Nagase, 2003). Corticosteroids inhibit several inflammatory activation signals, such as phospholipase A_2, and corticosteroid support of endothelial integrity, along with its other physiologic functions, is well known (Thompson, 2003). Whereas, high dose, short-term therapy has not proved useful in reducing mortality or risk for development of ARDS in several trials, there is debate on potential usefulness of lower dose therapy for longer periods, especially when the hypothalamic-pituitary-adrenal response to stress is compromised and trials are ongoing (Chadda and Annane, 2002; Luce, 2002; MacLaren and Jung, 2002; Marshall, 2003; Thompson, 2003). Late corticosteroid treatment may also be useful for suppressing irreversible fibrosis (Groeneveld *et al.*, 2003). Specific effects on endothelial permeability were not addressed in these trials. Numerous studies have reported beneficial effects of cyclooxygenase inhibitors in various models of endothelial and lung injury, but clinical results have been less encouraging. Ibuprofen did not improve incidence of ARDS or survival in trials on patients with sepsis (Bernard *et al.*, 1997b), but it could be hypothesized that both beneficial and pathologic eicosanoids were reduced; thus more specific blockers of detrimental eicosanoids would probably be more advantageous for clinical use.

PKA Inactivation

PKA activation is involved in establishing normal barrier properties and a number of permeability-inducing agonists decrease PKA (Rippe and Grega, 1978; Stelzner *et al.*, 1989; Ogawa *et al.*, 1992; Koga *et al.*, 1995; Manolopoulos *et al.*, 1997; Groeneveld, 1999; Patterson and Lum, 2001). Many agonists elevate Ca^{++}, which has been shown to mediate a decrease in cAMP that is essential to the permeability effect (Cioffi *et al.*, 2002; Creighton *et al.*, 2003), as discussed in Chapter 5. In preceding chapters, reduced PKA was shown to signal disassembly of the dense peripheral band and cell-cell adhesions, formation of focal adhesions, and motility, (Patterson *et al.*, 2000). Signaling of altered cAMP and PKA activation was described in detail in Chapter5, and this correlation of mediator effects on PKA and permeability suggests that blocking the signaled loss of PKA activity might be upstream targets for intervention. Nevertheless, means to directly elevate cAMP are probably more direct avenues and will be discussed below in the section on preventing and reversing endothelial monolayer permeability.

Kinases and Small GTPases

Plasma from post-burn injured rats induced hyperpermeability and actin stress fiber formation in rat lung microvascular endothelium, which was greatly attenuated by MLCK pharmacologic inhibition, transfection with an MLCK-inhibiting peptide, and pharmacological inhibition of Rho kinase (Tinsley *et al.*, 2004a). Similarly, MLCK knock-out or small molecule inhibition attenuated pulmonary edema and improved survival in mice injected with endotoxin (Tinsley *et al.*, 2004b). These studies indicate the importance of endothelial MLCK activation as a common downstream effector in pulmonary edema. Growth factors (e.g. VEGF and TGFβ1), cytokines (e.g. TNFα and IFNγ and agonists (e.g. thrombin and histamine) activate tyrosine kinases, which initiate many pathways that are ultimately edemagenic, such as phosphorylation of components of adherens junctions resulting in disassembly. Inhibition of tyrosine kinases attenuates permeability in a number of settings (Esser *et al.*, 1998; Shi *et al.*, 1998; Andriopoulou *et al.*, 1999; Cohen *et al.*, 1999; Baldwin and Thurston, 2001; Nakajima *et al.*, 2001; Bates *et al.*, 2002). For instance, genestein prevented H_2O_2-mediated tyrosine phosphorylation, gap formation, and permeability (Carbajal and Schaeffer, 1998). Activation of Rho GTPases, downstream from activation of various receptors, relays signals to the actin cytoskeleton involved in a multitude of processes such as cell adhesion, motility, and polarity, cell cycle progression, and apoptosis (Flinn and Ridley, 1996; Lim *et al.*, 1996; Schulze *et al.*, 1997). One reported effect of Rho activation is adenylate cyclase inhibition and, hence, reduced PKA activity (Chrzanowskawodnicka and Burridge, 1996; Tigyi *et al.*, 1996; Takaishi *et al.*, 1997). Thrombin challenge results in a rapid, transient RhoA activation, but not Rac1 or Cdc42, and either RhoA inactivation or Rho kinase inhibition reduced the barrier dysfunction (Essler *et al.*, 1998; van Nieuw Amerongen *et al.*, 1998; Carbajal and Schaeffer, 1999; van Nieuw Amerongen *et al.*, 2000a; Qiao *et al.*, 2003). Likewise, inhibition of a Rho kinase, p60 ROCK, attenuated thrombin-induced barrier dysfunction in pulmonary endothelial cells (Carbajal *et al.*, 2000) and in H_2O_2-induced edema in isolated perfused lungs (Chiba *et al.*, 2001). How interruption of endothelial Rho-signaling might prevent barrier disruption in clinical settings is unknown. An inhibitor of p38 MAPK ameliorated lung injury in rats induced with sterile, cytokine-free ascites fluid (Denham *et al.*, 2000).

Nuclear Signals

Oxidant stress resulting in DNA damage leads to activation of poly (ADP-ribose) polymerase, which can deplete NAD^+ content, inhibit metabolism, deplete ATP, and trigger necrosis (Watts and Kline, 2003). Inhibitors of the polymerase, such as 3-aminobenzamide and 1,5-dihydroxyisoquinoline, reduced neutrophil accumulation in lung, protect other organ failure, and improve survival in a variety of models for ischemia-reperfusion injury, sepsis, inflammation, shock, and traumatic brain injury (Watts and Kline, 2003). Mononuclear cell infiltration and lung histological damage were markedly reduced by the polymerase inhibitors in rodent models of intraperitoneal and intrathoracic zymosan-induced injury (Szabo *et al.*, 1997b; Cuzzocrea *et al.*, 2002). These inhibitors also reduced the upregulation of the leukocyte adhesion molecules, the loss of endothelial derived relaxing factors, suppression of mitochondrial respiration, and damage to the inter-endothelial tight junctions in cultured endothelium and in vascular rings challenged with peroxynitrite, TNFα, or IL-1β (Sharp *et al.*, 2001; Mazzon *et al.*, 2002).

Maximizing Intrinsic Protective Pathways

Feedback Inhibition and Autocrine Signaling

In many conditions, endothelium exhibits intrinsic intracellular homeostatic feedback mechanisms to dampen activation signaling and return to baseline function with time (Kozasa and Gilman, 1996; Yan *et al.*, 1998). Additionally, mediator degradation by surface peptidases and phosphorylation and internalization of receptors attenuate ongoing activation. Pathways that normally promote barrier formation, described in the first section, come into play as contacts are restored. Additionally, release of autocrines, such as PGI_2 and adrenomedullin with feedback receptor-mediated cAMP synthesis and PKA activation, may limit the response and restore the barrier. Adrenomedullin peptide, released primarily from the endothelium, has vasorelaxant properties and is increased in animal models of sepsis and in patients (Wang, 1998; Nikitenko *et al.*, 2002). PKA activation by adrenomedullin has a protective autocrine effect and exogenous addition has been shown in cultured endothelium to attenuate barrier dysfunction induced by thrombin, H_2O_2, or *E.coli* hemolysin (Hippenstiel *et al.*, 2002). Observations of intercellular gap formation *in vivo* in a variety of endothelia in response to a variety of inflammatory stimuli indicate that closure often occurs within 20 minutes (Baldwin and Thurston, 2001).

PKA Activation

Correspondingly, treatments to reverse activation signaling or accelerate normal barrier formation would be supportive, if it is not possible to block activation in the first place, as occurs with delayed presentation and diagnosis of pulmonary involvement in illness. Thus, various means to promote PKA activity have been employed to promote barrier function. In cultured endothelial cells, PKA activation provides barrier protection against endothelial activation by a wide variety of mediators (Demling *et al.*, 1981; Laposata *et al.*, 1983; Carson *et al.*, 1989; Minnear *et al.*, 1989; Sato *et al.*, 1991; Sheldon *et al.*, 1993; Siflinger-Birnboim *et al.*, 1993; Suttorp *et al.*, 1993; Patterson *et al.*, 1994a; Patterson *et al.*, 1994b; Ochoa *et al.*, 1997; Polte and Schroeder, 1998; Lum *et al.*, 1999; Patterson *et al.*, 2000; Hippenstiel *et al.*, 2002; Birukova *et al.*, 2004). Conversely, inhibition of PKA or overexpression of PKI (Figure 3) exacerbates the permeability response in cultured endothelium to agonist-stimulation (Lum *et al.*, 1991; Patterson *et al.*, 2000). PKA activation protected against several causes of experimental lung edema, such as air embolism, *tert*-butyl hydroperoxide, bacteremia, endotoxin, and hydrogen peroxide (Foy *et al.*, 1979; Demiling *et al.*, 1981; Minnear *et al.*, 1986; Farrukh *et al.*, 1987; Kobayashi *et al.*, 1987; Stelzner *et al.*, 1989; Adkins *et al.*, 1992; He and Curry, 1993b; Turner *et al.*, 1993; Baluk and McDonald, 1994; Barnard *et al.*, 1994; Seeger *et al.*, 1995; Waypa *et al.*, 2000). In intact vascular systems, cAMP benefits may be *via* reduced perfusion pressures as well as integrity of the barrier. Compounds that increase cAMP, such as long acting PGI_2 analog - Iloprost, diminished neutrophil adhesion and albumin permeability in isolated perfused lungs (Riva *et al.*, 1990; Hsu *et al.*, 1996; Schermuly *et al.*, 2003). Administration of the β-agonists in an intratracheal endotoxin-lung injury mouse model reduced cytokine release, neutrophil sequestration, and injury (Dhingra *et al.*, 2001; Maris *et al.*, 2004), but terbutaline administered after septic shock in a porcine model failed to reduce pulmonary permeability or mortality, suggesting that timing is critical (Groeneveld *et al.*, 1990). Inclusion of dibutyryl cAMP in perfusate did improve post-transplant lung function related to endothelial health (Kayano *et al.*, 1999). Adrenomedullin attenuated H_2O_2-induced permeability in isolated perfused rabbit lungs (Hippenstiel *et al.*, 2002), protected mice from LPS-induced septic shock (Shindo *et al.*, 2000), and inhibited endothelial apoptosis (Sata *et al.*, 2000). Pentoxifylline, an inhibitor of phosphodiesterases, raises cAMP. In a rat model of lung injury, due to intestinal ischemia-reperfusion, treatment with pentoxifylline 30 minutes after ischemia attenuated the albumin leak into the alveoli (Figure 4) (Carter *et al.*, 1995). Addition of roflumilast, a phosphodiesterase-4 inhibitor, improved oxygenation, neutrophil infiltration, and hyaline membrane formation (alveolar protein) in a rat model of repeated lavage lung injury, replenished with surfactant (Hafner and Germann, 1999). Importantly, PKA activation also stimulates the epithelial removal of alveolar fluid *via* increased sodium transport from the apical to basal surfaces; thus, aerosol delivery of cAMP-coupled agonists can further aid in resolution of alveolar edema if the epithelial and endothelial layers are reasonably intact or can be restored (Matthay *et al.*, 2000; Ware *et al.*, 2002). Therefore, these positive *in vivo* effects may occur by attenuating leukocyte adhesion and activation, smooth muscle relaxation and improved hemodynamics, platelet stability, epithelial sodium/water transport, or directly on endothelial barrier function, or by combinations. However, in severe injury, epithelial and endothelial proliferation is first required and other approaches may be called for, as described later.

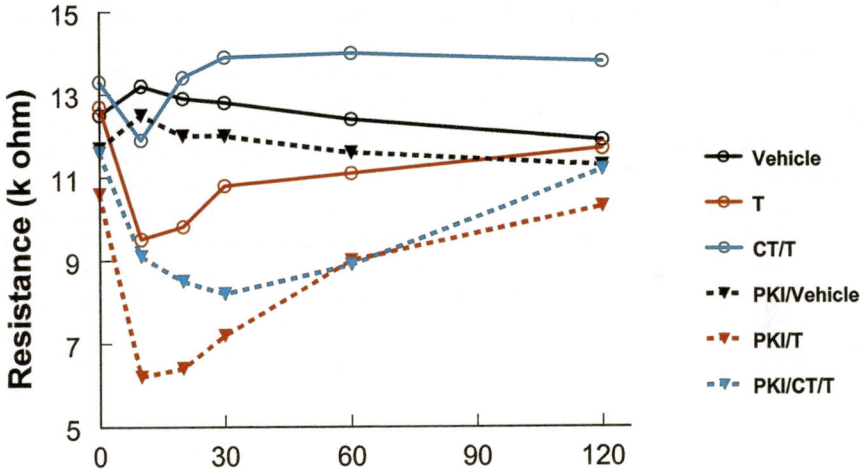

Figure 3: PKA activation attenuates thrombin-induced endothelial permeability, while PKA inhibition exacerbates dysfunction and slows recovery. SV40- transformed human dermal microvascular endothelial cells were grown to confluence and transfected with PKI (complete sequence DNA) for 24 h (*triangle and dashed lines*) or subjected to killed virus (*open circles and solid lines*). Monolayers were then treated with vehicle or cholera toxin (1ug/ml) for an additional 1 h to activate PKA (blue lines). Subsequently, electrical cell-impedance was determined as a measure of barrier integrity. PKA activation significantly increases basal barrier resistance and PKI significantly impairs basal resistance (see Patterson *et al.*, 2000). At this point (graph time 0), 200 nM α-human thrombin was added to some wells (*red lines* = *thrombin; blue lines* = *cholera toxin, then thrombin*) and others received vehicle (*black*). Thrombin-challenged endothelium showed the typical decreased impedance (indicative of lost cell-cell junctional integrity) with slow recovery. PKA activation attenuated, but did not totally prevent, the initial decline and accelerated the recovery. PKA inhibition with PKI worsened the thrombin-induced dysfunction and delayed recovery. PKI totally abrogated the protective effect of cholera toxin.

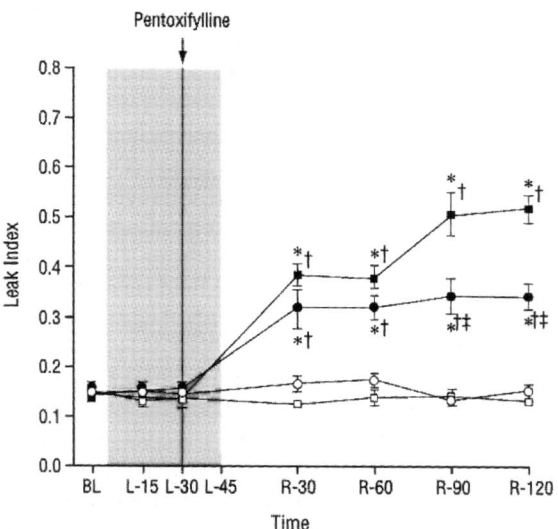

Figure 4: PKA activation attenuates alveolar-capillary membrane protein permeability after injury. Rats were subjected to 45 min. intestinal ischemia (*shadowed area*) followed by reperfusion and remote lung injury assessed by leak of intravascular fluorescently labeled albumin into alveoli, measured by epi-illumination fluorescent videomicroscopy. BL = baseline measure; L = ligation; R= reperfusion. Systemic mean arterial pressure was maintained by volume replacement to avert rapid death. Protein continued to accumulate in lung alveoli during the 90 minute period of intestinal reperfusion (*solid squares*), compared to no accumulation in sham-operation controls (*open squares*). After 30 minutes of the ischemia, some rats were treated with pentoxifylline, administered as an i.v. bolus to inhibit phosphodiesterases and activate PKA. Pentoxifylline significantly decreased the protein leak, which nearly flattened in the final 60 minutes (*closed circles*), but had no effect on sham control (*open circles*). Reproduced from *Carter, M.B., Wilson, M., Wead, W., and Garrison R.N. (1995) Pentoxifylline attenuates pulmonary macromolecular leakage after intestinal ischemia-reperfusion. Archives of Surgery. 130(12):1337-1344, 1995.* by kind permission of Dr. Mary Carter and The American Medical Association. Copyright © 1995. All rights reserved.

Pentoxifylline reduced TNF, improved clinical and radiologic signs of ARDS, and increased survival in a small study of cancer patients with ARDS (Ardizzoia *et al.*, 1993). Oxygenation was not improved, but was in another pentoxifylline study in ARDS patients (Montravers *et al.*, 1993). A study with a similar inhibitor, lisofylline, in ARDS failed to show clear benefit (ARDS Clinical Network, 2002). Terbutaline significantly reduced transferrin exudation into lung, determined by radiographic probing, within 30 minutes in the patients that later survived ARDS, but the non-survivors were non-responders (Basran *et al.*, 1986). Clinical benefits of intravenous PGI_2 for sepsis are attributed to systemic vasorelaxation and platelet stability (Scheeren and Radermacher, 1997), but effects on barrier function were usually not measured. In ARDS, PGI_2 decreased mean pulmonary pressure, but increased shunting of blood to poorly ventilated areas (Radermacher *et al.*, 1990) and dilation of systemic vessels resulting in problematic hypotension. To improve delivery to ventilated alveoli and reduce systemic hypotension, aerosolized PGI_2 was studied. Pulmonary pressures were decreased without systemic effects and arterial oxygenation was improved (Walmrath *et al.*, 1995, Bein *et al.*, 1996, Walmrath *et al.*, 1996). In a small study in ARDS patients given aerosolized PGI_2, oxygenation was improved despite no effect on mean pulmonary artery pressure, indicating either there was redistribution of blood flow to ventilated areas (i.e. reduced shunting) or reduced diffusion barrier (i.e. reduced alveolar and /or interstitial edema) (van Heerden *et al.*, 2000). Oxygenation was also improved in children with acute lung injury after aerosolized PGI_2 (Dahlem *et al.*, 2004). Similarly, aerosolized PGE_1 improved pulmonary resistance, admixture shunt, and oxygenation compared to intravenous infused PGE_1, which improved pulmonary resistance but worsened shunting and oxygenation (Putensen *et al.*, 1998). But in another, nebulized PGE_1 did not improve oxygenation, days on ventilation, or survival despite reduced pulmonary pressure (Vincent *et al.*, 2001). Permeability and endothelial function were not directly determined in these trials, but endothelium responds to PGI_2 and PGE_1 with cAMP generation, so it may be that some benefit was derived at the cellular level as well.

In smooth muscle, cAMP-induced relaxation is mediated *via* a decrease in $[Ca^{++}]_i$; but, in endothelium, PKA does not decrease basal or agonist-induced Ca^{++} mobilization and may even augment a secondary, late phase rise in agonist-induced Ca^{++} (Carson *et al.*, 1989; Buchan and Martin, 1991; Patterson *et al.*, 1994a; Patterson *et al.*, 2000; Hippenstiel *et al.*, 2002). The phosphorylation targets, and hence the biologic effects, of PKA activation are highly variable amongst different types of cells (Cass and Meinkoth, 1998; Miller *et al.*, 1998). Part of the PKA-mediated effect in endothelium was shown to be due to PKA-mediated phosphorylation of MLCK, which attenuates MLCK-phosphorylation of MLC induced by thrombin and other agonists (Moy *et al.*, 1993; Patterson *et al.*, 1994a; Garcia *et al.*, 1995a). However, PKA activation abrogated barrier dysfunction induced by agents, such as phorbol myristate acetate or pertussis toxin, which do not cause MLC phosphorylation, suggesting that other PKA targets are important in preventing the barrier dysfunction (Patterson *et al.*, 1994*a*; Patterson *et al.*, 1994*b*; Garcia *et al.*, 1995a; Patterson *et al.*, 1995; Moy *et al.*, 1998). As mentioned above, PKA activity may be decreased by Rho activation (Chrzanowskawodnicka and Burridge, 1996; Tigyi *et al.*, 1996; Takaishi *et al.*, 1997). Conversely, Rho proteins may themselves be targets for PKA-mediated phosphorylation (Flinn and Ridley, 1996; Schulze *et al.*, 1997). Pretreatment of endothelium to activate PKA lessened RhoA activation and translocation in response to thrombin (Qiao *et al.*, 2003). Moreover, downstream Rho family protein targets may also be subject to PKA counter-regulation (Burgering and Bos, 1995; Tigyi *et al.*, 1996; Arimura, *et al.*, 1997; Miller *et al.*, 1998).

Possible mechanisms were suggested above by which PKA would interfere with activation signaling, but there is strong evidence that rapid reversal of barrier dysfunction is a primary means for PKA promotion of barrier integrity after activation. For instance, when thrombin-dysfunction was fully established, subsequent cAMP elevation accelerated reversal of the dysfunction (Patterson *et al.*, 1994a). Moreover, in isolated lungs permeability established with hydrogen peroxide was reversed by isoproterenol-stimulated cAMP production or by phosphodiesterase inhibition of cAMP metabolism (Waypa *et al.*, 2000). The importance of PKA in initial establishment of the tightly confluent endothelium, the reduction of basal permeability in cultured endothelium, the prevention of agonists-induced dysfunction, and the rapid reversal of established dysfunction would suggest that protocols to activate PKA would be beneficial in lung edema and injury, involving endothelial dysfunction. On the other hand, PKA was reported to inhibit endothelial cell survival and angiogenesis with long-term administration (Kim *et al.*, 2002), so timing is critical, as well as assessment of whether proliferation or sealing of the barrier by the attached cells is most needed for return of endothelial barrier integrity.

PKG Activation and Nitric Oxide

In contrast to the direct benefit of cAMP on endothelial function, the reported benefits of NO and subsequent cGMP elevation are controversial. *In vitro* and *in vivo* studies of NO and cGMP have shown beneficial, mixed, and untoward effects. In some cases, cGMP elevation *per se* attenuated mediator-and hydrogen peroxide-induced permeability (Lofton, 1990; Westendorp, 1994; Suttorp 1996; Holschermann 1997; Lopez-Ongil *et al.*, 2001). In contrast, in other studies direct increases in cGMP with cell permeable analogs either failed to protect isolated endothelium from a variety of permeability-inducing mediators or even increased the permeability (Laposota *et al.*, 1983; Boulanger *et al.*, 1990; Schnittler *et al.*, 1990; Buchan and Martin, 1992; Patterson *et al.*, 1994a; Holschermann *et al.*, 1997). Whereas, cGMP reduced thrombin-induced permeability in aortic endothelium, it was without effect in freshly isolated human umbilical vein endothelium (Draijer *et al.*, 1995). This variation in response was explained by the differential expression of PKG in various endothelia (Draijer *et al.*, 1995). In endothelial cells, which are responsive to cGMP-barrier promotion, the effect appears to be related to either decreased mediator-induced Ca^{++} elevation or *via* cGMP-sensitive phosphodiesterase III and indirect increases in cAMP and PKA activity. This is similar to its effect in smooth muscle and in contrast to the effects of PKA, which does not alter mediator-induced Ca^{++} elevation. PKG activation increased untoward leukocyte adhesion, increased thromboxane to prostacyclin levels, and accompanied VEGF-induced permeability (Leszczynski, 1994; Fischer *et al.*, 1999). Moreover, histamine-induced endothelial permeability was dependent on NO and cGMP (Yuan, 1993). Thus, PKA activation, even subsequent to indirect cGMP activation, appears to be barrier protective, while PKG activation appears to have other signaling effects.

NO may stimulate endothelial cGMP production, but it also can result in increased cAMP (Polte and Schroeder, 1998). Moreover, it exhibits both pro-and anti-oxidant effects as will be explained below. Thus, interpretation of the available research, the mechanisms involved, and its usefulness in treating lung injury is even more complicated than with direct cGMP elevation. NO therapy has generally been applied in lung injury with the intent to correct perfusion abnormalities *via* its vasodilatory effects, but NO may actually benefit endothelial function and repair under some conditions. NO protected against thrombin-induced permeability in pulmonary endothelium (Westendorp *et al.*, 1994) and against phorbol ester-induced- permeability in lung microvasculature (Sprague *et al.*, 1998). NO prevented oxidant-induced increases in permeability in isolated endothelial cells and in pulmonary arteries (Lofton *et al.*, 1991; Suttorp *et al.*, 1996). Importantly, NO protection against TNF-induced endothelial cytotoxicity was due to PKA activation, rather than *via* PKG activation (Polte and Schroeder, 1998). In hyperoxic-lung injury in rats an intermediate level of inhaled NO (20ppm) attenuated both endothelial and alveolar epithelial type I cell injury (McElroy *et al.*, 1997). This same level of inhaled prophylactic NO attenuated pulmonary permeability and neutrophil migration and improved oxygenation in a septic pig model of lung failure (Figure 5) (Bloomfield *et al.*, 1997). In this study neutrophil-superoxide production was also suppressed, which benefited the lung in the short course of the experiment; but it should be noted that this could have detrimental impact if the ability to fight the sepsis is impaired. NO ventilation prevented neutrophil migration, reduced lavage myeloperoxidase levels, and prevented the macromolecular leak in isolated perfused IL-1-challenged rat lungs despite the potential for oxidant effects with higher levels (50 ppm) (Guidot *et al.*, 1996). It is unclear in ARDS models treated with NO whether improvement in arterial oxygenation was due to direct effects on smooth muscle vasodilation or due to improved release of endothelial-derived vasorelaxing mediators, decreased endothelial-derived vasoconstrictors, decreased neutrophil recruitment and activation, or directly on endothelial barrier function.

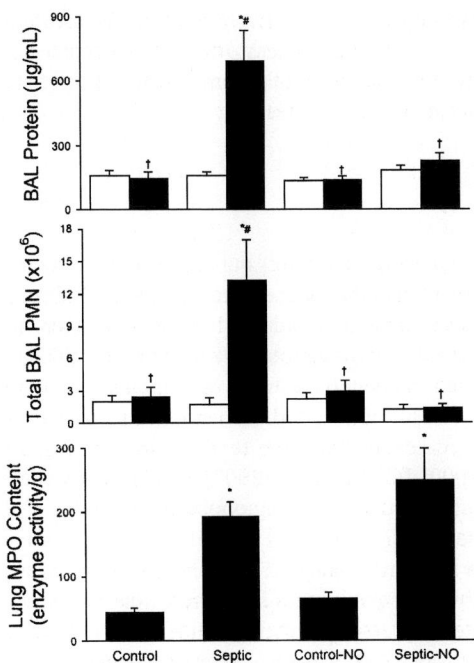

Figure 5: Inhaled NO inhibits sepsis-induced neutrophil accumulation and lung injury. Swine were ventilated with either air or air supplemented with NO (20 ppm), and then received a 1h i.v. infusion of either saline or live *Pseudomonas aeruginosa*. NO treatment prevented the increase in bronchoalveolar lavage protein in the septic animals, indicating reduction of alveolar-capillary membrane permeability. *Open bars are measures at time 0 and black bars at 5h.* Bronchoalveolar lavage neutrophil counts were also suppressed by NO treatment, indicating decrease transmigration; however, total lung myeloperoxidase increased in both septic groups, indicating the sequestration in the pulmonary vasculature had occurred, without abatement by the NO treatment. From *Bloomfield, G.L., Holloway, S., Ridings, P., Fisher, B., Blocher, C., Sholley, M., Bunch, T., Sugerman, H., and Fowler, A. (1997) Pretreatment with inhaled NO inhibits neutrophil migration and oxidative activity resulting in attenuated sepsis-induced acute lung injury. Crit. Care Med. 25:584-593.* by permission of Lippincott, Williams, and Wilkins, publishers.

In one clinical study, NO ameliorated permeability in acute lung injury (Benzing *et al.*, 1995) and in another it reversed the pulmonary endothelial dysfunction after heart surgery (Schulze-Neick *et al.*, 1999). However, in a study of cardiopulmonary bypass NO ventilation attenuated hemodynamic alterations but did not prevent endothelial dysfunction (Serraf *et al.*, 1997). In ARDS patients, NO improved arterial oxygenation without untoward systemic side effects of vasodilation and hypotension, but did not improve mortality (Pappert *et al.*, 1995; Rossaint *et al.*, 1995). In other studies, inhaled NO and aerosolized PGI_2 had similar pulmonary benefits, but the time course of PGI_2 was extended compared to NO (Pappert *et al.*, 1995; Walmrath *et al.*, 1996; Scheeren and Radermacker, 1997). Moreover, as discussed below, NO can contribute to free-radical damage depending on level and conditions, whereas PGI_2 does not have the oxidant effects. Five other clinical studies of inhaled NO in children and adults with ARDS conducted in 1998 and 1999 were re-analyzed together and it was concluded that no particular dose (1-80 ppm) showed specific advantage, that improvement in oxygenation was only seen initially, and that there was no improvement in ventilation therapy required, hospital stay, or mortality (Sokol *et al.*, 2003). In a recent study, the dose response to inhaled NO was examined for improved oxygenation in patients with and without continuous 10 ppm NO ventilation (Gerlach *et al.*, 2003). Initially, both groups responded best to 10 ppm. By the fourth day, the patients on NO had a heightened NO response, such that more responded to lower, 1ppm, doses; while the dose response curve in the control patient was unchanged. Additionally, at higher NO, the patients showed deterioration of oxygenation indicating sensitization. These studies suggest that the previous use of continuous NO may have resulted in overdosing and an unintended detrimental effect with long term NO at high doses, reducing benefit and complicating interpretation of the studies. Although, there was no difference in mortality between the two groups in this last study, the NO patients required extracorporeal membrane oxygenation less often. It is unclear how much of any beneficial effect might be due to direct effects on endothelium and how much to the vasoactive, inflammatory suppression, or platelet stabilization effects.

Angiopoietin-1

Not only is angiopoietin-1 essential for vascular development, but it also is important in maintaining endothelium in the adult through activation of the Tie2 receptor (Jones, 2003; Pizurki *et al.*, 2003). As mentioned above megacaryocytes express angiopoietin-1 and platelets store and release angiopoietin-1 upon stimulation, which could serve as a protective feedback mechanism. This suggests that exogenous angiopoietin-1 might be beneficial when endothelial function is compromised. Supporting this, pre-treatment of lung and umbilical vein endothelium with angiopoietin-1 attenuated the permeability increases in response to thrombin, bradykinin, histamine, PAF, and VEGF (Gamble *et al.*, 2000; Pizurki *et al.*, 2003). Importantly, basal permeability in unstimulated endothelium was also reduced with angiopoietin-1 (Gamble *et al.*, 2000). In a study of transgenic mice with overexpression of VEGF, double overexpression of angiopoietin-1 attenuated the VEGF-effects on albumin leakage (Thurston, 2002). Similarly, when wild-type mice were injected with adenovirus containing VEGF there was widespread plasma leakage, but transient adenovirus-mediated expression of angiopoietin-1 prior to injection of VEGF reduced the permeability response (Thurston, 2002). In the cell studies, angiopoietin-1 also reduced the thrombin-induced secretion of IL-8 and neutrophil adhesion and TNF-induced neutrophil transmigration (Gamble *et al.*, 2000; Pizurki *et al.*, 2003). While promotion of survival involves PI_3 kinase/Akt signaling, promotion of endothelial barrier function is independent of this pathway (Gamble *et al.*, 2000). This protective anti-permeability and anti-inflammatory effect of angiopoietin was not due to inhibition of thrombin-induced ROCK activation (Pizurki *et al.*, 2003). Curiously, study of potential activation of PKA has not been reported.

PPARγ

A recent, exciting experimental study examined the PPARγ-receptor agonist, rosiglitazone, normally used to control hyperglycemia and dyslipidemia in diabetic patients, but also suppresses cytokine expression by leukocytes. In a mouse model of non-septic zymosan-induced lung injury, rosiglitazone reduced mortality from 90% to 10% (Cuzzocrea *et al.*, 2004). Importantly, increases in plasma TNF, IL-1, and nitrite + nitrate and in lung myeloperoxidase, iNOS, leukocyte infiltration, TBARS, and nitrotyrosine were reduced, and lung alveolar macrophage NFκB nuclear translocation was prevented. These favorable effects on inflammation, oxidant stress, and function were shown in other organs as well. Direct or indirect effects on the endothelium and lung edema were not evaluated, but improvement might be anticipated as PPARγ agonists inhibit endothelial Akt activation, expression of leukocyte adhesion molecules, selected chemokines, and endothelin-1, and migration (Plutzky, 2001; Goetze et al., 2002; Hannan et al., 2003). High doses of rosiglitazone can cause transient permeability and apoptosis (Hannan et al., 2003; Idris et al., 2003). Thus, it is interesting to speculate about the potential benefits in other conditions leading to ARDS, such as sepsis, and on other potential direct effects in the endothelium.

Antioxidants

Decreasing Oxidant Effects and Oxidant Generation

Reactive oxidant species and reactive nitrogen species exert both signaling and cytotoxic effects and may be derived from the endothelium *per se*, from surrounding tissues, or from invading or resident leucocytes. A variety of anti-oxidant treatments provided protection against oxidant-mediated permeability or lung damage. Superoxide dismutase and catalase, or small mimetics, that catalyze degradation of superoxide and H_2O_2, have proved beneficial in animal models of ARDS (Milligan et al., 1988; Seekamp et al., 1988; Amari et al., 1993; Gonzalez et al., 1996; McClintock et al., 2002). Non-lethal oxidant exposure induces resistance to lethal oxidant insult by upregulating cellular defense and repair mechanisms, such as superoxide dismutase and glutathione peroxidase. Protection against H_2O_2-mediated increases in permeability and pressure in perfused lungs was provided by immunotargeting of catalase to the endothelium (Atochina et al., 1998). Gene transfer of heme oxygenase-1 stress protein into microvascular endothelial cells increased survival from H_2O_2 toxicity (Yang et al., 1999). Red blood cells protected endothelia against H_2O_2-mediated damage (Tsan and White, 1988) and, as discussed below, albumin exhibits protective effects that may be related to antioxidant properties (Power et al., 2003).

Continuous intravenous infusion of the antioxidants, cysteamine and N-acetylcysteine, improved pulmonary edema, lung glutathione state, and survival (Figure 6) in rats exposed to 100% oxygen (Patterson et al., 1985; Patterson and Rhoades, 1988). N-acetylcysteine also protected isolated perfused lungs from damage by activated neutrophils (Simon and Suttorp, 1985). Other studies of hyperoxic lung injury in dogs, guinea pigs, and rats confirmed the benefits of N-acetylcysteine treatment on neutrophil sequestration and function-structural damage to the lungs (Jamieson et al., 1987; Wagner et al., 1989; Langley and Kelly, 1993; Lucas et al., 1995). But, untoward effects on type II epithelial cells were noted with once daily intraperitoneal N-acetylcysteine at lower levels of oxygen stress (van Klavern et al., 1997), indicating the importance of dosing and targeting drug delivery. In other animal models of ARDS, N-acetylcysteine also improved pulmonary gas exchange and reduced lung edema, neutrophil influx, and reduced mortality from respiratory failure (Bernard et al., 1984; Bernard, 1990; Feddersen et al., 1993; Leff et al., 1993; Fan et al., 2000; McClintock et al., 2002; Ritter et al., 2004; Timlin et al., 2004). The beneficial effects of one-time intratracheal administration were prolonged when the N-acetylcysteine was administered in liposomes (Fan et al., 2000). In contrast, post-treatment with N-acetylcysteine after septic shock or endotoxemia was induced in pigs, reduce pulmonary microvascular permeability determined by post-mortem inspection, but did not alter hemodynamics or mortality (Groeneveld et al., 1990; Vassilev et al., 2004), indicating the importance of early intervention.

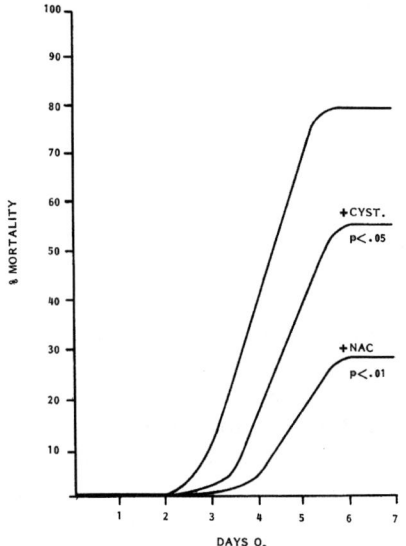

Figure 6: **Sulfhydryl antioxidants improve survival in hyperoxic lung injury.** Rat were continuously infused with vehicle, cysteamine (150mg/kg/day), or N-acetylcysteine (390 mg/kg/day) and exposed to continuous 99-100% O_2 for 7 days. Cysteamine reduced mortality by 30%, restored lung non-protein sulfhydryl levels, and attenuated lung edema. N-acetylcysteine reduced mortality a remarkable 50% (Patterson et al., 1985). Figure from Patterson and Rhoades, 1998. Copyright (1988) from *Protective role of sulfhydryl reagents in oxidant lung injury.* by C.E. Patterson and R.A, Rhoades. Reproduced by permission of Taylor & Francis, Inc., http://www.taylorandfrancis.com

Treatment of patients with moderate lung injury with N-acetylcysteine to directly scavenge oxidants or to increase cell cysteine, and hence glutathione, improved oxygenation and reduced need for ventilatory support, although development of ARDS and mortality was not altered (Suter et al., 1994). N-acetylcysteine administration in ARDS patients resulted in improved oxygen delivery, lung compliance, edema, and resolution of the edema (Bernard, 1990; Bernard, 1991). In subsequent small clinical trials, N-acetylcysteine or oxothiozolidine administration after established ARDS, reduced days of lung injury, but did not alter mortality (Jepsen et al., 1992; Bernard et al., 1997a; Domenighetti et al., 1997). It seems likely that the failure could be due to treatment that is simply too late in the process and high-dose N-acetylcysteine can upregulate iNOS leading to NO overproduction (Van Klaveren et al., 1997; Groeneveld and Sipkema, 2000). Early treatment of trauma patients with high risk for ARDS with α-tocopherol and ascorbic acid did not alter the incidence of ARDS, but did reduce the alveolar protein levels and mortality from multiple organ failure (Nathens et al., 2002). The small numbers in the clinical studies, the complex nature of ARDS, and the presence of co-morbidities make full assessment problematic. Moreover, effects of these treatments on endothelial injury and function are unknown.

In addition to the ability of NO to increase cGMP and activate PKG, discussed above, NO is a free radical itself, which can either protect from oxidant injury by scavenging superoxide or contribute to oxidant stress *via* peroxynitrite and nitrogen dioxide production, depending on the immediate environment. This can explain problems encountered with high levels of therapeutic NO and altered sensitivity with time. In a study of hyperoxic-lung injury in rats, inhibition of endogenous NO production increased TBARS, indicative of oxidant stress, and worsened survival, suggesting the importance of low level, endogenous NO in reducing oxidant stress (Garat *et al.*, 1997). In this study, added inhalation therapy with low, 10 ppm NO reduced the hyperoxic-increases in TBARS and lung edema, yet did not alter protein leak or survival. However, when the inhaled NO was increased to 100 ppm, rather than producing a further benefit, there was an increase in vascular protein permeability (Garat *et al.*, 1997). Thus, dosage is extremely critical. When there is excessive NO production, as with induction of iNOS and there is concomitant production of superoxide, as occurs with activation and upregulation of NADPH oxidase in the leukocytes or in the endothelium itself, peroxynitrite generation results.

Peroxynitrite can alter tyrosine-phosphorylation signaling by nitrosylation of protein-tyrosine residues. Furthermore, peroxynitrite activates the nuclear enzyme poly (ADP-ribose) polymerase. When tissues from knock-out mice for this enzyme or pharmacologic inhibitors were used, endotoxin-mediated and low level peroxynitrite-mediated endothelial dysfunction, characterized by reduced vasodilator release and decreased mitochondrial respiration, was attenuated, indicating this signaling pathway was involved in dysfunction (Szabo *et al.*, 1997a). NO enhanced hydrogen peroxide-mediated neutrophil adhesion and endothelial injury (Okayama *et al.*, 1998). Specific inhibition of iNOS to block NO overproduction reduced infused endotoxin-induced acute lung injury and proteinaceous edema in rabbits, if given within 2 hrs after insult but not at later times (Mikawa *et al.*, 2003). Propofol, a peroxynitrite radical scavenger similarly attenuated lung injury and mortality in rats infused with LPS, with diminished effectiveness administered after the insult (Gao *et al.*, 2004). Peroxynitrite caused depolarization of the mitochondrial membrane, redox dysregulation, DNA fragmentation, and death in pulmonary endothelial cells (Gow *et al.*, 1998). Pre-treating the cells with N-acetylcysteine or cysteine attenuated the cellular injury and permeability (Gow *et al.*, 1998; Knepler *et al.*, 2001). Type II pulmonary epithelial cells were even more sensitive than the endothelium to the cytotoxic effects of both NO and peroxynitrite (Gow *et al.*, 1998). VEGF-induced permeability was ascribed to the combined effects of NO and PGI$_2$, although PGI$_2$ is generally considered protective as discussed above (Murohara *et al.*, 1998; Fischer *et al.*, 1999). Similarly, the combination of NO and prostaglandins were implicated in increased permeability due to endotoxin in skin (Fujii *et al.*, 1996). Paradoxically, inhaled NO was found to reduced tyrosine nitration after pulmonary instillation of LPS in rats (Honda *et al.*, 1999). Thus, it is not surprising that controversy persists regarding administration of blood borne or ventilatory NO, during edemagenic conditions to protect the endothelium or preserve other NO-mediated functions. Obviously, timing and dosage and control of tissue and leukocyte generation of NO to prevent increased oxidant load is extremely critical.

In addition to blocking oxidant signaling and toxicity with scavengers and metabolizing enzymes, new strategies are coming to light to prevent pathologic oxidant generation. Agents, such as apocynin, are effective blockers of NADPH oxidase (Cotter and Cameron, 2003). Statins, widely used to lower cholesterol associated with atherosclerosis, are now recognized to reduce mediator-induced oxidant stress. Non-lipid-related benefits of statins in endothelium include improved NO production, anticoagulant properties, antiproliferative effects, plaque stabilization, and prevention of tumor growth (Altieri, 2001; Siegel-Axel, 2003). These effects could be attributed to decreased oxidant signaling *via* inhibition of the activation of NAD(P)H oxidase and also to interfering with other RhoA-directed events such as cytoskeletal reorganization. By statin's interference with isoprenylin production, mediator-induced activation of rac1 is prevented, which prevents rac1 participation in assembly of cytosolic components of NADPH oxidase on the membrane to produce a functional oxidase (Dinauer, 2003). Pre-treatment of cultured endothelium with simvastatin, reduced Rho-A translocation to the membrane and thrombin-induced permeability without a change in cAMP level (van Nieuw Amerongen et al., 2000b). In vivo statin treatment, without a change in the vascular wall fat disposition, decreased permeability in the aortic vascular wall (van Nieuw Amerongen et al., 2000b). Similarly, statin pretreatment greatly attenuated ischemia/reperfusion lung injury (Naidu et al., 2003). Given the importance of Rac to barrier integrity (Waschke et al., 2004), it is not clear how the protective effects of statins relate to specific rac-signaling regarding the quiescent cytoskeleton.

Albumin

Salutary effects of high albumin in lung injury have also been attributed to its antioxidant properties, which account for the vast majority of free thiols in plasma. Albumin-infusion attenuated elevation of 8-isoprostane, whereas albumin with depleted antioxidant capacity failed to protect rodent model of lung injury from shock and endotoxin, (Powers et al., 2002; Powers et al., 2003). Protection of endothelial cells from HOCl-induced injury was attributed to antioxidant properties of albumin apart from the thiol status (Lang et al., 2004). In this study, albumin also attenuated binding of neutrophils and neutrophil-derived myeloperoxidase, whereas hydroxyethyl starch enhanced binding (Lang et al., 2004). In a small study of patient with acute lung injury, albumin improved plasma anti-oxidant status and protein oxidative damage (Quinlan et al., 2004). Additional, direct benefits of albumin-binding to endothelial cells were mentioned above in the section on normal endothelium- autocrine and bloodborne signals, but the benefit of using albumin in patients with critical illnesses is controversial and the focus has been primarily on the osmotic effects of the albumin. Although low serum albumin correlates with mortality (Vincent et al., 2003), in patients with hypovolemia due to surgery or injury, hypoproteinemia, or burns, albumin administration either had no effect (Wilkes and Navickis, 2001) or slightly increased the overall low mortality (Cochrane Injuries Group, 1998). On the other hand, the superiority of albumin to artificial colloids and over crystalloids in restoring fluid balance and reducing morbidity and mortality was noted in other trials (Groeneveld, 2000). In a more specific study of patients with acute lung injury and hypoproteinemia, combined therapy with albumin and a loop diuretic, improved fluid balance, arterial oxygenation, and hemodynamics indicating positive effects, although there was no difference mortality (Martin et al., 2002). Thus, albumin supplementation is also no "magic bullet," but its potential oncotic, buffering, anti-oxidant, and endothelial barrier promoting effects warrant further consideration in combination therapies.

Prevention and Reversal of Injury

Cell Death and Replacement

Endothelial injury, extending beyond mere activation and temporarily loss of barrier function, presents with increased plasma and mitochondrial membrane permeability, altered distribution of ions, impaired energy metabolism, activation of cellular proteases, and possibly DNA damage. These processes potentially lead to cytotoxic death or activation of apoptotic or proliferative responses. As mentioned above in blocking activation signaling, oxidant-induced DNA damage-activation of poly (ADP-ribose) polymerase can initiate necrosis and its inhibition can protect from various organ damage (Watts and Kline, 2003). Specific endothelial function was not examined in most of the *in vivo* studies, but *in vitro* cytotoxicity in endothelium induced with exogenous peroxynitrite was reduced by polymerase inhibition (Zingarelli *et al.*, 1997).

Recovery of the cell from injury is more complex than recovery from activation and if loss of the cell due to death or detachment occurs, re-endothelialization of the vascular surface must occur. Temporarily platelets and leukocytes may attach to the interstitium forming a quasi barrier. Normally, dead cells remain attached, frank denudation is rare, and growing new cells crawl under the dead cells and pushes them off. In either case, endothelial activation, including oxidant signaling, promotes endothelial proliferation and motility for wound repair. Additionally, there is new evidence that circulating endothelial precursor cells can be recruited to promote re-endothelialization. Most circulating endothelial-like cells are likely derived from the vascular wall and have limited proliferative capacity, but the small number of bone-marrow-derive endothelial precursor cells have high capacity for proliferation and appear to be more important for re-endothelialization and angiogenesis (Shi, 1998; Yi *et al.*, 2000). Pretreatment of grafts with bone marrow CD34+ cells prior to implantation in the descending thoracic aortas of dogs yielded improved endothelialization (Bhattacharya *et al.*, 2000). As expected, these precursor cells are more numerous in cord blood (Murohara *et al.*, 2000) and are responsive to VEGF (Reyes *et al.*, 2002). But it is surprising that statins also promote proliferation of the endothelial progenitor cells (Dimmeler *et al.*, 2001; Llevador, 2001). Selective application of specific endothelial and epithelial growth factors may be useful to promote re-establishment of the intact alveolar capillary membrane (Ware and Matthay, 2001), but this must be carefully balance with potential permeability effects noted above. The corollary to this is that PKA activation, shown above to be beneficial in promoting barrier function and reducing pulmonary hypertension, at the point of severe injury may be counterproductive for growth and spreading of the endothelium, due to its signaling of contact inhibition and suppression of cell proliferation and motility.

Matrix Destruction, Prevention, and Repair

In addition to prevention and reversal of endothelial injury, prevention of damage to the matrix or reversal of matrix disruption is important to restoration of endothelial barrier function due to the important interactions and intimate association of endothelium and the basal lamina. New modes of low volume ventilation and efforts to control hemodynamic stress on the lungs have reduced mechanical damage to the interstitium. In acute lung injury, disruption of the basal lamina and the interstitium is due, not only to mechanical stress, but also due to protease digestion. Normal leukocyte diapedesis across the endothelial layer and into tissue is aided by activation of matrix metalloproteinases in a precisely localized and tightly regulated manner that generally results in no barrier dysfunction. The matrix metalloproteinases are endopeptidases that function in matrix degradation and remodeling and are controlled by expression, binding the surface receptors, proteolytic activation, and extracellular inhibitors. With overactivation of endothelial matrix metalloproteinases and injection of matrix metalloproteinases and elastases from the leukocytes, there is uncontrolled matrix protein digestion. In addition, this proteolytic activity degrades occludin and VE-cadherin, thereby increasing permeability directly at the endothelial cell level (Carden *et al.*, 1998; Herren *et al.*, 1998; Gibbs *et al.*, 1999; Wachtel, 1999). As in investigation of growth factors, clinical interests in inhibition of proteases has been directed at arresting angiogenesis to prevent or regress tumors, but the interactions of endothelial cells and matrix elements is highly relevant to pulmonary edema. Inhibition of neutrophil elastase prevented inflammatory and edematous changes and improved survival in several models of acute lung injury and in patients (Kawabata *et al.*, 2000; Hagio *et al.*, 2001; Eaton and Martin, 2002; Zeiher *et al.*, 2002; Kawabata *et al.*, 2003). Likewise, acute lung injury was reduced and survival improved by inhibition of neutrophil-metalloproteinases with COL-3 in a rodent model of sepsis (Steinberg *et al.*, 2003) and by batimistat in a rodent model of pancreatitis (Muhs *et al.*, 2003). Additionally, there are ways in which the interstitium is repaired and in which the matrix feeds back to benefit the endothelium. For instance, matrix proteins *via* integrin signaling attenuated endotoxin-lung endothelial apoptosis (Hoyt *et al.*, 1996). In a bacteremia sheep model, plasma fibronectin was incorporated into lung matrix, suggesting this may be a repair mechanism (Charash *et al.*, 1993). Similarly, in cultured endothelium, soluble fibronectin incorporated into matrix and lessened TNFα-induced permeability (Wheatley *et al.*, 1993; Curtis *et al.*, 1995).

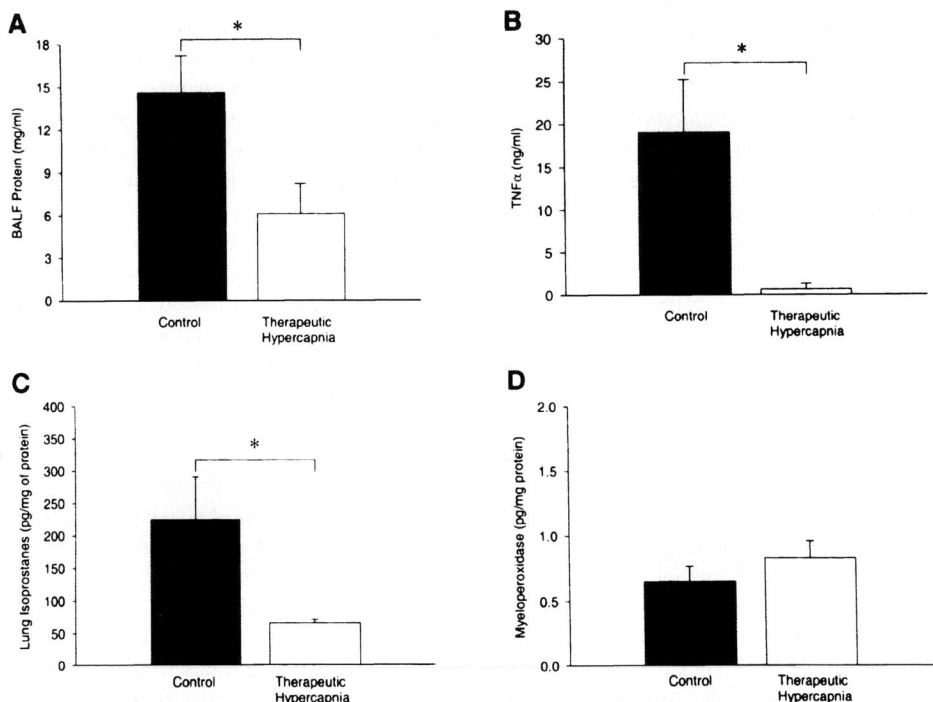

Figure 7: Hypercapnia attenuates lung reperfusion injury. Rabbits ventilated with (FI_{CO_2} 0.12) resulted in pH, 7.1 acidosis. Lung ischemia was initiated for 75 min, followed by 90 min reperfusion. The therapeutic hypercapnia reduced lung wet weight gain, BAL protein, and TNF, indicative of reduced alveolar-capillary permeability and inflammation. Lung tissue nitrotyrosine and 8-isoprostane levels were reduced in the hypercapnia group, but myeloperoxidase levels were similar in both groups, indicating that oxidant stress was reduced, although neutrophil sequestration was comparable. Reproduced from *Laffey, J.G., Tanaka, M., Engelberts, D., Luo, X., Yuan, S., Tanswell, A., Post, M., Lindsay, T., and Kavanagh, B.P. (2000b) Therapeutic hypercapnia reduces pulmonary and systemic injury following in vivo lung reperfusion. Amer. J. Resp. Crit. Care Med. 162:2287-2294.* By kind permission of Dr. Brian Kavanagh and The American Journal of Respiratory and Critical Care Medicine and The American Thoracic Society.

Surfactant and Acidosis

Other therapies with only indirect endothelial effects, such as surfactant replacement therapy to reduce mechanical stress on the alveolar capillary wall and aid in restoration of lung fluid balance, have had limited success in ARDS (Eaton and Martin, 2002; Spragg et al., 2003). One rather interesting study and promising treatment addressed the basis for beneficial effects of permissive hypercapnia (Laffey et al., 2000b). It had been reasonably assumed that reduced ventilation volume was the basis of protection afforded by allowing arterial CO_2 levels to modestly increase (Hickling and Joyce, 1995; Gillette and Hess, 2001; Hickling, 2002). Yet therapeutic hypercapnia (FI_{CO2} = 0.12) in ischemia-reperfusion rabbit lung injury reduced protein leakage and edema and improved oxygenation independent of ventilation (Fig. 7). Importantly, 8-isoprostane levels were reduced 70% and TNF levels were reduced 95%, indicating attenuation of oxidant stress and inflammation. Suggestions from earlier studies in isolated perfused lungs indicated that acidosis, due to hypercapnia, may be important in the protective effect. Whereas acidosis attenuated permeability due to either ischemia/reperfusion or to exogenous superoxide generation (Shibata et al., 1998), buffering of the hypercapnia-related acidosis nullified the protection (Figure 8) (Laffey et al., 2000a). Similarly, increased alveolar to arterial oxygen gradient and peak airway pressures due to 4h hyperventilation (12 ml/kg vs. 7.5 ml kg) were reduced by hypercapnia and exacerbated by hypocapnia, independent from changes in broncho-alveolar lavage surfactant recovery (Laffey et al., 2003).

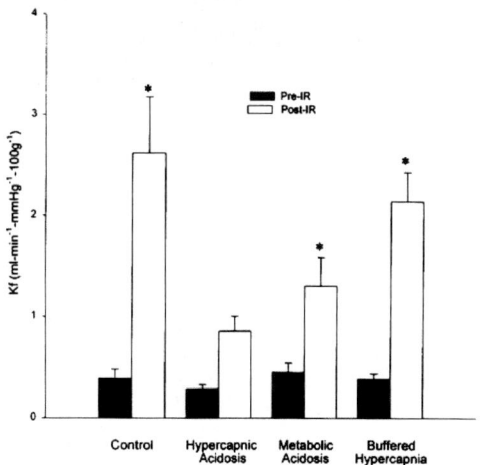

Figure 8: Buffering attenuates hypercapnic-protection in lung injury. Isolated perfused rabbit lungs were stabilized; then ventilation and perfusion were stopped 45 min. and recommenced for 30 min. Lungs, ventilated with standard 5% CO_2, had increased wet weight and capillary pressures and a 6.5 fold increase in specific capillary filtration at controlled capillary pressure, whereas lungs receiving supplemental 25% CO_2 (pH = 6.9) were protected with significant attenuation of wet weight gain, capillary pressure, and specific capillary permeability. Ventilation with 5% CO_2 and perfusion with comparable pH 6.9 metabolic acidosis had the same attenuation of the increased capillary pressure and tended toward protection of weight gain and capillary filtration, but was much less effective than hypercapnia. Buffering of the acidosis, while ventilating with the 25% CO_2, did not alter the reduction in the ischemia/reperfusion-induced rise of capillary pressure, but it blocked the hypercapnic protection against increase capillary filtration, indicating that CO_2 ventilation and acidosis reduces ischemia/reperfusion lung injury. Reproduced from *Laffey, J.G., Engelberts, D., and Kavanagh, B.P. (2000a) Buffering hypercapnic acidosis worsens acute lung injury. Amer. J. Respir. Crit. Care Med. 161:141-146* by kind permission of Dr. Brian Kavanagh and The American Journal of Respiratory and Critical Care Medicine and The American Thoracic Society.

And more recently, pre- or post-hypercapnic acidosis improved lung compliance and arterial oxygenation and reduced neutrophil influx and lung injury histology changes due to endotoxin intratracheal endotoxin instillation in rats (Laffey et al., 2004). Endothelial function and permeability was not directly examined, but it would not be surprising that endothelial function was better preserved by acidosis relative to alkalosis, as has been observed by the authors and others with regard to both basal and agonists-altered barrier function in cultured cells. Caution must be taken with hypercapnia, however, due to potential untoward effects on cardiac contractility and on over-dilation and over-perfusion of brain vessels resulting in increased intracranial pressure (Malarkkan et al, 2003).

Protecting the Dam on Multiple Fronts

In summary, despite clear benefits of the numerous treatments in specific experimental studies, no single therapy to reduce lung failure has translated to provide clear, across the board improvement in survival in ARDS (Artigas et al., 1998; Ware and Matthay, 2000; Groeneveld, 2003; Matthay et al., 2003). This is due to inherent differences in the many etiologies leading to lung dysfunction and injury, the varied contribution of the underlying pathologies or injuries to death, the pattern of lung ventilation and perfusion, the presence of inflammation stresses including oxidant stress in the lung, and the interactions among the endothelium, epithelium, and matrix in the injury response. This complexity and the likelihood that the mechanisms and mediators have interactive, synergistic pathologic effects, means that no one agent can stand alone against the flood and the fire. Improved understanding of the participation of the endothelium of the lung and of other organs in the integrated pathophysiologic processes and the best means to protect and restore endothelial function, with its important influence over lung fluid balance, vascular function, inflammation, and coagulation will be critical to further advances.

ACKNOWLEDGEMENTS:

This work was supported by a Merit Grant from the Veterans Administration Medical Research Service (CEP) and NIH HL-071081 (HL).

REFERENCES:

Abraham, E., Reinhart, K., Opal, S., Demeyer, I., Doig, C., Rodriguez, A., Beale, R., Svoboda, P., Laterre, P., Simon, S., Light, B., Spapen, H., Stone, J., Seibert, A., Peckelsen, C., De Deyne, C., Postier, R., Pettila, V., Artigas, A., Percell, S., Shu, V., Zwingelstein, C., Tobias, J., Poole, L., Stolzenbach, J., and Creasey, A. (2003) OPTIMIST Trial Study Group. Efficacy and safety of tifacogin (recombinant tissue factor pathway inhibitor) in severe sepsis. JAMA 290:238-247.

Acosta, J., Qin, X., and Halperin, J. (2004) Complement and complement regulatory proteins as potential molecular targets for vascular diseases. Curr. Pharm. Design 10:203-211.

Adamson, R.H., Liu, B., Nilson, G. Rubin, L., and Curry, F.E. (1998) Microvascular permeability and number of tight junctions are modulated by cAMP. Amer. J. Physiol. 247:H1885-H1894.

Adkins, W., Barnard, J., May, S., Seibert, A., Haynes, J., and Taylor, A.E. (1992) Compounds that

cells. J. Biol. Chem. 273:21867-21874.
Ettenson, D.S. and Gotlieb, A.I. (1992) Centrosomes, microtubules, and microfilaments in the reendothelialization and remodeling of double-sided in vitro wounds. Lab. Invest. 66:722-733.
Fahy, R.J., Lichtenberger, F., McKeegan, C., Nuovo, G., Marsh, C., and Wewers, M.D. (2003) The acute respiratory distress syndrome: a role for TGF-β1. Amer. J. Respir. Cell Molec. Biol. 28:499-503.
Fan, J., Shek, P., Suntres, Z., Li, Y., Oreopoulos, G., and Rotstein, O.D. (2000) Liposomal antioxidants provide prolonged protection against acute respiratory distress syndrome. Surgery. 128:332-338.
Farmer, P.J., Girardot, D., Lepage, A., Regoli, D., and Sirois, P. (2002) Inhibition of prostaglandin G/H synthase unveils a potent effect of platelet activating factor on the permeability of bovine aortic endothelial cells to albumin. Inflamm. 26:253-258.
Farrukh, I.S., Gurtner, G.H., and Michael,J.R. (1987) Pharmacological modification of pulmonary vascular injury: possible role of cAMP. J. Appl. Physiol. 62:47-54.
Feddersen, C.O., Barth, P., Puchner, A., and von Wichert, P. (1993) N-acetylcysteine decreases functional and structural, ARDS-typical lung changes in endotoxin-treated rats. Medizin. Klinik 88:197-206.
Fischer, S., Clauss, M., Wiesnet, M., Renz, D., Schaper, W., and Karliczek, G.F. (1999) Hypoxia induces permeability in brain microvessel endothelial cells via VEGF and NO. Amer. J. Physiol. 276:C812-C820.
Flick, M.R., Webster, R., Hoeffel, J., Julien, M., Milligan, S. Kent, B., and Lesser, M. (1993) Effect of phenytoin on acute lung injuries in unanesthetized sheep. Crit. Care Med. 21:1563-1571.
Flinn, H.M. and Ridley, A.J. (1996) Rho stimulates tyrosine phosphorylation of focal adhesion kinase, p130 and paxillin. J. Cell. Sci. 109:1133-1141.
Folkesson, H.G., Matthay, M., Hebert, C., and Broaddus, V.C. (1995) Acid aspiration-induced lung injury in rabbits is mediated by interleukin-8-dependent mechanisms. J. Clin. Invest. 96:107-116.
Foy, T., Marion, J., Brigham, K.L., and Harris, T.R. (1979) Isoproterenol and aminophylline reduce lung capillary filtration during high permeability. J. Appl. Physiol. 46:146-151.
Gainor, J.P., Morton, C., Roberts, J., Vincent, P., and Minnear, F.L. (2001) Platelet-conditioned medium increases endothelial electrical resistance independently of cAMP/PKA and cGMP/PKG. Amer. J. Physiol. 281:H1992-2001.
Galloway, S.W. and Kingsnorth, A. (1996) Lung injury in the microembolic model of acute pancreatitis and amelioration by lexipafant, a platelet-activating factor antagonist. Pancreas 13:140-146.
Gamble, J.R., Drew, J., Trezise, L., Underwood, A., Parsons, M., Kasminkas, L., Rudge, J., Yancopoulos, G., and Vadas, M.A. (2000) Angiopoietin-1 is an antipermeability and anti-inflammatory agent in vitro and targets cell junctions. Circ. Res. 87:603-607.
Gao, J., Zeng, B., Zhou, L., and Yuan. S.Y. (2004) Protective effects of early treatment with propofol on endotoxin-induced acute lung injury in rats. Brit. J. Anaesth. 92:277-279.
Garat, C., Jayr, C., Eddahibi, S., Laffon, M., Meignan, M., and Adnot, S. (1997) Effects of inhaled nitric oxide or inhibition of endogenous nitric oxide formation on hyperoxic lung injury. Amer. J. Respir. Crit. Care Med. 155:1957-1964.
Garcia, J.G.N., Davis, H.W., and Patterson, C.E. (1995a) Regulation of endothelial cell gap formation and barrier dysfunction: role of myosin light chain phosphorylation. J. Cell. Physiol. 163:510-522.
Garcia, J.G.N., Pavalko, F.M., and Patterson, C.E. (1995b) Vascular endothelial cell permeability responses to thrombin. Blood Coag. Fibrin. 6:609-626.
Garcia, J.G.N., Siflinger-Birnboim, A., Bizios, R., DelVecchio, P., Fenton, J., and Malik, A.B. (1986) Thrombin-induced increase in albumin permeaility across endothelium. J. Cell. Physiol. 128:96-104.
Gerlach, H., Keh, D., Semmerow, A., Busch, T., Lewandowski, K., Pappert, D., Rossaint, R., and Falke, K. (2003) Dose-response characteristics during long-term inhalation of nitric oxide in patients with severe acute respiratory distress syndrome. Amer. J. Respir. Crit. Care Med. 167:1008-1015.
Gibbs, D.F., Shanley, T., Warner, R., Murphy, H., Varani, J., and Johnson, K.J. (1999) Role of matix metalloproteinases in models of macrophage-dependent acute lung injury. Amer. J. Respir. Cell Molec. Biol. 20:1145-1154.
Gillette, M.A. and Hess, DR. (2001) Ventilator-induced lung injury and the evolution of lung-protective strategies in acute respiratory distress syndrome. Respir. Care 46:130-148.
Goetze, S., Bungenstock, A., Czupalla, C., Eilers, F., Stawowy, P., Kintscher, U., Spencer-Hansch, C.,

Graf, K., Nurnberg, B., Law, R., Fleck, E., and Grafe, M. (2002) Leptin induces endothelial cell migration through Akt, which is inhibited by PPARgamma-ligands. Hyperten. 40:748-754.

Goldberg, P.L., MacNaughton, D., Clements, R., Minnear, F., and Vincent, P.A. (2002) p38 MAPK activation by TGF-β1 increases MLC phosphorylation and endothelial monolayer permeability. Amer. J. Physiol. 282:L146-L154.

Gonzalez, P.K., Zhuang, J., Doctrow, S., Malfroy, B., Benson, P., Menconi, M., and Fink M.P. (1996) Role of oxidant stress in ARDS: evaluation of a novel antioxidant strategy in a porcine model of endotoxin-induced acute lung injury. Shock. 6:S23-S26.

Gow, A., Thom, S., and Ischiropoulos, H. (1998) Nitric oxide and peroxynitrite-mediated pulmonary cell death. Amer. J. Physiol. 274:L112-L118.

Grazia-Lampugnani, M., Zanetti, A., Breviario, F., Balconi, G., Orsenigo, F., Corada, M., Spagnuolo, R., Betson, M., Braga, V., and Dejana, E. (2002) VE-cadherin regulates endothelial actin activating Rac and increasing membrane association of Tiam. Molec. Biol. Cell 13:1175-1189.

Grazia-Lampugnani, M., Zanetti, A., Corada, M., Takahashi, T., Balconi, G., Breviario, F., Orsenigo, F., Cattelino, A., Kemler, R., Daniel, T., and Dejana, E. (2003) Contact inhibition of VEGF-induced proliferation requires vascular endothelial cadherin, beta-catenin, and the phosphatase DEP-1/CD148. J. Cell Biol. 161:793-804.

Greenman, R.L., Schein, R., Martin, M., Wenzel, R., MacIntyre, N., Emmanuel, G., Chmel, H., Kohler, R., McCarthy, M., and Plouffe J. (1991) A controlled clinical trial of E5 murine monoclonal IgM antibody to endotoxin in the treatment of gram-negative sepsis. The XOMA Sepsis Study Group. JAMA. 266:1097-1102.

Groeneveld, A.B.J. (1999) Treatment of increased microvascular permeability following inflammation? Crit. Care Med. 27:23-24.

Groeneveld, A.B.J. (2000) Albumin and artificial colloids in fluid management: where does the clinical evidence of their utility stand? Crit. Care (London) 4:S16-S20.

Groeneveld, A.B.J. (2003) Vascular pharmacology of acute lung injury and acute respiratory distress syndrome. Vasc. Pharm. 39:247-256.

Groeneveld, A.B.J., den Hollander, W., Straub, J., Nauta, J., and Thijs, L.G. (1990) Effects of N-acetylcysteine and terbutaline treatment on hemodynamics and regional albumin extravasation in porcine septic shock. Circulatory Shock. 30:185-205.

Groeneveld, A.B.J. and Sipkema, P. (2000) Interaction of oxyradicals, antioxidants, and nitric oxide during sepsis. Crit. Care Med. 28:2161-2162.

Guidot, D.M., Hybertson, B., Kitlowski, R., and Repine, J.E. (1996) Inhaled NO prevents IL-2-induced neutrophil accumulation and associated acute edema in isolated rat lungs. Amer. J. Physiol. 271:L225-L229.

Hafner, D. and Germann, P.G. (2000) Additive effects of phosphodiesterase-4 inhibition on effects of rSP-C surfactant. Amer. J. Respir. Crit. Care Med. 161:1495-1500.

Hagio, T., Nakao, S., Matsuoka, H., Matsumoto, S., Kawabata, K., and Ohno, H. (2001) Inhibition of neutrophil elastase activity attenuates complement-mediated lung injury in the hamster. Euro. J. Pharm. 426:131-138.

Hallahan, D.E. and Virudachalam, S. (1997) Intercellular adhesion molecule 1 knockout abrogates radiation induced pulmonary inflammation. PNAS USA 94:6432-6437.

Hannan, K.M., Dilley, R., de Dios, S., and Little, P.J. (2003) Troglitazone stimulates repair of the endothelium and inhibits neointimal formation in denuded rat aorta. Arterioscl. Throm. Vasc. Biol. 23:762-768.

Haselton, F.R. and Alexander, J.S. (1992) Platelets and a platelet-released factor enhance endothelial barrier. Amer. J. Physiol. 263:L670-L678.

Haselton, F.R., Alexander, J.S., and Mueller, S.N. (1993) Adenosine decreases permeability of in vitro endothelial monolayers. J. Appl. Physiol. 74:1581-1590.

Hastie, L.E., Patton, W.F., Hechtman, H.B., and Shepro, D. (1997) H_2O_2-induced filamin redistribution in endothelial cells is modulated by the PKA pathway. J. Cell Physiol. 172:373-381.

He, P. and Curry, F.E. (1993a) Albumin modulation of capillary permeability: role of endothelial cell $[Ca^{2+}]_i$. Amer. J. Physiol. 265:H74-H82.

He, P. and Curry, F.E. (1993b) Differential actions of cAMP on endothelial [Ca^{++}]$_i$ and permeability in microvessels exposed to ATP. Amer. J. Physiol. 265:H1019-H1023.
Herren, B., Levkau, B., Raines, E., and Ross, R. (1998) Cleavage of β-catenin and plakoglobin and shedding of VE-cadherin during endothelial apoptosis. Mol. Biol. Cell 9:1598-1601.
Hickling, K.G. (2002) Permissive hypercapnia. Respir. Care Clinics N. Amer. 8:155-169.
Hickling, K.G. and Joyce, C. (1995) Permissive hypercapnia in ARDS and its effect on tissue oxygenation. Acta Anaesth. Scand. 107:S201-S208.
Hippenstiel, S., Krull, M., Ikemann, A., Risau, W., Clauss, M., and Suttorp, N. (1998) VEGF induces hyperpermeability by a direct action on endothelial cells. Amer. J. Physiol. 274:L678-L684.
Hippenstiel, S., Witzenrath, M., Schmeck, B., Hocke, A., Krisp, M., Krull, M., Seybold, J., Seeger, W., Rascher, W., Schutte, H., and Suttorp, N. (2002) Adrenomedullin reduces endothelial hyperpermeability. Circ. Res. 91:618-625.
Hoffmann, J.N., Vollmar, B., Romisch, J., Inthorn, D., Schildberg, F., and Menger, M.D. (2002) Antithrombin effects on endotoxin-induced microcirculatory disorders are mediated mainly by its interaction with microvascular endothelium. Crit. Care Med. 30:218-225.
Hoffmann, J.N., Vollmar, B., Inthorn, D., Schildberg, F., and Menger, M.D. (2000) The thrombin antagonist hirudin fails to inhibit endotoxin-induced leukocyte/endothelial cell interaction and microvascular perfusion failure. Shock 14:528-534.
Holschermann, H., Noll, T., Hempel, A., and Piper, H.M. (1997) Dual role of cGMP in modulation of macromolecule permeability of aortic endothelial cells. Amer. J. Physiol. 272:H91-H98.
Honda, K., Kobayshi, H., Hitaishi, R., Hirano, S., Fukuyama, N., Nakazawa, H., and Tomita, T. (1999) Inhaled nitric oxide reduces tyrosine nitration after LPS instillation into lungs of rats. Amer. J. Respir. Crit. Care Med. 160:678-688.
Hoyt, D.G., Mannix, R., Gerritsen, M., Watkins, S., Lazo, J., and Pitt, B.R. (1996) Integrins inhibit LPS-induced DNA strand breakage in cultured lung endothelial cell. Amer. J. Physiol. 270:L689-L694.
Hsu, K., Wang, D., Chang, M., Wu, C., and Chen, H. (1996) Pulmonary edema induced by phorbol myristate acetate is attenuated by compounds that increase intracellular cAMP. Res. Exp. Med. 196:17-28.
Hybertson, B.M., Jepson, E., Allard, J., Cho, O., Lee, Y., Huddleston, J., Weinman, J., Oliva, A., and Repine, J.E. (2003) TGF-β contributes to lung leak in rats given interleukin-1 intratracheally. Exper. Lung Res. 29:361-373.
Idris, I., Gray, S., and Donnelly, R. (2003) Rosiglitazone and pulmonary oedema: an acute dose-dependent effect on human endothelial cell permeability. Diabetologia 46:288-290.
Iivanainen, E., Kaharl, V. Jeino, J., and Elenius, K. (2003) Endothelial cell-matrix interactions. Micros. Res. Techniq. 60:13-22.
Jamieson, D.D., Kerr, D., and Unsworth, D.R. (1987) Interaction of N-acetylcysteine and bleomycin on hyperbaric oxygen-induced lung dmaged in mice. Lung 165:239-247.
Jepsen, S., Herlevsen, P., Knudsen, P., Bud, M., and Klausen, N.O. (1992) Antioxidant treatment with N-acetylcysteine during ARDS. Crit. Care Med. 20:918-923.
Johnson, A., Tahamont, M., Kaplan, J., and Malik, A.B. (1982) Lung fluid balance after pulmonary embolization: effects of thrombin vs. fibrin aggregates. J. Appl. Physiol 52:1565-1570.
Jones, P.F. (2003) Not just angiogenesis--wider roles for the angiopoietins. J. Path. 201:515-527.
Joyce, D.E. and Grinnell, B.W. (2002) Recombinant human activated protein C attenuates the inflammatory response in endothelium and monocytes by modulating NFκB. Crit. Care Med. 30:S288-S293.
Juliano, R.L. (2002) Signal transduction by cell adhesion receptors and the cytoskeleton: functions of integrins, cadherins, selectins, and immunoglobulin-superfamily members. Ann. Rev. Pharm. Tox. 42:283-323.
Karzai, W., Cui, X., Mehlhorn, B., Straube, E., Hartung, T., Gerstenberger, E., Banks, S., Natanson, C., Reinhart, K., and Eichacker, P. (2003) Protection with antibody to TNF differs with similarly lethal *Escherichia coli* versus *Staphylococcus aureus* pneumonia in rats. Anesthes. 99:81-89.
Kawabata, K., Hagio, T., and Matsuoka, S. (2003) Pharmacological profile of a specific neutrophil elastase inhibitor, Sivelestat sodium hydrate. Nippon Yak. Zasshi - Folia Pharm. Jap. 122:151-160.

Kawabata, K., Hagio, T., Matsumoto, S., Nakao, S., Orita, S., Aze, Y., and Ohno, H. (2000) Delayed neutrophil elastase inhibition prevents subsequent progression of acute lung injury induced by endotoxin inhalation in ha msters. Amer. J. Respir. Crit. Care Med. 161:2013-2018.

Kayano, K., Toda, K., Naka, Y., Okada, K., Oz, M., and Pinsky, D.J. (1999) Superior protection in orthotopic rat lung transplantation with cyclicAMP and nitroglycerin-containing preservation solution. J. Thorac. Cardiovasc. Surg. 118:135-144.

Kelly, J.J., Moore, T., Babal, P., Diwan, A., Stevens, T., and Thompson, W.J. (1998) Pulmonary microvascular and macrovascular endothelial cells: differential regulation of Ca^{2+} and permeability. Amer. J. Physiol. 274:L810-L819.

Kelly, R.M. (2000) Current strategies in lung preservation. J. Lab. Clin. Med. 136:427-440.

Kim, S., Manjiri Bakre, Hong Yin, and Varner, J.A. (2002) Inhibition of endothelial survival and angiogenesis by protein kinase A. J. Clin. Invest. 110:933-941.

Klabunde, R.E. and Cavello, C. (1995) Inhibition of endotoxin-induced microvascular peakage by a platelet-activating factor antagonist and 5-li[poxygenase inhibitor. Shock 45:368-372.

Knepler, J, Taher L., Gupta, M., Patterson, C., Ober, M., and C.M. Hart. (2001) Peroxynitrite causes endothelial cell monolayer barrier dysfunction. Amer. J. Physiol. 281:C1064-C1075.

Kobayashi, H., Kobayashi, T., and Fukushima, M. (1987) Effects of dibutyryl cAMP on pulmonary air embolism-induced lung injury in awake sheep. J. Appl. Physiol. 63:2201-2207.

Koga, S., Morris, S., Ogawa, S., Liao, H., Bilezikian, J.P., Chen, G., Thompson, W.J., Ashikaga, T., Brett, J., and Stern, D.M. (1995) TNF modulates endothelial properties by decreasing cAMP. Amer. J. Physiol. 268:C1104-C1113.

Kozasa, T. and Gilman, A.G. (1996) Protein kinase C phosphorylates Gα and inhibits its interaction with Gβγ. J. Biol. Chem. 271:12562-12567.

Kuklin, V.N., Kirov, M., Evgenov, O., Sovershaev, M., Sjöberg, J., Kirova, S., and Bjertnaes, L. (2004) Novel endothelin receptor antagonist attenuates endotoxin-induced lung injury in sheep. Crit. Care Med. 32: 766-773.

Laffey, J.G., Engelberts, D., Duggan, M., Veldhuizen, R., Lewis, J., and Kavanagh, B.P. (2003) CO_2 attenuates pulmonary impairment resulting from hyperventilation. Crit. Care Med. 31:2634-2640.

Laffey, J.G., Engelberts, D., and Kavanagh, B.P. (2000a) Buffering hypercapnic acidosis worsens acute lung injury. Amer. J. Respir. Crit. Care Med. 161:141-146.

Laffey, J.G., Honan, D., Hopkins, N., Hyvelin, J., Boylan, J., and McLoughlin, P. (2004) Hypercapnic acidosis attenuates endotoxin-induced acute lung injury. Amer. J. Respir. Crit. Care Med. 169:46-56.

Laffey, J.G., Tanaka, M., Engelberts, D., Luo, X., Yuan, S., Tanswell, A., Post, M., Lindsay, T., and Kavanagh, B.P. (2000b) Therapeutic hypercapnia reduces pulmonary and systemic injury following in vivo lung reperfusion. Amer. J. Resp. Crit. Care Med. 162:2287-2294.

Laffon, M., Pittet, J., Modelska, K., Matthay, M., and Young, D.M. (1999) Interleukin-8 mediates injury from smoke inhalation to both the lung endothelial and the alveolar epithelial barriers in rabbits. Amer. J. Respir. Crit. Care Med. 160:1443-1449.

Lampugnani, M., Zanetti, A., Corada, M., Takahashi, T., Balconi, G., Breviario, F., Orsenigo, F., Cattelino, A., Kemler, R., Daniel, T., and Dejana, E. (2003) Contact inhibition of VEGF-induced proliferation requires vascular endothelial cadherin, ☐-catenin, and the phosphatase DEP-1/CD148. J. Cell Biol. 161:793-804.

Lang, J.D., Figueroa, M., Chumley, P., Aslan, M., Hurt, J., Tarpey, M., Alvarez, B., Radi, R., and Freeman, B.A. (2004) Albumin and hydroxyethyl starch modulate oxidative inflammatory injury to vascular endothelium. Anesthesiol. 100:51-58.

Lang, P., Gesbert, F., Delespine-Carmagnat, M., Stancou, R., Pouchelet, M., and Bertoglio, J. (1996) Protein kinase A phosphorylation of RhoA mediates the morphological and functional effects of cyclic AMP in cytotoxic lymphocytes. EMBO J. 15:510-519.

Langeler, E.G., Fiers, W., and van Hinsbergh, V.W. (1991a) Effects of TNF on prostacyclin production and the barrier function of human endothelial cell monolayers. Arterioscl. Thromb. 11:872-81.

Langeler, E.G. and van Hinsbergh, V.W. (1991b) Norepinephrine and iloprost improve barrier function of human endothelial cell monolayers: role of cAMP. Amer. J. Physiol. 260:C1052-C1059.

Langley, S.C. and Kelly, F.J. (1993) N-acetylcysteine ameliorates hyperoxic lung injury in the preterm guinea pig. Biochem. Pharmacol. 45:841-846.
Laposata, M. Dovnarsky, D., and Shin, H.S. (1983) Thrombin-induced gap formation in confluent endothelial cell monolayers in vitro. Blood 62:549-556.
Leff, J.A., Bodman, M., Cho, O., Rohrbach, S., Reiss, O., Vannice, J., and Repine, J.E. (1994) Post-insult treatment with interleukin-1 receptor antagonist decreases oxidative lung injury in rats given intratracheal interleukin-1. Amer. J. Respir. Crit. Care Med. 150:109-112.
Leff, J.A., Wilke, C., Hybertson, B., Shanley, P., Beehler, C., and Repine, J.E. (1993) Post-insult treatment with N-acetyl-L-cysteine decreases IL-1-induced neutrophil influx and lung leak in rats. Amer. J. Physiol. 265:L501-L506.
Leszczynski, D., Josephs, M., and Foegh, M. (1994) IL-1 beta-stimulated leucocyte-endothelial adhesion is regulated, in part, by the cyclic-GMP-dependent signal transduction pathway. Scand. J. Immunol. 39:551-556.
Lim, L., Manser, E., Leung, T., and Hall, C. (1996) Regulation of phosphorylation pathways by p21 GTPases - The p21 Ras-related Rho subfamily and its role in phosphorylation signalling pathways. Euro. J. Biochem. 242:171-185.
Linder, R., Frebelius, S., Jansson, K., and Swedenborg, J. (2003) Inhibition of endothelial cell-mediated generation of activated protein C by direct and antithrombin-dependent thrombin inhibitors. Blood Coag. Fibrinol. 14:139-46.
Liu, F., Schaphorst, K., Verin, A., Jacobs, K., Birukova, A., Day, R., Bogatcheva, N., Bottaro, D., and Garcia, J. (2002) Hepatocyte growth factor enhances endothelial cell barrier function and cortical cytoskeletal rearrangement: potential role of glycogen synthase kinase-3beta. FASEB J. 16:950-962.
Liu, H. and Zhao, J. (2002) An experimental study of therapeutic effects of ACE inhibition on chemical-induced ARDS in rats. Chin. J. Prevent. Med. 36:93-96.
Llevadot, J., Murasawa, S., Kureishi, Y, Uchida, S., Masuda, H., Kawamoto, A., Walsh, K., Isner, J., and Asahara, T. (2001) HMG-CoA reductase inhibitor mobilizes bone-marrow-derived endothelial progenitor cells. J. Clin. Invest. 198:399-405.
Lo, S.K., Garcia-Szabo, R., and Malik, A.B. (1990) Leukocyte repletion reverses protective effect of neutropenia in thrombin-induced increase in lung vascular permeability. Amer. J. Physiol. 259):H149-H155.
Lofton, C.E., Baron, D.A., Heffner, J.E., Currie, M.G., and Newman, W.H. (1991) Atrial natriuretic peptide inhibits oxidant-induced increases in endothelial permeability. J. Molec. Cell. Cardiol. 23:919-927.
Lopez-Ongil, S., Gonzalez-Santiago, L., Griera, M., Molpeceres, J., Rodriguez-Puyol, M., and Rodriguez-Puyol D. (2001) Mechanisms involved in the relaxation of bovine aortic endothelial cells. Life Sci. 70:699-714.
Lucas, M.C., Ludena, M.A., Barbero, E.A., and Sanchez-Gascon, A. (1995) Effects of N-acetylcysteine on hyperoxic lung in the rat. Respir. Med. 89:311-314.
Luce, J.M. (2002) Corticosteroids in ARDS. An evidence-based review. Crit. Care Clin. 18:79-89.
Lum, H., Aschner, J.L., Phillips, P.G., Fletcher, P.W., and Malik, A.B. (1992) Time course of thrombin-induced increase in endothelial permeability: relationship to $Ca^{2+}{}_i$ and inositol polyphosphates. Amer. J. Physiol. 263:L219-L225.
Lum, H., Hao, Z., Gayle, D., Kumar, P., Patterson, C.E., and Uhler, M.D. (2002) Vascular endothelial cells express isoforms of protein kinase A inhibitor. Amer J. Physiol. 282:C59-C66.
Lum, H., Jaffe, H.A., Schulz, I.T., Masood, A., RayChaudhury, A., and Green, R.D. (1999) Expression of PKA inhibitor (PKI) gene abolishes cAMP-mediated protection to endothelial barrier dysfunction. Amer. J. Physiol. 46:C580-C588.
Lum, H. and Malik, A.B. (1994) Regulation of vascular endothelial barrier function. Amer. J. Physiol. 267:L223-L241.
Lum, H., Siflinger-Birnboim, A., Blumenstock, F., and Malik, AB. (1991) Serum albumin decreases transendothelial permeability to macromolecules. Microvasc. Res. 42:91-102.
MacArthur, R.D., Miller, M., Albertson, T., Panacek, E., Johnson, D., Teoh, L., and Barchuk, W. (2004) Adequacy of early empiric antibiotic treatment and survival in severe sepsis. Clin. Infect. Dis. 38:284-

288.
MacLaren, R. and Jung, R. (2002) Stress-dose corticosteroid therapy for sepsis and acute lung injury or acute respiratory distress syndrome in critically ill adults. Pharmacotherapy 22:1140-1156.
Madri JA. and Williams, S.K. (1983) Capillary endothelial cell cultures: Phenotypic modulation of matrix components. J. Cell Biol. 97:153-165.
Malarkkan, N., Snook, N., and Lumb, A.B. (2003) New aspects of ventilation in acute lung injury. Anaesthesia 58:647-667.
Manganello, J.M., Huang, J., Kozasa, T., Voyno-Yasenetskaya, T., and Le Breton, G.C. (2003) Protein kinase A-mediated phosphorylation of the $G\alpha 13$ switch I region alters the $G\alpha\beta\gamma 13$-G protein-coupled receptor complex and inhibits Rho activation. J. Biol. Chem. 278:124-130.
Manolopoulos, V.G., Fenton, J.W., and Lelkes, P.I. (1997) The thrombin receptor in adrenal medullary microvascular endothelial cells is negatively coupled to adenylyl cyclase through a Gi protein. Biochim. Biophys. Acta 1356:321-332.
Maris, N.A., van der Sluijs, K., Florquin, S., de Vos, A., Pater, J., Jansen, H., and van der Poll, T. (2004) Salmeterol, a β_2-receptor agonist, attenuates lipopolysaccharide-induced lung inflammation in mice. Amer. J. Physiol. 286:L1122-L1128.
Marshall, J.C. (2003) Such stuff as dreams are made on: mediators directed therapy in sepsis. Nature Rev. 2:391-405.
Martin, G.S., Mangialardi, R., Wheeler, A., Dupont, W., Morris, J., and Bernard, G.R. (2002) Albumin and furosemide therapy in hypoproteinemic patients with acute lung injury. Crit. Care Med. 30:2175-2182.
Matthay, M.A., Fukuda, N., Frank, J., Kallet, R., Daniel, B., and Sakuma, T. (2000) Alveolar epithelial barrier. Role in lung fluid balance in clinical lung injury. Clin. Chest Med. 21:477-490.
Matthay, M.A., Zimmerman, G., Esmon, C., Bhattacharya, J., Coller, B., Doerschuk, C., Floros, J., Gimbrone, M., Hoffman, E., Hubmayr, R., Leppert, M., Matalon, S., Munford, R., Parsons, P., Slutsky, A., Tracey, K., Ward, P., Gail, D., and Harabin, A.L. (2003) Future research directions in acute lung injury: Summary of a National Heart, Lung, and Blood Institute working group. Amer. J. Respir. Crit. Care Med. 167:1027-1035.
Mazzon, E., De Sarro, A., Caputi, A., and Cuzzocrea, S. (2002) Role of tight junction derangement in the endothelial dysfunction elicited by exogenous and endogenous peroxynitrite and poly (ADP-ribose) synthetase. Shock 18:434-439.
McClintock, S.D., Till, G., Smith, M., and Ward, P.A. (2002) Protection from half-mustard-gas-induced acute lung injury in the rat. J. Appl. Tox. 22:257-262.
McElroy, M.C., Wiener-Kronish, J., Miyazaki, H., Sawa, T., Modelska, K., Dobbs, L., and Pittet, J.F. (1997) Nitric oxide attenuates lung endothelial injury caused by sublethal hyperoxia in rats. Amer. J. Physiol. 272:L631-L638.
Meduri, G.U., Kohler, G., Headley, S., Tolley, E., Stentz, F., and Postlethwaite, A. (1995) Inflammatory cytokines in the BAL of patients with ARDS. Persistent elevation over time predicts poor outcome. Chest 108:1303-1314.
Mikawa, K., Nishina, K., Takao, Y., and Obara, H. (2003) ONO-1714, a NOS inhibitor, attenuates endotoxin-induced acute lung injury in rabbits. Anesthesia & Analgesia. 97:1751-1755.
Millauer, B.S., Wizigmann-Voos, H., Schnurch, R., Martinez, N., Moeller, W., Risau, W., and Ulrich, A. (1993) High affinity VEGF binding and developmental expression suggest Flk-1 as a major regulator of vasculogenesis and angiogenesis. Cell 15:1020-1027.
Miller M.J., Rioux, L., Prendergast, G., Cannon, S., White, M., and Meinkoth, J.L. (1998) differential effects of PKA on Ras effector pathways. Molec. Cell. Biol. 18:3718-1726.
Milligan, S.A, Hoeffel, J., Goldstein, I., and Flick, M.R. (1988) Effect of catalase on endotoxin-induced acute lung injury in unanesthetized sheep. Amer. Rev. Respir. Dis. 137:420–428.
Minnear, F.L., DeMichele, M.A., Moon, D.G., Rieder, C.L., and Fenton, J.W. (1989) Isoproterenol reduces thrombin-induced pulmonary endothelial permeability. Amer. J. Physiol. 257:H1613-H1623.
Minnear, F.L., Johnson, A., and Malik, A.B. (1986) Beta-adrenergic modulation of pulmonary transvascular fluid and protein exchange. J. Appl. Physiol. 60:266-274.
Minnear, F.L., Patil, S., Bell, D., Gainor, J., and Morton, C.A. (2001) Platelet lipid(s) bound to albumin

increases endothelial electrical resistance: mimicked by LPA. Amer. J. Physiol. 281:L1337-L1344.
Minnema, M.C., Chang, A., Jansen, P., Lubbers, Y., Pratt, B., Whittaker, B., Taylor, F., Hack, C., and Friedman, B. (2000) Recombinant human antithrombin III improves survival and attenuates inflammatory responses in baboons lethally challenged with Escherichia coli. Blood 95:1117-1123.
Modelska, K., Pittet, J., Folkesson, H., Broaddus, V., and Matthay, M.A. (1999) Acid-induced lung injury. Protective effect of anti-interleukin-8 pretreatment on alveolar epithelial barrier function in rabbits. Amer. J. Respir. Crit. Care Med. 160:1450-1456.
Montravers, P., Fagon, J., Gilbert, C., Blanchet, F., Novara, A., and Chastre, J. (1993) Pilot study of cardiopulmonary risk from pentoxifylline in ARDS. Chest 103:1017-1022.
Moodie, S.A. and Martin, W. (1991) Effects of cyclic nucleotides and PMA on proliferation of pig aortic endothelial cells. Br. J. Pharmacol. 102:101-106.
Moore, T.M., Chetham, P.M., Kelly, J.J., and Stevens, T. (1998) Signal transduction and regulation of lung endothelial cell permeability. Interaction between calcium and cAMP. Amer. J. Physiol. 275:L203-L222.
Morgan, B., Thomas, A., Drevs, J., Hennig, J., Buchert, M., Jivan, A., Horsfield, M., Mross, K., Ball, H., Lee, L., Mietlowski, W., Fuxuis, S., Unger, C., O'Byrne, K., Henry, A., Cherryman, G., Laurent, D., Dugan, M., Marme, D., and Steward, W.P. (2003) Dynamic contrast-enhanced magnetic resonance imaging as a biomarker for the pharmacological response of PTK787/ZK 222584, an inhibitor of the VEGF receptor tyrosine kinases, in patients with advanced colorectal cancer and liver metastases. J. Clin. Oncol. 21:3955-3964.
Moy, A.B., Bodmer, J.E., Blackwell, K., Shasby, S., and Shasby, D.M. (1998) cAMP protects endothelial barrier function independent of inhibiting MLC20-dependent tension development. Amer. J. Physiol. 274:L1024-L1029.
Moy, A.B., Shasby, S.S., Scott, B.D., and Shasby, D.M. (1993) The effect of histamine and cyclic adenosine monophosphate on myosin light chain phosphorylation in human umbilical vein endothelial cells. J. Clin. Invest. 92:1198-1206.
Muhs, B.E., Patel, S., Yee, H., Marcus, S., and Shamamian, P. (2003) Inhibition of matrix metalloproteinases reduces local and distant organ injury following experimental acute pancreatitis. J. Surg. Res. 109:110-117.
Murakami, K., Enkhbaatar, P., Shimoda, K., Mizutani, A., Cox, R., Schmalstieg, F., Jodoin, J., Hawkins, H., Traber, L., and Traber, D. (2003) High-dose heparin fails to improve acute lung injury following smoke inhalation in sheep. Clin. Sci. 104:349-356.
Murakami, K., McGuire, R., Cox, R., Jodoin, J., Bjertnaes, L., Katahira, J., Traber, L., Schmalstieg, F., Hawkins, H., Herndon, D., and Traber, D. (2002) Heparin nebulization attenuates acute lung injury in sepsis following smoke inhalation in sheep. Shock 18:236-241.
Murohara, T., Horowitz, J.R., Silver, M., Tsurumi, Y., Chen, D., Sullivan, A., and Isner, J.M. (1998) Vascular endothelial growth factor/vascular permeability factor enhances vascular permeability via nitric oxide and prostacyclin. Circ. 97:99-107.
Murohara, T., Ikeda, H., Duan, J., shintani, S., Sasaki, K., Egulchi, H., Onitsuka, I., matsui, K., and Imaizumi, T. (2000) Transplanted cord blood-derived endothelial precursor cells augment postnatal neovascularization. J. Clin. Invest. 105:1527-1536.
Nadel, J.A. (1992) Biologic effects of mast cell enzymes. Amer. Rev. Respir. Dis. 145:S37-S41.
Nagase, T., Uozumi, N., Aoki-Nagase, T., Terawaki, K., Ishii, S., Tomita, T., Yamamoto, H., Hashizume, K., Ouchi, Y., and Shimizu, T. (2003) A potent inhibitor of cytosolic PLA2, arachidonyl trifluoromethyl ketone, attenuates LPS- lung injury in mice. Amer. J. Physiol. 284:L720-L726.
Naidu, B.V., Farivar, A., Woolley, S., Grainger, D., Verrier, E., and Mulligan, M.S. (2004) Novel broad-spectrum chemokine inhibitor protects against lung ischemia-reperfusion injury. J. Heart Lung Transpl. 23:128-134.
Naidu, B.V., Woolley, S., Farivar, A., Thomas, R., Fraga, C., and Mulligan, M.S. (2003) Simvastatin ameliorates injury in an experimental model of lung ischemia-reperfusion. J. Thorac. Cardiovasc. Surg. 126:482-489.
Nakajima, M., Cooney, M., Tu, A., Chang, K., Cao, J., Ando, A., An, G., Melia, M., and deJuan, E.J. (2001) Normalization of retinal vascular permeability in experimental diabetes with genestein. Invest.

Opthal. Vis. Sci. 42:2110-2114.
Narravula, S., Lennon, P., Mueller, B., and Colgan, S.P. (2000) Regulation of endothelial CD73 by adenosine: paracrine pathway for enhanced endothelial barrier function. J. Immunol. 165:5262-5268.
Nathens, A.B., Neff, M., Jurkovich, G., Klotz, P., Farver, K., Ruzinski, J., Radella, F., Garcia, I., and Maier, R.V. (2002) Randomized, prospective trial of antioxidant supplementation in critically ill surgical patients. Ann. Surg. 236:814-822.
Nikitenko, L.L., Smith, D., Hague, S., Wilson, C., Bicknell, R., and Rees, M.C. (2002) Adrenomedullin and the microvasculature. Trends Pharm. Sci 23:101-103.
Noll, T., Holschermann, H., Koprek, K., Gunduz, D., Haberbosch, W., Tillmanns, H., and Piper, H.M. (1999) ATP reduces macromolecule permeability of endothelial monolayers despite increasing [Ca^{2+}]i. Amer. J. Physiol. 276:H1892-H1901.
Ochoa, L., Waypa, G., Mahoney, J.R., Rodriguez, L., and Minnear, F.L. (1997) Contrasting effects of hypochlorous acid and hydrogen peroxide on endothelial permeability: Prevention with cAMP drugs. Amer. J. Resp. Crit. Care Med. 156:1247-1255.
Ogawa, S., Koga, S., Kuwabara, K., Brett, J., Morrow, B., Morris, S.A., Bilezikian, J.P., Silverstein, S.C., and Stern, D. (1992) Hypoxia-induced increased permeability of endothelial monolayers occurs through lowering of cellular cAMP levels. Amer. J. Physiol. 262:C546-C554.
Okayama, N., Ichikawa, H., Coe, L., Itoh, M., and Alexander, J.S. (1998) Exogenous NO enhances hydrogen peroxide-mediated neutrophil adherence to cultured endothelial cells. Amer. J. Physiol. 274:L820-L826.
Opal, S.M. (2003) Clinical trial design and outcomes in patients with severe sepsis. Shock. 20:295-302.
Opal, S.M., Fisher, C., Dhainaut, J., Vincent, J., Brase, R., Lowry, S., Sadoff, J., Slotman, G., Levy, H., Balk, R., Shelly, M., Pribble, J., LaBrecque, J., Lookabaugh, J., Donovan, H., Dubin, H., Baughman, R., Norman, J., DeMaria, E., Matzel, K., Abraham, E., and Seneff, M. (1997) Confirmatory interleukin-1 receptor antagonist trial in severe sepsis. Crit. Care Med. 25:1115-1124.
Opal, S.M. and Gluck, T. (2003) Endotoxin as a drug target. Crit. Care Med. 31:S57-S64.
Ostermann, H. (2002) Antithrombin III in Sepsis. New evidences and open questions. Minerva Anest. 68:445-448.
Pappert, D., Busch, T., Gerlach, H., Lewandowski, K., Radermacher, P., and Rossaint, R. (1995) Aerosolized prostacyclin vs. inhaled NO in children with severe ARDS. Anesth. 82:1507-1511.
Patterson, C.E., Davis, H., Schaphorst, K., and Garcia, J.G.N. (1994a) Mechanisms of cholera toxin prevention of thrombin- and PMA-induced endothelial cell barrier dysfunction. Microvasc. Res. 48:212-235.
Patterson, C.E. and Garcia, J.G.N. (1994b) Regulation of thrombin-induced endothelial cell activation by bacterial toxins. Blood Coag. Fibrinol. 5:63-72.
Patterson, C.E., Lum, H., Verin, A.D., Schaphorst, K.L., and Garcia, J.G.N. (2000) Regulation of endothelial barrier function by the cAMP-dependent protein kinase. Endothelium 7:287-308.
Patterson, C.E. and Rhoades, R.A. (1988) Protective role of sulfhydryl reagents in oxidant lung injury. Exper. Lung Res. 14:1005-1019.
Patterson, C.E., Butler, J.A., Byrne, F., and Rhodes, M.L. (1985) Oxidant lung injury: Intervention with sulfhydryl reagents. Lung 163:23-32.
Patterson, C.E., Stasek, J.E., Schaphorst, K.L., Davis, H.W., and Garcia, J.G.N. (1995) Mechanisms of pertussis toxin-induced barrier dysfunction in bovine pulmonary artery endothelial cell monolayers. Amer. J. Physiol. 268:L926-L934.
Patton, W.F., Alexander, J.S., Dodge, A.B., Patton, R.J., Hechtman, H.B., and Shepro, D. (1991) Mercury-arc photolysis: a method for examining second messenger regulation of endothelial cell monolayer integrity. Analyt. Biochem. 196:31-38.
Peters, K., Unger, R., Brunner, J., and Kirkpatrick, C. (2003) Molecular basis of endothelial dysfunction in sepsis. Card. Res. 60:49-57.
Pittet, J.F., Griffiths, M., Geiser, T., Kaminski, N., Dalton, S., Huang, X., Brown, L., Gotwals, P., Koteliansky, V. Matthay, M., and Sheppard, D. (2001) TGF-β is a critical mediator of acute lung injury. J. Clin. Invest. 107:1537-1544.
Pizurki, L., Zhou, Z., Glynos, K., Roussos, C., and Papapetropoulos, A. (2003) Angiopoietin-1 inhibits

endothelial permeability, neutrophil adherence and IL-8 production. Brit. J. Pharmacol. 139:329-336.
Platts, S.H. and Duling, B.R. (2004) Adenosine A3 receptor activation modulates the capillary endothelial glycocalyx. Circ. Res. 94:77-82.
Plutzky, J. (2001) Peroxisome proliferator-activated receptors in endothelial cell biology. Curr. Opin. Lipidol. 12:511-518.
Polte, T. and Schroder, H. (1998) Cyclic AMP mediates endothelial protection by nitric oxide. Biochem. Biophys. Res. Comm. 251:460-465.
Powers, K.A., Kapus, A., Khadaroo, R., He, R., Marshall, J., Lindsay, T., and Rotstein, O.D. (2003) Twenty-five percent albumin prevents lung injury following shock/resuscitation. Crit. Care Med. 31:2355-2363.
Powers, K.A., Kapus, A., Khadaroo, R., Papia, G., and Rotstein, O.D. (2002) 25% Albumin modulates adhesive interactions between neutrophils and the endothelium following shock/resuscitation. Surgery. 132:391-398.
Pratt, B.M., Form, D., and Madri, J.A. (1985) Endothelial cell-extracellular matrix interactions. Ann. NY Acad. Sci. 460:274-288.
Putensen, C., Hormann, C., Kleinsasser, A., and Putensen-Himmer, G., (1998) Cardiopulmonary effects of aerosolized prostaglandin E_1 and nitric oxide inhalation in patients with ARDS. Amer. J. Respir. Crit. Care Med. 157:1743-1747.
Qiao, J., Huang, F., and Lum, H. (2003) PKA inhibits RhoA activation: a protective mechanism against endothelial barrier dysfunction. Amer. J. Physiol. 284:L972-L980.
Quinlan, M.P. and Hyatt, J.L. (1999) Establishment of the circumferential actin filament network is a prerequisite for localization of the cadherin-catenin complex in epithelial cells. Cell Growth Diff. 10:839-854.
Quinlan, G.J., Mumby, S., Martin, G., Bernard, G.R., Gutteridge, J., and Evans, T. (2004) Albumin influences total plasma antioxidant capacity favorably in patients with acute lung injury. Crit. Care Med. 32:755-759.
Rabbani, L.E., Johnstone, M., Rudd, A., Devine, P., George, D., and Loscalzo, J. (1993) PPACK attenuates plasmin-induced changes in endothelial integrity. Thromb. Res. 70:425-436.
Radermacher, P., Santak, B., Wust, H., Tarnow, J., and Falke, K.J. (1990) Prostacyclin and right ventricular function in patients with pulmonary hypertension associated with ARDS. Intens. Care Med. 16:227-232.
Reimold, A..M. (2003) New indications fot the treatment of chronic inflammation by TNFα blockade. Amer. J. Med. Sci. 325:75-92.
Reinhart, K., Bayer, O., Brunkhorst, F., and Meisner, M. (2002) Markers of endothelial damage in organ dysfunction and sepsis. Crit. Care Med. 30:S302-S312.
Reinhart, K., Menges, T., Gardlund, B., Zwaveling, J., Smithes, M., Vincent, J., Tellado, J., Salgado-Remigio, A., Zimlichman, R., Withington, S., Tschaikowsky, K., Brase, R., Damas, P., Kupper, H., Kempeni, J., Eiselstein, J., and Kaul, M. (2001) Randomized, placebo-controlled trial of the anti-TNF antibody fragment afelimomab in hyperinflammatory response during severe sepsis: The RAMSES Study. Crit. Care Med. 29:765-769.
Remick, D.G. (2003) Cytokine therapeutics for the treatment of sepsis: why has nothing worked? Curr. Pharm. Design 9:75-82.
Reyes, M., Dudek, A., Jahagirdar, B., Koodie, L., Marker, P., and Verfaille, C.M. (2002) Origin of endothelial progenitors in human postnatal bone marrow. J. Clin. Invest. 109:337-346.
Richard, L.F., Dahms, T., and Webster, R.O. (1998) Adenosine prevents permeability increase in oxidant-injured endothelial monolayers. Amer. J. Physiol. 274:H35-H42.
Riedemann, N.C., Neff, T., Guo, R., Bernacki, K., Laudes, I., Sarma, J., Lambris, J., and Ward, P.A. (2003) Protective effects of IL-6 blockade in sepsis are linked to reduced C5a receptor expression. J. Immun. 170:503-507.
Rippe, B. and Grega, G.J. (1978) Effects of isoprenaline and cooling on histamine-induced changes in capillary permeability in the rat hindquaterter vascular bed. Acta Phsiol. Scand. 103:252-262.
Ritter, C., Andrades, M., Reinke, A., Menna-Barreto, S., Moreira, J., and Dal-Pizzol, F. (2004) Treatment with N-acetylcysteine plus deferoxamine protects rats against oxidative stress and improves survival in

sepsis. Crit. Care Med. 32:342-349.
Riva, C.M., Morganroth, M., Ljungman, A., Schoeneich, S., Marks, R., Todd, R., Ward, P.A., and Boxer, L.A. (1990) Iloprost inhibits neutrophil-induced lung injury and neutrophil adherence to endothelial monolayers. Amer. J. Respir. Cell Molec. Biol. 3:301-309.
Roemisch, J., Gray, E., Hoffmann, J., and Wiedermann, C.J. (2002) Antithrombin: a new look at the actions of a serine protease inhibitor. Blood Coag. Fibrinol. 13:657-670.
Roepke, J.E., Patterson, C., Packer, S., and Rhoades, R.A. (1991) Responses of perfused lung and isloated pulmonary artery to adenosine. Exp. Lung Res. 17:25-37.
Rossaint, R., Gerlatch, H., Schmidt-Ruhnke, H., Pappert, D., Lewandowski, K., Steudel, W., and Falke, K. (1995) Efficacy of inhaled NO in patients with severe ARDS. Chest 107:1107-1115.
Sata, M., Kakoki, M., Nagata, D., Nishimatsu, H., Suzuki, E., Aoyagi, T., Sugiura, S., Kojima, H., Nagano, T., Kangawa, K., Matsuo, H., Omata, M., Nagai, R., and Hirata, Y. (2000) Adrenomedullin and nitric oxide inhibit human endothelial cell apoptosis via a cyclic GMP-independent mechanism. Hyperten. 36:83-88.
Sato, K, Stelzner, T., O'Brien, R., Weil, J., and Welsh, C.H. (1991) Pentoxifilline lessens the endothoxin-induced increase in albumin clearance across pulmonary artery endothelial monolayers with and without neutrophils. Amer. J. Respir. Cell Molec. Biol. 4:219-227.
Schaeffer, R.C., Barnhart, M., and Carlson, R.W. (1987) Pulmonary fibrin deposition and microvascular permeability to protein following fibrin microembolism in dogs. Microvasc. Res. 33:327-352.
Schaphorst, K.L., Chiang, E., Jacobs, K., Zaiman, A., Natarajan, V., Wigley, F., and Garcia, J.G. (2003) Role of sphingosine-1 phosphate in enhancement of endothelial barrier integrity by platelet-released products. Amer. J. Physiol. 285:L258-L267.
Scheeren, T. and Radermacher, P. (1997) Prostacyclin (PGI2): new aspects of an old substance in the treatment of critically ill patients. Inten. Care Med. 23:146-158.
Schermuly, R.T., Leuchte, H., Ghofrani, H., Weissmann, N., Rose, F., Kohstall, M., Olschewski, H., Schudt, C., Grimminger, F., Seeger, W., and Walmrath, D. (2003) Zardaverine and aerosolised iloprost in a model of acute respiratory failure. Euro. Resp. J. 22:342-347.
Schilling, M.K., Eichenberger, M., Maurer, C., Sigurdsson, G., and Buchler, M.W. (2001) Ketoconazole and pulmonary failure after esophagectomy: a prospective clinical trial. Diseases Esoph. 14:37-40.
Schnittler, H.J., Wilke, A., Gress, T., Suttorp, N., and Drenckhahn, D. (1990) Role of actin and myosin in the control of paracellular permeability in pig, rat and human vascular endothelium. J. Physiol. 431:379-401.
Schnitzer, J.E. (1992) Gp60 is an albumin-binding glycoprotein expressed by continuous endothelium involved in albumin transcytosis. Amer. J. Physiol. 262:H246-H254.
Schraufstatter, I.U., Chung, J., and Burger, M. (2001) IL-8 activates endothelial cell CXCR1 and CXCR2 through Rho and Rac signaling pathways. Amer. J. Physiol. 280:L1094-L1103.
Schulze, C., Smales, C., Rubin, L., and Staddon, J.M. (1997) Lysophosphatidic acid increases tight junction permeability in cultured brain endothelial cells. J. Neurochem. 68:991-1000.
Schulze-Neick, I., Penny, D.J., Rigby, M.L., Morgan, C., Kelleher, A., Collins, P., Li, J., Bush, A., Shinebourne, E.A., and Redington, A.N. (1999) L-arginine and substance-P reverse the pulmonary endothelial dysfunction caused by congenital heart surgery. Circ. 100:749-755.
Schuster, D.P., Metzler, M., Opal, S., Lowry, S., Balk, R., Abraham, E., Levy, H., Slotman, G., Coyne, E., Souza, S., and Pribble, J. (2003) ARDS Prevention Study Group. Recombinant platelet-activating factor acetylhydrolase to prevent acute respiratory distress syndrome and mortality in severe sepsis. Crit. Care Med. 31:1612-1619.
Seeger, W., Hansen, T., Rossig, R., Schmehl, T., Schutte, H., Kramer, H.J., Walmrath, D., Weissmann, N., Grimminger, F., and Suttorp, N. (1995) Hydrogen peroxide-induced increase in lung endothelial and epithelial permeability--effect of adenylate cyclase stimulation and phosphodiesterase inhibition. Microvasc. Res. 50:1-17.
Seekamp A, Lalonde C, Zhu, D.G., and Demling, R. (1988) Catalase prevents prostanoid release and lung lipid peroxidation after endotoxemia in sheep. J. Appl. Physiol. 65:1210–1216.
Serraf, A., Robotin, M., Bonnet, N., Detruit, H., Baudet, B., Mazmanian, M.G., Herve, P., and Planche, C. (1997) Alteration of the neonatal pulmonary physiology after total cardiopulmonary bypass. J. Thor.

Cardiovasc. Surg. 114:1061-1069.
Sharp, C., Warren, A., Oshima, T., Williams, L., Li, J., and Alexander, J.S. (2001) Poly ADP ribose-polymerase inhibitors prevent the upregulation of ICAM-1 and E-selectin in response to Th1 cytokine stimulation. Inflamm. 25:157-163.
Shay-Salit, A., Shushy, M., Wolfovitz, E., Yahav, H., Breviario, F., Dejana, E., and Resnick, N. (2002) VEGF receptor 2 and the adherens junction as a mechanical transducer in vascular endothelial cells. PNAS USA 99:9462-9467.
Sheldon, R., Moy, A., Lindsley, K., Shasby, S., and Shasby, D.M. (1993) Role of myosin light-chain phosphorylation in endothelial cell retraction. Amer. J. Physiol. 265:L606-L612.
Shenkar, R., Coulson, W., and Abraham, E. (1994) Anti- TGF-β monoclonal antibodies prevent lung injury in hemorrhaged mice. Amer. J. Respir. Cell Molec. Biol. 11:351-357.
Shi, Q., Rafii, S., Wu, M., Wijelath, E., Yu, C., Ishida, A., Fujita, Y., Kothari, S., Mohle, R., Sauvage, L., Moore, M., Storb, R., and Hammond, W.P. (1998) Evidence for circulating bone marrow-derived endothelial cells. Blood 92:362-367.
Shi, S., Verin, K.L., Gilbert-McClain, L., Patterson, C.E., Irwin, R.P., Natarajan, V., and Garcia, J.G.N. (1998) Role of tyrosine phosphorylation in thrombin-induced endothelial cell contraction and barrier function. Endothelium 6:153-171.
Shibata, K., Cregg, N., Engelberts, D., Takeuchi, A., Fedorko, L., and Kavanagh, B.P. (1998) Hypercapnic acidosis may attenuate acute lung injury by inhibition of endogenous xanthine oxidase. Amer. J. Respir. Crit. Care Med. 158:1578-1584.
Shindo, T., Kurihara, H., Maemura, K., Kurihara, Y., Kuwaki, T., Izumida, T., Minamino, N., Ju, K., Morita, H., Oh-hashi, Y., Kumada, M., Kangawa, K., Nagai, R., and Yazaki, Y. (2000) Hypotension and resistance to lipopolysaccharide-induced shock in transgenic mice overexpressing adrenomedullin in their vasculature. Circ. 101:2309-2316.
Siegel-Axel, D.L. (2003) Ceristatin: a cellular and molecular drug for the future? Cell. Molec. Life Sci. 60:144-164.
Siflinger-Birnboim, A., Bode, D.C., and Malik, A.B. (1993) Cyclic-AMP attenuates neutrophil-mediated increase in endothelial permeability. Amer. J. Physiol. 264:H370-H375.
Siflinger-Birnboim, A., Lum, H., del Vecchio, P., and Malik, A. (1996) Involvement of Ca^{2+} in the H_2O_2-induced increase in endothelial permeability. Amer. J. Physiol. 270:L973-L978.
Simon, L.M. and Suttorp, N. (1985) Lung cell oxidant injury: decrease in oxidant mediated cytotoxicity by N-acetylcysteine. Euro. J. Respir. Dis. 139:132-135.
Sokol, J., Jacobs, S.E., and Bohn, D. (2003) Inhaled nitric oxide for acute hypoxic respiratory failure in children and adults: a meta-analysis. Anesth. Analges. 97:989-998.
Spragg, R.G., Lewis, J., Wurst, W., Hafner, D., Baughman, R., Wewers, M., and Marsh J.J. (2003) Treatment of acute respiratory distress syndrome with recombinant surfactant protein C surfactant. Amer. J. Respir. Crit. Care Med. 167:1562-1566.
Sprague, R.S., Stephenson, A., McMurdo, L., and Lonigro, A. (1998) Nitric oxide opposes phorbol ester-induced increases in pulmonary microvascular permeability. J. Pharm. Exp. Therap. 284:443-448.
Steinberg, J., Halter, J., Schiller, H., Dasilva, M., Landas, S., Gatto, L., Maisi, P., Sorsa, T., Rajamaki, M., Lee, H., and Nieman, G. (2003) Metalloproteinase inhibition reduces lung injury and improves survival after cecal ligation and puncture in rats. J. Surg. Res. 111:185-195.
Stelzner, T.J., Weil, J.V., and O'Brien, R.F. (1989) Role of cyclic adenosine monophosphate in the induction of endothelial barrier properties. J. Cell. Physiol. 139:157-166.
Stevens, T., Fouty, B., Hepler, L., Richardson, D., Brough, G., McMurtry, I.F., and Rodman, D.M. (1997) Cytosolic Ca2+ and adenylyl cyclase responses ia phenotypically distinct pulmonary endothelial cells. Amer. J. Physiol. 16:L51-L59.
Sudo, T., Ito, H., and Kimura, Y. (2003) Phosphorylation of the vasodilator-stimulated phosphoprotein (VASP) by the anti-platelet drug, cilostazol, in platelets. Platelets. 14:381-390.
Suter, P.M., Domenighetti, G., Schaller, M., Laverriere, M., Ritz, R., and Perret, C. (1994) N-acetylcysteine enhances recovery from acute lung injury in man. A randomized, double-blind, placebo-controlled clinical study. Chest. 105:190-194.
Suttorp, N., Hippenstiel, S., Fuhrmann, M., Krull, M., and Podzuweit. T. (1996) Role of nitric oxide and

phosphodiesterase isoenzyme II for reduction of endothelial hyperpermeability. Amer. J. Physiol. 270:C778-C785.

Suttorp, N., Weber, U., Welsch, T., and Schudt, C. (1993) Role of phosphodiesterases in the regulation of endothelial permeability in vitro. J. Clin. Invest. 91:1421-1428.

Szabo, C., Cuzzocrea, S., Zingarelli, B., O'Connor, M., and Salzman, A.L. (1997a) Endothelial dysfunction in a rat model of endotoxic shock . J. Clin. Invest. 100:723-735.

Szabo, C., Lim, L., Cuzzocrea, S., Getting, S., Zingarelli, B., Flower, R., Salzman, A., and Perretti, M. (1997b) Inhibition of poly (ADP-ribose) synthetase attenuates neutrophil recruitment and exerts antiinflammatory effects. J. Exp. Med. 186:1041-1049.

Taggart, D.P. (2001) Effects of a platelet-activating factor antagonist on lung injury and ventilation after cardiac operation. Ann. Thorac. Surg. 71:238-242.

Taichman, D.B., Schachtner, S., Li, Y., Puri, M., Bernstein, A., and Baldwin, H.S. (2003) A unique pattern of Tie1 expression in the developing murine lung. Exp. Lung Res. 29:113-122.

Takaishi, K., Sakaki, T., Kotani, H., and Takai, Y. (1997) Regulation of cell-cell adhesion by Rac and Rho small G proteins in MDCK cells. J. Cell Biol. 139:1047-1059.

Tedesco, F., Fischetti, F., Pausa, M., Dobrina, A., Sim, R., and Daha, M. (1999) Complement-endothelial cell interactions: pathophysiological implications. Molec. Immun. 36:261-268.

Terramani, T.T., Eton, D., Bui, P., Wang, Y., Weaver, F., and Yu, H. (2000) Human macrovascular endothelial cells: optimization of culture conditions. In Vitro Cell. Devel. Biol. Animal 36:125-132.

Thompson, B.T. (2003) Glucocorticoids and acute lung injury. Crit. Care Med. 31:S253-S257.

Thurston, G. (2002) Complementary actions of VEGF and angiopoietin-1 on blood vessel growth and leakage. J. Anat. 200:575-580.

Tigyi, G., Fischer, D.J., Sebok, A., Yang, C., Dyer, D.L., and Miledi, R. (1996) Lysophosphatidic acid-induced neurite retraction in PC12 cells: control by phosphoinositide-Ca^{2+} signaling and Rho. J. Neurochem. 66:537-548.

Timlin, M., Condron, C., Toomey, D., Power, C., Thornes, B., Kearns, S., Street, J., Murray, P., and Bouchier-Hayes, D. (2004) N-acetylcysteine attenuates lung injury in a rodent model of fracture. Acta Orthopaed. Scand. 75:61-65.

Tinsley, J.H., Teasdale, N.R., and Yuan, S.Y. (2004a) Myosin light chain phosphorylation and pulmonary endothelial cell hyperpermeability in burns. Amer. J. Physiol. 286:L841-L847.

Tinsley, J.H., Yuan, S.Y., and Wilson, E. (2004b) Isoform-specific knockout of endothelial myosin light chain kinase: closing the gap on inflammatory lung disease. Trends Pharm. Sci. 25:64-66.

Toda, K., Kayano, K., Karimova, A., Naka, Y., Fujita, T., Minamoto, K., Wang, C., and Pinsky, D. (2000) Antisense ICAM-1 oligodeoxyribonucleotide delivered during organ preservation inhibits posttransplant ICAM-1 expression and reduces primary lung isograft failure. Circ. Res. 86:166-174.

Toh, C., Ticknor, L., Downey, C., Giles, A., Paton, R., and Wenstone, R. (2003) Early identification of sepsis and mortality risks through simple, rapid clot-waveform analysis. Inten. Care Med. 29:55-61.

Tsan, M. and White, J.E. (1988). Red blood cells protect endothelial cells against H_2O_2-mediated but not hyperoxia-induced damage. Proc. Soc. Exp. Biol. Med. 188:323-327.

Turner, C.R., Esser, K., and Weeldon, E.B. (1993) Therapeutic intervention in a rat model of ARDS: phosphodisetarase IV inhibition. Circ. Shock 39:237-245.

Usui, H., Inoue, R., Tanabe, O., Nishito, Y., Shimizu, M., Hayashi, H., Kagamiyama, H., and Takeda, M. (1998) Activation of protein phosphatase 2A by PKA-catalyzed phosphorylation of the 74-kDa B" regulatory subunit *in vitro* and identification of the phosphorylation sites. FEBS Letters 430:312-316.

Van Heerden, P.V., Barden, A., Michalopoulos, N., Bulsara, M., and Roberts, B.L. (2000) Dose-response to inhaled aerosolized prostacyclin for hypoxemia due to ARDS. Chest. 117:819-827.

Van Klaveren, R.J., Dinsdale, D., Pype, J., Demedts, M., and Nemery, B. (1997) N-acetylcysteamine does not protect against type II cell injury after prolonged hyperoxia. Amer. J. Physiol. 273:L548-L555.

Van Nieuw Amerongen, G.P., Delft, S., Vermeer, M.A., Collard, J., and van Hinsbergh, V.M. (2000a) Activation of RhoA by thrombin in endothelial hyperpermeability. Circ. Res. 87:335-340.

Van Nieuw Amerongen, G.P., Draijer, R., Vermeer, M.A., and van Hinsbergh, V.M. (1998) Transient and prolonged increase in endothelial permeability induced by histamine and thrombin - Role of protein kinases, calcium, and RhoA. Circ. Res. 83:1115-1123.

van Nieuw Amerongen, G.P. and van Hinsbergh, V.W. (20020 Targets for pharmacological intervention of endothelial hyperpermeability and barrier function. Vasc. Pharmacol. 39:257-272.

Van Nieuw Amerongen, G.P., Vermeer, M., Negre-Aminou, P., Lankelma, J., Emieis, J., and van Hinsbergh, V.W. (2000b) Simvastatin improves endothelial barrier function. Circ. 102:2803-2809.

Vassilev, D., Hauser, B., Bracht, H., Ivanyi, Z., Schoaff, M., Asfar, P., Vogt, J., Wachter, U., Schelzig, H., Georgieff, M., Bruckner, U., Radermacher, P., and Froba, G. (2004) Systemic, pulmonary, and heptaosplancnic effects of N-acetylcysteine during porcine endotoxemia. Crit. Care Med. 32:525-532.

Vestweber, D. (2000) Molecular mechanisms that control endothelial cell contacts. J. Path. 190:281-291.

Victorino, G.P., Newton, C., and Curran, B. (2003) The impact of albumin on hydraulic permeability: comparison of isotonic and hypertonic solutions. Shock. 20:171-175.

Victorino, G.P., Newton, C., and Curran, B. (2004) Modulation of microvascular hydraulic permeability by platelet-activating factor. J. Trauma-Injury Infect. Crit. Care 56:379-384.

Victorino, G.P., Wisner, D., and Tucker, V.L. (2001) Basal release of nitric oxide and its interaction with endothelin-1 on single vessel hydraulic permeability. J. Trauma-Injury Infect. Crit. Care. 50:535-539.

Vincent, J.L., Brase, R., Santman, F., Suter, P., McLuckie, A., Dhainaut, J., Park, Y., and Karmel, J. (2001) A multi-center, double-blind, placebo- controlled study of liposomal prostaglandin E_1 in patients with ARDS. Inten. Care Med. 27:1578-1583.

Vincent, J.L., Wilkes, M., and Navickis, R.J. (2003) Safety of human albumin--serious adverse events reported worldwide in 1998-2000. Brit. J. Anaesth. 91:625-630.

Von Dobschuetz, E., Bleiziffer, O., Pahernik, S., Dellian, M., Hoffmann, T., and Messmer, K. (2004) Soluble complement receptor 1 preserves endothelial barrier function and microcirculation in postischemic pancreatitis in the rat. Amer. J. Phsysiol. 286:G791-G796.

Wachtel, M., Frei, K., Ehler, E., Fontana, A., Winterhalter, K., and Gloor, S.M. (1999) Occludin proteolysis and increased permeability in endothelial cells through tyrosine phosphatase inhibition. J. Cell Sci. 112:4347-4356.

Wagner, P.D., Mathieu-Costello, O., Bebout, D., Gray, A., Natterson, P., and Glennow, C. (1989) Protection against pulmonary O2 toxicity by N-acetylcysteine. Euro. Resp. J.. 2:116-126.

Walmrath, D., Schneider, T., Pilch, J., Schermuly, R., Grimminger, F., and Seeger, W. (1995) Effects of aerosolized PGI_2 in severe pneumonia. Amer. J. Resp. Crit. Care Med. 151:724-730.

Walmrath, D., Schneider, T., Schermuly, R., Olschewski, H., Grimminger, F., and Seeger, W. (1996) Direct comparison if inhaled NO and aerosolized prostacyclin in ARDS. Amer. J. Resp. Crit. Care Med. 153:991-996.

Wang, P. (1998) Adrenomedullin in sepsis and septic shock. Shock 10:383-384.

Wang, Y., Gu, Y., Zhang, Y., and Lewis, D.F. (2004) Evidence of endothelial dysfunction in preeclampsia: decreased endothelial nitric oxide synthase expression is associated with increased cell permeability in endothelial cells from preeclampsia. Amer. J. Obstet. Gyn. 190:817-824.

Ward, P.A. (1996) Role of complement, chemokines, and regulatory cytokines in acute lung injury. Ann. NY Acad. Sci. 796:104-112.

Ward, P.A., Riedemann, N., Guo, R., Huber-Lang, M., Sarma, J., and Zetoune F.S. (2003) Anti-complement strategies in experimental sepsis. Scand. J. Infec. Dis. 35:601-603.

Ward, W.F., Molteni, A., and Ts'ao, C.H. (1989) Radiation-induced endothelial dysfunction and fibrosis in rat lung: modification by the ACE inhibitor CL242817. Rad. Res. 117:342-350.

Ware, L.B., Fang, X., Wang, Y., Sakuma, T., Hall, T., and Matthay, M. (2002) Lung Edema Clearance:20 years of progress. Mechanisms that may stimulate resolution of alveolar edema in the transplanted lung. J. Appl. Physiol. 93:1869-1874.

Ware, L.B. and Matthay, M.A. (2000) The acute respiratory distress syndrome. New Eng. J. Med. 342:1334-1349.

Ware, L.B. and Matthay, M.A. (2001) Keratinocyte and heptaocyte growth factor in the lung: roles in lung development, inflammation, and repair. Amer. J. Physiol. 282:L924-L940.

Waschke, J., Baumgartner, W., Adamson, R., Zeng, M., Aktories, K., Barth, H., Wilde, C., Curry, F., and Drenckhahn, D. (2004) Requirement of Rac activity for maintenance of capillary endothelial barrier properties. Amer. J. Physiol. 286:H394-H401.

Watts, J.A. and Kline, J.A. (2003) Bench to bedside: the role of mitochondrial medicine in the

pathogenesis and treatment of cellular injury. 10:985-997.
Waypa, G.B., Morton, C., Vincent, P., Mahoney, J., Johnston, W., and Minnear, F.L. (2000) Oxidant-increased endothelial permeability: prevention with phosphodiesterase inhibition vs. cAMP production. J. Appl. Physiol. 88:835-842.
Waypa, G.B., Vincent, P., Morton, C., and Minnear, F.L. (1996) Thrombin increases fluid flux in isolated rat lungs by a hemodynamic and not a permeability mechanism. J. Appl. Physiol. 80:1197-1204.
Westendorp, R.G., Draijer, R., Meinders, A., and van Hinsbergh, V.W. (1994) Cyclic-GMP-mediated decrease in permeability of human umbilical and pulmonary artery endothelial cell monolayers. J. Vasc. Res. 31:42-51.
Wheatley, E.M., McKeown, P., Vincent, P., and Saba, T. (1993) Incorporation of fibronectin into matrix decreases TNF-induced endothelial monolayer permeability. Amer. J. Physiol. 265:L146-L157.
Whittles, C.E., Pocock, T., Wedge, S., Kendrew, J., Hennequin, L., Harper, S., and Bates, D.O. (2002) ZM323881, a novel inhibitor of VEGFR2 tyrosine kinase activity. Microcirc. 9:513-522.
Wiedemann, H.P., Arroliga, A.C., and Komara, J.J. (2003) Emerging systemic pharmacologic approaches in ARDS. Respir. Clin. North Amer. 9:419-435.
Wiedemann, H.P. and Gillis, C.N. (1986) Altered metabolic function of the pulmonary microcirculation. Early detection of lung injury and possible functional significance. Crit. Care Clin. 2:497-509.
Wilkes, M.M. and Navickis, R.J. (2001) Patient survival after human albumin administration. A meta-analysis of randomized, controlled trials. Ann. Interr. Med. 135:149-164.
Wojcikiewicz, R. and Luo, S. (1998) Phosphorylation of inositol 1,4,5-trisphosphate receptors by cAMP-dependent protein kinase. J. Biol. Chem. 273:5670-5677.
Wu, H.M., Yuan, Y., Zawieja, D., Tinsley, D., and Granger, H.LJ. (1999) Role of phospholipase C, PKC, and calcium in VEGF-induced venular permeability. Amer. J. Physiol.:H535-H542.
Yan, W., Tiruppathi, C., Lum, H., Qiao, R., and Malik, A.B. (1998) PKCβ regulates heterologous desensitization of thrombin receptor (PAR-1) in endothelial cells. Amer. J. Physiol. 274:C387-C395.
Yang, L., Quan, S., and Abraham, N.G. (1999) Retrovirus-mediated HO gene transfer into endothelial cells protects against oxidant-induced injury. Amer. J. Physiol. 277:L127-L133.
Yatomi, Y., Ohmori, T., Rile, G., Kazama, F., Okamoto, H., Sano, T., Satoh, K., Kume, S., Tigyi. G., Igarashi, Y., and Ozaki, Y. (2000) Sphingosine 1-phosphate as a major bioactive lysophospholipid that is released from platelets and interacts with endothelial cells. Blood 96: 3431-3438.
Yi, L., Weisdorf, D., Solovey, A., and Hebbel, R.P. (2000) Origins of circulating endothelial cells and endothelial outgrowth from blood. J. Clin. Invest. 105:71-77.
Yu, M. and Tomasa, G. (1993) A double-blind, prospective, randomized trial of ketoconazole, a thromboxane synthetase inhibitor, in the prophylaxis of ARDS. Crit. Care Med. 21:1635-1642.
Yuan, Y., Granger, H., Zawieja, D., DeFily, D., and Chilian, W.M. (1993) Histamine increases venular permeability via PLC-NO synthase-guanylate cyclase cascade. Amer. J. Physiol. 264:H1734-H1739.
Zeiher, B.G., Matsuoka, S., Kawabata, K., and Repine, J.E. (2002) Neutrophil elastase and acute lung injury: prospects for sivelestat and other neutrophil elastase inhibitors. Crit. Care Med. 30:S281-S287.
Zhang, C., Hein, T., Wang, W., and Kuo, L. (2003) Divergent roles of angiotensin II AT1 and AT2 receptors in modulating coronary microvascular function. Circ. Res. 92:322-329.
Zhang, H., Schmeisser, A., Garlichs, C., Plotze, K., Damme, U., Mugge, A., Daniel, W. (1999) Angiotensin II-induced superoxide anion generation in human vascular endothelial cells: role of membrane-bound NADH-/NADPH-oxidases. Cardiovasc. Res. 44:215-222.
Zhang, Y., Gu, Y., Lucas, M., and Wang, Y. (2003) Antioxidant superoxide dismutase attenuates increased endothelial permeability induced by platelet-activating factor. J. Soc. Gyn. Invest. 10:5-10.
Zimmerman, J.L., Dellinger, R., Straube, R., and Levin, J.L. (2000) Phase I trialof soluble recombinant receptor 1 in ALI and ARDS. Crit. Care Med. 28:3149-3154.
Zingarelli, B., Cuzzocrea, S., Zsengeller, Z., Salzman, A., and Szabo, C. (1997) Protection against myocardial ischemia and reperfusion injury by 3-aminobenzamide, an inhibitor of poly (ADP-ribose) synthetase. Cardiovasc. Res. 36:205-215.
Zoellner, H., Hou, J., Lovery, M., Kingham, J., Srivastava, M., Bielek, E., Vanyek, E., and Binder, B.R. (1999) Inhibition of microvascular endothelial apoptosis in tissue explants by albumin. Microvasc. Res. 57:162-73.

Chapter 16.
Looking to the Future as Keepers of the Dam

Carolyn E. Patterson, Ph.D.[1]* & Michael A. Matthay, M.D.[2]

[1]Departments of Medicine & Physiology, Indiana University School of Medicine & Roudebush VA Med. Center, Indianapolis IN., USA 46202, & [2]Cardiovascular Research Institute, Univ. of California, San Francisco CA, USA 94143

CONTENTS:

Introduction

Basic construction and function of the endothelial dam
Advances in understanding
Gaps and new directions

Basic mechanisms of activation
Experimental advances
Gaps and new directions

Basic mechanisms promoting integrity
Experimental advances
Gaps and new directions

Clinical pathophysiology
Simple concepts to complex realities
Advances in inflammation, hemostasis, and endothelial function
Pathogens and pulmonary endothelium
Definitions and diagnosis
Potential predictors of permeability
Progress in understanding the physiology of resolution

Protection and restoration in the clinical setting
First line support
Advances and retreats in blocking activation
Controversies over albumin and anti-oxidants
Progress in protective mechanisms
Combinations and considerations

Charge to the keeper

Introduction

Though we are as vulnerable to inundation and inflammation as those hapless victims of the Johnstown flood were to the consequences of rupture of the Southfork Dam, we have come to understand that our endothelial dam is "fearfully and wonderfully made." Despite the extreme delicacy of the alveolar capillary membrane, the pulmonary microvascular endothelium normally functions faithfully as an effective and selective barrier, protecting the lung parenchyma and the alveoli from excesses of the blood fluid, bioactive molecules, and cellular components, while allowing optimal exchange of oxygen and carbon dioxide, critical to life. Advances in cellular, molecular, and biomechanical sciences have enabled us to more fully appreciate and understand the construction and function of the endothelial barrier, as well as the many important regulatory functions of the endothelium. Nevertheless, if severe, acute failure of the pulmonary endothelium occurs, the clinical consequences are serious as described in Chapters 1, 12, and 15. It has become increasingly apparent that to make advances in our treatment of acute lung injury, we must improve our ability to provide support for safeguarding and restoration of endothelial integrity. Thus, as keepers of the dam, we need to have an intimate acquaintance with its composition, maintenance, and regulation. Furthermore, there is growing recognition that even moderate endothelial dysfunction in any anatomical site contributes to the pathological processes leading to diverse acute and chronic, vascular and organ dysfunction.

Thus, this volume has been dedicated to an integrated presentation of the advancements in endothelial biology and to development of a view toward expansion of our comprehension of endothelial functional integrity. As the current state of knowledge has been thoroughly expounded in the preceding chapters, in this final chapter we will only summarize the advances, beginning with basic laboratory investigation and concluding with clinical application. The principal focus will be to identify those areas within our grasp, which require more intensive basic and clinical investigation, and to project approaches needed to apply this knowledge, particularly in the setting of respiratory failure from acute lung injury.

Basic construction and function of the endothelial dam

Advances in understanding

With the development of isolated perfused lung preparations, endothelial cell culture, and isolation of specific types of endothelium, including human pulmonary microvascular endothelium, we have been able to define the fundamental importance of endothelial integrity in maintenance of lung fluid balance and to relate barrier function to architectural integrity of the endothelial cell, as explored in Chapters 1 and 2. Advances in biochemistry and molecular biology have allowed definition of mediators and mechanisms governing cell function, as will be explored in the following section. With advances in electron microscopy and confocal, fluorescence imaging we have been able to link the biochemical events to the effectors responsible for the integration of the structural and functional effects. Besides improved understanding of the specific role of endothelial continuity in prevention of pulmonary edema, progress has been made in defining the complementary roles of interstitial integrity, epithelial function, ventilation and perfusion characteristics, as well as the interplay between the endothelium and these other components. Finally, we have expanded our view of the endothelium in regulation of vascular smooth muscle contractile function and phenotype expression, regulation of coagulation and fibrinolysis, regulation of inflammatory processes, and the critical interactions of these functions with barrier function.

Chapters 2 and 7 detailed identification of cytoskeletal components in endothelium and their control. These form the molecular basis for relating cellular structure to its function as a semi-permeable, selective barrier. New information conveyed in Chapter 7 on actin and the microtubule cytoskeletal system has allowed understanding of the polarity of endothelium, the organization of enzyme complexes, and the importance of these cytoskeletal systems to cellular biochemical function. Importantly, these advances have permitted development of models, described in Chapter 2, which further the understanding and testing of the mechanical properties of the endothelium and the responsiveness of endothelium to its external environment and mechanical stress. Reorganization of the actin cytoskeleton and its associated proteins with cellular activation has now been described. Focal adhesions at the endothelial basal surface were identified as sites of transmembrane integrin binding to specific extracellular proteins. These integrins were shown to be coupled to the actin stress fibers via specific intracellular molecules. Importantly, a number of signaling molecules have been linked to this focal adhesion complex and to the transduction of mechanical stress into cellular biochemical responses as defined in Chapter 8. Moreover, our understanding of the role of focal adhesions to cell spreading and motility has been advanced in endothelial and other cells. The significance of cell-cell tethering to barrier function has been clarified and several different adhesion complexes or molecules defined as reviewed in Chapter 9. Regulation of these junctional complexes and the participation of particular junctional proteins in signaling have also been discovered. Finally, Chapter 10 displayed the heterogeneity of endothelium, rooted in dedicated gene expression that confers specific site and organ function and permits adaptation to pathophysiologic stimuli.

Gaps and new directions

For clarity and simplicity of presentation, the endothelium has been represented as existing in two basic states: 1) the normal, barrier-functioning, quiescent state, which signals quiescence to vascular smooth muscle, platelets, and leukocytes, and 2) the activated, leaky endothelium, which expresses molecules for leukocyte interaction and releases molecules that activate many other cells. But this, like most simplifications, is inadequate to describe the endothelial diversity. Even in reasonably quiescent endothelium, a few of the cells appear to be somewhat activated, in that some stress fibers are observed. Physical and temporal variability in response to activators results in a complex continuum with variable patterns of transition and recovery. Activators differ in intercellular signals initiated. Activation may involve different degrees of injury, necrosis, and apoptosis. Some cells may be dividing, some spreading, some growing, some moving. Moreover, there are overt differences in form and expression of markers in different vascular beds and in different organs. There are marrow-derived progenitor endothelial cells with high capacity for proliferation. There are transformed cells studied in culture that may or may not reflect normal endothelium. Effects of culture conditions and regression with passage alter endothelial characteristics, just as age alters endothelial characteristics *in vivo*. Even endothelial cells that classically express either the activated or the quiescent types, exhibit continuous dynamic reactions with shifting and reformation of molecular complexes. This genetic, physiologic, structural, and state diversity must be appreciated to gain a better understanding of the endothelium in health or in disease.

Although development of models has been useful in advancing our understanding, there is a need to redefine, expand, and resolve the two models now commonly used to describe the endothelium: the contractile/tethered model and the prestressed, tensegrity model. This is especially desirable in light of the diversity discussed above. A sampling of such questions was raised in Chapter 2 regarding the potential role of actin microfilament as scaffolding elements in quiescent endothelium, the actual role of actin/myosin interactions in activated endothelium, the transitions and interactions among the various cytoskeletal elements and the junctional complexes, and the biomechanical properties of confluent, quiescent endothelium. There is still much to learn with regard to the coupling of mechanical and biochemical events, both cell to environment and environment to cell. Likewise, we need improved understanding of the coupling of biochemical signals in response to mediators, oxidants, chemical stress, or the binding of endothelial ligands by leukocytes or pathogens to the end effectors governing the physiologic, functional response of the cell.

Basic mechanisms of activation and injury

Experimental advances

Many of the mediators, receptors, and physiologic responses of endothelium related to pathophysiology have been identified in the last 25 years. Such agents of activation outlined in Chapter 3 include cell-derived mediators, vasoactive mediators, cytokines, chemokines, growth factors, lysophospholipids, oxidants, microorganisms, and mechanical stimuli. Methods have been developed to quantitate barrier dysfunction *in vitro* and *in vivo*. Markers of endothelial activation have been described, identified in plasma, and used in patient studies. We have come to appreciate that activation involves a spectrum of responses as described in Chapter 3. Although effects of stimuli on endothelium are generally direct, we are well aware of the dance between the endothelium and leukocytes, which amplifies the primary response. Though absolutely necessary for defense of the organism against attack by foreign agents, the dance becomes deadly when the endothelium is injured and when edema impairs organ function, particularly in the lung. The endothelium is not merely a victim of inflammation gone awry, but is an important player as well. Altered endothelial control over leukocyte and platelet function, coagulation, fibrinolysis, and smooth muscle function can trigger and/or exacerbate a pathologic sequence of events. Moreover, the endothelium can become a source of oxidants. This Janus role for endothelium as victim and participant is exemplified in the studies of *E. coli* sepsis in baboons, where endothelium was identified as the primary target but was shown to regulate both the inflammatory and coagulopathic aspects of the sepsis and become a primary generator of oxidant radicals (Taylor, 1994). Thus loss of pulmonary endothelial barrier function may be a primary event, a secondary result of injurious products that reach the endothelium, a secondary event due to remote activation of signaling that reaches the endothelium, or a result of a vicious amplification cycle involving endothelial dysfunction.

Armed with new information on endothelial function, mechanisms involved in activation of inflammation, leukocyte biology, and the pathophysiology of lung edema, numerous protocols to block activation have been successfully tried in *in vitro* and *in vivo* models of endothelial dysfunction and permeability edema, as examined in Chapters 12 and 15. Table 1 offers a summary list of blockers that may be protective for endothelium and other cells from inflammatory and oxidant activation and may prevent participation of endothelium in the exacerbation of pathologic processes leading to tissue damage, but much work is needed to move from speculation to usefulness. Detailed information may be found in Chapter 15.

Table 1
Potential Endothelial Protective Strategies – Blockers

Factor	Rationale/Value	Reference
ACE inhibitors (fosinopril) and AT1 blockers (losartan)	Prevent angiotensin activation of EC and upregulation of oxidant producing systems	Zhang et al., 1999; Liu and Zhou, 2002; Zhang et al., 2003
Endothelin-1 receptor antagonists (tezosentan)	Prevent endothelial activation & vaso-constrictive state	Kuklin et al., 2004
Thromboxane-blockers	Prevent EC activation; also attenuate vasoconstriction and platelet activation	Morel et al., 1990; Teixeira et al., 1995
Anti-oxidants (NAC, NAL, apocynin, vitamins C & E); anti 8-isoprostane; targeted increase in cell catalase and SOD	Counter oxidant stress – helpful only if given early (Groeneveld et al., 1990; Vassilev et al., 2004)	Patterson et al., 1985; Milligan et al., 1988; Seekamp et al., 1988; Amari et al., 1993; Lucas et al., 1995; Gonzalez et al., 1996; Atochina et al., 1998; Gow et al., 1998; Yang et al., 1999; McClintock et al., 2002; Mikawa et al., 2003; Zhang, Y. et al., 2003; Antonicelli et al., 2004; Gao et al., 2004; Gertzberg et al., 2004; Ritter et al., 2004; Timlin et al., 2004
PARP inhibitors (3-aminobenzamide; 1,5-dihydroxyisoquinoline)	Decreases EC upregulation of adhesion molecules and loss of tight junctions	Szabo et al., 1997b; Sharp et al., 2001; Cuzzocrea et al., 2002; Mazzon et al., 2002; Watts and Kline, 2003
Cytokine neutralizing antibodies or receptor antagonists (anti-TNF, IL-1, IL-6, IL-8, TGF)	Prevent EC activation, adhesion molecule expression; systemic inflammatory stress	Laffon et al., 1999; Modeleska et al., 1999; Pittet et al., 2001; Hybertson; et al., 2003; Karzai et al., 2003; Tiedermann et al., 2003; Wang et al., 2004a
C5a - neutralizing antibodies or soluble receptor fragment	Prevent EC activation; inhibit systemic inflammation	Biffl et al., 1995; Basta et al., 2003; Ward et al., 2003; von Dobschuetz et al., 2004;
PAF – receptor antagonist (lexipafant)	Prevent EC activation	Galloway and Kinsnorth, 1996; Kelly, 2000; Zhang et al., 2003; Bhatia & Moochhala, 2004; Victorino et al., 2004
Neutrophil elastase inhibitors (sivelestat)	Prevent intercellular junction digestion; also protect matrix proteins	Carden et al., 1998; Herren et al., 1998; Gibbs et al., 1999; Wachtel, 1999; Zeiher et al., 2002
Neutrophil azurocidin inhibitor	Prevent EC activation	Gautam et al., 2001

ACE – angiotensin converting enzyme; AT-1 - receptor for angiotensin II; EC – endothelial cell(s); NAC – Nacetylcysteine; NAL – Nacystelyn; PAF – platelet activating factor; PARP – poly (ADP-ribose) polymerase; SOD- superoxide dismutase

Since most inflammatory mediators have dual activating effects on endothelium and on other inflammatory cells and processes, it has not always been clear what target was most important in the *in vivo* inhibition studies, but the benefits for preventing permeability have been decisive. As specifics have been related and referenced in earlier chapters, only a few examples will be briefly listed here to demonstrate the range of effective protocols and different points in the pathophysiologic mechanisms targeted. Tezosentan, an endothelin-1 antagonist, reduced lung edema in endotoxin-infused sheep (Kuklin *et al.*, 2004). ACE inhibitors and AT1 blockers prevented hemodynamic stress, endothelial dysfunction, and endothelial oxidant production (Zhang *et al.*, 1999; Liu and Zhou, 2002; Zhang, C. *et al.*, 2003). Survival from sepsis was improved in baboons by antithrombin III (Minnema *et al.*, 2000). Yet, antithrombin III also increased endothelin release and exacerbated the increase caused by endotoxin; thus, dual therapy was recommended (Dschietzig *et al.*, 2000). Of note, activated protein C directly suppressed ICAM expression, apoptosis, and thrombin- and IL-1-induced permeability in endothelial cells and lung injury in a sepsis model (Taylor *et al.*, 1987; Joyce and Grinnell, 2002; Zeng *et al.*, 2004). Nebulized, but not intravenous, heparin improved pulmonary functions in a lung injury model, indicating that delivery mode is important in anti-coagulant therapy (Murakami *et al.*, 2002; Murakami *et al.*, 2003). Antibodies to inflammatory cytokines (e.g. IL-6, IL-8, TNF; TGFβ) prevented lung endothelial permeability and improved survival in various models of lung injury (Leff *et al.*, 1994; Shenkar *et al.*, 1994; Folkesson *et al.*, 1995; Laffon *et al.*, 1999; Modelska *et al.*, 1999; Hybertson *et al.*, 2003; Karzai *et al.*, 2003; Riedermann *et al.*, 2003; Wang *et al.*, 2004a). Antibodies to or knockout of complement, cytokines, and chemokine receptors prevented permeability in both isolated endothelium and in animals and leukocyte infiltration and circulatory collapse in animal models (Biffl *et al.*, 1995; Gerard *et al.*, 1997; Bhatia *et al.*, 2000; Pittet *et al.*, 2001; Schraufstatter *et al.*, 2001; Eaton and Martin, 2002a; Basta *et al.*, 2003; Bhatia *et al.*, 2003; Ward *et al.*, 2003; von Dobschuetz *et al.*, 2004). Inhibitors of neutrophil elastase and metalloproteinases prevented inflammatory and edematous changes and improved survival in several models of acute lung injury and in patients (Kawabata *et al.*, 2000; Hagio *et al.*, 2001; Zeiher *et al.*, 2002; Eaton and Martin, 2002a; Kawabata *et al.*, 2003; Muhs *et al.*, 2003; Steinberg *et al.*, 2003). Such inhibition would not only protect extracellular matrix from digestion, but also endothelial and epithelial tight junctions and adherens junctions from proteolytic degradation (Carden *et al.*, 1998; Herren *et al.*, 1998; Gibbs *et al.*, 1999; Wachtel, 1999). Selective inhibition of neutrophil-derived azurocidin (CAP37) reduced endothelial activation and monolayer permeability (Gautam *et al.*, 2001).

There has been an explosion of information about receptor-membrane signaling events, second messenger release, subsequent kinase and phosphatase cascades, and activation or inactivation of intracellular effector molecules. The application of this knowledge to signaling and altered function in endothelial cells, particularly as related to acute barrier dysfunction has been spelled out in Chapters 3 and 4. For instance, endothelial activation and permeability was ameliorated *in vitro* and *in vivo* by blocking MLCK or Rho kinase (Garcia *et al.*, 1995; Essler *et al.*, 1998; van Nieuw Amerongen *et al.*, 1998; Carbajal and Schaeffer, 1999; Carbajal *et al.*, 2000; van Nieuw Amerongen *et al.*, 2000a; Chiba *et al.*, 2001; Qiao *et al.*, 2003; Tinsley *et al.*, 2004a; Tinsley *et al.*, 2004b). Tyrosine kinase inhibitors attenuated permeability (Carbajal and Schaeffer, 1998; Esser *et al.*, 1998; Shi, S. *et al.*, 1998; Andriopoulou *et al.*, 1999; Cohen *et al.*, 1999; Baldwin and Thurson, 2001; Bates *et al.*, 2002). While PPARγ-receptor agonists prevented inflammation, oxidation, endothelial activation, lung injury, and multi-organ failure (Plutzky, 2001; Goetze et al., 2002; Hannan et al., 2003; Collin et al., 2004; Cuzzocrea et al., 2004; Verrier et al., 2004); high levels of the agonists induce endothelial permeability and apoptosis (Hannan et al., 2003; Idris et al., 2003).

Oxidants may be produced by endothelial cells themselves or by other cells with both activating and injurious effects on endothelium. Moreover, oxidants have a synergistic action with neutrophil elastase in induction of lung injury and oxidants destroy the anti-proteases intended to block the elastases (Zeiher et al., 2002). Various treatments to prevent oxidant generation or to scavenge oxidants, such as specific inhibition of iNOS to block NO overproduction, have been shown to protect from endothelial dysfunction and lung injury in animal models (Bernard et al., 1984; Patterson et al., 1985; Simon and Suttorp, 1985; Milligan et al., 1988; Patterson and Rhoades, 1988; Seekamp et al., 1988; Amari et al., 1993; Lucas et al., 1995; Gonzalez et al., 1996; Atochina et al., 1998; Fan et al., 2000; Knepler et al., 2001; and others in Table1). Recently, a new thiol antioxidant was used to reduce leukocyte recruitment to lungs after endotoxin instillation (Antonicelli et al., 2004). An advantage of this anti-oxidant, Nacystelyn, is that it can be administered by inhalation. A related advancement was the revelation that inhibition of peroxynitrite activation of poly (ADP-ribose) polymerase reduced upregulation of the leukocyte adhesion molecules, leukocyte infiltration, loss of endothelial tight junctions, and lung damage in several models of injury (Szabo et al., 1997b; Sharp et al., 2001; Cuzzocrea et al., 2002; Mazzon et al., 2002; Watts and Kline, 2003). However, anti-oxidant treatment initiated late in the injury process is of limited benefit (Groeneveld et al., 1990; Vassilev et al., 2004). Statins have recently been shown to provide direct benefits to endothelium, such as reduced oxidant generation via NAD(P)H oxidase, improved endothelial-relaxant NO production, reduced Rho-A activation, reduced permeability, improved anticoagulant properties, decreased cytokine-induced adhesion molecule expression, and antiproliferative effects (van Nieuw Amerongen et al., 2000b; Altieri, 2001; Greenwood et al., 2003; Siegel-Axel, 2003; Nubel et al., 2004). Moreover, simvastatin pretreatment attenuated leukocyte accumulation and lung permeability in an *in vivo* model of ischemic lung injury (Naidu et al., 2003).

An important step along the path to providing specific protection against oxidant injury is represented by the immunotargeting of catalase and SOD to endothelium and by gene transfer of heme oxygenase-1 stress protein into microvascular endothelium (Muzykantov et al., 1996; Atochina et al., 1998; Yang et al., 1999). Importantly, catalase-immunotargeting protected against edema in H_2O_2-challenged lungs (Atochina et al., 1998). Heme-oxygenase upregulation increases production of bilirubin, which acts as a peroxyl radical scavenger and decreases leukocyte adhesion molecule expression (Carter et al., 2004). Airway instillation of adenovirus to transfect lungs with 1-cys peroxiredoxin also attenuated hyperoxic-induced lung injury (Wang et al., 2004b). Depending on the antibody chosen, proteins can be directed to particular endothelium. Injected Thy-1 antibodies directed selective uptake to capillaries> microvessels, but not to large vessels (Danilov et al., 2001). ACE antibodies directed linked protein expression mainly to pulmonary endothelium (Danilov et al., 2001). Coupling to ICAM-1 or to E-selectin antibodies direct the protein to expression in activated endothelium (Spragg et al., 1997; Everts et al., 2003; Muro et al., 2003). Proteins can be directed to the surface or can be internalized depending on the antibodies, the conjugates, and the conjugate size (Wiewrodt et al., 2002). Up to 66 % internalization of catalase was achieved in isolated endothelium with biotinylated-SA conjugated PECAM antibodies and 60% with ICAM and ACE antibodies (Danilov et al., 2001; Muzykantov et al., 1995; Sweitzer et al., 2003). Internalization of the chosen protein can also be aided by the use of immunoliposome delivery and by coupling to nanoparticle for amiloride-sensitive endocytosis (Spragg et al., 2001; Muro et al., 2003). In contrast, antibodies to a novel 85kD glycoprotein directed preferentially to the surface of pulmonary endothelium with minimal internalization (Murcianao et al., 2001). Immunotargeting provided rapid increases in the coupled proteins, while adenovirus-protein delivery was slower but sustained; simultaneous administration of adenovirus-protein with immunotargeted delivery gave both rapid and sustained increases in endothelial catalase (Sweitzer et al., 2003). With a different approach, adenovirus engineered with endothelial-targeting peptides, improved viral uptake by endothelium and decreased uptake by the liver (White et al., 2004). Additionally, complexing complement receptor-1 inhibitor with sialyl Lewisx enhanced endothelial binding and attenuated inflammatory and lung injury effects of cobra venom (Mulligan et al., 1999). Thus, organ, vascular segment, timing, internalization vs. surface expression, and protein choice can all be engineered with the right combination.

More recently, advances in molecular biology have been applied to altered endothelial functions involving altered expression of receptors, leukocyte binding molecules, junctional molecules, signaling molecules, and oxidant generating systems as outlined in Chapter 6. For instance, the receptors and signals for leukocyte-independent activation of endothelium mediated by endotoxin and other bacterial products have been discovered (Bannerman and Goldblum, 1999; Bochud and Calandra, 2003; Peters et al., 2003). There has especially been fervent interest in understanding the signaling involved in angiogenesis related to restenosis and to tumor development. Endothelial paracrine signaling involved in smooth muscle proliferation and in atherosclerotic plaque development is coming under increased scrutiny.

Gaps and new directions

Blockers of cytokines and other inflammatory mediators are effective in preventing lung injury in different experimental models, but poor translation into clinical usefulness indicates that more work is needed. The attenuation of neutrophil-induced permeability with the inhibitor of azurocidin, suggests that this might be a new tool to prevent the barrier dysfunction and lung injury without interfering with the ability of the neutrophils to combat infection (Gautam et al., 2001). The development of a low molecular weight inhibitor of neutrophil elastase that is also resistant to oxidant degradation, sivelestat, could be a significant development for preventing protease destruction of intracellular junctions and matrix damage (Zeiher et al., 2002).

Although many intracellular pathways that signal endothelial activation are known, there are major gaps in our understanding of the sequence and interaction of specific signals with respect to the variety of activators. Even larger gaps occur in knowing precisely how these signals are translated into structural and functional alteration. For instance, there is good evidence for rhoA involvement in actin polymerization and stress fiber formation, but the role of Rac is more confused. Whereas Rac was implicated in formation of adherens junctions and maintenance of the DPB (Braga et al., 1999; Grazia-Lampugnani et al., 2002; Waschke et al., 2004); Rac was essential for VEGF-induced permeability (Eriksson et al., 2003), upregulation of ICAM in response to TNF (Chen et al., 2003), and Rac activation stimulates assembly and activation of NAD(P)H oxidase, a primary source of oxidants in activated endothelium (Dinauer et al., 2003; Gertzberg et al., 2004). Thus it is not clear under what circumstances statins provide net benefit to the endothelium, if Rac availability is limited by statin treatment and if Rac is important to promote barrier integrity (Waschke et al., 2004). Likely, both timing and compartmentation are critical to which effects are expressed for Rac and for other signals; thus experimentation revealing the spatio-temporal nature of endothelial cell signaling will be important in the future. Blocking signals, such as inhibition of Ca^{++} release, Ca^{++} chelation, MLCK inhibition, and PKC inhibition, attenuate permeability in isolated endothelium (Chetham *et al.*, 1997; Kelly *et al.*, 1998), but application of these inhibitors in a multicellular *in vivo* setting is complicated by the commonality of signaling and the high potential for undesirable non-endothelial effects. However, inhibition of intracellular activation of phospholipase A_2 has shown potential benefit for endothelial integrity and for reducing edema in a lung injury model (Nagase, 2003; Thompson, 2003). Similarly, the range of beneficial effects shown in the few experiments with PPARγ-receptor agonists indicate this could be an advancement in intracellular modification of injury that should be further explored (Plutzky, 2001; Goetze *et al.*, 2002; Hannan *et al.*, 2003; Collin *et al.*, 2004; Cuzzocrea *et al.*, 2004). In addition to the signals related to reversible activation, there is much to be learned regarding the signaling of necrosis and apoptosis in the pathologic process and the balance of apoptosis and proliferation in repair.

Oxidant challenge can cause upregulation of cellular defense and repair mechanisms. If this natural protective mechanism could be initiated quickly by pharmacologic means without the need for oxidant priming, endothelium and epithelium could be made more resistant to oxidant damage. The immunotargeting of anti-oxidant treatments to specific cells and to modification of intracellular signaling as described above holds significant promise for future improved drug administration. This technique could be developed to target virtually any desired protein to specific vascular beds, with or without endothelial protein internalization.

Basic mechanisms promoting integrity

Experimental advances

Signaling involved in the transition to a quiescent, barrier competent monolayer following growth has been partially elucidated (Shay-Salit *et al.*, 2002; Grazia-Lampugnani *et al.*, 2003). For example, recently the role of cadherin ligation in signaling growth arrest and different roles for RhoA and Rac in signaling alterations in the actin cytoskeleton typical of activation and quiescence were revealed (Braga *et al.*, 1999; Grazia-Lampugnani *et al.*, 2002; Grazia-Lampugnani *et al.*, 2003; Waschke *et al.*, 2004). The importance of cAMP and PKA to endothelial barrier integrity as described in Chapters 5 and 15 has long been known. *In vitro* and *in vivo* studies have thoroughly documented the ability of elevated cAMP to prevent and even reverse barrier dysfunction due to a variety of mediators. For example, PKA activation with NECA, a stable adenosine analog, resulted in formation of tight junctions and improved barrier integrity (Comerford *et al*, 2002). Likewise, release of both prostacyclin and adrenomedullin peptide by activated endothelium result in cAMP/PKA-related autocrine protection of endothelium and paracrine vasorelaxation and exogenous addition has been used to attenuate barrier dysfunction (Wang, 1998; Hippenstiel *et al.*, 2002; Nikitenko *et al.*, 2002). Other means to raise cAMP and prevent pulmonary edema include the long acting PGI_2 analog - iloprost, β-agonists, dobutamine and dopexamine, phosphodiesterase inhibitors, such as pentoxifylline, and cholera toxin stimulation of adenylyl cyclase (Riva *et al.*, 1990; Patterson *et al.*, 1994; Carter *et al.*, 1995; Hsu *et al.*, 1996; Suttorp *et al.*, 1996; Sata *et al.*, 2000; Shindo *et al.*, 2000; Dhingra *et al.*, 2001; Schermuly *et al.*, 2003; Maris *et al.*, 2004; Perkins *et al.*, 2004). Co-administration of PGI_2 with phosphodiesterase inhibitors prolonged the beneficial effects on vasodilation and should prolong effect on epithelium and endothelium (Schermuly et al., 2001). One recent study showed that treatment with a β-2 adrenergic agonist partially reversed acid-induced lung injury and improved endothelial reflection of albumin (McAuley et al., 2004).

Moreover certain mediators exercise their barrier disruption via decreased cAMP (Ogawa et al., 1992; Koga et al., 1995). Studies disclosed the importance of cAMP and PKA compartmentation to the effects produced, other cAMP-effectors besides PKA, the potentially important role of PKI in regulation of PKA phosphorylation, and a few possible barrier-related PKA targets (Arimura et al., 1997; Kawasaki et al., 1998; Lum et al., 1999; Patterson et al., 2000; Comerford et al., 2002; Lum et al., 2002; Brunton, 2003; Qiao et al., 2003; Bundey and Insel, 2004). Importantly Chapter 5 detailed current knowledge on regulation of adenylate cyclase-mediated synthesis of cAMP, phosphodiesterase cAMP degradation, and the interaction of calcium influx to focal alterations of cAMP levels. Interestingly there are unique differences in the calcium/cAMP regulation in endothelium from different pulmonary vascular segments appropriate to their unique functions (Creighton et al., 2003; King et al., 2004). Finally, in addition to beneficial effects on pulmonary endothelium, PKA also promotes clearance of alveolar edema fluid by increasing sodium transport from the alveoli across the lung epithelia (Wang et al., 1999; Matthay et al., 2000; Matthay et al., 2002; Ware et al., 2002; Sartori, et al., 2002; McAuley et al., 2004).

Inhaled NO has been used with various intentions: smooth muscle relaxation and improved perfusion, direct benefit to endothelial function, and as a superoxide scavenger. In some cases improvement was noted, but effects are variable depending on the mode and dose of NO administration, the model, and the outcome parameter measured (Benzing et al., 1995; Guidot et al., 1996; Suttorp et al., 1996; Bloomfield et al., 1997; Garat et al., 1997; Okayama et al., 1998; Honda et al., 1999). Importantly NO protection in endothelium was ascribed to PKA activation, rather than PKG as in smooth muscle (Polte and Schroeder, 1998).

Other factors are now identified, independent of cAMP and PKA, which promote endothelial barrier integrity, such as sphingosine-1 phosphate and ATP (Noll et al., 1999; Yatomi et al., 2000; Gainor et al., 2001; Minnear et al., 2001; Schaphorst et al., 2003). Our understanding of the influence of the matrix on endothelial signaling and function and of the endothelium on matrix composition has improved (Vestweber, 2000; Juliano, 2002; Iivanainen et al., 2003). Permissive hypercapnia or direct acidosis reduced permeability edema and oxidant generation, and improved oxygenation independent of ventilation in several in vitro and in vivo experimental models of lung injury (Shibata et al., 1998; Laffey et al., 2000a; Laffey et al., 2000b; Laffey et al., 2004b). TNF levels and neutrophil influx were decreased, indicating suppressed inflammation, yet mechanisms will need to be determined for applicability in the complex clinical setting (Laffey et al., 2004b). In addition to improving lung fluid balance by restoring or elevating plasma protein osmotic pressure, albumin benefits may also be ascribed to its thiol-related and thiol-independent antioxidant properties (Powers et al., 2002; Powers et al., 2003; Lang et al., 2004).

In the last decade much has been learned about endothelial cell growth factors, their receptors, and their influence on barrier function (Gamble et al., 2000; Liu et al., 2002; Thurston, 2002; Jones, 2003; Pizurki et al., 2003; Taichman et al., 2003). There is a major current effort to understand the growth factors and physiology of angiogenesis. On one side, there is desire to control unwanted vascular development, as in tumors; but on the other side, revascularization is desirable in conditions like diabetic neuropathy and ischemic heart damage. How endothelial growth factors affect lung function or dysfunction and recovery dysfunction may forestall lung injury and/or impending dysfunction in multiple other systems, which occur secondarily to the inciting trauma or disease.

Although many of these suggested markers are associated with acute respiratory dysfunction, inadequate study in patients, non-specificity for lung injury, poor availability of assays, slow laboratory turn around, and costs have unfortunately hampered their use. Importantly, elevated vWF in non-pulmonary sepsis has predictive value for identifying patients who will develop acute lung injury (Rubin et al., 1990), although multiple other biochemical and clinical indices failed to predict lung injury with high probability. New data show it also has prognostic value for mortality and other outcomes in early acute lung injury based on a large multicenter study (Ware et al., 2004). This new multicenter study was prompted by an earlier study showing elevated plasma vWF predicted mortality (Ware et al., 2001). Some prior smaller studies had not found that vWF was of predictive value (Moss et al., 1996; Bajaj and Triconomi, 1999; Reinhart et al., 2002), but the new multicenter study confirms the prognostic value of elevated plasma vWF for early lung injury (Ware et al., 2004). Although vWF and these other markers are not specific for only lung endothelium or impending lung injury, they do indicate endothelial stress responses. Importantly, prophylactic endothelial protection when vWF is elevated could improve the overall morbidity and mortality regardless of the particular organ involved, as endothelial dysfunction is central to the systemic inflammatory response, as well as to pulmonary permeability.

Perhaps more directly specific for lung injury, is the dramatic 16 fold difference in the level of IL-8 -autoantibody complexes in BAL fluid of at risk patients, who later developed ARDS compared to comparably ill patients, who did not develop ARDS; whereas the ratio of free IL-8 in these two groups was only 2.4 (Kurdowska et al., 2002). Interestingly, the patients, who did not develop ARDS, actually had higher BAL protein than those who did, further emphasizing the potential value of this IL-8 -autoantibody marker for predicting progress to lung injury. Moreover, this complex possessed the same proinflammatory properties as free IL-8, the major chemotactic factor for neutrophils (Krupa et al., 2004). Recently, a plasma marker of alveolar epithelial injury, KL-6 was also correlated with severity of respiratory illness and survival outcome (Ishizaka et al., 2004; Sato et al., 2004).

Early indication of oxidant stress and effective early intervention in very ill patients could even conceivably reduce the numbers requiring eventually admittance to ICU for a variety of reasons, if a reliable, easily administered, and reasonably priced test were developed. In one study, all patients admitted to the ICU were given serial tests to measure urine albumin. A positive correlation between urinary microalbuminuria in the first 48 hours and progression to low PaO_2/FIO_2 ratio was found, suggesting the predictive power of this measure for acute respiratory failure (Abid et al., 2001). The sensitivity and specificity of this indicator of generalized endothelial barrier dysfunction was also good for predicting development of multiple organ failure, thus administration of a protective intervention would be warranted, if a specific agent or set of agents could be identified without side-effects on other cell types.

Given the central importance of capillary leak in lung failure, improved diagnostic imaging techniques for assessing pulmonary vascular permeability could be highly useful to detect early changes. As described in Chapter 1, bedside test using ^{67}Ga and ^{99m}Tc was shown to measure lung permeability (Groeneveld et al., 1996). Both CAT and PET scans are capable of resolving the sites and magnitude of edema (Schuster, 1998, Ketai and Godwin, 1998). Unfortunately, these are expensive tests and are generally available STAT. Thus, even if a good predictive marker is identified that would help guide prophylactic treatment, testing must be available and cost reasonable for practical application in the critical care setting.

Table 4. Potential Markers of Early Endothelial Activation and Predictive and Prognostic Biologic Markers of Acute Lung Injury

Factor/Test	Rationale/Value	Reference
Von Willebrand factor,* plasma	indicative of endothelial activation, whether due to trauma or sepsis, pro-coagulant state; predicts lung injury	Rubin et al., 1990; Bajaj and Tricomi, 1999; Ware et al., 2001; Ware et al., 2004
Endothelin-1, plasma	indicative of endothelial activation & vaso-constrictive state	Dschietzig et al., 2001; Kuklin et al., 2004
Thromboxane, plasma, particularly a high TXA_2/PGI_2 ratio	endothelial and inflammatory cell activation, pro-coagulant & vaso-constrictive state	Schilling et al., 2001; Dandona et al., 2003
sICAM-1, plasma	indicative of endothelial activation & inflammatory state	Reinhart et al., 2002; Flori et al., 2003; Bhatia & Moochhala, 2004
8-isoprostane, plasma, urine, and breath	oxidant stress	Carpenter et al., 1998; Lum & Roebuck, 2001; Del Rio et al., 2002;
Nitrite/nitrate, plasma & breath	oxidant stress	Lee et al., 2001; Zhu et al., 2001; Gessner et al., 2003; Gao et al., 2004
TBARS, plasma	oxidant stress	Del Rio et al., 2002
4-hydroxy-2-nonenol	oxidative stress	Quinlan et al., 1995
Circulating, non-p34 endothelial cells	endothelial injury; late marker; hard to quantitate but specific	Reinhart et al., 2002
Cytokines (TNF, IL-1, IL-6, IL-8, TGF), plasma & AF	inflammatory stress; correlation with severity of disease; mediators of endothelial activation	Miller et al., 1992; Pugin et al., 1999; Ranieri et al., 1999; Bauer et al., 2000; Reinhart et al., 2002; Bhatia & Moochhala, 2004
C5a, plasma	complement factor, chemoattractant, endothelial cell activator; both pro- and anti-inflammatory	Weinberg et al., 1984; Parsons et al., 1990; Bhatia & Moochhala, 2004
PAF, in AF	pro-inflammatory bioactive phospholipid	Bhatia & Moochhala, 2004
IL-8/autoantibody complex, AF	inflammation in lung; major neutrophil chemotaxis factor	Kurdowska et al., 2002
Thrombomodulin & Protein C, plasma & AF	Increased thrombomodulin indicates endothelial shedding and epithelial synthesis, decreased Protein C correlates with coagulation abnormality	Ware et al., 2003
Von Willebrand's screen of 10 clotting indices*	pro-coagulant state	Reinhart et al., 2002
Microalbuminuria,* urine	increased systemic endothelial permeability	Abid et al., 2001
Albumin,* plasma and AF	plasma - low protein osmotic pressure, low anti-oxidant protection; AF – if elevated > 0.65 in edema fluid indicates alveolar-capillary permeability	Matthay & Wiener-Kronish, 1990; Pittet et al., 1997; Groeneveld, 2000; Martin et al., 2002; Vincent et al., 2003b; Bowler et al., 2004; Quinlan et al., 2004
PET scan, CAT scan, "bed side" Ga/Tc screening	pulmonary permeability imaging	Groeneveld et al., 1996; Schuster et al., 1998; Ketai & Godwin, 1998.

Table 4. These factors all have proven associations with endothelial dysfunction, ARDS, or major risk conditions. A few (e.g. vWF) have proven prognostic value for outcome in ARDS, but most are not proven predictors of impending lung injury. Importantly, even when a marker is not specifically predictive of lung injury, elevation of the marker may indicate a need for treatment of endothelial injury in the overall critical care context. * Most of these factors are current performed as standard clinical laboratory assays; however, only plasma albumin generally has a turn around < 1 day. To be useful for indicating early intervention, test and results must be more readily available. AF –refers to alveolar fluid, whether obtained by broncho-alveolar lavage or undiluted edema fluid from intubated patients obtained by catheter suction without lavage.

The exponential growth of genomics and proteomics offer new opportunities for discovery of factors related to endothelial, lung, and multiorgan/multisystem dysfunction. Advances in techniques, application to analysis of BAL fluid, plasma, and lung proteins, and the potential usefulness for the study of lung disease were recently reviewed (Hirsch et al., 2004). Proteomics was used to compare plasma and alveolar fluid from normal subjects and ALI patients (Bowler et al., 2004). Of the proteins analyzed, plasma levels of β2-glycoprotein 1, hemopexin, serum amyloid, and oromucoid increased, while haptoglobin and α1-antitrypsin decreased in ALI. In the alveolar fluid, albumin, β-globin, and oromucoid increased, while surfactant protein A decreased. This technique and marker identification will potentially lead the way to new early markers for impending lung injury. Further study is necessary to judge the predictive value of these markers in at-risk patients, prior to development of ALI. Proteomics was also recently used to demonstrate programmed differences in endothelium from different vascular beds, even when maintained under the same culture conditions (Chi et al., 2003). Factors unique to genetic background, which render certain persons more susceptible to development of ALI and ARDS, can also be identified with these new tools. Polymorphisms in TNFα, Il-6, ACE, and TLR receptors for pathogen products are examples of complex genetic make-up that increase such susceptibility to sepsis and to lung involvement (Marshall et al., 2002; Bochud and Calandra, 2003). This kind of information is useful to insure that the optimal cell type and state is used for experimental investigation of relevant endothelium and drug responsiveness. Advances in targeted delivery of drugs and genetic material to selected cells will enable more effective utilization of our knowledge of such mediators and mechanisms. For instance, coupling adenovirus vectors to polycations resulted in 10 to 100 fold increased transduction into mouse lung epithelial cells without evidence of inflammatory activation (Kaplan et al., 1998).

Progress in understanding the physiology of resolution

We have expended most effort at defining the downward processes: activation of inflammation, activation of endothelial cells, and initiation of barrier dysfunction, with the intent to block these processes. But have paid less attention to understanding the natural processes involved in termination of inflammation, reversal of coagulation, and restoration endothelial integrity. These reversible states permit temporary, focused activation to deal with a disturbance that is normally balanced by full and rapid return to a physiologic functional status. Capitalizing on mechanisms leading to reversal of injury would offer new opportunities for treating lung injury. Normal turnover of endothelium is very low, but following injury, it is not well known how repair occurs. Recent reports on angiogenesis, growth factors, and endothelial progenitor cells have begun investigations that will help fill this gap (Reyes et al., 2002; Aicher et al., 2003; Hibbert et al., 2003; Jones et al., 2003).

Resolution of alveolar edema involves the reabsorption of water driven by the active vectorial Na^+ transport from the alveoli to the interstitial surface, as noted above, can be activated by cAMP elevation (Matthay et al., 1996; Wang et al., 1999; Matthay et al., 2002; Ware et al., 2002; Sartori, et al., 2002). Note that proteins may actually be concentrated when the active Na^+ transport and water removal is operating effectively. Proteins are then removed by diffusion, endocytosis, transcytosis, and by macrophage phagocytosis (Folkesson et al., 1996). Re-epithelialization is controlled in part by keratinocyte and hepatocyte growth factors (Ware and Matthay, 2000). While some patients seem to recover completely, others develop fibrosing alveolitis that may itself be fatal (Martin et al., 1995; Ware and Matthay, 2000). This condition is initiated early in the course of lung failure as evidence by procollagen III peptide in alveoli even at the time of initial diagnosis of ARDS (Clark et al., 1995; Chesnutt et al., 1997). By 5-7 days it is characterized by: invasion of the alveoli with mesenchymal cells; development of new vessels and destruction of some original vessels; and accumulation of collagen and fibronectin (Bachofen and Weibel, 1982; Fukuda et al., 1987; Artigas et al., 1998). The specific conditions and mechanisms that differentiate between this fibrotic response and more normal resolution are not yet clear, but the presence of alveolar protein, a role for inflammation with cytokine activation of fibroblast, and epithelial damage have been identified (Matthay et al., 1990; Martinet et al., 1996; Olman et al., 2004).

Protection and restoration in the clinical setting

First line support

Improved treatment of the initiating trauma or disease that leads to endothelial damage is fundamental for preventing and dealing with permeability edema (Ware and Matthay, 2000; Marshall et al., 2003; Toh et al., 2003; Wiedemann et al., 2003; Dellinger et al., 2004; MacArthur et al., 2004). When acute lung injury does occur, therapy to provide optimal ventilatory, circulatory, and nutritional support is important in reducing mortality (Artigas et al., 1998; Eisner et al., 2001; Malarkkan et al., 2003; Brower et al., 2004). Though epithelial injury and alveolar flooding contribute to lack or poor surfactant performance in ARDS, efforts at supplementation have not proven to be of benefit, as in IRDS (Spragg et al., 2003).

In addition to these protective ventilatory strategies, a recent paper on supportive nutrition is of interest. Parenteral lipid emulsion feeding with medium and long chain triglycerides to patients with acute lung injury resulted in deterioration of arterial oxygenation, lung compliance, and lung vascular resistance (Lekka et al., 2004). Lavage fluid protein, phospholipid, phospholipase activity, PAF, and neutrophil counts also increased, suggesting exacerbation of inflammation and endothelial barrier dysfunction. Depending on content, excessive serum triglycerides and free fatty acids are known to have inflammatory consequences and untoward direct effects of oleic acid (but not triolein emboli) on pulmonary vascular and barrier function have long been known (Selig et al., 1986; Selig et al., 1987; Steinberg et al., 1997). However, serum lipids can bind and reduce toxic effects of LPS. One explanation of bone fracture trauma is release of fat emboli, pulmonary capillary trapping, metabolism of the lipids by endothelial phospholipases, local increases in free fatty acids, resulting in secondary lung injury; but this remains unproven as the connection between multiple fracture and ARDS (Muller et al., 1992). Thus, exacerbating effects of lipid feeding may be explained by lipid-induced inflammation or by direct effects of fatty acids on lung endothelium. Regardless, this is important to know in order to improve patient support.

Advances and retreats in blocking activation

Blocking mediators of lung injury, which are common to the variety of initiating events, has been a primary strategy in experimental and clinical studies. A few blockers that were successful in preventing or treating endothelial permeability in culture or experimental models of lung edema, as described above and in Chapter 15, showed promise in patients. PAF antagonism or catabolism attenuated lung injury and improved survival in specific patients (Galloway and Kingsworth, 1996; Schuster et al., 2003). Some clinical studies of thromboxane inhibition yielded improved outcomes (Yu and Tomasa, 1993, Schilling et al., 2001) and ibuprofen COX inhibition was beneficial in a sub-set of ARDS patients with hypothermia (Bernard et al., 1997b). Although the effect was modest, the use of activated protein C is a recent advance in treating sepsis, a leading cause of ARDS (Bernard et al., 2001; Joyce and Grinnell, 2002; Vincent et al., 2003a; Opal, 2003). Activated protein C not only attenuates clotting by inhibiting plasminogen activator inhibitor and thrombin activation, but is also anti-inflammatory and has direct beneficial effects on the endothelium. Importantly, reduced protein C levels in ALI/ARDS patients was recently shown to be prognostic for clinical outcome even when data from patients without sepsis was analyzed separately (Matthay and Ware, 2004). Sivelestat, an oxidant resistant inhibitor of neutrophil elastase, tended to reduced time on mechanical ventilation and time in ICU, but did not result in a significant improvement in overall mortality in a trial of ALI patients with systemic inflammatory syndrome (Zeiher et al., 2002). Complement blocker was deemed safe, but promised efficacy trials have not been forthcoming (Zimmerman et al., 2000).

Despite these studies and positive experimental results in blocking inflammation, endothelial cell activation, and lung edema, successful application of many of the blockers and protocols in patients have not yielded the hoped for clinical benefits in improving survival in severe lung edema (Artigas et al., 1998; Ware and Matthay, 2000; Groeneveld, 2003; Matthay et al., 2003). Clinical trials using various approaches to block TNF or IL-1 were not effective against sepsis or ARDS (Opal et al., 1997; Reinhart et al., 2001; Butty et al., 2003; Remick, 2003; Reimold, 2003). Similarly, trials to oppose endotoxins did not show benefit in spite of the known role of endotoxin in certain types of sepsis (Opal and Gluck, 2003; Marshall, 2003). Clinical trials for anti-TLR4 for Gram negative bacterial sepsis are currently underway and anti-TLR6 for Gram positive sepsis is proposed, but care must be taken not to block the defensive responses necessary to fight infection, so patients must be carefully selected (Bochud and Calandra, 2003). Although positive effects were reported above in human trials to block PAF and TxA_2, such strategies failed in other cases to ameliorate the pulmonary dysfunction (ARDS Network, 2000a; Taggart, 2001) or outcome in sepsis and pancreatitis (Bhatia and Moochala, 2004). Importantly, the studies showing effectiveness of the ketoconozol involved prophylactic administration, but this approach is regretfully not well practiced (Slotman *et al.*, 1988; Yu and Tomasa, 1993). Overall results with ibuprofen in ARDS or sepsis were disappointing (Bernard *et al.*, 1997b), but this is not surprising considering the non-specific nature of ibuprofen leading to decreases in beneficial, physiologic metabolites as well as inflammatory metabolites.

Treatments, such as high dose cortisone, which attenuated experimental permeability, have not attenuated mortality from ARDS (Chadda and Annane, 2002; Luce, 2002; MacLaren and Jung, 2002; Marshall, 2003; Thompson, 2003). Longer-term use in ALI and ARDS with extended disease course using lower doses has been proposed in certain situations where the stress response is compromised or where fibrosis may result after acute injury and such trials are ongoing (Groeneveld *et al.*, 2003; Meduri *et al.*, 2003). In this same light, long-term low dose corticosteroids reduced mortality in septic shock, where high doses were not beneficial (Bochud and Calandra, 2003). Neither anti-thrombin III nor tissue factor pathway inhibitor was protective in ARDS, despite benefits in experimental animal models (Osterman, 2002; Abraham, *et al.*, 2003; Eichacker and Natanson, 2003; Opal, 2003). Possibly the combination of antithrombin III with anti-endothelin might be beneficial, based on the experimental results showing antithrombin III increased endothelin release (Dschietzig *et al.*, 2000). In many of these trials, the timing of the administration of the blockers by treating diagnosed ALI or ARDS may have been the very factor that prevented their potential effectiveness. To block, they need to be present before the mediators have already had their effect. Thus, early identification of true risk would be critical.

Controversies over albumin and anti-oxidants

Potential benefits of albumin supplementation in ALI/ARDS treatment have several physiologic bases: its contribution to protein osmotic pressure necessary for lung fluid balance, its role as the major plasma buffer, its antioxidant potential as shown in preventing lung injury in several models, and its stabilizing effects on endothelial cells. Moreover, low albumin correlated with mortality in ARDS, while adverse effects are low (Vincent *et al.*, 2003b). In patients with acute lung injury, albumin helped restore fluid balance and oxygenation (Groeneveld, 2000; Martin *et al.*, 2002a) and attenuated oxidative damage (Quinlan *et al.*, 2004). Yet, it had either no effect or slightly increased overall mortality in other clinical studies (Cochrane Injuries Group, 1998; Wilkes and Navickis, 2001; The SAFE Study Investigators, 2004). This may be due to variability in lipid and anti-oxidant character of the albumin, appropriateness for particular patients, that a single treatment cannot be expected to address the multiple factors involved in lung barrier failure, or that, in advanced permeability with significant protein loss into the alveoli, albumin adds to the milieu. Thus, use of albumin remains controversial and is not standard care in ARDS.

Regardless of the accepted role of oxidants in pathology and protection afforded by antioxidants in models of ARDS, their clinical usefulness has been confined to improvements in oxygen delivery, lung compliance, alveolar protein levels, and need for ventilatory support (Bernard, 1990; Bernard, 1991; Jepsen *et al.*, 1992; Suter *et al.*, 1994; Bernard *et al.*, 1997a; Domenighetti *et al.*, 1997; Walsh and Lee, 1999; Nathens *et al.*, 2002). Probably, earlier antioxidant therapy may have helped prevent development of ARDS and improved mortality, as delayed anti-oxidant treatment has been shown experimentally to be much less effective (Van Klaveren et al., 1997; Groeneveld and Sipkema, 2000). Possibly, the newly developed Nacystelyn, which can be administered by inhalation, may prove beneficial for lung targeting, but this is now only at the experimental stage (Antonicelli et al., 2004).

Progress in cellular protective mechanisms

Studies have partly elucidated pathways by which endothelial cells normally establish the barrier or recover from activation and/or injury. These pathways offer a second approach to providing protection from the consequences of lung edema, besides blocking pathologic changes. As in the case of blocking drugs, clinical benefits of protective drugs have had mixed reviews. Agents that raise cAMP, such as the phosphodiesterase inhibitor - pentoxifylline and the β-adrenergic agonist - terbutaline, improved some related conditions, such as platelet stability and TNF levels, and several indices of lung failure in patients, such as permeability (Basran et al., 1986; Ardizzoia et al., 1993; Montravers et al., 1993; Scheeren and Radermacher, 1997; Perkins et al., 2004). However, lisofylline, an analog of pentoxyfylline with several pharmacologic effects, failed to show benefit (ARDS Clinical Network, 2002) and there were non-responders in the terbutaline trial (Basran et al., 1986). Many experimental studies have shown that phosphodiesterase blockers alone have little effect on cell cAMP levels and that activators of adenylyl cyclase are far more effective, thus combinations should be tested.

The need to improve targeting was highlighted by improved results with aerosolized PGI_2 vs. systemic PGI_2 administration. Whereas systemic PGI_2 delivery, with the aim of producing vasorelaxation and alleviating elevated pulmonary pressures, worsened patient condition due to shunting and systemic hypotension (Radermacher et al., 1990; Putensen et al., 1998), aerosolized delivery improved pulmonary hypertension, shunting, and arterial oxygenation, without the undesirable systemic hypotension (Pappert et al., 1995; Walmrath et al., 1995; Bein et al., 1996, Walmrath et al., 1996; Putensen et al., 1998; van Heerden et al., 2000; Dahlem et al., 2004). Effects on endothelial permeability and epithelial clearance were not assessed and studies were too small to assess survival, but cAMP elevation in these lung cells should be of benefit. Possibly, use of long acting analogs (e.g. iloprost) or co-administration with phosphodiesterase inhibitors (e.g. pentoxyfylline) would improve effectiveness (Schermuly et al., 2001). In one small study nebulized PGE_1 improved oxygenation (Meyer et al., 1998); yet in another trial nebulized PGE_1 did not improve oxygenation, ventilation days, or survival despite reduced pulmonary pressure (Vincent et al., 2001). Nevertheless, it is apparent that mode of delivery to target functional alveoli and their associated microvessels was an important step forward. The potential benefits of this protocol to lung endothelial and epithelial function should be considered. Possibly, targeting of cAMP elevation to endothelium and epithelium, without effect on vascular smooth muscle, may be desirable in some cases. Interestingly, in ARDS patients given intravenous PGE_1, cardiac and hepatic function was improved (Slotman, et al., 1992).

NO is a simple molecule with very complex effects. On one hand, it is physiologically important in vascular smooth muscle relaxation for control of blood flow and reduction of the driving pressure for edema, has direct beneficial effects on other tissues such as cardiomyocytes and even some endothelium, and can scavenge superoxide. While on the other hand, it is a free radical itself, contributes to peroxynitrite generation, and high levels as may occur with iNOS upregulation are detrimental. Thus it can act as an anti-oxidant or pro-oxidant as described in Chapter 15. It is not surprising that exogenous administration would be fraught with controversy. Even experimental results with administration or manipulation of NO have been highly variable due to these differences, so variability in the clinical benefits of supplemental NO could be foreseen. Overall, there appears to be an initial improvement in oxygenation in ARDS patients, but long-term benefits and better survival have not materialized (Pappert *et al.*, 1995; Gerlach *et al.*, 2003; Sokol *et al.*, 2003). We have learned that sensitization occurs and that higher doses result in deterioration of oxygenation (Gerlach *et al.*, 2003). For these reasons, dosage, duration of treatment, and consideration of the oxidant environment is extremely critical in NO therapy.

Thus, none of the "beneficial" treatments alone have exhibited dramatic results or are without controversy. Possibly, many approaches tried with limited clinical success may yet be valuable if applied early, for specifically defined conditions, and in combination with other treatments. Thus, despite disappointing trials, improved specificity, targeting, timing, and combinations might yield positive results. Newer means to elevate cAMP, such as the natural peptide, adrenomedullin, are of particular interest for development and testing. Likewise, as more is learned about the beneficial mechanisms for angiopoietin and S1P on endothelial barrier function, versions of these agents may be possibilities for future clinical development.

Combinations and considerations

Obviously, both earlier treatment and an approach to address the complex, multifactoral nature of the syndrome are required. While it is necessary to focus experimental studies, the integration of the multiple recognized endothelial functions must be fully appreciated to translate investigational results to clinical results. Barrier dysfunction cannot be severed from endothelial reactivity to leukocytes or presentation of a pro-coagulant influence. This is highlighted by recent successful trials, noted above, with activated protein C, with its multiple actions on coagulation, inflammation, and endothelial survival. Another problem in successful translation is that many agents, which effectively block activation of endothelial cell signaling, block processes in other cells that may be necessary for their normal function. Difficulty in specific organ-directed or cell-directed delivery of inhibitors, which have undesirable side effects in other systems, has been a major stumbling block. The success of experimental immunotargeted delivery of agents to particular vascular beds, especially to lung, as described above, holds extraordinary promise for improved treatment of lung injury in humans. For instance, thrombomodulin could be targeted to the lumen or Rho kinase inhibitors targeted for internalization by pulmonary endothelium. It may be desirable to direct drugs to particular types of cells within the lung and with further development and molecular engineering this should be possible. Further identification of unique markers for endothelial heterogeneity, see Chapter 10, will be useful to this end. Finally, it will be critical to provide direct protection and support for matrix and epithelial integrity in treating lung edema.

Timing is critical. Prevention of leukocyte over-activation is important to prevent lung permeability injury, but without leukocyte activation infection can worsen or arise and, thus, long-lasting or ill-timed intervention could exacerbate failure. Specific timing, targeting to the correct cells in need, and customized combinations of therapy modalities to address the multiple causes and pathologies will be required for further advancements. Perhaps most important, is the ability to deliver blocking drugs before the mediators have exerted their initial effects. A very recent study of prophylaxis in cardiopulmonary bypass surgery demonstrates the likely benefit of early treatment (Wei *et al.*, 2004). Four drugs were added to the cryoperfusion media instilled into the pulmonary artery: anisodamine to inhibit leukocyte and platelet aggregation, arginine as an NO precursor, aprotinin to inhibit serine proteases, and methyprednisolone as an anti-inflammatory agent. This protective cocktail resulted in reduced plasma TNFα, TBARS, vWF, and ET-1 after weaning from bypass, indicating decreased endothelial activation and oxidant generation. Alveolar edema, IL-6, IL-8, and leukocyte count was also reduced and arterial oxygenation was improved. It is not known how important each component was to improvement, but use of combinations and early administration is a model to improve treatment in ALI and ARDS from various causes.

In a report of ARDS patients given immediate combined therapy including optimized ventilation, permissive hypercapnia, fluid restriction, furosemide, and intermittent prone positioning, high survival was noted compared to historical data of apparently comparable patients (Lewandowski *et al.*, 1997). With specific indication, optional treatments included supplemental red cells, plasma, or albumin, and aerosolized NO. It should be noted, there are controversies regarding some of these protocols, such as transfusion in critically ill patients (Shoemacher and Wo, 1998; Wu *et al.*, 2001; Taylor *et al.*, 2002; Vincent *et al.*, 2002) and use of permissive hypercapnia (Swenson, 2004; Thompson *et al.*, 2001). Despite concerns regarding interpretation and controversies over specific protocols, use of a data-based algorithm and combined therapy is a model for improved practice. Ideally such a framework would incorporate predictive risks and a stratified definition of lung failure including "at risk" patients, so that certain treatments could be instituted before ARDS is a fact. With this in mind, advancements discussed above, accepted and new diagnostic possibilities noted in Table 4, a "working model" paradigm for recognition, prevention, and combination treatment of permeability-lung failure is presented in Figure 1 to clarify the need for investigation.

Importantly, survival often reflects survival from the underlying illness (Ware and Matthay, 2000). Indeed, Vincent noted, "most patients die with ARDS, rather than from ARDS" and concluded that "the approach to treatment must not be lung-limited but must take into account the systemic effects of the inflammatory response" (Vincent, 2002; Vincent *et al.*, 2003a). Moreover, "the endothelium is a major target of sepsis-induced events and endothelial damage accounts for much of the pathology of septic shock" (Peters *et al.*, 2003). Yet, treatments that might benefit endothelial dysfunction associated with systemic inflammation, such as ketoconozol, were not mentioned in the recent detailed survival guidelines for sepsis (Dellinger *et al.*, 2004), probably because more study is needed and because of the variable reports of benefits, or the lack thereof, partly depending on timing of administration (Yu and Tomasa, 1993, Bernard *et al.*, 1997b; ARDS Network, 2000a; Schilling *et al.*, 2001). Thus, protection of endothelium from inappropriate activation and injury, whether pulmonary or systemic, should be a major goal to improve overall survival. Given the contribution of endothelial dysfunction to activation of coagulation and multiorgan failure, early protection of endothelium when markers of endothelial activation, such as vWF (Table 4), are first elevated would seem be important (Bhatia and Moochala, 2004).

Several of the more recent experimental results demonstrating direct benefits on endothelial function may be of considerable clinical interest, especially as several of these drugs are approved drugs for other purposes. For example, cholesterol-lowering statins, such as lovastatin, stabilized endothelial function and prevention of excessive endogenous oxidant generation. And as also noted above, PPARα-agonists, such as rosiglitazone for treatment of diabetes, improved endothelial function in several studies and dramatically improved survival in an ARDS model. The recently defined role of angiopoietin in both barrier function and endothelial growth may be beneficial in patients. Use of endothelial progenitor cells for repair of injury may be more ambitious, but may be a valid strategy to test experimentally. Specific targeting will certainly advance treatment. As we improve prediction of impending endothelial activation or of early phase activation, it will desirable to advance the use of particular protocols to a prophylactic status, rather than wait for appearance of the pathologic markers, which they are intended to abrogate. Thus, to improve overall mortality we must address the problems at the cellular level as well as at the physiologic level and by preventing dysfunction, rather than merely trying to patch the dam once it has already been sabotaged.

Figure 1. Paradigm for preventing activation and injury and promoting endothelial barrier function. Primary pathologies and trauma leading to loss of lung endothelial integrity and pulmonary dysfunction may be either remote or pulmonary in origin and may arise from a wide variety of etiologies that trigger inflammation, coagulation, and hemodynamic changes. These untoward effects on lung endothelium may be direct or secondary to leukocyte sequestration and activation, airway/ alveolar or vascular mechanical trauma, perfusion abnormality, or injury to other lung tissues. Treating initiating events is critical to ultimate endothelial integrity and to the outcome in patients with ALI and ARDS, but here we highlight factors that have direct bearing on the endothelium, which are potential or tested points for intervention explored in this chapter. These protective and restorative treatments, naturally, will have crossover benefits on leukocytes, vascular smooth muscle, platelets, and other tissues because there is considerable commonality in mechanisms and mediators and also because the endothelium contributes to control of leukocytes, vascular smooth muscle, and platelets. Nevertheless, in some cases non-specificity of treatment or inopportune timing can produce undesirable effects. Thus, targeted delivery of treatment may be desirable. In many cases single treatments have not proved to be effective in preventing incidence of ARDS or improving survival in the complex clinical setting, as suggested by the multifaceted and interactive pathophysiology in the schema. See Tables 3 and 4 for addition details on diagnostic parameters. Tables 1 & 2 are speculative tables of some agents which have been shown to protect or promote endothelial function and may at some point be useful for clinical development for early treatments aimed to prevent pulmonary permeability related to specific inciting factors. Importantly, many of the proposed indices and treatments may not be specific for lung injury. This is not undesirable, but indicates the need for a comprehensive critical care treatment of endothelial dysfunction and injury that plays a role in multiple organ failure, disseminated intravascular coagulation, shock, and acute lung injury.

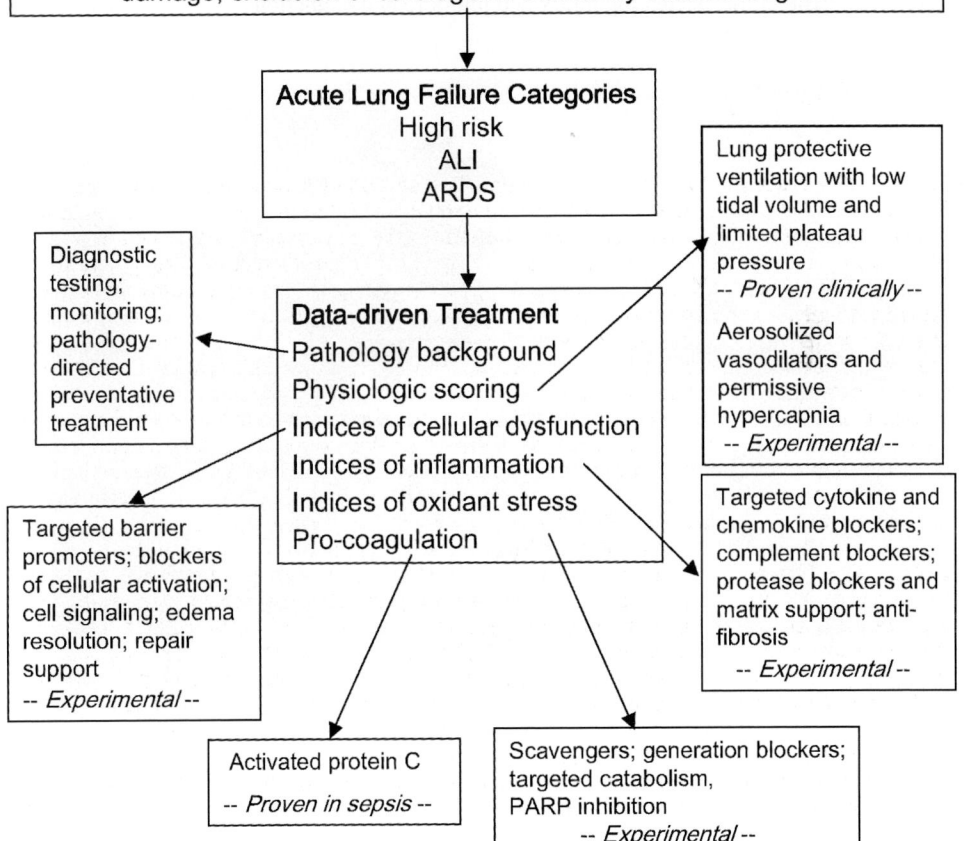

Charge to the keeper

Our knowledge of the role of endothelium in maintenance of the functional lung has expanded exponentially. Many mediators, signaling mechanisms, and cellular responses involved in assault and destruction of the integrity of the endothelial dam have been identified. Numerous strategies have been used to reduce injury in experimental models of lung permeability edema and injury with fair success; however, the transfer of this knowledge to effective intervention in clinical practice has been hampered by our inability to appropriately time, target, customize, and combine treatment options. Continued experimental refinement of the complex interplay of factors involved in lung barrier failure will provide the sound basis upon which innovations are made for the early diagnosis and effective prevention and treatment of acute lung failure and ARDS. Moreover, we recognize that endothelium is a key participant in multiple pathophysiologic disease processes, such as inflammation, atherosclerosis, thrombosis, heart and kidney failure, organ transplant failure, and in metastasis and tumor growth, and that advances in endothelial biology will have benefits for the overall heath of individuals. Our continued commitment to the partnership of basic research and clinical science will usher in the fulfillment of our goal to more effectively prevent and treat human acute lung injury and multiorgan dysfunction.

Figure 2: Harmony on the River. Early effective treatment of endothelium in trauma and in conditions leading to acute activation could prevent the havoc wrought by the flood and the fire and maintain the river within, securely contained inside the sound walls of its levees, for the harmonious conduct of its vital, normal commerce. Reproduction from The Granger Collection, New York.

ACKNOWLEDGEMENTS:

This work was supported by a Merit Award from the Veteran's Administration Medical Research Service (CEP), and by RO1 HL51856 (MAM), RO1 HL51854 (MAM), and P50 HL740005 (MAM).

REFERENCES:

Abid, O., Sun, Q., Sugimoto, K., Mercan, D., and Vincent, J.L. (2001) Predictive value of microalbuminuria in medical ICU patients: results of a pilot study. Chest. 120:1984-1988.

Abraham, E., Reinhart, K., Opal, S., Demeyer, I., Doig, C., Rodriguez, A., Beale, R., Svoboda, P., Laterre, P., Simon, S., Light, B., Spapen, H., Stone, J., Seibert, A., Peckelsen, C., De Deyne, C., Postier, R., Pettila, V., Artigas, A., Percell, S., Shu, V., Zwingelstein, C., Tobias, J., Poole, L., Stolzenbach, J., and Creasey, A. (2003) OPTIMIST Trial Study Group. Efficacy and safety of tifacogin (recombinant tissue factor pathway inhibitor) in severe sepsis. JAMA 290:238-247.

Aicher, A., Heeschen, C., Mildner-Rihm, C., Urbich, C., Ihling, C., Technau-Ihling, K., Zeiher, A., and Dimmeler, S. (2003) Essential role of endothelial nitric oxide synthase for mobilization of stem and progenitor cells. Nat. Med. 9:1370-1376.

Aird, W.C. (2004) Endothelium as an organ system. Crit. Care Med. 32:S271-S279.

Altieri, D.C. (2001) Statins' benefits begin to sprout. J. Clin. Invest. 108:365-366.

Antonicelli, F., Brown, D., Parmentier, M., Drost, E., Hirani, N., Rahman, I., Donaldson, K., and MacNee, W. (2004) Regulation of LPS-mediated inflammation in vivo and in vitro by the thiol antioxidant Nacystelyn. Amer. J. Physiol. 286:L1319-L1327.

Amari, T., Kubo, K., Kobayashi, T., and Sekiguchi, M. (1993) Effects of recombinant human superoxide dismutase on tumor necrosis factor-induced lung injury in awake sheep. J. Appl. Physiol. 74:2641–2648.

Andriopoulou, P., Navarro, P., Zanetti, A., Lampugnani, M.G., and Dejana, E. (1999) Histamine induces tyrosine phosphorylation of endothelial cell-to-cell adherens junctions. Arterioscl. Throm. Vasc. Biol. 19:2286-2297.

Ardizzoia, A., Lissoni, P., Tancini, G., Paolorossi, F., Crispino, S., Villa, S., and Barni S. (1993) Respiratory distress syndrome in patients with advanced cancer treated with pentoxifylline: a randomized study. Support. Care in Cancer. 1:331-333.

ARDS Network. (2000a) Ketoconazole for early treatment of acute lung injury and acute respiratory distress syndrome: a randomized controlled trial. JAMA. 283:1995-2002.

ARDS Network. (2000b) Ventilation with lower tidal volumes as compared with traditional tidal volumes for acute lung injury and ARDS. N. Eng. J. Med. 342:1301-1308.

ARDS Network. (2002) Randomized, placebo controlled trial of lisofylline for early treatment of acute lung injury and ARDS. Crit. Care. Med. 30:1-6.

Arimura, S., Nakata, H., Tomiyama, K., and Watanabe, Y. (1997) Phosphorylation of H-ras proteins by protein kinase A. Cell. Signal. 9:37-40.

Artigas, A., Bernard, G., Carlet, J., Dreyfuss, D., Gattinoni, L., Hudson, L., Lamy, M., Marini, J., Matthay, M., Pinsky, M., Spragg, R., and Suter, P.M. (1998) The American-European Consensus Conference on ARDS, part 2. Ventilatory, pharmacologic, supportive therapy, study design strategies and issues related to recovery and remodeling. Inten. Care Med. 24:378-398.

Ashbaugh, D, Bigelow, D., Petty, T., and Levine, B. (1967) Acute respiratory distress in adults. Lancet 2:319-232.

Atabai, K. and Matthay M.A. (2002) The pulmonary physician in critical care: Acute lung injury and the acute respiratory distress syndrome: definitions and epidemiology. Thorax 57:452-458.

Atochina, E.N., Balyasnikova, I., Danilov, S., Granger, D., Fisher, A., and Muzykantov, V.R. (1998) Immunotargeting of catalase to ACE or ICAM-1 protects perfused rat lungs against oxidative stress. Amer. J. Physiol. 275:L806-L817.

Bachofen, M. and Weibel, E.R. (1982) Structural alterations of lung parenchyma in the adult respiratory distress syndrome. Clin. Chest Med. 3:35-56.
Bajaj, M.S. and Tricomi, S.M. (1999) Plasma levels of the three endothelial-specific proteins von Willebrand factor, tissue factor pathway inhibitor, and thrombomodulin do not predict the development of ARDS. Intens. Care Med. 25:1259-1266.
Balwin, A.L. and Thurson, G. (2001) Mechanics of endothelial cell architecture and vascular permeability. Crit. Rev. in Biomed. Eng. 29:247-278.
Bannerman, D.D. and Goldblum, S.E. (1999) Direct effects of endotoxin on the endothelium: barrier function and injury. Lab. Invest. 79:1181-1199.
Basran, G.S., Hardy, J., Woo, S., Ramasubramanian, R., and Byrne, A.J. (1986) Beta-2-adrenoceptor agonists as inhibitors of lung vascular permeability to radiolabelled transferrin in the adult respiratory distress syndrome in man. Euro. J. Nucl. Med. 12:381-384.
Basta, M., Van Goor, F., Luccioli, S., Billings, E., Vortmeyer, A., Baranyi, L., Szebeni, J., Alving, C., Carroll, M., Berkower, I., Stojilkovic, S., and Metcalfe D.D. (2003) F(ab)'2-mediated neutralization of C3a and C5a anaphylatoxins: a novel effector function of immunoglobulins. Nat. Med. 9:431-438.
Bates, D.O., Hillman, N., Williams, B., Neal, C., and Pocock, T.M. (2002) Regulation of microvascular permeability by vascular endothelial growth factor. J. Anat. 200:581-597.
Bauer, T.T., Monton, C., Torres, A., Cabello, H., Fillela, X., Maldonado, A., Nicolas, J., and Zavala, E. (2000) Comparison of systemic cytokine levels in patients with acute respiratory distress syndrome, severe pneumonia, and controls. Thorax 55:46-52.
Bein, T., Metz, C., Keyl, C., Sendtner, E., and Pfeifer, M. (1996) Cardiovascular and pulmonary effects of aerosolized prostacyclin adminstration in severe respiratory failure using a ventilator nebulization system. J. Cardiovasc. Pharmacol. 27:583-586.
Benzing, A., Brautigam, P., Geiger, K., Loop, T., Beyer, U., and Moser, E. (1995) Inhaled nitric oxide reduces pulmonary transvascular albumin flux in patients with acute lung injury. Anesth. 83:1153-1161.
Bernard, G.R. (1990) Potential of N-acetylcysteine as treatment for the adult respiratory distress syndrome. Euro. Resp. J. 11:496S-498S.
Bernard, G.R. (1991) N-acetylcysteine in experimental and clinical ALI. Amer. J. Med. 91:54S-59S.
Bernard, G.R., Artigas, A., Brigham, K., Carlet, J., Falke, K., Hudson, L., Lamy, M., LeGall, J., Morris, A., Spragg, R. (1994) Report of the American-European consensus conference on ARDS: definitions, mechanisms, relevant outcomes and clinical trial coordination. Intensive Care Med. 20:225-32.
Bernard, G.R., Lucht, W., Niedermeyer. M., Snapper, J., Ogletree, M., and Brigham, K.L. (1984) Effect of N-acetylcysteine on the pulmonary response to endotoxin in the awake sheep and upon in vitro granulocyte function. J. Clin. Invest. 73:1772–1784.
Bernard, G.R., Vincent, J., Laterre, P., LaRosa, S., Dhainaut, J., Lopez-Rodriguez, A., Steingrub, J., Garber, G., Helterbrand, J., Ely, E., and Fisher, C. (2001) Efficacy and safety of recombinant human activated protein C for severe sepsis. N. Eng. J. Med. 344:699-709.
Bernard, G.R., Wheeler, A., Arons, M., Morris, P., Paz, H., Russell, J., and Wright, P.E. (1997a) A trial of antioxidants N-acetylcysteine and procysteine in ARDS. Chest. 112:164-172.
Bernard, G.R., Wheeler, A., Russell, J., Schein, R., Summer, W., Steinberg, K., Fulkerson, W., Wright, P., Christman, B., Dupont, W., Higgins, S., and Swindell, B. (1997b) The effects of ibuprofen on the physiology and survival of patients with sepsis. New Eng. J. Med. 336:912-918.
Bhatia, M., Brady, M., Zagorski, J., Christmas, S., Campbell, F., Neoptolemos, J., and Slavin, J. (2000) Treatment with neutralising antibody against cytokine induced neutrophil chemoattractant (CINC) protects rats against acute pancreatitis associated lung injury. Gut 47:838-844.
Bhatia, M. and Moochhala, S. (2004) Role of inflammatory mediators in the pathophysiology of acute respiratory distress syndrome. J. Pathol. 202:145-156.
Bhatia, M., Proudfoot, A., Wells, T., Christmas, S., Neoptolemos, J., and Slavin, J. (2003) Treatment with Met-RANTES reduces lung injury in caerulein-induced pancreatitis. Brit. J. Surg. 90:698-704.

Bhattacharya, V., McSweeney, P., Shi, Q., Bruno, B., Ishida, A., Nash, R., Storb, R., Sauvage, L., Hammond, W., and Wu, M. (2000) Enhanced endothelialization and microvessel formation in polyester grafts seeded with CD34(+) bone marrow cells. Blood. 95:581-585.

Biffl, W.L., Moore, E., Moore, F., Carl, V., Franciose, R., and Banerjee, A. (1995) Interleukin-8 increases endothelial permeability independent of neutrophils. J. Trauma Inj. Inf. Crit. Care 39:98-102.

Bloomfield, G.L., Holloway, S., Ridings, P., Fisher, B., Blocher, C., Sholley, M., Bunch, T., Sugerman, H., and Fowler, A. (1997) Pretreatment with inhaled NO inhibits neutrophil migration and oxidative activity resulting in attenuated sepsis-induced acute lung injury. Crit. Care Med. 25:584-593.

Bochud, P. and Calandra, T. (2003) Pathogenesis of sepsis: new concepts and implications for future treatment. Brit. Med. J. 326:262-266.

Bowler, R.P., Duda, B., Chan, E., Enghild, J., Ware, L., Matthay, M.A., and Duncan, M.W. (2004) Proteomic analysis of pulmonary edema fluid and plasma in patients with acute lung injury. Amer. J. Physiol. 286:L1095-L1104.

Braga, V.M., Del Maschio, A., Machesky, L., and Dejana, E. (1999) Regulation of cadherin function by Rho and Rac: modulation by junction maturation and cellular context. Molec. Biol. Cell 10:9-22.

Brower, R.G., Lanken, P., MacIntyre, N., Matthay, M., Morris, A., Ancukiewicz, M., Schoenfeld, D., Thompson, B.T.: NHLBI ARDS Clinical Trials Network. (2004) Higher versus lower PEEP in patients with ARDS. N. Eng. J. Med. 351:327-336.

Brunton, L.L. (2003) PDE4: arrested at the border. Sci. STKE 204:PE44.

Bundey, R.A. and Insel, P.A. (2004) Discrete intracellular signaling domains of soluble adenylyl cyclase: camps of cAMP? Sci. STKE 231:PE19.

Butty, V., Roux-Lombard, P., Garbino, J., Dayer, J., and Ricou, B. (2003) Anti-inflammatory response after infusion of p55 soluble tumor necrosis factor receptor fusion protein for severe sepsis. Euro. Cytok. Network. 14:15-19.

Carbajal, J.M., Gratix, M., Yu, C., and Schaeffer, R.C. (2000) ROCK mediates thrombin's endothelial barrier dysfunction. Amer. J. Physiol. 279:C195-C204.

Carbajal, J.M. and Schaeffer, R.C. (1999) RhoA inactivation enhances endothelial barrier function. Amer. J. Physiol. 277:C955-C964.

Carden, D., Xiao, F., Moak, C., Willis, B.H., Robinson-Jackson, S., and Alesander, S. (1998) Neutrophil elastase promotes lung microvascular injury and proteolysis of endothelial cadherins. Amer. J. Physiol. 275:H385-H392.

Carpenter, C.T., Price, P.V. and Christman, B.W. (1998) Exhaled breath condensate isoprostanes are elevated in patients with acute lung injury or ARDS. Chest 114:1653-1659.

Carter, M.B., Wilson, M., Wead, W., and Garrison R.N. (1995) Pentoxifylline attenuates pulmonary macromolecular leakage after intestinal ischemia-reperfusion. Arch. Surg. 130:1337-1344.

Carter, E.P., Garat, C., and Imamura, M. (2004) Continual emerging roles of HO-1: protection against airway inflammation. Amer. J. Physiol. 287:L24-L25.

Chadda, K. and Annane, D. (2002) The use of corticosteroids in severe sepsis and acute respiratory distress syndrome. Ann. Med. 34:582-589.

Chen, X.L., Zhang, Q., Zhao, R., Ding, X., Tummala, P., and Medford, R.M. (2003) Rac1 and superoxide are required for the expression of cell adhesion molecules induced by tumor necrosis factor-alpha in endothelial cells. J. Pharm. Exp. Therap. 305:573-580.

Chesnutt, A.N., Matthay, M.A., Tibayan, F.A., and Clark, J.G. (1997) Early detection of type III procollagen peptide in acute lung injury: pathogenetic and prognostic significance. Amer. J. Respir. Crit. Care Med. 156:840-845.

Chetham, P., Guldemeester, H., Mons, N., Brough, G., Bridges, J., Thompson, W., and Stevens, T. (1997) Ca^{++}-inhibitable adenylyl cyclase and pulmonary microvascular permeability. Amer. J. Physiol. 273:L22-L30.

Chi, J.T., Chang, H., Haraldsen, G., Jahnsen, F., Troyanskaya, O., Chang, D., Wang, Z., Rockson, S., van de Rijn, M., Botstein, D., and Brown, P.O. (2003) Endothelial cell diversity revealed by global expression profiling. PNAS USA 100:10623-10628.

Chiba, Y., Ishii, Y., Kitamura, S., and Sugiyama, Y. (2001) Activation of Rho is involved in the mechanism of H_2O_2-induced lung edema in isolated perfused rabbit lungs. Microvasc. Res. 62:164-171.

Clark, J.G., Milberg, J.A., Steinberg, K.P., and Hudson, L.D. (1995) Type III procollagen peptide in the ARDS: association of increased peptide levels in bronchoalveolar lavage fluid with increased risk for death. Ann. Intern. Med. 122:17-23.

Cochrane Injuries Group (1998) Human albumin administration in critically ill patients: systematic review of randomised controlled trials. BMJ. 317:235-240.

Cohen, A.W., Carbajal, J.M., and Schaeffer, R.C.J. (1999) VEGF stimulates tyrosine phosphorylation of beta-catenin and small-pore endothelial barrier dysfunction. Amer. J. Physiol. 277:H2038-H2049.

Collin, M., Patel, N., Dugo, L., and Thiemermann, C. (2004) Role of peroxidsom proliferator-activated receptor-γ in the protection afforded by 15-deoxy$\Delta^{12,14}$ prostaglandin J_2 against the multiple organ failure caused by endotoxin. Crit. Care Med. 32:826-831.

Comerford, K.M., Lawrence, D., Synnestvedt, K., Levi, B.,and Colgan, S.P. (2002) Role of vasodilator-stimulated phosphoprotein in PKA-induced changes in endothelial junctional permeability. FASEB J. 16:583-585.

Creighton, J.R., Masada, N., Cooper, D., and Stevens, T. (2003) Coordinate regulation of membrane cAMP by Ca2+-inhibited adenylyl cyclase and phosphodiesterase activities. Amer. J. Physiol. 284:L100-L107.

Cuzzocrea, S., McDonald, M., Mazzon, E., Dugo, L., Serraino, I., Threadgill, M., Caputi, A., and Thiemermann, C. (2002) Effects of 5-aminoisoquinolinone, a water-soluble, potent inhibitor of the activity of PARP, in a rodent model of lung injury. Biochem. Pharmacol. 63:293-304.

Cuzzocrea, S., Pisano, B., Dugo, L., Ianaro, A., Patel, N., Di Paola, R., Genovese, T., Chatterjee, P., Fulia, F., Cuzzocrea, E., Di Rosa, M., Caputi, A., and Thiemermann, C. (2004) Rosiglitazone, a ligand of the peroxisome proliferator-activated receptor-gamma, reduces the development of nonseptic shock induced by zymosan in mice. Critical Care Medicine. 32:457-66.

Dahlem, P., van Aalderen, W., de Neef, M., Dijkgraaf, M., and Bos, A. (2004) Randomized controlled trial of aerosolized prostacyclin in children with acute lung injury. Crit. Care Med. 32:1055-1060.

Dandona, P., Aljada, A., and Chaudhuri, A. (2003) Vascular reactivity and thiazolidinediones. Amer. J. Med. 115:81S-86S.

Danilov, S.M., Gavrilyuk, V., Franke, F., Pauls, K., Harshaw, D., McDonald, T., Miletich, D., and Muzykantov, V.R. (2001) Lung uptake of antibodies to endothelial antigens: key determinants of vascular immunotargeting. Amer. J. Physiol. 280:L1335-L1347.

Dellinger, R.P., Carlet, J., Masur, H., Gerlach, H., Calandra, T., Cohen, J., Gea-Banacloche, J., Keh, D., Marshall, J., Parker, M., Ramsay, G., Zimmerman, J., Vincent, J., and Levy, M. (2004) Surviving sepsis campaign guidelines for management of severe sepsis and septic shock. Crit. Care Med. 32:858-873.

Del Rio, D., Serafini, M., and Pellegrini, N. (2002) Selected methodologies to assess oxidative/antioxidant status *in vivo*: a critical review. Nutr. Metab. Cardiovasc. Dis. 12:343-351.

Dhingra, V.K., Usaro, A., Holmes, C., and Walley, K. (2001) Attenuation of lung inflammation by adrenergic agonists in murine acute lung injury. Anesthes. 95:947-953.

Dimmeler, S., Aicher, A., Vasa, M., Mildner-Rihm, C., Adler, K., Tiemann, M., Rutten, H., Fichtlscherer, S., Martin, H., and Zeiher, A.M. (2001) HMG-CoA reductase inhibitors (statins) increase endothelial progenitor cells via the PI3-kinase/Akt pathway. J. Clin. Invest. 108:391-397.

Dinauer, M.C. (2003) Regulation of neutrophil function by Rac GTPases. Curr. Opin. Hematol. 10:8-15.

Domenighetti, G., Suter, P., Schaller, M., Ritz, R., and Perret, C. (1997) Treatment with N-acetylcysteine during ARDS: a randomized, double-blind, placebo-controlled clinical study. J. Crit. Care 12:177-182.

Dschietzig, T., Alexiuo, K., Laule, M., Becker, R., Schror, K., Baumann, G., Brunner, F., and Stangl, K. (2000) Stimulation of pulmonary big ET-1 and ET-1 by antithrombin III: a rationale for

combined application of antithrombin II and ET antagonists in sepsis related ARDS. Crit. Care Med. 28:2445-2449.
Dschietzig, T., Richter, C., Pfannenschmidt, G., Bartsch, C., Laule, M., Baumann, G., and Stangl, K. (2001) Dexamethasone inhibits stimulation of pulmonary endothelins by proinflammatory cytokines: possible involvement of NFκB dependent mechanism. Inten. Care Med. 27:751-756.
Eaton, S. and Martin, G. (2002) Clinical developments for treating ARDS. Exp. Opin. Inv. Drugs 11:37-48.
Eichacker, P.Q. and Natanson, C. (2003) Recombinant human activated protein C in sepsis: inconsistent trial results, an unclear mechanism of action, and safety concerns resulted in labeling restrictions and the need for phase IV trials. Crit. Care Med. 31:S94-S96.
Eisner, M.D., Thompson, T., Hudson, L., Luce, J., Hayden, D., Schoenfeld, D., and Matthay, M.A. (2001) ARDS Network. Efficacy of low tidal volume ventilation in patients with different clinical risk factors for ALI and ARDS. Amer. J. Respir. Crit. Care Med. 164:231-236.
Eriksson, A., Cao, R., Roy, J., Tritsaris, K., Wahlestedt, C., Dissing, S., Thyberg, J., and Cao, Y. (2003) Small GTP-binding protein Rac is an essential mediator of VEGF-induced endothelial fenestrations and vascular permeability. Circ. 107:1532-1538.
Esser, S., Lampugnani, M.G., Corada, M., Dejana, E., and Risau, W. (1998) Vascular endothelial growth factor induces VE-cadherin tyrosine phosphorylation in endothelial cells. J. Cell Sci. 111:1853-1865.
Essler, M., Amano, M., Kruse, H.J., Kaibuchi, K., Weber, P.C., and Aepfelbacher, M. (1998) Thrombin inactivates myosin light chain phosphatase via Rho and its target Rho kinase in human endothelial cells. J. Biol. Chem. 273:21867-21874.
Everts, M., Koning, G., Kok, R., Asgeirsdottir, S., Vestweber, D., Meijer, D., Storm, G., and Molema, G. (2003) In vitro cellular handling and in vivo targeting of E-selectin-directed immunoconjugates and immunoliposomes used for drug delivery to inflamed endothelium. Pharm. Res. 20:64-72.
Fan, J., Shek, P., Suntres, Z., Li, Y., Oreopoulos, G., and Rotstein, O.D. (2000) Liposomal antioxidants provide prolonged protection against acute respiratory distress syndrome. Surgery. 128:332-338.
Fink, M.P. (2002) Role of reactive oxygen and nitrogen species in ARDS. Curr. Opin. 8:6-11.
Flori, H.R., Ware, L., Glidden, D., and Matthay, M.A. (2003) Early elevation of plasma soluble ICAM-1 in pediatric acute lung injury identifies patients at increased risk of death and prolonged mechanical ventilation. Ped. Crit. Care Med. 4:315-321.
Folkesson, H.G., Matthay, M., Hebert, C., and Broaddus, V.C. (1995) Acid aspiration-induced lung injury in rabbits is mediated by interleukin-8-dependent mechanisms. J. Clin. Invest. 96:107-116.
Folkesson, H.G., Matthay, M.A., Westrom, B.R., Kim, K.J., Karlsson, B.W., and Hastings, R.H. (1996) Alveolar epithelial clearance of protein. J. Appl. Physiol. 80:1431-1435.
Fukuda, Y., Ishizaki, M., Masuda, Y., Kimura, G., Kawanami, O., and Masugi, Y. (1987) The role of intraalveolar fibrosis in the process of pulmonary structural remodeling in patients with diffuse alveolar damage. Amer. J. Pathol. 126:171-182.
Gainor, J.P., Morton, C., Roberts, J., Vincent, P., and Minnear, F.L. (2001) Platelet-conditioned medium increases endothelial electrical resistance independently of cAMP/PKA and cGMP/PKG. Amer. J. Physiol. 281:H1992-2001.
Galloway, S.W. and Kingsnorth, A. (1996) Lung injury in the microembolic model of acute pancreatitis and amelioration by lexipafant, a platelet-activating factor antagonist. Pancreas 13:140-146.
Gamble, J.R., Drew, J., Trezise, L., Underwood, A., Parsons, M., Kasminkas, L., Rudge, J., Yancopoulos, G., and Vadas, M.A. (2000) Angiopoietin-1 is an antipermeability and anti-inflammatory agent in vitro and targets cell junctions. Circ. Res. 87:603-607.
Gao, J., Zeng, B., Zhou, L., and Yuan. S.Y. (2004) Protective effects of early treatment with propofol on endotoxin-induced acute lung injury in rats. Brit. J. Anaesth. 92:277-279.
Garat, C., Jayr, C., Eddahibi, S., Laffon, M., Meignan, M., and Adnot, S. (1997) Effects of inhaled nitric oxide or inhibition of endogenous nitric oxide formation on hyperoxic lung injury. Amer. J. Respir. Crit. Care Med. 155:1957-1964.

Garcia, J.G.N., Davis, H.W., and Patterson, C.E. (1995a) Regulation of endothelial cell gap formation and barrier dysfunction: role of myosin light chain phosphorylation. J. Cell. Physiol. 163:510-522.

Gautam, N., Olofsson, A., Herwald, H, Iversen, L., Lungren, E., Hedqvist, P., Arfors, K., Flodgaard, H., and Lindbom, L. (2001) Heparin-binding protein (HBP?CAP37): a missing link in neutrophil-evoked alteration of vascular permeability. Nature Med. 7:1123-1127.

Gerard, C., Frossard, J., Bhatia, M., Saluja, A., Gerard, N., Lu, B., and Steer, M. (1997) Targeted disruption of the β-chemokine receptor CCR1 protects against pancreatitis-associated lung injury. J. Clin. Invest. 100:2022-2027.

Gerlach, H., Keh, D., Semmerow, A., Busch, T., Lewandowski, K., Pappert, D., Rossaint, R., and Falke, K. (2003) Dose-response characteristics during long-term inhalation of nitric oxide in patients with severe acute respiratory distress syndrome. Amer. J. Respir. Crit. Care Med. 167:1008-1015.

Gertzberg, N., Neumann, P., Rizzo, V., Johnson, A. (2004) NAD(P)H oxidase mediates the endothelial barrier dysfunction induced bt TNFα. Amer. J. Phsiol. 286:L37-L48.

Gessner, C., Hammerschmidt, S., Kuhn, H., Lange, T., Engelmann, L., Schauer, J., and Wirtz, H. (2003) Exhaled breath condensate nitrite and its relation to tidal volume in ALI. Chest 124:1046-1052.

Gibbs, D.F., Shanley, T., Warner, R., Murphy, H., Varani, J., and Johnson, K.J. (1999) Role of MMPs in models of macrophage-dependent acute lung injury. Amer. J. Respir. Cell Molec. Biol. 20:1145-1154.

Goetze, S., Bungenstock, A., Czupalla, C., Eilers, F., Stawowy, P., Kintscher, U., Spencer-Hansch, C., Graf, K., Nurnberg, B., Law, R., Fleck, E., and Grafe, M. (2002) Leptin induces endothelial cell migration through Akt, which is inhibited by PPARgamma-ligands. Hyperten. 40:748-754.

Gonzalez, P.K., Zhuang, J., Doctrow, S., Malfroy, B., Benson, P., Menconi, M., and Fink M.P. (1996) Role of oxidant stress in ARDS: evaluation of a novel antioxidant strategy in a porcine model of endotoxin-induced acute lung injury. Shock. 6:S23-S26.

Gow, A., Thom, S., and Ischiropoulos, H. (1998) Nitric oxide and peroxynitrite-mediated pulmonary cell death. Amer. J. Physiol. 274:L112-L118.

Grazia-Lampugnani, M., Zanetti, A., Breviario, F., Balconi, G., Orsenigo, F., Corada, M., Spagnuolo, R., Betson, M., Braga, V., and Dejana, E. (2002) VE-cadherin regulates endothelial actin activating Rac and increasing membrane association of Tiam. Molec. Biol. Cell 13:1175-1189.

Grazia-Lampugnani, M., Zanetti, A., Corada, M., Takahashi, T., Balconi, G., Breviario, F., Orsenigo, F., Cattelino, A., Kemler, R., Daniel, T., and Dejana, E. (2003) Contact inhibition of VEGF-induced proliferation requires vascular endothelial cadherin, beta-catenin, and the phosphatase DEP-1/CD148. J. Cell Biol. 161:793-804.

Greenwood, J., Walers, C., Pryce, G., Kanuga, N., Beraud, E., Baker, D., and Adamson, P. (2003) Lovastatin inhibits brain endothelial cell Rho-mediated lymphocyte migration and attenuates experimental autoimmune encephalomyelitis. FASEB J. 17:905-907.

Groeneveld, A.B.J. (2000) Albumin and artificial colloids in fluid management: where does the clinical evidence of their utility stand? Crit. Care (London) 4:S16-S20.

Groeneveld, A.B.J. (2003) Vascular pharmacology of acute lung injury and acute respiratory distress syndrome. Vasc. Pharm. 39:247-256.

Groeneveld, A.B.J., den Hollander, W., Straub, J., Nauta, J., and Thijs, L.G. (1990) Effects of N-acetylcysteine and terbutaline treatment on hemodynamics and regional albumin extravasation in porcine septic shock. Circulatory Shock. 30:185-205.

Groeneveld, A.B.J., Raijmakers, P., Teule, G., and Thijs, L.G. (1996) The ^{67}gallium pulmonary leak index in assessing the severity and course of ARDS. Crit. Care Med. 24:1467-1472.

Guidot, D.M., Hybertson, B., Kitlowski, R., and Repine, J.E. (1996) Inhaled NO prevents IL-2-induced neutrophil accumulation and acute edema in isolated rat lungs. Amer. J. Physiol. 271:L225-L229.

Hagio, T., Nakao, S., Matsuoka, H., Matsumoto, S., Kawabata, K., and Ohno, H. (2001) Inhibition of neutrophil elastase attenuates complement-mediated lung injury in the hamster. Euro. J. Pharm. 426:131-138.

Hannan, K.M., Dilley, R., de Dios, S., and Little, P.J. (2003) Troglitazone stimulates repair of the endothelium and inhibits neointimal formation in denuded rat aorta. Arterioscl. Throm. Vasc. Biol. 23:762-768.

Herren, B., Levkau, B., Raines, E., and Ross, R. (1998) Cleavage of β-catenin and plakoglobin and shedding of VE-cadherin during endothelial apoptosis. Mol. Biol. Cell 9:1598-1601.

Hibbert, B., Olsen, S., and O'Brien, E. (2003) Involvement of progenitor cells in vascular repair. Trends Cardiovasc. Med. 13:322-326.

Hippenstiel, S., Witzenrath, M., Schmeck, B., Hocke, A., Krisp, M., Krull, M., Seybold, J., Seeger, W., Rascher, W., Schutte, H., and Suttorp, N. (2002) Adrenomedullin reduces endothelial hyperpermeability. Circ. Res. 91:618-625.

Hirsch, J., Hansen, K., Burlingame, A., and Matthay, M.A. (2004) Proteomics: current techniques and potential applications to lung disease. Amer. J. Physiol. 287:L1-L23.

Honda, K., Kobayshi, H., Hitaishi, R., Hirano, S., Fukuyama, N., Nakazawa, H., and Tomita, T. (1999) Inhaled nitric oxide reduces tyrosine nitration after LPS instillation into lungs of rats. Amer. J. Respir. Crit. Care Med. 160:678-688.

Hsu, K., Wang, D., Chang, M., Wu, C., and Chen, H. (1996) Pulmonary edema induced by phorbol myristate acetate is attenuated by compounds that increase intracellular cAMP. Res. Exp. Med. 196:17-28.

Hybertson, B.M., Jepson, E., Allard, J., Cho, O., Lee, Y., Huddleston, J., Weinman, J., Oliva, A., and Repine, J.E. (2003) TGF-β contributes to lung leak in rats given interleukin-1 intratracheally. Exper. Lung Res. 29:361-373.

Idris, I., Gray, S., and Donnelly, R. (2003) Rosiglitazone and pulmonary oedema: an acute dose-dependent effect on human endothelial cell permeability. Diabetologia 46:288-290.

Iivanainen, E., Kaharl, V. Jeino, J., and Elenius, K. (2003) Endothelial cell-matrix interactions. Micros. Res. Techniq. 60:13-22.

Ishizaka, A., Matsuda, T., Albertine, K., Koh, H., Tasaka, S., Hasegawa, N., Kohno, N., Kotani, T., Morisaki, H., Takeda, J., Nakaamura, M., Fang, X., Martin, T., Matthay, M.A., and Hashimoto, S. (2004) Translational physiology. Elevation of KL-6, a lung epithelial cell marker, in plasma and epithelial lining fluid in acute respiratory distress syndrome. Amer. J. Physiol. 286:L1088-L1094.

Jepsen, S., Herlevsen, P., Knudsen, P., Bud, M., and Klausen, N.O. (1992) Antioxidant treatment with N-acetylcysteine during ARDS. Crit. Care Med. 20:918-923.

Jones, P.F. (2003) Not just angiogenesis--wider roles for the angiopoietins. J. Path. 201:515-527.

Joyce, D.E. and Grinnell, B.W. (2002) Recombinant human activated protein C attenuates the inflammatory response in endothelium and monocytes by modulating NFκB. Crit. Care Med. 30:S288-S293.

Juliano, R.L. (2002) Signal transduction by cell adhesion receptors and the cytoskeleton: functions of integrins, cadherins, selectins, and immunoglobulin-superfamily members. Ann. Rev. Pharm. Tox. 42:283-323.

Kaplan, J.M., Pennington, S., St. George, J., Woodworth, L., Fasbender, A., Marshall, J., Cheng, S., Wadsworth, S., Gregory, R., and Smith, A. (1998) Potentiation of gene transfer to the mouse lung by complexes of adenovirus vector and polycations improves therapeutic potential. Human Gene Therapy 9:1469-1479.

Karzai, W., Cui, X., Mehlhorn, B., Straube, E., Hartung, T., Gerstenberger, E., Banks, S., Natanson, C., Reinhart, K., and Eichacker, P. (2003) Protection with antibody to TNF differs with similarly lethal *Escherichia coli* versus *Staphylococcus aureus* pneumonia in rats. Anesthes. 99:81-89.

Kawabata, K., Hagio, T., and Matsuoka, S. (2003) Pharmacological profile of a specific neutrophil elastase inhibitor, Sivelestat sodium hydrate. Nippon Yak. Zasshi - Folia Pharm. Jap. 122:151-160.

Kawabata, K., Hagio, T., Matsumoto, S., Nakao, S., Orita, S., Aze, Y., and Ohno, H. (2000) Delayed neutrophil elastase inhibition prevents subsequent progression of acute lung injury induced by endotoxin inhalation in ha msters. Amer. J. Respir. Crit. Care Med. 161:2013-2018.

Kawasaki, H., Springett, G., Mochizuki, N., Toki, S., Nakaya, M., Matsuda, M., Housman, D., and Graybiel, A.M. (1998) A family of cAMP-binding proteins that directly activate Rap1. Science 282:2275-2279.

Keitzmann, D., Kahl, R., Muller, M., Burchardi, H., and Kettler, D. (1993) H_2O_2 in expired breath condensate of patients with acute respiratory failure and ARDS. Int. Care Med. 19:78-81.

Kelly, J.J., Moore, T., Babal, P., Diwan, A., Stevens, T., and Thompson, W.J. (1998) Pulmonary microvascular and macrovascular endothelial cells: differential regulation of Ca^{2+} and permeability. Amer. J. Physiol. 274:L810-L819.

Kelly, R.M. (2000) Current strategies in lung preservation. J. Lab. Clin. Med. 136:427-440.

Ketai, L.H. and Godwin, J.D. (1998) A new view of pulmonary edema and acute respiratory distress syndrome. J. Thor. Imag. 13:147-171.

Kim, S., Manjiri Bakre, Hong Yin, and Varner, J.A. (2002) Inhibition of endothelial survival and angiogenesis by protein kinase A. J. Clin. Invest. 110:933-941.

King, J., Hamil, T., Creighton, J., Wu, S., Bhat, P., McDonald, F., and Stevens, T. (2004) Structural and functional characteristics of lung macro- and microvascular endothelial cell phenotypes. Microvasc. Res. 67:139-151.

Knepler, J, Taher L., Gupta, M., Patterson, C., Ober, M., and C.M. Hart. (2001) Peroxynitrite causes endothelial cell monolayer barrier dysfunction. Amer. J. Physiol. 281:C1064-C1075.

Koga, S., Morris, S., Ogawa, S., Liao, H., Bilezikian, J.P., Chen, G., Thompson, W.J., Ashikaga, T., Brett, J., and Stern, D.M. (1995) TNF modulates endothelial properties by decreasing cAMP. Amer. J. Physiol. 268:C1104-C1113.

Krupa, A., Kato, H., Matthay, M.A., and Kurdowska, A.K. (2004) Translational Physiology Proinflammatory activity of anti-IL-8 autoantibody:IL-8 complexes in alveolar edema fluid from patients with acute lung injur. Amer. J. Physiol. 286:L1105-L1113.

Kuklin, V.N., Kirov, M., Evgenov, O., Sovershaev, M., Sjöberg, J., Kirova, S., and Bjertnaes, L. (2004) Novel endothelin receptor antagonist attenuates endotoxin-induced lung injury in sheep. Crit. Care Med. 32: 766-773.

Kurdowska, A., Noble, J., Grant, I., Robertson, C., Haslett, C., and Donnelly, S.C. (2002) Anti-interleukin-8 autoantibodies in patients at risk for ARDS. Crit. Care Med. 30:2335-2337.

Laffey, J.G., Engelberts, D., and Kavanagh, B.P. (2000a) Buffering hypercapnic acidosis worsens acute lung injury. Amer. J. Respir. Crit. Care Med. 161:141-146.

Laffey, J.G., Honan, D., Hopkins, N., Hyvelin, J., Boylan, J., and McLoughlin, P. (2004a) Hypercapnic acidosis attenuates endotoxin-induced acute lung injury. Amer. J. Respir. Crit. Care Med. 169:46-56.

Laffey, J,G., O'Croinin, D., McLoughlin, P., Kavanagh, B.P. (2004b) Permissive hypercapnia--role in protective lung ventilatory strategies. Inten. Care Med. 30:347-356.

Laffey, J.G., Tanaka, M., Engelberts, D., Luo, X., Yuan, S., Tanswell, A., Post, M., Lindsay, T., and Kavanagh, B.P. (2000b) Therapeutic hypercapnia reduces pulmonary and systemic injury following in vivo lung reperfusion. Amer. J. Resp. Crit. Care Med. 162:2287-2294.

Laffon, M., Pittet, J., Modelska, K., Matthay, M., and Young, D.M. (1999) Interleukin-8 mediates injury from smoke inhalation to both the lung endothelial and the alveolar epithelial barriers in rabbits. Amer. J. Respir. Crit. Care Med. 160:1443-1449.

Lang, J.D., Figueroa, M., Chumley, P., Aslan, M., Hurt, J., Tarpey, M., Alvarez, B., Radi, R., and Freeman, B.A. (2004) Albumin and hydroxyethyl starch modulate oxidative inflammatory injury to vascular endothelium. Anesthesiol. 100:51-58.

Lee, R.P., Wang, D, Kao, S., and Chen, H.I. (2001) The lung is the major site that produces nitric oxide to induce acute pumonary edema in endotoxic shock. Clin. And Exp. Pharm. Phsiol. 28:315-320.

Leff, J.A., Bodman, M., Cho, O., Rohrbach, S., Reiss, O., Vannice, J., and Repine, J.E. (1994) Post-insult treatment with interleukin-1 receptor antagonist decreases oxidative lung injury in rats given intratracheal interleukin-1. Amer. J. Respir. Crit. Care Med. 150:109-112.

Lekka, M.E., Liokatis, S., Nathanail, C., Galani, V., and Nakos, G. (2004) The impact of intravenous fat emulsion administration in acute lung injury. Amer. J. Respir. Crit. Care Med. 169:638-644.

Lenz, A.G., Jorens, P., Meyer, B. De Backer, W., Van Overveld, F., Bossaert, L., and Maier, K.L. (1999) Oxidatively modified proteins in BAL of patients with ARDS and patients at risk for ARDS. Eur. Respir. J. 13:169-174.

Lewandowski, K., Rossaint, R., Pappert, D., Gerlach, H., Slama, K., Weidemann, H., Frey, D., Hoffmann, O., Keske, U., and Falke, K.J. (1997) High survival rate in 122 ARDS patients managed according to a clinical algorithm including extracorporeal membrane oxygenation. Inten. Care Med. 23:819-835.

Liu, F., Schaphorst, K., Verin, A., Jacobs, K., Birukova, A., Day, R., Bogatcheva, N., Bottaro, D., and Garcia, J. (2002) Hepatocyte growth factor enhances endothelial cell barrier function and cortical cytoskeletal rearrangement: role of glycogen synthase kinase-3β. FASEB J. 16:950-962.

Liu, H. and Zhao, J. (2002) An experimental study of therapeutic effects of ACE inhibition on chemical-induced ARDS in rats. Chin. J. Prevent. Med. 36:93-96.

Llevadot, J., Murasawa, S., Kureishi, Y, Uchida, S., Masuda, H., Kawamoto, A., Walsh, K., Isner, J., and Asahara, T. (2001) HMG-CoA reductase inhibitor mobilizes bone-marrow-derived endothelial progenitor cells. J. Clin. Invest. 198:399-405.

Lucas, M.C., Ludena, M.A., Barbero, E.A., and Sanchez-Gascon, A. (1995) Effects of N-acetylcysteine on hyperoxic lung in the rat. Respir. Med. 89:311-314.

Luce, J.M. (2002) Corticosteroids in ARDS. An evidence-based review. Crit. Care Clin. 18:79-89.

Lum, H., Hao, Z., Gayle, D., Kumar, P., Patterson, C.E., and Uhler, M.D. (2002) Vascular endothelial cells express isoforms of protein kinase A inhibitor. Amer J. Physiol. 282:C59-C66.

Lum, H., Jaffe, H.A., Schulz, I.T., Masood, A., RayChaudhury, A., and Green, R.D. (1999) Expression of PKA inhibitor (PKI) gene abolishes cAMP-mediated protection to endothelial barrier dysfunction. Amer. J. Physiol. 46:C580-C588.

Lum, H. and Roebuck, K.A. (2001) Oxidant stress and endothelial function. Amer. J. Physiol. 280:C719-C741.

MacArthur, R.D., Miller, M., Albertson, T., Panacek, E., Johnson, D., Teoh, L., and Barchuk, W. (2004) Adequacy of early empiric antibiotic treatment and survival in severe sepsis. Clin. Infect. Dis. 38:284-288.

MacLaren, R. and Jung, R. (2002) Stress-dose corticosteroid therapy for sepsis and acute lung injury or acute respiratory distress syndrome in critically ill adults. Pharmacotherapy 22:1140-1156.

Malarkkan, N., Snook, N., and Lumb, A.B. (2003) New aspects of ventilation in acute lung injury. Anaesthesia 58:647-667.

Maris, N.A., van der Sluijs, K., Florquin, S., de Vos, A., Pater, J., Jansen, H., and van der Poll, T. (2004) Salmeterol, a $β_2$-receptor agonist, attenuates lipopolysaccharide-induced lung inflammation in mice. Amer. J. Physiol. 286:L1122-L1128.

Marshall, J.C. (2003) Such stuff as dreams are made on: mediators directed therapy in sepsis. Nature Rev. 2:391-405.

Marshall, R.P., Webb, S., Hill, M., Humphreis, S., and Laurent, G. (2002) Genetic polymorphisms associated with susceptibility and outcome in ARDS. Chest:68S-69S

Martin, C., Papazian, L., Payan, M.J., Saux, P., and Gouin, F. (1995) Pulmonary fibrosis correlates with outcome in ARDS: a study in mechanically ventilated patients. Chest 107:196-200.

Martin, G.S., Mangialardi, R., Wheeler, A., Dupont, W., Morris, J., and Bernard, G.R. (2002a) Albumin and furosemide therapy in hypoproteinemic patients with acute lung injury. Crit. Care Med. 30:2175-2182.

Martin, T.A., Mansel, R.E., and Jiang, W.G. (2002b) Antagonistic effect of NK4 on HGF/SF induced changes in the transendothelial resistance and paracellular permeability of human vascular endothelial cells. J. Cell Physiol. 192:268-275.

Martinet, Y., Menard, O., Vaillant, P., Vignaud, J.M., and Martinet, N. (1996) Cytokines in human lung fibrosis. Arch. Toxicol. 18:127S-139S.

Mathru, M., Rooney, M., Dries, D., Hirsch, L., Barnes, L., and Tobin, M.J. (1994) Urine H_2O_2 during ARDS in patients with and without sepsis. Chest 105:232-236.

Matthay, M.A., Bhattacharya, S., Gaver, D., Ware, L., Lim, L., Syrkina, O., Eyal, F., and Hubmayr, R. (2002) Ventilator-induced lung injury: in vivo and in vitro mechanisms. Amer. J. Physiol. 283:L678-L682.

Matthay, M.A., Folkesson, H., and Verkman, A.S. (1996) Salt and water transport across alveolar and distal airway epithelia in adult lung. Amer. J. Physiol. 270:L487-L503.

Matthay, M.A., Fukuda, N., Frank, J., Kallet, R., Daniel, B., and Sakuma, T. (2000) Alveolar epithelial barrier. Role in lung fluid balance in clinical lung injury. Clin. Chest Med. 21:477-490.

Matthay, M.A. and Ware, L.B. (2004) Plasma protein C levels in patients with acute lung injury: Prognostic significance. Crit. Care Med. 32:S229-S232.

Matthay, M.A. and Wiener-Kronish, J.P. (1990) Intact epithelial barrier function is critical for the resolution of alveolar edema in humans. Amer. Rev. Respir. Dis. 142:1250-1257.

Matthay, M.A., Zimmerman, G., Esmon, C., Bhattacharya, J., Coller, B., Doerschuk, C., Floros, J., Gimbrone, M., Hoffman, E., Hubmayr, R., Leppert, M., Matalon, S., Munford, R., Parsons, P., Slutsky, A., Tracey, K., Ward, P., Gail, D., and Harabin, A.L. (2003) Future research directions in acute lung injury. Amer. J. Respir. Crit. Care Med. 167:1027-1035.

Mazzon, E., De Sarro, A., Caputi, A., and Cuzzocrea, S. (2002) Role of tight junction derangement in the endothelial dysfunction elicited by exogenous and endogenous peroxynitrite and poly(ADP-ribose) synthetase. Shock 18:434-439.

McAuley, D.F., Frank, J., Fang, X., and Matthay, M.A (2004) Clinically relevant concentrations of β2-adrenergic agonists stimulate maximal cAMP-dependent airspace fluid clearance and decrease pulmonary edema in experimental acid-induced lung injury. Crit. Care Med. 32:1470-1476.

McClintock, S.D., Till, G., Smith, M., and Ward, P.A. (2002) Protection from half-mustard-gas-induced acute lung injury in the rat. J. Appl. Tox. 22:257-262.

Meduri, G.U., Carratu, P., and Freire, A.X. (2003) Evidence of biological efficacy for prolonged glucocorticoid treatment in patients with unresolving ARDS. Eur. Respir. J. 42:57s-64s.

Meduri, G.U., Kohler, G., Headley, S., Tolley, E., Stentz, F., and Postlethwaite, A. (1995) Inflammatory cytokines in the BAL of patients with ARDS. Persistent elevation over time predicts poor outcome. Chest 108:1303-1314.

Metnitz, P.G., Bartens, C., Fischer, M., Fridrich, P., Steltzer, H., and Druml, W. (1999) Antioxidant status in patients with ARDS. Inten. Care Med. 25:180-185.

Meyer, J., Theilmeier, G., Van Aken, H., Bone, H., Busse, H., Waurick, R., Hinder, F., and Booke, M. (1998) Inhaled prostaglandin E1 for treatment of acute lung injury in severe multiple organ failure. Anesth. Analg. 86:753-758.

Mikawa, K., Nishina, K., Takao, Y., and Obara, H. (2003) ONO-1714, a NOS inhibitor, attenuates endotoxin-induced acute lung injury in rabbits. Anesth. Analg. 97:1751-1755.

Miller, E.J., Cohen, A., Nagao, S., Griffith, D., Maunder, R., Martin, T., Weiner-Kronish, J., Sticherling, M., Christophers, E., and Matthay, M.A. (1992) Elevated levels of NAP-1/interleukin-8 are present in the airspaces of patients with ARDS and are associated with increased mortality. Amer. Rev. Respir. Dis. 146:427-432.

Milligan, S.A, Hoeffel, J., Goldstein, I., and Flick, M.R. (1988) Effect of catalase on endotoxin-induced acute lung injury in unanesthetized sheep. Amer. Rev. Respir. Dis. 137:420–428.

Minnear, F.L., Patil, S., Bell, D., Gainor, J., and Morton, C.A. (2001) Platelet lipid(s) bound to albumin increases endothelial electrical resistance: mimicked by LPA. Amer. J. Physiol. 281:L1337-L1344.

Minnema, M.C., Chang, A., Jansen, P., Lubbers, Y., Pratt, B., Whittaker, B., Taylor, F., Hack, C., and Friedman, B. (2000) Recombinant human antithrombin III improves survival and attenuates inflammatory responses in baboons lethally challenged with Escherichia coli. Blood 95:1117-1123.

Modelska, K., Pittet, J., Folkesson, H., Broaddus, V., and Matthay, M.A. (1999) Acid-induced lung injury. Protective effect of anti-interleukin-8 pretreatment on alveolar epithelial barrier function in rabbits. Amer. J. Respir. Crit. Care Med. 160:1450-1456.

Montravers, P., Fagon, J., Gilbert, C., Blanchet, F., Novara, A., and Chastre, J. (1993) Pilot study of cardiopulmonary risk from pentoxifylline in ARDS. Chest 103:1017-1022.

Morel, N.M., Petruzzo, P., Hechtman, H., and Shepro, D. (1990) Inflammatory agonists that increase microvascular permeability in vivo stimulate cultured pulmonary microvessel endothelial cell contraction. Inflamm. 14:571-583.

Mosnier, L.O. and Griffin, J.H. (2003) Inhibition of staurosporine-induced apoptosis of endothelial cells by activated protein C requires protease-activated receptor-1 and endothelial cell protein C receptor. Biochem. J. 373:65-70.

Moss, M., Gillespie, M., Ackerson, L., Moore, F., Moore, E., and Parsons, P.E. (1996) Endothelial cell activity varies in patients at risk for ARDS. Crit. Care Med. 24:1782–1786.

Muhs, B.E., Patel, S., Yee, H., Marcus, S., and Shamamian, P. (2003) Inhibition of MMP reduces local and distant organ injury following experimental acute pancreatitis. J. Surg. Res. 109:110-117.

Muller, C., Rahn, B., and Pfister, U. (1992) Fat embolism and fracture, a review of the literature. Aktuelle Traum. 22:104-113.

Mulligan, M.S., Warner, R., Rittershaus, C., Thomas, L., Ryan, U., Foreman, K., Crouch, L., Till, G., and Ward, P.A. (1999) Endothelial targeting and enhanced antiinflammatory effects of complement inhibitors possessing sialyl Lewisx moieties. J. Imm. 162:4952-4959.

Murakami, K., Enkhbaatar, P., Shimoda, K., Mizutani, A., Cox, R., Schmalstieg, F., Jodoin, J., Hawkins, H., Traber, L., and Traber, D. (2003) High-dose heparin fails to improve acute lung injury following smoke inhalation in sheep. Clin. Sci. 104:349-356.

Murakami, K., McGuire, R., Cox, R., Jodoin, J., Bjertnaes, L., Katahira, J., Traber, L., Schmalstieg, F., Hawkins, H., Herndon, D., and Traber, D. (2002) Heparin nebulization attenuates ALI in sepsis following smoke inhalation in sheep. Shock 18:236-241.

Murciano, J.C., Harshaw, D., Ghitescu, L., Danilov, S., and Muzykantov, V.R. (2001) Vascular immunotargeting to endothelial surface in a specific macrodomain in alveolar capillaries. Amer. J. Respir. Crit. Care Med. 164:1295-1302.

Muro, S., Cui, X., Gajewski, C., Murciano, J., Muzykantov, V.R., and Koval, M. (2003) Slow intracellular trafficking of catalase nanoparticles targeted to ICAM-1 protects endothelial cells from oxidative stress. Amer. J. Physiol. 285:C1339-C1347.

Murohara, T., Ikeda, H., Duan, J., Shintani, S., Sasaki, K., Egulchi, H., Onitsuka, I., matsui, K., and Imaizumi, T. (2000) Transplanted cord blood-derived endothelial precursor cells augment postnatal neovascularization. J. Clin. Invest. 105:1527-1536.

Murray, J.F., Matthay, M.A., Luce, J., Flick, M.R. (1988) An expanded definition of the adult respiratory distress syndrome. Amer. Rev. Respir. Dis. 138:720-723.

Muzykantov, V.R., Atochina, E., Ischiropoulos, H., Danilov, S., and Fisher, A.B. (1996) Immunotargeting of antioxidant enzyme to the pulmonary endothelium. PNAS USA 93:5213-5218.

Muzykantov, V.R., Gavriluk, V., Reinecke, A., Atochina, E., Kuo, A., Barnathan, E., and Fisher, A.B. (1995) The functional effects of biotinylation of anti-angiotensin-converting enzyme monoclonal antibody in terms of targeting in vivo. Analyt. Biochem. 226:279-287.

Nagase, T., Uozumi, N., Aoki-Nagase, T., Terawaki, K., Ishii, S., Tomita, T., Yamamoto, H., Hashizume, K., Ouchi, Y., and Shimizu, T. (2003) A potent inhibitor of cytosolic PLA2, arachidonyl trifluoromethyl ketone, attenuates LPS- lung injury in mice. Amer. J. Physiol. 284:L720-L726.

Naidu, B.V., Woolley, S., Farivar, A., Thomas, R., Fraga, C., and Mulligan, M.S. (2003) Simvastatin ameliorates injury in an experimental model of lung ischemia-reperfusion. J. Thor. Cardiovasc. Surg. 126:482-489.

Nathens, A.B., Neff, M., Jurkovich, G., Klotz, P., Farver, K., Ruzinski, J., Radella, F., Garcia, I., and Maier, R.V. (2002) Randomized, prospective trial of antioxidant supplementation in critically ill surgical patients. Ann. Surg. 236:814-822.

Nikitenko, L.L., Smith, D., Hague, S., Wilson, C., Bicknell, R., and Rees, M.C. (2002) Adrenomedullin and the microvasculature. Trends Pharm. Sci 23:101-103.

Noll, T., Holschermann, H., Koprek, K., Gunduz, D., Haberbosch, W., Tillmanns, H., and Piper, H.M. (1999) ATP reduces macromolecule permeability of endothelial monolayers despite increasing [Ca2+]i. Amer. J. Physiol. 276:H1892-H1901.

Nubel, T., Dippold, W., Kleinert, H., Kaina, B., and Fritz G. (2004) Lovastatin inhibits Rho-regulated expression of E-selectin by TNFalpha and attenuates tumor cell adhesion. FASEB J. 18:140-142.

Ogawa, S., Koga, S., Kuwabara, K., Brett, J., Morrow, B., Morris, S.A., Bilezikian, J.P., Silverstein, S.C., and Stern, D. (1992) Hypoxia-induced increased permeability of endothelial monolayers occurs through lowering of cellular cAMP levels. Amer. J. Physiol. 262:C546-C554.

Okayama, N., Ichikawa, H., Coe, L., Itoh, M., and Alexander, J.S. (1998) Exogenous NO enhances hydrogen peroxide-mediated neutrophil adherence to cultured endothelial cells. Amer. J. Physiol. 274:L820-L826.

Olman, M.A., White, K., Ware, L., Simmons, W., Benveniste, E., Zhu, S., Pugin, J., and Matthay, M.A. (2004) Pulmonary edema fluid from patients with early lung injury stimulates fibroblast proliferation through IL-1 beta-induced IL-6 expression. J. Immunol. 172:2668-2677.

Opal, S.M. (2003) Clinical trial design and outcomes in patients with severe sepsis. Shock. 20:295-302.

Opal, S.M., Fisher, C., Dhainaut, J., Vincent, J., Brase, R., Lowry, S., Sadoff, J., Slotman, G., Levy, H., Balk, R., Shelly, M., Pribble, J., LaBrecque, J., Lookabaugh, J., Donovan, H., Dubin, H., Baughman, R., Norman, J., DeMaria, E., Matzel, K., Abraham, E., and Seneff, M. (1997) Confirmatory interleukin-1 receptor antagonist trial in severe sepsis. Crit. Care Med. 25:1115-1124.

Opal, S.M. and Gluck, T. (2003) Endotoxin as a drug target. Crit. Care Med. 31:S57-S64.

Ostermann, H. (2002) Antithrombin III in Sepsis. New evidences and open questions. Minerva Anest. 68:445-448.

Pappert, D., Busch, T., Gerlach, H., Lewandowski, K., Radermacher, P., and Rossaint, R. (1995) Aerosolized prostacyclin vs. inhaled nitric oxide in children with severe ARDS. Anesth. 82:1507-1511.

Parsons, P.E. and Giclas, P.C. (1990) The terminal complement complex (sC5b-9) is not specifically associated with the development of the adult respiratory distress syndrome. Amer. Rev. Respir. Dis. 141:98-103.

Patterson, C.E., Davis, H., Schaphorst, K., and Garcia, J.G.N. (1994) Mechanisms of cholera toxin prevention of thrombin- and PMA-induced endothelial cell barrier dysfunction. Microvasc. Res. 48:212-235.

Patterson, C.E., Lum, H., Verin, A., Schaphorst, K., and Garcia, J.G. (2000) Regulation of endothelial barrier function by cAMP-dependent protein kinase. Endothelium 7:287-308.

Patterson, C.E. and Rhoades, R.A. (1988) Protective role of sulfhydryl reagents in oxidant lung injury. Exper. Lung Res. 14:1005-1019.

Patterson, C.E., Butler, J.A., Byrne, F., and Rhodes, M.L. (1985) Oxidant lung injury: Intervention with sulfhydryl reagents. Lung 163:23-32.

Peters, K., Unger, R., Brunner, J., and Kirkpatrick, C. (2003) Molecular basis of endothelial dysfunction in sepsis. Card. Res. 60:49-57.

Perkins, G.D., McAuley, D., Richter, A., Thickett, D., and Gao, F. (2004) Bench-to-bedside review: β2-agonists and the ARDS. Crit. Care London 8:25-32.

Petty, T.L. and Ashbaugh, D.G. (1971) The adult respiratory distress syndrome, clinical features, factors influencing prognosis, and principals of management. Chest 60:233-239.

Pittet, J.F., Griffiths, M., Geiser, T., Kaminski, N., Dalton, S., Huang, X., Brown, L., Gotwals, P., Koteliansky, V. Matthay, M., and Sheppard, D. (2001) TGF-β is a critical mediator of acute lung injury. J. Clin. Invest. 107:1537-1544.

Pittet, J.F., Mackersie, R., Martin, T., and Matthay, M.A. (1997) Biological markers of acute lung injury: prognostic and pathogenetic significance. Amer. J. Respir. Crit. Care Med. 155:1187-1205.

Pizurki, L., Zhou, Z., Glynos, K., Roussos, C., and Papapetropoulos, A. (2003) Angiopoietin-1 inhibits endothelial permeability, neutrophil adherence and IL-8 production. Brit. J. Pharmacol. 139:329-336.

Plutzky, J. (2001) PPARs in endothelial cell biology. Curr. Opin. Lipidol. 12:511-518.

Polte, T. and Schroder, H. (1998) Cyclic AMP mediates endothelial protection by nitric oxide. Biochem. Biophys. Res. Comm. 251:460-465.

Powers, K.A., Kapus, A., Khadaroo, R., He, R., Marshall, J., Lindsay, T., and Rotstein, O.D. (2003) Twenty-five percent albumin prevents lung injury following shock/resuscitation. Crit. Care Med. 31:2355-2363.

Powers, K.A., Kapus, A., Khadaroo, R., Papia, G., and Rotstein, O.D. (2002) 25% Albumin modulates adhesive interactions between neutrophils and the endothelium following shock/resuscitation. Surgery. 132:391-398.

Pugin, J., Verghese, G., Widmer, M., and Matthay, M.A. (1999) The alveolar space is the site of intense inflammatory and profibrotic reactions in the early phase of ARDS. Crit. Care Med. 27:304-312.

Putensen, C., Hormann, C., Kleinsasser, A., and Putensen-Himmer, G., (1998) Cardiopulmonary effects of aerosolized prostaglandin E_1 and nitric oxide inhalation in patients with ARDS. Amer. J. Respir. Crit. Care Med. 157:1743-1747.

Qiao, J., Huang, F., and Lum, H. (2003) PKA inhibits RhoA activation: a protective mechanism against endothelial barrier dysfunction. Amer. J. Physiol. 284:L972-L980.

Quinlan, G.J., Evans, T., Gutteridge, J. (1995) 4-hydroxy-2-nonenol levels increase in the plasma of patients with ARDS as linoleic acid appears to fall. Free Rad. Res. 21:95-106.

Quinlan, G.J., Mumby, S., Martin, G., Bernard, G.R., Gutteridge, J., and Evans, T. (2004) Albumin influences total plasma antioxidant capacity favorably in patients with ALI. Crit. Care Med. 32:755-759.

Radermacher, P., Santak, B., Wust, H., Tarnow, J., and Falke, K.J. (1990) Prostacyclin and right ventricular function in patients with pulmonary hypertension associated with ARDS. Intens. Care Med. 16:227-232.

Ranieri, V.M., Suter, P., Tortorella, C., De Tullio, R., Dayer, J., Brienza, A., Bruno, F., and Slutsky. A.S. (1999) Effect of mechanical ventilation on inflammatory mediators in patients with acute respiratory distress syndrome: a randomized controlled trial. JAMA 282:54-61.

Reimold, A..M. (2003) New indications for the treatment of chronic inflammation by TNFα blockade. Amer. J. Med. Sci. 325:75-92.

Reinhart, K., Bayer, O., Brunkhorst, F., and Meisner, M. (2002) Markers of endothelial damage in organ dysfunction and sepsis. Crit. Care Med. 30:S302-S312.

Reinhart, K., Menges, T., Gardlund, B., Zwaveling, J., Smithes, M., Vincent, J., Tellado, J., Salgado-Remigio, A., Zimlichman, R., Withington, S., Tschaikowsky, K., Brase, R., Damas, P., Kupper, H., Kempeni, J., Eiselstein, J., and Kaul, M. (2001) Randomized, placebo-controlled trial of the anti-TNF antibody fragment afelimomab in hyperinflammatory response during severe sepsis: The RAMSES Study. Crit. Care Med. 29:765-769.

Remick, D.G. (2003) Cytokine therapeutics for the treatment of sepsis: why has nothing worked? Curr. Pharm. Design 9:75-82.

Reyes, M., Dudek, A., Jahagirdar, B., Koodie, L., Marker, P., and Verfaille, C.M. (2002) Origin of endothelial progenitors in human postnatal bone marrow. J. Clin. Invest. 109:337-346.

Riedemann, N.C., Neff, T., Guo, R., Bernacki, K., Laudes, I., Sarma, J., Lambris, J., and Ward, P.A. (2003) Protective effects of IL-6 blockade in sepsis are linked to reduced C5a receptor expression. J. Immun. 170:503-507.

Ritter, C., Andrades, M., Reinke, A., Menna-Barreto, S., Moreira, J., and Dal-Pizzol, F. (2004) Treatment with N-acetylcysteine plus deferoxamine protects rats against oxidative stress and improves survival in sepsis. Crit. Care Med. 32:342-349.

Riva, C.M., Morganroth, M., Ljungman, A., Schoeneich, S., Marks, R., Todd, R., Ward, P.A., and Boxer, L.A. (1990) Iloprost inhibits neutrophil-induced lung injury and neutrophil adherence to endothelial monolayers. Amer. J. Respir. Cell Molec. Biol. 3:301-309.

Roumen, R.M., Hendrika, T., de Man, B., and Goris, R.J. (1994) Serum lipofusin as a prognostic indicator of ARDS and multiple organ failure. Br. J. Surg. 81:1300-1305.

Rubin, D.B., Wiener-Kronish, J., Murray, J., Green, D., Turner, J., Luce, J., Montgomery, A., Marks, J., Matthay, M.A. (1990) Elevated von Willebrand factor antigen is an early plasma predictor of acute lung injury in nonpulmonary sepsis syndrome. J. Clin. Invest. 86:474-80.

SAFE Study Investigators. (2004) A comparisom of albumin and saline for fluid resuscitation in the intensive care unit. New Eng. J. Med. 350:2247-2256.

Sartori, C., Fang, X., McGraw, D., Koch, P., Snider, M., Folkesson, H., and Matthay M.A. (2002) Selected contribution: long-term effects of beta(2)-adrenergic receptor stimulation on alveolar fluid clearance in mice. J. Appl. Physiol. 93:1875-1880.

Sata, M., Kakoki, M., Nagata, D., Nishimatsu, H., Suzuki, E., Aoyagi, T., Sugiura, S., Kojima, H., Nagano, T., Kangawa, K., Matsuo, H., Omata, M., Nagai, R., and Hirata, Y. (2000) Adrenomedullin and nitric oxide inhibit human endothelial cell apoptosis via a cyclic GMP-independent mechanism. Hyperten. 36:83-88.

Sato, H., Callister, M., Mumby, S., Quinlan, G., Welsh, K., duBois, R., and Evans, T.W. (20040 KL-6 levels are elevated in plasma from patients with ARDS. Eur. Respir. J. 23:142-145.

Sayner, S.L., Frank, D., King, J., Chen, H., VandeWa, J., and Stevens, T. (2004) Paradoxical cAMP-induced lung endothelial permeability revealed by *P. aeruginosa* ExoY. Circ. Res. 95:196 - 203.

Schaphorst, K.L., Chiang, E., Jacobs, K., Zaiman, A., Natarajan, V., Wigley, F., and Garcia, J.G. (2003) Role of sphingosine-1 phosphate in enhancement of endothelial barrier integrity by platelet-released products. Amer. J. Physiol. 285:L258-L267.

Scheeren, T. and Radermacher, P. (1997) Prostacyclin (PGI2): new aspects of an old substance in the treatment of critically ill patients. Inten. Care Med. 23:146-158.

Schermuly, R.T., Leuchte, H., Ghofrani, H., Weissmann, N., Rose, F., Kohstall, M., Olschewski, H., Schudt, C., Grimminger, F., Seeger, W., and Walmrath, D. (2003) Zardaverine and aerosolised iloprost in a model of acute respiratory failure. Euro. Resp. J. 22:342-347.

Schermuly, R.T., Roehl, A., Weissmann, N., Ghofrani, H., Leuchte, H., Grimminger, F., Seeger, W., and Walmrath, D. (2001) Combination of nonspecific PDE inhibitors with inhaled prostacyclin in experimental pulmonary hypertension. Amer. J. Physiol. 281:L1361-L1368.

Schilling, M.K., Eichenberger, M., Maurer, C., Sigurdsson, G., and Buchler, M.W. (2001) Ketoconazole and pulmonary failure after esophagectomy: a prospective clinical trial. Diseases Esoph. 14:37-40.

Schraufstatter, I.U., Chung, J., and Burger, M. (2001) IL-8 activates endothelial cell CXCR1 and CXCR2 through Rho and Rac signaling pathways. Amer. J. Physiol. 280:L1094-L1103.

Schuster, D. (1998) Evaluation of lung function with PET. Seminars in Nuclear Medicine 28:341-351.

Schuster, D.P., Metzler, M., Opal, S., Lowry, S., Balk, R., Abraham, E., Levy, H., Slotman, G., Coyne, E., Souza, S., and Pribble, J. (2003) ARDS Prevention Study Group. Recombinant PAF acetylhydrolase to prevent ARDS and mortality in severe sepsis. Crit. Care Med. 31:1612-1619.

Seekamp A, Lalonde C, Zhu, D.G., and Demling, R. (1988) Catalase prevents prostanoid release and lung lipid peroxidation after endotoxemia in sheep. J. Appl. Physiol. 65:1210–1216.

Selig, W.M., Patterson, C.E., Henry, D., and Rhoades, R. (1986) The role of histamine in acute oleic acid-induced injury. J. Appl. Physiol. 61:233-239.

Selig, W.M., Patterson, C.E., and Rhoades, R. (1987) Cyclooxygenase metabolites contribute to oleic acid-induced lung edema. Exp. Lung Res. 13:69-82.

Sharp, C., Warren, A., Oshima, T., Williams, L., Li, J., and Alexander, J.S. (2001) Poly ADP ribose-polymerase inhibitors prevent the upregulation of ICAM-1 and E-selectin in response to Th1 cytokine stimulation. Inflamm. 25:157-163.

Shay-Salit, A., Shushy, M., Wolfovitz, E., Yahav, H., Breviario, F., Dejana, E., and Resnick, N. (2002) VEGFR2 and the adherens junction as a mechanical transducer in vascular endothelial cells. PNAS USA 99:9462-9467.

Shenkar, R., Coulson, W., and Abraham, E. (1994) Anti- TGF-β monoclonal antibodies prevent lung injury in hemorrhaged mice. Amer. J. Respir. Cell Molec. Biol. 11:351-357.

Shi, Q., Rafii, S., Wu, M., Wijelath, E., Yu, C., Ishida, A., Fujita, Y., Kothari, S., Mohle, R., Sauvage, L., Moore, M., Storb, R., and Hammond, W.P. (1998) Evidence for circulating bone marrow-derived endothelial cells. Blood 92:362-367.

Shi, S., Verin, K.L., Gilbert-McClain, L., Patterson, C.E., Irwin, R.P., Natarajan, V., and Garcia, J.G.N. (1998) Role of tyrosine phosphorylation in thrombin-induced endothelial cell contraction and barrier function. Endothelium 6:153-171.

Shibata, K., Cregg, N., Engelberts, D., Takeuchi, A., Fedorko, L., and Kavanagh, B.P. (1998) Hypercapnic acidosis may attenuate acute lung injury by inhibition of endogenous xanthine oxidase. Amer. J. Respir. Crit. Care Med. 158:1578-1584.

Shindo, T., Kurihara, H., Maemura, K., Kurihara, Y., Kuwaki, T., Izumida, T., Minamino, N., Ju, K., Morita, H., Oh-hashi, Y., Kumada, M., Kangawa, K., Nagai, R., and Yazaki, Y. (2000) Hypotension and resistance to lipopolysaccharide-induced shock in transgenic mice overexpressing adrenomedullin in their vasculature. Circ. 101:2309-2316.
Shoemaker, W.C. and Wo, C.C. (1998) Circulatory effects of whole blood, packed red cells, albumin, starch, and crystalloids in resuscitation of shock and acute critical illness. Vox Sanguinis 74:69s-74s.
Siegel-Axel, D.L. (2003) Ceristatin: a cellular and molecular drug for the future? Cell. Molec. Life Sci. 60:144-164.
Sittipunt C., Steinberg, K., Ruzinski, J., Myles, C., Zhu, S., Goodman, R., Hudson, L., Matalon, S., and Martin, T.R (2001) Nitric oxide and nitotyrosine in the lungs of patients with ARDS. Amer. J. Respir. Crit. Care med. 163:503-510.
Simon, L.M. and Suttorp, N. (1985) Lung cell oxidant injury: decrease in oxidant mediated cytotoxicity by N-acetylcysteine. Euro. J. Respir. Dis. 139:132-135.
Slotman, G.J., Burchard, K., D'Arezzo, A., and Gann, D.S. (1988) Ketoconozol prevents acute respiratory failure in critically ill surgical patients. J. Trauma Injury Infect Crit. Care 28:648-654.
Slotman, G.J., Kerstein, M., Bone, R., Silverman, H., Maunder, R., Hyers, T., and Ursprung, J.J. (1992) The effects of prostaglandin E1 on non-pulmonary organ function during clinical acute respiratory failure. J. Trauma Injury Infect Crit. Care 32:480-488.
Smith, H., Keppie, J., Stanley, J., and Harris-Smith, P.W. (1955) The chemical basis of the virulence of Bacillus anthracis. IV. Secondary shock as the major factor in death of guinea-pigs from anthrax. Br. J. Exp. Pathol. 36:323-335.
Spragg, D.D., Alford, D., Greferath, R., Larsen, C., Lee, K., Gurtner, G., Cybulsky, M., Tosi, P., Nicolau, C., and Gimbrone, M.A. (1997) Immunotargeting of liposomes to activated vascular endothelial cells: a strategy for site-selective delivery in the cardiovascular system. PNAS USA 94:8795-8800.
Spragg, R.G., Lewis, J., Wurst, W., Hafner, D., Baughman, R., Wewers, M., and Marsh J.J. (2003) Treatment of ARDS with recombinant protein C surfactant. Amer. J. Respir. Crit. Care Med. 167:1562-1566.
Sokol, J., Jacobs, S.E., and Bohn, D. (2003) Inhaled nitric oxide for acute hypoxic respiratory failure in children and adults: a meta-analysis. Anesth. Analges. 97:989-998.
Stanley, J.L. and Smith, H. (1961) Purification of factor I and recognition of a third factor of the anthrax toxin. J. Gen. Microbiol. 26:49-63.
Steinberg, H.O., Tarshoby, M., Monestel, R., Hook, G., Cronin, J., Johnson, J., Johnson, A., Bayazeed, B., and Baron, A.D. (1997) Elevated circulating free fatty acids levels impair endothelium-dependent vasodilation. J. Clin. Invest. 100:1230-1239.
Steinberg, J., Halter, J., Schiller, H., Dasilva, M., Landas, S., Gatto, L., Maisi, P., Sorsa, T., Rajamaki, M., Lee, H., and Nieman, G. (2003) Metalloproteinase inhibition reduces lung injury and improves survival after cecal puncture in rats. J. Surg. Res. 111:185-195.
Sudo, T., Ito, H., and Kimura, Y. (2003) Phosphorylation of VASP by the anti-platelet drug, cilostazol, in platelets. Platelets. 14:381-390.
Suter, P.M., Domenighetti, G., Schaller, M., Laverriere, M., Ritz, R., and Perret, C. (1994) N-acetylcysteine enhances recovery from acute lung injury in man. A randomized, double-blind, placebo-controlled clinical study. Chest. 105:190-194.
Suttorp, N., Hippenstiel, S., Fuhrmann, M., Krull, M., and Podzuweit. T. (1996) Role of nitric oxide and phosphodiesterase isoenzyme II for reduction of endothelial hyperpermeability. Amer. J. Physiol. 270:C778-C785.
Suttorp, N., Weber, U., Welsch, T., and Schudt, C. (1993) Role of phosphodiesterases in the regulation of endothelial permeability in vitro. J. Clin. Invest. 91:1421-1428.
Sweitzer, T.D., Thomas, A., Wiewrodt, R., Nakada, M., Branco, F., and Muzykantov, V.R. (2003) PECAM-directed immunotargeting of catalase: specific, rapid and transient protection against hydrogen peroxide. Free Rad. Biol. Med. 34:1035-1046.
Swenson, E.R. (2004) Therapeutic hypercapic acidosis. Pusing the envelope. Amer. J. Respir. Crit. Care Med. 169:8-9.

Szabo, C., Cuzzocrea, S., Zingarelli, B., O'Connor, M., and Salzman, A.L. (1997a) Endothelial dysfunction in a rat model of endotoxic shock . J. Clin. Invest. 100:723-735.

Szabo, C., Lim, L., Cuzzocrea, S., Getting, S., Zingarelli, B., Flower, R., Salzman, A., and Perretti, M. (1997b) Inhibition of poly (ADP-ribose) synthetase attenuates neutrophil recruitment and exerts antiinflammatory effects. J. Exp. Med. 186:1041-1049.

Taggart, D.P. (2001) Effects of a platelet-activating factor antagonist on lung injury and ventilation after cardiac operation. Ann. Thorac. Surg. 71:238-242.

Taichman, D.B., Schachtner, S., Li, Y., Puri, M., Bernstein, A., and Baldwin, H.S. (2003) A unique pattern of Tie1 expression in developing murine lung. Exp. Lung Res. 29:113-122.

Taylor, F.B. Jr. (1994) Studies on the inflammatory-coagulant axis in the baboon response to E. coli: regulatory roles of proteins C, S, C4bBP and of inhibitors of tissue factor. Prog. Clin. Biol. Res. 388:175-194.

Taylor, F.B. Jr., Chang, A., Esmon, C., D'Angelo, A., Vigano-D'Angelo, S., and Blick, K.E. (1987) Protein C prevents the coagulopathic and lethal effects of Escherichia coli infusion in the baboon. J. Clin. Invest. 79:918-925.

Taylor, R.W., Manganaro, L., O'Brien, J., Trottier, S., Parkar, N., and Veremakis, C. (2002) Impact of allogenic packed red blood cell transfusion on nosocomial infection rates in the critically ill patient. Crit. Care Med. 30:2249-2254.

Teixeira, C.F., Farmer, P., Laporte, J., Jancar, S., and Sirois, P. (1995) Increased permeability of bovine aortic endothelial cell monolayers in response to a TxA2-mimetic. Agents & Actions 45s:47-52.

Thompson, B.T. (2003) Glucocorticoids and acute lung injury. Crit. Care Med. 31:S253-S257.

Thompson, B.T., Hayden, D., Matthay, M.A., Brower, R., Parsons, P.E. (2001) Clinicians' approaches to mechanical ventilation in acute lung injury and ARDS. Chest 120:1622-1627.

Thurston, G. (2002) Complementary actions of VEGF and angiopoietin-1 on blood vessel growth and leakage. J. Anat. 200:575-580.

Timlin, M., Condron, C., Toomey, D., Power, C., Thornes, B., Kearns, S., Street, J., Murray, P., and Bouchier-Hayes, D. (2004) N-acetylcysteine attenuates lung injury in a rodent model of fracture. Acta Orthopaed. Scand. 75:61-65.

Tinsley, J.H., Teasdale, N.R., and Yuan, S.Y. (2004a) Myosin light chain phosphorylation and pulmonary endothelial cell hyperpermeability in burns. Amer. J. Physiol. 286:L841-L847.

Tinsley, J.H., Yuan, S.Y., and Wilson, E. (2004b) Isoform-specific knockout of endothelial myosin light chain kinase: closing the gap on inflammatory lung disease. Trends Pharm. Sci. 25:64-66.

Toh, C., Ticknor, L., Downey, C., Giles, A., Paton, R., and Wenstone, R. (2003) Early identification of sepsis and mortality risks through simple, rapid clot-waveform analysis. Inten. Care Med. 29:55-61.

Van Heerden, P.V., Barden, A., Michalopoulos, N., Bulsara, M., and Roberts, B.L. (2000) Dose-response to inhaled aerosolized prostacyclin for hypoxemia due to ARDS. Chest. 117(3):819-827.

van Nieuw Amerongen, G.P., Delft, S., Vermeer, M.A., Collard, J., and van Hinsbergh, V.M. (2000a) Activation of RhoA by thrombin in endothelial hyperpermeability. Circ. Res. 87:335-340.

van Nieuw Amerongen, G.P., Draijer, R., Vermeer, M.A., and van Hinsbergh, V.M. (1998) Transient and prolonged increase in endothelial permeability induced by histamine and thrombin - Role of protein kinases, calcium, and RhoA. Circ. Res. 83:1115-1123.

van Nieuw Amerongen, G.P., Vermeer, M., Negre-Aminou, P., Lankelma, J., Emieis, J., and van Hinsbergh, V.W. (2000b) Simvastatin improves endothelial barrier function. Circ. 102:2803-2809.

Vassilev, D., Hauser, B., Bracht, H., Ivanyi, Z., Schoaff, M., Asfar, P., Vogt, J., Wachter, U., Schelzig, H., Georgieff, M., Bruckner, U., Radermacher, P., and Froba, G. (2004) Systemic, pulmonary, and heptaosplancnic effects of NAC during long-term porcine endotoxemia. Crit. Care Med. 32:525-532.

Verrier, E., Wang, L., Wadham, C., Albanese, N., Hahn, C., Gamble, H., Krishna, V., Chatterjee, K., Vadas, M., and Xia, P. (2004) PPARγ agonists ameliorate endothelial cell activation via inhibition of DAG–PKC signaling pathway: Role of DAG kinase. Circ. Res. 94:1515–1522.

Vestweber, D. (2000) Molecular mechanisms that control endothelial contacts. J. Path. 190:281-291.
Victorino, G.P., Newton, C.R., and Curran, B. (2004) Modulation of microvascular hydraulic permeability by PAF. J. Trauma-Injury Inf. & Crit. Care 56:379-384.
Vincent, J.L. (2002) New management strategies in ARDS. Crit. Care Clinics 18:69-78.
Vincent, J.L., Baron, J., Reinhart, K., Gattoni, L., Thijs, L., Webb, A., Meier, A., Nollet, G., Peres, D. (2002) Anemia and blood transfusion in critically ill patients. JAMA 288:1499–1507.
Vincent, J.L., Brase, R., Santman, F., Suter, P., McLuckie, A., Dhainaut, J., Park, Y., and Karmel, J. (2001) A multi-center, double-blind, placebo- controlled study of liposomal prostaglandin E_1 in patients with ARDS. Inten. Care Med. 27:1578-1583.
Vincent, J.L., Sakr, Y., and Ranieri, V.M. (2003a) Epidemiology and outcome of acute respiratory failure in intensive care unit patients. Crit. Care Med. 31:S296-S299.
Vincent, J.L., Wilkes, M., and Navickis, R.J. (2003b) Safety of human albumin--serious adverse events reported worldwide in 1998-2000. Brit. J. Anaesth. 91:625-630.
von Dobschuetz, E., Bleiziffer, O., Pahernik, S., Dellian, M., Hoffmann, T., and Messmer, K. (2004) Soluble complement receptor 1 preserves endothelial barrier function and microcirculation in postischemic pancreatitis in the rat. Amer. J. Physiol. 286:G791-G796.
Wachtel, M., Frei, K., Ehler, E., Fontana, A., Winterhalter, K.,and Gloor, S.M. (1999) Occludin proteolysis and increased permeability in endothelial cells through tyrosine phosphatase inhibition. J. Cell Sci. 112:4347-4356.
Walmrath, D., Schneider, T., Pilch, J., Schermuly, R., Grimminger, F., and Seeger, W. (1995) Effects of aerosolized prostacyclin in severe pneumonia – impact of fibrosis. Amer. J. Resp. Crit. Care Med. 151:724-730.
Walmrath, D., Schneider, T., Schermuly, R., Olschewski, H., Grimminger, F., and Seeger, W. (1996) Direct comparison of inhaled NO and aerosolized prostacyclin in ARDS. Amer. J. Resp. Crit. Care Med. 153:991-996.
Walsh, T.S. and Lee, A. (1999) N-acetylcysteine in the critically ill. Inten. Care Med. 25:432-434.
Wang, P. (1998) Adrenomedullin in sepsis and septic shock. Shock 10:383-384.
Wang, Y., Folkesson, H., Jayr, C., Ware, L., and Matthay, M.A. (1999) Alveolar epithelial fluid transport can be simultaneously upregulated by both KGF and beta-agonist therapy. J. Appl. Physiol. 87:1852-1860.
Wang, Y., Gu, Y., Zhang, Y., and Lewis, D.F. (2004a) Evidence of endothelial dysfunction in preeclampsia: decreased endothelial nitric oxide synthase expression is associated with increased cell permeability in endothelial cells from preeclampsia. Amer. J. Obstet. Gyn. 190:817-824.
Wang, Y., Manevich, Y., Feinstein, S., and Fisher, A.B. (2004b) Adenovirus-transfer of 1-cys peroxiredoxin gene to mouse lung protects against hyperoxic injury. Amer. J. Physiol. 286:L1188-L1193.
Ward, P.A., Riedemann, N., Guo, R., Huber-Lang, M., Sarma, J., and Zetoune F.S. (2003) Anti-complement strategies in experimental sepsis. Scand. J. Infec. Dis. 35:601-603.
Ware, L.B., Conner, E., and Matthay, M.A. (2001) Von Willebrand factor antigen is an independent marker of poor outcome in patients with early acute lung injury. Crit. Care Med. 29:2325-2331.
Ware, L.B., Eisner, M., Thompson, B., Parsons, P., Matthay, M.A., and The ARDS Network. (2004) Significance of von Willebrand factor in septic and non-septic patients with acute lung injury. Amer. J. Respir. Crit. Care Med. published ahead of print on June 16.
Ware, L.B., Fang, X., and Matthay, M.A. (2003) Protein C and thrombomodulin in human acute lung injury. Amer. J. Physiol. 285:L514-L521.
Ware, L.B., Fang, X., Wang, Y., Sakuma, T., Hall, T., and Matthay, M. (2002) Lung Edema Clearance:20 years of progress. Mechanisms that may stimulate resolution of alveolar edema in the transplanted lung. J. Appl. Physiol. 93:1869-1874.
Ware, L.B. and Matthay, M.A. (2000) The ARDS. New Eng. J. Med. 342:1334-1349.
Waschke, J., Baumgartner, W., Adamson, R., Zeng, M., Aktories, K., Barth, H., Wilde, C., Curry, F., and Drenckhahn, D. (2004) Requirement of Rac activity for maintenance of capillary endothelial barrier properties. Amer. J. Physiol. 286:H394-H401.
Watts, J.A. and Kline, J.A. (2003) Bench to bedside: the role of mitochondrial medicine in the pathogenesis and treatment of cellular injury. 10:985-997.

Wei, B., Liu, Y., Wang, Q., Yu, C., Long, C., Chang, Y., and Ruan, Y. (2004) Lung perfusion with protective solution relieves lung injury in Tetralogy of Fallot. Ann. Thor. Surg. 77:918-924.

Wiedemann, H.P., Arroliga, A.C., and Komara, J.J. (2003) Emerging systemic pharmacologic approaches in ARDS. Respir. Clin. North Amer. 9:419-435.

Weinberg, P.F., Matthay, M.A., Webster, R., Roskos, K., Goldstein, I., and Murray, J.F. (1984) Biologically active products of complement and acute lung injury in patients with the sepsis syndrome. Amer. Rev. Respir. Dis. 130:791-796.

White, S.J., Nicklin, S., Buning, H., Brosnan, M., Leike, K., Papadakis, E., Hallek, M., and Baker, A.H. (2004) Targeted gene delivery to vascular tissue in vivo by tropism-modified adeno-associated virus vectors. Circ. 109:513-519.

Wiewrodt, R., Thomas, A., Cipelletti, L., Christofidou, M., Weitz, D., Feinstein, S., Schaffer, D., Albelda, S., Koval, M., and Muzykantov, V.R. (2002) Size-dependent intracellular immunotargeting of therapeutic cargoes into endothelial cells. Blood 99:912-922.

Wilkes, M.M. and Navickis, R.J. (2001) Patient survival after human albumin administration. A meta-analysis of randomized, controlled trials. Ann. Interr. Med. 135:149-164.

Wu, W.C., Rathore, S., Wang, Y., Radford, M., and Krumholz, H. (2001) Blood transfusion in elderly patients with acute myocardial infarction. New Engl. J. Med. 345:1230–1236.

Xin, X., Yang, S., Ingle, G., Zlot, C., Rangell, L., Kowalski, J., Schwall, R., Ferrara, N., and Gerritsen, M.E. (2001) Hepatocyte growth factor enhances vascular endothelial growth factor-induced angiogenesis in vitro and in vivo. Amer. J. Pathol. 158:1111-1120.

Yahr, T.L., Vallis, A., Hancock, M., Barbieri, J., and Frank, D.W. (1998) ExoY, an adenylate cyclase secreted by Pseudomonas aeruginosa type III system. PNAS USA 95:3899-3904.

Yang, L., Quan, S., and Abraham, N.G. (1999) Retrovirus-mediated HO gene transfer into endothelial cells protects against oxidant-induced injury. Amer. J. Physiol. 277:L127-L133.

Yatomi, Y., Ohmori, T., Rile, G., Kazama, F., Okamoto, H., Sano, T., Satoh, K., Kume, S., Tigyi. G., Igarashi, Y., and Ozaki, Y. (2000) S1P as a major bioactive lysophospholipid that is released from platelets and interacts with endothelial cells. Blood 96: 3431-3438.

Yi, L., Weisdorf, D., Solovey, A., and Hebbel, R.P. (2000) Origins of circulating endothelial cells and endothelial outgrowth from blood. J. Clin. Invest. 105:71-77.

Yu, M. and Tomasa, G. (1993) A double-blind, prospective, randomized trial of ketoconazole, a TxA_2 synthetase inhibitor, in the prophylaxis of the ARDS. Crit. Care Med. 21:1635-1642.

Zeiher, B.G., Matsuoka, S., Kawabata, K., and Repine, J.E. (2002) Neutrophil elastase and ALI: prospects for sivelestat and other neutrophil elastase inhibitors as therapeutics. Crit. Care Med. 30:S281-S287.

Zeng, W., Matter, W., Yan, S., Um, S., Vlahos, C., and Liu, L. (2004) Effect of drotrecogin alfa (activated protein C) on endothelial cell permeability and Rho kinase signaling. Crit. Care Med. 32:S302-S308.

Zhang, C., Hein, T., Wang, W., and Kuo, L. (2003) Divergent roles of angiotensin II AT1 and AT2 receptors in modulating coronary microvascular function. Circ. Res. 92:322-329.

Zhang, H., Schmeisser, A., Garlichs, C., Plotze, K., Damme, U., Mugge, A., Daniel, W. (1999) Angiotensin II-induced superoxide anion generation in human vascular endothelial cells: role of membrane-bound NADH-/NADPH-oxidases. Cardiovasc. Res. 44:215-222.

Zhang, Y., Gu, Y., Lucas, M., and Wang, Y. (2003) Antioxidant SOD attenuates increased endothelial permeability induced by platelet-activating factor. J. Soc. Gyn. Invest. 10:5-10.

Zhu, S., Ware, L., Geiser, T., Matthay, M.A., and Matalon, S. (2001) Increased levels of nitrate and surfactant protein a nitration in the pulmonary edema fluid of patients with acute lung injury. Amer. J. Resp. Crit. Care Med. 163:166-172.

Zimmerman, J.L., Dellinger, R., Straube, R., and Levin, J.L. (2000) Phase I trial of the recombinant soluble complement receptor 1 in acute lung injury and acute respiratory distress syndrome. Crit. Care Med. 28:3149-3154.

Index of abbreviations and acronyms

AC - Adenylyl cyclase
ACE - angiotensin converting enzyme, expressed on cell surfaces, particular on pulmonary capillaries
AJ - adherens junction
AKAPs - A kinase (PKA) anchoring proteins
ALI - Acute Lung Injury, by specific clinical criteria
ANCA - antineutrophil cytoplasmic antibodies, mainly directed to leukocyte Proteinase 3
AP-1 - a transcription factor, activating protein-1
APB - actin binding proteins
APC - Activated Protein C
ARDS - Acute Respiratory Distress Syndrome, by specific clinical criteria
Arp2/3 - actin related proteins, involved in nucleation and brancing
ASK-1 - Apoptosis signal-regulating kinase
AT1 - angiotensin II receptor
ATP - adenosine 5'-triphosphate
Bad – apoptotic protein
BAEC - bronchial artery endothelial cells
BAL - broncho-alveolar lavage
Bax - apoptotic protein
Bcl2 - anti-apoptotic protein
bFGF - basic fibroblast growth factor
BHQ - inhibitor of sarcoplasmic/endoplasmic reticulum calcium ATPase
BMVEC- bronchial microvascular endothelial cells
CAM - nonspecific for cellular adhesion molecule (generally ICAM, VCAM, & ELAM)
cAMP - adenosine 3',5'-cyclic monophosphate
$[Ca^{++}]_i$ - intracellular calcium level – interacts with calmodulin (CaM)
cGMP - guanosine 3',5'-cyclic monophosphate
C5a - an activated component of the complement cascade
Cdc42 - a Rho family GTPase protein
CINC - cytokine-inducible neutrophil chemoattractant
CNF1 - Cytotoxic necrotizing factor 1
CPA - inhibitor of sarcoplasmic/endoplasmic reticulum calcium ATPase
COX – cyclooxygenase – two enzymes 1 & 2, which synthesize prostanoids
CXC - one of the two major families of chemokines
DAG - diacylglycerol
DPB - dense peripheral, cortical band of actin
EB - infectious elementary body from chlamydia
ECMO - extracorporeal membrane oxygenation
Edg – endothelial differentiation gene-encoded G protein-coupled receptors for LPA and S1P
EDHF - endothelial derived hyper-permeability factor
Egr1 - early growth response gene 1
ELAM - E-selectin, an adhesion molecule
eNOS - endothelial type nitric oxide sythase, generally constitutive, expressed in other cells
ERM - Ezrin-radixin-moesin family of proteins that link the membrane to the cytoskeleton
ET-1 - endothelin-1
FAK - p125 focal adhesion kinase
FGF - Fibroblast Growth Factor
FiO_2 – fraction of inspired oxygen
FKHD – forkhead transcripption factor for gene program related to apoptosis
FLIP - FLICE-like inhibitor protein, an anti-apoptotic protein with structural antagonism of caspase-8

Flt-1 - VEGF Receptor-1
FMLP - N-Formylmethionine Leucyl-Phenylalanine
FRET - fluorescence resonance energy transfer
GAG - glycosaminoglycan
$G_{\alpha i}$ - trimeric GTP binding proteins, α subunit with negative effect on adenylate cyclase
$G_{\alpha s}$ - trimeric GTP binding proteins, α subunit with positive effect on adenylate cyclase
GAP - GTPase activating protein, increases the GTPase activity of small G-proteins
GEF - guanine nucleotide exchange factor, which alters small G-proteins allowing activation
GPCR – GTPases coupled receptor
GSH/GSSG - reduced glutathione/oxidized glutathione
GTP - guanosine 5'-triphosphate
H_2O_2 - hydrogen peroxide
HA - Hyaluronic acid
HBP/CAP37 - heparin binding protein, azurocidin, secreted by leukocytes
HCMV - human cytomegalovirus
HGF - Hepatocyte Growth Factor
HIF-1 – hypoxia induced factor-1, a transcription factor
HIV - human immunodeficiency virus
HSP - heat shock protein, e.g. Hsp27
HUVEC – human umbilical vein endothelial cells
HlyA - *E. coli* hemolysin
ICAM-1 – intracellular adhesion molecule, expressed by certain cells, binds integrins on leukocytes
IFN – interferon
IκB - inhibitory binder of NFκB
IKK – kinase that phosphorylates IκB
INOS - inducible nitric oxide synthase
IP_3 - Inositol 1,4,5, trisphosphate
IL - one of a number of interleukin bioactive molecules release from leukocytes
JAK – janus kinases
JAM - one of several small Ig family glycoprotein adhesion molecules associated with tight junctions
JNK – Jun kinase
K - hydraulic conductance
KSHV - Kaposi´s sarcoma-associated herpesvirus
LBP - LPS-binding protein
LDL - low density lipoprotein
LLO - *L. monocytogenes* hemolysin, listeriolysin O
LP - lysophospholipids
LPA - lysophosphatidic acid
LPC - lysophosphatidylcholine
LPS - lipopolysaccharide from Gram negative bacterial wall, endotoxin
LTB_4 - leukotriene B4
MAPK - mitogen-activated protein kinase, e.g. p38 MAPK
MCP-1 - monocyte chemoattractant protein-1
MHC - Major Histocompatibility Complex
MIP-2 - macrophage inflammatory protein
MLC - myosin light chain
MLCK - myosin light chain kinase
MMP - matrix metalloproteinase, family of proteases secreted by various cells
NF-κB - nuclear factor
NO - nitric oxide free radical
NOS - nitric oxide synthase, iNOS-inducible, eNOS-endothelial (more constitutive)

oxLDL - oxidized low-density lipoproteins
π_i - protein osmotic pressure
π_{pl} - osmotic pressure in plasma
PAF - platelet-activated factor, lipid bioactive molecule
PAI - plasminogen activator inhibitor (1 & 2)
PAK - p21-activated kinase
PAR – protease activated receptor, family of receptors for thrombin, trypsin, tryptase
PDE – phosphodiesterases, catabolize cAMP and cGMP
PECAM - platelet-endothelial adhesion molecule, CD31
PEEP - positive end expiratory pressure ventilation
PGE_1 - prostaglandin E1
$PGF_{2\alpha}$ - prostaglandin F2 alpha
PGI_2 - prostaglandin I2, prostacyclin
P_i - hydrostatic pressure of the interstitial space
PI3K - phosphotidylinositol-3 kinase
PIP_2 - phosphatidylinositol 4,5-bisphosphate
PKA - cAMP-dependent kinase, kinase A
PKC - protein kinase C, one of several types
PKG – cyclic GMP-dependent kinase
PKI - endogenous PKA inhibitor protein
PLA_2 - phospholipase A2
PLC – phospholipase C
PLD – phospholipase D
PMN – neutrophil, polymorphonuclear neutrophil
PMP - polymorphic outer membrane protein from *C. pneumoniae*
PPARγ - peroxisome proliferator-activated receptor
Pyk2 - proline rich tyrosine kinase-2
Q - net fluid flux
Rac1 - a Rho family GTPase
RhoA - a Rho family GTPase associated with actin polymerization and organization
RhoK or ROCK – Rho kinase
ROC - receptor operated calcium entry channels
ROS - reactive oxygen species
RpcAMPS - Rp isomer adenosine-3', 5'-monophosphothionate, a PKA inhibitor;
σ - macromolecule reflection coefficient
SAPK – stress activated protein kinases, including p38- and JNK- MAPKs
SH2 & SH3 - src homology 2 and 3 recognition domains
SHP-2 - src homology domain (SH) 2-containing tyrosine phosphatase
S1P - sphinosine 1-phosphate
SLPI - secretory leukocyte protease inhibitor, an inhibitor of serine proteases
SMC – smooth muscle cells
SOC - store operated calcium entry channel
SOD - superoxide dismutase
STAT - Signal Transducers and Activators of Transcription
TBARS - thiobarbituric acid reating substances, index of oxidative stress
TER - transendothelial electrical resistance
TFPI - Tissue factor pathway inhibitor
TGFβ - transforming growth factors – family of 3 gene products with 5 common receptors
TIE - endothelial-specific angiopoeitin-linked receptors
TIMP - inhibitors of matrix metalloproteinase
TJ - tight junction

TLR - *toll*-like receptors, for recognition of extracellular microbial components, like LPS
TM - thrombomodulin
TNF - tumor necrosis factor, generally meaning TNFα
tPA - tissue plasminogen activator
TPEN - chelates endoplasmic reticulum calcium
TRAF - TNF receptor-associated factor
TRP - transient receptor potential protein, which encodes a calcium entry pathway
TTS - type III secretion system, for injection of Gram-negative bacteria proteins cells
TxA_2 - thromboxane
u-PA - urokinase-type plasminogen activator
VASP - vasodilator-stimulated phosphoprotein of focal adhesions, a nucleation promoter for actin polymerization
VCAM - an adhesion molecule expressed by certain cells that binds ligands on leukocytes
VE - vascular endothelial, as in VE cadherin
VEGF - vascular endothelial growth factor, also known as VPF-vascular permeability factor
VEGFR - VEGF receptor, VEGFR1 = Flt-1; VEGFR-2 = KDR = flk
VLA - very late antigen
vWF - von Willebrand factor
WASP - Wiskott-Aldrich syndrome protein, proteins involved in signaling
WAVE - WASP-family verprolin homologous proteins, involved in actin organization
XO – xanthine oxidase
ZO - zonula occludins protein, one of several proteins which complex with tight junctions

INDEX

* terms such as endothelium, permeability, inflammation are not included as they are central to the entire volume; many items are found under their common abbreviated names and full names are not repeated here (see previous list for abbrev., p 508-511).

ACE (see angiotensin II)
Actin (also see cytoskeleton) 27, 44, 206-227
    binding proteins 27, 30-31, 108, 113, 211-216, 430
    depolymerization 29, 31, 42-45, 47-48, 52, 209-210
    polymerization 27, 29-31, 42-48, 52, 182, 209-210, 374, 437
    organization 28, 31-46, 113, 122, 182, 206-217, 220-227, 473
        (see also DPB and stress fibers)
    actomyosin contraction (see myosin)
Activated Protein C (see Protein C)
Acute lung injury – 13, 14, 18, 20, 166-167, 193, 267, 341, 424, 426, 443, 450, 472, 476-478, 480, 484-498
Adenosine 431
Adenylyl cyclase 108, 110, 140-156, 177, 425, 429, 437, 480-480, 494
Adherens Junctions 44, 49-52, 255-257, 425, 427
    normal composition and decreased content 44, 49-51, 192, 268, 340-341, 380, 485
    function and regulation 49, 51, 110, 115, 124, 155-156, 222-223, 252-254, 258-260, 267, 286, 428-429, 435, 437, 473
    associated signaling molecules 107, 176-177, 189
Adrenomedullin 383-382, 413, 415, 438-439, 483, 494
Akt 114, 124, 170-177, 180-181, 184-189, 192-193, 260, 299, 444, 445, 483
Albumin 6-9, 14, 107, 190, 325, 430-431, 448, 487-488, 492, 495
ALI (see acute lung injury)
α-actinin 30, 44, 47-48, 114, 223-224
Alveolar-capillary membrane 5-6, 8-9, 11-18, 278, 472
Alveoli 3-6, 15-16, 312, 426, 472, 480, 490-490, 493
Angiogenesis 19, 71, 82, 107, 166, 186-189, 252, 260, 280, 286, 402-403, 409, 412, 414, 428
Angiopoietin 86, 110, 116, 122-123, 174, 177, 185-190, 261, 291-292, 406, 415, 427, 431, 444, 480-482, 496
Angiostatin 167, 414

Angiotensin II, ACE inhibitors, & AT1 blockers 73, 110, 120-122, 177-180, 189, 191, 286-288, 295, 342, 344, 355, 425, 435, 475-477, 488
Antioxidants (also see apocynin and N-acetylcysteine) 120, 190, 351-353, 448, 475, 477-478, 492-493
AP1 120, 172-173, 179-181, 183, 184-185, 191-192, 299-300, 320, 342, 411
Apocynin 447, 475
Apoptosis & anti-apop. (endothelium) 11-12, 15-17, 140, 166-171, 173-174, 176, 185-189, 336, 341, 344, 352-353, 378, 385, 425, 437, 441, 473, 476-477, 484
ARDS 12-20, 167, 174, 226, 267, 289-290, 296, 341-342, 352-354, 366, 372, 386, 402, 410-412, 415, 424, 426, 432-433, 435-436, 440-443, 445-446, 451-452, 484-487, 490-498
Arp 2/3 31, 212-213, 217
Arterial hypoxia – see blood gas
ASK-1 172-173, 183-184, 191, 343, 344
Atherosclerosis 19, 226-227, 447, 498
Bacteria and products (see also LPS) 83-82, 156-156, 239, 314-315, 343, 349-351, 366-388, 404-405, 415, 425, 438, 450, 478, 485, 497
Basal lamina –see matrix
bFGF (see FGF)
β-agonists 264, 415, 438, 480, 482, 493
Blood-brain barrier 5, 252-253, 260, 268, 278, 280, 386
Blood gas exchange 3, 13-18, 66, 312, 380, 424-425, 440-442, 445, 451-452, 472, 480, 484, 486, 493, 497
Bradykinin 10, 53, 72-73, 76, 110, 148, 151, 283, 403, 444
Bronchial circulation 3-5, 85, 278, 300-301, 414-415
C5a (see complement)
Cadherin (see adherens junctions and VE cadherin)
Calcium 35, 51, 70, 107-108, 11-113, 118-119, 121, 125, 128, 144-146, 149-156, 175, 177, 184, 225, 240, 255, 284, 293-294, 298, 316-317, 319-320, 322, 325, 339, 377, 383, 403-404, 425, 429-431, 436, 441-442, 480
Caldesmon 44
Calmodulin (CaM) 35, 44, 108, 111, 118, 144-145, 156, 183, 225
cAMP 29, 33, 39, 41, 48, 52-54, 86, 107, 108-109, 115, 122, 140-157, 192-193, 260-261, 264, 266, 284, 351, 415, 429-431, 436, 438-441, 448, 480-483, 490, 493-493
Cardiogenic edema (see also hydrostatic edema) 11-12
Caspase 49, 171-172, 176, 180-183, 186, 190-191, 344-345, 385-385
Catalase 338, 344, 349, 352-353, 445, 475, 477-478

Catenin (see also adherens junctions) 44, 115, 254-258
  alpha 49-51, 428
  beta 49, 115, 174-175, 192, 256, 258-260, 263, 410, 428
  gamma 49, 252, 429
  P122 51, 192, 405
Caveolae & caveolin 110, 125, 143, 147, 151-156, 176, 178, 187, 283, 407
CD34 49, 286-288, 286, 295, 449
CD36 369, 387, 382, 385, 390
Cdc42 116-118, 176, 216-217, 220, 374, 407, 437
Centrosome (see also cytoskeleton – microtubules) 32-33, 44
Chemokines 68, 72, 78, 82, 110, 166, 183, 191, 336-337, 344, 378, 425, 433, 445, 476, 485, 497
Claudin (see also tight junctions) 44, 51-52, 253, 263, 286
Clotting (see hemostasis)
Coagulation 10, 15, 19, 66, 71, 278, 354, 378, 382-384, 425, 433-434, 447, 473-474, 476, 486, 488, 494-497
Cofilin 31, 118, 212, 216-217
Colchicine 33, 37, 41, 43
Collagen 5, 8, 14, 47, 285, 342, 354, 428, 490
Complement 166, 337, 342, 344, 355, 378, 425, 432-434, 475, 478, 485, 488, 497
Connexin (see gap junctions)
Contractile model 33-35
Cortactin 31, 241, 299, 374
Corticosteroids 192, 414-415, 436, 492, 495
COX 111, 177, 181, 190, 339, 349, 383, 410, 491
Cytochalasin 29, 33, 37, 41-44, 51
Cytokeratins 31-32
Cytokines (see also IL and TNF) 10, 16-18, 66, 77-78, 172-173, 183-185, 187, 267, 280, 286, 336-337, 343-344, 353, 378, 415, 425, 432, 475-476, 478, 485, 497
Cytoskeleton 44 (see also actin)
  Endothelial shape/function 26-38, 85, 208, 320, 425, 427, 447, 473
  Actin cytoskeleton 26-30, 33-44, 48, 51-52, 55, 182, 113, 121, 206-227, 317, 373, 473
  Microtubule cytoskeleton 26, 32-33, 44, 207-208, 211, 217-222, 473
  Intermediate filaments 26, 31-32, 206
Cytotoxicity 11, 15, 18, 167, 336, 341, 349-350, 378, 384-386, 425, 448, 473, 484-485
DAG 111-112, 115, 121
Diperoxovanadate 118, 119
Disseminated intravascular coagulation 17
DPB (cortical actin) 28, 31, 33, 37-43, 51-53, 55, 115, 148, 206, 208, 223, 225, 244, 255, 263, 265, 299, 483

Edg receptors 116, 123, 180, 192-193, 286
EGF 120, 177-178
Elastase 167, 267, 337-338, 340-344, 349, 450, 475-478, 485, 492
Endothelial precursor cells (see progenitor)
Endothelin 70-73, 77, 82, 85, 180, 190-190, 324, 382, 425, 435, 444, 475-476, 488, 495
Endotoxin (see LPS)
Epithelial permeability/injury 9, 14-15, 355, 425, 485, 487, 490, 495
Epithelium (lung) 5-6, 15, 194, 278, 313, 337, 349, 355-354, 412, 432, 439, 445, 472, 476, 480-480, 487, 490
ERM 31, 118, 241-242, 321, 323, 380
E-selectin 80, 82, 172, 246, 286-289-290, 296, 313, 322, 337, 354, 377, 406, 434, 477
FAK 46-49, 110, 114-115, 119, 176-177, 188, 224, 241-243, 266, 375, 428
Fatty acid 9, 17, 325, 491
FGF 70, 75, 80, 264, 413
Fibrin & fibrinogen 71,75, 285, 319-324, 353-354, 369, 382, 409, 425, 490
Fibroblast 5, 32, 167, 349, 355-355, 490
Filamin 30-31, 44, 119, 430
FKHR 171-173, 175, 184-186
Focal Adhesion 27, 39, 44-48, 53, 425
  complex 45, 48, 192, 223, 241-242, 266, 428-429
  composition 44-48, 115
  adhesion plaques 39-40, 45-48, 167, 206, 219, 223, 241, 254, 298, 408, 473
Forskolin 108
Gallium 14, 487, 488
GAP 111, 116
Gap Junctions 44, 49, 252, 266, 283, 286
Gas exchange (see Blood gas)
GEF 111, 116, 220
Gelsolin 31, 44, 114, 215
Glucose 180, 185, 189-192, 258
Glycocalyx 283, 296-297, 340, 343, 368, 431
Glycosaminoglycans 72, 343, 485
G-protein coupled receptors 69, 72, 75, 110-111, 377
Heme oxygenase-1 445, 477
Hemostasis (see coagulation)
Heparan proteoglycans 45-47, 428
HGF 75, 80, 118, 189, 482, 490
HIF-1 174, 190, 407, 410-410, 413-414
Histamine 10, 31, 53, 69, 74, 76-77, 112-113, 140, 148, 225, 256, 267, 284, 338, 403-404, 415, 437, 442, 444
HSP27 170, 192, 319-320
HSP90 119, 125, 175, 352
Hyaluronic acid 286
Hydraulic conductivity 7-9, 239, 284

Hydrogen peroxide 81, 108, 118-120, 191-193, 338, 346-352, 404, 415, 425, 437-439, 442, 445
Hydrostatic edema 8-9, 11-14
Hydrostatic pressure 8-17
4-hydroxynonenol 119, 488
Hypercapnia 450-451, 480, 495, 497
Hyperoxia 191-192, 445-446, 486
Hypoxia 12, 16-17, 72, 174-175, 190, 193, 283-282, 286, 407, 411, 413, 424, 484, 486
ICAM 15, 31, 68-69, 85, 166, 172-174, 190, 226m 280, 286-290, 295, 314-324, 336-337, 340-342, 352-353, 355, 368, 370, 376-377, 407, 432, 434, 476-478, 488
IFN 73, 74, 115, 182, 256, 437
IL-1 69, 72, 74, 166, 174, 183-182, 336-337, 343, 354, 372, 382, 432-433, 437, 442, 444, 475-476, 488, 491
IL-8 71-72, 74, 177, 183, 192, 286, 336, 341-343, 354, 370-372, 432, 444, 475-476, 487-488, 495
Infection 10, 12, 15-16, 66, 336, 366-388, 425, 485, 497
Integrins (see also focal adhesion) 36, 41, 47, 244, 369, 408
    matrix 43, 45-48, 85, 223, 238-241
    cell-cell 44, 49, 239, 253, 266
    apical surface 45, 239
Interleukins (see also specific ones) 74, 77-78, 166, 177, 183, 192, 337, 342, 351, 354, 372, 377, 382, 411, 432, 475, 488, 495
Intermediate filaments (see cytoskeleton)
Intrinsic phenotype 279, 285, 473
IP$_3$ 111-112, 115, 240, 429
Isoprostane 181, 425, 435, 448, 451, 475, 488
JAM 44, 49, 51-52, 261, 264-265, 288-288, 313
Lactosyl ceramide 179
LDL & oxidized LDL 71, 80, 175, 179-183, 193, 227, 340, 343, 353, 425
Lectins 283, 296-297
Leukocyte 312-325, 336-338, 378, 425, 474
    adhesion 45, 66, 68, 80, 289, 322, 342, 369, 425, 439, 442, 445, 448, 474, 478
    lung sequestration 4, 15-18, 312, 336-338, 355, 372, 426, 432, 437, 477, 480, 495-496
    migration 49, 68, 226-227, 284, 312-317, 337, 341-342, 432, 476
    products 10, 167, 336, 337-338, 344, 403, 432, 484-485
LIM kinase 117-118
LPA 74, 80, 110, 114, 117, 180, 324, 404
LPC 75, 80
LPS 10, 16, 71-72, 82-83, 175, 180-182, 285, 338, 343, 350-354, 372-373, 377-385, 386, 425, 431-432, 435-436, 438-439, 447, 450, 353, 476, 478, 490-491
Lung fluid balance 6-9, 12, 284, 424, 452, 472
Lymphatics and lymph drainage (lung) 5, 8-9, 11-13
Macrophage 6
MAPK 44, 114-116, 120-122, 124, 168-183, 184-187, 190-193, 241, 260, 264, 266, 285, 298-299, 319-322, 344, 353, 407
Matrix (lung) 238-246
    components, content, and quality 5-8, 27, 69-70, 72, 285, 313, 339-340, 341, 355-355, 369, 425, 428, 432, 449-450, 472, 476, 478, 480, 485, 495, 497
    adhesions (see focal adhesions) 26, 27, 33-39, 41-48, 53-54, 85, 219, 223-224
MCP-1 71-72, 74, 78-80, 82, 177, 185, 192, 341, 353, 370, 384
Mechanical ventilation 12-16, 244, 341, 407, 424, 426, 449, 473, 486-486, 490-491, 495, 497
Mechanotransduction 36-38, 175-176, 268, 474
Metalloproteinases 72, 121, 167, 173, 188, 192, 340-344, 349, 355, 410, 449-450, 476
Microorganisms 83-84
Microtubules (see cytoskeleton)
MLCK (see also myosin) 111-112, 115, 123, 298-299, 316, 403-404, 406, 425, 429-430, 436-437, 441, 476, 480
Moesin (see ERM)
Myosin (see also cytoskeleton) 27, 29, 33, 35-39, 44, 69, 108, 110-113, 117-118, 123, 209, 224-225, 299, 380, 404, 406-407, 430, 441
Multi-organ failure 14, 476, 487, 495-498
N-acetylcysteine & Nacystelyn 120, 165, 191, 445-447, 475, 493
NADPH Oxidase 114, 118, 119-123, 174, 180-180, 184-185, 189-193, 300, 318-319, 322, 338-339, 345, 347-349, 352, 406, 425, 429, 435, 447, 477-478
Necrosis (see cytotoxicity)
Neoplasm 19-20, 167, 177, 186, 239, 286, 374, 449, 498
Neurogenic edema 12, 17
Neutrophils (see also leukocytes) 6, 16-19, 80, 284, 312-325, 336-338, 345-348, 415, 431, 442-443, 445, 447, 480, 485
Neutrophil cationic proteins 337, 341, 345, 349, 475-476, 478
Neutrophil elastase (see elastase)
NFκB 82, 169-173, 176-177, 183, 184-186, 191-193, 300, 315-317, 340, 353, 377-378, 386, 411, 434, 444, 485
Nitric oxide (see NO)

NO 70, 74, 76-77, 85, 125, 192, 266, 284, 299, 338, 340, 343, 349, 351-352, 380-383, 413, 425, 427, 429-430, 442-444, 446-447, 477, 480-483, 494-495
NOS 82, 120, 173-174, 176, 187, 192, 299-300, 338, 352, 383-382, 413, 444, 477, 483
Occludin (see also tight junctions) 44, 51-52, 242, 253, 255, 263-264, 268, 380, 449
Oleic Acid 17, 435, 491
Oncotic pressure –see osmotic
Osmotic pressure 7-9, 11-13
Osmotic stress 243-246
Oxidants (see also antioxidants)
    effect of 10, 15-16, 18, 72, 75, 80, 166, 190-190, 267, 284, 337-340, 343-353, 355, 371, 380, 386, 407, 414, 425, 431, 435, 437, 442, 444-447, 476-477, 493
    production by 72, 120, 185, 284, 300, 322, 338-339, 432, 474, 476, 478, 495
    markers 435, 444, 446, 447, 448, 450, 487-488, 495, 497
    signaling 80, 118-122, 183, 185
Oxygenation (see blood gas)
PAF 72, 81, 110, 113, 115, 166, 317, 336, 343, 382, 415, 425, 431, 435, 444, 475, 488, 490-491
PAI 71, 187, 285, 384
PARP 184, 437, 447-449, 475, 477, 497
Paxillin 31, 44, 47-49, 116, 224, 241, 244, 266, 428
PDGF 120, 173, 177-180, 187, 190, 355, 413
PECAM (CD31) 44, 49, 51, 110, 114, 115, 123, 176, 252-254, 264-267, 286-289, 286, 293, 295, 313-314, 317, 377, 429, 477
Pericyte 5, 283, 386
Permeability edema 8-17, 66, 69-70, 252, 259, 267-268, 372, 380, 386, 402, 412, 415, 424, 437-438, 440, 442-443, 450, 474, 476, 484, 486, 487-497
Pertussis toxin 441
PGI_2 (see prostanoids)
Phallicidin 29
Phosphatases 111-113, 116-118, 122, 225, 258, 321, 404, 425, 428-430
Phosphodiesterases 108-109, 141-145, 154-156, 425, 429, 439-440, 442, 480-480, 482, 493
PI3K 114, 168, 170-172, 176-177, 184-188, 224, 260, 321, 375, 444, 483
PKA 31,33, 39, 52-54, 108-109, 123, 124, 145-148, 171, 180, 192-193, 266, 425, 429-431, 436-441, 444, 449, 480-483
PKC & phorbol esters 31, 42, 47-48, 70, 110-114, 121, 176-177, 224, 266, 315, 416, 436, 441-442
PKI 108-109, 425, 430, 438-439, 480
PLA_2 82, 85, 110-111, 115, 118, 124, 177, 189, 240, 372, 375, 436
Plakoglobin (see catenins-gamma)

Platelets 109, 122, 324, 380, 426, 431, 439, 444, 496
PLC 81, 110-111, 113, 118, 170, 175, 177, 241
PLD 81, 111, 113-115, 118-118, 121, 177
PPARγ 86, 168, 173-174, 191, 444-445, 476, 482, 496
Profilin 31, 44, 114
Progenitor (circulating endothelial) 11, 425, 449, 483, 496
Proliferation 66-67, 70-71, 120-122, 166-167, 169, 172-177, 180-180, 186-188, 193-194, 264, 278, 286, 301, 352, 378, 386-386, 425-426, 437, 447-449, 473, 483
Prostacyclin (see prostanoids)
Prostanoids
    synthesis and release 111, 113, 115, 123-125, 190, 324, 343, 351, 413, 425, 436, 438
    effect of 354, 383, 425, 427, 430, 435, 438, 440-443, 447, 480, 482, 493
Protease 10, 16, 337-344, 349-351, 353, 355, 372, 380, 425, 432, 448-449, 476, 478, 485, 497
Protein C 19, 71, 353, 382, 424, 434, 476, 482, 491, 494
P-selectin 287, 296, 313, 322, 380, 380
Pulmonary blood flow 3-4
Pulmonary circulation 3-4, 278, 312, 472
Pulmonary edema (see hydrostatic and permeability edema)
Pulmonary fibrosis 336, 354-355, 411, 433, 435-436, 490, 497
Pulmonary hypertension 8, 12, 18, 85, 283-282, 288, 402, 410, 412-414, 424
Rac 52, 107, 109-110, 114, 116-118, 120, 176, 216-217, 220, 225, 258, 280, 322, 374, 405, 425, 429, 437, 448, 478-480, 483
Raf 122, 169, 173, 180
Ras 108, 116, 122, 169, 173, 180-180, 192, 241, 321, 428
Re-endothelialization 26, 193-194, 449, 483, 485, 490
Reflection coefficient 7, 480
Rho GTPase 31, 48, 52, 82-83, 108, 110-111, 116-118, 122, 155, 176, 180, 216-220, 224-226, 258, 280, 320, 373, 380, 383, 404-405, 408, 411, 416, 425, 429, 437, 441, 447-448, 476-480
ROCK 111, 117, 120, 225-226, 299, 437, 444, 483, 495
ROS (see oxidants)
Rosiglitazone 444-445, 482, 496
S1P 51, 75, 80, 86, 107, 109-110, 114, 122, 180, 243, 324, 415, 480, 494
Sepsis/Septic shock 14, 16, 239, 290, 296, 336, 380-384, 386, 402, 424, 432, 432, 434, 439, 443, 445, 474, 476, 485, 487, 492, 496
Sivelestat 475, 478, 491

Shear stress 72, 84-86, 118, 176-177, 185, 192, 206-207, 227, 283, 298-300, 408, 449, 473
SH2 265, 321
Smooth muscle – see vascular smooth muscle
SOC 149-156, 294
SOD 338, 344, 347, 435, 475, 477
Starling-Landis equation 7, 12
STAT 169, 173, 178, 181, 183, 184, 192
Statin 179-181, 192-193, 447-449, 477-478, 483-482, 496
Stress fibers (see also cytoskeleton-actin) 29-33, 35-48, 52-53, 109, 116, 206, 225-226, 298, 425, 429, 437
Surfactant 341, 439, 451, 489-490
Talin (see also focal adhesion) 31, 44, 47-48, 241
Tensegrity 33-44
TGF 69, 71, 74, 80,85, 115, 172, 186, 187, 355, 425, 432-433, 437, 475-476, 488
Thapsigargin 149-156
Thrombin 10, 19, 29, 43, 46-48, 50, 53, 69-70, 74-75, 118, 122, 148-151, 155-156, 177, 180, 193, 225, 239, 242-243, 256, 260, 267, 284, 301, 344, 353-354, 382, 402, 406, 425, 431, 433-434, 436-437, 441-442, 444, 476, 492-493
Thrombomodulin 19, 71, 188, 286-288, 291-294, 341, 353-354, 377, 382, 488, 494
Thrombosis (see coagulation)
Thromboxane 111, 324, 380, 382, 435, 442, 475, 488, 491, 495
Tie: angiopoietin receptors 122, 186-186, 188-189, 286-286, 290-292, 427, 444, 483-482
Tight junctions 6, 44-45, 51-52, 70, 155, 223, 252-255, 260-265, 267, 280-283, 380, 380, 435, 437, 476-477
TLR 83, 368, 372-373-374, 385, 431, 484, 488, 491
TNF 48, 70-71, 74, 78, 108, 115, 122, 166-167, 174, 180-183, 184-185, 191-193, 256, 284-285, 288, 290, 295, 298, 316-318, 336, 343, 378, 384, 402, 406, 415, 425, 427, 432-433, 437, 440, 442, 444, 450-451, 475-476, 478, 485, 488-489, 491-493, 495
tPA 71, 285-286, 301
Translation 167-168, 370
Troglitazone 174
Tyrosine kinases 112, 114-116, 119-122, 169-170, 176-178, 180, 186-187, 192-193, 224, 240-242, 258, 266, 299, 319-321, 342, 376, 376, 404, 416, 425, 437, 476
Tubulin (see also cytoskeleton microtubules) 26, 32, 44, 184, 217-218
Transcytosis 9, 325
Tumor – see neoplasm
Vasculogenesis 19, 107, 263, 280
VASP 31, 224, 430-431

Vascular smooth muscle 17-19, 71, 106, 125, 284, 353, 386, 413, 415, 424, 426, 429-430, 439, 443, 473-474, 480, 496
VCAM 31, 68-69, 80, 85, 172-174, 177, 227, 286-287, 289, 295, 313, 320, 322, 352-353, 368, 377, 407, 434
VE-cadherin (see also adherens junction) 49,51, 110, 115, 123, 177, 189, 211, 223, 243-244, 255-260, 263, 267-268, 286-287, 289, 341, 405, 428-429, 449
VEGF & VEGF receptors 69-70, 72, 74, 78-80, 85, 107, 110, 112, 115-118, 120, 124, 166, 173-177, 185-190, 193, 256, 258, 260, 265, 280, 286-286, 286-292, 371, 376, 380, 386, 386, 402, 404-415, 425-429, 435, 437, 444, 447, 449, 478, 480-483
Vimentin (see also cytoskeleton –intermediate filaments) 26, 31-32, 40, 42-44, 49, 51, 252
Vinculin 31, 114, 119, 244-246
Virus/Viral infection 83, 315, 370, 374, 375-378, 380, 382, 385-387
vWF – 14-15, 19, 85, 188, 280, 286-289, 292-294, 298, 300-302, 324, 355, 377, 382, 487-488, 495
Xanthine oxidase 120, 317-319, 339, 346-349
Xenobiotics 10, 14-15
ZO proteins (see also tight junctions) 44, 49, 51-52, 110, 223, 255, 263, 265, 380, 380